Polymers and Other Advanced Materials

Emerging Technologies and Business Opportunities

Edited by

Paras N. Prasad
State University of New York at Buffalo
Buffalo, New York

James E. Mark
University of Cincinnati
Cincinnati, Ohio

and

Ting Joo Fai
Standard and Industrial Research Institute of Malaysia
Kuala Lumpur, Malaysia

Plenum Press • New York and London

Library of Congress Cataloging-in-Publication Data

Polymers and other advanced materials : emerging technologies and business opportunities / edited by Paras N. Prasad, James E. Mark, and Ting Joo Fai.
p. cm.
"Proceedings of the Third International Conference on Frontiers of Polymers and Advanced Materials, held January 16-20, 1995, in Kuala Lumpur, Malaysia"--T.p. verso.
Includes bibliographical references and index.
ISBN 0-306-45210-3
1. Polymers. 2. Materials. I. Prasad, Paras N. II. Mark, James E., 1934- . III. Ting, Joo Fai. IV. International Conference on Frontiers of Polymers and Advanced Materials (3rd : 1995 : Kuala Lumpur, Malaysia)
TA455.P58P6954 1996
620.1'92--dc20 95-48884
CIP

Proceedings of the Third International Conference on Frontiers of Polymers and Advanced Materials, held January 16–20, 1995, in Kuala Lumpur, Malaysia

ISBN 0-306-45210-3

A Division of Plenum Publishing Corporation
233 Spring Street, New York, N. Y. 10013

10 9 8 7 6 5 4 3 2 1

Printed in the United States of America

COMMITTEE MEMBERS

Honorary Chairman

Datuk law Hieng Ding
Minister of Science, Technology
and Environment
MALAYSIA

Conference Chairmen

Professor Paras N. Prasad
Director, Photonics Research
Laboratory
Department of Chemistry
State University of New York
Buffalo, NY 14214

Dato' Dr. Ahmad Tajuddin Ali
Director General
Standards and Industrial Research
Institute of Malaysia
Persiaran Dato Menteri, Section 2
P.O.Box 7035, 40911
Shah Alam
Selangor, MALAYSIA

Co-Chairmen

Professor Frank E. Karasz
Distinguished Professor
Department of Polymer
Science and Engineering
University of Massachusettes
Amherst, MA 01003

Professor James E. Mark
Distinguished Research Professor
Department of Chemistry
University of Cincinnati
Cincinnati, OH 45221-0172

Mr. S. Y. Liew
President
Malaysian Plastics
Manufacturers Assoc.
MALAYSIA

International Coordinators

Professor D. Allan Cadenhead
Associate Dean
Faculty of Natural Sciences and
Mathematics
State University of New York
Buffalo, NY 14260

Dr. C. N. Chong
Director, Research & Technology
Development Division
SIRIM
Persiaran Dato Menteri
Section 2, P. O. Box 7035
Shah Alam, Selangor
MALAYSIA

ADVISORY COMMITTEE

PROGRAM COMMITTEE

Professor A. de Sousa
Department of Materials Engineering
Federal University of Sao Carlos
BRAZIL

Dr. O. H. Elbayoumi
United States Air Force
European Office of Aerospace Research and Development, UK

Dr. F. Kajzar
CEA/IRDI
Cen. Saclay
Dein/Lpem
FRANCE

Professor S. C. Kim
Korea Advanced Institute of Science and Technology
Seoul, KOREA

Professor I. C. Khoo
Department of Electrical Engineering
Pennsylvania State University
University Park, PA

Professor R. Laine
Department of Materials Science
University of Michigan
Ann Arbor, MI

Dr. C. Lee
Air Force Office of Scientific Research
Washington, DC

Professor S. Lefrant
Institut des Materiaux
Nantes, FRANCE

Professor A. MacDiarmid
Blanchard Professor of Chemistry
University of Pennsylvania
Philadelphia, PA

Professor S. Miyata
Tokyo University of Agriculture and Technology
Tokyo, JAPAN

Dr. J. K. Nigam
Director, Shriram Institute of Industrial Research
Delhi, INDIA

Professor N. Ogata
Department of Chemistry
Sophia University
Tokyo, JAPAN

Dr. J. Rajikan
Professor and Dean
School of Chemical Sciences
University of Science Malaysia
Pulau Pinang, MALAYSIA

Dr. C. Taliani
Instituto di Spettroscopia Molecolure
Consiglio Nazionale delle Ricerche
Bolognia, ITALY

Dr. J. F. Tung
Principal Research Officer
SIRIM
MALAYSIA

Professor G. G. Wallace
Intelligent Polymer Research Laboratory
Department of Chemistry
University of Wollongong
NSW, AUSTRALIA

PREFACE

This book continues the tradition of providing the scientific community with information on some of the most important advances reported at a series of conferences on Frontiers of Polymers and Advanced Materials. The particular meeting covered in this proceedings volume was held in Kuala Lumpur, Malaysia, from January 16th through the 20th, 1995. It follows earlier proceedings, also published by Plenum, for a conference in New Delhi in 1991, and another in Jakarta in 1993. All of these conferences focused on the most recent and important advances in a wide range of carefully chosen subject areas dealing with advanced materials and new technologies.

The Malaysia Conference was organized by the Malaysian Ministry of Science, Technology and Environment; Malaysian Industry-Government Group for Higher Technology; Standards and Industrial Research Institute of Malaysia; State University of New York at Buffalo; and Malaysian Plastic Manufacturers Association. The stated goals of the conference were:

- To highlight advances and new findings in Polymers and Advanced Materials
- To bring together leading international scientists, engineers and top level industrial management for discussions on the current status of advanced materials, new technologies and industrial opportunities
- To foster global communication in polymers and advanced materials technology.

The Malaysian conference covered by these proceedings emphasized "composites and blends", "high-performance materials", "materials for photonics", "materials for electronics", "biomaterials", "recycling of materials", "sol-gel and processed materials", "advanced materials from natural products", and "multifunctional and smart materials". There was also a separate symposium on "business opportunities".

The success of the conference can be gauged, in part, from the number of attendees and their origins: a total of approximately 350, with 200 having come from more than 30 countries other than the host, Malaysia. The Malaysian participants were well aware of the importance of new materials and of emerging new technologies in various high-technology areas such as microelectronics, aerospace, and telecommunications. Their stated interests were certainly consistent with the conference's goal of providing researchers and industrialists immediate access to advances and developments within the international research community.

The conference included general plenary lectures, given by high-level managers coming from important polymer industries, or from industries interested in the use of polymers in electronics or optoelectronics. Some are included here as part of this proceedings volume. The general emphasis was on the use of polymers and new organic materials over a very wide range of industrial applications.

The international sponsoring agencies for this conference were AKZO Nobel, Cheil Industries, Dow Corning, Hoechst-Celanese, International Institute of Theoretical Physics/UNESCO, Lucky, Ltd., US Office of Naval Research, TORAY Industries, US AID, US Army Research Office (Far East, ARO-FE), US Naval Regional Contracting Centre, and Unitika, Ltd. The Asian Office of Aerospace Research and Development, United States Air Force Office of Scientific Research sponsored the symposium "Materials for Photonics". There were also a number of local sponsoring agencies, and their support added greatly to the success of the conference. We wish to express our gratitude to the Honorable Datuk Law Hieng Ding, Minister of Science, Technology and Environment, Malaysia who played a valuable role as the honorary chairman and to Dato' Dr. Ahmad Tajuddin Ali as the conference chairman from Malaysia. Our gratitude is also to our co-chairman, Professor Frank Karasz, and Dr. Chong Chok Ngee of the National Organizing Committee for their effort in the success of this conference. Finally, we thank Professor D. A. Cadenhead, Professor A. J. Epstein, Dr. A. Guzdar, Professor S. J. Huang, Dr. F. Kajzar, Dr. D. Kaplan, Professor F. E. Karasz, Professor I. C. Khoo, Professor R. M. Laine, Professor J. E. Mark, Professor N. Ogata, and Professor G. G. Wallace for serving as the various symposium chairpersons and for their efforts in the selection of speakers.

The editors of this volume will consider themselves amply rewarded if the information provided not only brings interested readers up to date in these important subject areas, but encourages them to contribute to future advances as well.

P. N. Prasad
Buffalo, New York

J. E. Mark
Cincinnati, Ohio

J. F. Tung
Kuala Lumpur, Malaysia

CONTENTS

COMPOSITES AND BLENDS

HIGH PERFORMANCE MATERIALS

MATERIALS FOR ELECTRONICS

MATERIALS FOR PHOTONICS

BIOMATERIALS

RECYLING OF MATERIALS

SOL-GEL AND CHEMICALLY PROCESSED MATERIALS

MATERIALS FROM NATURAL PRODUCTS

MULTIFUNCTION AND SMART MATERIALS

BUSINESS OPPORTUNITIES

COMPOSITE APPLICATIONS IN COMMERCIAL TRANSPORT AIRCRAFT

John T. Quinlivan/Henry R. Fenbert

Boeing Commercial Airplanes
Seattle, Washington

ABSTRACT

Since the development of advanced fiber composite materials in the 1950s and 1960s, these materials, particularly carbon fiber/epoxy, have become increasingly important in commercial transport aircraft. These materials and the design/manufacturing technology for their application are now truly global.

To assure continued use of these materials to enhance aircraft performance, their life cycle costs must compete favorably with those of other materials. The two areas driving high composites costs are initial fabrication and repair. The composites industry must continue to demonstrate its commitment to provide value to its customers, the airlines, by aggressive improvements in those areas.

Significant use of advanced composites in commercial aircraft started when several new fibers with impressive structural properties were developed in the late 1950s and early 1960s. Primary among these were boron, graphite and carbon, aramid and S-glass. Resin matrix composites of these materials possess very high specific strength and/or modulus, making them attractive candidates for aircraft applications. By the late 1960s, testing and development had clearly identified carbon filaments as the fibers with the best overall balance of engineering properties, ease of manufacture, and cost. Industry interest in learning to use the materials led to the design, fabrication, and service evaluation of a number of commercial transport airplane components, such as the Boeing 727 elevators, Boeing 737 spoilers, Lockheed L-1011 inboard ailerons, and McDonnell-Douglas DC-10 rudders. Weight reductions averaging 25% were achieved.

New large commercial jet aircraft initiated in the late 1970s – the Boeing 757 and 767, and the Airbus Industries A310 – included the first widespread application of advanced composites to secondary structures. Subsequently, new and derivative models of existing aircraft, such as the MD-11, MD-80, B737, and A300, were introduced with similar composite components.

Polymers and Other Advanced Materials: Emerging Technologies and Business Opportunities. Edited by P. N. Prasad *et al.*, Plenum Press, New York, 1995

Pressure to reduce aircraft weight continued into the early 1980s. NASA sponsored programs aimed at the development of advanced composite primary structures for civil transport applications. The FAA and industry worked together to develop a means of showing compliance with the certification requirements. Certification of a composite Boeing 737 horizontal stabilizer was completed in 1982, and the McDonnell Douglas DC-10 composite vertical stabilizer was certified in 1984. A vertical fin was also developed for the Lockheed L1011, but was not introduced into commercial service. By the end of the 1980s, several new aircraft included advanced composite primary structures, including the Airbus Industries A320 and A330/A340, the Aerospatiale-Alenia ATR72, and the Boeing 777.

WHAT HAVE WE LEARNED?

Boeing placed its first composite components into commercial jet service 37 years ago with the B707. Twenty-six years ago we placed our first advanced composite component into commercial service. The service experience of our composite structures has been excellent. Millions of flight hours and years of successful service have demonstrated the corrosion and fatigue-free life we can expect from reinforced plastic materials. Our challenges–areas for improvement–are repair and initial fabrication cost.

Repair

Composite control surface and fairing structures, utilizing the superior strength and stiffness to weight ratios of composites use a minimum of material weight to generate efficient structures. This low weight, coupled with the fact that most of these structures, leading and trailing edges, cowlings, landing gear doors are in harm's way from hail, birds, lightning strike, and ground equipment can generate damage not always easily detected and severe repair problems.

Our airline customers are dissatisfied with the complexity of composites repairs which typically require careful removal of the damaged area and a considerable portion of the surrounding structure, removal of contaminates which have intruded into the damaged area and lengthy cures at elevated temperatures which can compromise the surrounding structure.

The repair materials themselves cause a problem in that they are usually original materials of manufacture. Thus, each structure has its own unique repair material. According to Air Transport World, both Airbus and Boeing repair manuals require inventory and control of large numbers of materials. These materials are often perishable, requiring low temperature storage and with limited shelf lives.

Boeing has joined with several original equipment manufactures, airlines, repair specialists, the FAA, and others in forming the Commercial Aircraft Composite Repair Committee (CACRC). The goals of this organization are standardization of inspection, materials, and repair techniques.

To satisfy our customers, repairs must not only be standardized, but also easier to effect. Our customers are clear that they need:

- Repair cure temperatures well below the initial manufacturing cure temperature.
- Designed-in repairability.
- Intelligent application of composites considering damage potential.

Fabrication Costs

Production costs of composite structures remain an issue. Treating composites as a simple extension of earlier, labor-intensive fiberglass designs led to some very expensive lessons learned. For example, the cost of one advanced composite 737-200 component is more than twice that of its aluminum/fiberglass counterpart.

Expanded application of composites for commercial transport structures will be based on cost effectiveness. The technology must provide the customer with net-added value. Fortunately, there are no insurmountable barriers to producing composite structures at costs which clearly provide value to airline customers. This value is the resultant of the operating cost benefits of composites weighed against the price premium generated by initial fabrication costs.

CURRENT MARKET CONDITIONS

After a long period of stability, jet fuel prices went through two turbulent increases in the 1970s and early 1980s. These increases placed a premium on the low weight components made possible by composites. More recently, energy prices have fallen. Jet fuel prices are forecast to reach 93 cents per gallon by the end of the century. This equates to 70 cents per gallon in 1992 dollars and is less than half the 1981 price of $1.59 per gallon (in 1992 dollars).

Meanwhile, deregulation of the airline industry has led to significant competitive pressures to reduce operating costs. Cost of ownership has replaced fuel costs as the major element of an airline DOC. Because of this, airlines are finding new ways of keeping their old airplanes viable through engine modifications, "hush kit" installations, new interiors, etc. The incentive to replace fleets of older, less fuel efficient airplanes is significantly decreased.

In the 1970s and 1980s, advanced composites were committed to production as the reduction in aircraft fuel burn was envisioned to offset the increased cost of component manufacture. In today's marketplace, new technology must demonstrate value-added benefits over its predecessors: it must "buy" its way onto our airplanes by providing some combination of enhanced safety, reliability, performance, and lower life cycle costs for our customers. Our goal is to understand what our customers need and to incorporate the available technologies into timely products which meet those needs.

The basic technological building blocks that form the foundation of the advanced composites industry are available to manufacturers worldwide. Inevitably, some of these firms have developed specialized expertise; the ability to make the most of this expertise--wherever it is found--can be critical to commercial success. This has led to an evolution of international business arrangements among aerospace firms beginning with subcontracting, and evolving into risk sharing agreements and long-term contracts. To implement the 777 composites structures, we selected eleven U.S. and international subcontractors for component fabrication and included them in the design process through design/build teams. The future will see more partnerships and consortia on a global scale. To not pursue such a global strategy, risks not only market share, but also our ability to launch new airplanes.

FUTURE DIRECTION

The future of commercial aircraft industry appears bright. Travel growth is forecast to average between 5 and 6 percent per year through 2010. At this rate, the market will be over 2 1/2 times greater by 2010.

The need for new airplanes is real; the question is, can the airlines afford new equipment? Replacement of the aging fleet and meeting the expanding market needs will require a net economic improvement. Achieving this net economic improvement will be a challenge for the manufacturers who are striving to reduce costs in an increasingly competitive environment. When applied to metal aircraft structures, this means we must reduce our costs to survive in the marketplace. As applied to advanced composites, this means the cost targets must be even more aggressive.

While composite technology alone is not likely to drive new product timing, composites are basic materials of consideration in the design of all new transport aircraft concepts. Advanced composites are enabling technologies for some vehicles (HSCT, tiltrotor) and design concepts (very high aspect ratio wings). Composites have proven their ability to perform as a primary structural material system while saving 15 to 25 percent weight when compared to modern aluminum designs. No other structural technology offers such promise of reduced weight combined with improved corrosion resistance and inherent fatigue resistance.

At Boeing, we are committed to composites technology as evidenced by the 777 empennage and floor beam applications, and the investment in the Composite Manufacturing Center at the Frederickson site, near Tacoma, Washington. The 777 empennage and floor beams are expected to be life cycle cost competitive with metal structures. We're excited about this opportunity to optimally apply the technology developed to date. Highly automated processes have reduced the cost of these components by more than half the estimates a decade ago.

We at Boeing feel that the fabrication of the 777 empennage, the vertical and horizontal stabilizers, at the Composite Manufacturing Center is the key to cost effective composite production. The Composite Manufacturing Center is a 432,000 sq ft facility designed for composites manufacture. Its current single product is the 777 empennage which is capable of producing at seven per month. The facility is also capable of modular expansion in three directions.

The facility is designed for optimal manufacturing flow with innovative double-doored autoclave separating clean room operations from post cure operations. The unidirectional graphite/epoxy tape material that comprises the stiffened skin is layed down by automated tape laying machines. Skin plies are layed into contoured cure tools. These tools are fabricated of specialized iron/nickel alloys to match the thermal expansion of the carbon fiber/epoxy and provide durable, leak-free tool surfaces for autoclave cure.

Stringer "charges" are layed up flat transfer to a numerically-controlled cutter and cut to shape. The individual stringer components are brought to the stringer fabrication cell where they are assembled into an "I" section.

The cured stringers are trimmed and thru transmission ultrasonic inspected and joined to the uncured skin. The skin stringer subassembly is "co-bonded" in an autoclave process. This process utilizing precured stringers enables a number of process efficiencies. The stringers can be fully trimmed in a feed-through operation, eliminating the need for the five axis machining which would be required if stringer trimming were done after attachment to the skin. Cure tooling requirements are reduced and the defects which can result from tools bearing on the skin during cure are eliminated. The 777 skin is inspected from a single side.

Skin peripheries are abrasive waterjet trimmed. The spars and ribs, which complete the structural boxes of the 777 main empennage components, are manually layed up assisted by laser ply location and mechanically deployed permanent bags for ply compaction. After autoclave cure, the spars and ribs are immersion through transmission ultrasonic inspected and abrasive waterjet trimmed.

The structural box, leading and trailing edges are mechanically assembled using titanium fasteners and moldable plastic shims. This material facilitates the match of bagside surfaces with complex ply dropoffs which enable minimum weight designs.

Three subassemblies are delivered to the 777 assembly plant at Everett, north of Seattle. In Everett, centerline joining of the horizontal stabilizer and attachment to the aft fuselage section are completed. We are committed to continuing improvements to meet our aggressive cost reduction targets. Our current focus in on:

- The automated tape laying-increasing laydown rate.
- Spar fabrication-eliminating hand layup
- Assembly-elimination of shims

This assembly improvement is a cornerstone of our manufacturing improvement efforts. Using the stability of composites components as an asset, shim-free composite assembly without design compromise will be a reality.

Our second major effort is the implementation of automated fiber placement for labor and material efficient fabrication of large complex components. This process combines the continuous process and severe contour aspects of filament winding with tailoring capabilities of tape layer. Boeing has purchased a machine which will be operational in April of this year. The machine will be used for process development and technology verification.

With demonstrated success, we can confidently apply composites to gradually expanded applications. With the technology and manufacturing tools in hand, major payoffs appear to be available for both wing and fuselage structures. The high loads in wing structures offer the highest potential to exploit composites for weight reduction. Insensitivity to corrosion may be the most compelling reason to consider fuselage structure. Future high speed civil transport aircraft will likely look to significant use of composites to meet performance and elevated temperature requirements. While the necessary materials and structural configurations may differ, the fundamental experience with subsonic structures will be absolutely crucial in developing technology for the future.

CONCLUSION

During the past 25 years, the composite industry has become truly global. The technology, manufacturers, and markets are worldwide. Some interesting insights can be inferred from the history of this industry:

Technology cannot be monopolized for any appreciable length of time. For example, carbon fibers were invented in England, but Japan leads in their manufacture. While the United States certified the first CFRP structures, Germany was the first to commit CFRP primary transport structures to production.

Technological innovation alone does not ensure success in the marketplace. The Learfan and Beech Starship are examples of aircraft which were the first of their types. One never made it into production, while the other has accumulated few sales after a $500 million investment.

In the long run, leadership comes from the will and ability to incorporate mature technology into safe, reliable, cost-effective and market-driven products. The next generation at composite designs will build on this history and focus on the major driver in our business today: COST. The future of our industry depends on our ability to design to carefully selected design and cost targets.

HIGH MECHANICAL PERFORMANCE COMPOSITES AND DESIGN OF COMPOSITE STRUCTURES

Jean-Marie Berthelot

Laboratoire de Mécanique Productique et Matériaux
Université du Maine
BP 535, 72017 LE MANS Cedex, France

HIGH MECHANICAL PERFORMANCE COMPOSITES

First it is important to remind the reasons for using composite materials in structural applications and the range of composites used.

Initially composite materials have been developed in the fields of aerospace and aircraft. The challenge was to create materials which at a given mass offered the best mechanical performances. In a first approach, the mechanical performances of materials can be characterized by : i) the rigidity and the strength at the structure level, ii) the tensile modulus and the ultimate tensile stress at the material level.

Thus, the mass of structure being given, the mechanical performances have been characterized by introducing the specific properties (ratio of the properties to the mass per unit of volume) : the specific Young's modulus and the specific ultimate stress.

So the most attractive materials must have a high tensile modulus, a high ultimate tensile stress and a low specific mass. The materials must be also cost-effective, the cost-effectiveness depending on the field of applications.

Material scientists are continously developing new matrix materials and reinforcements. Many remain in the laboratories. Only composite materials with high mechanical performances and of current use will be considered here (aerospace and aeronautic composites will not be considered).

Conventional bulk materials (steel, aluminium alloys, wood, glass, etc) have similar values of the specific modulus. Moreover, it is observed a specific stress of glass which is clearly higher than the specific stress of steel and aluminium alloys. Furthermore it is well known that material strength is related to defects in materials : imperfections, flaws, etc. Thus, increasing material strength can be obtained by elaboration processes which minimize material defects. This consideration leads to the elaboration of materials in fibre form with low diameter of about 10 µm.

Comparison of the mechanical properties of usual fibres is reported on figure 1, giving the specific tensile stress as a function of tensile strain. High specific modulus is associated with carbon fibres and aramid fibres. High specific strength is associated with high strength carbon fibres, aramid fibres, S-glass fibres and also with E-glass fibres which are of low cost.

On account of their very low diameters, fibres cannot be directly handled in mechanical applications, whence fibres are embedded in a polymer matrix to constitute a fibre composite. Polymer matrix has different functions, in particular matrix transfers mechanical loads to the fibres and the mechanical properties of the fibre composites are given, first, by the fibres.

Polymers and Other Advanced Materials: Emerging Technologies and Business Opportunities. Edited by P. N. Prasad *et al.*, Plenum Press, New York, 1995

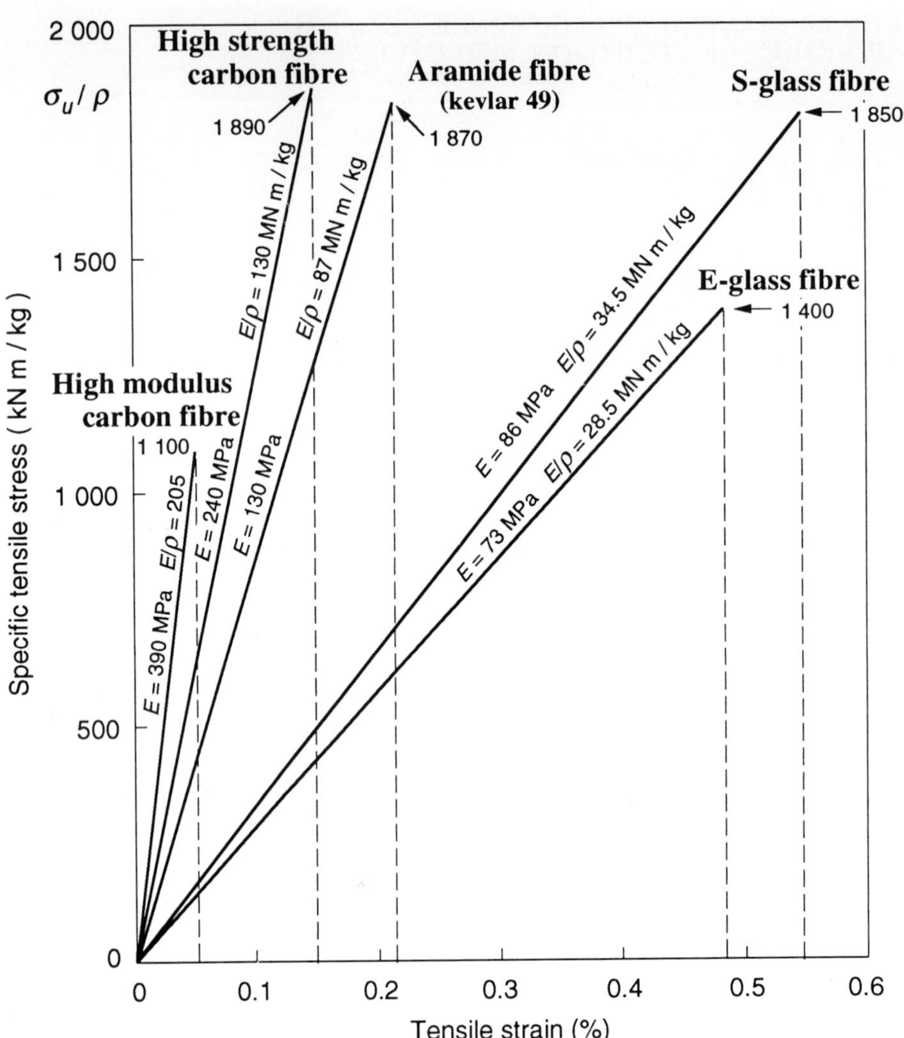

Figure 1. Mechanical properties of usual fibres.

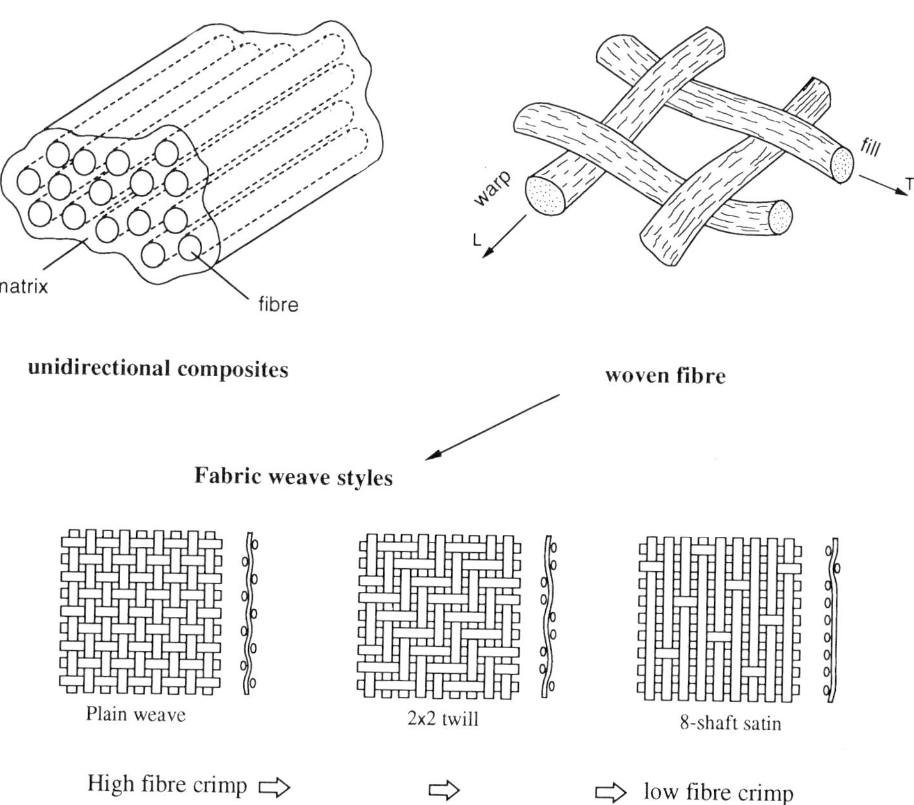

Figure 2. Fibre arrangement.

The principal matrix materials are organic polymers : polyester resins which are the most widely used ; high modulus epoxies for high performance applications ; multifunctional epoxies, polyimide and bismaleimide matrices for high temperature applications. Thermoplastic composites have emerged in recent years. However the processing of these composites requires high temperatures and large shearing forces, and cannot be done on equipement already in place for thermoset processing.

ARCHITECTURE OF COMPOSITE MATERIALS

Fibre arrangement

Fibrous reinforcement can be arranged according to two principal forms (figure 2). In unidirectional form, fibres are aligned in a parallel array. Fibre reinforcement can be also woven by interlacing fibre yarns in two directions : the warp direction and the fill direction. A wide variety of weave fabrics have been developed and the more common weave patterns are reported on the figure 2. These weave fabrics differ by the degree of crimp of fibre reinforcement. As the degree of fibre crimp decreases (from plain weave to satin weave styles), the mechanical properties of the resulting composites improve and it is easier to obtain complex shapes of composite structures.

Laminates and sandwich materials

Then composite materials are manufactured by stacking of successive layers with unidirectional fibres or woven fibres to obtain a *laminate* (figure 3).

Another architecture is obtained by introducing a material of low density (about 100 to 200 kg/m^3) between two laminates, to obtain a *sandwich material* (figure 3).

Nature, thickness, orientation, etc of layers can be varied according to the engineering requirements, thus assuring an optimised level of mechanical performances of the final composite material.

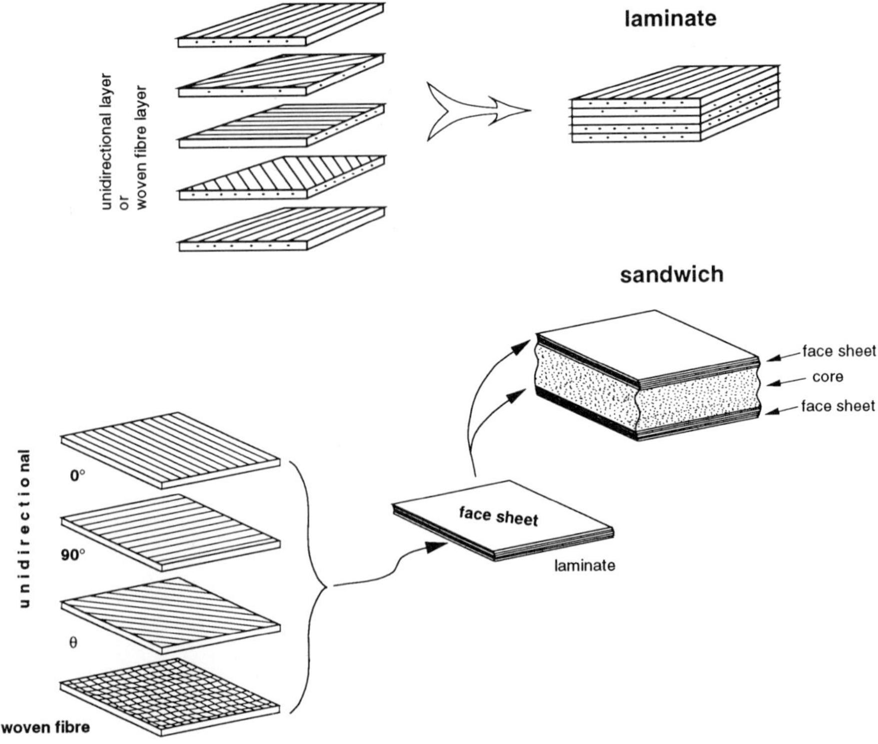

Figure 3. Laminate and sandwich material.

MECHANICAL PERFORMANCES OF COMPOSITES

Elastic behaviour

The elastic behaviour of unidirectional composite is characterized by the relation between stresses and strains, which introduce five independent stiffness constants. These constants are related to the engineering constants, measured in different tensile tests and shear tests :

E_L, ν_{LT} the Young's modulus and the Poisson's ratio, measured in a longitudinal tensile test ;

E_T the Young's modulus measured in a transverse tensile test ;

G_{LT}, $G_{TT'}$ the shear moduli measured respectively in longitudinal and transverse shear tests.

Engineering constants can be also estimated by analytical formulation from the mechanical characteristics of matrix and fibres, and from the fibre proportion. Unidirectional composites show a marked anisotropic behaviour with high mechanical performances along the fibre direction.

In the case of wowen reinforcement, composite material has 3 principal material directions : the warp direction, the fill direction and the direction transverse to the warp-fill plane. The mechanical behaviour of composite is characterized by nine independant engineering constants which can be measured in tensile and shear tests, or estimated by analytical formulation from the mechanical characteristics of fibres and matrix, from the characteristics of warp and fill yarns and from the mass per unit of area of the woven reinforcement.

Fracture properties

In the case of composite structures subjected to mechanical loading, the composite fracture results from the accumulation of different elementary mechanisms (figure 4) : fibre rupture, transverse cracking of matrix, longitudinal cracking of matrix, interfacial fibre matrix debonding and delamination resulting in the separation of layers from each other.

Fracture processes depend on constituants (fibres and matrix), architecture of laminates, mechanical loading, etc. Fracture mechanisms can be quantified by failure criteria such as stress criteria, quadratic criteria, etc.

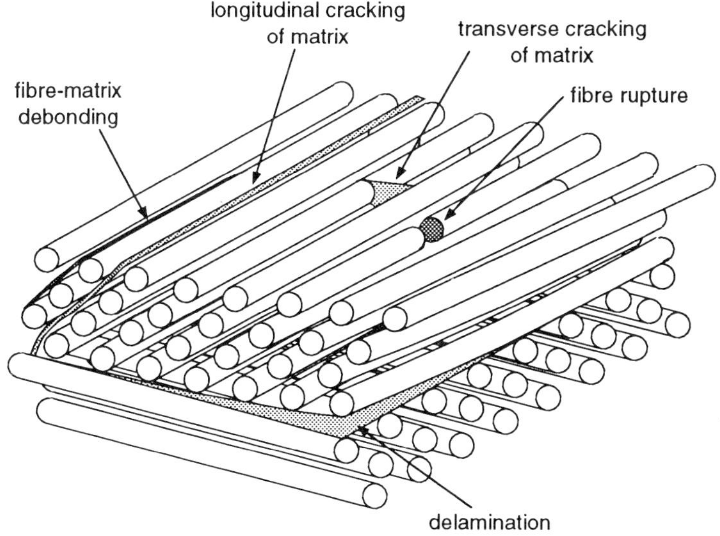

Figure 4. Fracture mechanisms in composites.

Tailoring composite materials

The previous considerations show clearly how engineer has the possibility of tailoring composite materials in order to obtain the desired engineering properties according to : constituants (fibre and matrix), fibrous reinforcement proportion, reinforcement form (unidirectional, woven form), nature of fibre matrix interface (high or poor adhesion), stacking sequence of layers, laminate or sandwich material, etc.

Two examples of tailoring composite materials can be given in the case of a plate with a central hole, subjected to mechanical loading in the 0° direction or 90° direction. Two stacking sequences are considered (figure 5) : 0° layers and [0°, ± 45°, 90°] laminate. The stress-strain curves and the crack patterns are reported on the figure 5. The results show a brittle behaviour in the case of 0° layers, with high performances in the 0° direction and low performances in the 90° direction. On the other hand the mechanical performances of [0°, ± 45°, 90°] are equivalent in the 0° and 90° directions, with a gradual failure of laminate, the stiffness and strength being intermediate between the previous ones.

The first behaviour is adapted to applications where a high stiffness and strength must be obtained. The ductile behaviour is needed in the case of fatigue application or impact, where a high energy must be absorbed with high deformation.

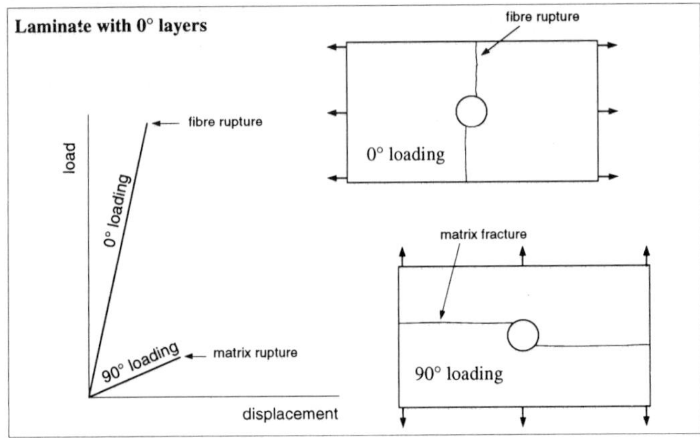

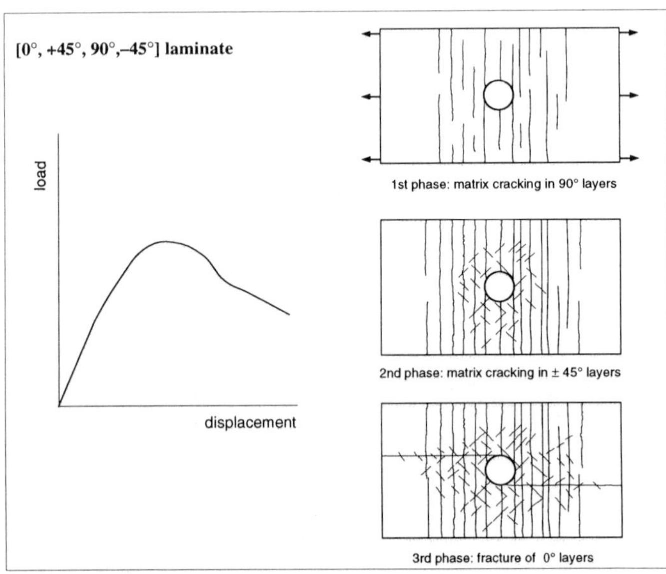

Figure 5. Influence of the nature of composite material.

DESIGN OF COMPOSITE STRUCTURES

Composite materials in structural applications

Composite materials are now widely used in many structural applications. The excellent fatigue resistance of composites has led to their widespread use in helicopter rotor blades. The use of composite materials has extended in transport for safety structures as spring and structural parts of truck, motorbike and french high speed train (TGV). Composite materials are also applied to the storage tank manufacturing, and used for boat construction, leisures and sports. Composite materials are also widely used in Civil Engineering, etc.

Design

We have considered the possibility of tailoring composite materials in order to obtain the desired engineering properties. The tailorability of composite materials considerably modifies the conventional approach to the structural design. In fact, the concepts of "material" and "structure", which are distinct and independent in the traditional design techniques, become here intimately joined.

To design a structure to be manufactured with composite materials involves design of the most suitable material, at ply and laminate levels, in close relation with the structure design (figure 6).

So in the process of design and optimization of composite structure, the design of material plays a fundamental role and consists of two steps :

— first step : design of the layers, by optimisation of their properties starting from the constituants : nature of fibre and matrix, fibre proportion, fibre matrix interface, fibre arrangement, etc ;

— second step : design of the laminate or sandwich, by optimisation of its properties from the layer properties and the stacking sequence.

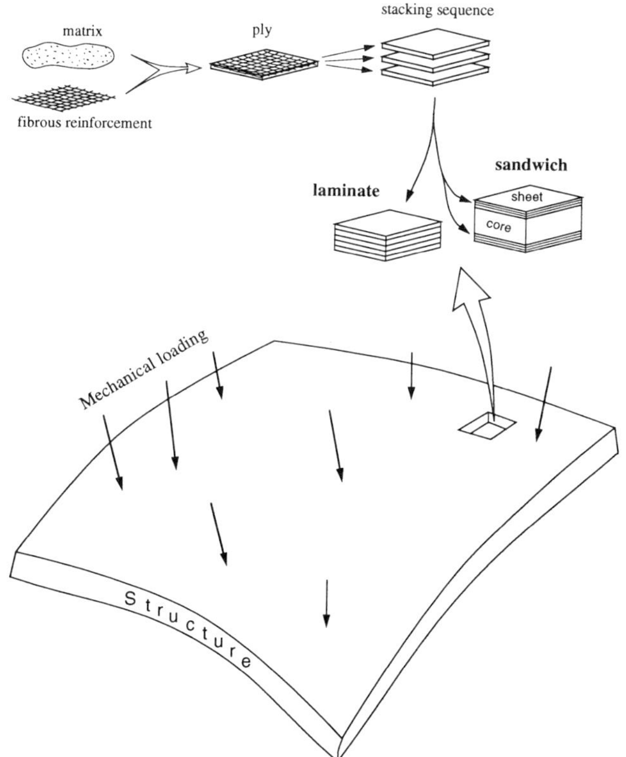

Figure 6. Design of composite structure.

Due to these peculiar characteristics of composites, no fixed entry point in the design process can be strickly defined. The structure performances depend on the composite material properties, which in turn can be properly tailored. So a "circular design process" mode can be defined from constituants to materials, from materials to structure and so on.

The tailorability of composite materials is the key to a non-trivial approach to composite structures, and constitutes a stimulating challenge for research and engineering.

By the way, the tailorability constitutes a considerable complication of the design process, because it requires a greater amount of knowledge of the analysts and designers. Design of composite structures involves a much deeper investigation of the local stresses. So design requires advanced analysis tools and needs to resort to computer in the very early stages of the design.

The design process can be sketched in the following way (figure 7). Mechanical properties of layers can be obtained by experimental testings or by analytical procedures from the constituant properties. The mechanical behaviour (elastic behaviour , damage, etc) of laminate or sandwich is evaluated, by different analytical or numerical models that we do not report here.

Then optimal design of large-scale composite structures combines Computer Aided Design : geometric modelling and meshing of the structure, with finite element analysis.

The figure 8 shows the modelling and meshing of the prototype of a car bonnet manufactured with a sandwich material : glass fibre-polyurethane and polyurethane foam. Finite element analysis calculates the stress and strain field at every point of the structure meshing. The figure 9 shows the stress distribution of the bonnet subjected to a concentrated load.

Then the results are compared with the specifications imposed. As function of this comparison, modifications can be reintroduced in the previous process at different levels : at level of constituants, at level of stacking sequence, at level of laminate or sandwich, at level of the models of the mechanical behaviour, at level of the structure.

Example of designing process

As an example, the designing process will be considered in the case of the elaboration of the prototype of an industrial ultra-light two-seater aircraft.

Initially the Company Ultravia Aero International of Montreal manufactured a light one-seater aircraft. This plane was constructed with one part of aluminium and one part of composite material. The composite part was elaborated with aeronautic pre-impregnated materials with epoxy resin, resulting in : a high cost of the pre-impregnated material, a high cure time (about 8 hours). Furthermore, the design was not optimised : the pre-impregnated materials were laid-up in an empirical arrangement.

The problem to be solved was : i) to convert the light one-seater aircraft into an ultra-light two-seater aircraft ; ii) to decrease the manufacturing cost ; iii) to optimize the design. The conception and manufacture of the prototype have been achieved by the Composite Materials Center of St Jérôme (Québec) Canada, the mechanical analysis of the design being implemented in partnership with our laboratory.

The first step was to substitute epoxy resin for another polymer matrix with lower cost and cure time. So polyester resins were considered. However polyester resins are sensitive to ultraviolet rays and to corrosion by hydrocarbons and solvents. Finally vinylester resin was chosen.

Then mechanical design of the aircraft structure was implemented by Computer Aided Design and Finite Element Analysis. The figure 10 shows the meshing of the aircraft structure with the wing frame, and the stress distributions. The mechanical analysis led to an optimised design with sandwich materials for the high loaded parts and laminate materials for the less loaded parts.

The figure 11 shows the aircraft structure after moulding. Pre-pregs of glass reinforcement and vinylester matrix were developed, leading to a cure time of half an hour. Tests (figure 12) were implemented to correlate experimental results with the mechanical analysis evaluated by finite element analysis. And figure 13 shows the ultra-light aircraft in flight.

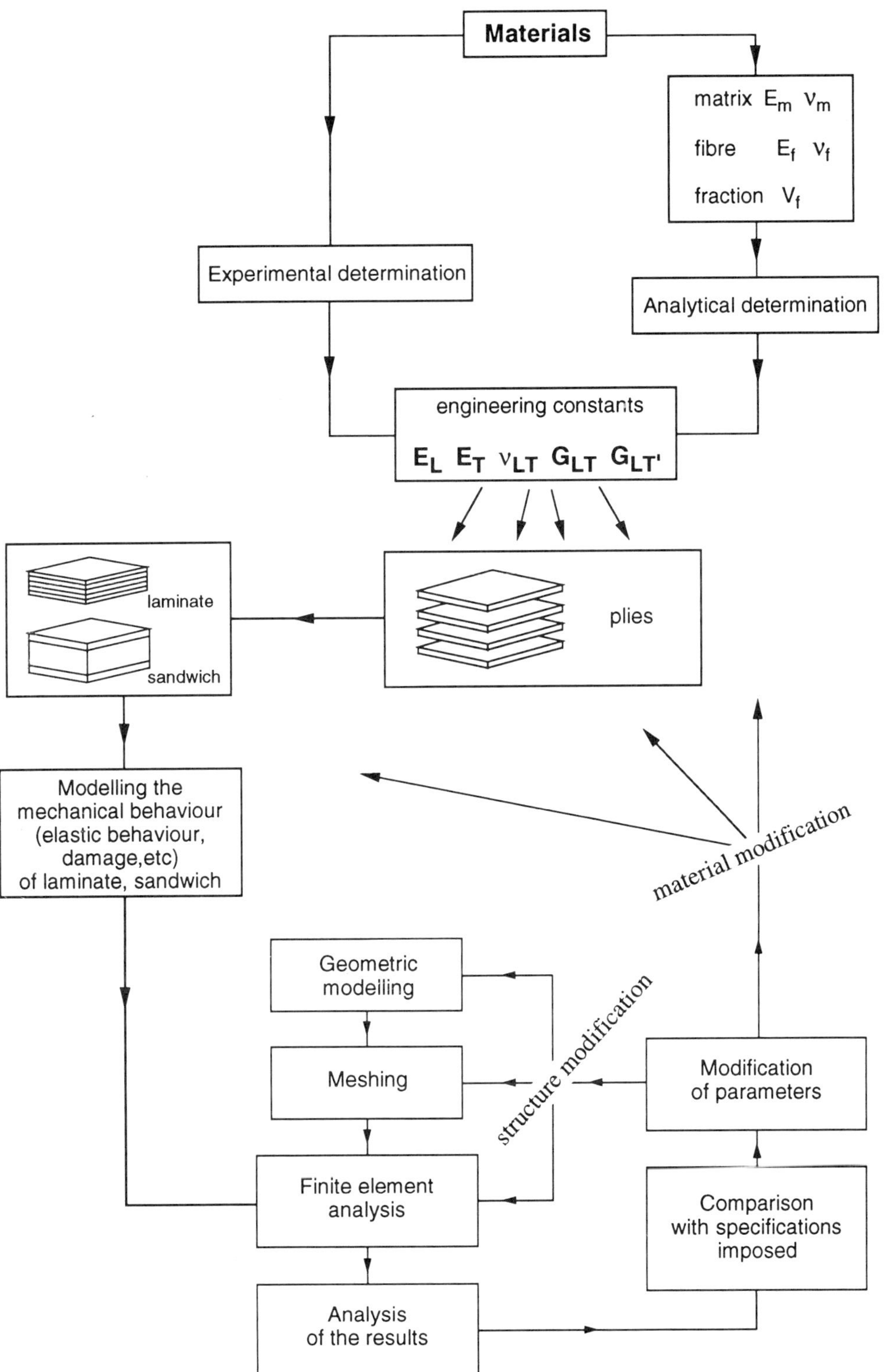

Figure 7. Design process.

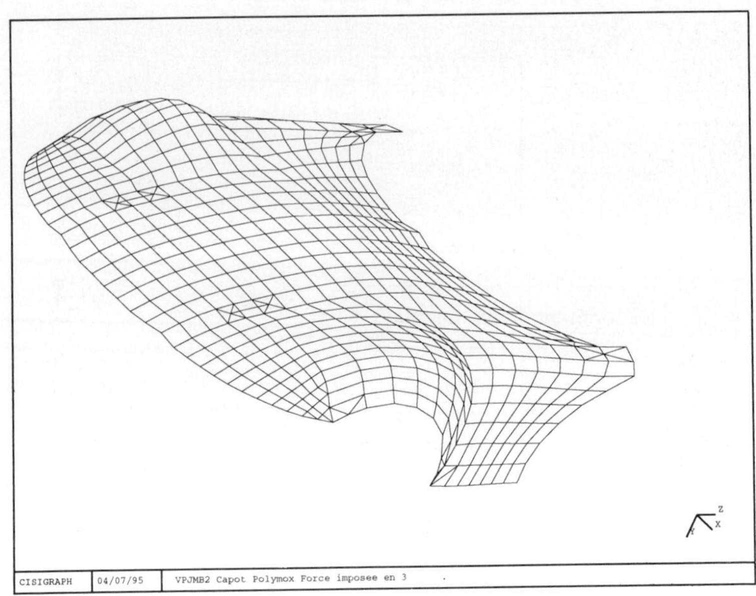

Figure 8. Modelling and meshing of the prototype of a car bonnet.

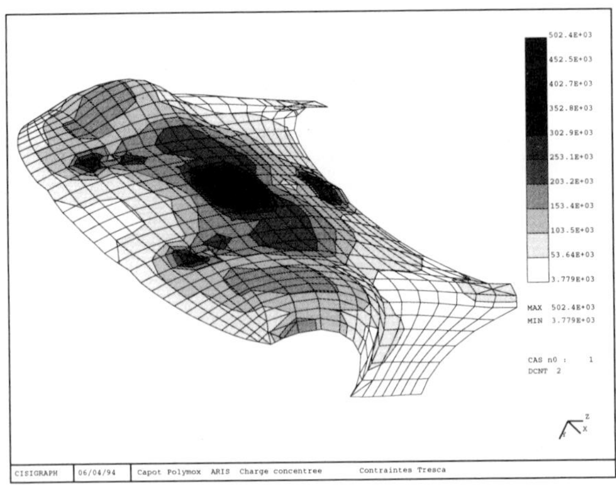

Figure 9. Stress distribution of the bonnet subjected to a concentrated load.

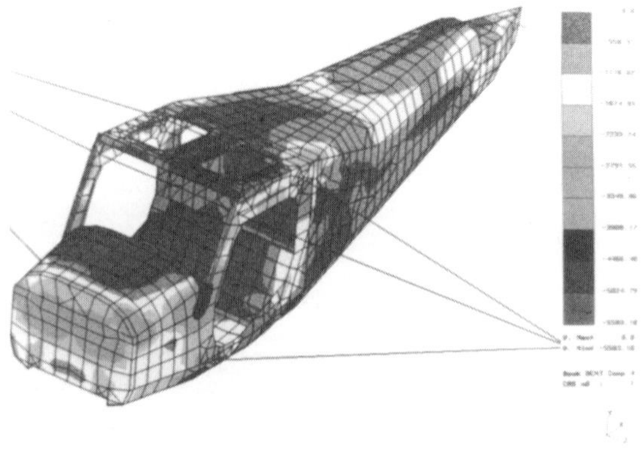

Figure 10. Meshing and stress distribution in the aircraft structure.

Figure 11. Aircraft structure after moulding.

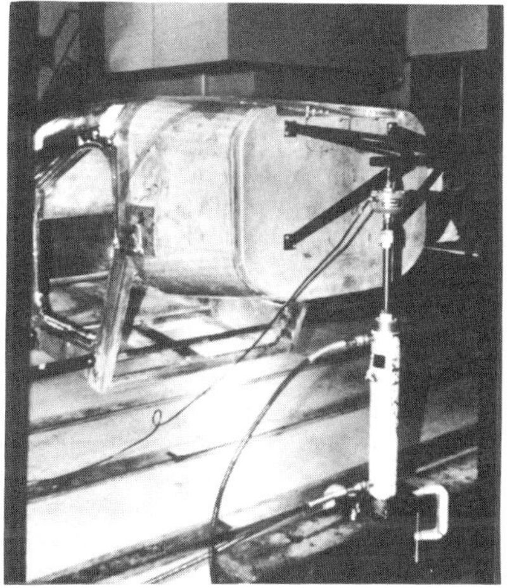

Figure 12. Torsion test on aircraft structure.

Figure 13. The ultra-light aircraft in flight.

CONCLUSIONS

The specific characteristics of fibre composites have been highlighted. The possibility of tailoring composite materials in order to obtain the desired engineering properties appears as the most impressive feature. The tailorability of composite materials considerably modifies the conventional approach to the structural design, since the concepts of material design and structure design become intimately joined.

Then to design a structure to be manufactured with composites involves design of the most suitable composite material : at ply level, at laminate or sandwich level, at structure level. At ply level, optimization of the material is obtained from : constituants (fibre , matrix), fibre matrix interface, fibre arrangement, fibre proportion, etc. At laminate or sandwich level, optimization depends on the ply properties and the stacking sequence. At structure level, optimization requires the necessity of a deep investigation of the local stresses, based on Computer Aided Design and Finite Element Analysis.

No fixed entry point in the design process of composite structures can be directly defined and optimization can extend to the constitution of the materials itself. Furthermore, design with composite materials requires a very high degree and level of interdisciplinary expertise, related to the fields of : matrix materials, fibres, weave fabrics, mechanics of materials, stress analysis, optimization, process and technology, etc.

PREPARATION OF NOVEL POLYAMIDE 6 BLENDS MADE BY REACTIVE EXTRUSION COMPOUNDING

J.F. Tung, P.R. Hornsby

Plastics Technology Group
Standards & Industrial Research Institute of Malaysia

INTRODUCTION

The preparation of polymer blends is normally undertaken by mixing polymers in the melt state, typically in a compounding extruder, whereby the applied shear can be varied and hence the blend microstructure controlled. In immisicible blends, which are thermodynamically unstable, phase morphology can be further modified through the use of compatibilising additives, such as block copolymers, which generally leads to improved physical properties. Compatibilisation can also be achieved by reactive modification of one or more of the polymer phases, including chemical changes to the polymer structure. This paper discusses a novel approach to the preparation of polymer blends, through the in-situ polymerisation of monomer in the presence of a secondary polymer phase during passage through a co-rotating twin-screw extruder. When prepared in this manner, not only can the molecular mass of the polymerising monomer be influenced, but there is the potential for improved compatibility in the blend, possibly arising from chemical interaction between the phases or the formation of interpenetrating network structures. Furthermore, the phase microstructure in the blend can be strongly affected by the mixing conditions employed inside the compounder, in particular the levels of shear generated, through adjustment of screw profile and operating conditions. This approach to reactive compounding will be illustrated with reference to the preparation and characterisation of polyamide 6 (PA6) blends, by polymerisation of $\in$-caprolactam in the presence of polypropylene (PP) and ethylene-propylene rubber (EPR). There is particular interest in the formation of PA6/PP and PA6/EPR blends on account of improvements in toughness which can result in these materials, together with reduced moisture sensitivity of the compositions relative to pure PA6.

Similar analysis of the 90/10 PA6/EPR reactive formulation also demonstrated that through reaction blending some unusual structural changes had occurred. WAXD confirmed the present of α-cyrstalline PA6 and surprisingly of α_1 and α_2 cyrsalline phases of PP ($2\theta = 14.2°$ and $16.8°$). However, EPR in isolation exhibited only a single broad reflection peak at $2\theta = 20.6°$, believed to be associated with polyethylene crystallinity. Further evidence of structural changes are provided by DSC, where two distinct endothermic melting and exothermic crystallisation peaks were observed during

Polymers and Other Advanced Materials: Emerging Technologies and Business Opportunities. Edited by P. N. Prasad *et al.*, Plenum Press, New York, 1995

Table 1 Mechanical properties of PA6 and selected reaction blend compositions

Composition	Blend Ratio	Youngs's Modulus (GPa)	Tensile yield Stress (MPa)	Elongation at break (%)	Impact Strength (Jm^{-1})	Flexural modulus (GPa)
	90/10	2.52	52.7	115.6	80.2	1.73
PA6/PP	70/30	1.43	30.1	18.6	61.8	1.52
	50/50	1.12	27.8	10.5	57.8	1.48
PA6/EPR	90/10	2.46	55.9	182	194	2.56
	90/10[(a)]	2.86	28.2	65	152	-
PA6 [(b)]	-	2.68	66.4	88.7	93.8	2.72

(a) melt blend of PA6/EPR after (3)

(b) PA6 made by the anionic polymerisation of $\in$ - caprolactam in a twin-screw extruder.
(Samples tested 'dry' as moulded)

both heating and cooling cycles, corresponding to transitions of PA6 and PP respectively, with no indication of crystalline PE.

Cryogenic fracture surfaces of PA6 modified with EPR by reaction compounding show the presence of spherical domains with approximate diameters between 1 and 3μm, together with ellipsoidal structures abotu 20-30μm in length and 3-5μm in thickness. There is also evidence for some improvement in interfacial adhesion between the PA6 and modifying phases. Ellipsoid formation is considered to result from differences in viscosity between the blend components and generation of elongational stresses during preparation of the compounds. Further information on the nature of possible structural changes occurring during formation of PA6/PP and PA6/EPR reaction blends has been gleaned from solubility studies combined with analysis by FTIR.

When immersed in 90% formic acid PA6/PP with 90/10 composition made by melt blending a clear solution was produced, corresponding to dissolution of the PA6, with white particles floating on the liquid. These were identified by FTIR as pure PP. The 90/10, PA6/PP reaction blend however produced a colloidal suspension in the upper portion of the solution, with no undissolved particles visible. Similar observations were recorded with the 90/10 PA6/EPR reaction blend, although in addition to the formation of a milky colloidal extracts, suspension, small pieces of insoluble material were seen floating on the solution. FTIR analysis of the colloidal extracts, suggests the likelihood of PA6-g-PP graft copolymer formation in the PA6/PP blend, containing C-N imide linkages formed as a result of mechano-chemical degradation reactions and the generation of chain radicals (Table2). Similarly, in the PA6/EPR system it can be postulated from the experimental results that a graft copolyme of PA6-g-PE, located at the PA6/EPR interface, may have been formed due to a combination of the high levels of the high levels of shear generated and reactive conditions encountered during blend preparation.

Co-polymer formation, as proposed above, would be expected to enhance compatibility between the phases and have an influence on the mechanical properties of the blends formed.

Table 2 FTIR characteristic absorption bands of PA6/EPR blend

Band assignment	PA6	PA6/EPR	PA6/EPR*	Literature Reference
NH stretching (H bonded)/cm^{-1}	3300 (s)+	3300 (vs)+	3298 (s)	3302 [159]
CH_2 assymm. (aliphatic)/cm^{-1}	2936 (s)	2937 (vs)	2923 (m)+	2940 [157]
CH_2 symm. (aliphatic)/cm^{-1}	2865 (s)	2867 (s)	2868 (m)	2865 [157]
Amide I, C=O/cm^{-1}	1641 (vs)	1642 (vs)	-	1642 [157]
Amide II, NH/cm^{-1}	1542 (vs)	1552 (vs)	1545 (vs)	1545 [157]
CH_2 deformation/cm^{-1}	1475 1462 1439 1418 (m)	1475 1462 - 1418 (m)	1477 1464 1436 1417 (w)+	1480 1465 1438 1417 [157]
CN stretching/cm^{-1}	-	-	1241 (w)	1238 [157,188]

+ Band intensity classification: >3(vs); >2 & <3(s); >1 & <2(m);<1 (w)

* Insoluble extract from formic acid solution

PHYSICAL PROPERTIES OF BLENDS

Selected mechanical properties of PA6/PP and PA6/EPR reaction blends are presented in Table 1, together with results for PA6 made by reactive polymerisation in a twin-screw extruder.

Relative to unmodified PA6, PA6/PP compositions made by reactive blending exhibit a progressive decline in most mechanical properties with increasing amounts of PP present. However, the 90/10 reaction blend shows some interesting effects which may

MICROSTRUCTURAL ANALYSIS

Various characterisation techniques were employed to investigate the composition, crystalline and blend morphology, in addition to the extent of compatibility between the phases present. These included WAXD on thin strips sectioned perpendicular to the long axis of the rod extrudate and scanning electron microscopy on low temperature fracture surfaces coated with gold palladium alloy. Samples were also examined by DSC in nitrogen, following a three-stage thermal cycling procedure, by initially heating to 250°C at a rate of 10°C min^{-1} (Stage 1), maintaining the temperature at 250°C for 15 minnutes before cooling to 40°C at a rate of 5°C min^{-1} (Stage II), then remelting again (Stage III) using the same conditions as the first heating run.

To examine blend compability, the relative solubility of the phases in 90 % formic acid was determined according to the procedure described by Molau (2).

Supporting information was also provided by Fourier transform infra-red spectroscopy on ~5μm thick cast films of blends and on extracts obtained from the solubility tests.

RESULTS AND DISCUSSION

MICROSTRUCTURE OF BLENDS

DSC thermograms for 90/10 PA6/PP compositions, made by reactive blending and melt mixing procedures are compared in Figure 1. For the conventional melt blend, distinct endothermic melting and exothermic crystallisation peaks were observed during heating and cooling cycles, corresponding to the PP and PA6 phases present. However, the reaction blend of the same composition displayed only one melting endotherm during both stage I and stage III heating cycles and only a single crystallisation exothermic peak in the cooling thermogram. Phase transitions are attributed solely to PA6, with no evidence for the presence of PP.

Complementary information was obtained from WAXD results, the mechanical melt blends yielding diffraction profiles consistent with the formation of γ-crystalline PA6 (peak at $2\theta = 21.3°$) and α-crystallinity in PP (reflections at $2\theta = 14°$ and 16.8°). The

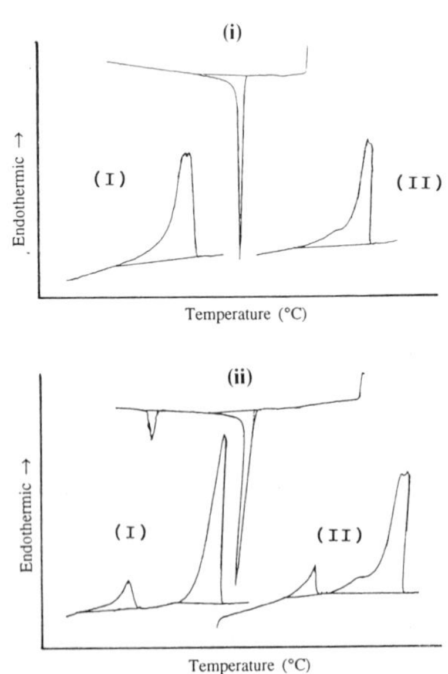

Figure 1 DSC thermogram of PA6/PP blends
(i) reaction 90/10 (ii) mechanical 90/10

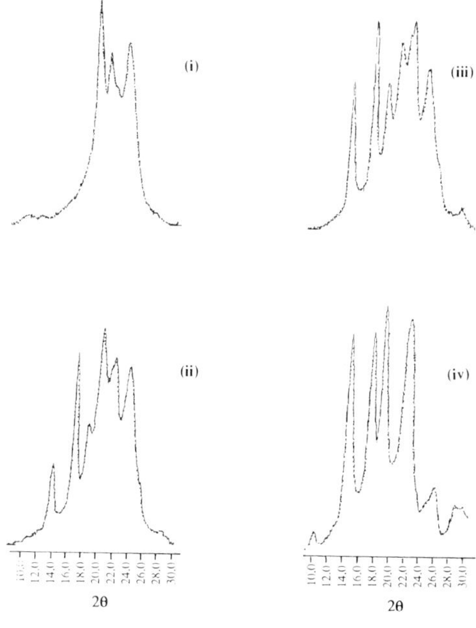

Figure 2 WAXD equitorial profile of reaction PA6/PP blends
(i) 90/10 (ii) 70/30 (iii) 50/50 (iv) 30/70

reaction blend gave a mixture of α ($2\theta = 23.5°$) and γ reflections, resulting from crystalline PA6, but no X-ray diffraction peaks which could be attributed to crystalline PP (Fig. 2).

It is interesting that corresponding analysis of PA6/PP reaction blends with 70/30 and 50/50 polymer ratios, clearly indicated crystalline forms of both PA6 and PP, as was also the case with melt blends of the same composition.

Examination by SEM of fracture surfaces from the PA6/PP blends showed clear phase separation and formation of PP droplets in a PA6 matrix with all melt blends and most reaction blends, other than the 90/10 formulation. In this material PP was not initially visible on the fracture surface but after etching with xylene, a microporous texture was observed which is attributed to dissolution of the PP present.

EXPERIMENTAL

PREPARATION OF COMPOUNDS

In previous work, procedures were established for the polymerisation of $\in$ - caprolactam to yeild PA6 inside a co-rotating intermeshing twin-screw extruder, using sodium caprolactamate catalyst and bis-acyllactam hexamethylenediamene activator (1). Through optimisation of processing parameters, in particular screw speed and barrel temperature profile, and use of a monomer/catalyst/activator ratio of 95:3:2 parts by weight, it was possible to extruder/PA6 with high molecular mass and low residual monomer content. For the preparation of PA6/PP and PA6/EPR blends, which are the subject of this communication, the modifying polymer was introduced at the extruder feedport as a premix with the $\in$ - caprolactam monomer and reactants. In this manner,

PA6 'reaction blends' with varying composition were prepared, although emphasis is given in this study to binary blends containing 90 % by weight of polyamide with 10 % of polypropylene homopolymer (MFI 9 g.10 min^{-1} - 2.16 kg/230°C) and ethylene-propylene rubber, (MR 5.1 - 10 kg/230°C) comprising 75 % by weight of ethylene.

The extruder screw geometry was configured to incorporate zones for feeding, reaction, mixing, venting and melt delivery through a single strand die. Melt extrudate was water cooled and pelletised. The barrel temperature profile was maintained between 80 and 220°C to accomodate the changing rheology of the compositions during extrusion and the screw speed set at 150rpm.

For comparison, selected blends of PA6 and PP were also prepared by melt mixing of commercial polymers using PA6 made by hydrolytic polymerisation (Mw ≈ 32,000) combined with polypropylene homopolymer. These are designated 'melt blends'.

DETERMINATION OF MECHANICAL PROPERTIES

In order to extract any residual monomer, which might complicate interpretation of results, compositions made by reactive extrusion were maintained in boiling distilled water for 12h, then dried under vacuum at 110° C for 24h.

Tensile, flexural modulus and impact test specimens were injection moulded according to ISO 1874-2 (1987), and tested in the 'dry' as moulded state.

Tensile strength and modulus were determined according to ASTM D638 at a strain rate of 50mm min^{-1}. Specimen elongation was measured using an infra-red non-contacting extensometer. Flexural modulus was measured following ASTM D790 (Method 1) and Izod impact strength assessed using ASTM D256, with a Type A notch.

Some three-point bending dynamic mechanical measurements were also undertaken on 4mm diameter rods of extrudate (after monomer extraction and sample conditioning), over a temperature range of - 100 to 130°C at a frequency of 6.28 rads sec^{-1} (1Hz).

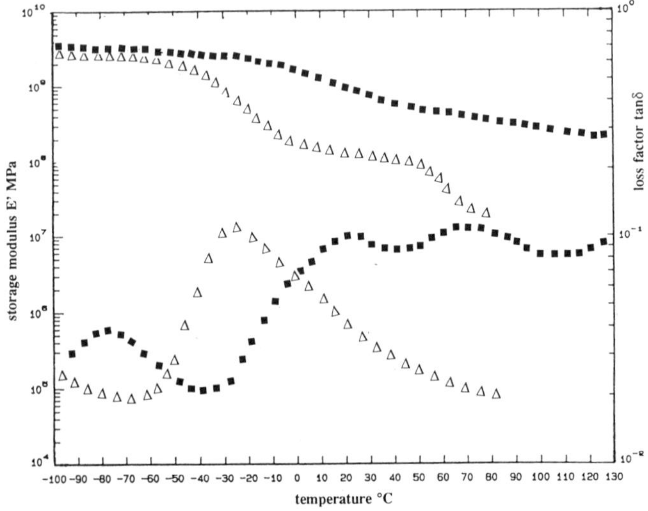

Figure 3 Dynamic mechanical spectrum of reaction PA6/EPR 90/10 blend and EPR

(■ PA6/EPR △ EPR)

result from increased compatibility between the phases, as disucssed earlier. In particular there is a substantial increase in elongation at break measured with this material a high value for Young's modulus, and an impact strength close to results obtained for PA6. From dynamic mechanical data, it is apparent that the 30/10 PA6/PP reaction blend displays only a single broad relaxation peak at about 51°C, corresponding to the glass transition temperature, or α- transition, of PA6. A further sharp relaxation peak is seen at ~77°C (β-transition) arising from crank-shaft motion of an unbonded amide group and several methylene carbon groups of PA6. α- and β-transition and at ~ 7°C and -80°C normally associated with amorphous PP were not detected, in contrast to results obtained for the PA6/PP melt blend of a equivalent composition, which displayed two distinct α-transition peaks at 10 and 42°C attributed to the amorphous content of the PP and PA6 phases respectively.

Mechanical properties obtained for the PA6/EPR reaction blend (Table) demonstrates that the presence of only 10 % EPR can give a very marked improvement in elongation at break and impact resistance, with only a small drop in tensile yield strength, relative to unmodified PA6, and a significantly better balance of mechanical properties compared to reported values from a 90/10 melt blend of PA6 and EPR.

This enhancement in properties for the reaction variant is directly attributed to the co-polymer formation at the rubber/PA6 interface discussed earlier, leading to enhanced phase compatibility and a more effective crack arresting mechanism.

Comparing dynamic mechanical spectra for the 90/10 PA6/EPR reaction blend and as received EPR it is apparent that two loss factor peaks were observed with the blend attributed to the α and β transition of PA6 at 55°C and -78°C respectively. However, a further peak is present at 9°C which is presumably associated with the PP identified in the composition. The distinct transition at -28°C noted for EPR in isolation was not detected in the blend (Fig. 3).

CONCLUSION

Novel PA6 blends have been made by polymerisation of ∊ - caprolactam in the presence of a secondary polymeric phase through a reactive extrusion compounding procedure.

With refence to PA6/PP and PA6/EPR compositions it has been shown that, particularly at a compositional ratio of 90/10, marked differences in phase morphology can exist between reaction blends and more conventional blends produced by melt mixing of the polymers. The former exhibit greatly enhanced phase compatibility, probably through copolymer formation, resulting from mechano-chemical radical reactions, which occur during blend preparation.

These changes in microsctructure are also conducive to enhanced mechanical properties, notably toughness and tensile elongation at break, the effects being particularly significant with the PA6/EPR reaction blend composition.

ACKNOWLEDGEMENT

The authors are grateful to the Standards and Industrial Research Institute of Malaysia for financial support for JFT during the course of this investigation.

REFERENCES

1. P.R. Hornsby, J. F. Tung and K. Tarverdi
Proceedings from SPE-ANTEC Conference, New Orleans, May, 1993, p470

2. G.E. Molau, J. Appl. Polym. Sci. Part A, 3, (1965), 1267

3. S. Cimmino et al, Polymer, 26 (1988), 1874

4. J. F. Tung, PhD thesis, Brunel Univesity, United Kingdom, 1993

PREPARATION AND PROPERTIES OF HAIRY-ROD MOLECULAR COMPOSITES

Jin Chul Jung*, Sang-Bong Park, Heesub Kim, Sung Hwan Ji and Wang-Cheol Zin

Department of Materials Science & Engineering, Pohang University of Science & Technology, San 31, Hyoja-dong, Pohang, 790-784 KOREA

INTRODUCTION

Molecular composites are polymer blends, in which rigid rod-like "fiber" polymers are molecularly dispersed in flexible polymer matrix[1,2]. To obtain excellent mechanical properties of the composites, fiber polymers must be so isotropically dispersed in the blend system that no phase separation can be observed[3,4]. Once phase separation has taken place, the molecular composite transforms to a macroscopic composite, and this phase separation is a serious problem in most of molecular composites. In polymer systems containing rigid-rod polymers a phase separation is particularly facile to occurr, since between the rigid backbones a strong van der Waals attraction is present to form a separate phase. To overcome this phase separation problem a number of novel methods has been introduced for making molecular composites with satisfactory properties.

Hairy-rod polymers are the polymers of rigid backbones substituted by long, flexible side chains. Although these polymers belong to a multigrafted copolymer, they are crystalline to form a layered crystal structure[5], in which the rigid main chains are stacked over each other to form layer domains and the side chains emanating from neighboring backbones are interdigitated with each other to form side chain domain. For making molecular composites hairy-rod polymers are particularly interesting[6]. Because their rigid backbones are interactive with the fiber polymers and their side chains tend to be miscible with the matrix polymers[7-9], they can be utilized as compatibilizers for molecular composites. In addition, when these polymers are mixed with coil polymers to form an isotropic, amorphous phase, these polymer systems form also a molecular composite, in which the rigid main chains function as fiber polymers.

On mixing coil polymers to the hairy-rod polymers having layered crystalline structures, one can obtain true molecular composites when the polymer blends become amorphous. To make the blend systems amorphous, the coil polymers must penetrate into both backbone layers and side chain domains to completely disrupt the layered crystal structures of the hairy-rod polymers (cf Figure 1). When they penetrate only into either main chain layers or side chain domains, we obtain a "pseudo-molecular composite", since in these blend systems coil polymers are molecularly dispersed, but the whole systems are not completely amorphous. The penetrability of coil polymers must be affected by various factors such as chemical structure of coil polymers, backbone structure of hairy-rod polymers, mixing temperature, and structure and length of side chains of the hairy-rod polymers.

To study these effects on molecular composites based on hairy-rod polymers, three series of rigid-rod polymers having linear (n-alkyloxy)methyl ($-CH_2O\text{-n-}C_mH_{2m+1}$) side chains with varying lengths have been prepared and mixed with n-alkanes (C_k), polystyrenes (PSs)

Polymers and Other Advanced Materials: Emerging Technologies and Business Opportunities. Edited by P. N. Prasad *et al.*, Plenum Press, New York, 1995

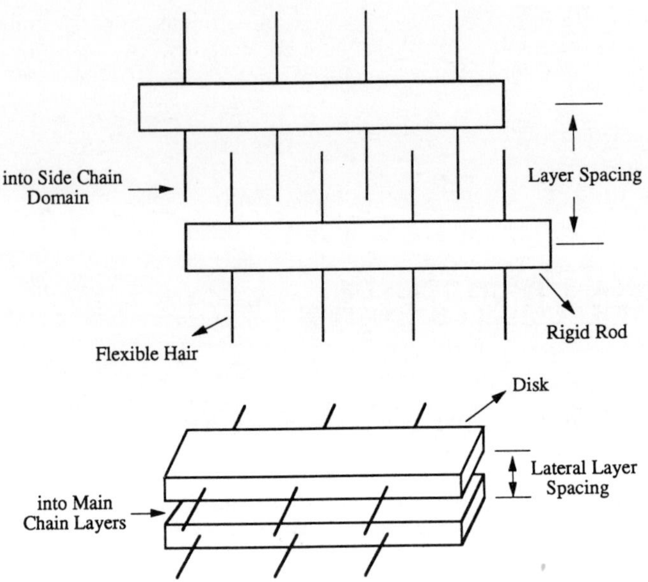

Figure 1. Schematic representation of layered crystal structure.

, poly(oxyethylene)s (PEGs) or poly(oxypropylene)s (PPGs) having low, but well-defined molecular weights and phase behaviors of the blends have been investigated. Coil polymers are designated with abbreviation and number, i.e., the poly(oxypropylene) with molecular weight 4000 g/mol was designated as PPG-4000 and the other coil polymers were designated in likewise manner.

SYNTHESIS

As the hairy-rod polymers poly(azomethine)s (C_m-PAMs), poly(pyro-mellitimide)s (C_m-PY-PIs) and poly(1,2,4-oxadiazole)s (C_m-PAOs) having linear (n-alkyoxy)methyl side chains with varying lengths were used. In their repeat units they all have at least one p-phenylene unit which is, directly or via bridges, linked to the other aromatic or heterocyclic aromatic rings. Over these functional groups the π-electrons of the p-phenylene groups can come into conjugation with those of the other aromatic rings.

C_m-PAMs were prepared by the condensation of 2,5-bis((n-alkyl-oxy)methyl)-p-benzenedialdehydes (C_m-TPAs) with p-phenylenediamine, as depicted in Scheme 1. The synthetic method for monomeric C_m-TPAs and detailed conditions of the polymerization reactions and polymers purification were described in our publication[10].

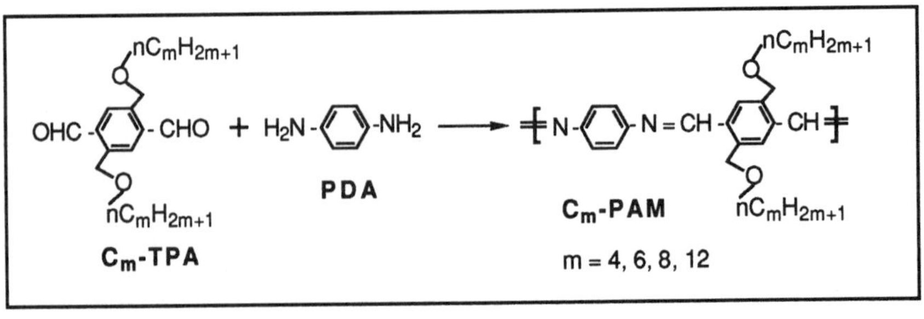

nC_mH_{2m+1} … OHC- … -CHO + H_2N- … -NH_2 ⟶ [N- … -N=CH- … -CH]

C_m-TPA **PDA** **C_m-PAM** m = 4, 6, 8, 12

Scheme 1. Preparation of C_m-PAMs.

C_m-PY-PIs were prepared by the one-step condensation of imidazole-blocked 2,5-bis((n-alkyloxy)methyl)-1,4-benzenediisocyanates with pyromellitic dianhydride, as shown in Scheme 2. The synthetic method for the monomeric BADIs and the detailed

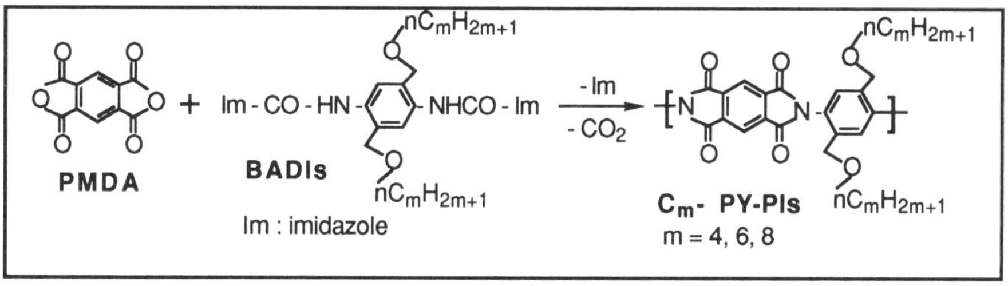

Scheme 2. Preparation of C_m-PY-PIs.

conditions of polymerization and polymers purification will be reported in a separate paper.

C_m-PAOs were prepaed by the condensation of 2,5-bis((n-alkyloxy)-methyl)-1,4-benzenediamideoximes (BABAs) with terephthaloyl chloride followed by the thermal cyclodehydration of precursors poly(tere-phthaloyl-2,5-bis((n-alkyloxy)methyl)-1,4-benzenediamideoxime)s, as shown in Scheme 3. The synthetic method for the monomeric BABAs and the detailed conditions of polymerization, cyclodehydration and polymers purification were reported in our earlier publications[11-15].

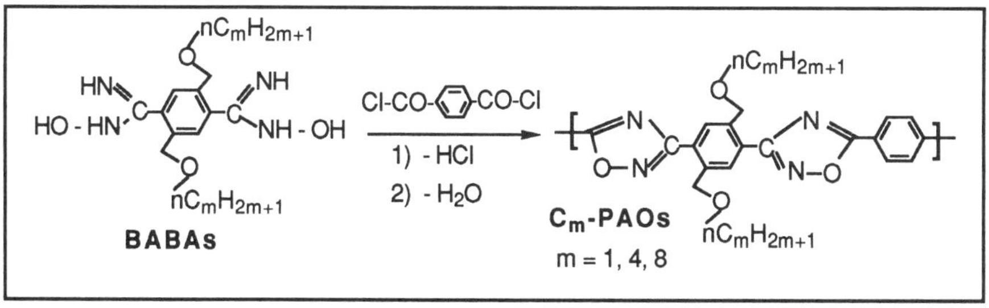

Scheme 3. Preparation of C_m-PAOs.

The composites were prepared by bulk blending of the component polymers at temperatures higher than melting points of the hairy-rod polymers in nitrogen atmosphere.

C_m-PAM COMPOSITES

It was reported earlier[10] that the layered structures of C_m-PAMs (m = 4, 6, 8, 12) are well developed in their crystalline state, and the layer spacing between main chain layers could be clearly elucidated by wide-angle X-ray scattering(WAXS). In C_m-PAMs the attractive interaction between rigid main chains are so strong that only the mesophase of C_{12}-PAM was completely fusible to form an isotropic melt at 255 °C, although, as DSC studies revealed, all C_m-PAMs exhibited a smectic-like liquid crystalline phase at temperatures higher than their crystal-mesophase transition points. These phase transitions could be also confirmed by WAXS measurements at varying temperatures, as shown in Figure 2.

Figure 2 shows that at certain temperatures the layer spacing suddenly jumps and this jump temperature is dependent on side chain length of C_m-PAMs. The shorter the side chain is, the higher is the jump temperature. These jump temperatures are well coincident with the crystal-mesophase transition temperatures measured by DSC. This behavior indicates

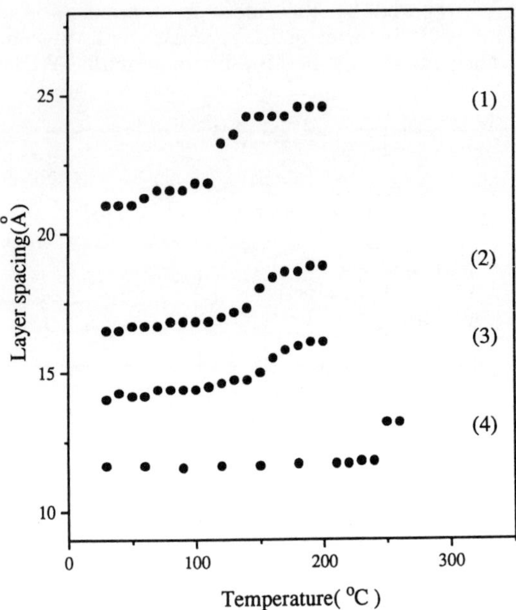

Figure2. Temperature dependence of layer spacing of C_m-PAMs. (1) C_{12}-PAM (2)C_8-PAM (3)C_6-PAM (4)C_4-PAM

that the penetration of coil polymers into hairy-rod poymers having layered crystal structure can be traced by WAXS technique.

To make a molecular composite of C_m-PAMs, we mixed equal weight of C_{20}-eicosane to C_{12}-PAM at temperature above crystal-mesophase transition of C_{12}-PAM under vaccum and measured layer spacing at varying temperatures. These results are plotted in Figure 3. It shows that both layer spacings of C_{12}-PAM and C_{12}-PAM/C_{20}-eicosane blend jump around 120°C. About the same jump temperature indicates that no attractive interaction exists between the side chain crystal and the molten eicosane. Figure 3 demonstrates also that the jump in layer spacing of C_{12}-PAM/C_{20}-eicosane blend is much larger than that of pure C_{12}-PAM. This indicates that C_{20}-eicosane molecules have penetrated into the side chain domain of C_{12}-PAM mesophase, in which side chain crystals are molten, but the layered structure is still maintained. In a separate study it was found that the lateral layer spacing does not change on blending of C_{20}-eicosane. This means that at temperatures higher than the jump temperature a pseudo molecular composite has been formed, because the backbone layers stacked over each other are not disrupted yet, and the blend system is not completely amorphous, but C_{20}-eicosane has been penetrated to and molecularly dispersed in C_{12}-PAM mesophase.

Figure 3 shows also that in 1/1(wt.) C_{12}-PAM/C_{20}-eicosane blend layer spacing does not increase any more when it has reached a maximum value. This asymptotic behavior could be confirmed to be present also in 1/2(wt.) blend, but with the same numerical value. This result indicates that both in 1/1 and 1/2 blends only a portion of C_{20}-eicosane has penetrated into side chain domain of C_{12}-PAM. This portion is equivalent to the solubility limit.

In Figure 4 are plotted the measured layer spacing against number of carbon atoms in C_k(n-alkanes). It shows that with increasing chain length of C_k solubility limit decreases. This behavior results from an entropy effect that with increasing molecular weight of C_k the entropy of mixing decreases. From Figure 4 it is also to see that with increasing side chain length solubility limit increases, indicating that higher side chain content can hold larger quantity of C_k.

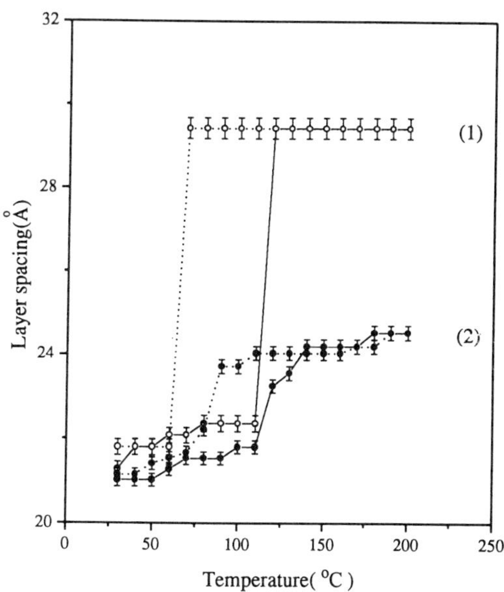

Figure 3. Temperature dependence of layer spacing of 1/1 C_{12}-PAM/C_{20}-eicosane blend(1) and C_{12}-PAM(2). Solid line denotes heating and dotted line does cooling .

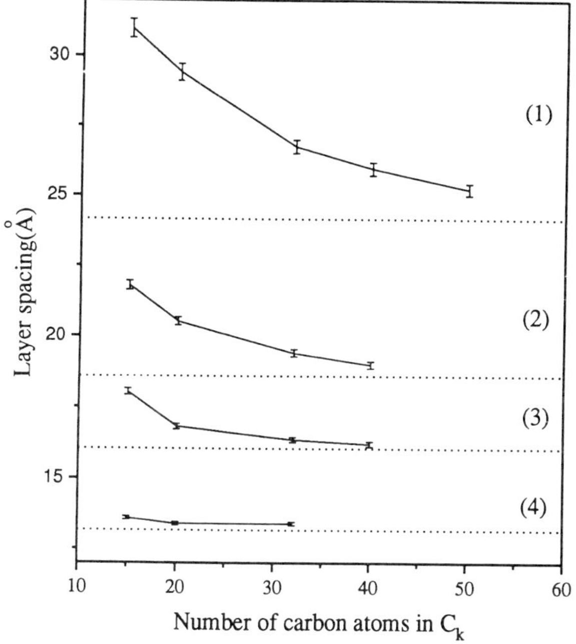

Figure 4. Equilibrium layer spacing of C_k/C_m-PAM blends plotted against number of carbon atoms in C_k-alkanes. The solid lines denote (1) C_k/C_{12}-PAM at 150°C (2) C_k/C_8-PAM at 180°C (3) C_k/C_6-PAM at 200°C (4) C_k/C_4-PAM at 260°C. The dotted lines denote C_m-PAM; C_{12}-PAM(top) at 150°C, C_8-PAM at 180°C, C_6-PAM at 200°C and C_4-PAM(bottom) at 260°C.

When we put equal weights of PPG-425 to C_8-PAM and measured changes in layer spacing, eventually the same behaviors as C_k blends were observed, i.e., asymptotic increases in layer spacings, but no changes in lateral layer spacings.

From the equilibrium layer spacing experiments described above for C_m-PAM blends with coil polymers such as C_k or PPG-425, it can be summarized that due to a strong backbone-backbone attraction of C_m-PAMs no true but pseudo molecular composites can be obtained only in such high temperature ranges that mesophase structures of C_m-PAMs are maintained. In equilibrium states the restricted quantities of coil polymers are penetrated into side chain domains of C_m-PAMs and increases in layer spacings are observed, but no changes in lateral layer spacings can take place.

C_m-PY-PI COMPOSITES

Taking into account the chemical structure of repeat unit of C_m-PY-PIs (m = 4, 6, 8), it is readily to expect that main chain of C_m-PY-PIs must be much more rigid than that of C_m-PAMs, since three aromatic coplanar rings are fused to form the pyromellitimide ring and the π-electrons are delocalized along the main chain[16]. Therefore, backbone-backbone attraction in C_m-PY-PIs must be much stronger than that of C_m-PAMs.

When we put equal weights of C_{20}-eicosane to C_4- and C_8-PY-PIs and measure layer spacing at various temperatures by WAXS, we obtain a result as reproduced in Figure 5. It shows that in neither of C_4- nor C_8-PY-PI there is a change in layer spacing up to 200 °C. In a separate experiment it was confirmed that no change takes place also in lateral layer spacing.

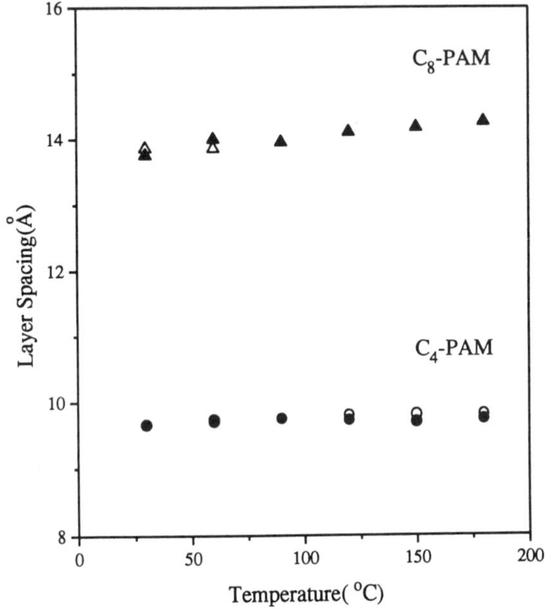

Figure 5. Layer spacing changes on temperature for C_m-PY-PIs and their blends with C_{20}-eicosane. Open triangle or circle denote pure C_m-PY-PIs and solid ones do the blends.

These experimental results indicate that in C_m-PY-PIs the backbone-bakcbone interaction not only to the lateral direction but also to the parallel direction is so strong that up to 200 °C the side chains are frozen and immobile and any coil polymers can penetrate into neither main chain layers nor side chain domains. Therefore, no successful true or pseudo molecular composites can be made from C_m-PY-PIs and the coil polymers used at all.

C_m-PAO COMPOSITES

As reported earlier[11-15], C_m-PAOs (m = 0, 1, 4, 8) have relatively low molecular weights and their layered crystalline structures are only loosely developed, as confirmed by WAXS investigations. This loose structure may be ascribed to the nonlinear structure of (1,2,4-oxadiazole)-3,5-diyl link[11]. C_m-PAOs exhibit crystal-liquid phase transition at eleverated temperature and isotropization points of C_m-PAOs are depressed as m increases.

When we add C_m-PAOs to C_{20}-eicosane less than 1 wt.%, transparency could not be observed up to 200°C. This immiscibility was contrary to our expectation, but we came later to a belief that this immiscibility between C_m-PAOs and C_{20}-eicosane up to 200 °C might be caused by a too large difference in polarity of the two polymers. C_{20} -eicosane is an aliphatic hydrocarbon, and is very hydrophobic, but in the repeat units C_m-PAOs have an oxadiazole ring, which must be quite polar on account of presence of two electronegative nitrogen atoms and one oxygen atom.

When we add any kind of PSs, PEGs or PPGs to C_0-PAO, no miscibility could be observed up to 240 °C. The immiscibility of C_0-PAO blends is natural to understand, since it does not have any side chains to make the blends miscible. In C_0-PAO, its blending behavior is the same as C_m-PY-PIs due to the very strong backbone-backbone interaction.

When we add 1wt.% of C_4-PAO to PPG-4000 and measured light scattered intensities at various temperatures, we obtain a result as reproduced in Figure 6. The heating curve of Figure 6 shows that the light scattered intensity suddenly falls at a temperature around 160 °C and the blend becomes transparent. This critical temperature is called isotropization temperature. In Figure 6 the cooling curve is also reproduced to show that this process is reversible. This isotropization process could also be confirmed by polarized microscopic investigations. The result obtained from Figure 6 indicates that a true molecular composite is obtained from the 1wt.% C_4-PAO/PPG-4000 blend at temperatures higher than 160 °C upon heating.

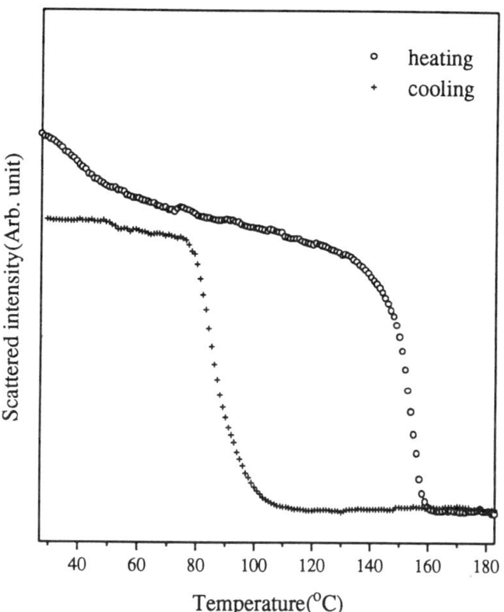

Figure 6. Intensity of scattered light for 1wt% C_4-PAO/PPG-4000 blend as a function of temperature.

If we measure isotropization temperatures for C_4-PAO blends with PPGs having various molecular weights and plot them against C_4-PAO weight fractions, Figure 7 is obtained. The curves of Figure 7 are nothing else than phase diagrams of C_4-PAO/PPG-4000 blends of low concentration. From Figure 7 it is to see that with increasing C_4-PAO content the isotropization temperature increases. Figure 7 also shows that isotropization temperature increases with increasing PPG molecular weights. This behavior resulting from an entropy effect indicates that C_4-PAO forms true molecular composites at lower temperatures with lower molecular weight PPGs and at higher temperatures with higher molecular weight PPGs.

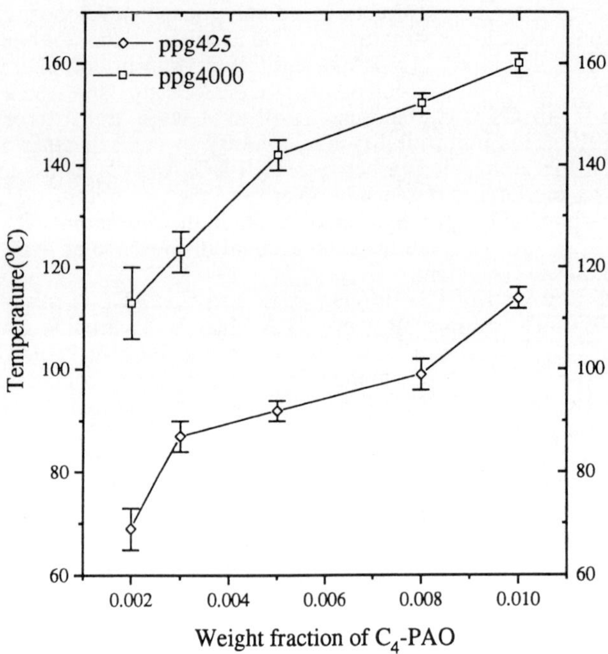

Figure 7. Changes in isotropization point against C_4-PAO wt. fraction in PPGs/C_4-PAO blends.

In Figure 8 are reproduced DSC thermograms of C_4-PAO/PPG-4000 blends. Figure 8 shows that with decreasing wt % of C_4-PAO in the blend its melting point decreases. This indicates that at melting temperature PPG-4000 penetrates not only into side chain domain but also into main chain layers, disrupting the layered crystalline structure.

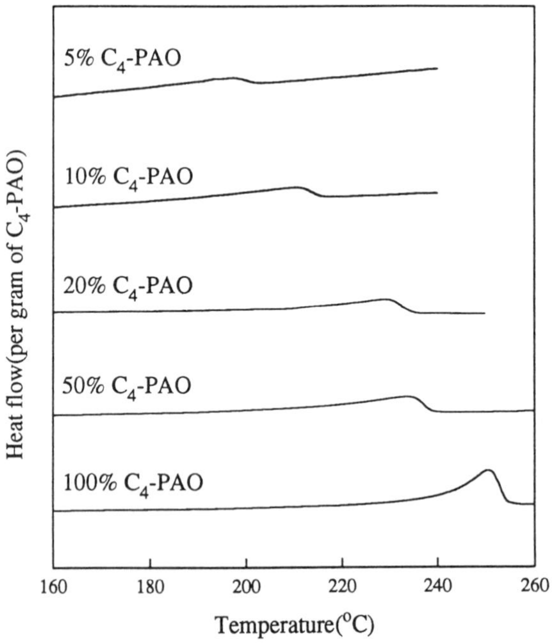

Figure 8. DSC thermograms of PPG-4000/C_4-PAO blends. Heat flow is normalized by the weight of C_4-PAO in blends.

The formation of true molecular composites could also be confirmed by WAXS studies. At lower temperatures than isotropization points the presence of layered structure of C_4-PAO could be detected, but at higher temperatures than isotropization points no layers were present at all, and only amorphous halo was observed.

In forming true molecular composites PEGs were found to be more effective than PPGs. It is evidenced by the fact that, as to see from Figure 8, PPG-4000 exhibits an isotropization temperature at 160 °C in 1wt.% C_4-PAO/PPG-4000 composite, but PEG-4600 brought an isotropization temperature at 110 °C for the same 1wt%, as measured in a separate experiment. Although the PEG has a bit higher molecular weight than the PPG, the PEG has an isotropization point 50 °C lower than the PPG. This result means that PEG is more compatible than PPG. The greater compatibility of PEG might arise from the fact that PEG has more polarity than PPG.

When we mix PEGs and PPGs with smaller quantities of C_8-PAO than 1 wt %, it was found from light scattering experiments that all the blend systems are completely isotropic even at room temperature. This means that side chain length of C_8-PAO is enough long and side chain content is accordingly enough high to make the blend systems completely miscible on molecular level and true molecular composites are obtainable from C_8-PAO blends with PPGs or PEGs even at room temperature. If this room temperature isotropization of the C_8-PAO composites is compared with elevated isotropization points of C_4-PAO composites, it is to see how noticeable the effect of side chain length is.

CONCLUSIONS

Formation of hairy-rod molecular composites could be satisfactorily characterized by WAXS and light scattering techniques.

In C_m-PY-PI blends with some coil polymers investigated the backbone-backbone attraction is so strong that independent of side chain length of C_m-PY-PIs, neither true nor pseudo molecular composites were obtainable up to 200 °C. In C_m-PAM blends the attraction is so lowered that the coil polymers can penetrate only into side chain domain to form pseudo molecular composites. In C_m-PAO blends the attraction is far much lowered that true molecular composites are obtainable, if side chain lengths of C_m-PAOs are not too short, and the isotropization depends much on their side chain length.

In conclusion, to obtain successful molecular composites from mixtures of hairy-rod polymers with coil poymers, the rod-rod attraction and the side chain length should be in the proper range and the structure of the coil polymers should be properly designed.

Acknowledgments: The authors are thankful for the kind financial support from Korea Science and Engineering Foundation (92-48-00-01)

REFERENCES

1. W.-F. Whang, D.R. Wiff, and T.E. Helminiak, *J. Macromol. Sci.-Phys.* B22:231 (1983)
2. S. Takayanagi, *Polym. Prepr. Jpn.* 36:3060 (1987)
3. S.J. Kruase and T.E. Helminiak, *J. Polym. Sci. Polym. Phys. Ed.* 84:1991 (1986)
4. D.R. Wiff, S. Timms, T.E. Helminiak, and W.-F. Whang, *Polym. Eng. Sci.* 27:424 (1987)
5. M. Ballauff, *Angew. Chem. Int. Ed. Engl.* 28:253 (1989)
6. J.F. Wolfe, B.H. Loo, and F.E. Arnold, *Macromolecules* 4:915 (1981)
7. C. Tanaka, *Kobunshi* 38:893 (1989)
8. J.F. Wolfe and F.E. Arnold, *Macromolecules* 4:909 (1981)
9. J. Burkett, Polym. Prepr. 28(2):278 (1987)
10. S.-B. Park, H. Kim, W.-C. Zin, and J.C. Jung, *Macromolecules* 26:1627 (1993)
11. S.-B. Park and J.C. Jung, *Makromol. Chem.* 193:2519 (1992)
12. E.-J. Choi and J.C. Jung, *Polym. J.* 24:121 (1992)
13. J.C. Jung and E.-J. Choi, *Angew. Makromol. Chem.* 197:72 (1992)
14. S.-B. Park and J.C. Jung, *Angew. Makromol. Chem.* 208:97 (1993)
15. S.-B. Park and J.C.Jung, *J. Polym. Sci. Polym.Chem. Ed.* 31:3351 (1993)
16. J.C. Jung and S.-B. Park, *J. Polym. Sci. Polym. Phys. Ed.* submitted

VIBRATION CONTROL IN FIBRE-REINFORCED COMPOSITE STRUCTURES

Roslan,A.R., and Dr. M.S. Leong

Faculty of Mechanical Engineering
Universiti Teknologi Malaysia
Locked Bag 791
Sekudai, 80990 Johor Bharu
Malaysia

INTRODUCTION

Machines inevitably generate unwanted vibration which are then transmitted into the substructures via their mountings. The resulting vibration subsequently transmits via various propagation waves to adjoining structures or regions where structure-borne noise radiation occurs. It is usually necessary to control excessive vibration such as to avoid potential fatigue failure to structures. Attempts can been made to control vibration at the point of vibration generation to the substructure to reduce the transmission of vibrational power flow to remote areas.

Previous studies had been undertaken on steel, alluminium and concrete structures by others to control the vibration transmission resulting from forces, moments and torque excitations[1,2,3,4]. The study of vibration control using power flow concept has been commonly used. Review of this technique is well documented by White[5].

This paper reports on the application of power flow concept to substructure of fibre-reinforced composite materials. The objective of the study was to evaluate potential control of vibrational power input into the structure by the control of material properties and structural configuration. Consideration is limited to flexural type of waves motion in beam structures.

VIBRATIONAL POWER INPUT TO A BEAM

A structure subjected by harmonic force, **F**, will have input power in proportion to the real part of the structure's mobility[1], β,

Polymers and Other Advanced Materials: Emerging Technologies and Business Opportunities. Edited by P. N. Prasad *et al.*, Plenum Press, New York, 1995

$$P = 1/2\ |F^2|\ \mathrm{Re}\ (\beta). \tag{1}$$

If the structure is excited by a random force of spectral density G_{FF}, the power per unit frequency is thus obtained as [2],

$$P/Hz = G_{FF}\ \mathrm{Re}\ (\beta). \tag{2}$$

Skudryzk[6] has shown that a finite structure may be approximated as an equivalent structure of infinite length with no reflecting devices, thereby allowing a convenient description of the average behaviour of the structure. Based on similar simplification, and using Euler-Bernouli beam theory, the power input to an infinite fibre-reinforced composite beam subjected by harmonic bending force and moment excitations can be respectively expressed as :

$$P_F = |F|^2 / [\ 8\ \omega^{1/2}\ (E_x I)^{1/4}\ (\rho A)^{3/4}\ ], \tag{3}$$

$$P_M = |M|^2\ \omega^{1/2} / [\ 8\ (E_x I)^{3/4}\ (\rho A)^{1/4}\ ], \tag{4}$$

where A is the beam's cross-sectional area, ρ the density, I the second moment of area, and E_x the bending equivalent Young's modulus elastic constant in the x-direction. Based on ply stiffness analysis in transfomation of elastic constants, the equivalent Young's modulus E_x for a singly oriented ply laminate, is derived as[7],

$$\frac{1}{E_x} = \frac{m^4}{E_1} + \frac{n^4}{E_2} + m^2 n^2 \left(\frac{1}{G_{12}} - 2\frac{\nu_{12}}{E_1}\right), \tag{5}$$

where m = cos(θ) and n = sin(θ), with θ as the angle of fibre direction with respect to beam coordinate axis. For a multioriented ply laminate, such as cross-ply and angly ply laminates, the equivalent Young's modulus is derived as [8],

$$E_x = 12/(h^3\ D'11), \tag{6}$$

with D'11 as laminate compliance. For a symmetric laminate, in which all coupling stiffness terms B_{ij} are zero, the compliance is given as $D'11 = D_{22} / [D_{11}\ D_{22} - D_{12}]$, where D_{ij} are bending stiffness terms. This is derived generally as,

$$D_{ij} = \sum_{p=1}^{N} 1/3(z_p^3 - z_{p-1}^3)(Q'_{ij})_p \tag{7}$$

For a p-layered laminate, ply p has cordinates z_p for its top surface and z_{p-1} for its bottom surface measured from the laminate midplane and Q'_{ij} is the ply transformed reduced stiffness for a given ply angle. The Q'_{ij} expression as generally derived in standard text books (for example, Refs. (7) and (8)) is a function of ply angle and reduced stiffness terms Q_{ij}, and are expressed as :

$Q_{11} = E_1 / (1 - \nu_{12}\ \nu_{21})$ $\qquad$ $Q_{22} = E_2 / (1 - \nu_{12}\ \nu_{21})$

$Q_{33} = G_{12}$ $\qquad$ $Q_{12} = \nu_{12}\ E_1 / (1 - \nu_{12}\ \nu_{21})$

Table 1. Property of E-Glass fibre and Polyester resin[9]

Material	E_f (GPa)	E_m (GPa)	ν_f	ν_m	γ_f	γ_m	C	K
E-Glass/ Polyester Resin	69	3.45	0.22	0.33	2.5	1.2	0.2	1.0

E =Young's modulus, ν = Poisson's ratio, γ = Specific gravity, C = Fibre contiguity K = Straightness factor. Subscript f and m refer to fibre and matrix respectively.

MATERIAL AND LAMINATION

An analytical study undertaken for typical laminates to compute vibrational power input to infinite beam with a given excitation is reported herewith. The variables of the laminates examined include type of ply, fibres orientation angle, resin weight fraction, cross-ply ratio, fibre contiguity and straightness, influence of short-fibre composite, and influence of beam configuration.

The material chosen for this analysis was a typical E-Glass fibre in polyester resin commonly used in boat building[9].The material property is listed in Table 1. The beam length was considered infinite, with a width of 24 mm. All laminates were assumed as symmetrical in the midplane; so as to eliminate bending-membrane coupling. The laminate details are shown in Table 2. The general plies arrangement of $(0/-60/60)_S$ and $(0/30/45)_S$ have been used by Schultz and Tsai[10] in studying the dynamic moduli of fibre reinforced composite material. The type of cross-ply laminates of (0/90/0) and (0/90/0/90/0) plies and general ply of $(0/90/45/-45)_S$ have been used by Adams and Bacon[11] in investigating the effect of fibre orientation and the dynamic properties of carbon fibre-reinforced plastic.

Exact solutions of the elasticity approach to the micromechanics of composite materials were used for the determination of moduli and poisson's ratio. The moduli and Poisson's ratio were computed with Tsia's expression[12] for unidirectional and cross-ply, and Manera's expression[13] for randomly oriented short fibres.

UNIDIRECTIONAL PLIES

Analytical results for the variation of vibrational power input associated with flexural force and moment excitations with frequency as computed from Eqs. (3) and (4), are plotted in Fig. (1) for singly oriented, unidirectional ply. The power input associated with force excitation decreased linearly and associated with moment excitation increased linearly on a logarithmic scale with increasing frequency.[1]

Results of the effect of variation in material properties and beam configuration on the vibrational power input to singly oriented, unidirectional ply, laminated beam are discussed as follows.

Effect of Fibre Orientation Angle

It was observed that increasing the ply fibre orientation angles resulted in an increase in vibrational power input for both force and moment excitations; as illustrated in Fig.(1). Equations (3) and (4) suggest that for a constant cross-sectional area and density, the

Table 2. Details of beam laminate

Lay-up	Thickness of each layer (mm)	Total thickness (mm)
Unidirectional (+θ)	-	3
Cross-plies (CP)		
i) (0/90/0)	See Table 3. for details	2.5
ii)(0/90/0/90/0)	Same as above	3.0
General Ply		
i) (θ/-θ/θ)	(0.5/2.0/0.5)	3.0
ii) $(0/-60/60)_S$	$(0.5/0.5/0.5)_S$	3.0
iii) $(0/30/45)_S$	$(0.5/0.5/0.5)_S$	3.0
iv) $(0/90/45/-45)_S$	$(0.5/0.5/0.25/0.25)_S$	3.0
Chop Strand Mat (CSM)		
i) CSM	-	3.0
ii) CSM + CP (0/90/CSM/90/0)	$(0.25/0.25/2.0/0.25/0.25)_T$	3.0
iii) CSM + UD (0/CSM/0)	$(0.5/2.0/0.5)_T$	3.0

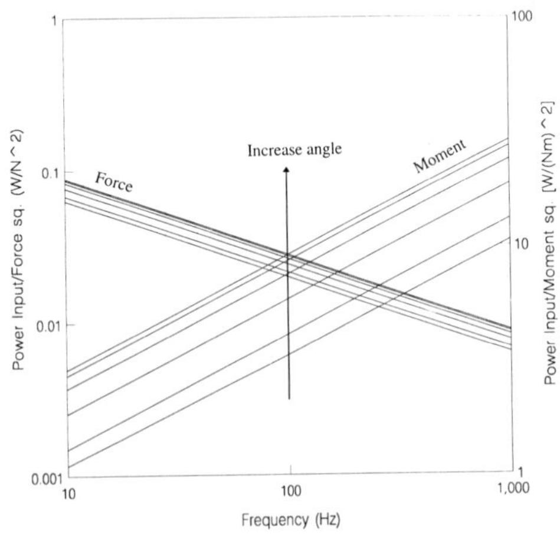

Figure 1. Vibrational power input associated with bending force and moment on infinite beam of singly oriented unidirectional ply for the frequency range of 10 Hz to 1 kHz. (θ = 0^0, 15^0, 30^0, 45^0, 60^0, 90^0).

vibrational power input is inversely proportional to the equivalent Young's modulus. To ensure minimum power input, the structure should have a high equivalent modulus. Figure (2) shows the variation of equivalent Young's modulus in the x-direction with fibre orientation angle. The shape of the curves is similar to that obtained by Ni and Adams[14]. It was thus observed that high equivalent Young's modulus can be obtained by employing lower orientation of fibre angle.

Changes in power input associated with bending force excitation were relatively small compared to power input associated with moment excitation as fibre orientation angles were increased from 0° to 90°. Numerical calculations showed that the power input associated with force increased by a factor of 1.4 as compared to a factor of 2.75 for a moment excitation whose fibre orientation angles were increased from 0^0 to 90^0. The equivalent Young's modulus raised to the power of quarter for force excitation, compared to three-quarter for moment excitation, are evident from Eqs.(3) and (4). High changes in equivalent Young's modulus accordingly resulted in small changes in the power input associated with force excitation due to the equivalent modulus being numerically reduced significantly. It was noted that the changes in power input with frequency for the angles of 60^0, and 90^0 were almost the same due to similar values of equivalent Young's modulus.

Effect of Resin Weight Fraction, R.

A high equivalent Young's modulus can be further obtained by controlling the resin weight fraction, R. The vibrational power input increased as resin weight fractions were increased, as illustrated in Fig.(3). The power input can thus be further minimised by having low resin weight fraction resulting in a high equivalent modulus.

Effect of Fibre Contiguity

In the fabrication of fibrous composites the fibres are contiguous, i.e. fibres touch each other rather than being entirely surrounded by matrix material. In many instances however, the fibres do not touch. Determination of moduli based on Tsai expression[12], shows that the fibre contiguity, denoted by C, is independent of Young's modulus in the fibre direction. The power input associated with flexural wave motion thus has small influence on C as shown in Fig.(4) for fibre orientation angle of 30°. Fibre contiguity has no influence on power input associated with flexural wave motion when θ was set to 0^0.

Effect of Fibre Straightness or Alingment

Tsai[12] has modified the rule of mixtures to account for imperfections in fibre alignment. The alignment factor, K, strongly influence the value of moduli in the fibre direction and thus on the values of laminate compliance, D'ij. Wide separation of power input with straightness factor hence occured for flexural wave motion, as shown in Fig.(5). The power input decreased as fibre straightness factor is increased, thereby suggesting that fibres should be properly aligned for minimum power input

Effect of Beam Configuration

The power input associated with bending force and moment excitations is given by Eqs.(3.0) and (4.0). These equations can be written in term of laminate compliance to give,

$$P_F = |F|^2 (D'11)^{1/4}/[8\ \omega^{1/2}\ b\ (\rho h)^{3/4}] \qquad (8)$$

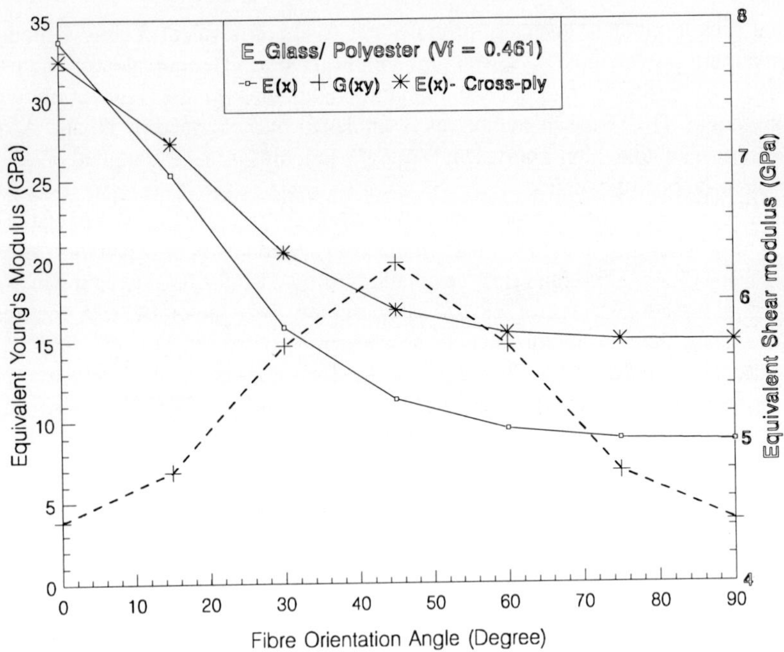

Figure 2. Variation of equivalent Young's modulus with fibre orientation angles.

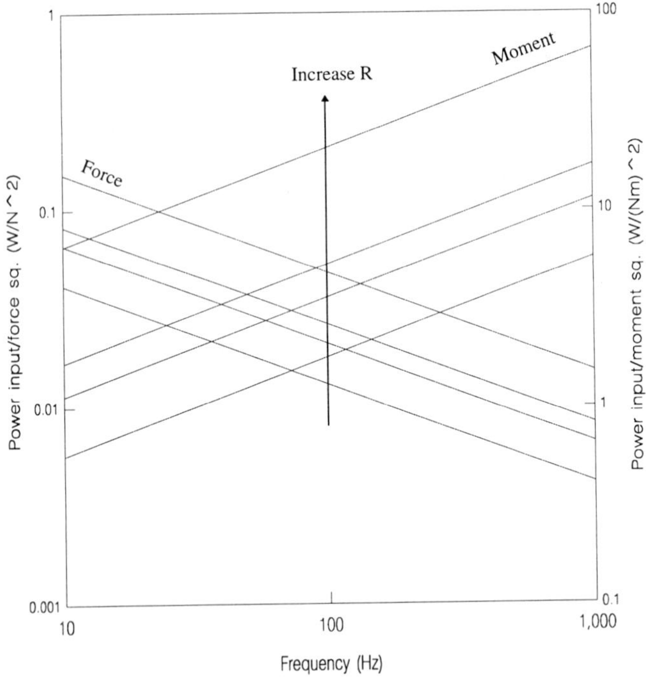

Figure 3. Vibrational power input per unit bending force and moment squares on infinite beam of 0^0 unidirectional ply at various resin weight fraction.(R= 0, 0.4, 0.6, 1.0)

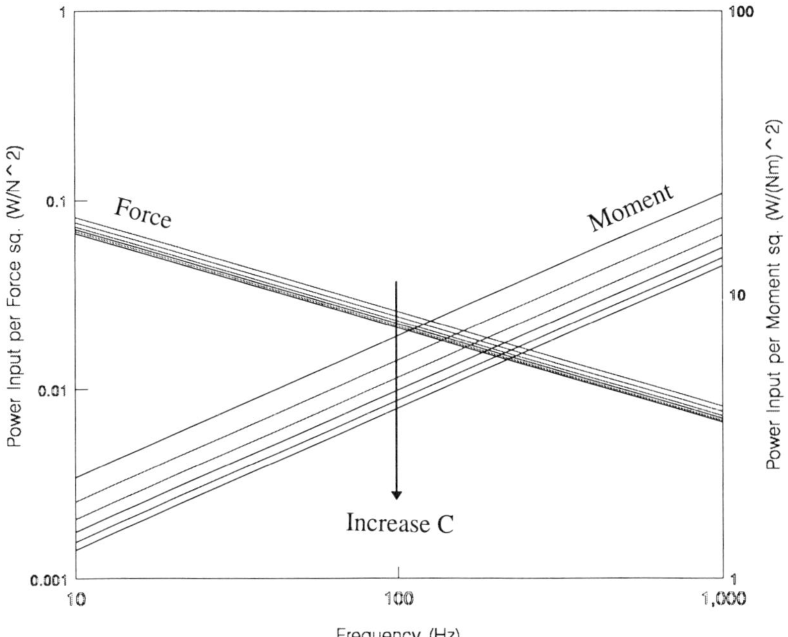

Figure 4. Vibrational power input per unit bending force and moment squares on infinite beam of 30^0 unidirectional ply at various fibre contiguity factor. (C= 0.0, 0.2, 0.4, 0.6, 0.8, 1.0)

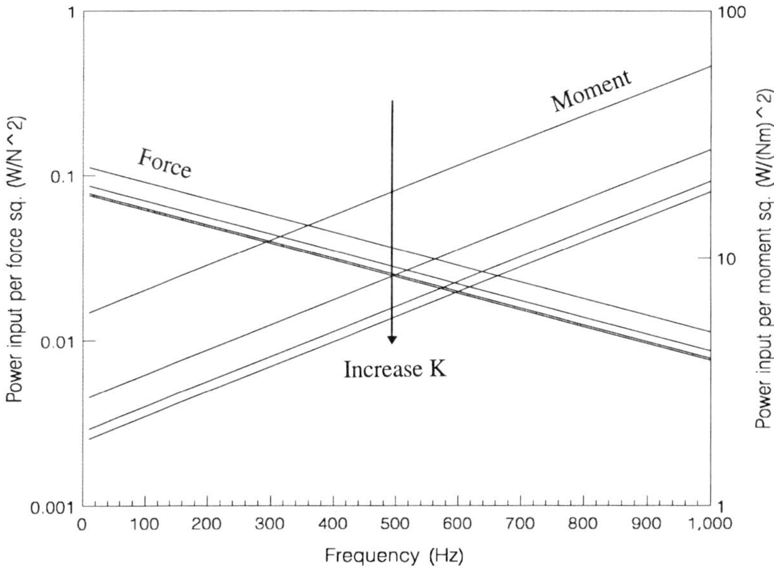

Figure5. Vibrational power input per unit bending force and moment squares on infinite beam of 30^0 unidirectional ply at various fibre straightness factor.. (K= 0.0, 0.2, 0.6, 1.0)

$$P_M = |T|^2\, \omega^{1/2}\, (D_{11}')^{3/4} / [8\, b\, (\rho h)^{1/4}].$$

The power input is therefore a complex function of beam thickness. The effect of beam thickness on the power input associated with bending force and moment excitations is shown in Fig.(6), for the case of 0^0 fibre orientation angle. The power inputs were seen to be linearly dependent to the beam thickness on a semi-log scale, which reduces as the beams thickness were increased. Comparatively, the beam thickness has a more significant effect on the moment component of power input. This could be further reduced by increasing the thickness of the beam structure. Equations (7) and (8) show that the bending force and moment component of power input are inversely proportional to beam width. Increase in beam width resulted in reduction in power input, as evident in Fig.(7).

The variation in beam configuration hence showed that minimum power input to an infinite beam of singly oriented unidirectional ply could be achieved by having thicker and wider beam for a given type of excitation. Numerical computations showed that with the same ratio of increase in beam thickness and width, thickness gave the highest reduction in power input. This suggests that thickness of the beam is a control parameter for a given beam configuration.

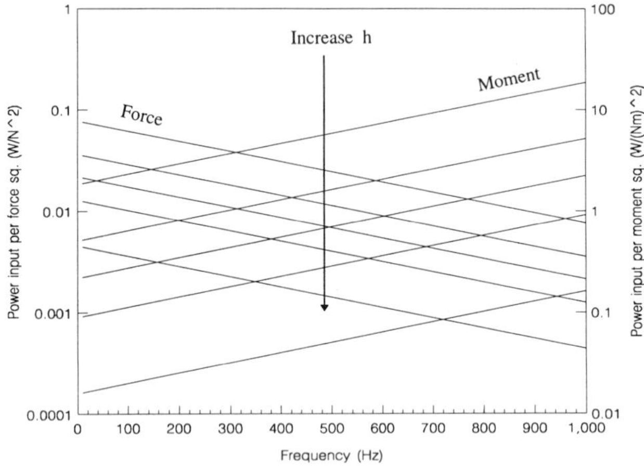

Figure 6. Vibrational power input per unit bending force and moment squares on infinite beam of 0^0 unidirectional ply at various beam thickness. (h= 3, 5, 10, 15, 20 mm)

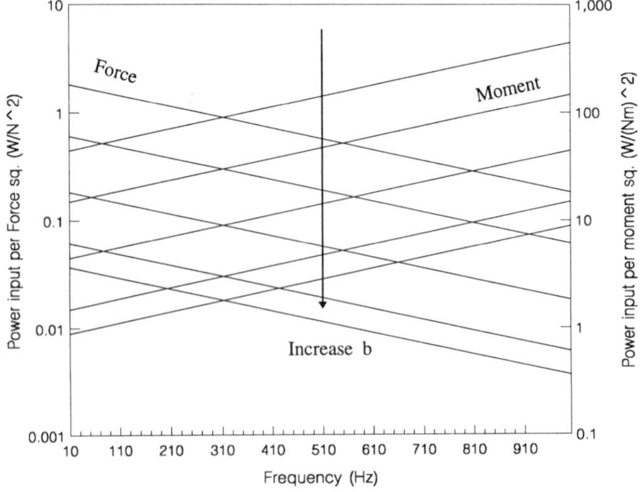

Figure 7. Vibrational power input per unit bending force and moment squares on infinite beam of 0^0 unidirectional ply at various beamwidth. (b = 1, 3, 10, 30, 50 mm)

CROSS PLIES

Vibrational power input into structures of cross-ply laminates are similarly computed as above. The dynamic properties of the cross-ply laminates are determined using Tsai expression. Two types of ply were considered : 3 layers of (0/90/0) ply and 5 layers of (0/90/0/90/0) ply in different lay-up of cross-ply ratios, as shown in Table 3.0.

The cross-ply ratio is defined as the ratio of total thickness of odd layers (0°) to the total thickness of even layers (90°). There are no bending/twisting coupling and streching /shearing coupling. Figures (8) and (9) show the vibrational power input associated with the bending force and moment, as computed from Eqs (3) and (4), for 3 and 5 number of layers respectively. It was observed that, the power input reduces as the cross-ply ratio increased for a given number of layer, thus implying that the total thickness of odd (0°) increased. This is consistent with the observation that an unidirectional ply of 0° angle gave the minimum power input. It was also observed that as the number of layers were increased from 3 to 5, with increase in thickness, the power input reduced accordingly. This may be due to the fact that the power input is inversely proportional to the thickness as evident from Eqs. (8) and (9).

Table 3. Cross-ply lay-up

Number of layer N	Cross-ply Ratio M	Layer Thickness (mm)	Total Thickness (mm)
3	0.25 (1/4)	(0.25/2.0/0.25)	2.5
(0/90/0)	0.666 (2/3)	(0.5/1.5/0.5)	2.5
	1.5 (3/2)	(0.75/1.0/0.75)	2.5
	4.0 (4/1)	(1.0/0.5/1.0)	2.5
5	0.5	(0.25/1.0/0.5/1.0/0.25)	3.0
(0/90/0/90/0)	1.0	(0.5/0.75/0.5/0.75/0.5)	3.0
	2.0	(0.75/0.5/0.5/0.5/0.75)	3.0
	5.0	(1.0/0.25/0.5/0.25/0.75)	3.0

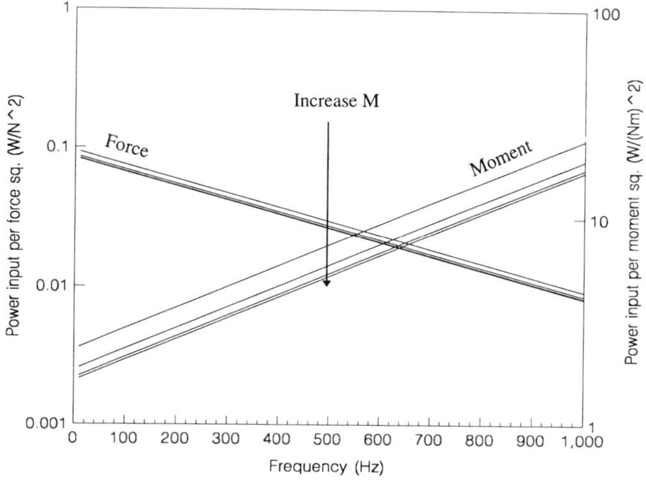

Figure 8. Vibrational power input associated with bending force and moment on infinite, cross-ply, laminated beam for N = 3, and M = 1/4, 2/3, 3/2, 4/1.

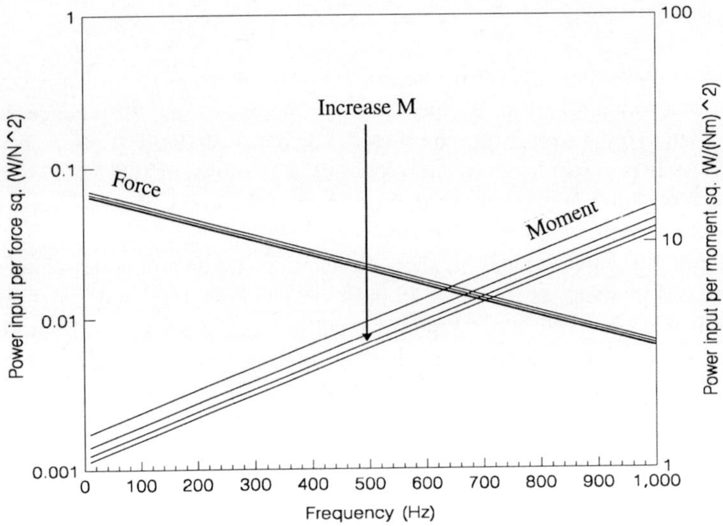

Figure 9. Vibrational power input associated with bending force and moment on infinite, cross-ply, laminated beam for N = 5, and M =0.5, 1.0, 2.0, 5.0

GENERAL PLIES

The results of symmetric general ply of (+θ/-θ/+θ) lay-up, with details given in Table 2, are shown in Figs. (10) and (11) . Both flexural force and moment excitations resulted in an increase in power input as outer layer fibre orientation angles were increased. The results of equal total thickness of $(0/-60/60)_S$, $(0/30/45)_S$ and $(0/90/45/-45)_S$ plies, which were used by Schultz and Tsai[10] , and by Adams and Bacon[11] respectively, are also shown. These plies gave almost similar values of power input at all frequencies. It was also noted that these plies were almost identical with the (+15/-15/+15) ply.

7.0 CHOP STRAND MAT (CSM)

The properties of the E-Glass CSM fibre and resin used are similar to those given in Table 1. The moduli were estimated using the expression proposed by Manera.[13] The computed power input was plotted as shown in Fig. (13). Results of cross-ply laminate (of Fig.(9)) and 0^0 unidirectional laminate (of Fig.(1)) which had equal total thickness as CSM, are similarly plotted in Fig.(13). The power input for CSM laminate gave the highest power input. The power could however be reduced by combining CSM with cross-ply and with 0^0 unidirectional ply, as illustrated in Fig.(14). The lowest power input ws obtained from structures having the combination of CSM and 0^0 unidirectional ply. This was due to the unidirectional ply having a higher equivalent Young's modulus as compared to other plies

EXPERIMENTAL STUDY

It has been established by Goyder and White[1] that the time averaged vibrational power input to infinite beam was dependent on the real component of point mobility. An

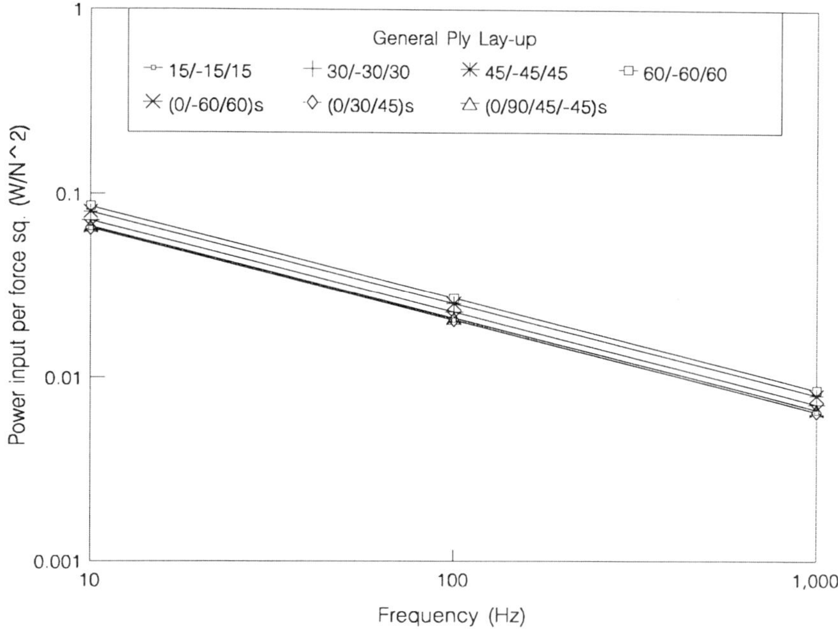

Figure 10. Vibrational power input per unit bending force square on infinite beam of general ply laminate

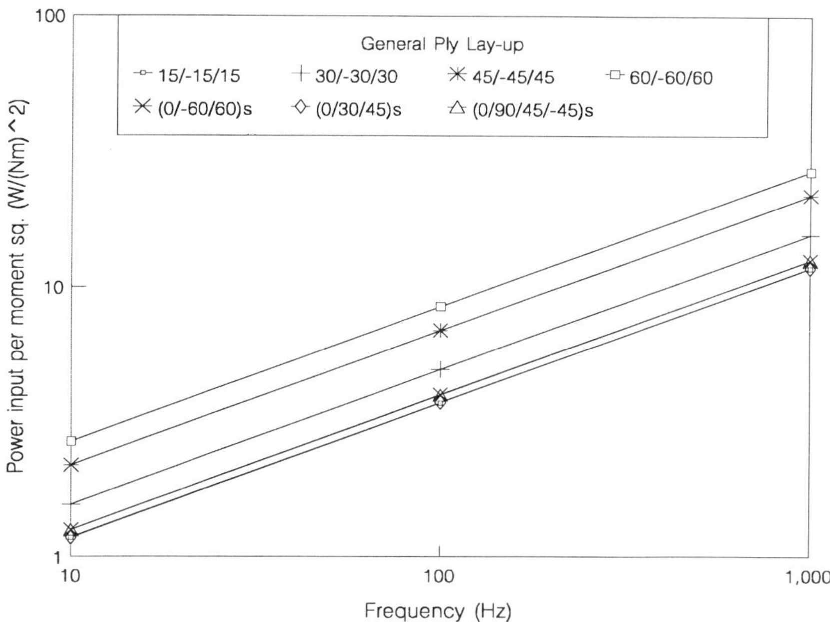

Figure 11. Vibrational power input per unit bending moment square on infinite beam of general ply laminate.

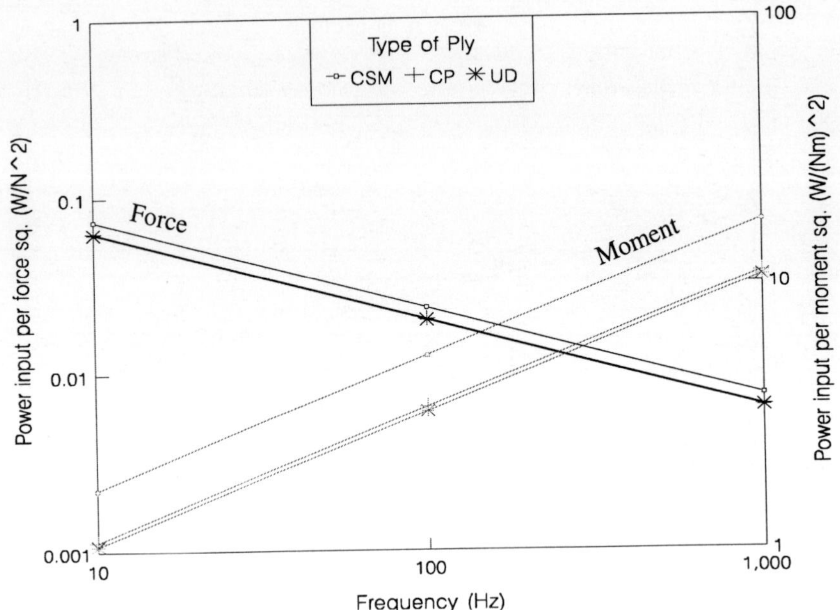

Figure 12. Vibrational power input associted with bending on infinite beams of CSM, cross-ply, 0^0 unidirectional ply.

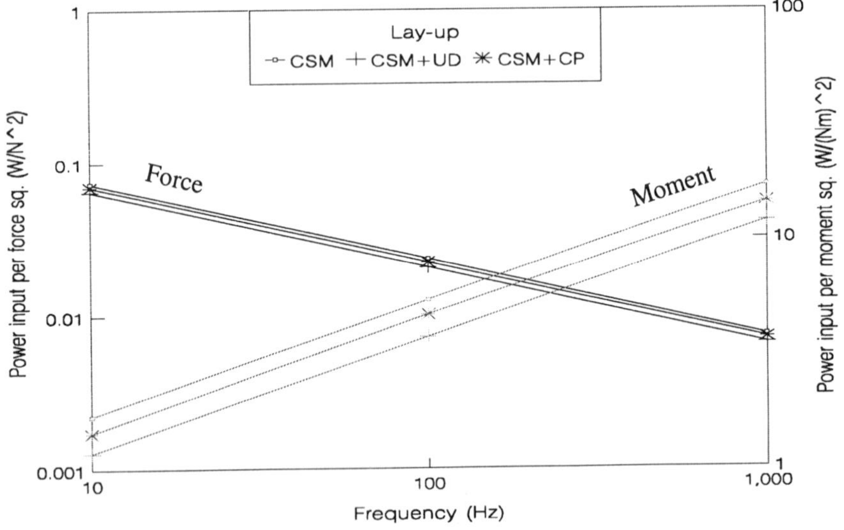

Figure 13. Vibrational power input associated with bending on infinite beam of CSM, CP+CSM, and UD+CSM.

experimental study was carried out to measure the point mobility, and to verify its variation with changes in fibre orientation angle.

Test specimens used consisted of a combination of E-Glass fibre of unidirectional woven roving (ERW-430-551C, 650 g/m^2) and polyester resin (268 BQT). The beam specimens were cut at various orientation of fibre angle, as tabulated in Table 4. The beams were subjected with random force signal at one end and the resulting wave motions in beam were attenuated at the other end by embedding this other end in a tapering aerated rubber sponge. The response was measured by a miniature accelerometer. Figures (14) and (15) illustrate typical results of modulus point mobility for various fibre orientation angles. Both groups of specimen gave similar results. The averaged behaviour of point mobility at various fibre orientation angle were almost identical. This correlates well with the analytical predictions as discussed above. These results have a significant practical implication which suggest that there is no need for fibre orientation control in the manufacture of engineering structures for a minimal power input.

Table 4. Dimensions and physical constant of beam specimens
(Material : E-Glass Unidirectional woven roving mat/Polyester resin)

Plate No.	Specimen No.	Orientation Angle	Resin Wt. Fraction	Dimensions in mm	Mass per length (kg/m)
1	A1	0.75^0	0.387	981 x 24.845 x 5.843	0.2547
	A2	26.75^0		981 x 24.867 x 5.827	0.2542
2	A3	58.9^0	0.367	981 x 24.737 x 5.886	0.2584
	A4	87.75^0		981 x 24.98 x 5.8380	0.2600
1	B1	1.5^0	0.387	920 x 14.99 x 5.772	0.1559
	B2	26.87^0		920 x 15.06 x 5.932	0.1585
	B3	47.4^0		920 x 15.00 x 5.868	0.1565
2	B4	58.83^0	0.367	920 x 14.93 x 5.832	0.1570
	B5	89.5^0		920 x 14.98 x 5.844	0.1568

CONCLUSION

The vibrational power input to fibre-reinforced composite beams subjected to bending type of excitation was studied, and influences of fibre orientation angle, resin weight fraction, cross-ply ratio and short fibre were examined analytically and experimentally. This study therefore confirms the following :

i) The vibrational power input to infinite beam structures may be controlled by using unidirectional off-axis laminates of high equivalent Young's moduli. This can be achieved by setting the fibre orientation angle at 0^0. The equivalent Young's modulus may be further increased by using a low resin weight fraction.

ii) The vibrational power input to cross-ply laminated beam could be reduced by employing a high cross-ply ratio.

iii) The use of 'pure' chop strand mat gave a larger vibrational power input as compared to cross-ply and unidirectional ply. The power input to the 'pure' chop strand mat laminate may be reduced by combining with unidirectional ply.

iv) The type of ply that gave a minimum power input was found to be 0^0 unidirectional ply.

v) Experimental results on variation of fibre orientation angles with point mobility showed that there were minimal differences in the average behaviour of mobility. This therefore significantly suggest that there is no constraint imposed on fibre orientation angle of laminated structures for vibration control.

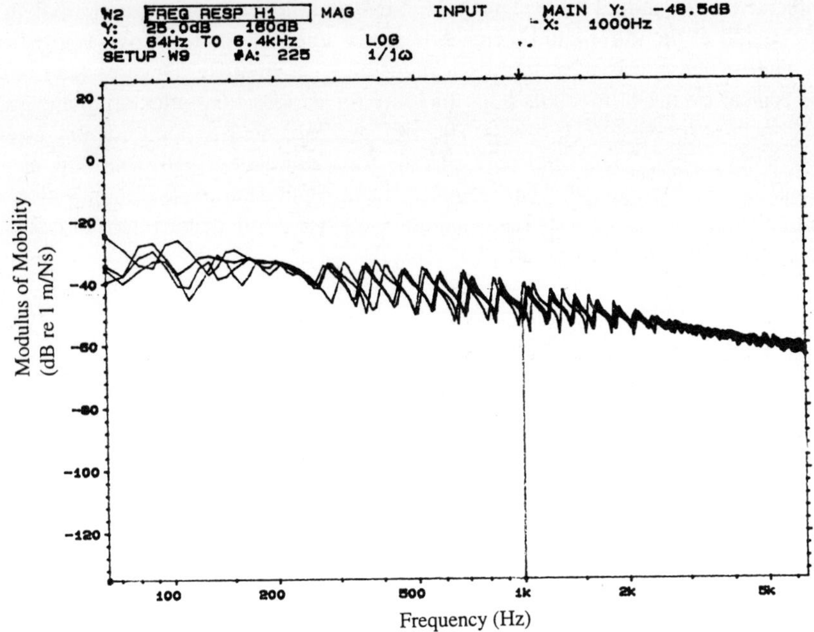

Figure 14. Measured plot of point mobility modulus of beams A1, A2, A3, A4.

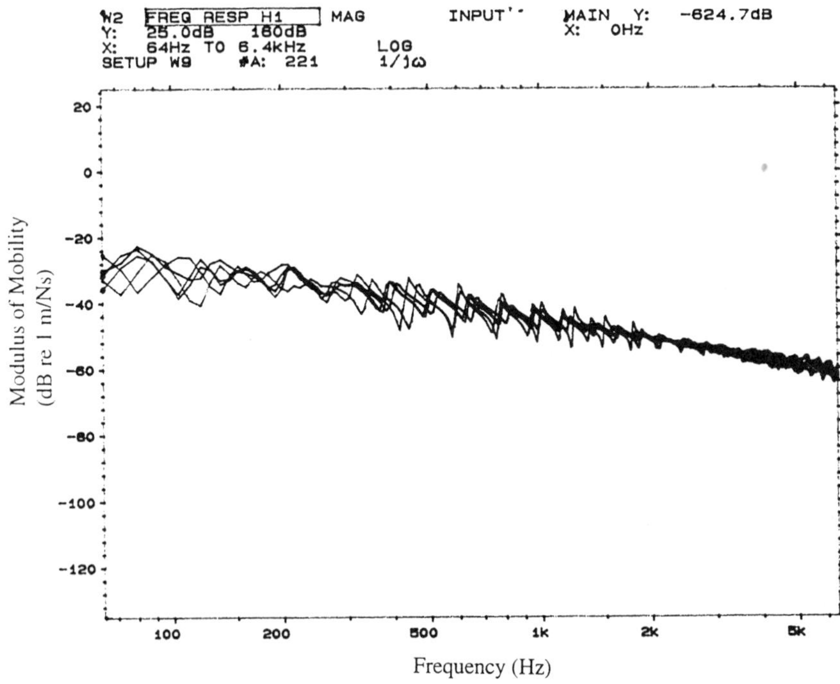

Figure 15. Measured plot of point mobility of beam B1, B2, B3, B4, and B5.

REFERENCES

1. H.D.Goyder and R.G.White, Vibrational power flow from machines into built-up structures. Part I:Introduction and approximate analyses of beam and plate-like foundations, *Journal of Sound and Vibration*, 68(1):59-75 (1980).
2. R.J. Pinnington and R.G. White, Power flow through machine isolators to resonant and non-resonant beams, *Journal of Sound and Vibration*, 75(2):179-197 (1981).
3. B. Petersson and J. Plunt, On effective mobilities in the prediction of structure-borne sound transmission between a source structure and a receiving structure, Part I: Theoretical background and basic experimental studies, *Journal of Sound and Vibration*, 82(4):517-529 (1982).
4. M.F. White and K.H. Liasjo, Measurement of mobility and damping of floors, *Journal of Sound and Vibration*, 81(4):535-547 (1982).
5. R.G. White, Vibration control of machinery installations and structures : Some design procedures and experimental diagnostic techniques, *Vibration and Noise Conference*, Melbourne, :1-9 (1990)
6. E. Skudrzyk, Vibrations of a system with a finite or an infinite number of resonances, *Journal of the Acoustical Society of America*, 30:1140-1152 (1958).
7. R.M. Jones, "Mechanics of composite materials", McGraw Hill, U.S.A. (1975).
8. M.H. Datoo, "Mechanics of fibrous composites", Elsevier Science Publishers Ltd., England. (1991).
9. C.S. Smith, "Design of Marine Structures in Composite Materials", Elsevier Science Publishers Ltd., England. (1990).
10. A.B. Schultz and S.W.Tsai, Measurement of Complex Moduli for laminated fibre-reinforced composites, *Journal of Composite Materials*, 3:434 (1969).
11. R.D. Adams and G.G.C. Bacon, Effect of fibre-orientation and laminated geometry on the dynamic properties of CFRP, *Journal of Composite Materials*, 7:402-428(1093).
12. S.W. Tsai, Structural behaviour of Composite Materials, *NASA, CR71*.(1964).
13. M. Manera, Elastic properties of randomly oriented short fibre-glass composites, *Journal of Composite Materials*, 11:235 (1977)
14. R.G. Ni and R.D. Adam, Damping and Dynamic Moduli of Symmetric Laminated Composite Beams-Theoretical and Experimental Results, *Journal of Composite Materials*, 18:104-121 (1984).

FATIGUE BEHAVIOR OF LONG FIBER REINFORCED POLYAMIDE AND POLYCARBONATE COMPOSITES UNDER TENSION-TENSION LOADING

C. C. M. MA[1], S. H. LIN[1] and N. H. TAI[2]

[1]INSTITUTE OF CHEMICAL ENGINEERING
[2]MATERIALS SCIENCE CENTER
NATIONAL TSING HUA UNIVERSITY
HSIN-CHU, TAIWAN, R. O. C. 30043

ABSTRACT

Static tensile strength and fatigue behavior of long glass fiber reinforced semicrystalline polyamide(Nylon6,6) and amorphous polycarbonate(PC) composites were investigated. The static tensile measurement at various temperatures and tension-tension fatigue loading tests at various levels of stress amplitudes were studied. Two-parameters Weibull distribution function and pooled Weibull distribution function were applied to obtain the statistical probability distribution of experimental data of static tensile strength and fatigue life under different stress amplitude tests. The dynamic creep property and the temperature increasing phenomenon under tension-tension fatigue loading were compared between semicrystalline and amorphous composites. Results showed that the static tensile strength of polyamide composites is higher than that of polycarbonate composites, with lower fatigue life and more sensitive to temperature. The slope of S-N_0 curves of long glass fiber reinforced semicrystalline polyamide and amorphous polycarbonate composites were almost identical.

KEYWORDS: Fatigue behavior, Long glass fiber, Polyamide, Polycarbonate, Pooled Weibull distribution function, Dynamic creep.

INTRODUCTION

Recent development of long fiber reinforced composites has been attracted by the composite industry for the longer fiber length and excellent property. The static mechanical properties of long fiber reinforced composites are much better than that of short fiber reinforced composites[(1-4)]. However, the investigation on the fatigue behavior of long fiber reinforced composites is still ambiguous.

For metallic materials, the fatigue damage can be characterized by the growth of a predominant crack which can be detected by using various nondestructive testing techniques[5]. For discontinuous fiber reinforced polymer composites, the damage is initiated at the interfacial of matrix and fibers aligned transverse to the loading direction[6]. The crack growth is due to fiber break, matrix crack and debonding between matrix and fibers.

Polymer materials show a temperature increase under cyclic loading due to the internal friction and viscoelastic characteristics[7]. J.W.Dally et. al[7,8] investigated the surface temperature increase of the specimens under fatigue loading as a function of frequency. The magnitude of the heat generation is different between semicrystalline and amorphous matrix resulted in the decreasing of fatigue life.

The test results in strength and fatigue life of polymer composites show larger scatter than that of metal[9]. The statistical Weibull distribution function can be applied to analyze the experimental data[10-12].

The objective of this research is intended to study the static tensile strength at various temperatures and the fatigue life of long glass fiber reinforced semicrystalline polyamide and amorphous polycarbonate at various stress amplitudes. The experimental data were analyzed by both the median rank function and the Weibull distribution function. The dynamic creep and temperature effect were discussed. In this study, the S-N_0 curves of both PA/LGF and PC/LGF were obtained by temperature correction.

EXPERIMENTAL

The 10mm long glass fiber reinforced semicrystalline polyamide(PA/LGF) and amorphous polycarbonate(PC/LGF) pellets were supplied by the Imperial Chemical Industry Co. (I.C.I.) U.K. and BAYCOMP Co. Canada, respectively. The test specimens were prepared to the ASTM D638 I-type specimen by injection molding.

All static tensile tests at various temperatures were conducted by using a Testometric MICRO 500, U.K.. Crosshead speed was 2mm/min. All fatigue tests were conducted by using a MTS 810.22 servo-hydraulic test machine under load control mode. Fatigue tests were conducted at room temperature with constant amplitude under tension-tension triangular load cycles. For all tests, the r-value ($r=\sigma_{min}/\sigma_{max}$) of 0.1 with a frequency of 10 Hz was adopted. Fatigue strain was recorded by using extensormeter. The surface temperature of testing specimen was measured by using a digital type thermocouple detector.

THEORETICAL ANALYSIS

Strength and Fatigue Life Distributions

The static strength data are analyzed by the two-parameter Weibull distribution function which is given in the following equation[10]:

$$R_s(X_i) = \exp\left[-\left(\frac{X_i}{X_0}\right)^{\alpha_s}\right] \quad (1)$$

where R_S represents the survival probability when specimen subjected to a X_i static load. α_S and X_0 denote the shape parameter and characteristic strength of the distribution, respectively. X_i is experimental datum. The shape parameter α_S and scale parameter X_0 can be determined by using the method of maximum likelihood estimation (MLE)[10].

The survival probability $R(X_i)$ at X_i, $1 \leq i \leq m$, of the static strength distribution or fatigue life is given by the median rank described as follows[13]:

$$R(X_i) = 1 - \frac{i - 0.3}{m + 0.4} \quad (2)$$

X_i denotes the test strength of i th sample. The same equation was also applied to the analysis of the fatigue life.

The two-parameter Weibull distribution is utilized to analyze the fatigue life data and to associate the censoring procedure to the data pooling procedure. In this study, the shape parameter and scale parameter are determined by maximum likelihood estimator (MLE). In order to prevent the inherent bias property of the MLE for small specimen sizes, a data pooling technique was used to determine the value of shape parameter for all stress amplitude by utilizing the normalized data. The form of Weibull distribution for survivor probability of fatigue life can be expressed as follows[10,12]:

$$R_f(N_i) = \exp\left[-\left(\frac{N_i}{N_{0i}}\right)^{\alpha_s}\right] \quad (3)$$

where $R_f(N_i)$ denotes the survivor probability of the ith specimen after being subjected to N_i fatigue cycles, α_{fi} and N_{0i} denote the shape and scale parameter of the distribution,respectively. The MLE equations for data censoring was proposed by Awerbuck[10], Bain[11] and Whitney[12].

Again, in order to prevent the inherent bias property of the MLE for small specimen sizes,a data pooling technique was used to determinate a single value of shape parameter for all stress amplitude by utilizing the normalized data. This procedure can determine the global shape parameter by using a few replicates of stress amplitude. The MLE associated with the data pooling procedure in conjunction with censored specimens was introduced by Whitney[12].

RESULTS AND DISCUSSION

Static Tensile Test

Tables 1 and 2 summarize the static tensile strength of PA/LGF and PC/LGF, respectively. The shape parameter α_S=24.09 and characteristic strength X_0= 186.38 MPa of PA/LGF can be obtained by MLE method. The shape parameter α_S=72.4 and characteristic strength X_0=89.97 MPa of PC/LGF also can be found. The reinforced fiber should increase the crystalillity of crystalline matrix. The semicrystalline PA/LGF shows a higher static tensile strength than that of the amorphous PC/LGF, with a larger scatter. Figures 1 and 2 show the Weibull distribution function curves of static ultimate tensile strength in X_0 of PA/LGF and PC /LGF, respectively. Figure 3 shows the static tensile strengths of PA/LGF and PC/LGF at various temperatures. The ultimate tensile strength was presented as a function of temperature from room temperature to 45^{0}C. It can be seen that the static tensile strength of PA/LGF depends on the temperature due to the characteristic of semicrystalline polyamide matrix.

Table 1. Static tensile strength of long glass fiber reinforced polyamide(Ny66) composite. testing rate: 2 mm/min, (20 Test Specimens).

Static Tensile Strength (MPa)						
172.32	173.08	174.34	174.97	175.60	178.20	178.75
179.94	180.06	181.20	183.01	183.06	183.47	187.76
188.17	188.37	188.66	189.01	195.04	200.06	
Average	182.75 MPa					
α_s	24.09					
X_0	186.38 MPa					

Table 2. Static tensile strength of long glass fiber reinforced polycarbonate(PC) composite. testing rate: 2 mm/min,(20 Test Specimens).

Static Tensile Strength (MPa)						
86.11	86.60	87.29	87.33	88.01	88.17	89.05
89.08	89.79	89.83	90.01	90.05	90.28	90.32
90.73	90.80	91.01	91.22	91.26		
Average	182.75 MPa					
α_s	24.09					
X_0	186.38 MPa					

Tension-Tension Fatigue Life and Reliability

Tables 3 and 4 list the tension-tension fatigue data at various stress amplitudes with a stress ratio r=0.1 and a loading frequency of 10 Hz of PA/LGF and PC/LGF, respectively. The fatigue limit of PA/LGF is 29.8% of X_0 less than PC/LGF with 34.34% of X_0. The

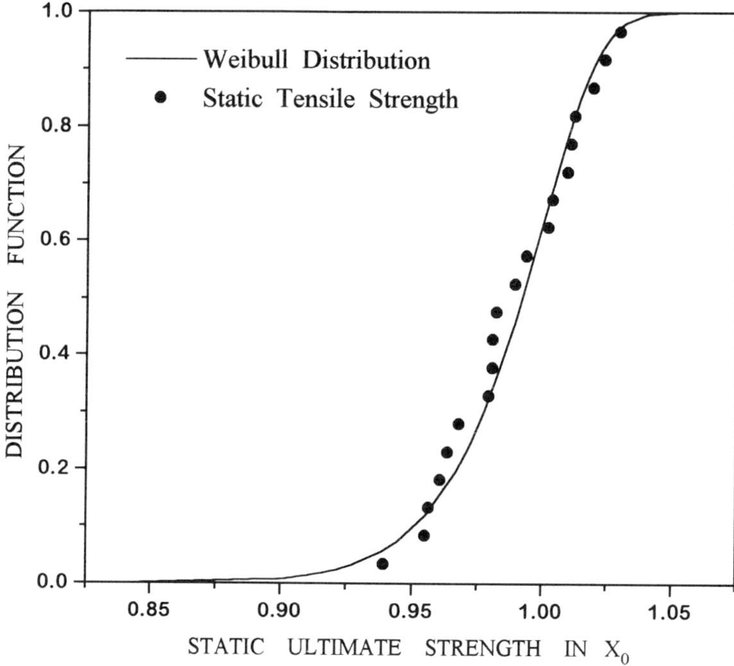

Figure 1. Comparison between the theoretical Weibull prediction curve and experimental data of static tensile strength of long glass fiber reinforced polyamide composite.

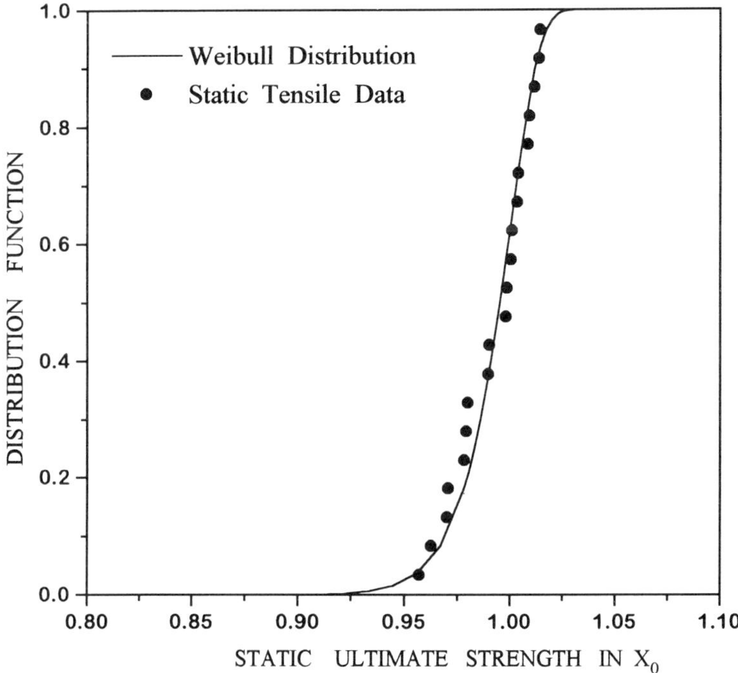

Figure 2. Comparison between the theoretical Weibull prediction curve and experimental data of static tensile strength of long glass fiber reinforced polycarbonate composite.

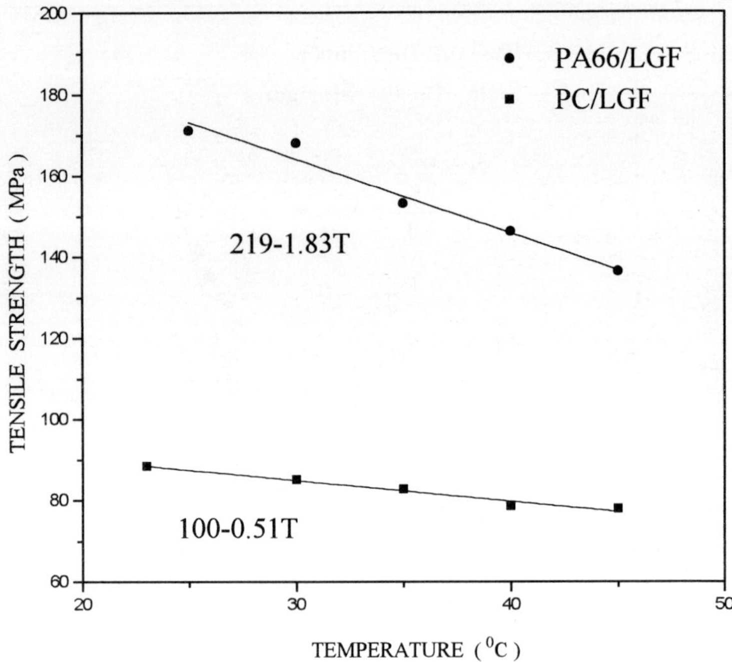

Figure 3. Static tensile strength of long glass fiber reinforced polyamide and polycarbonate composite at various temperature.

Table 3. Tension-tension fatigue data of long glass fiber reinforced polyamide composite under various stress amplitudes, (10 Hz, stress ratio = 0.1).

Stress (MPa) (% of X_0)	Fatigue Cycles (10 Hz, Stress Ratio = 0.1)
108.4 58.2%	698,710,722,746,780,793,822,862,880,890,929,964
100.9 54.1%	1213,1216,1227,1272,1312,1346, 1411,1460,1478,1622,1774,1819
93.3 50.1%	2245,2387,2590,2659,3570,3833 4121,4332,4484,4501,4683,5412
82.2 44.1%	5863,6797,6922,7680,8936,9766 9785,10766,10783,12379,16433,22219
75.7 40.6%	9036,15212,15891,19371,22416,24183 25520,39358,39635,50793,97258,107209
68.1 36.5%	44564,57480,91238,105551,106867 108258,141184,144612,148992, 205869,359619,528687,944851
60.5 32.5%	79220,101134,155243,196243,220554,240977 258498,277635,636979,639293,860477,10^6*
55.5 29.8%	10^6*,10^6*,10^6*,10^6*,10^6*,10^6*

*: Represented the fatigue test is over 10^6 cycles and no failure occurred.

Table 4. Tension-tension fatigue data of long glass fiber reinforced polycarbonate composite under various stress amplitudes, (10 Hz, stress ratio = 0.1).

Stress (MPa) (% of X_0)	Fatigue Cycles (10 Hz, Stress Ratio = 0.1)
67.46 75.0%	859,1295,1315,1396,1427,1437 1447,1448,1575,1631,1770,1824
62.97 70.0%	2550,2613,2700,3123,3315,3321 3465,3612,3663,4000,4012
58.46 65.0%	6560,7014,7123,7206,7494 7796,8065,8852,9437,9563
52.35 58.2%	15895,20712,21686,24164,26101 26564,28331,29426,35213,53071
48.69 54.1%	23780,28063,28961,34985,38098 38988,46690,50046,53390,59228
45.04 50.1.%	66511,94876,99803,102288,105731 112466,117198,121813,123614,137640
40.18 44.7.%	100001,148438,215194,297599,338766 340066,371849,446788,507434,642790
34.34 38.2%	10^6*,10^6*,10^6*,10^6*,10^6*,10^6*

*: Represented the fatigue test is over 10^6 cycles and no failure occurred.

stress level-fatigue life results were plotted in Figures 4 and 5. Figures 6 and 7 depict the change of surface temperature of specimen of both PA/LGF and PC/LGF verse fatigue cycles. The temperature change of PA/LGF was higher than PC/LGF, due to the semicrystalline PA/LGF is more sensitive to temperature than the amorphous PC/LGF. This result affects the fatigue limit and fatigue life distribution. PA/LGF shows a larger hysteresis loop in comparing with PC/LGF, as illustrated in Figure 8. The enveloped area of hysteresis loop is an indicator of the energy released during fatigue test.

The strain of specimen is increasing with fatigue cycles, which can be considered as dynamic creep phenomenon, owing to viscoelastisity of matrix and production of microcracks. Figures 9 and 10 show the strain change of PA/LGF and PC/LGF verse fatigue cycles, respectively. The stress relaxation has been found followed the strain control mode(15).

The linear regression lines may be represented by the following equation(7):

$$S/\sigma_{UTS} = A + m\text{Log } N \tag{4}$$

where S is the stress range, σ_{UTS} is the ultimate tensile strength, N is the number of cycles to failure, A and m are constants.

The temperature increase affects the ultimate tensile strength under tension-tension fatigue loading. From the results of Figures 3, 6 and 7, the σ_{UTS} can be represented as a function of temperature and the temperature of specimen can be indicated at cycles N. The temperature of specimen can be measured when the specimen was broken, then the σ_{UTS}

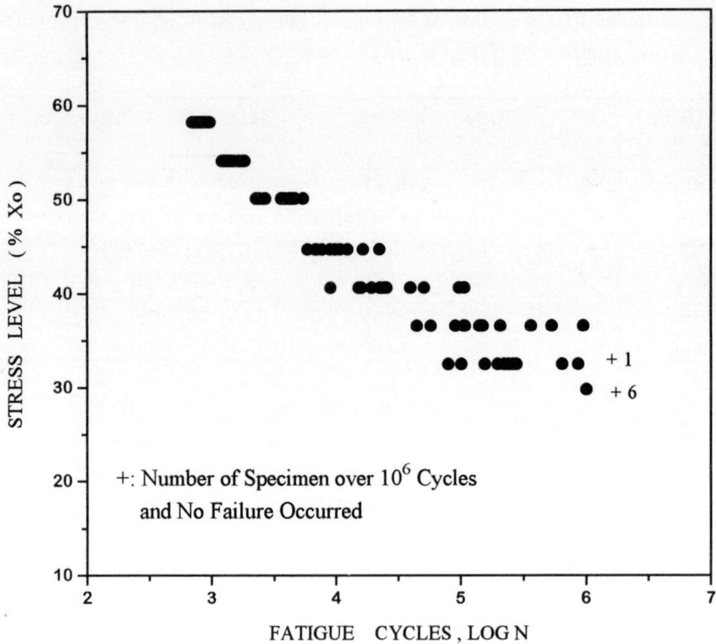

Figure 4. Experimental data of fatigue stress-life of long glass fiber reinforced polyamide composite under tension-tension fatigue loading (X_0 is the characteristic strength of Weibull function).

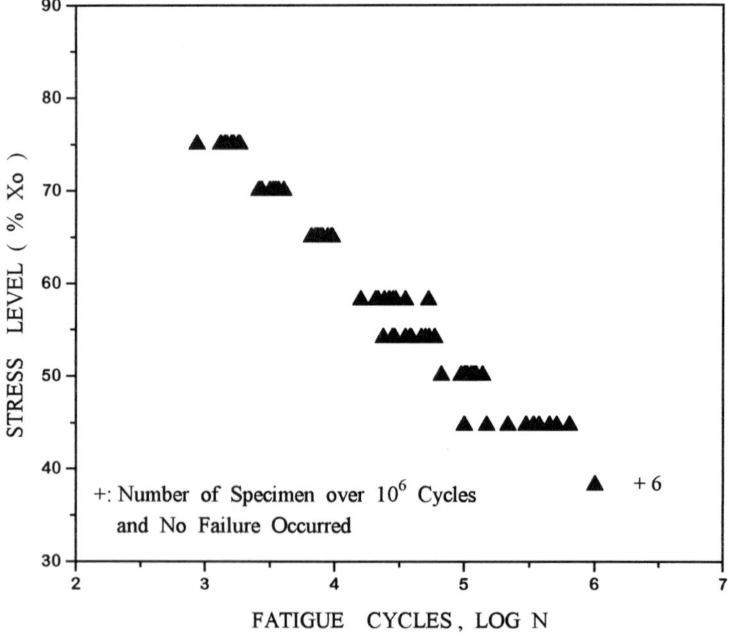

Figure 5. Experimental data of fatigue stress-life of long glass fiber reinforced polycarbonate composite under tension-tension fatigue loading (X_0 is the characteristic strength of Weibull function).

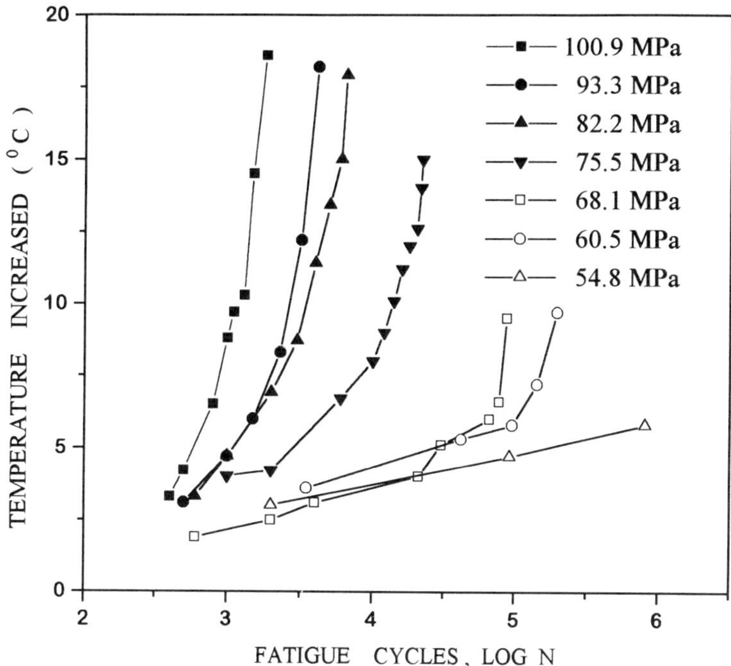

Figure 6. Surface temperature change of specimen vs. fatigue cycles of long glass fiber reinforced polyamide composite at 10Hz fatigue frequency under tension-tension loading.

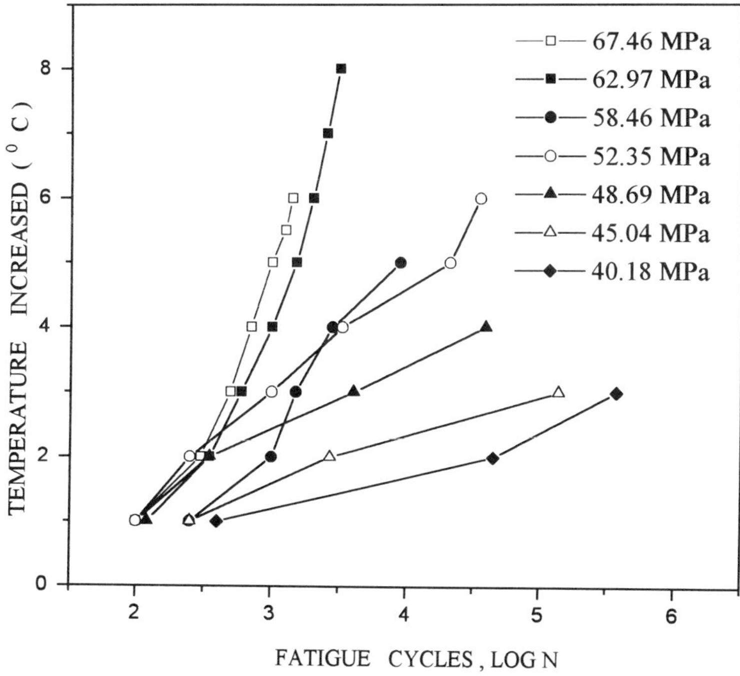

Figure 7. Surface temperature change of specimen vs. fatigue cycles of long glass fiber reinforced polycarbonate composite at 10Hz fatigue frequency under tension-tension loading.

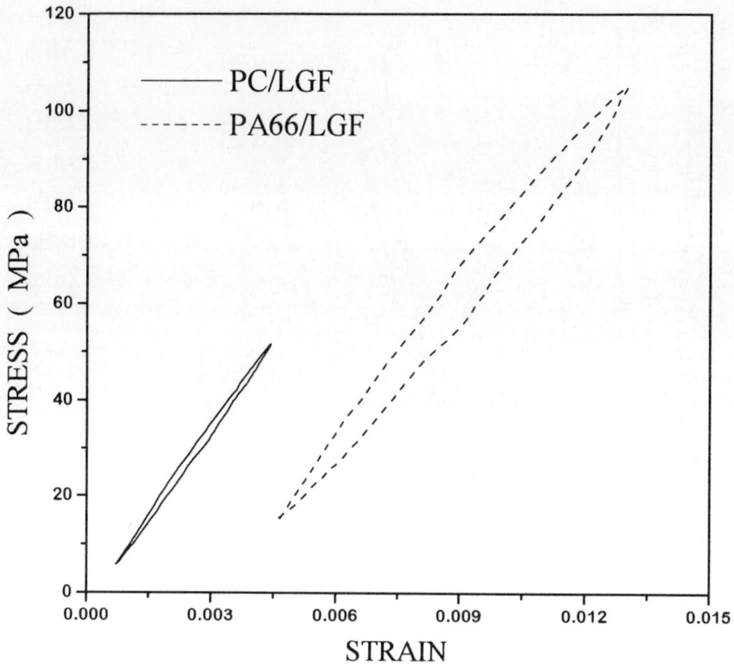

Figure 8. Stress-strain curve of long glass fiber reinforced polyamide and polycaıbonate composites under 58.1% stress level tension-tension fatigue loading .

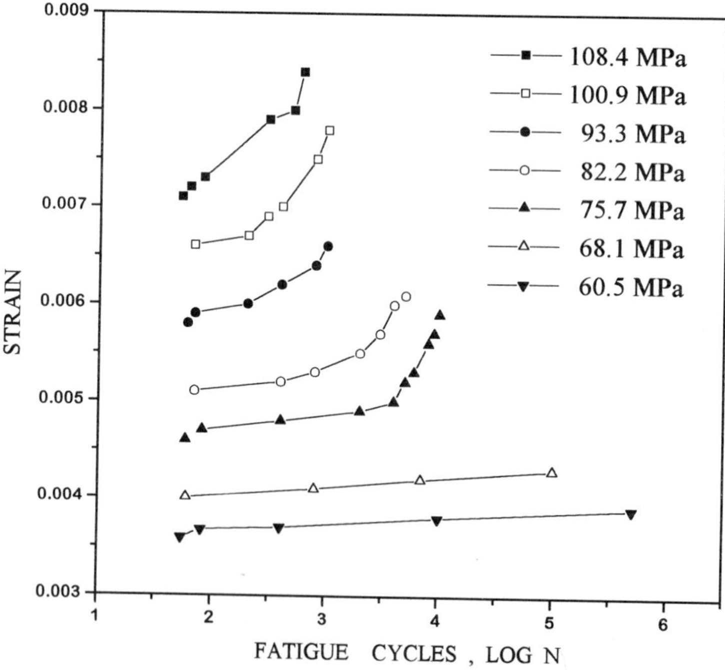

Figure 9. Strain change of specimen vs. fatigue cycle of long glass fiber reinforced polyamide composite under tension-tension fatigue loading .

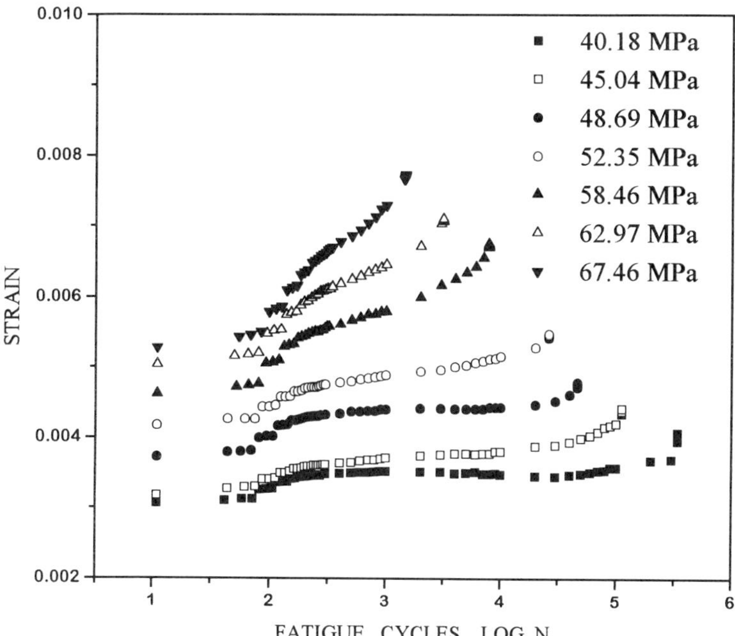

Figure 10. Strain change of specimen vs. fatigue cycle of long glass fiber reinforced polycarbonate composite under tension-tension fatigue loading.

of these specimens can be corrected from Figure 3. Consequently, different values of σ_{UTS} and stress ratio(S/σ_{UTS}) can be obtained at various stress amplitudes. Considering the temperature effect under tension-tension fatigue, the S-N_0 curves were plotted as shown in Figure 11. The slopes of S-N_0 curves of both PA/LGF and PC/LGF are almost identical and their degradation rates are similar.

Table 5 and 6 summarize the characteristic life of Weibull distribution function and Pooled Weibull distribution function of PA/LGF and PC/LGF,respectively. Figure 12 shows the Weibull distribution for fatigue life under different cyclic stress levels of PA/LGF and Figure 13 shows that of the PC/LGF. The pooled Weibull distribution of PA/LGF and PC/LGF are shown in Figures 14 and 15, respectively. The correlation between the experimental data and analyzed distribution function of PC/LGF is better than that of PA/LGF.

CONCULSIONS

1. The static tensile strength of semicrystalline PA/LGF is higher than that of amorphous PC/LGF, and depend on temperature tested.

2. The change of surface temperature of PA/LGF depends on the stress applied and exhibits a higher surface temperature in comparing with PC/LGF during fatigue test.

3. Dynamic creep phenomenon is observed by the tension-tension fatigue test of PA/LGF and PC/LGF. It was found that strains are increasing with fatigue cycles.

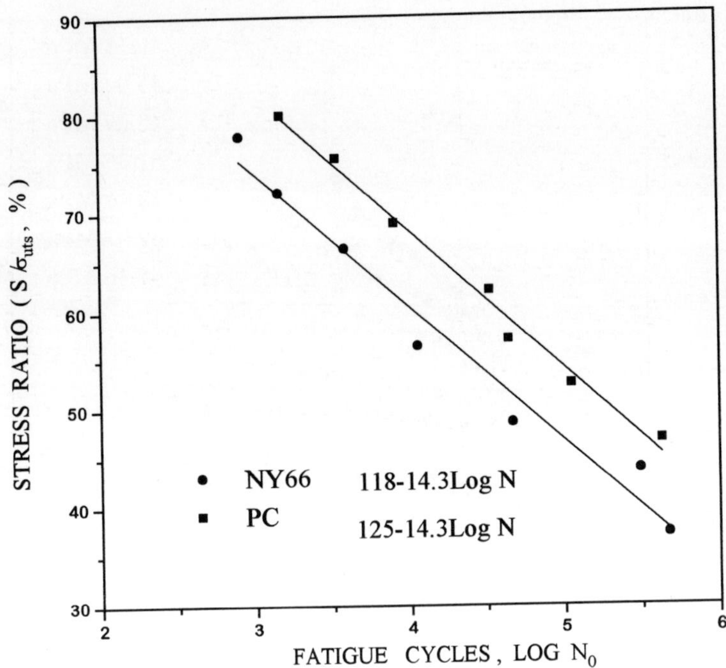

Figure 11. The S-N_0 curves of long glass fiber reinforced polyamide and polycarbonate composites

Table 5. Parameters of Weibull function of of long glass fiber reinforced polyamide composite under various stress amplitudes tension-tension fatigue test(10 Hz, stress ratio = 0.1)

Stress (MPa) (% of X_0)	Weibull Scale Parameter	Weibull Shape Parameter	Pooled Weibull Scale Parameter	Pooled Weibull Shape Parameter
108.4 58.2%	855	10.50	820	1.75
100.9 54.19%	1,521	7.14	1,440	1.75
93.3 50.1%	4,108	4.38	3,830	1.75
82.2 44.1%	12,073	2.52	11,150	1.75
75.7 40.6%	43,047	1.40	46,802	1.75
68.1 36.5%	241,438	1.12	307,588	1.75
60.5 32.5%	434,768	1.31	469,451	1.75

Table 6. Parameters of Weibull function of of long glass fiber reinforced polycarbonate composite under various stress amplitudes tension-tension fatigue test(10 Hz, stress ratio = 0.1)

Stress (MPa) (% of X_0)	Weibull Scale Parameter	Weibull Shape Parameter	Pooled Weibull Scale Parameter	Pooled Weibull Shape Parameter
108.4 58.2%	1,548	7.34	1,564	4.0
100.9 54.19%	3,518	7.87	3.555	4.0
93.3 50.1%	8,365	8.35	8,453	4.0
82.2 44.1%	31,451	2.93	31,782	4.0
75.7 40.6%	44,468	4.02	44,936	4.0
68.1 36.5%	115,645	7.14	116,863	4.0
60.5 32.5%	385,422	2.35	389,482	4.0

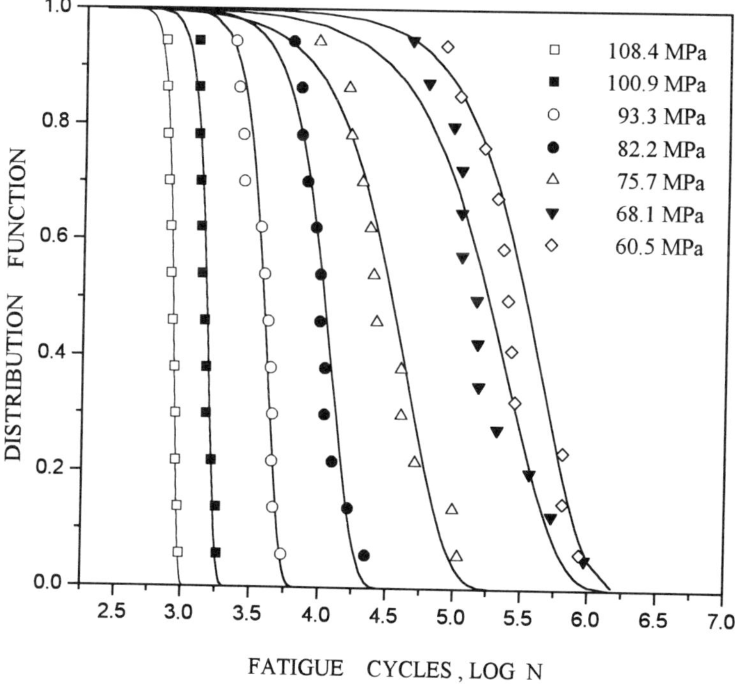

Figure 12. Comparison between the theoretical curves and experimental data of the fatigue life of long glass fiber reinforced polyamide composite

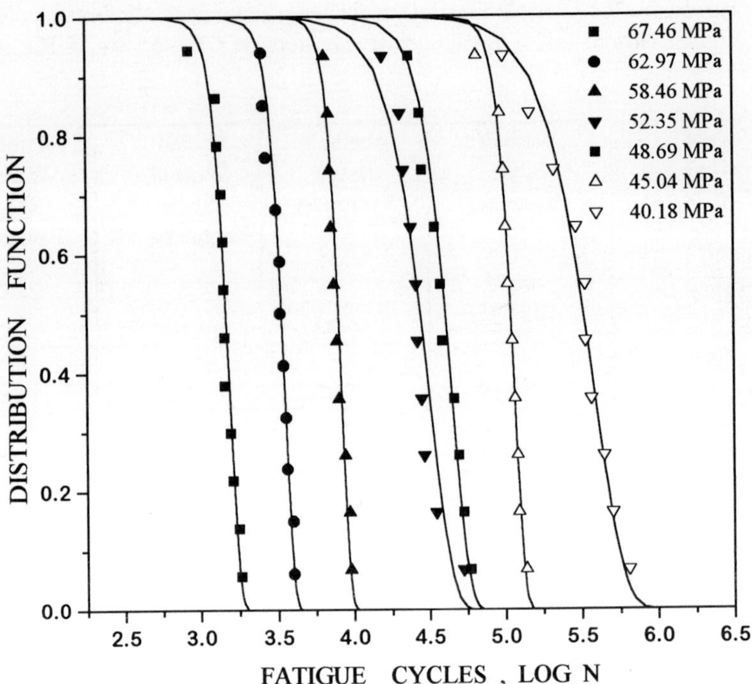

Figure 13. Comparison between the theoretical curves and experimental data of the fatigue life of long glass fiber reinforced polycarbonate composite.

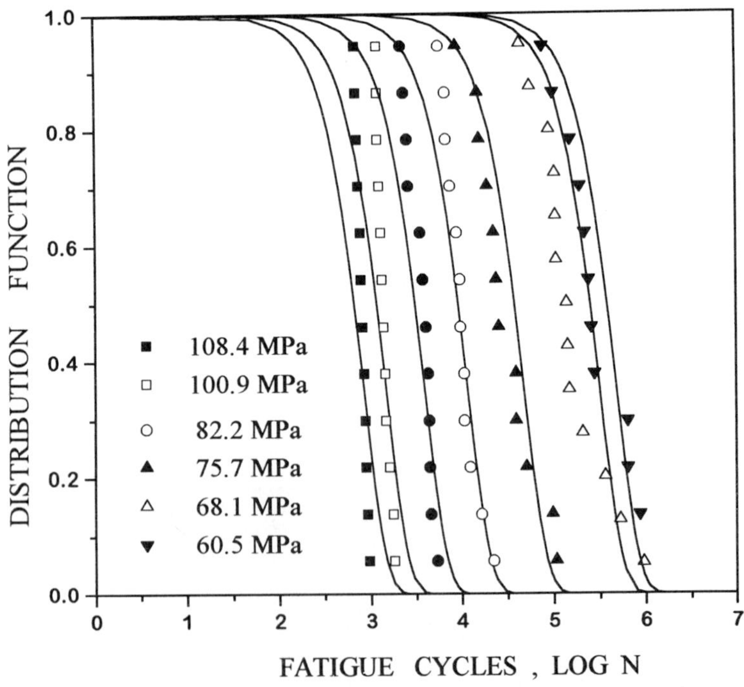

Figure 14. Comparison between the Pooled Weibull theoretical prediction curves and experimental data of the fatigue life of long glass fiber reinforced polyamide composite.

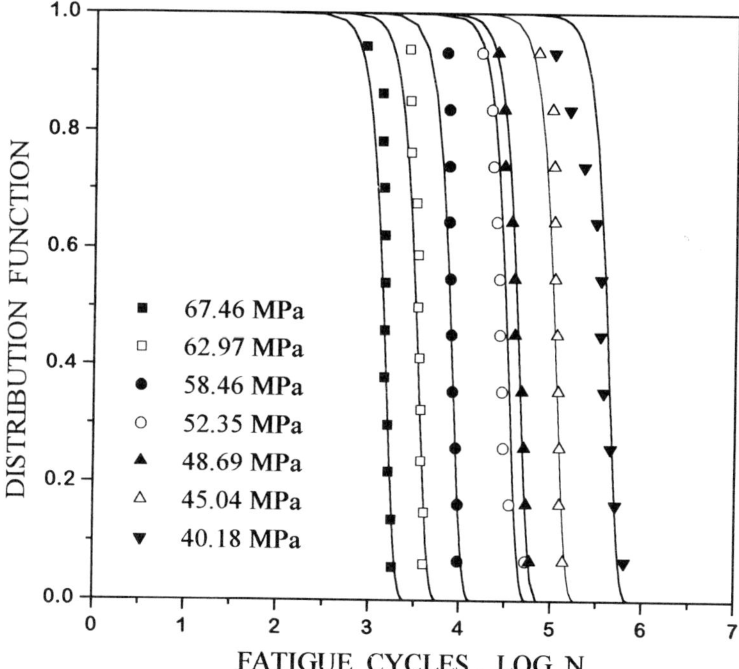

Figure 15. Comparison between the Pooled Weibull theoretical prediction curves and experimental data of the fatigue life of long glass fiber reinforced polycarbonate composite.

4. Considering the temperature effect and by data correction, the S-N curves of both PA/LGF and PC/LGF composite obtained are linear and the slopes of both material are almost identical.

ACKNOWLEDGMENTS

This research was supported by the National Science Council of the Republic of China, under the Contract No. NCS-83-0405-E-007-065.

REFERENCES

1. Staffan Toll and Carl-Gustaf Aronsson," Notched Strength of Long- and Short-Fibre Reinforced Polyamide", J. of Comp. Sci. and Technol., 45:43 (1992).

2. T. Vu-Khanh, J. Denault, P. Habib and A. Low," The Effects of Injection Molding on the Mechanical Behavior of Long-Fiber Reinforced PBT/PET Blends", J. of Comp. Sci. and Technol., 40:423 (1991).

3. J. Denault, T. Vu-Khanh and B. Foster, " Tensile Properties of Injection Molded Long Fiber Thermoplastic Composites", J. of Polym. Compsites.,10:313 (1989).

4. V. B. Gupta, R. K. Mittal, P. K. Sharma, G. Mennig and J. Wolters," Some Studies on Glass Fiber-Reinforced Polypropylene. Part II: Mechanical Properties and Their

Dependence on Fiber Length, Interfacial Adhesion, and Fiber Dispersion", J. of Polym. Composites, 10:16 (1989).

5. F. H. Chang, D. E. Gordon and A. H. Gardner," A Study of Fatigue Damage in Composites by Nondestrustive Testing Techniques", ASTM STP 636, 57 (1977).

6. S. A. Hitchen and S. L. Ogin," Damage Accumulation During the Fatigue of an Injection Moulded Glass/Nylon Composite", J. of Comp. Sci. and Technol., 47:83 (1993).

7. H. Neubert , H. Harig and K.Schutle," Monitoring of Fatigue Induced Damage Processes in CFRP by Means of Thermometric Methods", the Sixth International Conference on Composite Materials and Second European Conference on Composite Materials (ICCM 6 and ECCM 2), 1:1359 (1987).

8. J. W. Dally and L. J. Broutman," Frequency Effects on the Fatigue of Glass Reinforced Plastics", J. Comp. Mater., 1:424 (1967).

9. P.C.Chou and H.Miller,"Maximum likelihood Estimation of a Two-Segment Weibull Distribution for Fatigue Life", ASTM STP 744, 114 (1981).

10. J. Awerbuck and H. T. Hahn," Fatigue and Proof-Testing of Unidirectional Graphite/Epoxy Composites", ASTM STP 636, 248 (1977).

11. Lee J. Bain ," Statistical Analysis of Reliability and Life-Testing Models", Marcel Dekker Inc., New York (1978).

12. J. M. Whitney," Fatigue Characteristic of Composite Materials", ASTM STP 723, 133 (1981).

13. A. Rotem and H. G. Nelson," Residual Strength of Composite Laminates Subjected to Tension-Compressive Fatigue Loading", J. Comp. Technol. and Res., 12:76 (1990).

14. C. C. Osgood, " Fatigue Design", Wiley-Interscience, New York (1970).

15. P. K. Mallick," Fiber-Reinforced Composites: Materials, Manufacturing and Design", Marcel Dekker Inc., New York (1988).

POLY(ARYLENE ETHER KETONE) CHEMISTRY:

RECENT ADVANCES IN SYNTHESIS AND APPLICATIONS

K. J. Dahl and V. Jansons

Raychem Corporation
Menlo Park, CA, USA

INTRODUCTION

Many poly(arylene ether ketone)s (PAEKs) are high performance engineering thermoplastics with outstanding properties. The semicrystalline members are tough materials that show excellent thermooxidative, hydrolytic, and solvent resistance. The crystalline nature of these polymers contributes to their toughness and stems from similar bond angles of the aryl ether and ketone linkages. The presence of *para*-arylene linkages in the polymer chains causes these polymers to crystallize from the melt. Drawbacks include high cost and high melt processing temperature. In some cases, the glass transition temperature is relatively low, compared to that of poly(arylene ether sulfone)s and poly(arylene ether imide)s.

This survey outlines recent work on the preparation of PAEKs and copolymers, with emphasis on low cost and high performance. Attempts to develop low cost routes to PAEKs often lead to inferior materials, that are of low molecular weight, or contain structural defects, or difficult to remove impurities. This results in low melt stability, brittle resins, and lowered thermooxidative stability. Expensive fluoro-monomers previously used in nucleophilic routes are being replaced by activated chloro-monomers. Electrophilic routes tend to use low cost acid chlorides, such as phosgene, terephthaloyl chloride and isophthaloyl chloride. Buffered $AlCl_3$ solvent systems permit effective low temperature polymerization. Various routes to PAEKs utilize a soluble precursor approach that requires the removal of solubilizing groups in the work-up stage. Considerable work has been done towards the preparation of molecular composites using PAEK side chains attached to rigid rod polymer backbones. Various PAEKs and copolymers have found a broad range of applications.

Polymers and Other Advanced Materials: Emerging Technologies and Business Opportunities. Edited by P. N. Prasad *et al.*, Plenum Press, New York, 1995

LOW COST ELECTROPHILIC ROUTES

Poly(ether ketone ketone) (PEKK)

Aluminum trichloride catalyzed polymerization of phenyl ether with mixtures of terephthaloyl chloride and isophthaloyl chloride is a low cost route to semicrystalline copolymers containing *meta-* and *para*-phenylene units (PEKKs):

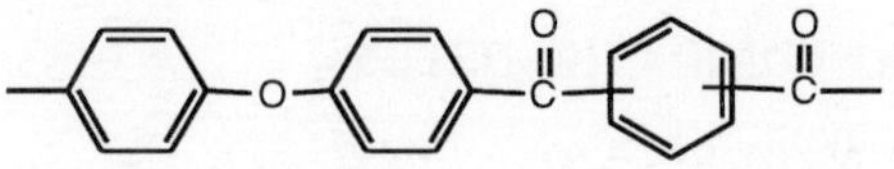

(meta- and para-)

The reactions are customarily run in the presence of an excess of the $AlCl_3$ in a non-basic solvent such as methylene chloride, 1,2-dichloroethane, or *o*-dichlorobenzene.[1] Under such conditions the reaction mixture contains undissolved $AlCl_3$ and results in phase separation to form an intractable layer of $AlCl_3$-complexed polymer. Occasionally, the separated viscous or rubbery phase can granulate (crystallize). Generally, a melt unstable product is obtained that requires special treatment to improve melt processability. Xanthylium ion impurities often result from *ortho*-acylation of the phenyl ether moiety, resulting in a capping effect to afford polymers of reduced molecular weight. Tris-aryl carbonium ion impurities are formed from the interaction of aryl groups with activated carbonyl sites. The carbonium ion sites tend to anchor catalyst residues that are difficult to remove. Some acylation of the reaction solvent *o*-dichlorobenzene can also limit the formation of high molecular weight resins.[2] The use of methylene chloride or 1,2-dichloroethane as polymerization solvent can cause alkylation of the aromatic polymer chains.[3] *Para*-alkylation of the chain ends also promotes *ortho*-acylation of the phenyl ether units and formation of xanthylium ions. Crosslinked polymers are often obtained upon synthesis or after melt processing because of the presence of alkyl linkages.[4] The terephthaloyl chloride and isophthaloyl chloride are of weak basicity towards $AlCl_3$ and are more difficult to activate than the more basic acid chlorides, such as 4,4'-oxydibenzoyl chloride. Furthermore, the resulting acylcarbonium ion intermediates are more electrophilic and correspondingly less selective towards aromatic sites, leading to increased *ortho*-acylation.[5]

Polymerization of phenyl ether with isophthaloyl chloride in 1,2-dichloroethane was shown by Sakaguchi et al. to be promoted by replacing the used 1,2-dichloroethane from a partially polymerized reaction mixture with fresh 1,2-dichloroethane.[6] This replacement apparently removed the HCl byproduct from the reaction mixture and caused the inherent viscosity to increase from 0.68 to 0.92 dL/g after 21-23 hr reaction time at room temperature.[6] Marvel et al. had previously polymerized phenyl ether with isophthaloyl chloride and terephthaloyl chloride with $AlCl_3$ in 1,2-dichloroethane to give polymers with an inherent viscosity of up to 0.96 dL/g,[7] but the resin was of unknown melt stability. Others also obtained similar viscosities, but the polymers were also unstable in the melt.[8] Heating of such polymers with formic acid improved their melt stability.[9] Formic acid tends to reduce carbonium ion content and facilitates the removal of catalyst residues. However, the hydrogenated defects are easily reoxidized in air. A more recent procedure for improving melt stability is to treat the crude polymer from reactions in *o*-dichlorobenzene with aqueous organic hydroperoxides at 180-200 °C for several hours.[10]

Polymerizations in the presence of $AlCl_3$ complexed with a base, such as DMF, benzophenone, $DMSO_2$, LiCl, pyridine, N-methylpyridinium chloride, or triethylamine hydrochloride, give cleaner and higher molecular weight polymers.[11] The basic additives promote the dissolution of excess $AlCl_3$ and buffer the acidity of the $AlCl_3$ and $[HAlCl_4]$. The resulting polar complexes or salts promote charge separation of the acylcarbonium ion intermediates. The adducts are dissolved in a non-basic diluent, such as methylene chloride or 1,2-dichloroethane. The polar adduct solutions also promote dissolution of the poorly soluble polymer complexes and salts (with $AlCl_3$ and $[HAlCl_4]$). The moderated reactions can be run at or below room temperature which further minimizes side reactions. The more highly solvated polymer chains are also easier to decomplex and to separate from the acidic catalyst residues.

The above-mentioned $AlCl_3$ complexes, as well as related salts containing HCl, when dissolved in methylene chloride or 1,2-dichloroethane, are good solvents for various poly(arylene ether ketones) that are at least partially complexed with $AlCl_3$. Dissolution of the excess $AlCl_3$ or its dimer (Al_2Cl_6) and buffering of the acidity of the reaction medium appear to be due to the weak residual basicity of the 1:1 complexes and salts of the above bases with $AlCl_3$ or [$HAlCl_4$]. For example, the solid and poorly soluble 1:1 complex of $AlCl_3$ with DMF can undergo a weak interaction with additional $AlCl_3$ or Al_2Cl_6 to form liquid complexes. Similarly the 1:1 $AlCl_3$ complexes of chloride salts, such as the liquid tetrachloroaluminate of 1-butylpyridinium chloride,[12] can accommodate additional $AlCl_3$ to form $Al_2Cl_7^-$ counterions that can reversibly release $AlCl_3$ to activate the weakly basic carboxylic acid chloride groups.

As an example of the use of basic co-agents, Hashimoto and Kobayashi (Toyobo Co.) polymerized phenyl ether (2.83 parts) with terephthaloyl chloride (2.53 parts) and isophthaloyl chloride (0.84 parts) in a solution of excess $AlCl_3$ / Et_3NH^+ Cl^- / methylene chloride (5 hr / 20 °C) to give PEKK of 1.44 dL/g inherent viscosity and low $AlCl_3$ residue content.[13] Similarly, *p*-phenoxybenzoyl chloride was polymerized by Fukawa and Tanabe (Asahi Chem. Ind.) in the presence of excess $AlCl_3$ and phenyl ether capping agent in a solution of DMF in methylene chloride (at -30 to 25 °C) to give PEK of 1.37 dL/g inherent viscosity.[14]

Alternatively, the PEKK polymers may be obtained by reaction of isophthaloyl chloride and terephthaloyl chloride with isolated intermediates. These intermediates are 4,4'-diphenoxyisophthalophenone and 4,4'-diphenoxyterephthalophenone. Such polymerizations yield cleaner polymers, particularly in the presence of $AlCl_3$-complex co-agents. There is also more control over the comonomer distribution in the chain. For example, polymerization of 1,4-bis(*p*-phenoxybenzoyl)benzene (4,4'-diphenoxyterephthalophenone) with isophthaloyl chloride proceeds rapidly to yield a perfectly alternating *meta*- and *para*-phenylene group containing crystalline copolymer (T_g 166 °C, T_m 332 °C).[15] The high T_g and low T_m are desirable for high heat deformation resistance and facile melt processing. The intermediates may also be generated in situ at a lower temperature and optimum acidity for mono-acylation of phenyl ether, followed by reaction with the required amount of additional diacid chloride and increased amount of $AlCl_3$, to form the polymer.

Such sequential addition of reactants was carried out by Du Pont, to prepare more ordered (blocky and alternating) copolymers showing higher crystallinity than obtainable from premixed monomers.[16] The perfectly alternating copolymer was also prepared from the isolated intermediates, both from the *meta*-isomer and the *para*-isomer, that is 1,4-bis(*p*-phenoxybenzoyl)benzene and isophthaloyl chloride, and 1,3-bis(*p*-phenoxybenzoyl)-benzene and terephthaloyl chloride.[16] The conditions used for the polymerizations can be represented by the following example:[16] To a solution of phenyl ether (0.306 mole) and isophthaloyl chloride (0.150 mole) in 150 g *o*-dichlorobenzene at 0 °C was slowly added $AlCl_3$ (0.65 mole) in 30 min, with the temperature held at 0 °C. Then terephthaloyl chloride (0.150 mole) was added, followed by more $AlCl_3$ (0.30 mole) at 0 °C. The reaction mixture was held an additional 10 min at 0 °C, and then *o*-dichlorobenzene (500 g) preheated to 160 °C was added with vigorous stirring. The jacketed reactor was then heated with steam. Particle formation was observed in the red-brown reaction mixture in about 1 min after the addition of hot *o*-dichlorobenzene. The stirred reaction mixture was held at 100 °C for 30 min, followed by cooling to 30 °C over a period of 1 hr. The polymer was then decomplexed by addition of 1 liter of MeOH that had been pre-chilled to -40 °C. Stirring was continued for an additional 1 hr, until all color was gone. The crude polymer was then collected by filtration, washed with methanol, heated in water to 98 °C, and filtered. The polymer was then soaked in 500 g of glacial formic acid for 30 min, followed by filtration and drying overnight at 170 °C under vacuum. The partially alternating copolymer (with a 1:1 terephthalophenone to isophthalophenone ratio) had an inherent viscosity of 0.63 dL/g (0.5 wt. % in conc. H_2SO_4) and showed a melting point of 319 °C. Other PEKK copolymers were obtained of inherent viscosities of up to 0.90 dL/g. They were exposed to formic acid during the work-up stage to reduce carbonium ion sites and to facilitate removal of catalyst residues. The products are suitable for melt extrusion to form films and fibers.[17]

Phenyl ether was also polymerized with terephthaloyl chloride in CF_3SO_3H containing 0.35 moles P_2O_5 per mole of CF_3SO_3H (48 hr / 25 °C) to give high molecular weight PEKK (inherent viscosity: 1.02 dL/g in CH_3SO_3H).[18] However, a 3 mole % excess of phenyl ether was needed to compensate for xanthylium ion formation from *ortho*-acylation.[18] The ^{13}C NMR spectrum indicated the presence of predominantly *para*-linkages.[18] Replacement of CF_3SO_3H with the less costly CH_3SO_3H gave inadequate results. Earlier, it was shown that phenyl ether polymerizes with terephthalic acid in CF_3SO_3H containing PCl_5 to give high molecular weight polymer (inherent viscosity: 1.19 dL/g in 95% H_2SO_4).[19]

Other Poly(arylene ether ketone)s

Another low cost system is based on phenyl ether / oxalyl chloride / $AlCl_3$ and co-agents, to give poly(*p*-phenyleneoxy-*p*-phenylenecarbonyl) (PEK).[20] Phenyl ether reacts with oxalyl chloride in the presence of $AlCl_3$ and an $AlCl_3$ complex of dimethyl sulfone in methylene chloride, to give a polymer of 0.92 dL/g inherent viscosity.[20]

Very pure PEK of controlled molecular weight is obtainable on polymerization of *p*-phenoxybenzoyl chloride in the presence of $AlCl_3$ complexes of LiCl and DMF in 1,2-dichloroethane solution at -5 to 0 °C.[21] Double capping with benzoyl chloride and *p*-phenoxybenzophenone controlled the molecular weight and deactivated all reactive end groups to give a melt stable polymer of 1.1 dL/g inherent viscosity (T_g 165 °C, T_m 365 °C).[21]

Earlier, phenyl ether was polymerized with phosgene in the presence of excess $AlCl_3$ in carbon disulfide, a non-alkylating solvent, to form PEK of 0.68 dL/g inherent viscosity (0.1 g/ml in 97% H_2SO_4).[22] More recently, excess phosgene has been reacted with phenyl ether in 1,2-dichloroethane to form PEK of higher molecular weight (1.15 dL/g inherent viscosity).[23] However, the melt stability of the products is uncertain. The polymerization was also carried out in the presence of a Lewis base, such as LiCl, in 1,2-dichloroethane.[24] After reductive treatment with Et_3SiH, this PEK showed no gel formation on heating at 400 °C for 5 min.[24]

Phosgene is also reported to polymerize with phenyl ether in the presence of $AlCl_3$ in CF_3SO_3H to form PEK of 0.95 dL/g inherent viscosity.[25] Certain phosgene derivatives, such as MeSC(=O)Cl and EtSC(=O)Cl, required the use of HF/BF_3 to produce high molecular weight PEK (1.24 dL/g inherent viscosity) and PEEK(1.54 dL/g inherent viscosity).[26] However, CF_3SO_3H or CH_3SO_3H in the presence of P_2O_5 or PCl_5 may also be suitable for polymerization of these phosgene derivatives with phenyl ether and other aromatics.

Phenyl ether reacts readily with phosgene / $AlCl_3$ to form 4,4'-diphenoxybenzophenone. A less pure product is obtained using carbon dioxide as the acylating agent.[27] This intermediate polymerizes very well with terephthaloyl chloride and/or isophthaloyl chloride to form high molecular weight PEKEKK using $AlCl_3$ in solutions of $AlCl_3$-Lewis base complexes in methylene chloride or 1,2-dichloroethane.[28] The high molecular weight achieved allows capping of the chain ends during the polymerization. This permits achievement of controlled melt viscosity and results in improved melt stability. Addition of small amounts of $POCl_3$ to $AlCl_3$ complexes in methylene chloride or 1,2-dichloroethane increases the dehydrating ability of the reaction medium and reduces the effects of partial hydrolysis of the acid chloride monomers. As an example, terephthaloyl chloride (0.200 mole), benzoyl chloride capping agent (7 mmole), and 4,4'-diphenoxybenzophenone (0.204 mole) were added to $AlCl_3$ (1.105 mole), dimethyl sulfone (28.239 g), and $POCl_3$ (10 mmole) at -15 °C, followed by a 4 hr reaction at 25 °C, to give end-capped PEKEKK of 1.12 dL/g inherent viscosity.[29] The $POCl_3$ increased control and reproducibility of the polymerizations and afforded polymers of improved properties.

1,4-Diphenoxybenzene similarly reacts with terephthaloyl chloride and isophthaloyl chloride in the presence of Lewis base buffered $AlCl_3$, to form high molecular weight PEEKKs.[30] 1,4-Diphenoxybenzene can also be polymerized with terephthalic acid in CF_3SO_3H / P_2O_5 to afford PEEKK of 1.43 dL/g inherent viscosity.[31]

LOW COST NUCLEOPHILIC ROUTES

4,4'-Dichlorobenzophenone was polymerized by reaction with sodium carbonate and SiO_2 / $CuCl_2$ catalyst in Ph_2SO_2 (2 hr at 280 °C, 1 hr at 300 °C, and an additional 1 hr at 320 °C).[32] PEK of 1.15 dL/g inherent viscosity was obtained. The SiO_2 was removed by leaching with 4% aqueous NaOH at 60 °C.[32] Copper also accelerated the reaction of chloro-monomers with phenolic monomers in the absence of SiO_2.[33] However, there could result considerable copper entrapment. Copper contamination reduces the thermooxidative stability of poly(arylene ether ketone)s, as shown by accelerated degradation of the solid polymer in air at elevated temperatures and crosslinking during melt processing.[34] Procedures to deal with Cu catalyst residues are described. For example, 4-(*p*-chlorobenzoyl)phenol was polymerized in the presence of K_2CO_3 and Cu powder in Ph_2SO_2 at 250-350 °C. The cooled reaction mixture was ground and then extracted with acetone to remove the Ph_2SO_2, followed by extraction with aqueous HCl containing 1 wt. % $FeCl_3{\cdot}6H_2O$ to remove the Cu, to afford PEK with good melt processability.[33]

Low cost dichloro-monomers were also polymerized with aryldiols in the presence of alkali metal nitrites,[35] sulfates,[36] or fluorides,[37] as promoting agents. For example, heating of 4,4'-dichlorobenzophenone and hydroquinone in the presence of K_2CO_3 and KF in diphenyl sulfone at 200-330 °C gave PEEK of 0.80 dL/g inherent viscosity (1 g/dL $CHCl_2$-CO_2H).[38] The fluoride ions promote displacement of the chloride to form more reactive aryl fluoride sites for reaction with the phenoxide anion.

The reactivity of the monomers is also increased when the activating effect of the two carbonyl groups is combined in a monomer, such as in 1,3-bis(*p*-chlorobenzoyl)benzene and 1,4-bis(*p*-chlorobenzoyl)benzene. These monomers have yielded amorphous PAEKs of inherent viscosities up to 1.27 dL/g (0.5 wt. % in $CHCl_3$) on reaction with difunctional phenols, in the absence of catalysts.[39] The dichloro-monomers are obtainable by Friedel-Crafts acylation of chlorobenzene with isophthaloyl or terephthaloyl chloride.

OTHER ROUTES

Silylated Phenol Reactions with Activated Aryl-Fluorides

Ketone group activated fluoro-aromatics, such as 4,4'-difluorobenzophenone, react with trimethylsilyl ethers of phenols in the presence of a catalytic amount of fluoride ions (CsF) to form silyl fluoride and poly(arylene ether ketone)s.[40] 2,6-Difluorobenzophenones react similarly with silyl ethers of bisphenols to give polyaryl ethers with pendent ketone groups.[41]

Ring Opening Polymerization

Ring opening polymerization of macrocyclic arylene ether ketones has been examined.[42] The polymerization is initiated by a catalytic amounts of nucleophiles, such as CsF or potassium salt of 4-hydroxybenzophenone.[42] The process could provide a means of direct conversion of low viscosity cyclic monomers to final, shaped, polymer articles.

Acylation and Arylation of Tin-Organic Compounds

Aromatic diacid chlorides have been coupled with trimethyltin substituted aromatic compounds in the presence of $PdCl_2(AsPh_3)_2$ catalyst in 1,2-dichloroethane at 70 °C to yield amorphous PAEKs.[43] Ketals of 4,4'-dibromobenzophenone were similarly coupled with bis(trialkylstannane) aromatic compounds in the presence of $PdCl_2(PPh_3)_2/2PPh_3$ in DMAc at 165-200 °C.[44] The solubilizing ketal groups were subsequently removed by hydrolysis.

Aryl-Aryl Coupling Reactions of Aryl-Halides

Soluble *meta*-phenylene containing poly(biphenylene ether ketones) can be prepared by coupling of chlorophenyl groups with Zn / $NiCl_2$ complexes of Ph_3P and 2,2'-bipyridine.[45] For example, 1,1'-(*p*-phenylenedioxy)bis[4-(3-chlorobenzoyl)benzene] was polymerized in DMAc at 90 °C to give a soluble polymer with an inherent viscosity of 0.87 dL/g in NMP.[45] High molecular weight all-*para* crystalline PAEKs could not be prepared, due to insufficient solubility in the solvents examined. The optimum reaction temperature in DMAc, based on the viscosity of polymer obtained, was shown to be around 90 °C. Unless a better solvent is found, temporary derivatization of the monomers with solubilizing groups appears required for achieving adequate molecular weight. The soluble precursor polymers could then be appropriately treated to remove the solubilizing groups. Such reversible derivatization may include Schiff base formation,[46] ketal formation,[47] or *tert*-butylation of the aromatic rings;[48] however, the extra steps required are economically unattractive.

Soluble Precursor Approaches

The Schiff base (ketimine) of 4,4'-dichlorobenzophenone with aniline was polymerized in the presence of zinc dust and complexes of $NiCl_2$ with PPh_3 and 2,2'-bipyridine in NMP at 80 °C to form a soluble ketimine polymer of 0.36 dL/g inherent viscosity in NMP.[49] This was hydrolyzed with dilute HCl to give crystalline poly(4,4'-biphenylenecarbonyl). Similar derivatization was used previously to prepare high molecular weight semicrystalline PAEKs by nucleophilic polymerizations.[50]

Previously, the cyclic ethylene glycol ketal of 4,4'-dihydroxybenzophenone was used in reactions with activated aryl halides, to form soluble amorphous polyketals.[51] These were converted to the poly(ether ketones) by acid catalyzed hydrolysis.

Another approach used *tert*-butylated monomers, to form soluble amorphous *tert*-butylated PAEKs.[52] Acid catalyzed cleavage of the *tert*-butyl substituents afforded the final PAEK.

More recently, a new polyketone synthesis was developed based on the reaction of dialkylaminonitrile derivatives of aromatic aldehydes with activated aromatic fluorides.[53] The conjugate bases of the *alpha*-aminonitriles act as selective and powerful nucleophiles towards displacement of the activated aryl halides to form a carbon-carbon bond. Hydrolytic removal of the solubilizing protecting groups of the resulting amorphous polymers produces free ketone groups and afforded polyarylene ketones not previously accessible. For example, Strecker reaction of isophthalaldehyde with aqueous morpholine and NaCN in the presence of $NaHSO_3$ gave the corresponding bis(*alpha*-aminonitrile). This reacted at room temperature under anhydrous conditions with 4,4'-bis(*p*-fluorophenyl) sulfone in DMF in the presence of NaH, to give a soluble derivatized polymer of high molecular weight. Acid hydrolysis yielded the corresponding poly(ketone sulfone) (T_g 208 °C, T_m 458 °C).[54]

Friedel-Crafts polymerization of acid chloride monomers in the presence of $AlCl_3$ can also be considered as a soluble precursor approach. The ketone groups formed are complexed with $AlCl_3$ and to some extent also protonated with [$HAlCl_4$]. These complexed polymers are generally amorphous, but they are not readily soluble in the non-basic and relatively inert solvents such as methylene chloride, 1,2-dichloroethane, carbon disulfide, or *o*-dichlorobenzene. Such solvents are insufficiently polar to dissolve the highly associated polymeric complexes. Excess $AlCl_3$ tends to interfere with crystallization or interchain association, but also tends to promote side-reactions. A solvent that is non-basic and highly polar, as well as inert towards $AlCl_3$, is not generally accessible. However, it was found that polar complexes of $AlCl_3$ with various bases or their HCl or [$HAlCl_4$] salts, when molten or dissolved in the inert non-basic diluents, result in suitable solvents for the complexed and protonated PAEKs.[55] Liquid HCl containing $AlCl_3$ was also found to be a suitable solvent, but required pressure equipment.[56]

NEW PAEKs AND POLY(ARYLENE KETONES)

Ortho-Linked Copolymers

2,2'-bis(*p*-fluorobenzoyl)biphenyl was reacted with various biphenols in the presence of K_2CO_3 in NMP. Hydroquinone gave an amorphous polymer of 0.41 dL/g inherent viscosity (0.5% in $CHCl_3$) and a T_g of 175 °C.[57] 4,4'-Biphenyldiol gave a polymer of 0.83 dL/g inherent viscosity and a T_g of 193 °C.[57] Other *ortho*-linked PAEKs have also been prepared.[58]

Polycyclic Copolymers

Naphthalene analogs of isophthaloyl and terephthaloyl units were prepared by reaction of 1,5-naphthalenedicarboxylic acid chloride and 2,6-naphthalenedicarboxylic acid chloride with fluorobenzene in the presence of $FeCl_3$ catalyst. This yielded 1,5-bis(*p*-fluorobenzoyl)naphthalene and 2,6-bis(*p*-fluorobenzoyl)naphthalene. The difluoromonomers were polymerized with hydroquinone using K_2CO_3 in diphenyl sulfone. The 1,5-isomer yielded a polymer of 1.0 inherent viscosity with a T_g of 192 °C and a T_m of 291 °C.[59] This represents a very high ratio of T_g to T_m. The degree of crystallinity of the polymer and whether it would crystallize from the melt was not discussed. The 2,6-isomer yielded a polymer of 1.08 inherent viscosity with a T_g of 178 °C and T_m of 367 °C.[59] Copolymers were also made with 4,4'-dihydroxybiphenyl. Naphthalenedicarboxylic acid chlorides were also directly copolymerized with phenyl ether and terephthaloyl chloride using $AlCl_3$ in 1,2-dichloroethane.[60]

Heteroatom-Linked Copolymers

N-Cyclohexylpyrolidone (bp 285 °C) was found to be the solvent of choice for the preparation of PEEK from 4,4-difluorobenzophenone and hydroquinone using K_2CO_3 as base.[61] It was used for the reaction of 4,4'-difluorobenzophenone with Na_2S to afford poly(*p*-phenylenecarbonyl-*p*-phenylenesulfide) of 0.95 inherent viscosity after 1 hr / 290 °C.[61] This previously made crystalline polymer has a T_g of 152 °C and T_m of 335 °C. 4,4'-Dichlorobenzophenone was also used to make this polymer in NMP,[62] showing a T_m of 365 °C and T_c (crystallization temperature) of 295 °C.[63] The more reactive 1,4-bis(*p*-chlorobenzoyl)benzene was similarly polymerized.[64] Phenyl ketimine intermediates from 4,4'-dichlorobenzophenone and 1,4-bis(*p*-chlorobenzoyl)benzene were polymerized with Na_2S at reduced temperatures.[65] Hydrolysis with dilute HCl yielded the semicrystalline ketone sulfides.

A phenylphosphine oxide bisphenol [PhP(=O)(Ph'OPh'OH)$_2$, where Ph' = *p*-phenylene] was polymerized with 4,4'-difluorobenzophenone, 1,3-bis(*p*-fluorobenzoyl)benzene, and 1,4-bis(*p*-fluorobenzoyl)benzene / K_2CO_3 in DMAc or sulfolane to give amorphous polymers of 0.62-0.67 inherent viscosity.[66] Various activated fluorine containing monomers were reacted with silylated intermediates in the presence of CsF catalyst to form phosphine oxide, sulfone, and ketone arylene ether copolymers.[67]

Heterocyclic Copolymers

Ether ketone copolymers have been prepared containing various heterocyclic rings, including imide,[68] imidazole,[69] benzimidazole,[70] benzobisoxazole,[71] benzobisthiazole,[71] oxadiazole,[72] pyrazole,[73] quinoxaline,[74] thiophene,[75] triazole,[76] and 1,2-dicarbadodecaborane.[77]

For example, bis(5-chlorothienyl-2)ketone, 1,3-bis(2-chlorothiophenoyl)benzene, and 1,4-bis(2-chlorothiophenoyl)benzene were polymerized with 4,4'-isopropylidenediphenol in the presence of K_2CO_3 in DMAc.[75]

Poly(arylene ether ketone 1,2-dicarbadodecaborane)s were prepared in CF_3SO_3H.[77] The polymer from 1,2-bis(4-phenoxyphenyl)-1,2-dicarbadodecaborane and biphenyl-4,4'-

dicarboxylic acid was amorphous and showed a weight loss of only 11% on pyrolysis at up to 1000 °C / N_2 (TGA heating rate 10 °C / min). Pyrolysis in air, under the same conditions, resulted in a weight loss of only 5%, indicating formation of boron oxides.[77]

Block Poly(arylene ether ketones)

An acid chloride terminated block was prepared from terephthaloyl chloride and 4,4'-diphenoxybenzophenone in methylene chloride containing $AlCl_3$-complexed dimethyl sulfone (at -13 °C to +20 °C). This was copolymerized with 4,4'-bis(*p*-phenylphenoxy)-diphenyl sulfone and 4-phenoxybenzophenone as end-capping agent to give a crystalline copolymer.[78]

p-Dichlorobenzene was polymerized with Na_2S in NMP and the block was reacted with 4,4'-dichlorobenzophenone and additional Na_2S to give a crystalline block copolymer.[79] 1,4-Bis(*p*-chlorobenzoyl)benzene was similarly polymerized with Na_2S in NMP (0.5 hr / 265 °C), and the block was combined with a block from *p*-dichlorobenzene and Na_2S to give a crystalline block copolymer.[80]

Liquid crystalline polyester blocks have been attached to PAEK blocks.[81] Random PAEK copolymers containing high levels of *p*-biphenylene or *p*-terphenylene units (30-80 mole %) also show thermotropic liquid crystalline behavior.[82]

Molecular Rod PAEKs

A low cost monomer for the preparation of molecular fiber composites is obtained by mono-acylation of *p*-dichlorobenzene with benzoyl chloride in the presence of $AlCl_3$ to yield 2,5-dichlorobenzophenone.[83] Other acid chlorides may also be used. The same monomer is also obtainable on Friedel-Crafts reaction of 2,5-dichlorobenzoyl chloride with benzene.[84] The latter reaction permits introduction of other aromatic compounds, such as phenyl ether or 4-phenoxybenzophenone, or PAEK oligomers. The dichloro-monomers can be polymerized by nickel-catalyzed dehalogenative aryl-aryl coupling, in the presence of zinc dust and complexes of $NiCl_2$ with PPh_3 in DMF or NMP at 65-140 °C.[84] Addition of 2,2'-bipyridine promotes the polymerization at the early stages.[85] 2,5-Dichlorobenzophenone yielded high molecular weight poly(benzoyl-1,4-phenylene) with an inherent viscosity of 1.1 dL/g (tetrahydrofuran as solvent), and a weight average molecular weight of 57,000 g / mole (polydispersion: 2.3).[84] 2,5-Dichloro-4'-phenoxybenzophenone yielded the corresponding poly(*p*-phenoxybenzoyl-1,4-phenylene). The polymers have the structure:

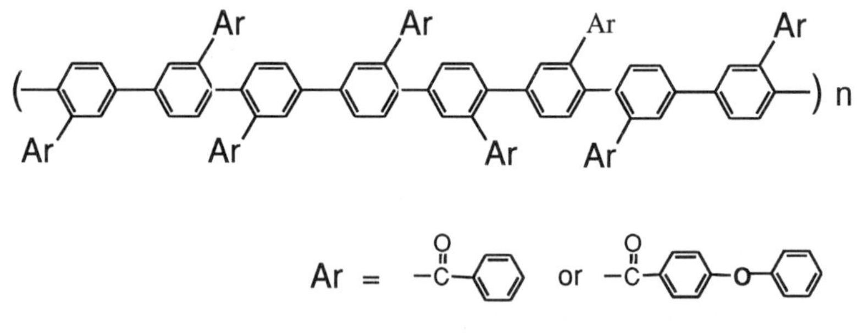

Under certain polymerization conditions, the repeat units were regularly connected in a head-to-tail manner, with the benzoyl groups pointing in one direction along the axis of the polyphenylene backbone.[86] The stereoregular polymer from 2,5-dichlorobenzophenone shows a T_g of 205-206 °C.[86]

Maxdem Inc. (San Dimas, CA) is in the process of developing poly-*p*-phenylenes with benzoyl, *p*-methylbenzoyl, and *p*-phenoxybenzoyl side groups.[87,88] These self-reinforcing polymers are collectively trade named Poly-X. They show a T_g in the range of 140-160 °C

(DSC), and are stable in air up to 500 °C (TGA).[87] The polymers are amorphous and soluble in DMF, chlorinated alkanes, and aromatic alcohols. They are prepared from 2,5-dichlorobenzophenones by Ni-catalyzed aryl coupling polymerization. End capped versions containing reactive ester end groups have been made.[89] Certain of the polymers are also melt processable. Unfilled Poly-X resins have mechanical properties comparable to fiber-reinforced conventional thermoplastics.[87] Tensile moduli of the Poly-X polymers range from 1 to 2.5 million psi.[87] These polymers can also serve as matrix resins for conventional high modulus fibers for high load bearing applications.[87]

Acylated poly-*p*-phenylenes with longer arylene ether ketone side chains may serve as compatible molecular fibers for reinforcement of PAEKs. PAEK side chains have also been attached to other rigid-rod polymer backbones, such as poly(*p*-phenylenebenzobisthiazole):[90]

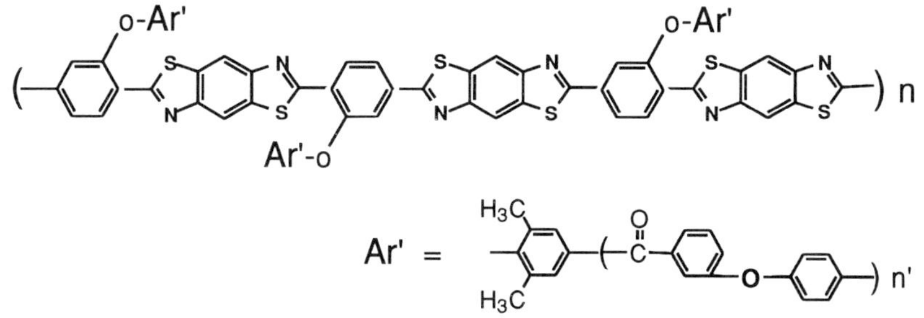

o-Ar' o-Ar'

S N S N S N

N S N S N S) n

Ar'-o

H_3C O

Ar' = C) n'

H_3C O

Such "hairy rod" PAEKs can act as self-reinforcing polymers, and may serve as compatibilizers for blends of rigid-rod polymers and PAEKs.[91] They are also being examined as compatible molecular fibers for reinforcement of PAEKs, to form molecular composites.[92] The flexible side chains effectively suppress thermally induced rod aggregation. Increased frequency of graft sites on the rigid rod backbone is more effective than increased side chain length in reducing rod aggregation.[93]

COMMERCIAL POLY(ARYLENE ETHER KETONE)S AND THEIR APPLICATIONS

PEK is no longer manufactured by ICI, but PEEK (Victrex PEEK) (T_g 144 °C, T_m 334 °C) is still available through Victrex USA. PEEKK (Hostatec) is no longer available from Hoechst. PEKK is available under the trade name of Declar from Du Pont. Declar may be manufactured in the following compositions (mole % terephthalate to mole % isophthalate ratio): 80 / 20 (T_g 165 °C, T_m 360 °C), 70 / 30 (T_g 165 °C, T_m 360 °C), and 60 / 40 (T_g 165 °C, T_m 360 °C). Declar is available in filled and unfilled grades. It is also available in the form of sheets. PEKK with a high content of isophthaloyl moiety may crystallize only slowly, if at all, from the melt. Such polymers appear to be preferred for vacuum drawing processes. Kadel is an unspecified poly(arylene ether ketone) (PAEK) available from Amoco (Atlanta, GA). It may be a PEEK / PEEKK copolymer. It is offered in several grades: (1) unfilled, (2) carbon fiber filled, (3) glass fiber filled, and (4) bearing grade (apparently containing a perfluoropolymer). PEKEKK (T_g 173 °C, T_m 371 °C) is available in various grades from BASF under the trade name of Ultrapek.

Desirable characteristics of PAEK's for applications include the following: (1) high level of mechanical properties over a wide temperature range (rigidity, strength, dimensional stability, low creep, and resistance to heat deformation), (2) favorable tribological properties (high abrasion resistance and low coefficient of friction), (3) good dielectric properties, (4) very good fire resistance, with low evolution of toxic or corrosive gases and very low smoke formation, (5) extremely good resistance to hydrolysis, (6) very good chemical resistance and biocompatibility, and (7) extremely good resistance to high-energy radiation.

PAEKs are gaining access to applications reserved for glass, ceramics, or metals, and have opened up entirely new uses for thermoplastics. They are increasingly used where conventional polymers are inadequate, and offer advantages of more economic fabrication processes when compared to thermosets, ceramics and metals.

Electrical and electronics applications of PAEKs include: (1) wire, cable, and flat conductor insulation coating (geothermal, oil well, mining, nuclear power plants, and thin wall wire products for aircraft and spacecraft), (2) fiber optic jacketing, (3) insulating film, (4) coil formers, (5) connectors (plug-and-socket, cable and hose), (6) carbon brush holders, (7) parts for circuit breakers, (8) lamp bases and holders, (9) ablation shields, (10) flexible printed circuit boards, (11) protective envelopes for sensors and electrodes.

Process engineering applications include: (1) parts for automatic hot water installations, (2) heat exchangers, (3) packing for absorption and distillation columns, (4) seals for valves, and compressors, and valve linings, (5) conveyor belt idlers, and flux links, (6) roller clutches, and guard plates, (7) bearing races, and friction bearings, (8) gears, sleeves, and clamps, (9) pump casings and rotors, (10) ultrafiltration and separation membranes, (12) filter fabrics from monofilaments (for gas and liquids), (13) tubings and discharge nozzles.

Medical engineering applications include: (1) laparoscopy equipment, (2) bone implants, and other orthopedic devices, (3) parts for medical instruments, (4) dialysis membranes and supports, (5) biocompatible tubing and other constructions.

Automotive applications include: (1) parts in the engine periphery, (2) parts in the vicinity of brakes, (3) seals and bushings, (4) bearing races, friction bearings, (5) piston skirts and oil rings, (6) sensor housing, (7) water pump components, (8) transmission bearing cages, (9) carbon fiber reinforced matrix for energy storing flywheels in hybrid electric vehicles.

Aircraft / aerospace applications include: (1) thermoplastic matrix for advanced composites, (2) carbon fiber woven fabric or prepregs, (3) hot air duct fiber composites, (4) jet engine components, (5) radomes (resistant to rain erosion), (6) seating shells, and other light weight interiors of low fire hazard, (7) hot melt structural adhesives, (8) rigid foamed shapes and composites.

Miscellaneous other applications of PAEKs are: (1) fasteners (rivets and screws), (2) parts for printers (ink-jet print heads, nozzle plates, ink channels), (3) amorphous PAEKs for optical components, (4) temperature resistant parts for cookers, (5) coated fibers and powder coating material, (6) nuclear reactor components, (7) biaxially oriented windows for ovens, (8) high temperature conveyor belts, (9) carbon filled conductive heating elements, and (10) heat shrinkable (elastic memory) tubing and couplers.

CONCLUSIONS

Lower cost routes to PAEKs are accessible. New copolymers are being explored. Molecular fiber ether ketones are in the process of development. Three basic types of PAEKs are commercially available (PEEK, PEKK, PEKEKK). The polymers are suitable for harsh environments or premium applications justifying the relatively high cost. They are gaining access to applications previously reserved for glass, ceramics, or metals. Thermoplastic PAEKs offer the advantage of more economical fabrication of precision shaped parts when compared with the use metals or thermosets. The trends in PAEK technology optimization include: (1) reduction of monomer cost, (2) simplification of polymerization processes, (3) high speed, well controlled precision melt processing to final shapes, and (4) broadening of the service temperature range. Further reduction in materials and polymerization costs should lead to expanded applications.

REFERENCES

1. E. Brugel (to Du Pont), US 4,987,171 (1991); I. Goodman, J. McIntyre, and W. Russell, Brit. Pat. 971,227 (1964).
2. T. Buckley III and H. Rapoport, *J. Am Chem. Soc.* 102:3056 (1980); Ref. 3.
3. R. Clendinning, D. Kelsey, J. Botkin, P. Winslow, M. Yousefi, R. Cotter, R. Matzner, and G. Kwiatowski, *Macromolecules* 26:2361 (1993).
4. Y. Sakaguchi, M. Tokai, and Y. Kato, *Polymer* 34:1512 (1993); Ref. 3.
5. G. Olah and S. Kobayashi, *J. Am. Chem. Soc.* 93:6964 (1971).
6. Y. Sakaguchi, M. Tokai, and Y. Kato, *Polymer* 34:1512 (1993).
7. A. Sutter, P. Schmutz, and C. Marvel, *J. Polym. Sci.: Polym. Chem. Ed.* 20:609 (1982).
8. E. Brugel (to Du Pont), US 4,987,171 (1991); Ref. 9.
9. S. Nozawa, M. Nakada, and J. Mihara (to Mitsubishi Kasei Corp.), Jpn. Kokai Tokkyo Koho JP 01161019 (1989); M. Tokai, E. Endo, N. Nishimori, and Y. Kato (to Toybo Co.), Jpn. Kokai Tokkyo Koho JP 01318035 (1989); S. Nozawa, M. Nakada, and Y. Mihara (to Mitsubishi Chem. Ind. Co.), Jpn. Kokai Tokkyo Koho JP 63 00317 (1988).
10. E. Brugel (to Du Pont), US 4,987,171 (1991).
11. V. Jansons and H. Gors (to Raychem Corp.), US 4,912,195 (1990); US 4,709,007 (1987); PCT Int. Appl. WO 84 / 03891 (1984); V. Jansons, H. Gors, S. Moore, R. Reamey, and P. Becker (Raychem Corp.), US 4,698,393 (1987).
12. R. Gale, B. Gilbert, and R. Osteryoung, *Inorg. Chem.* 18:2723 (1979).
13. H. Hashimoto and T. Kobayashi (to Toyobo Co.), Jpn. Kokai Tokkyo Koho JP 03149223 (1991).
14. I. Fukawa and T. Tanabe, *J. Polym. Sci.: Part A: Polym. Chem.* 31:535 (1993).
15. Unpublished Raychem data.
16. F. Gay and C. Brunette, US 4,816,556 (1989).
17. J. Bloom (to Du Pont), PCT Int. Appl. WO 90 / 06957 (1990); US 4,937,135 (1990); A. Moss and G. Oatapchenko (to Du Pont), Eur. Pat. Appl. EP 0 434 323 A2 (1991).
18. W. Risse, D. Sogah, and F. Boettcher, *Macromol. Chem. Macromol. Symp.* 44, 185 (1991).
19. J. Rose (Imp. Chem. Ind.), US 4,396,755 (1983).
20. J. Koch and G. Heinz (to BASF A.-G.), DE 3906178 (1990); Eur. Pat. Appl. EP 386,582 (1990).
21. J. Koch, W. Stegmaier, and G. Heinz, Ger. Offen. DE 3829520 (1990).
22. S. Nozawa and M. Nakada (to Mitsubishi Chem. Ind. Co.), Eur. Pat. Appl. EP 135 938 (1985); Jpn. Kokai Tokkyo Koho JP 61221228 (1986).
23. S. Nozawa, M. Nakada, and Y. Mihara (to Mitsubishi Chem. Ind. Co.), Jpn. Kokai Tokkyo Koho JP 62 119230 (1987).
24. S. Nozawa, M. Nakada, and Y. Mihara (to Mitsubishi Chem. Ind. Co.), Jpn. Kokai Tokkyo Koho JP 63 00316 (1988); S. Nozawa, M. Nakada, F. Sakaguchi, and Y. Mihara (to Mitsubishi Chem. Ind. Co.), Jpn. Kokai Tokkyo Koho JP 62 241 922 (1987).
25. P. Staniland, Poly(ether ketone)s, *in*: "Comprehensive Polymer Science," Vol. 5, G. Allen and J. Bevington, eds., Pergamon Press, New York (1989).
26. V. Jansons (to Raychem Corp.), US 4,361,693 (1982).
27. V. Jansons and H. Gors (to Raychem Corp.), US 4,843,179 (1989).
28. K. Dahl, V. Jansons, and S. Moore (to Raychem Corp.), US 4,808,693 (1989); Ref. 11.
29. J. Koch and G. Heinz (to BASF A.-G.), Ger. Offen. DE 3906178 A1 (1990).
30. Ref. 28.
31. W. Darnell and W, Jackson (to Eastman Kodak Co.), PCT Int. Appl. WO 88 0810 (1988).
32. I. Fukawa, T. Tanabe, and T. Dozono, *Macromolecules* 24:3838 (1991).
33. S. Matsumura and H. Inada (to Teijin Ltd.), Jpn. Kokai Tokkyo Koho JP 03192119 A2 (1991).
34. R. Bruce and J. Seferis, *J. Polym. Sci., Part C: Polym. Lett.* 24:641 (1986); R. McElhaney, D. Castner, and B. Ratner, *ACS Symp. Ser.* 440:370 (1990).
35. A. Aumueller, W. Eberle, G. Heinz, B. Hupfeld, C. Husemann, J. Koch, and H. Reihelt (to BASF A.-G.), Eur. Pat. Appl. EP 528266 A2 (1993).
36. S. Nozawa, H. Noguchi, and F. Sakaguchi (to Mitsubishi Chem. Ind.), Jpn. Kokai Tokkyo Koho JP 05178983 A2 (1993).
37. S. Matsuo (to Idemitsu Kosan Co.), Jpn. Kokai Tokkyo Koho JP 01178520 (1989); U. Hoffmann, F. Helmer-Metzmann, M. Klapper, and K. Mullen, *Macromolecules* 27:3575 (1994); D. Delsey (Union Carbide Corp.), Eur. Pat. Appl. EP 195448 (1986); S. Ebata and Y. Higuchi (to Mitsubishi Gas Chem Co.), Eur. Pat. Appl. EP 306051 A1 (1989); S. Matsuo (to Idemitsu Kosan Co.), Jpn. Kokai Tokkyo Koho JP 01178520 (1989).
38. S. Ebata and Y. Higuchi (to Mitsubishi Gas Chem. Co.), Eur. Pat. Appl. EP 88 11407 (1988).
39. P. Hergenrother, J. Jensen, and S. Havens, *Polym. Prepr.* 26:174 (1985).
40. H. Kricheldorf and G. Bier, *Polymer* 25:1151 (1984); H. Kricheldorf, U. Delius, and K.-U. Tonnes, *New Polym. Mater.* 1:127 (1988); K. Sugie, H. Inada, and S. Matsumura (to Teijin Ltd.), Jpn. Kokai Tokkyo Koho JP 05043680 (1993).

41. H. Kricheldorf and U. Delius, *Macromol. Chem. Rapid Commun.* 10:41 (1989); H. Kricheldorf and U. Delius, *Macromol. Chem.* 190:1277 (1989).
42. H. Colquhoun, C. Dudman, M. Thomas, C. O'Mahoney, and D. Williams, *J. Chem. Soc., Chem. Commun.* 1990:336; M. Mullins and E. Woo (to Dow Chem. Co.), Eur. Pat. Appl. EP 0 413 257 A2 (1991); see also: M. Mullins, R. Galvan, M. Bishop, E. Woo, D. Gorman, and T. Chamberlin, Polym. *Prepr.* 33 (1):414 (1992).
43. G. Deeter and J. Moore, *Macromolecules* 26:2535 (1993).
44. M. Bochmann and J. Lu, *J. Polym. Sci., Part A: Polym. Chem.* 32:2493 (1994).
45. M. Ueda and F. Ichikawa, Macromolecules 23:926 (1990).
46. R. Phillips, V. Sheares, E. Samulski, and J. DeSimone, *Macromolecules* 27:2354 (1994); R. Phillips, V. Sheares, E. Samulski, and J. DeSimone, *Polym. Prepr.* 35 (1):367 (1994).
47. K. Lyon, J. McGrath, and J. Geibel, *Polym. Mater. Sci. Eng.* 65:249 (1991); Ref. 50; Ref. 44.
48. M. Ueda, Y. Seino, Y. Haneda, M. Yoneda, and J.-I. Sugiyama, J. Polym. Sci., Part A: Polymer Chem. 32:675 (1994); S. Matsumura, S. Itoh, H. Imata, and J. Sadanobu (to Teijin Ltd.), US 5,290,906 (1994); Ref. 52.
49. Ref. 46.
50. D. Mohanty, R. Lowery, R. Lyle, and G. McGarth, *Int. SAMPE Symp. Exp.* 32:408 (1987); A. Brink, S. Gutzeit, T. Lin, H. Marand, K. Lyon, J. McGrath, and J. Riffle, *Polym. Prepr.* 33 (1):402 (1992); J. Roovers, J. Cooney, and P. Toporowski, *Macromolecules* 23:1611 (1990).
51. H. Kricheldorf and G. Bier, *Polymer* 25:1151 (1984); D. Kelsey, L. Robeson, R. Clendinning, and C. Blackwell, *Macromolecules* 20:1204 (1987); D. Kelsey, EP 148,633 (1985); Ref. 47.
52. D. Mohanty, T. Lin, T. Ward, and J. McGrath, *Int. SAMPE Symp. Exp.* 31:945 (1986); W. Risse and D. Sogah, *Polym. Prepr.* 31 (1):616 (1990); W. Risse and D. Sogah, *Macromolecules* 23:4029 (1990); Ref. 48.
53. J. Mikroyannidis, *J. Polym. Sci., Part A: Polym. Chem.* 28:669 (1990); A. Pandya, J. Yang, and H. Gibson, *Macromolecules* 27:1367 (1994).
54. A. Pandya, J. Yang, and H. Gibson, *Macromolecules* 27:1367 (1994); H. Gibson, J. Wang, and A. Pandya, *Polym. Prepr.* 34 (1):570 (1993)
55. Ref. 11.
56. R. Reamey (to Raychem Corp.), US 4,665,151 (1987).
57. C. Zhang and Z. Wang, *Macromolecules* 26:3330 (1993); Z. Wang and C. Zhang, *Polym. Prepr.* 33 (2):351 (1992).
58. R. Singh and A. Hay, *Macromolecules* 25:1017, 1025, 1033 (1992); *Polym. Prepr.* 32 (1):393 (1991); V. Percec, R. Clough, J. Fanjul, and M. Grigoras, *Polym. Prepr.* 34 (1):162 (1993); S. Matsuo, N. Yamukai, and C. Kayano (to Idemitsu Kosan Co.), Jpn. Kokai Tokkyo Koho JP 01201332 (1989).
59. S. Matsumura, S. Itoh, H. Inata, and J. Sadanobu (to Teijin Ltd.), US 5,290,906 (1994).
60. Y. Sakaguchi and Y. Kato (to Toyobo Co.), Jpn. Kokai Tokkyo Koho JP 01 29427 (1989).
61. V. Durvasula, F. Stuber, and D. Bhattacharjee, *J. Polym. Sci., Part A: Polym. Chem.* 27:661 (1989).
62. M. Cliffton and J. Geibel (to Phillips Petr. Co.), Eur. Pat. Appl. EP 385299 A1 (1990); J. Geibel (to Phillips Petroleum Co.), US 5,109,102 (1992); A. Nesheiwat and J. Geilbel (to Phillips Petr. Co.), US 5,089,597 (1992); K. Kashiwadate, et al. (to Kureha Kagaku Kagyo Kabushiki Kaisha), Eur. Pat. Appl. 0 369 653 A2 (1990).
63. K. Kashiwadata, Y. Satake, T. Kaneko, M. Tada, and T. Katto (to Kureha Chem. Ind. Co.), Eur. Pat. Appl. EP 0 369 653 A2 (1990).
64. Y. Satake, Y. Inaguma, and S. Yamomoyo (to Kureha Chem. Ind. Co.), Eur. Pat. Appl. EP 435674 A2 (1991); Y. Satake, Y. Inaguma, and J. Masuko (to Kureha Chem. Ind. Co.), Eur. Pat. Appl. EP 459621 A2 (1991).
65. K. Lyon, J. McGrath, and J. Geibel, *Polym. Mater. Sci. Eng.* 65:249 (1991).
66. J. Smith Jr., J. Connell, and P. Hergenrother, *Polym. Prepr.* 33 (2):241 (1992).
67. I.-Y. Wan, D Priddy, G. Lyle, and J. McGrath, *Polym. Prepr.* 34 (1):806 (1993).
68. D. White, et al., *J. Polym. Sci., Polym. Chem.* 19:1635 (1981); B. Jensen, P. Hergenrother, and R. Bass, *Polym. Prepr.* 31 (1):618 (1990); *Polym. Prepr.* 30 (2):132 (1989); C. Borrill and R. Whiteley, *J. Mater. Chem.* 1 (4): 655 (1991); A. Parthiban and N. Sundaram, *J. Polym. Sci., Part A: Polym. Chem.* 31:1233 (1993); N. Bicak, *Angew. Macromol. Chem.* 217:71 (1994); Ref. 71.
69. P. Roberts-McDaniel, J. Connell, and R. Orwoll, *Polym. Prepr.* 35 (1):543 (1994); J. Connell and P. Hergenrother, *J. Polym. Sci., Part A: Polym. Chem.* 29:1667 (1991).
70. J. Smith, Jr., J. Connell, and P. Hergenrother, *Polym. Prepr.* 33 (1):1098 (1992); *Polym. Prepr.* 32 (3):193 (1991); Ref. 71.
71. K. Dahl, P. Horner, H. Gors, V. Jansons, and R. Whiteley (to Raychem Corp.), US 4,868, 271 (1989); PCT Int. Appl. WO 86 02368 (1986).
72. J. Connell, P. Hergenrother, and P. Wolf, *Polym. Matl. Sci. and Eng. Proc.* 63:366 (1990); K. Carter, H. Jonsson, R. Twieg, R. Miller, and J. Hendrick, *Polym. Prepr.* 33:388 (1992).
73. R. Bass and K. Srinivasan, *Polym. Prepr.* 32 (1):619 (1991); R. Bass, K. Srinivasan, and J. Smith, *Polym. Prepr.* 32 (2):160 (1991).
74. J. Hedrick and J. Labadie, *Macromolecules* 23:1561 (1990); Ref. 71.

75. V. Sheares, S. Berndt, and J. DeSimone, *Polym. Prepr.* 33 (2):170 (1992); J. DeSimone, V. Sheares, and E. Samulski, *Polym. Prepr.* 33 (1):418 (1992); J. DeSimone, S. Stompel, and E. Samulski, *Polym. Prepr.* 32 (2):172 (1991); R. Bass and K. Srinivasan, *Polym. Prepr.* 33 (1):485 (1992).
76. K. Carter, R. Miller, and J. Hendrick, *Macromolecules* 26:2209 (1993); J. Connell, P. Hergenrother, and P. Wolf, *Polym. Matl. Sci. and Eng. Proc.* 63:366 (1990); K. Carter, H. Jonsson, R. Twieg, R. Miller, and J. Hendrick, *Polym. Prepr.* 33:388 (1992).
77. H. Calquhoun, J. Daniels, I. Stephenson, and K. Wade, *Polym. Commun.* 32:272 (1991).
78. I. Towle, R. Whiteley, and C. Borrill (to Raychem Ltd.), PCT Int. Appl. WO 9207894 A1 (1992).
79. Y. Sataka, Y. Inaguma, and Y. Suzuki (to Kureka Chem. Ind. Co.), Eur. Pat Appl. EP 459619 A2 (1991).
80. Ref. 64; Y. Satake, Y. Imaguma, and J. Masuko (to Kureha Chem. Ind. Co.), Jpn. Kokai Tokkyo Koho J P 01131236 (1989).
81. R. Kumpf, H. Pielartzek, D. Wicks, D. Nerger, and R. Wehrmann, *Polym. Prepr.* 32 (2):168 (1991); R. Kumpf, R. Wehrmann, H. Pielartzek, and D. Nerger (to Bayer A.-G.), Eur. Pat. Appl. EP 445, 409 (1991).
82. G. Bennett and R. Farris, *Polym. Eng. and Sci.* 34:781 (1994); M. Tokai, S. Endo, and Y. Kato (to Toyobo Co.), Jpn. Kokai Tokkyo Koho JP 01131236 (1989).
83. R. Phillips, V. Sheares, E. Samulski, and J. Simone, *Macromolecules* 27:2354 (1994); *Polym. Prepr.* 35:367 (1994).
84. G. Wang and R. Quirk, *Polym. Prepr.* 35 (1):712 (1994).
85. I. Colon and D. Kelsey, *J. Org. Chem.* 51:2627 (1986).
86. G. Wang and R. Quirk, *Polym. Prepr.* 35 (1):712 (1994); R. Phillips, V. Sheares, E. Samulski, and J. Simone, *Macromolecules* 27:2354 (1994); *Polym. Prepr.* 35:367 (1994).
87. R. Baum, *Chem. and Eng. News* 71 (7):27 (1993); M. Marrocco, M. Trimmer, L.-C. Hsu, and R. Gagne, *SAMPE Symp.* 39:1063 (1994).
88. M. Marrocco, R. Gagne, and M. Trimmer (to Maxdem Inc.), US 5,227,457 (1993).
89. R. Gagne, M. Marrocco, M. Trimmer, and N. Hendricks (to Maxdem Inc.), PCT Int. Appl., WO 9304099 A1 (1993).
90. M. Dotrong, M. H. Dotrong, and R. Evers, *Polymer* 34:727 (1993); U. Vakil, C. Wang, M. H. Dotrong, C. Lee, and R. Evers, *Polymer* 34:731 (1993); M. Dotrong, R. Evers, and G. Moore (to US / Air Force) US 5,194,519 (1993); M. Dotrong, M. H. Dotrong, and R. Evers, *Polym. Prepr.* 33 (1):477 (1992).
91. Y. Chen, W. Stevenson, E. Soloski, T. Vickstrom, C. Lee, and R. Evers, *J. Mater. Sci.* 27:6171 (1971).
92. U. Vakil, C. Wang, Y.-C. Lee, M. H. Dotrong, M. Dotrong, and R. Evers, *Polym. Prepr.* 33 (1):479 (1992); S. Trokhalaki and D. Dudis, *Polym. Prepr.* 33 (1):704 (1992); Y. Chen, W. Stevenson, and E. Soloski, *J. Materials Sci.* 27:6171 (1992); see also: S. Palsule, *ESA Journal* 17:133 (1993).
93. H. Song, M. Dotrong, G. Price, M. H. Dotrong, U. Vakil, U. Santhost, and R. Evers, *Polymer* 35:675 (1994); H. Song, G. Price, U. Vakil, and M. H. Dotrong, *Polym. Prepr.* 33 (1):319 (1992).

MORPHOLOGY MODIFICATION OF POLYMER BLEND FILMS BY ELECTRIC FIELD

Jong-Wook Ha and Seung-Man Yang

Department of Chemical Engineering
Korea Advanced Institute of Science and Technology
Taejon 305-701, KOREA

ABSTRACT

In this study, we consider the morphology evolution of immiscible blend of polystyrene (PS) and poly(vinylidene fluoride) (PVDF) in solution under uniform electric field. When the electric field strength is weak, the dispersed phase deforms slightly from spherical shape and the degree of deformation is proportional to the electrical Weber number. At higher electric field, the dispersed phase is highly elongated and finally fragmented. The linear stability analysis shows that the presence of copolymer made dispersed phase more deformable and unstable.

INTRODUCTION

When two immiscible fluids have different electrical properties, the externally applied electric field induces the electric stress at interface. Furthermore, when the electric conductivity of each phase is not negligible, the antisymmetric surface charge distribution is produced. This surface charge initiates the motion of fluid in order to match the tangential components of electric with hydrodynamic stresses. Since there is an additional electric stress, the normal stress is discontinuous at interface so that the initial steady-shape determined by interfacial tension must be deformed to remove normal stress imbalance[1]. This phenomenon has been studied extensively since it is encountered in various important industrial applications such as electrospraying, electroatomization and enhanced solvent extraction.

The motivation of the present study is twofold. First, the electric field can be used for control of polymer blend morphology. The properties of polymer blend depend on its morphology considerably and anisotropic morphology yields anisotropic mechanical and electrical properties. During the last decade, the new method for enhancing polymer blend film morphology by electric field has been suggested. It has been demonstrated that the

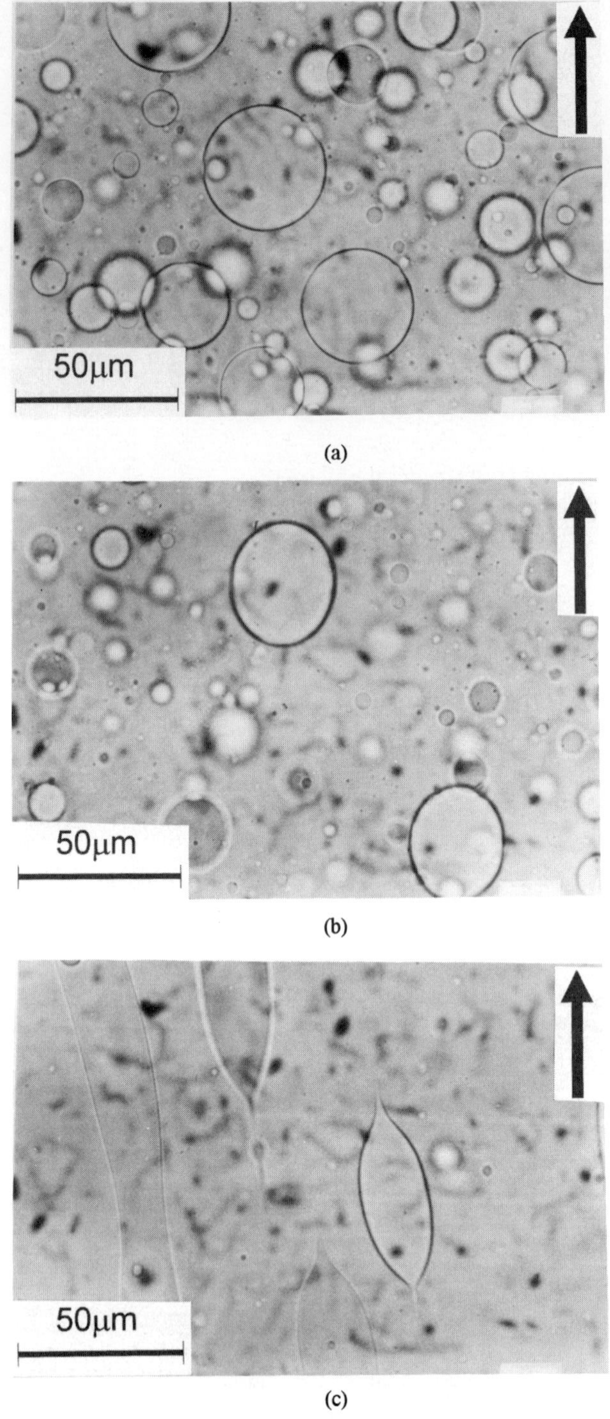

(a)

(b)

(c)

Figure 1. Electric field-induced morphology of (10wt% PVDF / 15wt% PS) blend. The electric field is applied in vertical direction ; (a) $E = 0V/cm$: (b) $E = 260V/cm$: (c) $E = 480V/cm$

electric field has an advantage of easy and exact manipulation in developing polymer film morphology oriented to the desired direction[2,3]. In general, the viscous polymeric droplet is elongated and aligned parallel to the electric field in a viscous medium, as shown in Figure 1. The degree of deformation is increased and the droplet is finally fragmented as the electric field strength increases and exceeds a certain critical value. Secondly, many practical polymer blending processes utilize the compatibilizers as surface active agents to reduce the interfacial tension between two immiscible polymers and thereby to enhance the phase adhesion. In this case, the interfacial tension gradient due to non-uniform distribution of surfactants induces additional flow via the so-called Marangoni effect and thus influences the response of droplet. In the present study, we consider the deformation and stability of a droplet in the applied electric field and determine the critical electric field strength above which the drop cannot sustain the steady-state shape and eventually breaks up. We include here the effects of surfactant on the deformation and stability of a droplet.

EXPERIMENTAL

The commercial grade PS ($\overline{M}_w$= 350000, Lucky Co.) and PVDF ($\overline{M}_w$= 520000, Aldrich) were used as a matrix and a dispersed phase, respectively. The block copolymer of polystyrene and poly (methyl methacrylate) was polymerized anionically. The molar ratio of PS to PMMA was nearly unity and the weight-averaged molecular weight was 75,000 as determined by GPC. The analyzed grade N,N-dimethylformamide was used as a solvent for two polymers. Each polymer was dissolved by rigorous stirring at room temperature in solvent individually and then PVDF solution added dropwisely to PS solution with stirring in order to prepare blends. The fraction of minor component in blend was fixed at as low as 5wt% to prevent the severe coalescence of dispersed phase. When the copolymer was added to blend, it was dissolved in PVDF solution. The electric properties such as dielectric constant and electric conductivity and viscosity, density and interfacial tension were measured. Camera-equipped optical microscope was used to observe the morphology of blends. We used the high voltage DC power supply (Glassman Co., EH series) to generate electric field.

DEFORMATION AND STABILITY OF A DROP IN ELECTRIC FIELD

In this subsection, we report, briefly, our previous study for the effect of surfactant on the drop deformation and stability under the action of a uniform electric field. We employed a purely analytical method of domain perturbation to investigate the deformation and stability of drop shape. Then, the steady-state shape and the concentration distribution of surfactant were determined up to second order in the electrical Weber number to investigate the role of surfactant on the shape deformation and stability of drop as below :

$$\begin{aligned} r &= 1+f \\ &= 1+We_0\delta_{12}P_2(\eta)+We_0^2(\delta_{20}+\delta_{22}P_2(\eta)+\delta_{24}P_4(\eta)) \end{aligned} \tag{1}$$

$$\begin{aligned} \Gamma &= 1+\varpi \\ &= 1+We_0 b_{12}P_2(\eta)+We_0^2(b_{20}+b_{22}P_2(\eta)+b_{24}P_4(\eta)) \end{aligned} \tag{2}$$

In (1) and (2), the electrical Weber number, We_0 $(=\varepsilon_0\varepsilon a E^2/\gamma_0)$, is a dimensionless group defined as a ratio of electric force to interfacial restoring force and $P_n(\eta)$ denotes the n-th degree of Legendre polynomial with $\eta = \cos\theta$. ε_0, ε, a, E and γ_0 stand for the dielectric constant of free space and suspending fluid, radius of undeformed drop, electric field strength, and interfacial tension for the uniform surfactant distribution at the initial state, respectively. When the concentration gradient of surfactant is moderate, We_0 can be related to We for a clean and completely mobile interface in terms of Gibbs elasticity.

In a linear stability analysis, we consider usually an arbitrarily small disturbance to the steady shape and examine whether the drop will return to the steady-state shape or continue to either deform or breakup by solving the corresponding unsteady problem for the disturbance. Recently, however, Kang and Leal[4] and Yang et al.[5] have shown that the critical Weber number which separates the stable and unstable steady-state solution branches can be determined without solving the unsteady disturbance problem. Indeed, following the same theory developed earlier, an estimate of the critical Weber number can be obtained by transforming the We_0-perturbation into a P_2-perturbation in which the small parameter is the magnitude of the $P_2(\eta)$ mode of deformation

$$\zeta \equiv \int_{-1}^{1} f P_2(\eta)\, d\eta \tag{3}$$

instead of We_0. Then, the Weber number as a function of ζ is given by

$$We_0 = \frac{5}{2\delta_{12}}\zeta - \frac{25\delta_{22}}{4\delta_{12}^3}\zeta^2 \tag{4}$$

The expansion in terms of ζ is equivalent to interchanging the dependent and independent variables from ζ and We_0, respectively, to We_0 and ζ. The most important feature of the present solution is that it exhibits multiple steady-state solutions for Weber number below the critical value, but no steady-state solution beyond the critical value. By the transformation, the limit point which appears as a singular point on the stable solution branch at a critical Weber number $(We_0)_c$ is converted to a regular point, and we can determine the Weber number We_0 as a function of ζ for both the unstable and stable branches of the solution curve. The critical Weber number, which occurs when $\partial We_0 / \partial \zeta = 0$, can thus be estimated as

$$(We_0)_c = \frac{\delta_{12}}{4\delta_{22}} \quad \text{at} \quad \zeta_c = \frac{\delta_{12}^2}{5\delta_{22}} \tag{5}$$

RESULTS AND DISCUSSION

For the small deformation analysis, the effects of surfactant convection appears via the parameters β and Ξ. The former represents the sensitivity of the interfacial tension to changes in surfactant concentration and the latter is inversely proportional to the surface diffusivity. It can be easily seen from (1) that when either $\beta \to 0$ or $\Xi \to 0$, the $O(1)$ solution reduces to the Taylor's result. These limits correspond, respectively, to the physical situations where either there is no effect of surfactant on the interfacial tension or where surfactant gradients do not arise due to the dominant surface diffusion. Furthermore, it should be noted that the special case of $SR = 1$ exhibits exactly the same dynamic responses to the electric field as the perfect dielectric materials. In both cases, the electric tangential stresses are continuous and at steady state the fluids are *motionless* in spite of the electric field. Thus, the surfactant concentration must be always uniform. When $\delta_{12} > 0$, the drop is deformed to a prolate and when $\delta_{12} < 0$, to an oblate spheroid. In the case of $\delta_{12} = 0$, the drop always remains spherical, which is independent of the electric field strength.

In Figure 2, the effects of ratios (R and S) of resistivities and dielectric constants on δ_{12} are demonstrated. In Regions I and III of Figure 2, the shape function δ_{12} is positive and thus the drop is a *prolate* spheroid. As noted from the present solution, for the perfect dielectric materials or the case of $SR = 1$ which separates Region I and III, the drop shape always prolate except when $S = R = 1$. In the latter case, the drop remains spherical at all times. However, when $SR \neq 1$ or at least one of the fluids is not perfect dielectric, the tangential components of electric stresses are not continuous and inevitably the boundary driven flow is generated. When $SR > 1$ (i.e., for Regions II and III), the boundary driven velocity field is an axisymmetric biaxial straining flow. On the other hand, when $SR < 1$ (i.e., for Region I), the fluid outside of the drop becomes uniaxially straining. Consequently, the fluid motion alone in Regions II and III would deform the drop into an oblate spheroid. However, the actual deformed shape in Region III is prolate due to the dominant normal stress imbalance associated with the electric field. It is worth pointing out that the electrical *normal* stress mismatch would deform always the drop into a prolate spheroid, which is independent of the fluid properties. Meanwhile, in Region II, the drop deformation is dominated by the flow induced by the tangential stress mismatch and thus the drop becomes oblate. It can be also noted from Figure 2 that the viscosity ratio λ has very weak effect on separating Regions II and III in which the drop deforms into either an oblate or a prolate spheroid. The presence of surfactant causes larger deformation than would occur for a drop with a constant interfacial tension equal to the initial equilibrium value. The physical mechanism for the increase in deformation is that the flow sweeps surfactant towards the ends of the drop, where it acts to locally lower the interfacial tension. This, in turn, requires larger deformation to balance viscous forces.

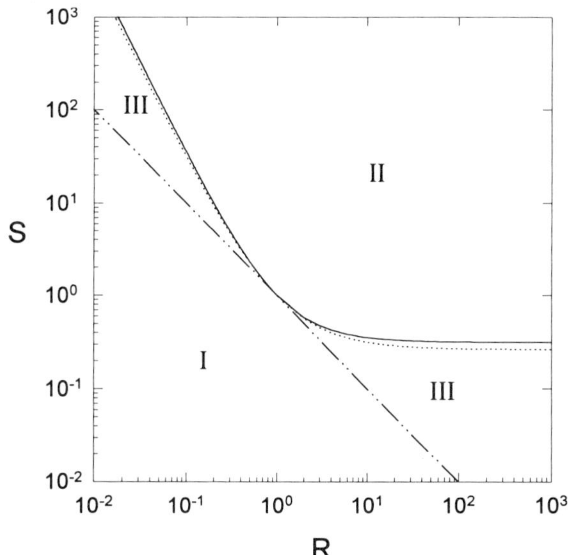

Figure. 2 Loci defined from $O(1)$ steady-state shape characterizing the steady deformation modes. Solid line : $\lambda=0$; dotted line : $\lambda \to \infty$.

In Figure 3, the Weber number We is plotted as a function of ζ for the prolate-type deformation in the absence of surfactant. Also included for comparison are the deformed drop shapes at the corresponding critical states. It can be seen that although the degree of deformation at the critical state does not change significantly as the permittivity ratio S increases, the critical value of the Weber number for the existence of a steady-state solution is monotonically increased.

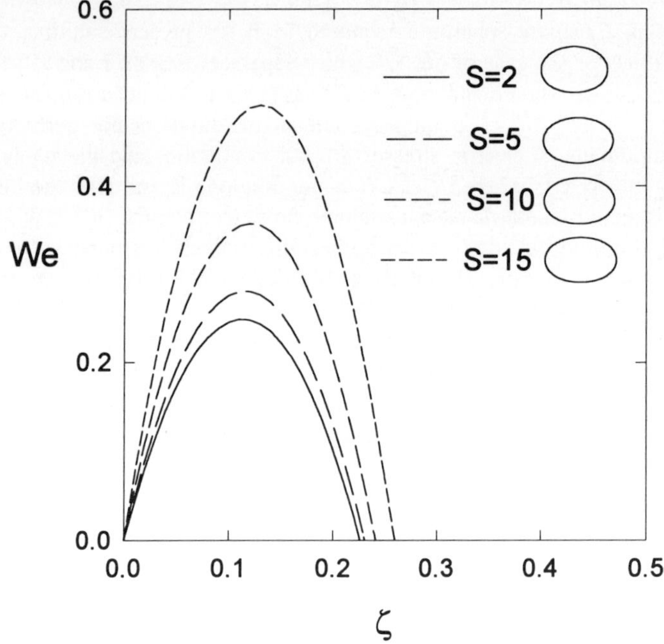

Figure. 3 Perturbation solutions for steady-state shape as a function of ζ for the prolate deformation with $R=0.1$ and $\lambda=1$. The drop shapes at the critical Weber numbers are also included.

Although the corresponding plots for the effects of surfactant on the stability of drop deformation are not reproduced here, our theory shows that as β increases (i.e., the sensitivity of the interfacial tension to changes in surfactant concentration increases), the drop becomes less stable. The stability is also reduced with increase in Ξ which is responsible for nonuniformity in surfactant concentration distribution. However, surfactant does not change the general features of stability, and in the presence of surfactant the oblate drop can also sustain a large deformation compared to the prolate drop. In addition, the effect of surfactant in reducing the critical Weber number is more significant for the oblate deformation than for the prolate one. It is perhaps worthwhile to repeat the statement that when $S=R=1$, the electric stresses are continuous and the fluids remain motionless at all times. As expected, very large Weber number is required to break up a drop near this region in the absence of surfactant. However, in this region surfactant is shown to be very effective in lowering the critical Weber number. The present theory clearly indicates that the addition of surfactant to an immiscible fluid mixture can create finely dispersed morphology and phase stability due to its emulsifying effect.

The immiscible polymer blends of PS and PVDF considered in the present study have the electric properties pertaining to the prolate deformation. In Figure 4, the degree of deformation in the absence of surfactant is illustrated as a function of the Weber number for various blend compositions. The experimental data contained in this plot were obtained up to the critical Weber numbers beyond which stable steady-state shapes did not exist. Thus, the last data point for each blend composition corresponds to the limit point. It can be noted easily that as PS content in the continuous phase increases the degree of deformation is decreased. This is due to the fact that increase in PS concentration not only lowers the conductivity of the continuous phase but also increases the interfacial tension. Thus, the parameter R as well as the Weber number are decreased with the content of PS in the continuous phase. As predicted from the theory, the general features of drop deformation is not changed appreciably by the viscosity ratio. The scattering of the experimental data may arise due either to the interaction between the neighboring drops or to the induced dipole or net charge which could not be eliminated completely. The presence of neighboring drops nearby one drop may cause coalescence and disturb the electric field which gives rise to a slight non-homogeneity of electric field. It is noteworthy that in the small Weber number limit the degree of drop deformation and the Weber number follow the theoretically predicted linear relationship.

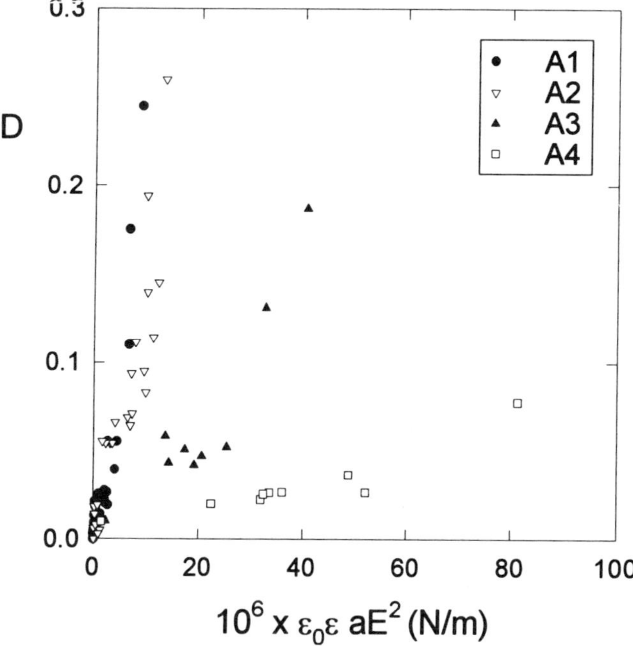

Figure. 4 Degree of deformation of dispersed phase as a function of Weber number in the absence of copolymer.
A1 : 5wt% PVDF / 10wt% PS
A2 : 5wt% PVDF / 15wt% PS
A3 : 5wt% PVDF / 20wt% PS
A4 : 5wt% PVDF / 30wt% PS

Table 1. Comparison of theoretically predicted critical Weber numbers with the experimental data

Sample	Predicted We_c	Experimental We_c
A1	0.29	0.36
A2	0.26	0.32
A3	0.23	0.26

From stability analysis as performed in the previous section, we can predict the critical Weber number and the corresponding degree of deformation. Table 1, the critical Weber numbers predicted from the theory are compared with the corresponding experimental data for various blend compositions. Considering that the experimental data have neither been corrected for the non-homogeneity of electric field nor for the presence of boundaries of the flow cell, and in view of the difficulties of preventing the drop migration induced by electrophoresis or dielectrophoresis and removing net charges, the agreement is quite good.

The effect of surfactant on he drop deformation is shown in Figure 5. A trace of copolymer produces a considerable change in the drop deformation. Owing to the addition of surface active copolymer, the drop becomes more deformable and unstable as predicted from the theory. The effect of copolymer is enhanced as the concentration of copolymer increases. In our experimental system, at least two kinds of breakup modes were observed. When the electric field exceeded a certain critical value, both of the poles of the drop became pointed and then, the small droplet or fluid jets were ejected from the tip of the cone, which is the so-called tip streaming (see, Figure 1(c)). At even higher electric field strength, the liquid jet was fragmented to small droplets. The tip streaming of drop and the fracture mode are observed for a large drop which can be highly elongated.

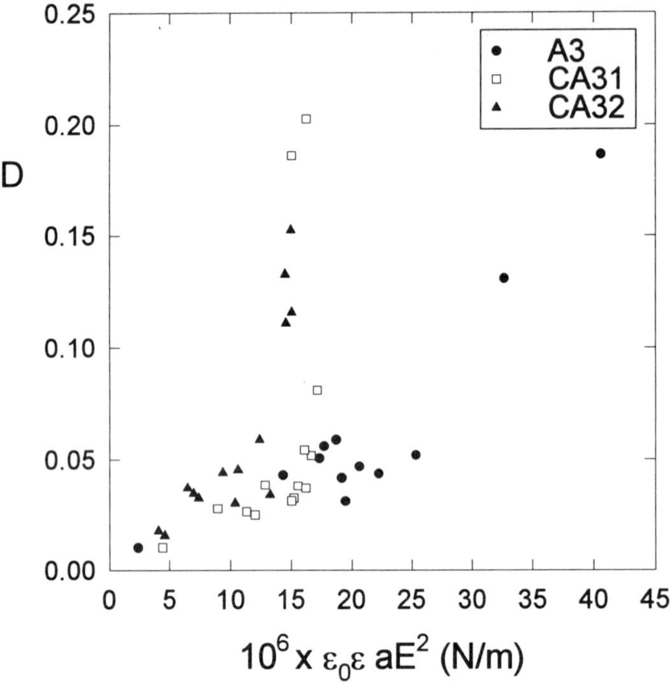

Figure. 5 Degree of deformation of dispersed phase as a function of Weber number in the presence of copolymer.
A3 : 5wt% PVDF / 20wt% PS
CA31 : (5wt% PVDF+0.05wt% PS-b-PMMA) / 20wt% PS
CA32 : (5wt% PVDF+0.15wt% PS-b-PMMA) / 20wt% PS

CONCLUSION

When an electric field is applied, the dispersed phase deforms and the degree of deformation is proportional to the electrical Weber number in the small deformation limit as predicted from the theory. When electric field is increased further, the dispersed phase becomes unstable, i.e., the drops experience time-dependent shape deformation. The stability analysis developed in this study predicts the morphology of polymer critical Weber number above which a steady-state drop shape cannot exist. The results agree fairly well with both the experimental data and the exact numerical predictions which are available. The addition of copolymer increases the degree of deformation and reduces the stability of drop. This is also consistent with the theory. Electric field is proved effective in aligning the dispersed phase to a desired direction and the morphology of polymer blend is easily controlled by the direction and strength of the applied electric field.

REFERENCES

1. Taylor, G. I., Studies in electrohydrodynamics. I. The circulation produced in a drop by an electric field, *Proc. R. Soc. Lond.*, A291 : 159 (1966)
2. Moriya, S., Adachi, K. and Kotaka, T., Deformation of droplets suspended in viscous media in an electric field : I. rate of deformation, II. Burst behaviour, *Langmuir*, 2 : 155(1986)
3. Venugopal, G. and Krause, S., Modification of polymer blend morphology using electric fields, *J. Polym. Sci. C : Polym. Lett.* 27 : 497 (1989)
4. Kang, I.S. and Leal, L.G., Small-amplitude perturbations of shape for a nearly spherical bubble in an inviscid straining flow (steady shapes and oscillatory motion), *J. Fluid Mech.* 187 : 231 (1988)
5. Yang, S.-M., Feng, Z.C. and Leal, L. G., Nonlinear effects in the dynamics of shape and volume oscillations for a gas bubble in an external flow, *J. Fluid Mech.* 247 : 417 (1993)

REACTIVE BLENDING OF PET AND LIQUID CRYSTALLINE POLYMER

C.I. Hwang and S.B. Kil and O O. Park

Department of Chemical Engineering
Korea Advanced Institute of Science & Technology,
373-1 Kusung-Dong, Yusung-ku, Taejon, 305-701, Korea

Copolyesters were prepared by reactive blending of poly(ethylene terephthalate)(PET) and a thermotropic liquid crystalline polymer, poly(p-hydroxybenzoic acid-co-ethylene terephthalate) under various operating conditions. In this investigation the blends of PET and LCP were analyzed by gel permeation chromatography(GPC), nuclear magnetic resonance spectroscopy(NMR), differential scanning calorimetry(DSC), and dynamic viscoelastic measurement.

GPC results and ^{1}H-NMR analysis revealed that PET/LCP blend were subjected to transesterification and chain scission reaction depending upon reaction conditions, resulting in reorganization of chain sequence. And acetate end group was found to be closely related to the final molecular weight of blends. From thermal analysis, the crystallization of PET were delayed and melting temperature were depressed with increasing LCP content due to exchange reactions. The rheological properties of reactive blends also showed delayed crystallization and rheologically homogeneous behavior.

INTRODUCTION

The blending of usual thermoplastic coil-shaped polymers with thermotropic liquid crystalline macromolecular material has been considerable interest in development of new advanced materials. Blending PHB/PET copolyester with various other polyesters represents an interesting procedure for application and fundamental studies of the properties of LCP/polymer blends.[1 – 4] Studies on the blends of this LCP with PET have been conducted by various research groups.[5 – 8] Blending of polymers is one of the effective means for improving their properties. However, in general, when a rigid rod polymer and a flexible polymer are simply mixed together, one would expect them to be immiscible in the first place.

Reactive blending may provide a useful way to improve physical properties during the preparation of new polymeric materials from already existing polymers. The reactive blending is particularly suitable for polycondensates due to the presence of many reactive groups along the polymer chains. The potential utilization of reactions between components of polyester blend has been found of interest. Friedrich and coworkers [9] reported that annealing the LCP/PET blend in the melt caused transesterification reactions that resulted in depression of melting temperature of the PET by 42 °C. Lee and Dibenedetto[10] suggested that chemical reactions can be used to improve the adhesion between aromatic fibers and thermoplastic matrix. The objectives of this present

Polymers and Other Advanced Materials: Emerging Technologies and Business Opportunities. Edited by P. N. Prasad *et al.*, Plenum Press, New York, 1995

investigation are to induce modification of the properties of PET by reactive blending with LCP, and to investigate the effects of chemical reaction on resulting properties of blends.

EXPERIMENTAL

The polymers used in this work were liquid crystalline polymer and PET. The LCP consists of a 60 mol % PHB and 40 mol % PET, obtained from Mistsubishi Kasei. PET was kindly supplied from Cheil Synthetics Inc. The weight average molecular weight of PET was 75,000 and that of LCP was 58,000. All materials were dried in a vacuum oven at 100 °C for 12 h to minimize moisture effects before use. Several reactive blends with two different blend composition were prepared by the melt-phase reaction of PET and LCP under various reaction conditions. All reactions were conducted in a cylindrical flask equipped with a vacuum distillation head, stainless steel agitator, and condensate trap. When the temperature of the mixture of PET and LCP reached to preset temperature, the agitator was engaged and vacuum applied during the blending. Moreover, simple blends of PET/LCP were prepared by using a Mini-Max Extruder. The processing temperature were ranged between 260 and 270 °C and mixing time was fixed within 2 minutes to minimize thermal degradation and transesterification.

Molecular weight and distribution were obtained by gel permeation chromatography(GPC). GPC measurements were carried out using a Waters 150C provided with three μ-stragel columns connected in series with porosity of 10^3, 10^4, 10^5 Å and OCP/$CHCl_3$(1:4) was used as the eluent at 35 °C. Calorimetric measurements were performed to evaluate thermal transitions of polymers in a Du Pont DSC 910. All samples were scanned from 30 °C to 300 °C at a heating rate of 10 °C /min . After first scan the samples were quickly quenched to room temperature and reheated using the same conditions. For monitoring the crystallization behavior from cooling experiments, the sample after second heating were scanned at a rate of 10 °C /min. NMR measurement were carried out using the Bruker AMX FT–NMR spectrometer operating at a resonance frequency of 500 MHz for protons. Rheological studies for all the materials were conducted on rotational rheometer (PHYSICA Rheo-Lab MC120) in the cone and plate configuration. Gaps employed were approximately 50 μm in all cases. Measurements were carried out within the range of linear viscoelastic response in a nitrogen atmosphere.

RESULTS AND DISCUSSION

Material Characterization

The detailed molecular weight distribution of LCP and PET are shown in Fig.1. It can be seen that some shoulder on GPC curve is shown in LCP. And it is likely that the shape of molar mass distribution is related to the copolymerization procedure of this copolyester. PET has higher fraction of molecular weight accompanying very low molecular weight of species.

Figure 2 shows the DSC thermograms of the second heating for PET and LCP. The glass transition temperature of PET is shown around at 86 °C. Moreover, a distinct exothermic crystallization peak is seen since quenched, amorphous PET is recrystallized during this heating scan. Further heating to higher temperature reveals typical endotherm of PET around 250 °C. The heating curve for LCP shows an increase of heat capacity at 87 °C due to the glass transition of the amorphous PET-rich phase and very weak endotherm between 170 and 250 °C due to a wide crystallite size distribution and/or defects. No glass transition for a PHB-rich amorphous phase was detected by DSC. Beyond 300 °C no further transition were seen until the decomposition of this copolyester.

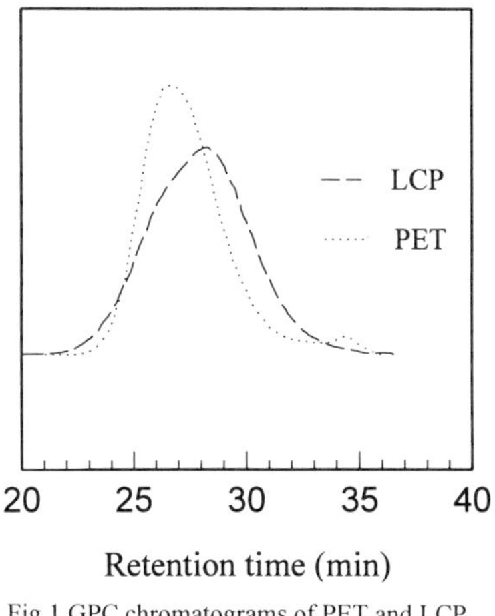

Fig.1 GPC chromatograms of PET and LCP

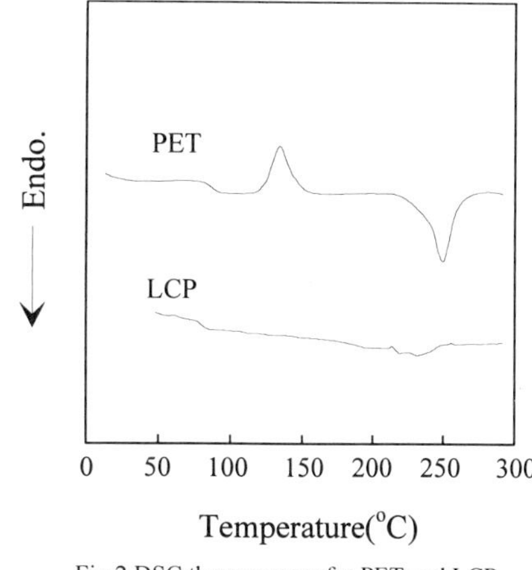

Fig.2 DSC thermograms for PET and LCP

Reactive Blend

Experimental conditions and GPC results for reactive blends are summarized in Table 1 and typical GPC curves for blends prepared from PET70/LCP30 are shown in Fig.3. The blends of A and B prepared at 280 °C under closed system yielded relatively low molecular weight polymers. This indicates that chain scission reaction occurred dominantly at 280 °C under closed system and it is essential to employ high vacuum to obtain high molecular weight polymers. In fact, since the low concentration of end group and the high concentration of ester group within the chains of two reactant polymers lead to many side or degradation reactions at high temperature, the removal of the by-products is necessary to maintain high molecular weight chains. In case of reaction temperature of 280 °C under reduced pressure, GPC curves of blend D exhibits higher molecular weight than B. When the reaction were performed at 270 °C, blend of K reveals the highest molecular weight of polymer prepared from PET70/LCP30 composition. Similar results were obtained for blends prepared from PET90/LCP10 composition. At high temperatures, many reactions can occur simultaneously, leading to chain scission or changing the chemical nature of end group, the relative role of these reactions is usually determined by the reaction conditions and by the catalysts employed.[11]

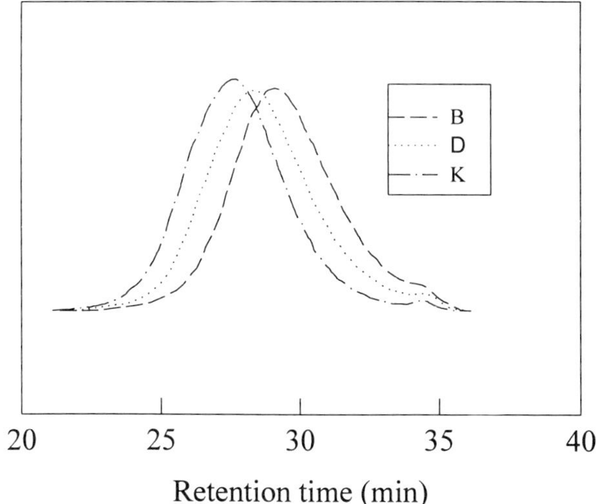

Fig.3 GPC curves for reactive blends of B, D, and K

Table 1. Experimental conditions and GPC results for reactive blend

Composition (wt. %)	Code	Temp. (°C)	Pressure (torr)	Time (min)	$\overline{M_n}$	$\overline{M_w}$
PET90/LCP10	A	280	No Vac.	60	4,070	13,120
	C	280	0.2	60	16,600	49,330
	H	270	0.2	60	12,650	50,540
PET70/LCP30	B	280	No Vac.	60	5,520	20,440
	D	280	0.2	60	8,680	38,850
	K	270	0.2	60	16,580	52,300

^{1}H-NMR Analysis

To investigate the reaction involved during blending, ^{1}H-NMR spectra can be used to obtain more information. Some of the typical proton NMR spectra for blends of B, D and K are shown from Fig.4.

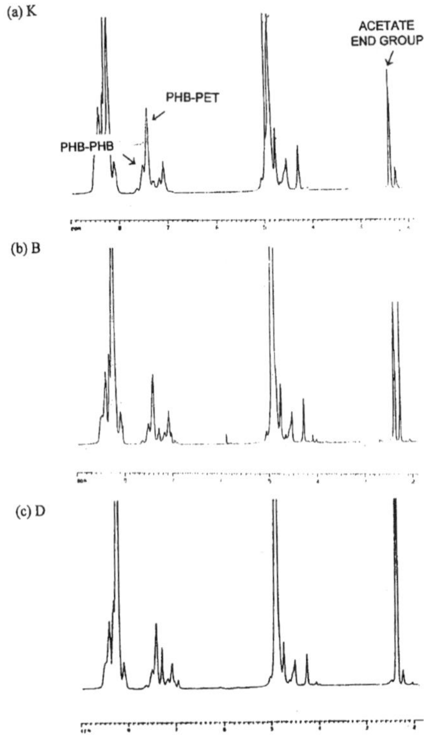

Fig.4 ^{1}H-NMR spectra of reactive blends of (a) K (b) B and D

From the NMR spectra, the most significant changes are increased peak intensity at 2.5 ppm based upon the ^{1}H-NMR spectra of LCP and PET. The peak at 2.5 ppm is attributed to the phenylene acetate end group and the peak at 2.3 ppm corresponds to ethylene acetate end group.[12] These two types of acetate end group can be found originally in small quantity both in PET and LCP.[13] The evolution of acetate end group is a byproduct of the complex degradation reactions.[14,15] The high reactivity of phenylene acetate group compared with ethylene acetate group is reflected higher peak intensity at 2.5 ppm than peak intensity at 2.3 ppm in Fig.4. Thus, high concentration of phenylene acetate as a residue within product as well as corresponding low molecular weight polymer are attributed to degradation reactions.

It can be shown that molecular exchange reaction between PET and LCP occurs. The direct evidence for transesterification of LCP/PET blends can be monitored by the relative amount of absorption peaks located in the region of $\delta = 7\sim7.8$ ppm, in which only LCP exhibits absorption in the ^{1}H-NMR spectra. If there were no transesterification reaction between LCP and PET, there should be no change in relative intensity in the region of $\delta = 7\sim7.8$ ppm. The peak area of PHB-PET bond relative to that of PHB-PHB bond for all blends in Fig.4 increased to above 3 from the initial ratio of 2.78 for pure LCP. The increasing value of dyad ratio represents actually that the peak at 7.42 ppm is increased while peak at 7.52 ppm is decreased. On the other hands, it indicates that the LCP have been subjected to transesterification resulting in randomization of chain sequence during the blending. It was confirmed that the LCP within the resulting polymer blend exhibited some loss of liquid crystallinity due to randomization reaction from the weaker depolarized light intensity compared with corresponding simple blends.

Thermal Studies

Fig.5 shows DSC thermograms for reactive blends of A, B, C, and D. The appreciable differences are observable with respect to depression of melting peak temperature of PET in blend with LCP content and vacuum applied as clearly seen in Fig.5. The melting temperature depression of PET in the blend of high LCP content were also reported by Joseph *et al.*[7] The reason for severe depression of melting temperature in reactive blend with high LCP content is largely attributed to the reduction of structural regularity of PET in blend due to high reactivity of phenol end group generated from the LCP in reactive blending.[16] Fig.6 shows the crystallization behavior of reactive blends from the melt state. The overall tendency of hindered crystallization is distinctly seen for blends with higher LCP content. Moreover, for the blends containing same LCP content, blends prepared under vacuum system exhibit more delayed crystallization behavior. This is possibly due to low crystallization rate for higher molecular weight of polymers such as C and D compared with polymers of A and B having low molecular weight of molecules even if the extent of transesterification is comparable.

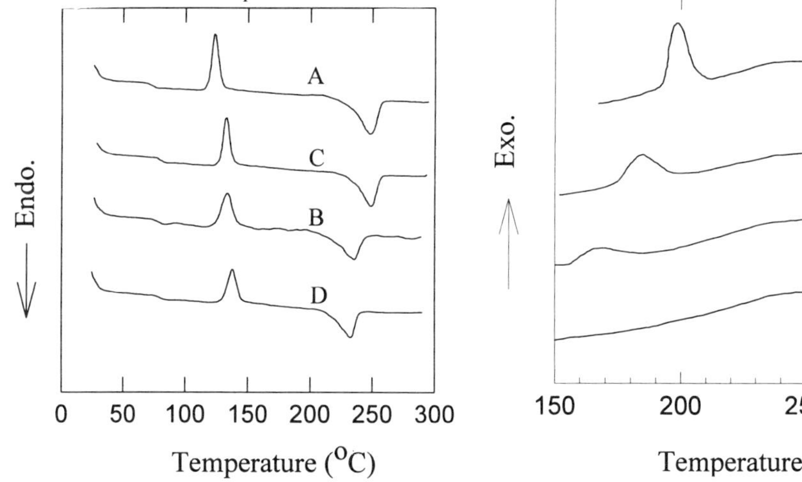

Fig.5 DSC thermograms of reactive blends of A, B, C, and D

Fig.6 DSC Cooling traces of reactive blends

Rheological Behaviors

Fig.7 shows the results of isothermal time sweep measurements of three blends with same LCP content of 30 wt. %, PET70/LCP30, D, and B at 200 °C(upon cooling from 270 °C), respectively. The simple blend(PET70/LCP30) is seen to solidify in about 100 sec, and it takes about 260 sec for solidification for B while it is more delayed in D. It is likely that relatively high molecular weight of D takes longer time in crystallization than low molecular weight of B. Moreover, the physical blend shows solidification behavior at 210 °C and it takes about 6 times longer compared with its solidification at 200 °C while D and B does not show the appreciable solidification.

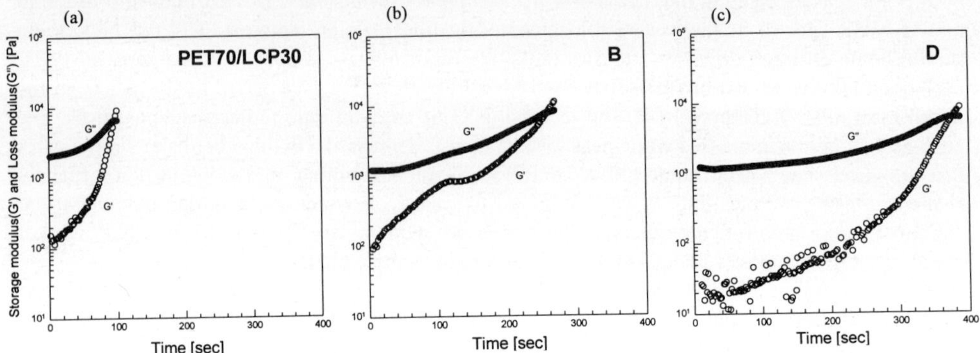

Fig.7 Isothermal time sweep of (a) PET70/LCP30 (b) B and (c) D at 200 °C(ω = 10 rad/sec)

Therefore, the crystallization/solidification of PET in reactive blend is delayed due to decreased structural irregularities as a result of exchange reaction.

The plots of logG' versus logG" [17,18] of PET and LCP at various temperatures are displayed in Fig.8. Note that LCP was cooled from 320 °C for 5 min for each frequency sweep to avoid the effect of thermal history. The plots of logG' versus logG" for PET in the isotropic region are fall into a single line, showing a temperature-independent correlation. The PET melt is a homogeneous polymer at the given temperatures and it is expected to have a slope of 2 in the terminal region.[19] However, logG' versus logG" for LCP show temperature dependence, indicating that the morphology of LCP varies with temperature.[18] As the temperature is increased from 250 to 290 °C, the slope of logG' versus logG" plots move towards the value of 2. Note that the LCP has biphasic behavior and the relative fraction of isotropic increases with temperature.[13] The plots of logG' versus logG" of PET, LCP and blends of LC10, LC30, C and D at 270 °C are shown in Fig.9.

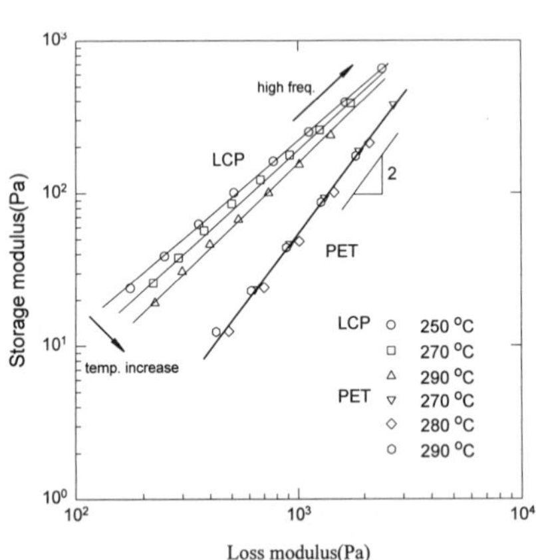

Fig.8 log G' vs. logG" plot for LCP and PET.

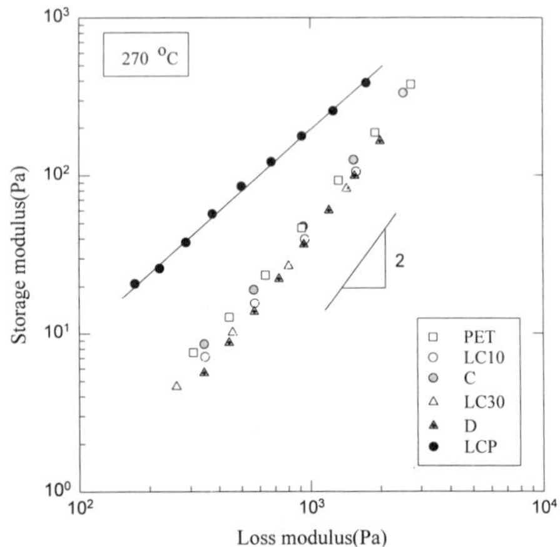

Fig.9 log G' vs. log G" plot for LCP and PET. and physical blends(LC10,LC30) and reactive blends(C, D)

The plot of logG' versus logG" for C is fall onto that of PET with a slope of 2 suggesting that blend of C is almost homogeneous melt at 270 °C. This may be homogenizing effect in blend of C due to randomization reaction. Blend of D also has similar slope parallel with that of PET and blend of C but they are located right, possibly due to low molecular weight compared with blend of C. Since the rheological behavior of physical blend is affected more or less by LCP, it is expected that the slope of plots of logG' versus logG" for LC10 and LC30 would falls into region between that of LCP and PET. But two physical blends are PET-rich blend and the slope is more likely to be shift toward that of PET. And LC30 would have relatively many LCP domain compared with LC10. It is thought that the minute differences in moduli between two types of blend is attributed to the almost comparable viscosity between two blend component at the low range of testing frequency. However, the frequency sweep at 270 °C shows that the reactive blend of C and D exhibits slightly higher complex viscosity than that of corresponding simple blends. Brostow[20] reported similar reports that longer mixing has resulted in higher moduli and higher viscosity in the blend system of PET and PHB60/PET. Our observation of the SEM morphology reveals that the reactive blend contains smaller, uniform domain size compared with the larger, irregular, and diverse-in-size domains in the simple blend. The fine LCP domain size in reactive blends is attributed to combined effect of mixing and chemical reactions. These effect cause to reduce the interfacial tension in the melt, leading to a finer dispersion of LCP phase in PET matrix. In addition, it results in large interfacial area in reactive blend with finer morphology. It is likely that the reactive blend is associated with the more miscible system and better adhesion than simple blend. Further investigation on this matter is in progress.

CONCLUSIONS

Reactive blending of PET and LCP were performed under various blending conditions and it gave some insight into modification of physical properties of polycondensation polymer. In addition, processing conditions can have a dramatic effect on the properties of the resulting blend. In this investigation, the high reactivity and concentration of phenylene acetate end group was responsible for reducing molecular weight of product. And the direct evidence for molecular exchange reaction between PET and LCP can be monitored by the reduced ratio of absorption peaks for PHB-PHB dyad and PHB-PET dyad.

Transesterification between the two component of the blends at high temperature has led to chemical modifications that influence markedly the blend behaviors. The effect of exchange reaction has led to some loss of mesomorphic order in the reactive blend based on the observation of weaker depolarized light intensity and smaller LCP domain size. Moreover, the melting temperature depression and hindered crystallization are attributed to the lowered structural regularity due to exchange reactions and are markedly affected by the LCP content and blending temperature. The reactive blends showed rheologically homogeneous behaviors due to randomizing reaction.

REFERENCES

1. J.F. Croteau and G.V. Laivins, Study and Control of Phase Morphology in Liquid Crystal Polyester-Poly(alkylene Terephthalate) Blends, *J. Appl. Polym. Sci.*, 39:2377 (1990).
2. A. Zaldua, G.Munoz, J.J. Pena, and A. Santamaria, Physical properties of blends of a liquid crystalline copolyester with poly(butylene terephthalate), *Polymer*, 32:682 (1991)
3. S.H. Jung, and S.C. Kim, Morphology and Mechanical Properties of Poly(ethylene terephthalate)-Poly(hydroxybenzoic acid) and Polycarbonate Blends, *Polym. J.*, 20:73 (1988).
4. T. Kyu, and P. Zhuang, Phase segregation in thermotropic liquid crystalline copolyester and polycarbonate blends, *Polym. Commun.*, 29:99 (1988).
5. S.K. Sharma, A. Tendokar, and A. Misra, Blends of liquid crystalline polyesters and polyethylene terephthalate-Morphological studies, *Mol. Cryst. Liq. Cryst.*, 157:597 (1988).
6. E.G. Joseph, G.L. Wilkes, and D.G. Baird, Preliminary thermal and structural studies of blends based on a thermotropic liquid crystalline copolyester and Poly(ethylene terephthalate), in: " Polymer Liquid Crystals," A. Blumstein Ed., Plenum Press, New York(1984).

7. E.G. Joseph, G.L. Wilkes, and D.G. Baird, Thermal and structural studies of flexible and semi-rigid polymeric blends, *Am. Chem. Soc., Div. Polym. Chem., Polym. Prepr.*, 24:304(1983)
8. A. Nakai, T. Shimaku, H. Hasigawa, and T. Hasimoto, Spinodal decomposition of polymer mixtures with a thermotropic liquid crystalline polymer as one component, *Macromolecules*, 19:3008 (1986).
9. K. Friedrich, M. Hess, and R. Kosfeld, Investigation of blends containing coil polymers and thermotropic main-chain polymers, *Makromol. Chem., Macromol. Symp.*, 16:251(1988)
10. W.C. Lee, and A.T. Dibenedetto, Processing of thermotropic liquid crystalline polymers and their blends-Analysis of an in-situ LCP composite system, *Polym. Eng. Sci.*, 32:400(1992)
11. F. Pilati, Solid-state polymerization, in: "Comprehensive Polymer Science," G. Allen, Pergamon Press, Oxford(1989).
12. E.E. Riecke, and F.L. Hamb, Copolyesters of Glycols and Bisphenols: Concurrent reactions of a new preparative process, *J. Polym. Sci.*, Polym. Chem., 15:593 (1977).
13. C.I. Hwang, *Ph.D.* Dissertation, KAIST(1995)
14. J.-M. Besnoin and K.Y.Choi, Identification and characterization of reaction byproducts in the polymerization of Polyethylene Terephthalate, *JMS-REV. Macromol. Chem. Phys.*, C29:55 (1989)
15. K. Ravindranath, and R.A. Mashelkar, PET-1. Chemistry, thermodynamics and transport properties, *Chem. Eng. Sci.*, 41:2197 (1986).
16. C.I. Hwang, and O.O. Park, Effects of chemical reactions on the properties of copolyesters made from Poly(ethylene terephthalate) prepolymer and liquid crystalline polymer, *Polym. J.*, 27:232(1995)
17. C.D. Han, and H.K. Chuang, Criteria for rheological compatibility of polymer blends, *J.Appl. Polym. Sci*, 30:4431 (1985).
18. S.S. Kim, and C.D. Han, *Polymer*, Oscillatory shear flow behaviour of a thermotropic liquid-crystalline polymer, *Polymer*, 35:93 (1994).
19. J.D. Ferry, Viscoelastic Properties of Polymers, 3rd Ed., Wiley, New York, (1980).
20. W. Brostow, T. Dziemianowicz, J. Romanski, and W. Werber, Transmission of mechanical energy through polymeric liquid crystals and their blends, *Polym. Eng. Sci.*, 28:785(1988)

MISCIBILITY OF POLYMER BLENDS AT THE AIR/WATER INTERFACE

Chang Sik Ha[1], Won-Ki Lee[1], Won-Jei Cho[1], and Hiroo Nakahara[2]

[1]Department of Polymer Science and Engineering, Pusan National University, Pusan 609-735, Korea
[2]Department of Chemistry, Faculty of Science, Saitama University, Urawa 338, Japan

INTRODUCTION

The miscibiity of polymer blends has been one of the main subjects in polymer science and technology[1-3]. A considerable amount of works have been reported on the miscibility of two polymers in the bulk state but there are only a few studies that reveal whether the miscibility of polymer mixtures in the bulk state correlates well with that in the two dimensional state or not. Investigation of polymer monomolecular films speread at the air/water interface leads to the interest of miscibility as well as static and dynamic properties in the two-dimensional system[4-8]. It is not easy, however, to choose a good pair to compare the miscibility in the bulk state with that in the film spread at the air/water interface, since almost polymers do not form stable monolayers at the air/water interface. Blends of poly(methyl acrylate)(PMA) and poly(vinyl acetate)(PVAc) as well as poly(ethylene oxide)(PEO) and poly(methyl methacrylate)(PMMA) are well-known to be miscible in the bulk state and do form stable monolayers at the air/water interface[5,7].

Monolayers of polymer mixtures at the air/water interface have been attracting much interest to study molecular nteractions between polymers[4-9]. Particularly, the surface pressure-area(π-A) isotherms of the monolayers strongly depend on the hydrophilic-hydrophobic balance, the chemical structures, orientation and packing of the molecules[9]. In general, the miscibility in monolayers of a binary polymer system at the air/water interface can be investigated mainly by two approaches, either following a change in the phase transition or plotting the mean area at a constant surface pressure versus the mole fraction of one component in the binary mixture[9-11]. If the plot is a linear relationship, i.e. the surface areas are additive, the mixed monolayers can be regarded as an ideal mixture or as a completely immiscible mixture in the entire mixture range. It has been established by Kawaguchi and Nishida that the deviation from the linear relation stems from the

Polymers and Other Advanced Materials: Emerging Technologies and Business Opportunities. Edited by P. N. Prasad *et al.*, Plenum Press, New York, 1995

contribution of intermolecular interaction between both substances[7]. A negative deviation means that the mixtures are considered to be stable and miscible, whereas a positive deviation indicates that the mixtures are less stable than the components alone at the interface.

In the present work, from measurements of the surface pressure-area isotherms for the binary mixed monolayers of polymers spread at the air/water interface, the authors have revealed the miscibility of the polymers having very similar molecular structures at the air/water interface to get the information concerning the general reasons for miscibility in the two-dimensional state. The polymers used in this study are PMMA and poly(isopropenyl acetate)(PIPA). And also, hydrolyzed PMMA with different degrees of hydrolysis were prepared to clarify the effects of the introduction of carboxylic groups to the PMMA on the miscibility at the air/water interface.

EXPERIMENTAL

Materials

Poly(methyl methacrylate)(PMMA, Mw.=120,000) was purchased from Aldrich Co. Poly(isopropenyl acetate)(PIPA) was prepared by polymerizing in toluene using 2-azobisisobutyronitrile at 65℃ for 72 hrs. The product was obtained by precipitating in n-hexane. Hydrolyzed PMMA(H-PMMA) with three different degrees of hydrolysis were obtained by dissolving PMMA in 96% sulfuric acid followed by stirring at room temperature. Each H-PMMA contains 4, 9, and 21 mol % of carbonyl groups, respectively and is designated by the degree of hydrolysis. For instance, H9-PMMA denotes that the carbonyl group in hydrolyzed PMMA is 9 mol %. The synthesized polymers were identified by Fourier-transform infrared and ^{1}H-NMR spectroscopy.

Measurements of surface pressure-area isotherms

A LAUDA film balance was used for measuring surface pressure(π) as a function of molecular area(A). The π-A isotherms were obtained at 25± 0.1℃ through a compression rate of 0.6 cm^2/sec. Blend systems containing H-PMMA were spread from N,N-dimethylacetoamide(DMAA) and benzene(1/1 by vol.) and for other systems, benzene was used as a spreading solvent.

RESULTS AND DISCUSSION

Effect of a spreading solvent

Figure 1 shows the π-A isotherms of PMMA monolayers at the air/water interface spread from two different spreading solvents, benzene and a mixture of 35/75(v./v.) dimethyl formamide(DMF) and chloroloform. The shapes of the surface pressure-area isotherms for the two different spreading solvents are comparable but the shift of the isotherm for the monolayer spread from DMF/chloroform mixed solvent should be noted. There could be several causes for this shift but it can be largely attributed to film loss due to the dissolving surface-active materials with DMF and chloroform into the aqueous subphase[6,13,14]. To achieve a well-defined state of the monolayer, the solvent must spread the surface-active material homogeneously over the whole water surface, disappear completely form

Table 1. Shapes for the π-A isotherms of PMMA and PIPA with different spreading solvents as compared with benzene as a spereading solvent.

Spreading solvent	PMMA	PIPA
Ethyl acetate	shift to smaller area	shift to smaller area
DMF/Chloroform (35/75 v./v.)	shift to smaller area	shift to smaller area
DMAA/Benzene	the same	the same

the surface layer and be chemically inert[15]. In this connection, the film loss into the aqueous subphase is one of the serious problems for the use of DMF/chloroform as a spreading solvent. For this reason, benzene was used as a spreading solvent for the homopolymer monolayers unless otherwise specified.

Table 1 shows the shapes of the π-A isotherms of the homopolymers, PMMA and PIPA, with three different spreading solvents, ethyl acetate, DMF/chloroform(35/75 by v./v.), and DMAA/benzene(5/5 by v./v.), as compared with benzene as a spreading solvent. Fairly reproducible π-A isotherms were obtained for the homopolymers when benzene was used as a spreading solvent. Similarly as in Figure 1, however, small shifts to lower area were observed for some of the homopolymer monolayers spread from some of the solvents, except the DMAA/benzene(5/5 by v./v.) mixture. The shift trend was more clearly observed in the case of the polymer containing carboxylic group, i.e., hydrolyzed

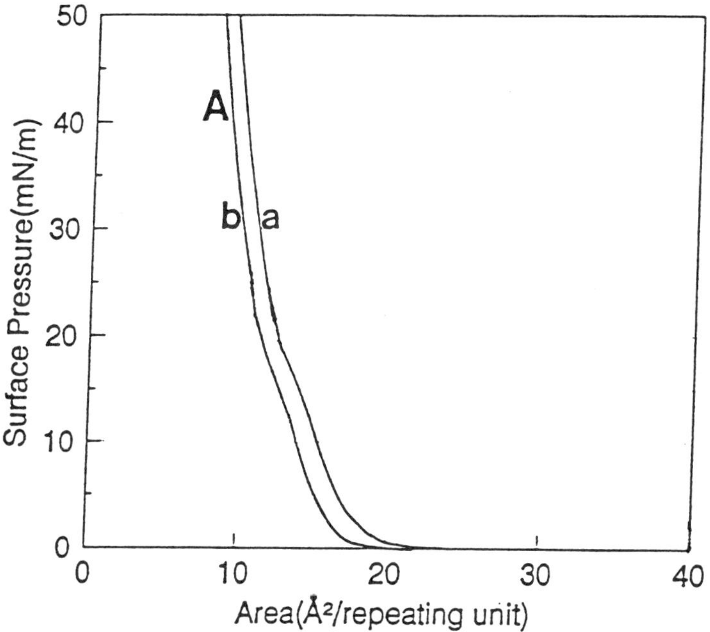

Figure 1. π-A isotherms of PMMA monolayers at the air/water interface spread from two different spreading solvents, benzene (a) and a mixture of 35/75(v./v.) dimethyl formamide(DMF) and chloroloform (b).

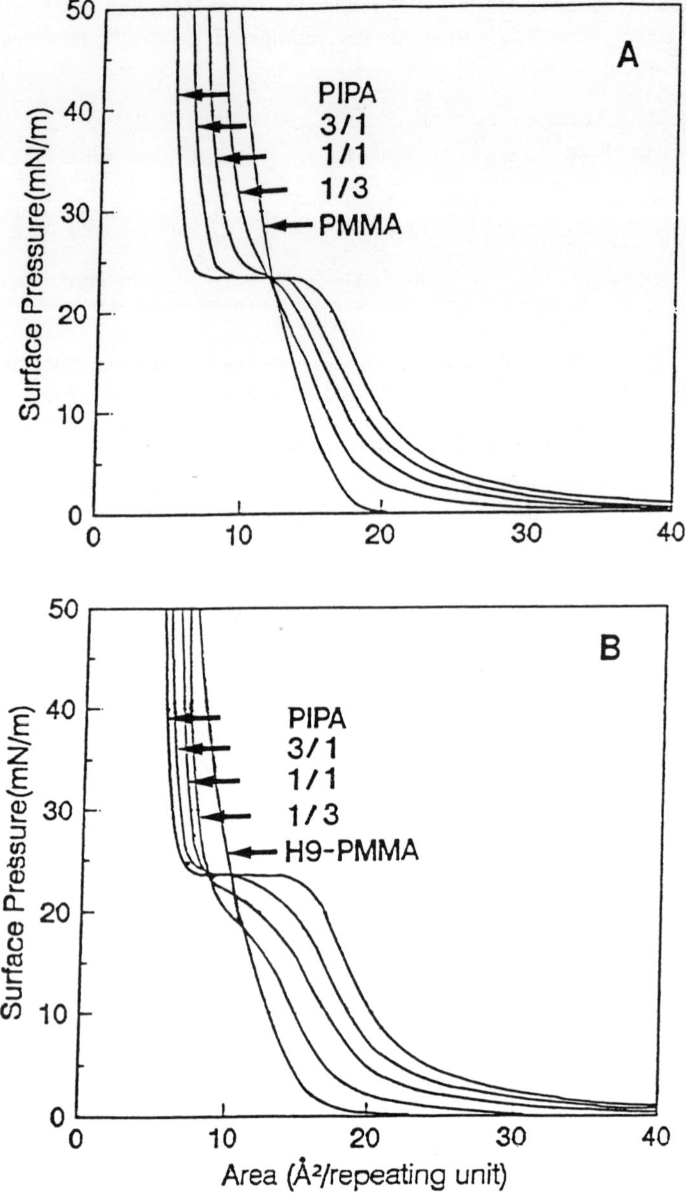

Figure 2. π-A isotherms of PMMA, PIPA, H9-PMMA, and their typical binary mixtures having different molar ratios; (A) PMMA/PIPA and (B)H9-PMMA/PIPA mixtures.

PMMA, even when benzene was used as a spreading solvent. This is reason why DMAA/benzene was used as a spreading solvent for the blends of the hydrolyzed PMMA.

The accurate π-A isotherms are very crucial in this work, since the determination of the miscibility of mixtures is purely based on the reproducible and exact π-A isotherm. Thus, the choice of a spreading solvent is very important. In order to verify conclusions from plots of the mean area-mole fraction of one component, the exact π-A isotherm is essential in this experiment. Thus, the isotherms were reproduced three times in order to rule out impurities and potential insufficient weighing accuracy.

PMMA-PIPA mixtures

Figure 2a shows the π-A isotherms of PMMA, PIPA and their typical binary mixtures having different molar ratios. PMMA was formed to be relatively condensed and compressible monolayers. The obtained π-A isotherm of PMMA is similar to that reported by other works[14-16], and the molecular area at 15 mN/m is about 14 $Å^2$/repeating unit. The collapse pressure and the corresponding area are 50 mN/m and 11 $Å^2$/repeating unit, respectively. The PIPA monolayer shows plateau regions around a surface pressure of 23 mN/m. The width of the plateau regions in terms of molecular areas ranges from 18 to 8 $Å^2$/repeating unit. On the other hand, the PIPA monolayer has plateau region around a surface pressure of 23 mN/m. The area region of the plateau decreases with increasing the mole fraction of PMMA. The surface pressure at the plateau depends on PMMA/PIPA molar ratios. Below the plateau, the surface pressure at the same area is lower than that of the PIPA film and decreases with an increase of the PMMA component, whereas above the plateau pressure the composition dependence of the surface pressure is quite the reverse, i.e. the π-A isotherm shows cross-over points at a certain surface pressure. However, when PMMA is hydrolyzed with 9 mol % of carbonyl group, it is noted that the H-9PMMA exhibits the condensed phase behavior at smaller areas in comparison to that of non-hydrolyzed PMMA, as shown in Figure 2b, and the crossover point on the π-A curve is not coincide with that of H9-PMMA homopolymer monolayers. The result suggests that the monolayer of H9-PMMA is more condensed by the presence of PIPA when they are mixing due to the molecular interaction between the two polymers at the air/water interface.

H-PMMA-PIPA mixtures

Plots of the mean surface areas at the surface pressures of 5, 10,20, and 30 mN/m versus the mole fraction of PIPA are shown in Figure 3. The mean area is expressed in terms of the area per unit mass of one repeating unit of component polymers used here.

A relation between the mean areas(A) at a given surface pressure and molar ratios of component 1(x_1) and component 2(X_2) in the mixture($x_1 + x_2=1$) is expressed by equation(1) if the two component are ideally miscible or completely immiscible[7].

$$A = x_1 A_1 + x_2 A_2 \qquad (1)$$

where A_1 and A_2 correspond to the molecular areas of components 1 and 2, respecively. This relation can be applied to the mixed polymer films by using the values of A, A_1, and A_2 as an area per unit mass of repeating unit instead of a molecular area.

The solid lines in Figure 3 represent the additivitiy ones according to the equation (1). In the case of the area plots at 20 mN/m, the negative deviation from the mean area was observed. These fugures show the effect of degree of the hydrolysis on the miscibility of PMMA and PIPA. The greater the degree of hydrolysis of PMMA, the larger the negative deviation behavior is. In the case of H-21PMMA/PIPA mixtures, the mean areas considerably deviate from the solid line. This negative deviation means that the H-21PMMA is nonideally miscible with PIPA at the air/water interface and the intemolecular interaction between H-21

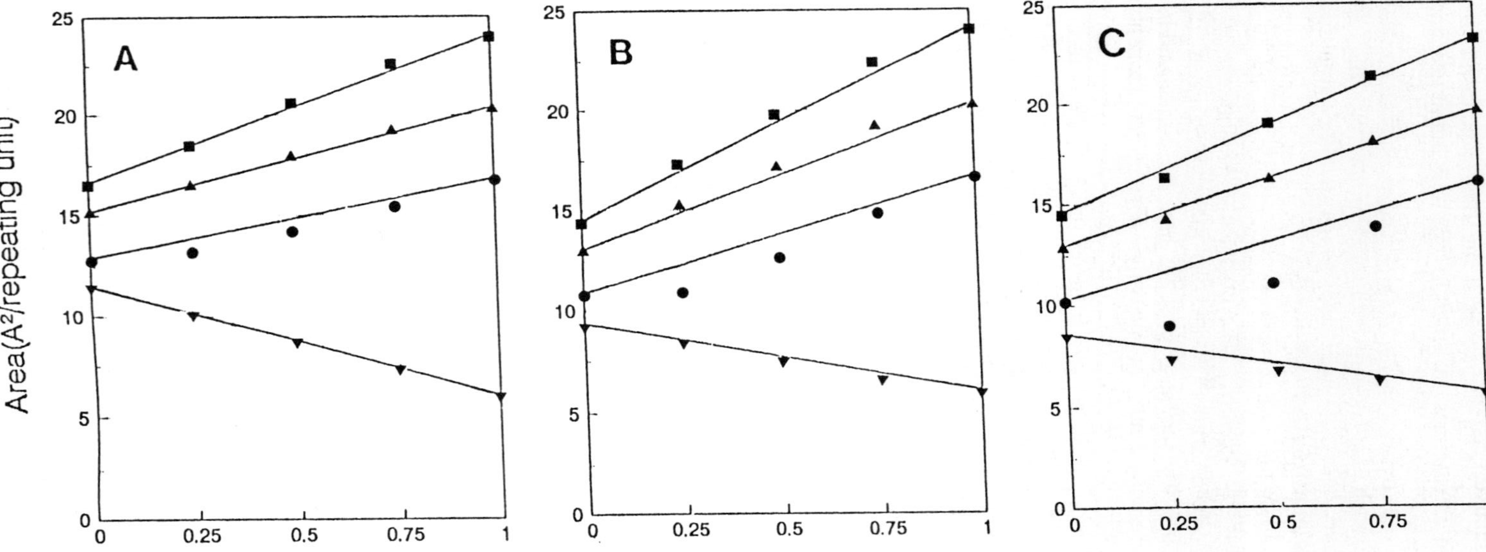

Figure 3. Plot of the mean surface areas at the surface pressures of 5(■), 10(▲), 20(●), and 30mN/m(▼) versus the mole fraction of PIPA for various mixtures (A: PMMA/PIPA, B:H4-PMMA/PIPA, C:H21-PMMA/PIPA). The numbers in the figures indicate the surface pressure(mN/m).

PMMA and PIPA is attractive. As the PMMA is hydrolyzed to more degree, the misciblility of PMMA and PIPA becomes enhanced due to the specific intermolecular interaction between carboxylic acid of hydrolyzed PMMA and PIPA. However, any mixtures did not show such clear negative deviation behaviors at lower surface pressures such as 5 and 10 mN/m. Even the H21-PMMA/PIPA mixtures exhibited additive behaviors at lower pressures. This result implies that the determination of miscibility from the π-A isotherm is not so clear at the lower surface pressure region, in which the polymer molecules do not pack enough to become condensed stable monolayers, and therefore the intermolecular interaction as well as the intramolecular one are relatively small to distinguish each other.

CONCLUSION

Miscibility behaviors of PMMA and PIPA in the monolayers at the air/water interface were investigated by the surface pressure measurements as a function of the mean area per unit mass of the repeating unit in the mxiture. The miscibility in the monolayer of polymer mixtures at the air/water interface was judged from the negative deviation behavior from the additivity rule in the plots of the mean surface area against the mole fraction of one component. The spreading solvent should be carefully choosed to obtain the reproducible surface pressure-area (π-A) isotherms for judgement of the miscibility of polymer mixtures at the air/water interface. PMMA was immiscible with PIPA, while hydrolyzed PMMA showed miscibility with PIPA at the air/water interface due to the presence of the molecular interaction between the carboxylic acid and the carbonyl groups.

REFERENCES

1. O. Olabishi, L.M. Robeson,, and M.T. Shaw, "Polymer-Polymer Miscibility", Academic Press, New York (1979).
2. A.C. Fernandes, J.W. Barlow, and D.R. Paul, *Polymer,* **27**:1799(1986).
3. W.K. Lee, C.S. Ha, C.S., and Cho, W.J., Polymer, **35**,10:2229 (1994).
4. D.J.Crisp, in "Surface Phenomena in Chemistry and Biology", F.Danielli, K.G.A.Pankhurst, and A.C. Riddiford, ed., Pergamon Press, London (1958), pp.23-54.
5. G.L.Gains,Jr., *Langmuir,* **7**:834(1991).
6. G.L.Gains,Jr., "Insoluble monolayers at Liquid-Gas Interface", Interscience, New York (1986).
7. M. Kawaguchi and R.Nishida, *Langmuir,* **6**:492 (1990).
8. G.Gabrielli, M. Puggelli, and R. Faccioli, *J.Colloid Interfice Sci.,* **37**:213(1971).
9. A.Es-Sounni, and R.M. Leblanc, *Langmuir,* **8**:1578 (1992).
10. G.Gabrielli, M.Puggelli, and R. Faccioli, *J.Colloid Interfice Sci.,* **44**:177(1973)
11. G.Gabrielli and P.Baglion, *J.Colloid Interfice Sci.,***73**:582(1980).
12. D.A.Cadenhcad and B.M.Kellner,Jr., *J. Colloid Interfice Sci.,***49**:143(1974).
13. A.Gericke, J.Simon-Kutscher, and H.Huhnerfuss,*Langmuir,***9**:2119(1993).
14. M.Kawaguchi, S.Komatsu, M.Matsuzumi, and A.Takahashi, *J. Colloid Interfice Sci.,* **102**:356(1984).
15. M.Kawaguchi, M.Tohyama,Y.Mutoh, and A.Takahashi, *Langmuir,***4**:407(1988).
16. M.Kawaguchi,M., B.B.Sauer, and H.Yu, *Macromolecules,* **22**:1735(1989).

OPTIMIZATION IN THERMOPLASTIC PREPREGGING SYSTEM

M. Nasir, H. Ishak and B. Azahari

School of Industrial Technology
Universiti Sains Malaysia
11800, Minden
Penang, Malaysia.

INTRODUCTION

Thermoplastic (TP) composites, particularly those of continuous and uniaxial types are gaining popularity, judging from the numerous reports available (Carlson,1991; Ford,1991). Amongst the various reasons for such a growth include better damage tolerance, longer pot-life, easier shaping and repair and of course recyclability, albeit still possessing inferior chemical resistance and lower elevated temperature performance.

The first step in any TP composites manufacturing, however involves the production of what is referred as prepreg or towpreg in which unlike the thermoset counterpart requires a more an elaborate impregnation sequence. This is to enable an optimum impregnation of fibre by the resin which is of higher viscosity to be attained. There are now numerous techniques, however, mostly proprietary being employed viz. solution-bond, powder coating/ fluidized commingled yarn and melt methods, to quote a few.(Wilson et al. 1989; Carlson, 1991)

Further, in view of their commercial implications, in depth discussions on these techniques are scarce, except those of Wilson et al. (1991) and Wilkinson et al (1993). In both cases using specifically designed and fabricated pultrusion machine, TP prepregs of PEI, PAS, PEEK etc are being produced in which the prepregs and the subsequent composite properties obtained are profoundly controlled by the processing variables such as temperature, pulling rate, the resin, doctor blade and so forth. However, that of Wilson et al. (1991) was only confined to solution technique of impregnation while that of Wilkinson et al. (1991) was more versatile which included powder and commingled approaches.

Polymers and Other Advanced Materials: Emerging Technologies and Business Opportunities. Edited by P. N. Prasad *et al.*, Plenum Press, New York, 1995

In our earlier report (Nasir et al., 1993), we have discussed some success stories of TP prepregging efforts based on a simple pultrusion system akin to that of a thermoset. Also, in view our logistic problems in trying to realize locally such a capability the prepreg is restricted to ABS matrix and glass-fibre reinforcement. Generally, the TP prepregs produced though consistent and reliable but had poorer prepregs' properties and numerous short-comings which required many major modifications to the system.

Here a report on the much improved prepreg quality is presented after major alterations were made to the old system, that included a better resin tank, a drying unit, a smaller die profile and also a more efficient cooling and prepreg puller. The influences of the numerous process variables and the subsequent used of a prepregging model in optimizing the process will be presented.

THE NEW PREPREGGING MACHINE

As reported previously (Nasir et al.,1993), there were numerous short-comings and limitations of the old prepregging machine which resulted in a much poorer quality prepregs being produced combined with lower output and handling problems. However, from the experiences in handling the prepregger plus numerous simulation studies have led to major modifications being done which include :

(a) a larger resin bath with a dispensing outlet plus impregnation rollers being attached to both the tank and the cover.

(b) a drying chamber of similar length to the tank and the compaction die. This is to serve as a predryer to the prepreg prior to entering the die which now serves as a consolidation module alone.

(c) a compaction die with a smaller exit die profile, to enable smaller prepreg being produced, hence enabling the prepreg to be rolled up.

(d) a cooling unit that cools the hot prepreg effectively after the consolidation die, hence prevents resin being squeezed by the puller rollers.

(e) a stronger puller system with rubber rather than steel rollers thus giving a much better grip plus not causing any marks on the prepreg.

The schematic diagram of the modified prepregger machine is shown in Fig. 1. These changes apart from producing a much enhanced prepregs viz. a four fold increase in prepreg strength, a more consistent and higher resin uptake, a much lower void content, analogous to that of a laminate, it also provides a much easier handling system. For example, the operation can be easily stopped at any juncture, the resin removed and the tank cleaned but with the fibre alone is left behind hence enabling the next operation to be commenced at any time. In other words the concept of 'flashing' in injection moulding is applied.

The presence of the drying module ensures prepreg to be completely dried prior to the die entrance, hence preventing excessive dripping of the resin at die mouth. Further, the new system enables the die serving to function as a compaction die with less chances of solvent entrapment, hence might result in disastrous effect to the prepreg. The more powerful puller enables continuous prepregging to be done for a longer period, albeit as shown later, limitations still exist particularly with a much thicker solution at a lower die temperature and a higher pulling rate.

In general, the various alterations being done to the system have enable a much better prepreg being produced at a much higher output will lesser physical problems, although the most glaring weakness of not having a measurable doctor blade's opening still prevails.

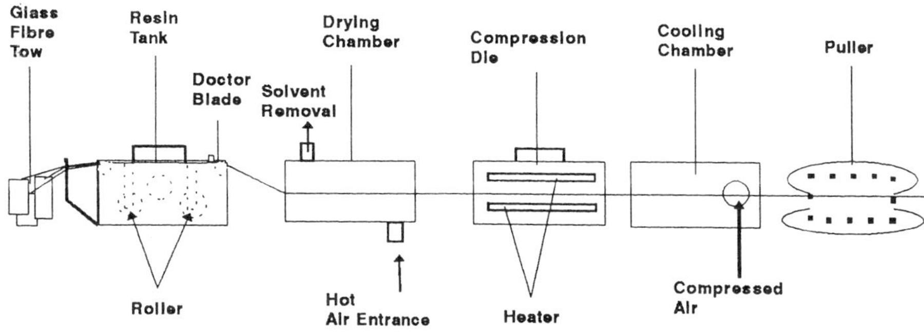

Figure 1. An overview of the new prepregger setup

PROCESSING VARIABLE INFLUENCES

Although not much work on these aspects were being reported, perhaps due to some commercial implications, it is apparent that the prepreg qualities are strongly controlled by the various processing variables which differ from system to system. From both the simulation studies and our earlier endeavour (Nasir et al., 1993) the important parameters are the ABS concentration, the die temperature, the pulling rate and the doctor blade's opening. Table 1 gives an overall view on the influences of the processing variables on the fibre weight fraction, W_f of the prepregs.

Table 1. The overall influences of the process variables on the fibre weight fractions of the prepreg.

Conc. % w/w	Pulling Speed m/min	W_f % Theory[a]	Measured Die Temp. °C 170	200	240
10	0.1	92	93	92	92
10	0.5	92	93	92	91
20	0.1	84	86	85	86
20	0.5	84	84	83	83
30	0.1	79	79	79	79
30	0.5	79	-	79	79
40	0.1	73	73	73	73
40	0.5	73	-	-	73

[a] theoretical values estimated using equation [2]

For both pulling rate, the W_f decreases with concentration while the effects of pulling rate and die temperature are less obvious or marginal. For example, for the 10 % w/w ABS solution, both 0.1 and 0.5 m/min pulling rates for all temperatures produced prepreg with more or less similar W_f i.e. 0.91 - 0.93 ; the same goes for 20 % w/w. Moreover, at higher concentrations, in which the resin uptake is correspondingly high, prepregs are not realised, shown by the (-) notation.

In other words, the influences of temperature and pulling rate are prominent when higher resin uptake occurs, here in the case of 30 and 40 % w/w solution.

The influences of the ABS concentration, the pulling rate and the die temperature on the prepreg strength are shown in Fig. 2. Unlike W_f, increasing the concentration increases initially the prepreg strength. However, further increase in concentration results in a subsequent drop; the maximum is noted to occur at 20 % w/w. Further, the prepreg strength is also affected by pulling rate, particularly at high concentration, which can be easily linked to the Non-Newtonian or the viscoelastic of the prepreg. The low prepreg strength observed at 10% w/w might be due to insufficient resin to hold the fibre, plus the high void content of the prepreg. While at 30% and 40% w/w, due to poor impregnation and also high void content. The die temperature effects are more obvious for the 40% w/w as compared to the others. The same reasons can be explained for such a trend in which the prepreg strength are controlled by other factors such as the void content, the interphases and the viscoelastic nature which in return are governed by the resin uptake, the impregnation and the consolidation process. This is clearly noted for example, the effect of die temperature on both the W_f and prepreg strength which shows different responses i.e. same W_f but different prepreg strengths.

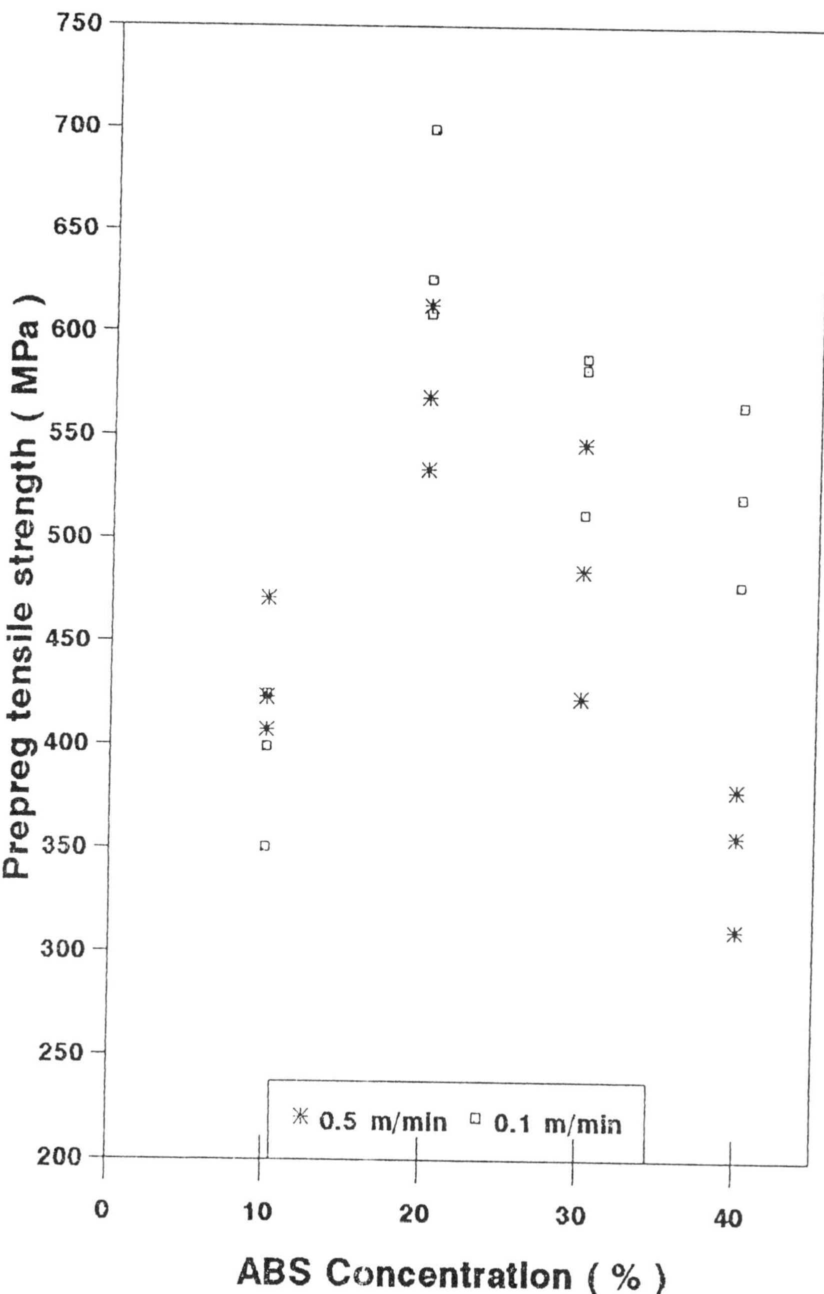

Figure 2. The influences of both the ABS concentration and pulling rate on the prepreg strength.

THE PREPREGGING MODEL

Generally, the use of model equation that correlates the composites properties to the processing variables such as pulling rates, the dwell time, the resin concentration will enable some form of control of eventual composite properties desired and as outlined by several workers such as Wilson et al. (1991) and Chandler et al. (1992). With the same purpose, here a simple prepregging model is introduced that enables processing variables to be set up so as to procure desirable prepegging level.

The model equations are easily derived by taking some assumptions viz. resin fully occupies the die, fibres are uniaxially aligned, complete wetting is realized with no void, the solution is a miscible and Newtonian in nature. They are as follows:

$$W_f = \frac{F}{A_d \rho_m + F(1 - \rho_f/\rho_m)} \qquad [1]$$

$$W_f = \frac{F}{A_{dr} \rho_{rs} + F(1 - \rho_f/\rho_{rs})} \qquad [2]$$

in which F is the fibre linear density ; A_d and A_{dr} are the die cross-sectional area

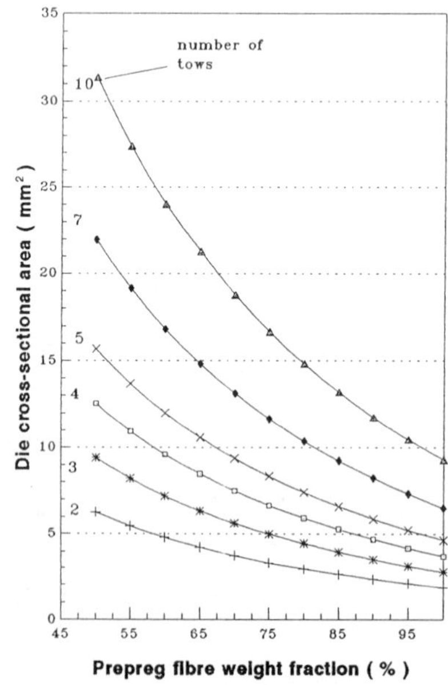

Figure 3. The theoretical plots of fibre weight fraction, W_f and the die area with varying number of tows derived using prepregging equation [1]

and the doctor blade opening, respectively; ρ_f, ρ_m and ρ_{rs} are densities of the fibre, the matrix and the resin solid in solution density.

It must be pointed that the ρ_{rs} is determined by assuming the solution is a miscible system. Further equation [1] represents event after prepreg is formed while equation [2] represents the resin uptake or the prepregging event, which is obviously affected by the solution's concentration.

Since the parameters in the above equations can be easily determined, theoretical plots correlating the composite properties, W_f to the A_d and A_{dr} can be derived at the different processing variables such as concentration and no. of tows. Typical plots of both are given in Fig. 3 and 4, respectively.

The usefulness of these plots in optimizing the prepregging under a particular conditions can be easily realized. For example , to procure optimum prepreg with say 10 mm^2 cross-sectional area if 5 tow of fibres are used would required a W_f about 67% while for a 7 tows, 78%. However from our experiment using 5 tows, the closest to that of 67% W_f, is attainable using the 40% w/w

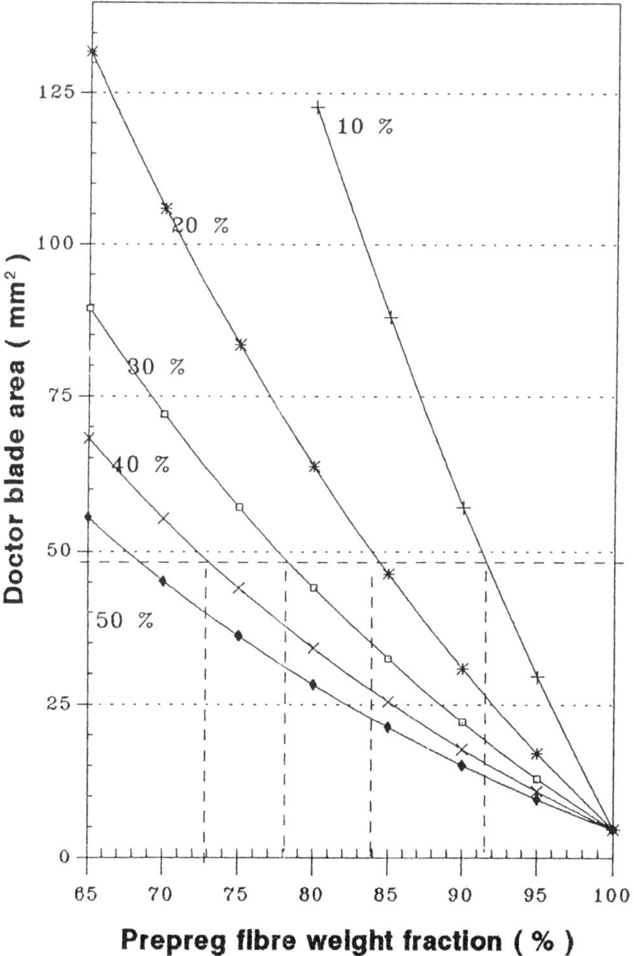

Figure 4. The theoretical plots of fibre weight fractions W_f and the doctor blade's opening at different concentrations derived from prepregging equation [2]

solution. However, one obvious weakness of the equation [1], that it does not able to account for the variation in the solution viscosity which our studies have found to be an important factor. Equation [2] looks at the impregnation occurring at the tank and controls to some extent by the blade's opening. Here the role of solution's concentration is taken into consideration. For example, if the doctor blade's opening is set at 50 mm^2, then for different concentrations the amount required to ensure optimum composite is different, i.e. 92, 84 and 78% W_f for 10, 20 and 30% w/w solutions. These estimated values are closed to those of the measured W_f, which are 91, 83 and 79% respectively, which strongly confirms the validity of the prepregging equations [2].

CONCLUSIONS

THe modifications of the old prepregger particularly the inclusion of the drying chamber, the more effective cooling and puller system clearly not only produces consistent prepregs but with property close to a laminate. The prepreg properties are governed by the process variables, especially the concentration and doctor blade's opening which are some what explained by the use of the prepregging model equation. The model also assist in determining the prepregging conditions that need to be used to ensure optimum prepreg.

Acknowledgement

This project on composite prepregging is supported by both the Malaysian Government via IRPA fund and the Universiti Sains Malaysia via the short-term grant.

REFERENCES

Carlson, L.A. (ed.), 1991, Thermoplastic Composite, Elsevier, New York.

Chandler, H.W., Devlin, B.J. and Gibson, A.G., 1992, A model for the continuous impregnation of fibre tows in resin baths with pins, Plastics Rubber and Composites Processing and Application, 18(4) : **215**

Ford, R., 1991, Thermoplastic Composite - A Time for Strategy, Reinforced Plastics, 35 : **26**

M. Nasir, H. Ishak, Z.A. Mohd. Ishak and B. Azahari, 1993, Production of thermoplastic Prepregs, Frontiers of Polymers and Advanced Material, Plenum Press, New York.

Wilkinson, S.P., Marchelio, J.M.,Dixon, D. and Johnston, N.J., 1993, A NASA LaRC multipurpose prepregging unit, 38th International SAMPE symposium, Anaheim.

Wilson, M.L., Buckley, Johnson, G.S., and Dickerson, G.F., 1991, Pultrusion Fabrication of Graphite Reinforced Thermoplastic Prepreg, Reinforced Plastics, 35 : **30**

THERMAL PROPERTIES OF NR-LLDPE THERMOPLASTIC BLENDS

lbrahim Abdullah, Sahrim Ahmad and Che Som Sulaiman
Faculty of Physical and Applied Sciences
Universiti Kebangsaan Malaysia
43600 Bangi, Selangor

ABSTRACT

Blends of NR-PP and NR-LLDPE innoculated with liquid natural rubber, LNR form thermoplastic material with improved physical properties. The glass transition temperatures, as shown by DMA measurement, of NR-PP and NR-LLDPE are -50°C and -55°C respectively. Addition of LNR into the blends causes the thermal expansion of the material to decrease by about 30% in both NR-PP and NR-LLDPE systems. In the NR-PP system for blends with and without LNR the expansion set in at 35°C and 42°C respectively. In NR-LLDPE blend addition of LNR causes the set in expansion to increase from 35°Cto55°C. The measured glass transition temperature sand the thermal expansibility of the blends denote the introduction of an interphase interaction on addition of LNR into the blend. The role of LNR in compatibilizing the NR-plastic phase will be discussed

INTRODUCTION

Polymer blending is a widely used method to produce new polymeric materials with the desired physical and chemical properties. These tailor-made materials are designed especially for certain specific applications. Incorporation of thermoplastic properties into natural rubber via melt blending of natural rubber (NR) with thermoplastics such polyethylene (PE) and polypropylene (PP), is very critical in determining the future industrial applications of NR as an engineering material. Blending of NR with PP or low polymeric materials with the linear low density PE (LLDPE) yields a thermoplastic natural rubber (TPNR) with diverging properties, ranging from a soft material to a semi rigid plastic.

These pyhsical properties depends on the composition of the material. Blends richer in NR are soft and the material becomes successively harder with the increase in the thermoplastic component[1]. Generally the NR-PP blend is harder than the NR-LLDPE blend forth samecomposition of the materials. Introducing cross-links into the NR phase of TPNR causes the material to become harder and leads to an improvement in the mechanical properties[2].

However, the main problem of compound NR with any thermoplastic is incompatibility which can result in the lost of the desired properties during blending. Dicumylperoxide was reported to have a positive effect on the NR-PP blends. While in our work liquid natural rubber (LNR) was found to compatibilise the NR-PP and NR-PE blends[3].

The applicability of the material industrially is also determined by the thermal properties of the material. Thermal expansion and softening of the blend will determine the suitability of the material for a certain application. In this paper a report on the thermal expansion and softening of the NR-PP and NR-LLDPE (linear low density PE) blends in relation to their tensile properties, morphology and glass transition temperature will be presented.

METHOD

Materials

NR of grade SMR-L was obtained from the Rubber Research Institute of Malaysia (RRIM), commercial grade linear low density polyethylene (LLDPE), obtained from Dowlex was used without any treatment and polypropylene (PP) was obtained from Shell (grade MT). The LNR was generated by photochemical oxidation of NR in our laboratory.

Blend Preparation

The blends were prepared by melt-blending of the components in a laboratory using intemal mixer at various temperatures and rates of mixing. Once homogeneous mixing was assumed the blend was removed and subsequently compressed at about 150°C and 7kN pressure for about 2 minutes into a thin sheet of about I mm thickness from which test specimen were cut.

Physical Measurement

i) Thermal Expansion

The expansion of the blend was determined by using a thermal mechanical analyzer, Shimadzu TMA-50. A sample size of 5 mm x 5 mm was used for the measurement. Average expansion and coefficient of expansion were calculated from the expansion.

ii) Dynamic Mechnical Thermal Analysis (DMTA)

The measurement were carried out, using DMTA - PL in a single cantilever deformation mode from - 100 to 150°C. The oscillating frequency was lHz and the scanning rate is 3°C/min.

Hardness

The hardness of the samples were determined by using a hardness tester. Durometer ASTM 2240, of Shore type A (Zwick Model) was used on a 3mm thickness sample. All measurement were performed at room temperature (25°C).

RESULTS AND DISCUSSION

A 50% NR-PP blends are generally light opague yellow in colour and of hardness about 90 Shore A. A crazing effect is clearly shown when the sample is folded. On the other hard, an equivalent NR-LLDPE blend is slighty darker, translucent, soft and craze-free on folding. The hardness of the material is about 85 shore A. The two blends are thus physically different in nature and properties. PP is a harder thermoplastic than LLDPE and thus the blends of the material with NR will be subsequently of different degree of hardness in accordant with the thermoplastic component. The general properties of thematerials are summarised in Table 1.

The thermal expansion of the NR-PP and NR-LLDPE with and without the addition of LNR as the compatibiliser as measured by Thermomechanical Analysis are depicted in Figures I and 2. Figure 1, shows that the NR-PP blends expands proportionally with heating from about 300°C to above 100°C with an average thermal coefficient of about 55.4 X 10^{-6} K^{-1}. The maximum expansion at IOO°C is about 18 µm With the incorporation of about 10% LNR into the NR-PP blend, the expansion coefficient decreases to 37.7 x 10^{-6} K^{-1}, that is about 30% decrement. The material also does not show any significant expansion until about 35°C and reaches a value of about 13 µm at IOO°C. The apparent effect of LNR is the reduction of the ability of the material, NR-PP, to expand.

The typical expansion curve of NR-LLDPE blends is as shown in Figure 2. Without LNR, the blend starts to expand at about 400°C till about 90°C with the maximum thermal expansion coefficient at about 200.6 K^{-1} between 600°C to 800°C. Addition of 10% LNR into the blend causes the material to become more resistant to heat expansion, where the expansion only set in at about 55°C and becomes soft at abnost IOO°C. The maximum thermal expansion between 60°C to 80°C is about 183.9 K^{-1}, a decrease of about 9%. The absolute expansion of the material up to 90°C are 47 Jim and 30 µm for the blend without and with LNR respectively.

In both systems the effect of LNR is similar i.e. binding the phases of the blend together so as to resist expansion of the material during heating. Apparently the LNR fimctions in a similar manner in these blends. The effect of LNR in NR-PP blends is more significant than in the NR-LLDPE blends where a 30% reduction in thermal expanson coefficient is observed in the former system in comparison to just about 9% in the later. The softening points of the material in both systems are also higher with the addition of LNR into the blends.

Figure 3 shows the variation of log (tan δ) with temperature in a DMTA on a single cantilever mode. For the NR-LLDPE (60 : 40) composition the Tg recorded is - 48.2°C and on addition of LNR into the blend with the NR:LLDPE composition at 60:40, the Tg is shifted to -50.2°C. The Tg for NR is about -60°C while in PE, the phase transition temperatures are not well defined. However for semicrystalline HDPE the reported values are 3 0 - 80°C, 0 to -60°C and - I 00 to -40°C for the ct, 0 and y relaxation peaks respectivelys[6]. As such the transition temperature in NR and LLDPE can be very close together and the shift in the Tg of the blend is toward that's of the pure components. In term of the morphology of the material the change indicates an increase in the compatibility or interaction between the rubber and plastic components.

In the NR-PP system as shown in Figure 4, the dynamic mechanical properties of the blend shows a shift in Tg from -55.70°C (O% LNR) to -53.9°C on addition of 10% LNR. The reported values of Tgs for PP are from -13 to O°C and -18 to -50°C for the isofactic and atactic polyrares respectively. The shift in the Tg of the blend is thus toward

Table 1: Hardness of NR-PP and NR-LLDPE with and without LNR

Blend Ratio (50:50)	Hardness		Features
	Shore A	Shore D	
NR - PP	93	36	Light Yellow Opague
NR - PP - LNR	95	39	Craze
NR - LLDPE	87	31	Yellow Translucent
NR - LLDPE - LNR	89	34	Craze-free

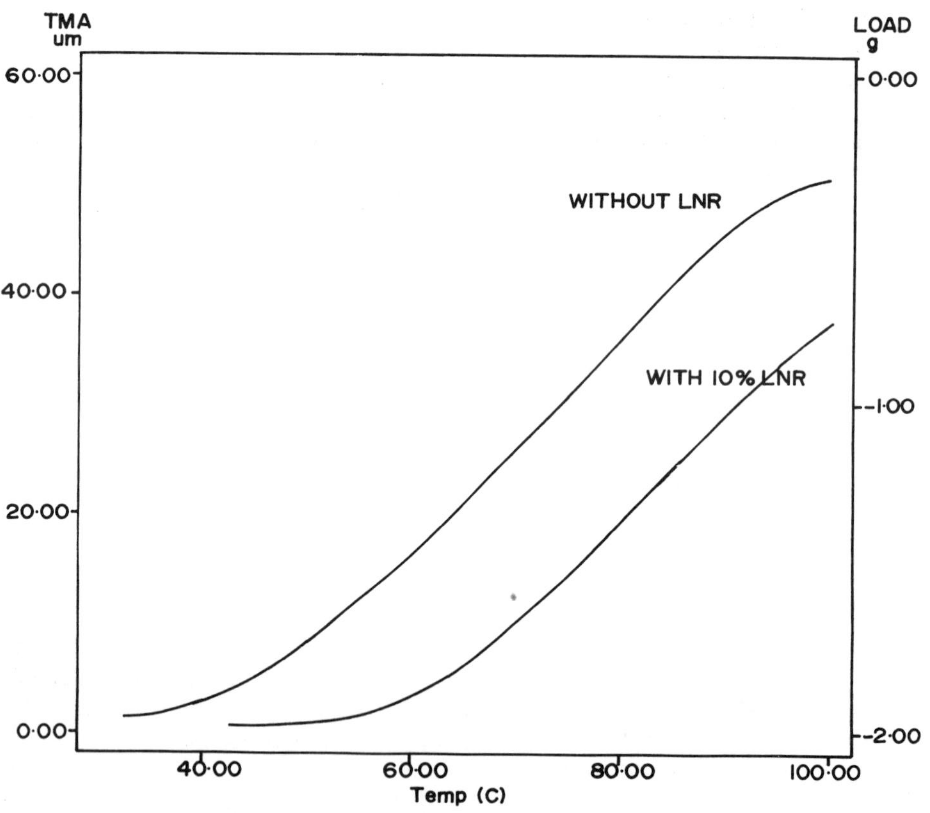

FIGURE 2: THERMAL EXPANSION OF NR-LLDPE

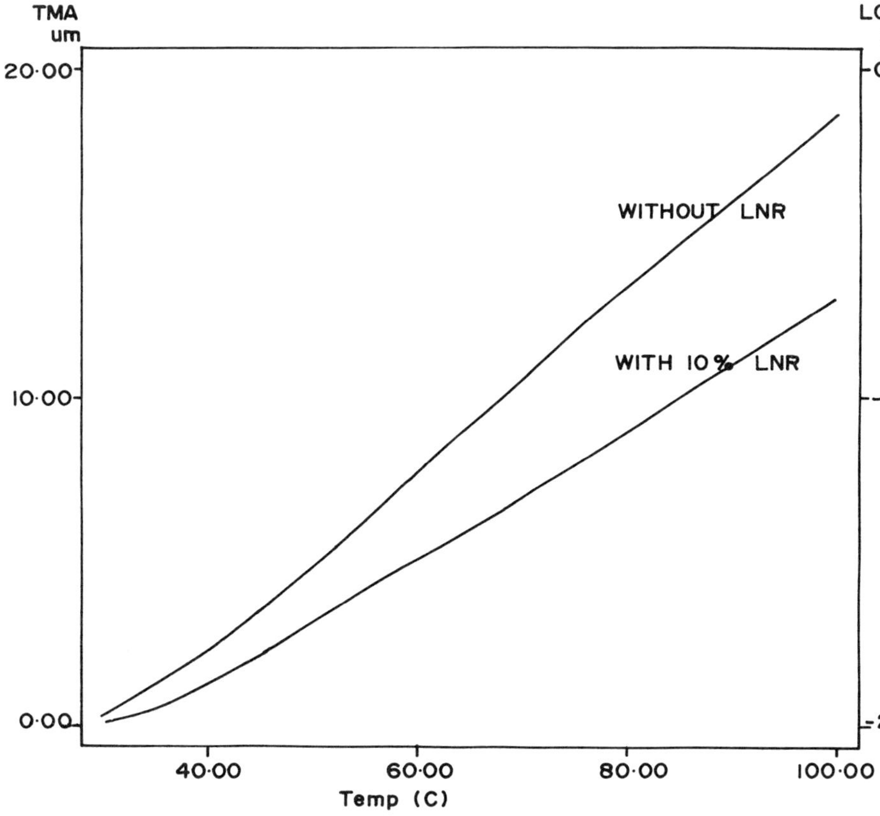

TMA
um
20·00
10·00
0·00
LOAD
g
0·00
-1·00
-2·00
WITHOUT LNR
WITH 10% LNR
40·00
60·00
80·00
100·00
Temp (C)

FIGURE 1 : THERMAL EXPANSION OF NR - PP

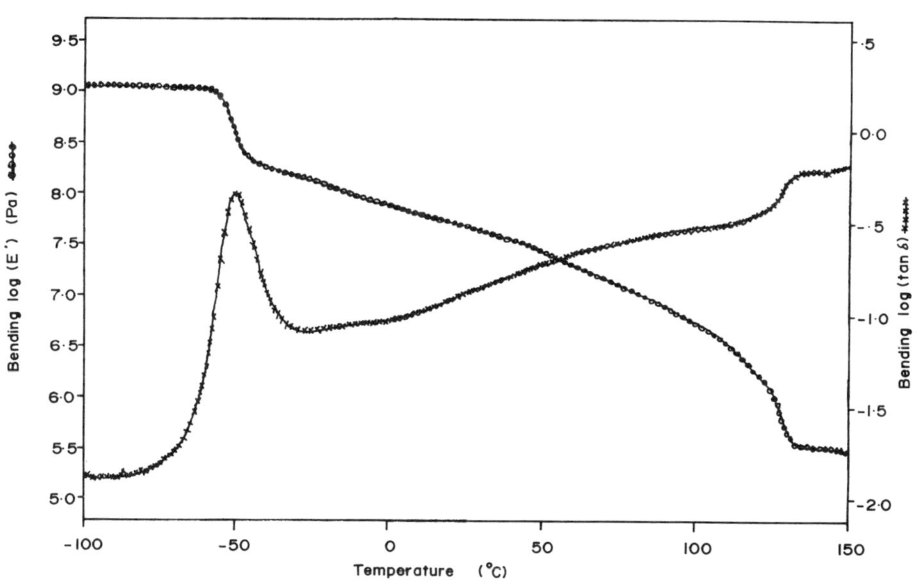

FIGURE 3: TYPICAL DMA TRACES OF NR-LLDPE-LNR

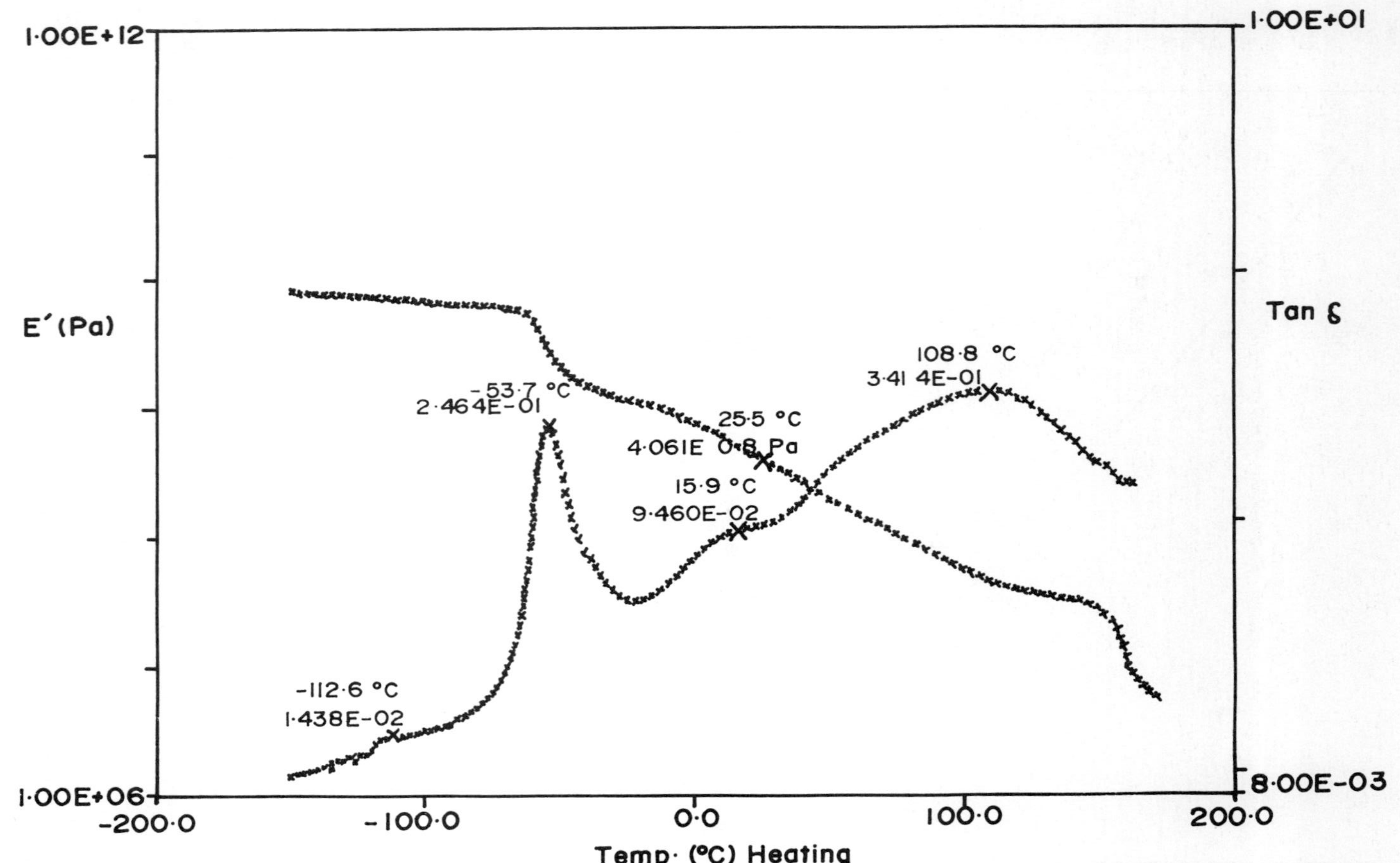

FIGURE 4: TYPICAL DMA TRACES OF NR-PP-LNR

the mid-point between the Tgs of the pure ocmponents. The single Tg and the direction of shift again strongly indicate an increase in the interaction between the PP and NR phases on addition of LNR.

In both systems, NR-PP and NR-LLDPE, the blends exhibit a Tg close to that's of the NR polymer and on addition of LNR into the blend, the Tg moves slightly towards the Tg of the plastic component. It is very obvious in NR-PP blend due the well defined Tg of PP but rather Vagne in the NR-LLDPE blend. The movement of Tg strongly suggests the existence of an interaction between the NR and plastic phases on addition of LNR into the blend. The mechanism of interaction is not well understood but it is believed as due the active sites or functional groups attached to the polyisoprene chain of LNR[7].

CONCLUSION

NR-PP and NR-LLDPE blends are relatively compatible systems. The compatibility of the blends is futher enhanced with the addition of LNR into the systems. The thermal expansion and variation of Tgs indicate an increased NR-plastic interphase interaction in the presence of LNR- Glass transition temperature, constitutes an accepted criterion which is helpfiil for the specification of the mutual interaction between constituents, that is the compatibility or incompatibility of polymer blends[4]. A single value for the Tg of the blends shows that the system is compatible. The compatibility in these blends is related to the parity between the polarities of the components and there is no finite interfacial tension resulting in a high work of adhesion[5]. This also implies that the blends are molecularly homogenous. Cross-linking is believed to take place in these systems.

ACKNOWLEDGEMENT

The authors would like to thank Dr. T.Y. Wong of Shimadzu Application Laboratory , Singapore for conying out the thermal expansion measurement, Universiti Kebangsaan Malaysia and the Malaysian Government for providing financial assistance and to everyone who helped in ensuring the completion of this manuscript.

REFERENCES

1. Elliot D.J., and Tinker, A.J., in Natural Rubber Science and Technology, A.P. Roberts, Eds. Oxford University Press, London, (1988).
2. Sahrim, A. Ibrahim A, S. Che Som, S. Kohjiya and J.K Yoon, J. App. Polym Sci, 51, 1357,(1994).
3. Mathew, N.M. and Tinker, A.J., J. Nat. Rubb. Ress, 1(4), 240, (1986)

4. Wetton, R.E., de Blok and Corish, D.J., Int. Poly. Science and Tech., 18, 5, (1991).
5. Olabis O.L., Robeson L.M. and Shaw, M.T., Polymer-polymer Miscability, New York, NY, Academic Press, (1979).
6. David L. Beach and Yung V. Kissin, Consice Encyclopedia of Polymer Science and Engineering, Wiley Interscience, pg. 355.
7. Ibrahim A. and Sahrim A., Material Forum 16, 353-357, (1992).

FLOW AND FRACTURE BEHAVIOR OF SAND-REINFORCED LOW DENSITY POLYETHYLENE DURING TENSILE TESTING

Abdelrazek Y. Kandeil[1] and Rouchdy R. Zahran[2]

[1]College of Engineering Studies and Technology
Arab Academy for Science and Technology
P.O. Box 1029, Miami
Alexandria, Egypt
[2]College of Engineering, Qatar University,
P.O. Box 2713, Doha, State of Qatar

ABSTRACT

Sand/low density polyethylene composite samples were prepared using injection molding. The effect of some processing and structural parameters on the flow and fracture behavior of the prepared samples was investigated. Typical processing parameters were melt and die temperatures. The investigated structural parameters were sand particle size and sand content.

Higher melt and die temperatures decrease the tensile strength and the modulus of elasticity for the various sand contents investigated, meanwhile, the ductility and toughness increase. Sand addition increases the tensile strength and the modulus of elasticity, as long as the sand size is smaller than 400 μm . Larger particle size weakens the structure. A maximum in the tensile strength, ductility and work done to fracture is observed for samples with a particle size of 400 μm. Adhesion and interlocking are responsible for the strengthening effect below 400 μm, while dewetting and interfacial friction are responsible for the reduced tensile strength and modulus of elasticity above 400 μm.

INTRODUCTION

Polymer composite systems are drawing much attention due to their broad involvement in many domestic and industrial applications. A diversity of reinforcements can be added selectively to the polymer matrix to enhance some mechanical, thermal and transport properties [1-5]. Particulate-filled polymers are among a special class of composite systems for several end uses, where mechanical properties are of primary interest [5,6]. In most of the

Polymers and Other Advanced Materials: Emerging Technologies and Business Opportunities. Edited by P. N. Prasad *et al.*, Plenum Press, New York, 1995

applications, the filler materials are expected to improve the mechanical properties of such systems. Furthermore, apart from the desired target transport and mechanical properties, the end product should be of a relatively low cost. For this reason, in the present study, attempts were made to develop a composite system using abundantly available sand in the State of Qatar as a filler reinforcing a matrix of low density polyethylene produced by Qatar Petrochemicals Co. Such sand-reinforced polyethylene (SRPE) systems may prove useful in domestic applications. Therefore, the present investigation evaluates the performance of injection-molded SRPE samples, during solid-state deformation, by reporting and analyzing the effect of some structural and processing parameters on their flow behavior and fracture characteristics.

EXPERIMENTAL

Sample Preparation

The materials utilized in this study were Lotrène FB 3003 low density polyethylene (LDPE) produced by Qatar Petrochemicals Co. (QAPCO) and beach sand (specific gravity = 2.7203), collected from the Ras-Laffan area, located on north-eastern coast of the State of Qatar. The properties of LDPE as supplied by QAPCO are listed in Table 1.

Table 1. Properties of Lotrène FB3003 LDPE*

Property	Value
Tensile strength at break	≥ 15 MPa
Density at 23°C	0.919 - 0.921 g/cm^3
Melting Point	115°C
Melt Index	0.25-0.23 g/10 min
Coefficient of friction	0.6

* Mechanical properties are measured on films 200 μm thick

Sand was first thoroughly washed, cleaned and soaked in distilled water for 24 hours and then dried at 105°C for five hours. The sand was sieved and four different cuts of average particle size 600, 425, 250 and 150 μm were obtained. Sand and LDPE were separately weighed to prepare composite mixtures bearing different sand content, namely 5, 10, 15, 20 and 25wt%. Ten milliliters of o-xylene were added onto LDPE pellets to promote wetting and homogeneous sand distribution. Sand was slowly added and intimately mixed with the polymer to form a homogeneous mix. A Unimolder injection molding machine of Tecquipment LTD, Model 10H Type C, was used to produce dumb-bell shaped tensile specimens with 65 mm gauge length, 12.6 mm width and a thickness of 3.1 mm. The injection pressure and dwell time were kept constant at 160 kg/cm^2 and 3 seconds, respectively. Composite SRPE samples were produced under different melt and die temperatures. Three melt temperatures were used: 140°C, 160°C and 180°C, and two die temperatures were used: 25°C and 40°C.

Mechanical Testing

Tensile testing was conducted at room temperature using a Lloyd Instruments materials testing machine, Model 'L 1000S', linked to a remote microcomputer for data acquisition and analysis, at a cross head speed of 100 mm/min. Five specimens were tested for each set of parameters involved and the average values of the properties are reported. The tensile strength, modulus of elasticity and strain at fracture were determined from the stress-strain diagrams, and the toughness -reported as work done- was determined from the load-extension diagrams.

RESULTS AND DISCUSSION

In order to compare the flow behavior of the composite samples with that of plain LDPE, some typical stress-strain diagrams are illustrated in Figure 1, for identical testing conditions. The modulus of elasticity and the peak flow stress increase upon sand addition. On the other hand, the ductility and the energy per unit volume to fracture the composite samples (modulus of toughness) are reduced. The difference in flow behavior between plain LDPE and SRPE may be rationalized in terms of the superposition of more than one stress component, in addition to the matrix tensile strength. These components are: the *interfacial strength* between the filler and the matrix, the *interlocking strength* resulting from the mechanical interaction between the irregularly-shaped filler and the polymer matrix, and the induced *matrix strengthening* because of relatively higher strain rates imposed on the matrix due to the presence of rigid sand particles. Figure 2(a) is a scanning electron micrograph of the fracture surface of a SRPE sample containing 25% sand of 250 μm, injection-molded at a melt temperature of 140°C and a die temperature of 25°C. It is clear that the matrix is not uniformly bonded to the sand phase. This is expected, however, as discussed by Shang et al.[2] and Chodak et al.[7]. Adhesion is observed in some areas, where considerable deformation occurred around the sand particle prior to fracture, as seen in the region labeled ① on the micrograph. Furthermore, the matrix phase is also interlocked within the sand particle as we can see from the region labeled ②, and enlarged in Figure 2(b), and from which the intervening matrix phase also undergone considerable

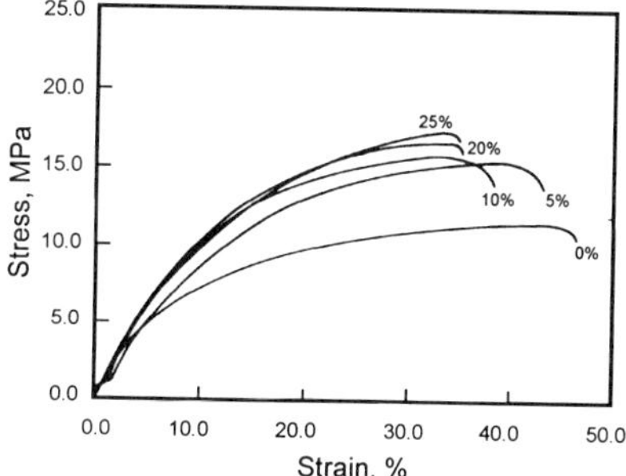

Figure 1 Stress-strain diagrams for plain LDPE and SRPE bearing different sand content. (Melt temperature=140°C, die temperature = 25°C, Sand Size = 250 μm)

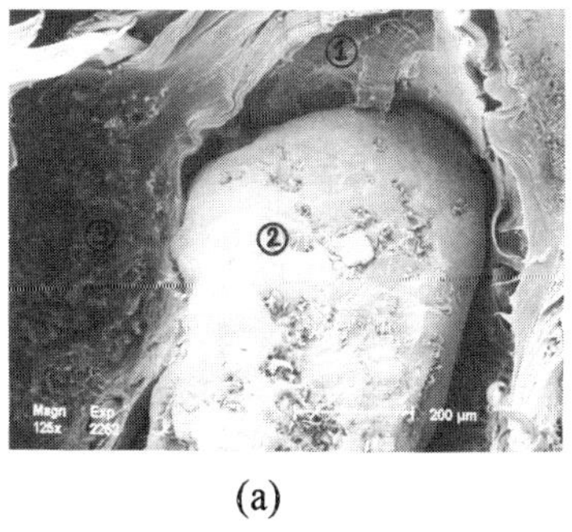

(a)

(b)

Figure 2 Scanning electron micrograph of a SRPE sample with 25% sand, 250 μm, mold temperature = 140°C and die temperature = 25°C; (a) fracture surface; (b) sand particle surface

deformation before fracture. Intuitively, some cohesion exists, obviously not on all the surface. Nevertheless, some sand particles were pulled out from the polymer matrix, still with some adhesion, as depicted from the replica in the region labeled ③. Therefore, it may be deduced that the sand particles act as resistors to flow, resulting in an increased interfacial shear resistance and mechanical interlocking. Figures 3 and 4 show the effect of sand content on the tensile strength and modulus of elasticity of SRPE samples, at different melt temperatures. Again it is clear that the presence of sand increases the tensile strength and modulus. These findings are in agreement with the results obtained for similar composite systems [1,6].

It is also observed from Figures 3 and 4 that -for plain LDPE or for SRPE at any given sand content- increasing the melt temperature is associated with a decrease in the tensile strength and the modulus. Therefore, a lower melt temperature will produce stronger and stiffer SRPE samples. At a constant die temperature, upon increasing the melt temperature, the degree of supercooling of the SRPE sample in the die will be higher when compared to a sample injection-molded at a lower melt temperature. The relatively higher degree of supercooling will a have a quench-like effect on the LDPE matrix, resulting in a relatively weaker and more ductile structure.

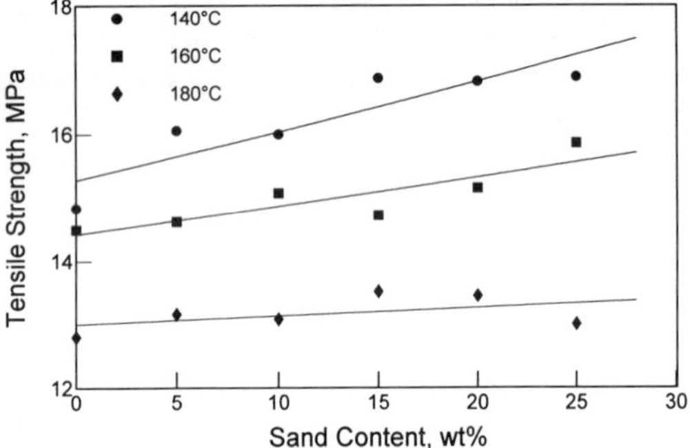

Figure 3 Variation of the tensile strength with sand content at different melt temperatures. (Die temperature = 25°C, sand size=250 μm)

Such an effect is also illustrated in Figure 5, which depicts the variation of the fracture strain with the sand content for SRPE samples, at three different melt temperatures. For a given sand content, the fracture strain increases as the melt temperature increases. In other words, increasing the melt temperature produces a more ductile structure, as mentioned earlier. It is anticipated, however, that for a given melt temperature and die temperature, sand addition will reduce the extent of deformation to fracture due to the restrained movement of the LDPE matrix phase and the accompanied increase in the number of possible void initiation sites, which tends to accelerate crack initiation and propagation. Accordingly, there will be a greater chance for the sand particles to dewet from the matrix and form ellipsoidic vacuoles as the amount of sand increases[5]. This explains the behavior illustrated in Figure 5, which exhibits a reduction in the fracture strain upon increasing the sand content. Figure 6 shows the variation of the work done to fracture, with the sand content of SRPE samples at different melt temperatures. The work done is taken as the area under the load-extension diagram; it also reflects the toughness of SRPE. Since sand addition has the effect of reducing the ductility (strain at fracture), therefore, it will reduce the area under the load-extension diagram; Figure 6 indicates that sand addition is, indeed, accompanied by a reduction in toughness.

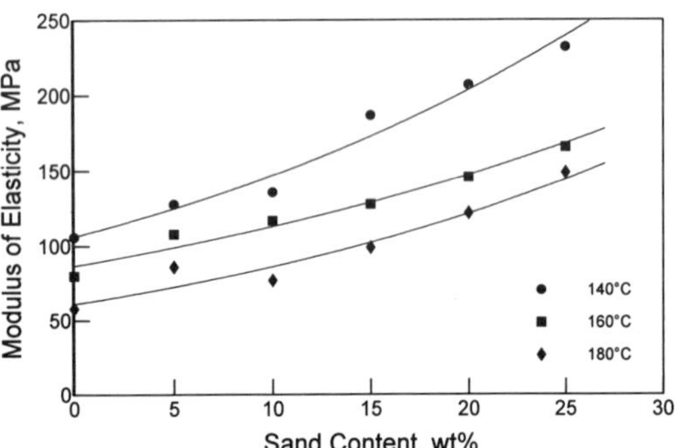

Figure 4 Effect of sand content on the modulus of elasticity at three different melt tempertures. (Die temperature=25°C, sand size=250 μm)

The effect of die temperature is depicted in Figure 7, which illustrates the variation of the tensile strength with sand content for two die temperatures of 25°C and 40°C. A lower die temperature yields a stronger composite system. This is related to the nature of the injection molding process. In other words, when the die temperature is 40°C, the specimen is ejected before it reaches room temperature, in which case the possible perturbations in the internal structure are relatively higher when compared to a specimen ejected from 25°C to room temperature. Conceivably, the time consumed for the sample to cool from the melt temperature to room temperature will depend on the die temperature; lower die temperatures will be associated with longer times. In this case, a more relaxed polymeric structure will result, giving rise to a relatively stiffer and stronger matrix. Therefore, SRPE samples produced at 25 °C are stiffer and stronger compared to those produced at a die temperature of 40°C. The latter samples, however, will be more ductile and will accommodate more deformation prior to fracture, as depicted from Figure 8, which shows that the fracture strains for SRPE specimens produced at a die temperature of 40°C are higher than those produced at a die temperature of 25°C, for any given sand

Figure 9 shows the variation of the tensile strength of SRPE samples with sand particle size at three different melt temperatures. A maximum tensile strength is exhibited for a particle size of about 400μm. In other words, a strengthening effect is observed to occur if the particle size of the sand does not exceed 400 μm, while upon further increase in the particle size, the tensile strength drops. Such a behavior may be attributed to the nature of interaction between the filler and the polymer matrix as well as the filler's geometry. For a given sand content, a smaller particle size will provide a relatively high available interfacial surface area per unit volume, and consequently a larger chance for physical adhesion and wetting,

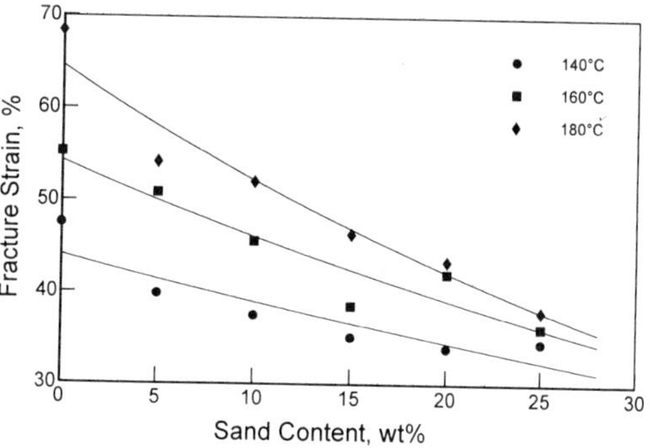

Figure 5 Plot of the fracture strain versus sand content at different melt temperatures.(Die temperature=25°C, sand size=250 μm)

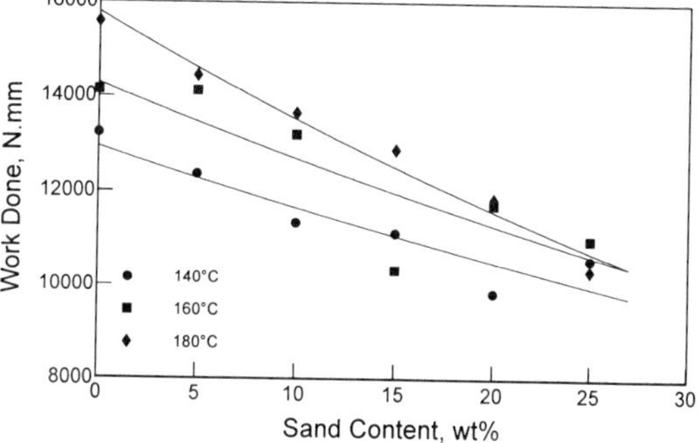

Figure 6 Effect of sand content on the work done to fracture, for samples produced at various melt temperatures. (Die temperature=25°C, sand size=250 μm)

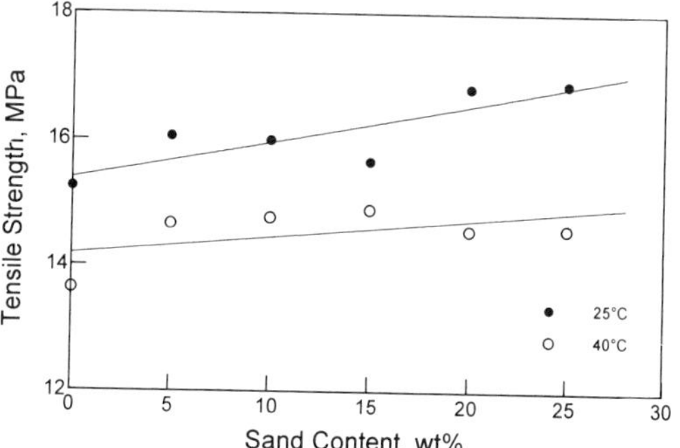

Figure 7 Plot of the tensile strength as a function of sand content at two die temperatures. (Melt temperature= 140°C, sand size = 250 μm)

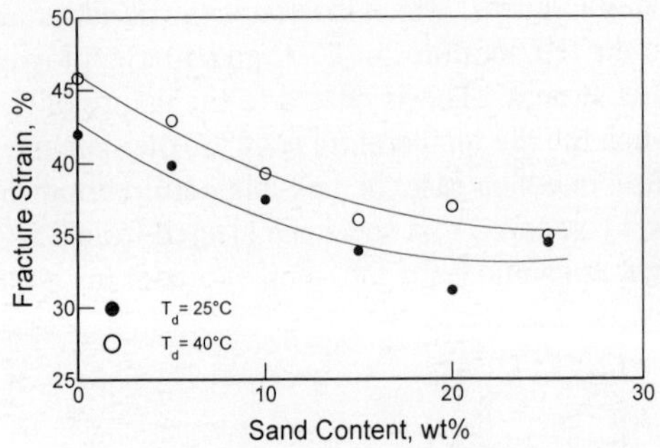

Figure 8 Effect of sand content on the fracture strain for samples injection-molded at two different die temperatures.(Melt temperature=140°C, sand size=250 μm)

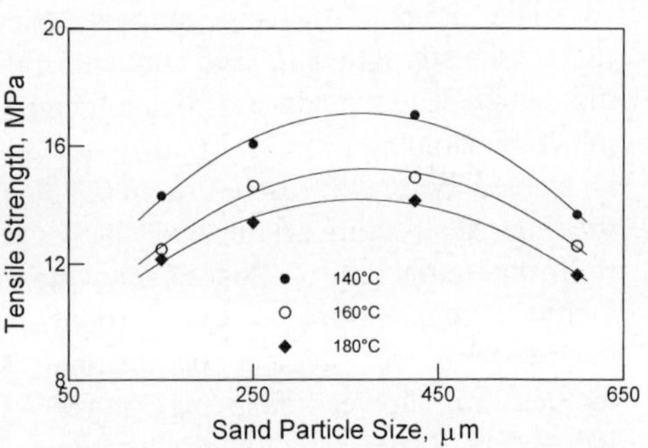

Figure 9 Variation of the tensile strength of SRPE samples with the sand particle size at various melt temperatures. (Die temperature=25°C, sand content=5wt%)

in spite of the possible irregularities in the shape of sand particles. In this range of particle size (up to 400 μm), as the particle size increases, the effect of mechanical interlocking also increases, and it overcomes the effect of internal friction and possible decohesion (dewetting) between the sand and the polymer. On the other hand, above 400 μm, the effect of internal friction and decohesion are more pronounced and dominant: as the particle size increases, although mechanical interlocking is also taking place, the internal friction decreases and it will reduce the strength of SRPE samples. Figure 10 shows a diagram illustrating the prevailing phenomena in each range of particle size. Again these findings are in line with previously published work on similar composite systems, and which reported that coarse particle-filled composites dewet easier than those with fine particles, and therefore, contribute to a loss in the modulus and the tensile strength[5].

At this point it is, therefore, important to realize that if the particle size of sand is greater than 400 μm, a strength reduction will be expected upon sand addition. The results shown in Figure 11 support this argument: for SRPE samples reinforced with 600 μm particles, as the sand content increases, the tensile strength is reduced regardless of the die temperature used.

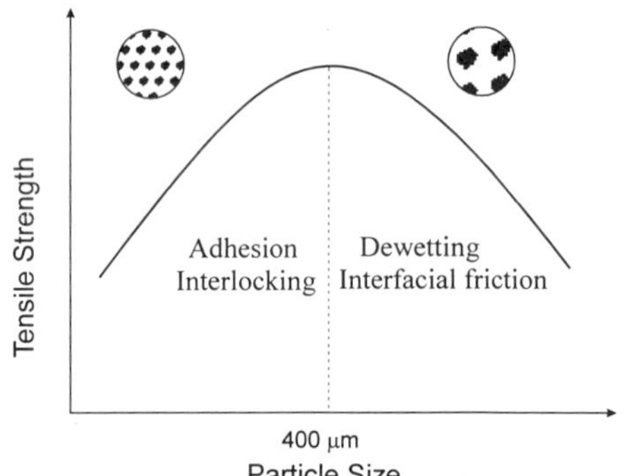

Figure 10 Dominating phenomena in the different size ranges

The modulus of elasticity is not very much affected by the sand particle size as depicted in Figure 12, which illustrates the variation of the modulus with the sand particle size at three different melt temperatures. Such an observation is in line with previous findings for similar systems[1,5,6]. Furthermore, the effect of sand size on the fracture strain and work done is similar to that on tensile strength, as seen from Figures 13 and 14, which show the variation of these properties with sand size at different sand content.

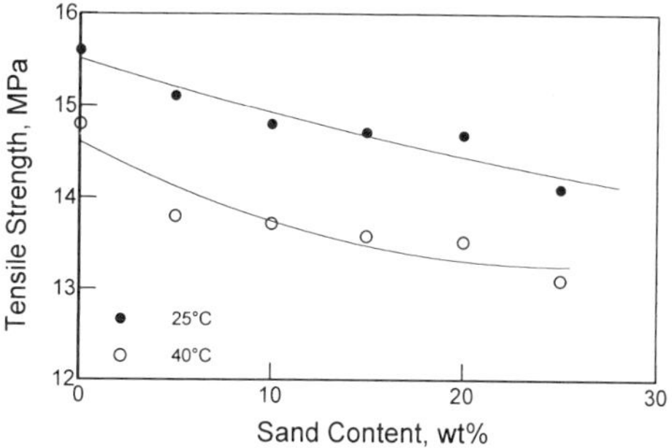

Figure 11 Plot of the tensile strength as a function of sand content at two die temperatures. (Melt temperature = 140°C, sand size = 600 μm)

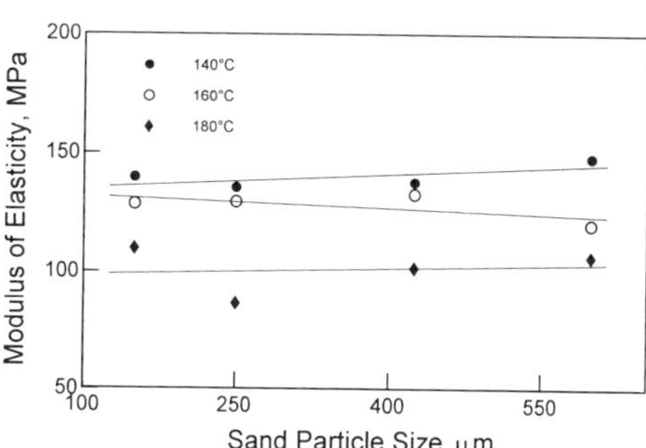

Figure 12 Effect of sand particle size on the elastic modulus of SRPE produced at different melt temperatures.(Die temperature=25°C, sand content=15wt%)

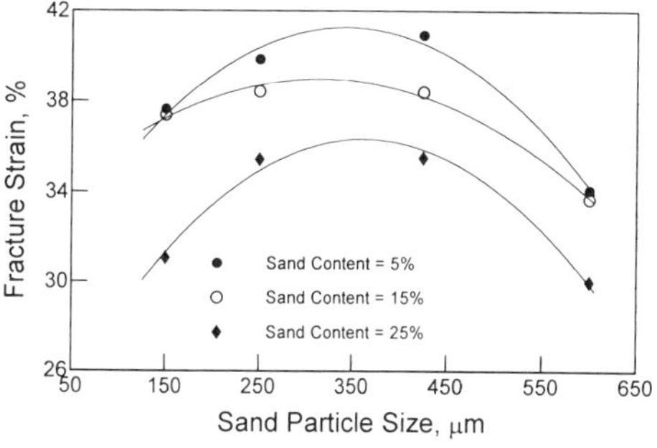

Figure 13 Effect of sand particle size on the fracture strain SRPE samples with different sand content. (Melt temperature = 140°C, die temperature = 25°C)

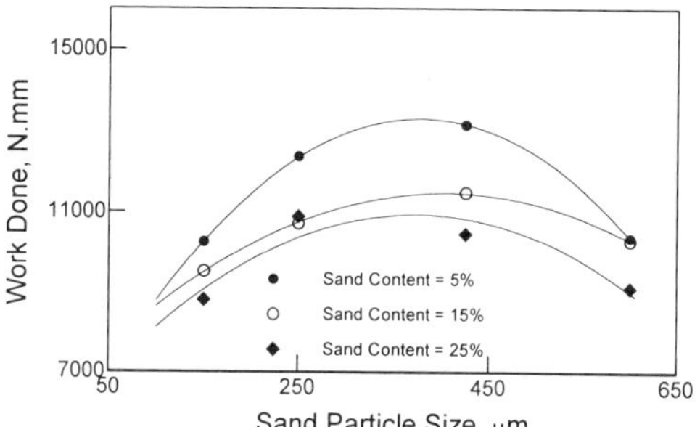

Figure 14 Plot of the work done versus the sand particle size for SRPE samples bearing different sand content. (Melt temperature = 140°C, die temperature = 25°C)

CONCLUSIONS

The flow behavior and fracture characteristics of SRPE are strongly dependent on the presence of sand and on the molding conditions. When the sand size is below 400 μm, its addition increases the modulus of elasticity and the tensile strength of the composite system. On the other hand, regardless of the sand size, it reduces the ductility and the toughness. The strengthening effect of sand addition may be attributed to several components, besides the nominal strength of the matrix. These components are: the interfacial strength, the interlocking strength and that resulting from the higher effective strain rate in the matrix phase of the reinforced material. Below a critical sand size of about 400 μm , adhesion and interlocking are prevailing. Dewetting and interfacial friction are dominating above 400 μm. A maximum in the tensile strength, ductility and toughness of SRPE occurs around a sand particle size of 400 μm and the modulus of elasticity is insensitive to sand particle size. Increasing the melt and the die temperature is associated with a decrease in both the tensile strength and the modulus of elasticity of SRPE and an increase in the ductility and the toughness.

ACKNOWLEDGMENTS

The authors are grateful to QAPCO for providing the LDPE used in this investigation. . The valuable discussions with Mr. H. Al-Rashid, Mr. M. Al-Daoudi and Mr. M. Abdulrahman of the LDPE plant, QAPCO, are highly acknowledged. Thanks are also due to Mr. B. Al-Jaidah, for his help in preparing the samples.

REFERENCES

1. C. Busigin, G.M. Martinez and R.T. Woodhams, Factors affecting the mechanical properties of mica-filled polypropylenes, *Polym. Eng. Sci.*, 23:766 (1983).

2. S.W. Shang, J.W. Williams, K-J.M. Söderholm, How the work of adhesion affects the mechanical properties of silica-filled polymer composites, *J. Mater. Sci.*, 29:2406 (1994).

3. A. Padilla, A.S. Solis and O. Manero, A note on the thermal conductivity of filled polymers, *J. Comp. Mater.*, 22:616 (1988).

4. A.P. Ramirez and A.S. Solis, Development of a new composite material from waste polymers, natural fiber and mineral fillers, *J. Appl. Polym. Sci.*, 29:2405 (1984).

5. U. Yilmazer and R.J. Farris, Mechanical behavior and dilatation of particulate-filled thermosets in the rubbery state, *J. Appl. Polym. Sci.*, 28:3369 (1983).

6. T.M. Malik, M.I. Farrooqi and C. Vachet, Mechanical and rheological properties of reinforced polyethylene, *Polym. Comp.* 13:174 (1992).

7. I. Chodack, I. Chorvath, I. Novack and K. Csmorova, Crosslinked low density polyethylene filled with silica-I: The effect of filler on crosslinking, *Eur. Polym. J.*, 28:107 (1992).

EFFECT OF HYGROTHERMAL AGEING ON THE TENSILE PROPERTIES OF SHORT GLASS FIBRE REINFORCED NYLON 6.6

Z. A. Mohd Ishak, U. S. Ishiaku and C. P. Tan

School of Industrial Technology
University Sains Malaysia
11800 Pulau Pinang, Malaysia

INTRODUCTION

In service, plastics and their corresponding composites are exposed to an environment in which the moisture and temperature vary in a prescribed manner. This combined change of moisture and temperature, which is known as the hygrothermal effect can be more damaging to the mechanical properties of the composites than to either of their constituent materials. The studies on the hygrothermal ageing of fibre reinforced plastics have been extensively reviewed by Springer (1982).

This paper presents the results of a study on the hygrothermal ageing of short glass fibre reinforced nylon 6.6. The effect of conditioning temperature and volume fraction of fibres on the diffusion coefficient or diffusivity and equilibrium or maximum moisture content is determined. In addition, the effect of such ageing process upon tensile properties is reported.

EXPERIMENTAL

Materials

The matrix material used was commercially available nylon 6.6 manufactured by Rhone-Poulene, and the material was in granular form. E-grade glass fibres of an average diameter of 12 µm and chopped length of 5 mm supplied by Eurochem Limited, was used as the reinforcement. 3-aminopropyltriethoxysilane supplied by Polyscience was used as the coupling agent.

Polymers and Other Advanced Materials: Emerging Technologies and Business Opportunities. Edited by P. N. Prasad *et al.*, Plenum Press, New York, 1995

Processing

Nylon 6.6 granules and various loadings of glass fibres were compounded in a Betol single screw extruder. The reinforced granules were then injection moulded in a Battenfeld Unilog 4000 reciprocating screw-type injection moulding machine, The nylon 6.6 and reinforced granules were dried overnight at 100^0C under vacuum prior to extrusion and injection moulding. A detailed description of the fabrication procedure is given by Tan (1993) elsewhere.

Kinetics of moisture absorption

The specimens were dried at 100^0C. The percentage moisture uptake at any time , t, M_t(%) was calculated by

$$M_t = 100(W_W - W_d)/W_d \quad (1)$$

Where W_d and W_W denote, respectively, weight of the dry material and weight of moist material. The equilibrium moisture content, M_m, was calculated as an average value of several consecutive measurements that showed no appreciable additional absorption.

Tensile test

The tensile test for both unaged and aged nylon 6.6 and its composites was conducted on an Instron Universal Testing Machine model 1114 at 30^0C, with a strain rate of 1.3 x 10^{-3} s^{-1}, using ASTM procedure D-1708.

Scanning Electron Microscopy

Examination of the fracture surface was carried out using the Cambridge Scanning Electron Microscope (SEM). The objective was to get some idea of the mode and the condition of the matrix and fibre surfaces.

RESULTS AND DISCUSSION

Figure 1 shows the moisture uptake of nylon 6.6 and its composites conditioned at 100^0C. In all cases M_t increases linearly with $t^{1/2}$ and then slows down until an equilibrium moisture content, M_m is attained. A good agreement is obtained between the theoretical prediction of equation 2 below (shown by the solid lines) and the experimental points. This indicates that the Fiction behaviour is observed. According to Bonniau and Bunsell (1981), the equation for single phase diffusion proposed originally by Crank (1956) can be modified as:

$$M_t = M_m \{1 - 8/\pi^2 \exp[-(Dt/h^2)\pi^2]\} \quad (2)$$

where D is the diffusivity of diffusion coefficient, t the time and h the thickness of the sample.

According to Shen and Springer (1979) the diffusivity can be calculated from the linear portion of the absorption curve as:

$$M_t = M_m\{ (4/h)\, (D/\pi)^{1/2} t^{1/2} \} \tag{3}$$

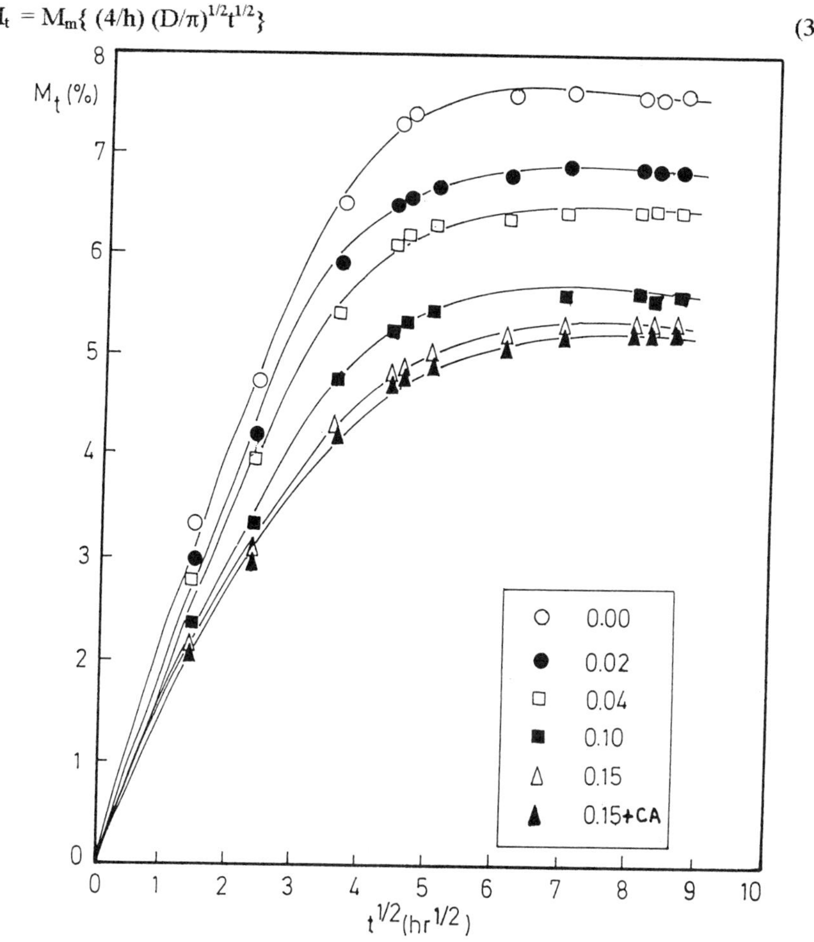

Figure 1: Plot of M_t vs $t^{1/2}$, showing the effect of volume fraction of fibres on the moisture uptake of nylon 6.6 at 100°C

Table 1. The effect of conditioning temperature on the diffusivity D and equlibrium moisture content, M_m, of nylon 6.6 ($V_f = 0.00$) and its composites ($V_f = 0.15$)

V_f	0.00		0.15		0.15 + C	
T(°C)	D x 10^{-12} (m^2sec^{-1})	M_m (%)	D x 10^{-12} (m^2sec^{-1})	M_m (%)	D x 10^{-12} (m^2sec^{-1})	M_m (%)
60	6.77	7.15	6.05	4.90	6.00	4.75
80	20.50	7.35	18.93	4.90	18.54	4.85
100	89.11	7.60	64.90	5.30	63.25	50.00

Table 1 summarises the values of diffusivity and equilibrium moisture content for nylon 6.6 and its composites aged under different conditioning temperatures. The values of diffusivity obtained are in agreement with the range of values reported by Loss and Springer (1979) and Mohd Ishak and Berry (1994). As expected, incorporation of glass fibres into nylon 6.6 matrix has led to a reduction in both the diffusivity and equilibrium moisture content. The reduction in diffusivity may be attributed to the presence complex fibre orientation distribution which hindered the direct moisture diffusion into the polymer matrix. A similar trend has been observed by Mohd Ishak et. al (1993) for short glass fibre reinforced poly(butylene terephthalate). The presence of 3-aminopropyltriethoxysilane in nylon 6.6 composites has resulted in a slight reduction in the values of diffusivity and maximum moisture content.. The improvement in the interfacial bonding between the nylon and the glass fibres is believed to give further hindrance on the moisture absorption into the composites. A similar situation has been encountered by Bian et al. (1991).

Table 2. The tensile Properties of unaged, aged and redried nylon 6.6 and its composites.

V_f	Condion	Tensile Strength (Mpa)	Tensile Modulus (Gpa)	Tensile Strain (%)
	Unaged	77.68	1.55	49.4
0.00	Aged at 100^{0}C, 100% RH	42.61 (54.2%)	0.47 (30.3%)	270.80
	Redried	74.46 (95.8%)	1.32 (85.2%)	
	Unaged	132.3	3.61	4.88
0.15	Aged at 100^{0}C, 100% RH	54.06 (40.9%)	1.80 (49.9%)	8.34
	Redried	89.74 (67.8%)	3.26 (90.3%)	-
	Unaged	102.31	3.44	4.77
0.15 + CA*	Aged at 100^{0}C, 100% RH	60.06 (58.7%)	2.20 (63.9%)	5.24
	Redried	102.22 (99.9%)	3.39 (98.6%)	-

-Parenthesis indicate the percentage retention or recovery of the tensile properties.
*CA indicate coupling agent

Table 2 illustrates the tensile properties i.e. the values of tensile modulus, tensile strength and elongation at break of the unaged and aged nylon 6.6 and its composites. In the case of the unaged samples, the incorporation of glass fibres into nylon 6.6 matrix has resulted in a significant improvement in the strength and stiffness properties. All the composites are seen to fail in a brittle manner with relatively low fracture strain value. The presence of coupling agent has no significant effect on the tensile properties of nylon 6.6 composites. This observation is in agreement with an earlier investigation carried out by Berry and Stanford (1977).

As for the samples aged at 100% RH and 100^{0}C, some interesting results were obtained. Hygrothermal ageing caused a dramatic reduction in both strength and stiffness of the composite. This may be attributed to the combined effect of plasticisation of the nylon 6.6 matrix, as well as the degradation of the polymer-polymer interface. The extent of the plasticisation and degradation processes seem to increase with increasing conditioning temperatures. The evidence of the plasticisation effect of moisture on nylon 6.6 has been proven via dynamic mechanical technique by Mohd Ishak and Berry (1994). From Table 2 also, it appears that the absorbed moisture has significantly changed the mode of fracture of the composite from being brittle to a ductile fracture. A similar observation has been reported by Mohd Ishak and Berry (1993) for the short carbon fibre reinforced nylon 6.6.

Further investigation on the extent of deterioration of tensile properties have been conducted by drying both unreinforced and reinforced nylon 6.6 samples. As expected, a good retention of tensile properties is observed in the case of the unreinforced nylon 6.6. This indicates that hygrothermal ageing in the samples is physical in nature i.e. the effect of moisture is merely that of a plasticiser. However, the tensile properties of nylon 6.6 composites are not fully recovered. This poor retentionability may be due to the fact that the action of moisture may have resulted in the partial disruption of the interfacial bonding which consequently leads to the formation of microcavities, which would be occupied with water. Upon drying, these cavities will act as stress concentrators, which can then result in the premature failure of the composite.

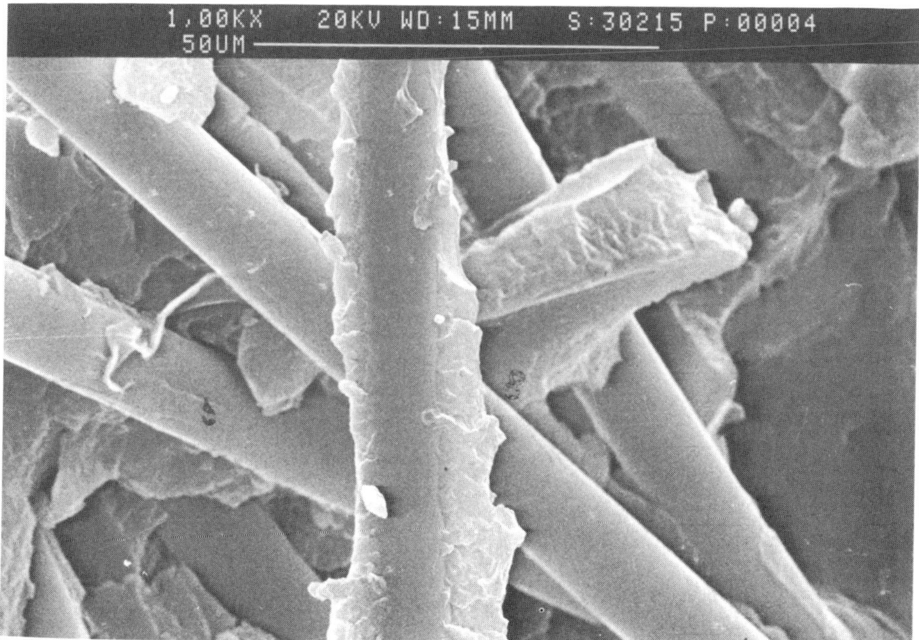

Figure 2: SEM micrograph of the fracture surface of unaged nylon 6.6 composites ($V_f = 0.15$)

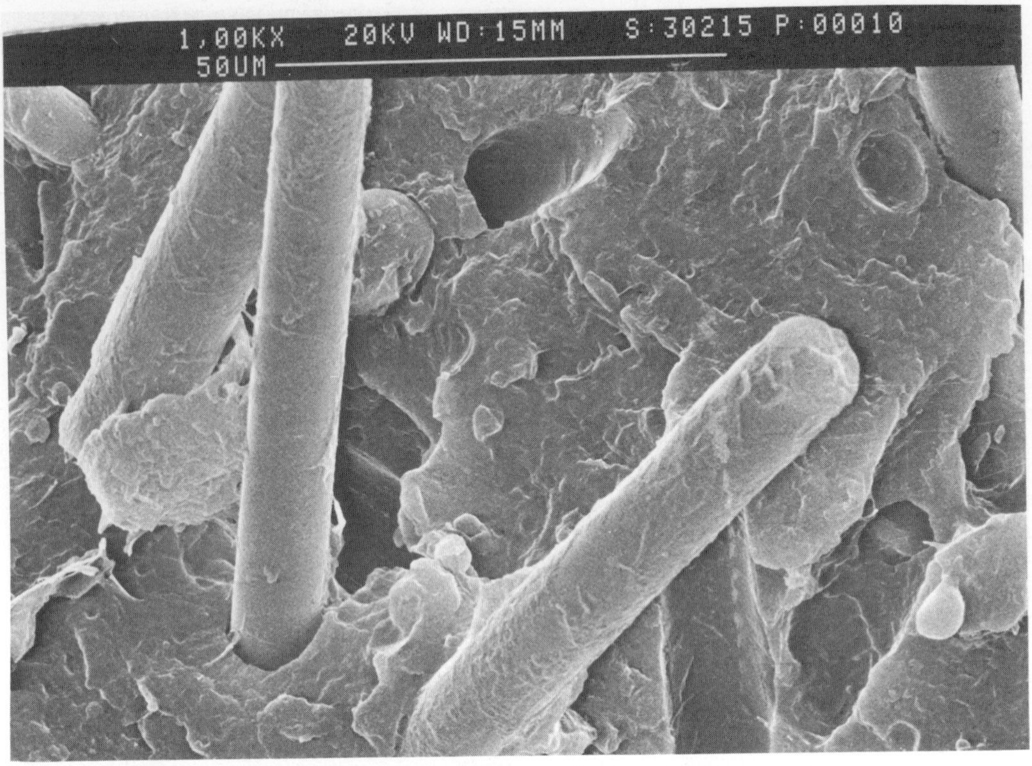

Figure 3: SEM micrograph of the fracture surface of unaged nylon 6.6 composites ($V_f = 0.15$) with silane coupling agent.

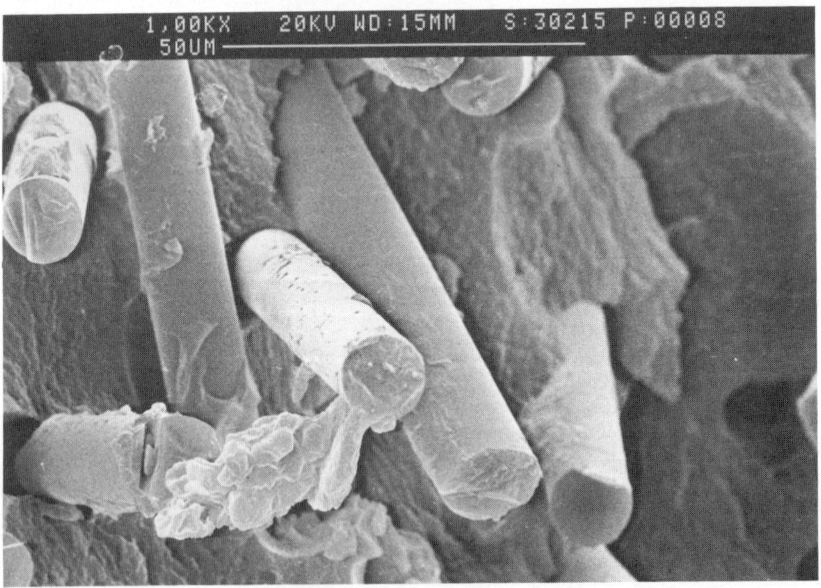

Figure 4: SEM micrograph of the fracture surface of nylon 6.6 composite ($V_f = 0.15$) aged at 100^{0}C and 100% RH.

Table 2 also highlights the significant role played by the silane coupling agent in maintaining the tensile properties of nylon 6.6 composites after the materials were subjected to hygrothermal ageing. The samples with coupling agent show a better strength and stiffness retention as compared to the uncoupled samples. This clearly indicates the capability of the silane in retaining a good interfacial bonding between the nylon 6.6 and glass fibres even when the composites were exposed to the most extreme conditioning temperature i.e. 100^0C . The excellent retention of the tensile properties upon drying further proves the fact that no significant interfacial damage has taken place

Figure 2 shows the fracture surface of unaged nylon 6.6 composites. The nylon 6.6 matrix indicates a predominantly brittle behaviour with extensive fibre pull-out in the crack plane. Apparently there is some evidence of nylon 6.6 matrix adherence to the fibre surface which indicate the existence of good interfacial bonding between the constituent materials. Some degree of matrix retention in the fibre surface can be seen more clearly in the case of nylon 6.6 with silane coupling agent. This is shown in Figure 3. Figure 4 shows the fracture surface of an aged nylon 6.6 composite. The role of moisture in reducing the effectiveness of the interfacial bonding is clearly indicated not only by extensive fibre pull-out but also by the absence of nylon 6.6 matrix on the surface of the fibres.

CONCLUSION

The single free-phase model of absorption has been applied successfully to the kinetic data and there after the values of diffusivity have been evaluated. The incorporation of short glass fibres into nylon 6.6 matrix has been observed to reduce the values of both diffusivity and the equilibrium moisture content. Hygrothermal ageing has reduced the tensile properties of both nylon 6.6 and its composites. This has been attributed to the plasticisation of the matrix and interfacial degradation. However, the extent of these plasticisation and degradation processes could be reduced by incorporating a silane coupling agent into the composites.

ACKNOWLEDGMENTS

The authors would like to thank Universiti Sains Malaysia and the Malaysia Government for providing financial assistance.

REFERENCES

Bian, X. N. S., Ambrosio, L., Kenny J. M. and Nicolais, L. 1991, Effect of water absorption on the behaviour of E-glass fibre/nylon-6 composites. Polym. Comp.,12:333

Berry, J. P. and Stanford J. L., 1977, Mecanical properties of short glass fibre reinforced nylon 6.6.Reinforced themoplastics II, *Plast. Rubb. Inst. Conf.*, Manchester.

Bonniau, P. and Bunsell, A. R., 1981. Comparative study of water absorption theories applied to glass epoxy composites , J. Compos. mater., 15:272.

Crank, J., 1956, The Mathematics of Diffusion, Oxford University Press, Claredon.

Loss, A. C. and Springer, G. S., 1979,Moisture absorption of graphite-epoxy composites immersed in liquid and in humid air, *J. Compos. Mater.*, 13:131

Mohd Ishak, Z. A., Mohd. Yusof, A. M., Senawi, R. and Abu Bakar, A. 1993, Effect of a silane coupling agent on the hygrothermal behaviour of short glass fibre reinforced polybutylene terephthalate, *Proceedings of the 3rd Pacific Polymer Conference,* Gold Coast, Australia.

Mohd Ishak, Z. A. and Berry, J. P., 1994,Effect of moisture absorption on the dynamic mechanical properties of short carbon fibre reinforced nylon 6.6, *J. Appl. Polym. Sci.,* 51:223

Shen, C. H. and Springer, G. S., 1976, Moisture absorption and desorption of composite materials *J. Compos. Mater., 10:2*

Springer, G. S., 1982, Moisture absorption in fibre-resin composites, in Developments in Reinforced Plastics - 2, 1982,G. Prichard, ed., Elsevier Applied Science, London

Applied Science, London.

Tan, C. P., 1993, Effect of hygrothermal ageing on short glass fibre reinforced nylon 6.6, *Final year project dissertation,* Universiti Sains Malaysia, Penang,

FINITE ELEMENT ANALYSIS OF PLASTIC YIELDING AT A CIRCULAR HOLE IN A LAMINATED COMPOSITE PLATE BASED ON REFINED PLATE THEORY

Tasneem Pervez
Faculty of Engineering
International Islamic University Malaysia

ABSTRACT

An elastic-plastic analysis of anisotropic laminated composite plate with a hole is carried out using a two dimensional finite element model based on higher order shear deformation theory (HSDT) of anisotropic laminated composite plates. The variation in material properties through thickness is made possible by discrete layer approach. The generalized Huber-Mises yield criterion, developed by Hill, for anisotropic material, is used to determine the onset of plastic flow. The inclusion of anisotropic parameters of plasticity generalizes the plastic yield function. These anisotropic parameters of plasticity are updated during the work hardening history of anisotropic laminated composite plate. The plastic potential, which is an anisotropic yield function, is used to determine an associated flow rule. Finally, an elastic-plastic incremental constitutive relation is obtained using finite element method. By means of an incremental and iterative procedure, the numerical solution of deformation problems in nonlinear anisotropic laminated plates is achieved based on first order shear deformation theory and higher order shear deformation theory. The transverse shear effects on the load deformation characteristics and on the spread of plastic zones around the circular hole are discussed. This method yields accurate values for transverse shear stresses as compared to other shear deformation theories and hence the growth of plastic zones.

INTRODUCTION

When studying the deformation of metallic solids due to applied external loads, it is common to assume that the material will behave in an isotropic manner. Typically, this assumption is extended to both the elastic and elastic-plastic responses. However, there are two categories of problems involving anisotropic behavior: one is the introduction of anisotropy due to a deformation mechanism into an initially isotropic material, most commonly observed in metal forming operations [1]. The second category involves those materials which are initially anisotropic such as composites. Unlike isotropic materials, for which some well defined theories of plasticity with all levels of sophistication exist [2-6], anisotropic materials lack of general theories for their inelastic behavior. Several investigators have attempted to describe the behavior of anisotropic metals. The concept of anisotropic plasticity was introduced into the Von-Mises theory of plasticity by Hill [7-8], who proposed an anisotropic yield criterion constituting a generalization of Mises criterion. Subsequently, there have been several modifications [9-14]. Yet, these theories do not properly account for all the effects associated with the orientation dependence of deformation mechanisms. In addition, some theories involve parameters that are difficult to be measured experimentally.

The complexity involved due to the material anisotropy, has restricted the research on the elastic-plastic analysis of fiber reinforced composites. Spencer [15] used a micromechanics approach and Lin, Salinis and Ito [16] used a micromechanical approach to model the nonlinear behavior of composites. Through the development of finite element method, considerable attention have been given to elastic-plastic plate problem, though limited to the isotropic case. Apparently, Whang [17] is the first one to consider the elastic-plastic analysis of orthotropic plates and shells. Owen and others [18-19] have extended the idea of Whang for the elastic-plastic analysis of plates and shells using FSDT theory.

In this paper, the elastic-plastic analysis of fiber reinforced laminated composite plates with and without a circular hole is carried out using a first order shear deformation theory and higher order shear deformation theory. Numerical results are obtained by adopting an anisotropic version of the Huber-Mises yield criterion. Nine noded Lagrangian and 8-noded serendipity elements are used in the analysis. The obtained solutions are compared with other available numerical results.

KINEMATICS OF LAMINATED COMPOSITE PLATE

Shear deformation theories for a laminated plate can be developed using a β displacement field of the following form

$$u(x,y,z,t) = u_o(x,y,t) + z\left(\beta_1\frac{\partial w}{\partial x} + \beta_2\theta_x \quad + \quad z^3\left(\beta_3\zeta_x + \beta_4\frac{4}{3h^2}(\theta_x + \frac{\partial w}{\partial x})\right)\right.$$

$$v(x,y,z,t) = v_o(x,y,t) + z\left(\beta_1\frac{\partial w}{\partial y} + \beta_2\theta_y\right) + z^3\left(\beta_3\zeta_y + \beta_4\frac{4}{3h^2}(\theta_y + \frac{\partial w}{\partial y})\right) \tag{1}$$

$$w(x,y,z,t) = w_o(x,y,t)$$

where (u,v,w) are the displacements of a generic point (x,y,z) in a laminated anisotropic plate at time t, (u_o,v_o,w_o) are the displacements at point (x,y,0) in the reference plane of the laminated plate, $\theta_x(x,y,t)$ and $\theta_y(x,y,t)$ are the rotations of the normals of the reference plane about y-axis and x-axis, and $\zeta_x(x,y,t)$ and $\zeta_y(x,y,t)$ are the so called warping functions. Different combinations of the parameters β_i, i=1,2,3,4 in equation (1) yield different plate theories [20]. The HSDT developed by the author [21] corresponds to $\beta_1 = 0$, $\beta_2 = -1$, $\beta_3 = 1$, and $\beta_4 = 0$. In this paper FSDT and HSDT are used. Using equation (1) the following in-plane and transverse shear strain components can be derived

$$\varepsilon^i = \varepsilon^p + \varepsilon^f + \alpha\varepsilon^w \tag{2a}$$

$$\varepsilon^s = \bar{\varepsilon}^s + \alpha\varepsilon^{sw} \tag{2b}$$

with ε^p, ε^f and ε^w representing the extensional, flexural, and warping parts of the strain, and ε^s, and ε^{sw} representing the shear, and shear-warp strains. The variable α describes the type of plate theory used. For FSDT $\alpha = 0$ and for HSDT $\alpha = 1$ is assumed.

LINEAR MATERIAL MODEL

The general relationship between the stresses and strains for an anisotropic elastic material involves twenty one independent elastic constants. However, this number is considerably reduced if the material possesses certain symmetry [22]. The type of material used in this research is the fiber reinforced composites, which are orthotropic in nature. Because of symmetry of the stress and strain tensors in orthotropic materials, and the assumption of plane stress, the number of constants reduces to nine. The constitutive equation for the inplane and transverse shear stresses of the k-th orthotropic lamina, under a plane stress state, can be written as

$$\begin{pmatrix} \sigma_x \\ \sigma_y \\ \sigma_{xy} \end{pmatrix}^{(k)} = \begin{pmatrix} \bar{Q}_{11} & \bar{Q}_{12} & \bar{Q}_{16} \\ \bar{Q}_{12} & \bar{Q}_{22} & \bar{Q}_{26} \\ \bar{Q}_{16} & \bar{Q}_{26} & \bar{Q}_{66} \end{pmatrix}^{(k)} \begin{pmatrix} \varepsilon_x \\ \varepsilon_y \\ \varepsilon_{xy} \end{pmatrix}^{(k)} \tag{3a}$$

$$\begin{Bmatrix} \sigma_{xz} \\ \sigma_{yz} \end{Bmatrix}^{(k)} = \begin{pmatrix} \bar{R}_{44} & \bar{R}_{45} \\ \bar{R}_{45} & \bar{R}_{55} \end{pmatrix}^{(k)} \begin{Bmatrix} \varepsilon_{xz} \\ \varepsilon_{yz} \end{Bmatrix}^{(k)} \tag{3b}$$

where the matrix $[\bar{Q}]$ and $[\bar{R}]$ are given as:

$$[\bar{\mathbf{Q}}]^{(k)} = [\mathbf{T}^{(k)}]^{-1}[\mathbf{Q}][\mathbf{T}^{(k)}]^{-T} \qquad [\bar{\mathbf{R}}]^{(k)} = [\hat{\mathbf{T}}^{(k)}]^{-1}[\mathbf{R}][\hat{\mathbf{T}}^{(k)}]^{-T}$$

where $[\mathbf{T}^{(k)}]$ and $[\hat{\mathbf{T}}^{(k)}]$ are the transformation matrices and are described in [23]. As in reference [24] the constitutive equation for the laminate based on HSDT is:

$$\begin{Bmatrix} \mathbf{N} \\ \mathbf{M} \\ \mathbf{P} \end{Bmatrix} = \begin{pmatrix} [\mathbf{A}] & [\mathbf{B}] & [\mathbf{E}] \\ [\mathbf{B}] & [\mathbf{D}] & [\mathbf{F}] \\ [\mathbf{E}] & [\mathbf{F}] & [\mathbf{G}] \end{pmatrix} \begin{Bmatrix} \varepsilon^{p} \\ \bar{\varepsilon}^{f} \\ \alpha\bar{\varepsilon}^{w} \end{Bmatrix} \tag{4a}$$

$$\begin{Bmatrix} \mathbf{V} \\ \mathbf{W} \end{Bmatrix} = \begin{pmatrix} [\mathbf{S}_1] & [\mathbf{S}_2] \\ [\mathbf{S}_2] & [\mathbf{S}_3] \end{pmatrix} \begin{Bmatrix} \bar{\varepsilon}^{s} \\ \alpha\bar{\varepsilon}^{sw} \end{Bmatrix} \tag{4b}$$

where

$$\{[\mathbf{A}], [\mathbf{B}], [\mathbf{D}], [\mathbf{E}], [\mathbf{F}], [\mathbf{G}]\} = \sum_k \int_{h_{k-1}}^{h_k} [\bar{\mathbf{Q}}]^{(k)} (1, z, z^2, z^3, z^4, z^6)dz \tag{5a}$$

$$\{[\mathbf{S_1}], [\mathbf{S_2}], [\mathbf{S_3}]\} = \sum_k \int_{h_{k-1}}^{h_k} [\bar{\mathbf{R}}]^{(k)} (1, z^2, z^4)dz. \tag{5b}$$

where h_k and h_{k-1} are the z-coordinates of the upper and lower surfaces of the k-th lamina.

NONLINEAR MATERIAL MODEL

Yield Criterion

The present formulation is limited to infinitesimal deformations and rate-independent plasticity is considered. All subsequent discussions and derivations are referenced to the material coordinate system. However, these equations can be transformed to an arbitrary Cartesian coordinate system. A law defining the onset of yielding under combined state of stress is known as the yield criterion, which for an anisotropic material should be able to account for different yielding conditions in different directions. With the assumption of no Bauschinger effect, a general yield function for an anisotropic material can be written as

$$f(\boldsymbol{\sigma}) - \sigma_Y = \bar{\sigma}(\boldsymbol{\sigma}) - \sigma_Y = 0 \tag{6}$$

where $\boldsymbol{\sigma}$ is a second order stress tensor, $\bar{\sigma}$ is the effective stress and σ_Y is the uniaxial yield stress. The effective stress, $\bar{\sigma}$, for a composite material is defined as

$$\bar{\sigma} = \sqrt{\boldsymbol{\sigma} . \mathbf{A} \boldsymbol{\sigma}} \tag{7}$$

with $\mathbf{A}$ a fourth order symmetric tensor of anisotropic parameters. The yielding of an anisotropic material is characterized by A_{ijkl} and σ_Y, which describe the shape and size of the yield surface, respectively. With the assumption that the normal stress $\sigma_{33} = 0$, equation (7) reduces to

$$\bar{\sigma}^2 = a_1\sigma_{11}^2 + a_2\sigma_{22}^2 + a_4\sigma_{11}\sigma_{22} + a_7\sigma_{12}^2 + a_8\sigma_{13}^2 + a_9\sigma_{23}^2 \tag{8}$$

where a's are the components of tensor $\mathbf{A}$ and defines the current state of anisotropy. This is the final form as it is required for the present analysis of a lamina. Using the plastic potential function, $g(\sigma)$, we can define a flow rule. In the present work, associated flow rule is assumed, therefore the yield and the plastic potential surfaces coincide i.e. $f = g$. Hence the plastic flow develops along the normal to the yield surface (the normality condition). The phenomena whereby yield stress increases with further plastic straining is known as strain hardening. In strain hardening theory of plasticity, one must relate the hardening parameters to the experimental uniaxial stress/strain curve. A stress variable called effective stress have been already defined. In order to define a corresponding strain variable, called effective plastic strain, which is a function of the plastic strains, the concept of plastic work i.e. work conjugate relation, is used. One can obtain the final expression of effective plastic strain increment, in terms of the components of plastic strains [20].

Elastoplastic Constitutive Relation

Once the material is yielded, its behavior will be partly elastic and partly plastic. During any incremental change of loading, the change in strain is assumed to be decomposed additively into elastic and plastic components.

$$d\varepsilon = d\varepsilon_e + d\varepsilon_p \tag{9}$$

The elastic strain increment can be related to the stress increment through an elastic anisotropic stress/strain relation. So far the yield criteria, flow rule and hardening law were defined. Using the consistency condition, which requires the state of stress to remain on the yield surface during plastic flow, we can derive the incremental stress/strain relationship, i.e.:

$$d\sigma = \left(\mathbf{D} - \frac{\left(\mathbf{D}\dfrac{\partial f}{\partial \sigma}\right) \otimes \left(\mathbf{D}\dfrac{\partial f}{\partial \sigma}\right)}{E_p + \left(\dfrac{\partial f}{\partial \sigma}\right) . \mathbf{D} \left(\dfrac{\partial f}{\partial \sigma}\right)} \right) d\varepsilon = \mathbf{D}_{ep}\, d\varepsilon \tag{10}$$

where $\mathbf{D}_{ep}$ is the elastic-plastic stiffness tensor. The matrix form of the above equation, is used in the computer implementation. In equation (10), the second term in the bracket is the incremental plastic stiffness. The incremental elastic-plastic stress/strain relation is valid only when there is a plastic deformation. Thus before using this relation, we have to check whether the material is in the plastic regime or not. If not, the elastic stress/strain relation should be used instead of equation (10).

Constitutive Representation of A Laminated Plate

The constitutive equation between incremental stresses and strains of the k-th orthotropic lamina, expressed in the global coordinate system, can be written as

$$\begin{pmatrix} d\hat{\sigma}^i \\ d\hat{\sigma}^s \end{pmatrix}^{(k)} = \begin{pmatrix} \hat{\mathbf{D}}_{ep}^{ii} & \hat{\mathbf{D}}_{ep}^{is} \\ \hat{\mathbf{D}}_{ep}^{si} & \hat{\mathbf{D}}_{ep}^{ss} \end{pmatrix}^{(k)} \begin{pmatrix} d\hat{\varepsilon}^i \\ d\hat{\varepsilon}^s \end{pmatrix}^{(k)} \tag{11}$$

where superscripts i and s denotes the inplane and transverse shear components of stress and strain respectively. $\hat{\mathbf{D}}_{ep}^{ii}$, $\hat{\mathbf{D}}_{ep}^{is}$, $\hat{\mathbf{D}}_{ep}^{si}$, and $\hat{\mathbf{D}}_{ep}^{ss}$ represent the planar-planar, planar-shear, shear-planar, and shear-shear elastoplastic stiffnesses respectively. The stress resultants are obtained by integrating the stresses with respect to 'z'

$$\left\{ d\mathbf{N}^T, d\mathbf{M}^T, d\mathbf{V}^T \right\} = \int_{-\frac{h}{2}}^{\frac{h}{2}} \left\{ \left(d\hat{\sigma}^i\right)^T, \left(d\hat{\sigma}^i\right)^T z, \left(d\hat{\sigma}^s\right)^T \right\} dz \tag{12}$$

where h is the thickness of the plate. After some algebra, it can be shown that

$$\begin{pmatrix} d\mathbf{N} \\ d\mathbf{M} \\ d\mathbf{V} \end{pmatrix} = \begin{pmatrix} [\mathbf{A}]_{ep} & [\mathbf{B}]_{ep} & [\mathbf{C}]_{ep} \\ [\mathbf{B}]_{ep} & [\mathbf{E}]_{ep} & [\mathbf{F}]_{ep} \\ [\mathbf{C}]_{ep} & [\mathbf{F}]_{ep} & [\mathbf{S}]_{ep} \end{pmatrix} \begin{pmatrix} d\varepsilon^{p} \\ d\bar{\varepsilon}^{f} \\ d\bar{\varepsilon}^{s} \end{pmatrix} \tag{13}$$

where $d\varepsilon^{p}$, $d\bar{\varepsilon}^{f}$ and $d\bar{\varepsilon}^{s}$ are the incremental planar, felxural and transverse shear component of strains, respectively. For the higher order shear deformation theory, which assumes a parabolic variation of the transverse shear strain through thickness and needs no shear correction factor. However, the constitutive equation between incremental stresses and strains of the k-th orthotropic lamina in global coordinate system require following additional stress resultants in addition to those given in equation (12).

$$\left\{ d\mathbf{P}^{T}, d\mathbf{W}^{T} \right\} = \int_{-\frac{h}{2}}^{\frac{h}{2}} \left\{ (d\hat{\sigma}^{i})^{T} z^{3}, \ (d\hat{\sigma}^{s})^{T} z^{2} \right\} dz \tag{14}$$

After some algebra, it can be shown that

$$\begin{pmatrix} d\mathbf{N} \\ d\mathbf{M} \\ d\mathbf{P} \\ d\mathbf{V} \\ d\mathbf{W} \end{pmatrix} = \begin{pmatrix} [\mathbf{A}]_{ep} & [\mathbf{B}]_{ep} & [\mathbf{G}]_{ep} & [\mathbf{C}]_{ep} & [\mathbf{C}_1]_{ep} \\ [\mathbf{B}]_{ep} & [\mathbf{E}]_{ep} & [\mathbf{J}]_{ep} & [\mathbf{F}]_{ep} & [\mathbf{F}_1]_{ep} \\ [\mathbf{G}]_{ep} & [\mathbf{J}]_{ep} & [\mathbf{L}]_{ep} & [\mathbf{F}_1]_{ep} & [\mathbf{F}_2]_{ep} \\ [\mathbf{C}]_{ep} & [\mathbf{F}]_{ep} & [\mathbf{F}_1]_{ep} & [\mathbf{S}]_{ep} & [\mathbf{S}_1]_{ep} \\ [\mathbf{C}]_{ep} & [\mathbf{F}_1]_{ep} & [\mathbf{F}_2]_{ep} & [\mathbf{S}_1]_{ep} & [\mathbf{S}_2]_{ep} \end{pmatrix} \begin{pmatrix} d\varepsilon^{p} \\ d\bar{\varepsilon}^{f} \\ d\bar{\varepsilon}^{w} \\ d\bar{\varepsilon}^{s} \\ d\bar{\varepsilon}^{sw} \end{pmatrix} \tag{15}$$

Finite Element Model

Considering transverse loads, the incremental principle of virtual work is written as

$$\int_{V_o} \delta d\hat{\varepsilon}^{i^T} d\hat{\sigma}^{i} \, dV + \int_{V_o} \delta d\hat{\varepsilon}^{s^T} d\hat{\sigma}^{s} \, dV = \int_{A_o} dq \delta dw \, dxdy \tag{16}$$

where A_o and V_o are the cross-sectional area and volume of the plate. The incremental principle of virtual work equation can be written in the form required for finite element calculation

$$\int_{A_o} \{\delta d\varepsilon^{p^T} [\mathbf{A}]_{ep} d\varepsilon_p + \delta d\varepsilon^{p^T} [\mathbf{B}]_{ep} d\bar{\varepsilon}^f + \delta d\varepsilon^{p^T} [\mathbf{C}]_{ep} d\bar{\varepsilon}^s + \delta d\bar{\varepsilon}^{f^T} [\mathbf{B}]_{ep} d\varepsilon^p$$

$$+ \delta d\bar{\varepsilon}^{f^T} [\mathbf{E}]_{ep} d\bar{\varepsilon}^f + \delta d\bar{\varepsilon}^{f^T} [\mathbf{F}]_{ep} d\bar{\varepsilon}^s + \delta d\bar{\varepsilon}^{s^T} [\mathbf{C}]_{ep} d\varepsilon^p + \delta d\bar{\varepsilon}^{s^T} [\mathbf{F}]_{ep} d\bar{\varepsilon}^f$$

$$+ \delta d\bar{\varepsilon}^{s^T} [\mathbf{S}]_{ep} d\bar{\varepsilon}^s \} dA = \int_{A_o} dq \delta dw \, dA \tag{17}$$

The general procedure for defining a finite element model from the principal of virtual work is well known and is only briefly stated here. The first step is to discretize the domain into a finite number of elements and define interpolation functions for each variable in the element. The second step is to derive the element strains in terms of the nodal degrees of freedom. The final step is to obtain stiffness matrix, and the load vector using the variational form of principle of virtual work. Using a discretization, the final assembled equations can be written as

$$[\mathbf{K}] \{d\mathbf{U}\} = \{d\mathbf{F}\} \tag{18}$$

where the unknown incremental vector $\{d\mathbf{U}\}$ is generated by the assemblage of element degrees of freedom $\{d\mathbf{u}\}^T_{(e)}$, e=1,2,....m. The assembled stiffness matrix and load vector for HSDT are given by equation (19a) and (19b) respectively.

$$[\mathbf{K}] = \sum_e \int_{A_e} [[\mathbf{B}^p]^T [\mathbf{A}]_{ep} [\mathbf{B}^p] + [\mathbf{B}^p]^T [\mathbf{B}]_{ep} [\mathbf{B}_f] + [\mathbf{B}^p]^T [\mathbf{G}]_{ep} [\mathbf{B}^w] +$$
$$[\mathbf{B}^p]^T [\mathbf{C}]_{ep} [\mathbf{B}^s] + [\mathbf{B}^p]^T [\mathbf{C}_1]_{ep} [\mathbf{B}^{sw}] + [\mathbf{B}^f]^T [\mathbf{B}]_{ep} [\mathbf{B}^p] + [\mathbf{B}^f]^T [\mathbf{E}]_{ep} [\mathbf{B}^f] +$$
$$[\mathbf{B}^f]^T [\mathbf{J}]_{ep} [\mathbf{B}^w] + [\mathbf{B}^f]^T [\mathbf{F}]_{ep} [\mathbf{B}^s] + [\mathbf{B}^f]^T [\mathbf{F}_1]_{ep} [\mathbf{B}^{sw}] + [\mathbf{B}^w]^T [\mathbf{G}]_{ep} [\mathbf{B}^p] +$$
$$[\mathbf{B}^w]^T [\mathbf{J}]_{ep} [\mathbf{B}^f] + [\mathbf{B}^w]^T [\mathbf{L}]_{ep} [\mathbf{B}^w] + [\mathbf{B}^w]^T [\mathbf{F}_1]_{ep} [\mathbf{B}^s] + [\mathbf{B}^w]^T [\mathbf{F}_2]_{ep} [\mathbf{B}^{sw}] +$$
$$[\mathbf{B}^s]^T [\mathbf{C}]_{ep} [\mathbf{B}^p] + [\mathbf{B}^s]^T [\mathbf{F}]_{ep} [\mathbf{B}^f] + [\mathbf{B}^s]^T [\mathbf{F}_1]_{ep} [\mathbf{B}^w] + [\mathbf{B}^s]^T [\mathbf{S}]_{ep} [\mathbf{B}^s] +$$
$$[\mathbf{B}^s]^T [\mathbf{S}_1]_{ep} [\mathbf{B}^{sw}] + [\mathbf{B}^{sw}]^T [\mathbf{C}_1]_{ep} [\mathbf{B}^p] + [\mathbf{B}^{sw}]^T [\mathbf{F}_1]_{ep} [\mathbf{B}^f] + [\mathbf{B}^{sw}]^T [\mathbf{F}_2]_{ep} [\mathbf{B}^w] +$$
$$[\mathbf{B}^{sw}]^T [\mathbf{S}_1]_{ep} [\mathbf{B}^s] + [\mathbf{B}^{sw}]^T [\mathbf{S}_2]_{ep} [\mathbf{B}^{sw}]] dxdy \tag{19a}$$

$$\{d\mathbf{F}\} = \sum_e \int_{A_e} [\mathbf{N}]^T \{d\mathbf{q}\} dxdy \tag{19b}$$

where the subscript 'ep' represents elastoplastic terms.

NUMERICAL EXAMPLES

An elastic-plastic analysis of a nonlaminated and laminated anisotropic composite plates with and without a circular hole are undertaken to show the applicability of the present model. The effects of transverse shear deformation and material anisotropy are investigated. All plates analyzed are square with planar dimensions. Numerical results presented here are obtained using the nine noded Lagrangian element. Each layer is a unidirectional fiber reinforced composite with the following material properties.

$E_1 = E_2 = 30000.0$ MN/m^2 $\qquad$ $G_{12} = G_{13} = G_{23} = 11540.0$ MN/m^2

$\sigma_0 = \sigma_{11_0} = 30.0$ MN/m^2 $\qquad$ $\sigma_{13_0} = \sigma_{23_0} = 17.32$ MN/m^2

$\sigma_{22_0} = 40.0$ MN/m^2 $\qquad$ $\sigma_{45_0} = 30.0$ MN/m^2

$\sigma_{12_0} = 20.2$ MN/m^2 $\qquad$ $E_p = 300.0$ MN/m^2

$G_p = \frac{E_p}{3}$ $\qquad$ $a = 6.0$ m $\qquad$ $\nu_{12} = 0.3$

Two types of loading are applied i.e. uniformly distributed load and inplane tensile load. Clamped and simply supported boundary conditions are used and they are as follows.

For a simply supported quarter plate model.

$$u(a/2,y,z) = v(0,y,z) = v(x,b/2,z) = 0$$

$$\theta_y(0,y,z) = \theta_y(x,b/2,z) = \theta_x(x,0,z) = \theta_x(a/2,y,z) = 0$$

$$\zeta_y(0,y,z) = \zeta_y(x,b/2,z) = \zeta_x(x,0,z) = \zeta_x(a/2,y,z) = 0$$

$$w(x,0,z) = 0$$

For a clamped quarter plate model.

$$u(0,y,z) = u(x,b/2,z) = v(x,0,z) = v(0,y,z) = v(a/2,y,z) = 0$$

$$\theta_y(x,0,z) = \theta_y(0,y,z) = \theta_y(x,b/2,z) = \theta_x(0,y,z) = \theta_x(x,0,z) = \theta_x(a/2,y,z) = 0$$

$$\zeta_y(x,0,z) = \zeta_y(0,y,z) = \zeta_y(x,b/2,z) = \zeta_x(0,y,z) = \zeta_x(x,0,z) = \zeta_x(a/2,y,z) = 0$$

$$w(x,0,z) = 0$$

Example 1: Effect of Transverse Shear Deformation on Growth of Plastic Zone.

This example investigates the thickness sensitivity of a laminated and nonlaminated composite plates. The effect of transverse shear deformation is demonstrated using a first order shear deformation (FSDT) and a higher order shear deformation (HSDT) theories. At first, an elastic-plastic analysis of a single layer clamped square plate is presented. Uniformly distributed load, q, is applied and a quarter of the plate is analyzed using a 3x3 mesh of elements. The fibers are oriented at an angle of zero degree. Stress sampling points are 2x2 Gauss points. Figure 1 shows the load deformation characteristic (normalized central deflection). To show the influence of the shear correction factor on the accuracy of the solutions obtained using FSDT, two different shear factors are used. The shear correction factor of 5/6 is used from Mindlin's plate theory, while the other is obtained based on the cross-section of the plate under consideration. From the results, it is clear that FSDT under predicts the deflection, which agrees with the observations made in static analysis [24] and in transient dynamic analysis. Also results indicate the determination

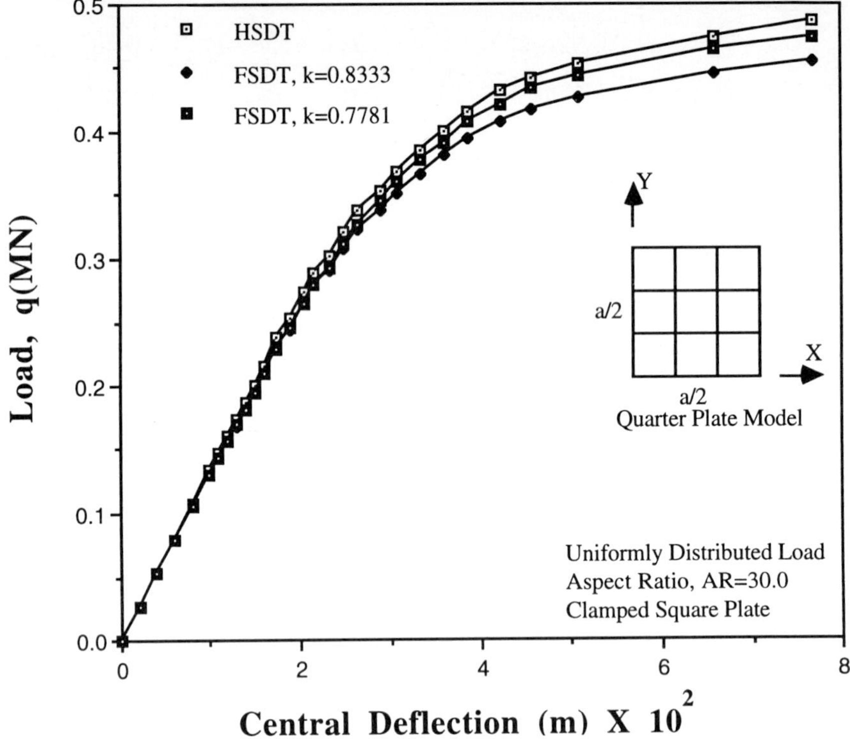

Figure 1: Elastic-plastic analysis of a clamped square plate using different theories.

of the shear correction factor is crucial to the accuracy of the solution obtained using FSDT. Figure 2 shows the spread of plastic zone using different theories along with the result of reference [19]. In single layer composite plates spread of plastic zone occurs similar to laminated composite plates i.e., successively from the center and the boundary. From the figure, it is also clear that the growth of plastic zone in a moderately thick non laminated plate using HSDT is almost identical to the one obtained using FSDT with the correct shear correction factor. The similar behavior is obtained for a moderately thick laminated plate. However, for thick laminated plate, the same conclusion is not true. For moderately thick plates, HSDT and FSDT with correct shear correction factor yield approximately the same solution. However, for thick plates only HSDT gives the adequate solution. Figure 4.9

shows the spread of plastic zone for different aspect ratios (AR = 50, 25, 5) at Gauss points 2 and 4 of elements 13, 14, 15 and 16. It can be seen that for thin to moderately thick plates, the center of the plate first becomes plastic with plastic zone spreading towards the boundary area. However, for thick plates the boundary area first becomes plastic, with the plastic zones

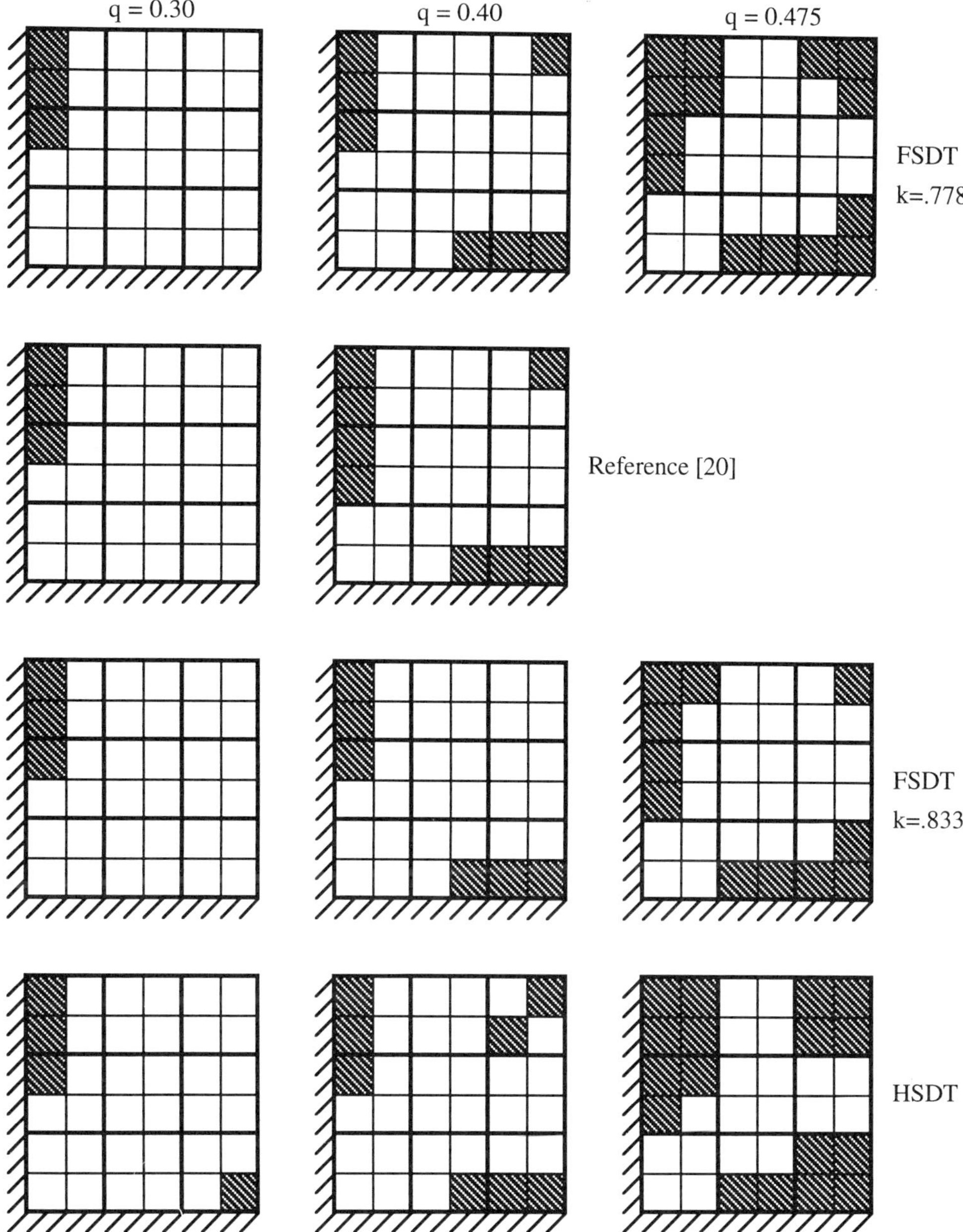

Figure 2: Spread of plastic zone in a clamped square plate using different theories

spreading towards the inside of the plate. This is because the transverse shear stress dominates the load carrying action of the plate. With an increase in the aspect ratio, the effect of transverse shear stresses decreases, while the effect of bending stress increases. It is also interesting to note that in anisotropic plates, the plastic zones tend to spread along the weaker layers.

EXAMPLE 2: Plastic deformation of an anisotropic laminated plate with a circular hole

In this example, the elasto-plastic model presented in previous sections, in conjunction with the finite element method, is used to examine the behavior of anisotropic laminated plate with a circular hole in the middle of the plate. The plate is loaded in x-directional in-plane tension. The elastic-plastic response of the plate is observed during the course of loading in terms of the development of plastic zones.

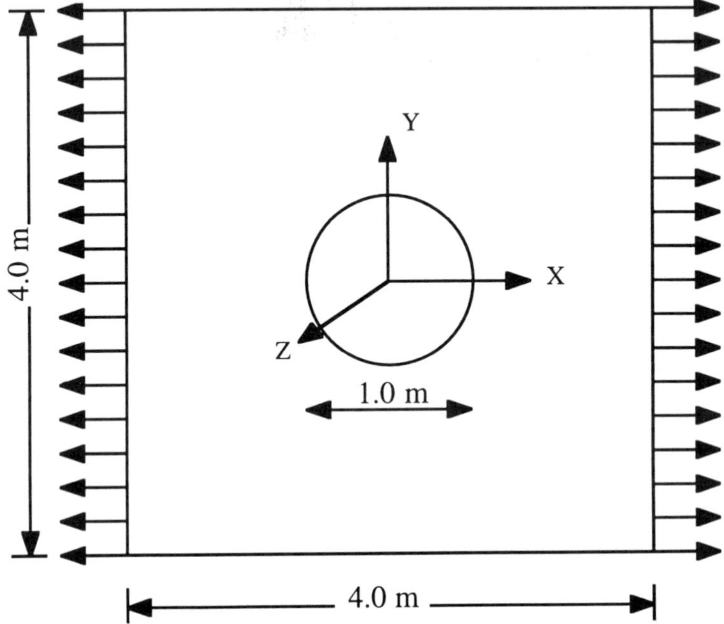

Figure 3: Geometry of an angle ply laminated plate with a circular hole

A (0°/45°) symmetric laminated plate is considered. Material properties used are same as before. The analysis of the plate is conducted using the PPLATE program developed in this research. The analyzed specimen is 4m. x 4m. and contains a circular hole of 1.0m diameter as shown in figure 3. The size of the elements are small around the holes, where stress concentrations are anticipated, and gets larger towards the edge of the plate as shown in figure 4. The fiber direction in the outer laminas coincide with the loading direction, while the fiber direction in the inner laminas are 45° to the loading direction. The plate is assumed to be free from initial stresses. This may be difficult to achieve in practice for composite materials, because of cooling from the high fabrication temperature.

The loading is applied uniformly to the edge of the plate and is increased gradually. First yielding is observed relatively early in the loading process at about 38.5 KN. This takes place in the 45° lamina at the edge of the circular hole as shown in figure 5. Figures 6 to 8 show the growth of plastic zones as loading continues. The plastic zone in the 45° lamina spread rapidly. It can be observed that the 45° lamina has yielded almost

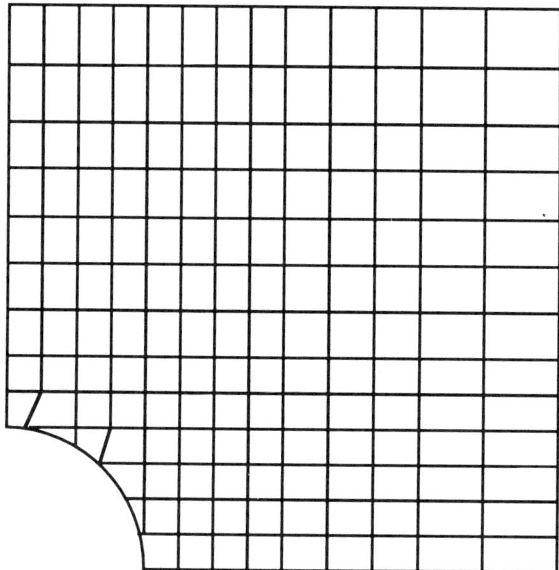

Figure 4: Finite element mesh for a quarter of plate shown in 3.

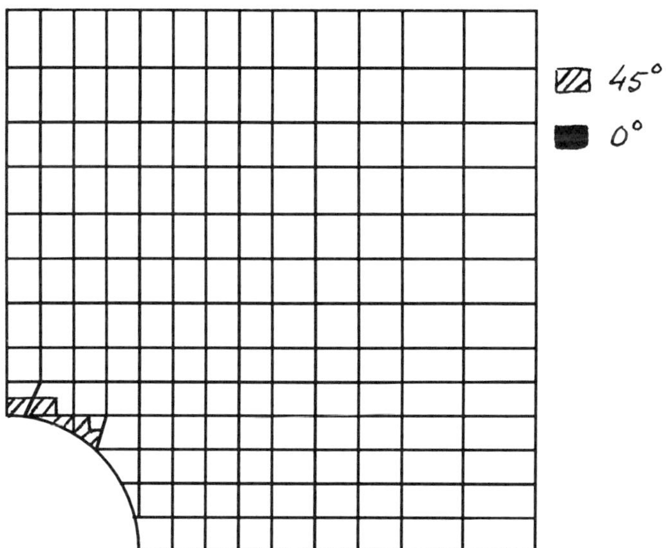

Figure 5: Plastic zones in an angle ply laminated square plate at 38.5 KN.

completely at about 150.0 KN. Yielding in 0° lamina on the other hand, is somewhat delayed and the growth of plastic zones remain comparatively small and contained to the immediate vicinity of the hole. For example, at 137 KN only 34 percent of the 0° lamina has yielded in comparison to about 82 percent yielding in 45° lamina at the same level of loading. It has been observed that the onset of plastic yielding takes place at a lower load level which is also the case for isotropic laminated plates. Therefore, the fiber reinforcement strengthen the matrix material substantially and are the principal source of composite stiffness, but they have a relatively small effect on the effective stress level which causes the onset of plastic flow.

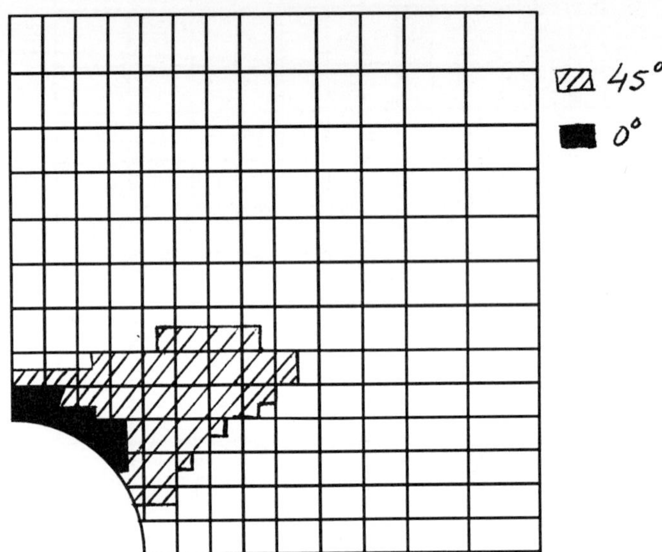

Figure 6: Plastic zones in an angle ply laminated square plate at 72.0 KN.

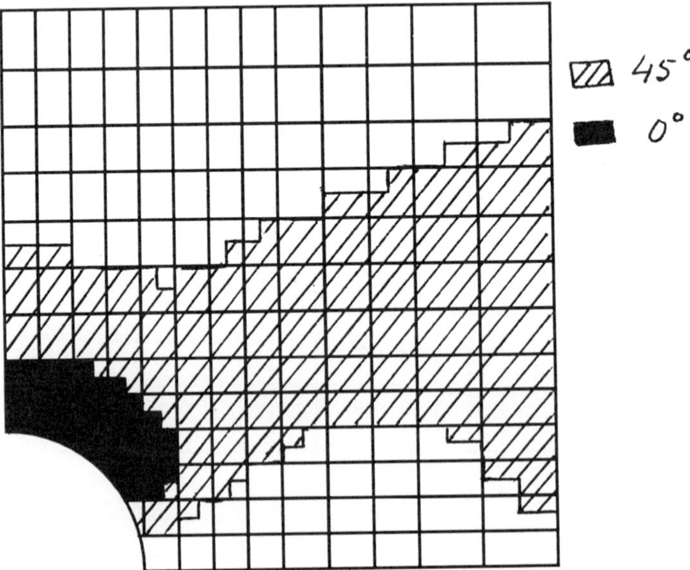

Figure 7: Plastic zones in an angle ply laminated square plate at 109.5 KN.

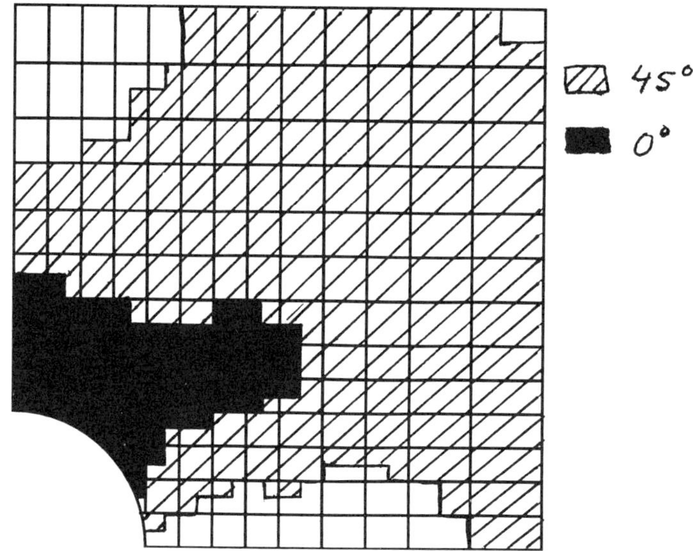

Figure 8: Plastic zones in an angle ply laminated square plate at 147.0 KN.

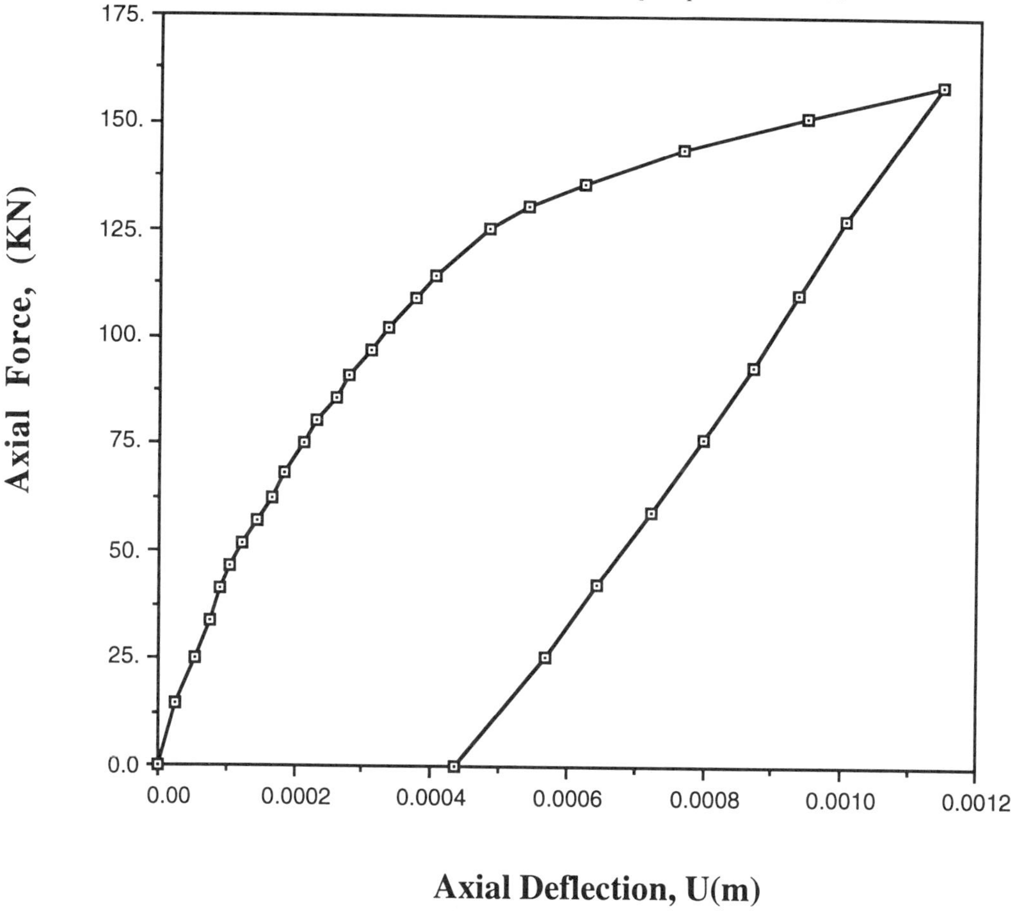

Axial Force, (KN)

175.
150.
125.
100.
75.
50.
25.
0.0

0.00 0.0002 0.0004 0.0006 0.0008 0.0010 0.0012

Axial Deflection, U(m)

Figure 9: Load deformation characteristic of a laminated angle ply square plate with a circular hole at point A as shown in figure 4

Unloading the laminated plate to complete removal of load results in permanent deformation as shown in figure 9. The laminated anisotropic plate is first unloaded by a small unloading step followed by 8 unloading increments. After the first unloading step, all laminas became elastic and remain elastic till the end of unloading. At the end of loading/unloading cycle, the plate is left with 30 percent of the total deformation as the permanent deformation. However, the magnitude of the permanent strain in 45° lamina is larger than the 0° lamina. This causes the induced residual stresses in each lamina as shown in figure 10. Figure 10 shows the residual stress σ_{xx} distributions along the vertical symmetry line. The compressive residual stress exists in the 45° lamina, but a tensile residual stress is present in 0° lamina. The tensile stress is probably caused by the permanent strains of larger magnitude in 45° lamina, which prevents the elastic strains from completely recovering in the 0° lamina. These results also suggest that crack initiation due to cyclic loading may be prevalent in 0° lamina.

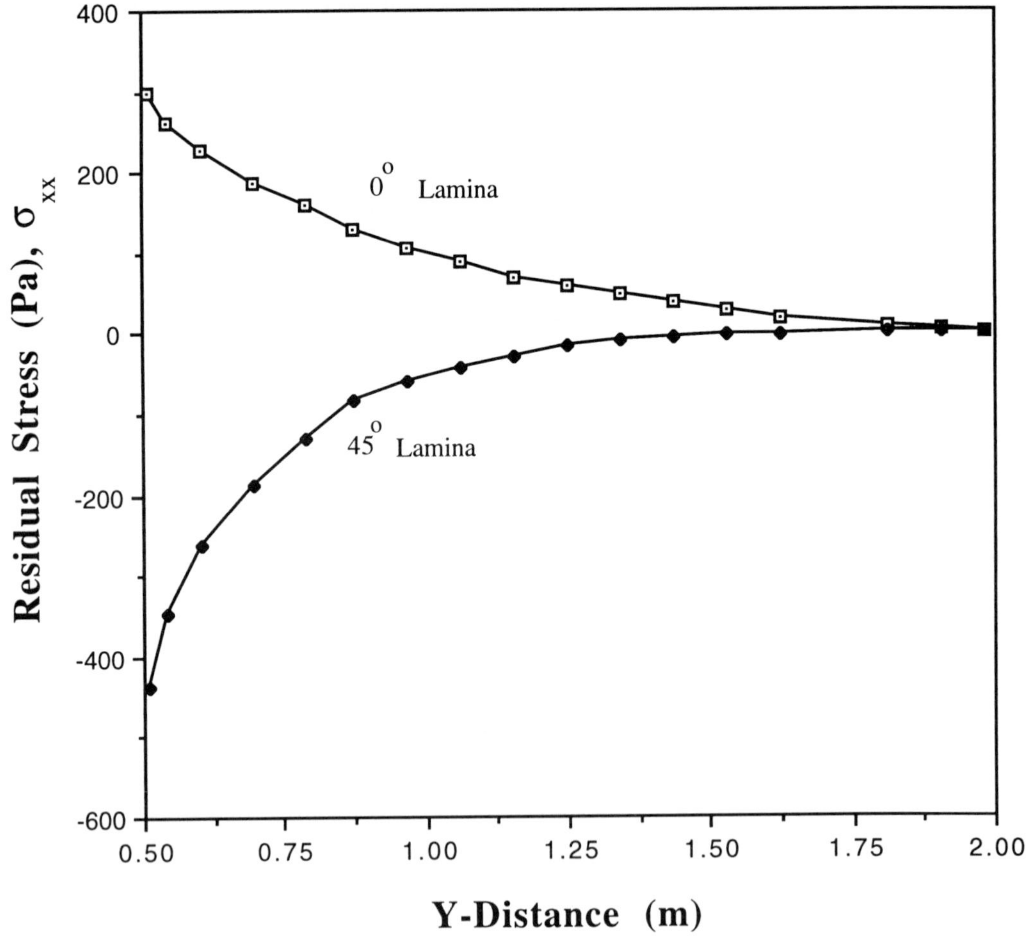

Figure 10: Residual stress, σ_{xx}, in each lamina for a laminated square plate with a circular hole subjected to unloading

CONCLUSION

This research is aimed for the finite element elastoplastic analysis of anisotropic laminated composite plates with and without a hole. A first order shear deformation theory and a new higher order shear deformation theory is employed herein to obtain the elastoplastic response of laminated fibrous composite plates. This new shear deformation theory contains two more independent variables (warp on xz-plane and yz-plane) as compared to previous and results in accurate prediction of deflections and stresses. It has been found that the different types of shear deformation theories have little effect on deflections, however, higher order shear deformation theory gives much accurate results for stresses than first order shear deformation theory. The numerical results exhibit some interesting characteristics, though no experimental verification is available. However, in view of the agreement obtained in predicting the growth of plastic zone using FSDT, it is reasonable to assume that the results obtained using HSDT are equally valid. Plastic zones have a tendency to spread along the weaker layer in both thick and thin laminates. For thick plates, the use of HSDT results in more reliable prediction of elastic-plastic response as compared to FSDT. This also demonstrates the significance of plasticity in fibrous composites for laminated plates. Elastic behavior is of limited significance in applications which hope to utilize the high strength of composite. Misleading results and predictions of stiffness and strength of laminated plates may occur if plastic deformation of the composite is disregarded. Apart from all of these, a plasticity model of a composite lamina needs to be developed based on a micromechanical model which accounts for fiber-matrix interaction, matrix or fiber cracking etc.

REFERENCES

1. I.L. Dillamore, A.J. Hazel and T.W. Watson, An experimental study of the mechanical anisotropy of some common metals, Interna. J. Mech. Sci. 13 (1971) 1049-1061.
2. V. Mises, Mechanik der plastischen formanderungen von kristallen, ZAMM, (1928).
3. W. Prager, A new method for analyzing stress-strain relations in work hardening plastic solids, J. Appl. Mech. 23 (1956) 493-501.
4. H. Ziegler, A modification of Prager's hardening rule, Quart. Appl. Math. 17(1) (1959) 55-66.
5. Z. Morz, An attempt to describe the behavior of metals under cyclic loads using a more general work hardening model, Acta Mechanica 7 (1969) 191-212.
6. R.D. Krieg, A practical two surface plasticity theory, J. Appl. Mech. 42, 641-646.
7. R. Hill, The mathematical theory of plasticity, Clarendon press, Oxford (1950).
8. R. Hill, A theory of yielding and plastic flow of anisotropic metals, Proc. Roy. Soc.

of London, Ser. A. 193 (1948) 281-298.

9. J.C. Fisher, Anisotropic plastic flow, ASME Trans. 71(4) (1949) 349-356.

10. F. Edelman and D.C. Drucker, Some extensions of elementary plasticity theory, J. Franklin Institute 251 (1951) 581-605.

11. L.W. Hu and J. Martin, Anisotopic lading function for combined stresses in the plastic range, J. Appl. Mech. (1955) 77-85.

12. Y. Hoshimura, Hypothetical theory of anisotropy and the Bauschinger effect due to plastic strain history, Aeronautical Research Institute, University of Tokyo, Report N0. 349 (1959) 221-247.

13. D.C. Bouge, The yield stress and plastic strain theory for anisotropic materials, Oak Ridge National Laboratory, Oak Ridge, Tennesse (1967) Rep. No. ORNL-TM-1869.

14. S.N. Shabashi and A. Shelton, The anisotropic yield flow and creep behavior of pre-strained En-24 steel, J. Mechanical Engrg. Sci. 17(2) (1975) 93-104.

15. A.J.M. Spencer, Deformations of fiber reinforced material, Clanderon press, (1972).

16. T.H. Lin, D. Salinas and Y.M. Ito, Elastic -plastic analysis of unidirectional composites, J. Composite Materials 6 (1972) 55-68.

17. B. Whang, Elasto-plastic orthotropic plates and shells, Proc. Symp. on Application of F.E.M. in Civil Engineering, Vanderbilt University, Tennessee, (1973) 481-515.

18. D.R.J. Owen and J.A. Figueras, Elastio-plastic analysis of anisotropic plates and shells by the Semiloof element, Internat. J. Numer. Meths. Engrg. 19 (1983) 521-539.

19. D.R.J. Owen and J.A. Figueras, Anisotropic elastic-plastic finite element analysis of thick and thin plates and shells, Internat. J. Numer. Meths. Engrg. 19 (1983) 541-556.

20. T. Pervez, "Transient Dynamic, Damping and Elasto-plastic Analysis of Higher Order Laminated Anisotropic Composite Plates Using Finite Element Method", Doctoral Thesis, University of Minnesota, USA, September 1991.

21. T. Pervez and N. Zabaras, Transient dynamic and damping analysis of laminated anisotropic plates using a refined theory, Internat. J. Numer. Meths. Engrg., Vol. 72, 1992.

23. T. Pervez, "A Higher Order Shear Deformation Theory of Anisotropic Laminated Composite Plates and Its Application in Defense Industry", Seminar on 'Research and Development in Support of Technologies with military Applications' at PNS Naval Engineering College, May 1992.

24. N. Zabaras and T. Pervez, A viscous damping approximation of laminated anisotropic composite plates using the finite element method, Comput. Meths. Appl. Mech. Engrg. 81(3) (1990) 291-316.

25. T. Pervez, "Finite Element Structural Analysis Via A New Integrated Force Method", Proceedings of 2nd International Naval Engineering Conference, Karachi, (1993).

HEAT TREATMENT BEHAVIOUR OF METAL MATRIX COMPOSITES

Azmi Rahmat, Shamsul Baharin and P. R. Khangaonkar

Universiti Sains Malaysia, Perak Branch Campus
Tronoh, 31750, Malaysia

ABSTRACT

High performance composites on the basis of aluminium alloy matrix and alumina-silica continuous fibres were studied from the point of view of their response to age hardening treatment as compared to the unreinforced matrix alloy. The changes were monitored by following the microhardness of the matrix and the electrical resistivity of the materials examined. It was observed that the matrix of the composites showed considerably more hardening effect than the matrix of the unreinforced alloy inspite of the fibre being innert to the matrix alloy. The resistivity changes in the composites during the ageing process indicated that appreciable internal stress continued to persist in the composite material well after overageing and hardness decline took place. EDX evaluation of regions close to the fibres indicated a higher magnesium content as compared to the regions away from the fibres.

INTRODUCTION

In the earlier stages of the development of metal matrix composites, it was presumed that if the reinforcement phase is innert to the matrix, the heat treatment behaviour may not be basically different from that of the unreinforced alloy material forming the matrix. However it was soon realized that in a composite, the matrix may show a different heat treatment response than the same material as an unreinforced alloy. An accelerated ageing response in the microhardness of the matrix of ceramic reinforced aluminium alloy composites has been acknowledged for quite some time[1], although the earlier studies, which measured primarily macrohardness, indicated no appreciable advantage in the composites over the unreinforced alloy. However recent studies by Song and Baker[2] have clearly brought out the superior age hardening response in the composites of 6061-SiC particles and 6061-alumina whiskers even in terms of macrohardness.

One of the causes advanced to explain this behaviour is the higher dislocation density in the composites, which in turn is claimed to be generated due to the mismatch between the coefficient of expansion the matrix and the reinforcement[3]. This approach has been examined critically particularly for discontinuously reinforced composites (DRA composites[4]). The composites reinforced by continuous fibres (CMMCs) have not been studied much from this point of view, except for the early work on continuous filaments of tungsten in copper[5]. Song and Baker[2] have, on the other hand, ascribed the better ageing response of the aluminium alloy composite to a higher density of precipitation, leading to depletion of Mg in the matrix and retarding the growth of β'-Mg_2Si phase.

Aluminium alloy based composites containing continuous alumina-silica fibres are reported to reach high performance levels in strength characteristics and retention of the same at elevated temperatures[6]. The present work studies the age hardening behaviour of squeeze cast composites based on aluminium alloy 6061 reinforced with continuous alumina-silica fibres in terms of the changes in matrix microhardness and resistivity changes.

The method of monitoring the structural changes during age-hardening by resistivity changes has been well known[7]. During precipitation from solid solution, the solute leaving the supersaturated solution will normally cause a fall in resistivity, as shown by the curve ab in fig.1, if the equilibrium phase precipitates directly without any precipitation hardening. However, if age hardening takes place the resistivity change during ageing may follow a curve of the type acb in fig.1, thus enabling a certain degree of monitoring of the structural and phase changes taking place during the age-hardening process.

MATERIALS AND METHODS

The alloy used for making the composites by squeeze casting was 6061 with the actual composition being stated in Table 1. The fibre used was alumina-silica fibre (brand name Altex) manufactured by Sumitomo Co. of Japan. The basic data about the fibres used are also stated in Table 1.

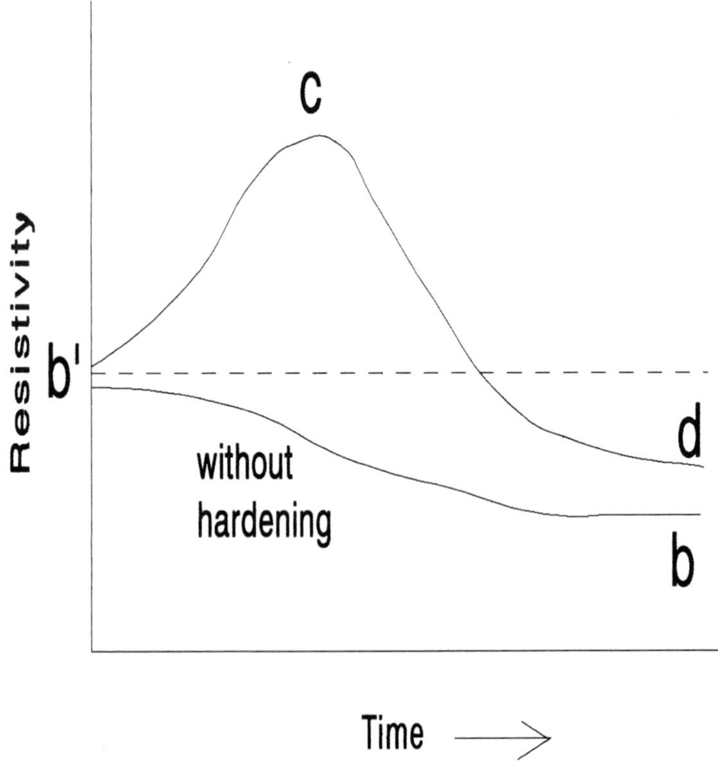

Fig. 1. Resistivity changes during precipitation hardening.

The composite was made by liquid infiltration of the fibre preform in a squeeze casting unit, with the alloy poured at 750°C and the die preheated to 350°C. The reinforcement was held in the mold on a preform made of steel plate and wire frame and preheated to 650°C before placing in the mold just before casting. A squeeze pressure of 25 MPa was applied by a hydraulic ram. Part of the cast billet contained no reinforcement, and this was retained as a suitable sample of unreinforced alloy 6061 for comparative ageing studies. Two levels of V_f (volume fraction), 0.50 (composite I) and 0.32 (composite II) were studied.

For comparison, one composite made by powder metallurgy route, based on 6061 Al alloy, was used. This material was SiC whisker reinforced (6061/SiC/20w) as supplied by Advanced Composite Materials Corporation (Greer SC USA) in non-heat-treated condition (composite III).

The age-hardening treatment of the samples consisted of solution treatment at 550°C for 8 h, followed by water quenching and ageing at room temperature, 120°C and 170°C for different times.

The composites were checked for microhardness of the matrix and resistivity during the ageing process. The microhardness indentations were done with 25 gf load, dwell time 15 s, on a specimen surface transverse to the fibres. The resistivity was monitored by eddy current method for conductivity, using Defectoscope SD 2.832 of Institut Dr. Forster (Germany), on the basis of calibration from standard samples of known conductivity and by placing the probe on the specimen surface parallel to the fibres. The effect of fibre orientation on the microhardness and resistivity was not studied. The composite samples were checked by optical microscopy and SEM, and the intermetallic phases were observed to be primarily Mg_2Si and (FeMn)SiAl with help of EDX analysis.

Table I

(a) Composition of the Matrix

Element	Weight %
Si	0.6
Fe	0.18
Cu	0.26
Mg	0.9
Cr	0.08
Other impurities	0.01 each

(b) Alumina Silica fiber (Altex - Sumitomo, Japan)

Composition range	>70% Al_2O_3
Actual Composition	85% Al_2O_3, 15% SiO_2
X ray analysis	Main component cubic phase, isomorphous with gamma alumina, along with amorphous phase.
Form used	fibre diameter about 13μm, with each tow containing several hundreds of fibres.
Density	3.3 g/cm^3
Tensile Strength	1.8 GPa

RESULTS AND DISCUSSION

Microstructural Observations

The unreinforced alloy showed the expected phase of Mg_2Si and also an intermetallic phase of the shape of chinese script, which was identified by EDX as possibly (FeMn)SiAl. The composite samples showed micrographs indicating that the penetration was generally satisfactory, although minute voids were seen, particularly in the regions where fibres happen to be closely packed. It was observed that the intermetallic phases in the matrix of the composites were smaller in size than those in the unreinforced alloy, due to the fast solidification of the melt in contact with fibres. The grain size of the matrix was smaller (about 15 µm) than the one in the reinforced alloy. Investigations with TEM, detailed separately earlier[8] indicated that within the matrix, dislocations were present, clustered mainly around the fibres also around small particles, identified separately as Mg_2Si by EDS. No evidence was found for the formation of additional phases at the fibre/matrix interface, although EDS did indicate segregation of some Mg to the regions near the fibres.

Fig.2 and fig.3 indicate the EDX spectrum of a spot near the fibre and a spot away from the fibre in the matrix of the above composite after ageing at 170°C. Apparently there is a migration of magnesium towards the regions near the fibre. It is possible that the fibre having substantial content of amorphous silica can influence this behaviour of magnesium. TEM studies mentioned above[8] also indicated migration of Mg to the regions near the fibre. Yang and Scott[9], in their study of Al- 7% Si alloy composite with alumina fibre (Safimax of ICI), also observed by selected area diffraction that Si and Mg were higher near the fibre, but undetectable in the main matrix phase.

Microhardness and Resistivity Changes

Fig.4 indicates the results of matrix microhardness changes and resistivity variation during ageing of the unreinforced cast alloy 6061 at 170°C. The peak hardness increase of 18 VPN (above the solution treated hardness of 65.8 VPN) is obtained in 4 h and declines rapidly after 6 h or so. The resistivity peak is attained at 2 hours followed by a rapid decline to the base value (zero increase over the solution treated state) in 4 h and decreases further on ageing later. The resistivity reaches its peak and starts declining by the time the hardness peak is reached. The behaviour is on the lines of known pattern for high temperature ageing, as schematically shown in fig.1.

Fig. 5 indicates the results of composite I (V_f 0.50) during ageing at 170°C. The hardness peak is attained in 3 h and the increase attained is over 40 VPN over the solution treated hardness of 85.2 VPN and this is appreciably higher than the results of the unreinforced alloy stated above. The presence of fibres definitely influences the hardening behaviour of the matrix considerably, thus enabling the composite to attain better overall properties. The resistivity does peak off slightly earlier than the hardness.

The results of the other temperatures, as well those of the composite II (V_f 0.32) and the commercial composite (composite III) are stated in table II, given below[10].

It is seen from the data of table II and the figures that the matrix in the composites attains a higher hardness than the unreinforced alloy, and in

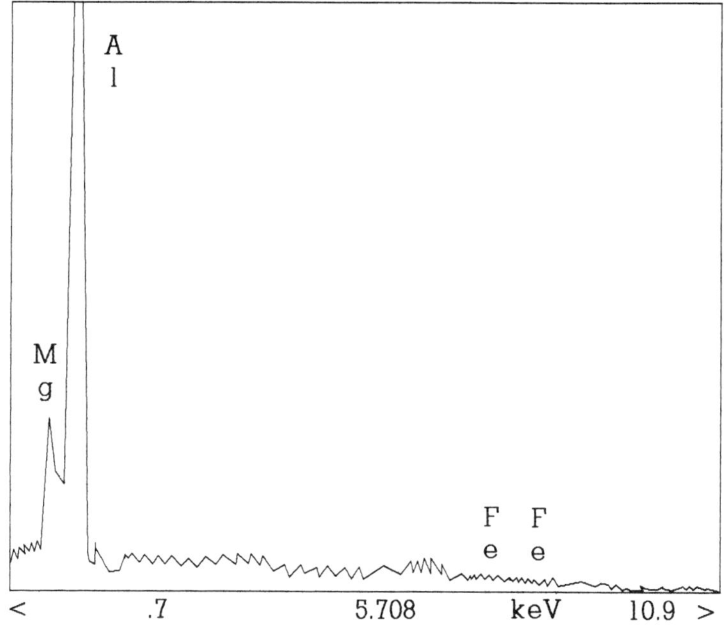

MEM1 : OVERAGED V_f=0.5 (matrix)

Fig. 2. EDX Matrix away from fibre Composite 1 overaged.

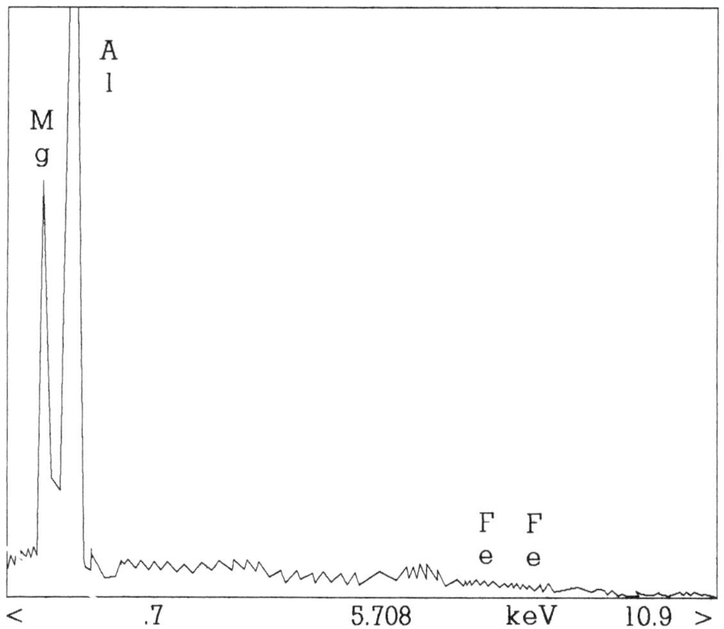

MEM1 : OVERAGED V_f=0.5 (matrix)

Fig. 3. EDX Matrix near fibre Composite 1 averaged.

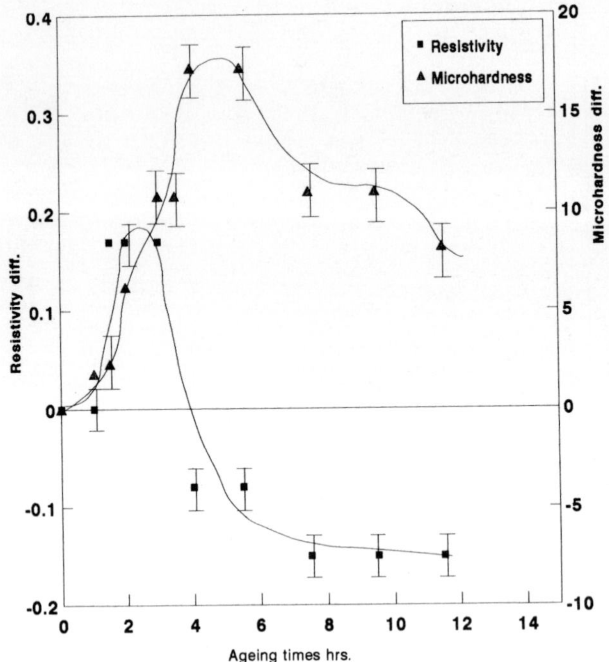

Fig.4. As cast 6061 , Ageing at 170°C

Solution treated hardness 65.8 VPN

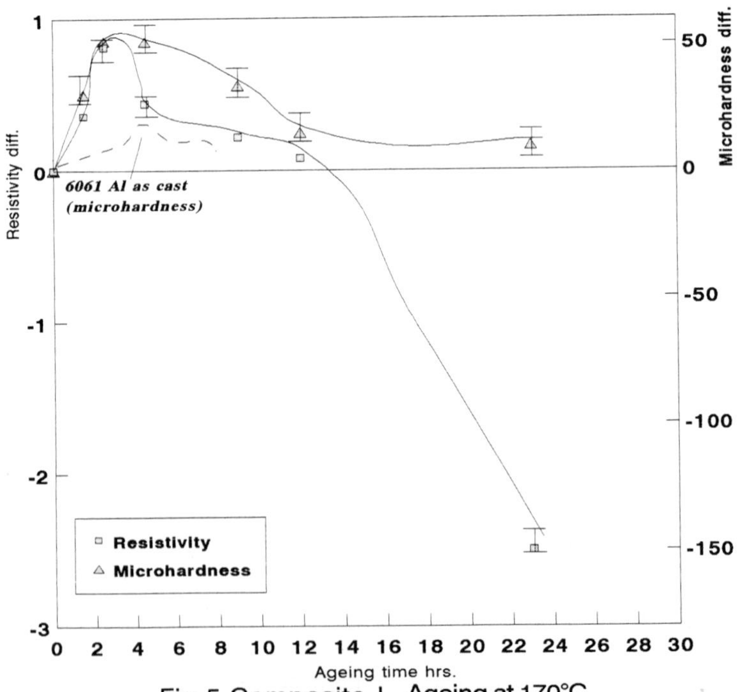

Fig.5.Composite I , Ageing at 170°C

Solution treated hardness 85.2 VPN

many cases the peak is attained faster. The coefficient of thermal expansion (CTE) of the matrix is about $23\text{-}24\times10^{-6}/°C$ while that of the fibre is $8.8\times10^{-6}/°C$, so that the CTE mismatch is substantial enough to create possibly a higher dislocation density, causing a higher matrix hardening. For discontinously reinforced composites, this phenomenon is supported by experimental evidence as well as theoretical models[1,2,11]. However, for continuously reinforced composites (CMMCs), the complete and satisfactory theoretical treatment for the generation of dislocations is not yet available. In fact the effect of possible dislocation generation due to CTE mismatch on the strength of a CMMC has not been determined and it has been stated that it is probably not large[11]. Higher dislocation density was observed in the investigations of Cu-W systems by Chawla and Metzger[3]. Simple thermal stress has been well studied for CMMCs due to their relatively simple geometry, but not the possibility stated above regarding higher dislocation density in the composite as compared to the unreinforced alloy.

Table II

Selected data from the age hardening experiments

Material	Ageing temp. °C	Peak hardness increase VPN	Time h	Peak Resistivity increase µohm cm	Time h
6061 as cast	room temp	45	60	0.8	80
Composite I	room temp	74	80	0.83	90
Composite II	room temp	75	65	1.5	60
Composite III	room temp	65	90	0.85	60
6061 as cast	120	27	8	0.25	3
Composite I	120	35	3	7.5	4
Composite II	120	65	4	2.5	5
Composite III	120	60	8	1.0	4
Composite II	170	55	3	12	3
CompositeIII	170	60	3	0.25	1.5

For the composite II, a comparison with the data of Friend et al.[12] on 6061/Alumina fibre composite (V_f 0.26) indicates that they reported a gain of hardness of 22 VPN in 22 hours and no increase in resistivity during natural ageing, while a much slower hardening at 140°C than the one observed in the present work for 120°C. A similar other study by Badini et al.[13] on SiC whiskers in 6061 alloy (V_f) also shows a slow hardening in natural ageing, but it may be noted that they measured the bulk hardness and not the matrix microhardness. On the other hand the data of earlier studies on 6061/SiC/20w of House et al.[4] and the data on 6061/SiC/20p of Srinivasan et al.[14] indicate several features similar to the data of the present work, viz. faster hardening of the matrix in the composites compared to the unreinforced alloy, and attainment of higher matrix microhardness in the composites under otherwise similar heat treating conditions.

It was also observed in the resistivity curves of the composites that after the peak of resistivity is attained it does not fall quickly as in the classical curves of age hardening alloys, particularly at room temperature and 120°C ageing. The resistivity continues to remain at peak well after the hardness fall due to overageing sets in. The behaviour suggests that the strains due to clustering (before GP zones) and other related strains in the transformation appear to be higher and they are relieved relatively more slowly compared to the unreinforced alloys.

CONCLUSIONS

The age hardening of the matrix in 6061- alumina silica fibre composites indicates a faster and stronger hardening effect than the unreinforced alloy treated under identical conditions. The material, as an example of continuously reinforced composite (CMMC), indicated that it shows a hardening effect of reinforcement on the matrix almost as intense as the one observed in discontinuously reinforced composites (DMMC). The resistivity changes indicated the existence of strains in the material that are relieved slowly compared to the softening in overageing.

REFERENCES

1. R. J. Arsenault, and R. M. Fisher, Microstructure of fiber and particulate SiC in 6061 Al composites, *Scripta Metallurgica.* 17:67-71 (1983).
2. Y. Song and T.N. Baker, Accelerated aging response in ceramic reinforced AA 6061 composites, *Materials Sci. Tech.* 10:406-413 (1994).
3. M. Vogelsang, R. J. Arsenault, and R. M. Fisher, An in situ HVEM study of dislocation generation at Al/SiC interface in metal matrix composites, *Met. Trans.A.* 17A:379-389 (1986).
4. M. B. House, K. C. Meinert, and R. B. Bhagat, The aging response and creep of DRA composites, *Journal of Metals.* 43(8): 24-28 (1991).
5. K. K. Chawla, and M. Metzger, Initial dislocation distributions in tungsten fibre/copper composites, *J. Mat. Sci.* 7: 34-39 (1972).
6. Sumitomo Metal Co, Tokyo, Japan. " Altex Fibres, Product information".
7. A.G. Guy, C.S. Barrett and R.F. Mehl, Age hardening in Copper Beryllium alloys, *Trans. A.I.M.E.* 175: 216 (1948).
8. J. B. Shamsul, N. Jomin, R. S. Bushby, and V. D. Scott, Fabrication, microstructure and ageing characteristics of aluminimum alloy (6061) reinforced with Altex fibre, *in*: " Proceedings of the International conference on Recent Advances in Materials and Mineral Resources," Universiti Sains Malaysia, Penang, Malaysia, 212-221 (1994).
9. M. Yang and V. D. Scott,, Microstructural studies of aluminium-silicon alloy reinforced with alumina fibres, *J. Mat. Sci.* 26: 2245-2254 (1991).
10. P.R. Khangaonkar, J.B. Shamsul, and R. Azmi, Age hardening of 6061/alumina silica fibre composites, *in*: " Symposium on High Performance

Composites: Commonality of Phenomena," K. K. Chawla, ed., TMS, Warrandale, PA USA, 435-443 (1994).

11. R. J. Arsenault, Strengthening of metal matrix composites due to dislocation generation through CTE mismatch, *in*:" Metal matrix composites: Mechanisms and Properties", R. K. Everett and R. J. Arsenault, eds., The Academic Press,Boston MA):79-100 (1991).

12. C. M. Friend, I. Horsfall, S. D. Luxon, and R. J. Young, The effect of fibre matrix interfaces on the age hardening characteristics of δ-alumina fibre reinforced AA6061, *in*: " Cast metal matrix composites", S. G. Fishman and A. K. Dhingra, eds., ASM International: 309-315 (1988).

13. A. Badini, F. Marino, and A. Tomasi, Natural aging characteristics of aluminium alloy 6061 reinforced with SiC whiskers and particles, *Material Science and Engineering*, 136A:99-107 (1991).

14. D. Srinivasan, and M. K. Surappa, Effect of iron impurity and thermomechanical processing on the age hardening behaviour of 6061 Al-SiC_p MMC produced by casting route, *Scripta Metallurgica et Materialia*, 27:1139-1144 (1992).

BIODEGRADABLE POLYMER BLENDS

Mohammed Yasin, Allan J. Amass and Brian J. Tighe

Biodegradable Polymers Unit, Speciality Materials Research Group, Department of Chemical Engineering & Applied Chemistry, Aston University, Aston Triangle, Birmingham B4 7ET, (UK).

INTRODUCTION

Several microbially produced or chemically synthesised polyesters are currently available commercially. Polyesters are finding use as biodegradable material not only because the chemical synthesis is simple, but also due to the fact that the raw materials are cheap, the ester group is easily hydrolysable, the rate of degradation is good and non-toxic products are formed. Microbially produced polyesters have unique characteristics which can be utilised to modulate price and performance. Their many useful and interesting properties can be extended and exploited by blending with synthetic polymers, which can allow both the performance and price of the material to be adjusted. Currently, there is great interest and awareness in 'environmentally friendly' packaging, brought on by either constantly updated legislation, or through environmental conscience. Our work on developing materials for such applications, is concerned with two aims; the scientific interest in the study of the factors affecting compatibility of polymer blends and the subsequent properties of these materials, particularly their degradation properties, and the importance of producing biodegradable materials that are economically viable [1, 2].

The bacterially produced biodegradable polyester, polyhydroxybutyrate (PHB) and its copolymers with hydroxyvalerate (HV), are receiving considerable interest. Applications of P(HB-HV) copolymers, such as for packaging has been limited because of their high cost of production in relation to commodity plastics. One possibility of producing commercially viable biodegradable materials for applications where price is of importance is by blending P(HB-HV) copolymers with other polymers. A number of easily available polymers (such as cellulose acetate butyrate, CAB), together with high molecular weight, non-crystallizable polyester plasticisers were identified as possible candidates in the production of compatible blends in which adequate performance could be linked with reduced cost. The extent to which the modulation of the physical properties in general, but the stability in various environments in particular, is possible by the formation of P(HB-HV) blends with cellulose acetate butyrate and polyester adipate plasticiser will be discussed in this paper.

Polymers and Other Advanced Materials: Emerging Technologies and Business Opportunities. Edited by P. N. Prasad *et al.*, Plenum Press, New York, 1995

EXPERIMENTAL

Materials

P(HB-HV) copolymers of 5 to 20% hydroxyvalerate content and weight average molecular weights (Mw) 350,000 Daltons used in these studies were obtained from ICI (now Zeneca), Teeside, UK, and are known under the trade name 'Biopol'. Cellulose acetate butyrate, (CAB, grade 531-1) was supplied by Eastman Chemical (UK) Ltd. The polymeric polyester-based plasticiser, polyester adipate (Paraplex G40, Mw = 23,500), was used as supplied by Croxton & Garry Ltd, Dorking, UK, (UK distributors for C P Hall & Co., Chicago, USA).

Preparation of samples

Typical solvent blending techniques were used to prepare film samples using 10% (w/v) solutions of the P(HB-HV) copolymers, CAB and the plasticiser in chloroform. Various compositions of the three component blends were mixed and poured on to glass petri dishes, allowing the solvent to evaporate slowly at room temperature before removing the films and annealing them at 90°C for 24 hours prior to testing.

Compatibility and Degradation studies

Polymer compatibility was assessed from both optical clarity of the films and differential scanning calorimetry (DSC) measurements (carried out on a Perkin-Elmer DSC 7 at scan rates of 10°C per minute). Tensile properties at 20°C were monitored using the Hounsfield tensometer, (model HTi 10, Hounsfield Ltd, Croydon, UK) at a cross-head speed of 5 mm/min and gauge length 10 mm. Hydrolytic degradation study was carried out in 50 mls of pH 7.4 and 10.6 buffer solution at both 25 and 70°C using 1 cm square films of 0.1 mm nominal thickness. Aerobic degradation was carried using 10% activated sludge (mixed inoculum) at 25°C and pH 7.

RESULTS AND DISCUSSION

Optical clarity of the films was used as a preliminary indicator for compatibility, with compatible blends giving rise to clear blends and incompatible blends to opaque films. Our results, illustrated in the three component phase diagram in Figure 1, indicate that blends containing high levels of CAB (>30%) and low levels of plasticiser (<40%) are compatible. DSC results used to correlate the compatibility of the multiphase crystalline blends of the P(HB-HV)/plasticiser, CAB/plasticiser and P(HB-HV)/CAB two component blends showed interesting trends in the compatibility, which is evident from the phase diagram and are in agreement with the work on binary blends of P(HB-HV) and CAB reported elsewhere [3-4]. CAB forms compatible blends (clear films) over the full range of compositions with the polyester adipate plasticiser. P(HB-HV) copolymers form immiscible blends with the plasticiser over the full range of compositions, whereas P(HB-HV) copolymers blends with CAB are compatible when the CAB content is >50%. With less than 50% CAB, the blends are incompatible. The general trend is that blends having high levels of CAB formed clear films and those containing high levels of plasticiser formed opaque films.

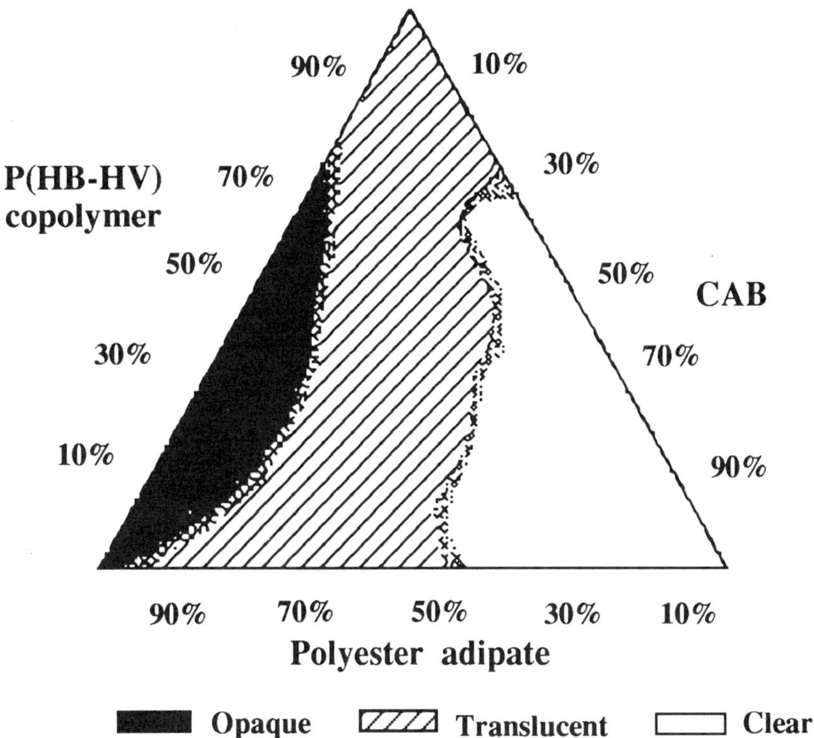

Figure 1. Phase compatibility diagram for P(HB-HV)/Cellulose acetate butyrate (CAB)/polyester adipate blends. Solvent cast films.

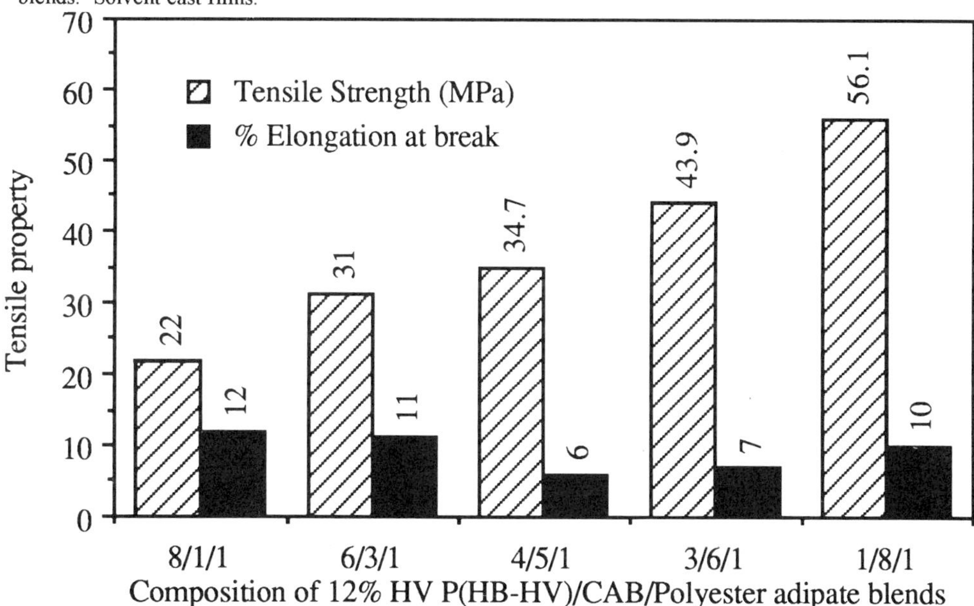

Figure 2. Tensile strength and % elongation at break for 12% HV P(HB-HV)/CAB/polyester blends. Solvent cast films.

The contribution of CAB and the plasticiser on the physical properties of two component blends with the P(HB-HV) copolymers was as expected, with the CAB imparting strength and the plasticiser flexibility. However, in the three component blends, certain compositions gave rise to blends which possessed both considerable strength and flexibility. Figures 2 and 3 illustrate the tensile strength and % elongation at break for 12% HV and 20% HV P(HB-HV)/CAB/polyester adipate blends, respectively. Our previous work showed the effect that increasing hydroxyvalerate content had on the mechanical properties of the P(HB-HV) copolymers [1]. Comparing Figure 2 and 3, the effect of increasing hydroxyvalerate content of the P(HB-HV) copolymers was again found to directly influence the physical properties of the blends, causing a decrease in the tensile strength and a corresponding increase in % elongation at break.

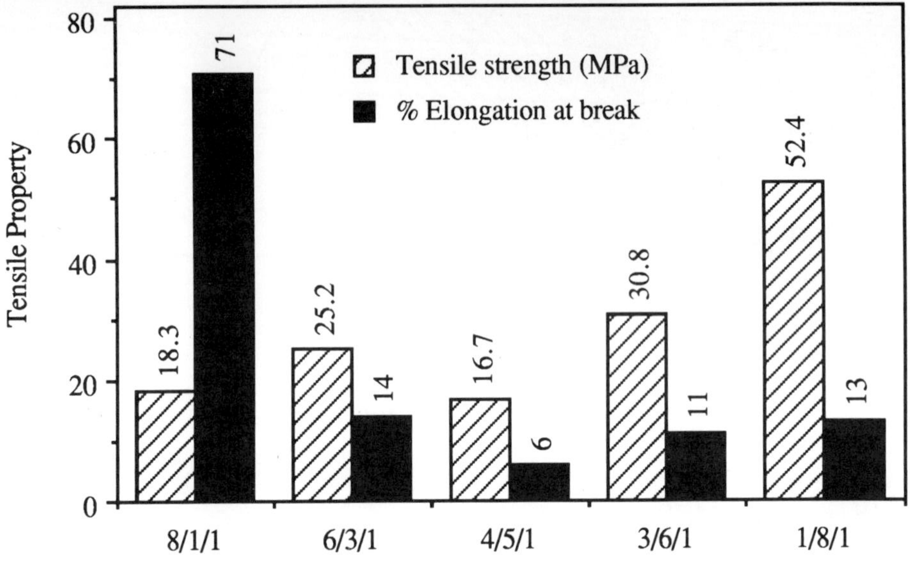

Figure 3. Tensile strength and % elongation at break for 20% HV P(HB-HV)/CAB/polyester blends. Solvent cast films.

The weight loss profiles for the 20% HV P(HB-HV) copolymer, CAB and 20% HV P(HB-HV)/CAB blends in pH 7.4 phosphate buffer at 25°C are illustrated in Figure 4. The degradation profiles are characterised by an initial slow weight loss, followed by a secondary enhanced phase. Although the weight loss is very small in the time period of this study, it can be seen from Figure 4 that the P(HB-HV) copolymer degrades faster than the CAB. The time taken for 10 and 50% loss of initial dry weight (t_{10} and t_{50}, respectively), for films of 12% HV P(HB-HV)/CAB blends in pH 10.6 buffer at 70°C are shown in Table 1. The effect of both high temperature and pH on the hydrolytic degradation rates of the P(HB-HV)/CAB blends is apparent when the t_{10} and t_{50} values in Table 1 are compared with the profiles in Figure 4.

The addition of the polymeric polyester adipate plasticiser, (weight average molecular weight 23,700), into the P(HB-HV) copolymer matrix had the effect of increasing the degradation rate of the blend in comparison to the unblended copolymer. Table 2 shows the t_{10} and t_{50} values for films of 20% HV P(HB-HV)/polyester adipate blends in pH 10.6 buffer at 70°C. The dramatic decrease in the t_{10} and t_{50} values with increasing plasticiser content suggests another method of controlling the degradation rate of the P(HB-HV) copolymers. This is illustrated quite well in Figure 5 with the weight loss profiles for the ternary blends of 20% HV P(HB-HV)/CAB/polyester adipate. With high compositions of CAB, the blend is more resistant to hydrolytic degradation and the contribution of the plasticiser in increasing the degradation rate is only evident with blends having either high P(HB-HV) copolymer or plasticiser content.

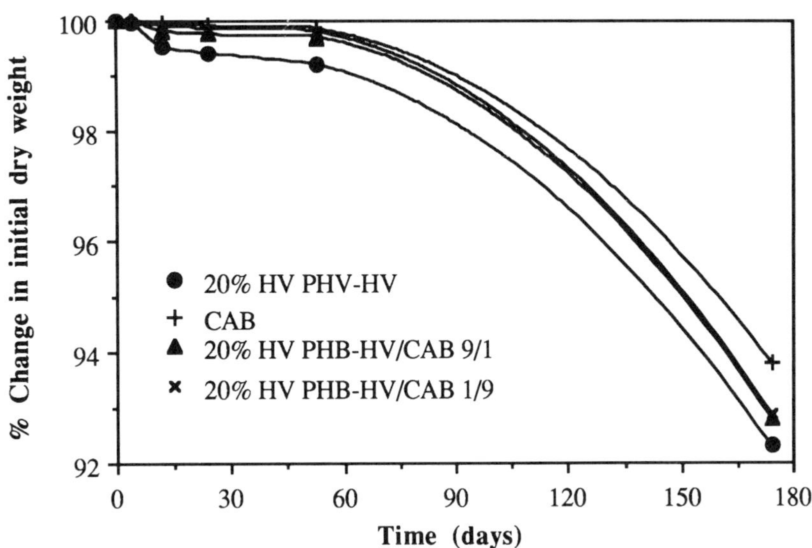

Figure 4. % Change in initial dry weight for solvent cast films of 20% HV P(HB-HV)/CAB blends in pH 7.4 buffer at 25°C.

Blend	t_{10} (days)	t_{50} (days)
CAB	46.0	-
12% HV P(HB-HV)	3.9	13.5
12% HV P(HB-HV)/CAB 9/1	8.1	15.6
12% HV P(HB-HV)/CAB 1/9	25.0	-

Table 1. 12% HV P(HB-HV)/CAB blends: Time for 10 and 50% loss of initial dry weight, (t_{10} and t_{50}, respectively). Solvent cast films in pH 10.6 buffer at 70°C.

Blend	t_{10} (days)	t_{50} (days)
20% HV P(HB-HV)	2.8	6.1
20% HV P(HB-HV)/G40 9/1	2.3	5.8
20% HV P(HB-HV)/G40 7/3	1.8	4.3
20% HV P(HB-HV)/G40 5/5	1.2	2.9

Table 2 20% HV P(HB-HV)/Polyester adipate (G40) blends: Time for 10 and 50% loss of initial dry weight, (t_{10} and t_{50}, respectively). Solvent cast films in pH 10.6 buffer at 70°C.

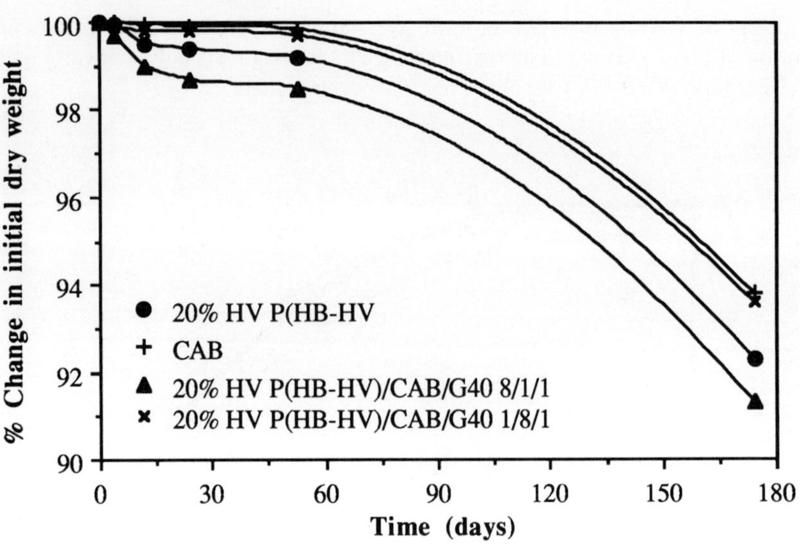

Figure 5. % Change in initial dry weight for solvent cast films of 20% HV P(HB-HV)/CAB/polyester adipate (G40) blends in pH 7.4 buffer at 25°C.

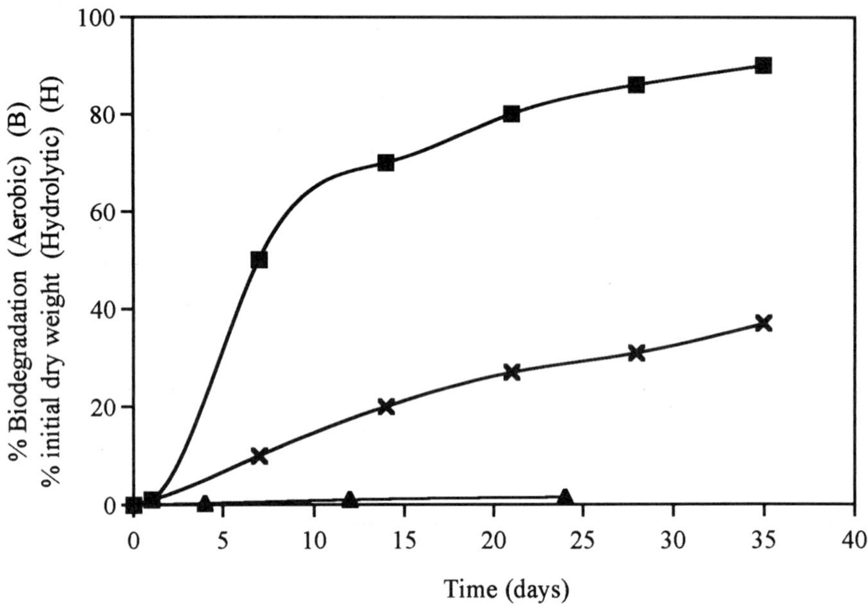

Figure 6. Solvent cast films of 20% HV P(HB-HV)/CAB/polyester adipate (G40) blends: % change in initial dry weight in pH 7.4 buffer at 25°C (hydrolytic degradation, H) and % biodegradation in activated sludge (B). (■) 20% HV P(HB-HV) copolymer (B), (✕) 20% HV P(HB-HV)/CAB/polyester adipate 8/1/1 blend (B) and (▲) 20% HV P(HB-HV)/CAB/polyester adipate 8/1/1 blend (H), respectively.

The aerobic degradation of the P(HB-HV)/CAB/polyester glutarate blends using 10% activated sludge (mixed inoculum) at 25°C has been reported elsewhere [2]. Results from the degradation of P(HB-HV)/CAB/polyester adipate blends under the same conditions suggest that these degrade slightly faster than the polyester glutarate blends. The marked effect on hydrolytic degradation rate of the blends shown in Figures 4 and 5 by the incorporation of 10% CAB into the P(HB-HV) matrix was also evident with the aerobic degradation of the three component blends. The % change in initial dry weight for solvent cast films of 20% HV P(HB-HV)/CAB/polyester adipate blends under hydrolytic degradation (H, pH 7.4 buffer at 25°C) and % biodegradation in activated sludge (B) are shown in Figure 6. The effect on the biodegradation rate with just 10% incorporation of CAB into the P(HB-HV) matrix is clearly seen in Figure 6. Also of interest the wide gulf between the hydrolytic and true biodegradation rates of the P(HB-HV)/CAB/polyester adipate 8/1/1 blend. This could be due to the rigid CAB molecule predominating the surface and thus shielding the P(HB-HV) copolymer from the micoorganisms in the activated sludge [5]. This shielding effect by the CAB could also be a reason as to the dramatic effects on the hydrolytic degradation rates even with 10% CAB addition into the blends of P(HB-HV)/CAB and P(HB-HV)/CAB/polyester adipate.

Our current work involves the degradation of these blends using several fungi, both in mono-cultures and mixed fungi cultures. Initial work has concentrated on the exposure of the samples to fungi growing on agar in petri dishes, followed by using liquid cultures of fungi containing powdered polymers and degradation in representative sterile and inoculated soils. Other polymeric polyester type plasticisers have been evaluated to see if the rate of degradation of the P(HB-HV)/CAB/plasticiser blends can be further manipulated.

CONCLUSION

A whole range of novel, biodegradable blends based on P(HB-HV) copolymers, CAB and polymeric plasticiser have been made. The physical properties and hydrolytic stability of the blends are affected by the CAB and plasticiser content. Applications for which there maybe price constraints in using expensive microbially produced material, such as the polyhydroxybutyrate and its copolymers with hydroxyvalerate, the technique of blending with synthetic material to enable adjustment of the price to be made is one technique that could potentially provide routes to commercially useful materials. The blending of natural and synthetic polymers provides the opportunity of increasing the range of materials that can be available for a wide spectrum of applications from medical to packaging, with both mechanical and degradative versatility, as well as price incentive.

ACKNOWLEDGEMENTS

We would like to thank Petra Püchner, (Institute for Water and Waste Water Engineering, University of Stuttgart, Germany), for the aerobic biodegradation work and the EC for funding of this project (Brite-EuRam II Project). We would also like to thank Dr Hänggi of Biomer for financial support.

REFERENCES

1 M. Yasin and B.J. Tighe, Strategies for the design of biodegradable polymer systems: Manipulation of polyhydroxybutyrate-based blends, *Plastics, Rubber and Composites: Processing and Applications*. 19:15-27 (1993).

2 M. Yasin, A.J. Amass, B.J. Tighe, P. Püchner, and W.R. Müller. New biodegradable three component polymer blends. *The 1994 IChemE Research Event,* 1: 368-370 (1994).

3 C.M. Buchanan, S.C. Gedon, A.W. White, M.D. Wood, Cellulose acetate butyrate and poly(hydroxybutyrate-co-valerate) copoymer blends. *Macromolecules*, 25:7373-7381 (1992).

4 N. Lotti and M. Scandola, Miscibility of bacterial poly(3-hydroxybutyrate-co-3-hydroxyvalerate) with ester substituted celluloses, *Polym. Bull.* 29:407-413, (1992).

5 D.F. Gilmore, R.C.Fuller, B. Scheneider, R.W. Lenz, N. Lotti and M. Scandola, Biodegradability of blends of poly(3-hydroxybutyrate-co-3-hydroxyvalerate) with cellulose acetate esters in activated sludge, *J. Environmental Polymer Degradation*, 2:49-57, (1994).

IMPROVED MECHANICAL PROPERTIES OF POLYMER FILMS DRIED FROM THE SWOLLEN STATE IN UNIAXIAL AND BIAXIAL STATES OF STRAIN

J. E. Mark

Department of Chemistry and the Polymer Research Center
The University of Cincinnati
Cincinnati, OH 45221-0172

INTRODUCTION

The technique employed in the investigations reported in this review emphasizes postponing the formation of anisotropic domains in semi-flexible polymers so as to better orient them along a common director. Most of the materials investigated were prepared by the following sequence of steps:

(i) Identifying polymer chains of sufficient stiffness to give liquid-crystalline, anisotropic phases (either homopolymers, or block copolymers consisting of stiff and flexible sequences)
(ii) Cross linking the chains, in the presence of solvent, thus conferring sufficient solidity for the polymer to remain in a deformed state for any length of time, with the solvent preventing the premature ordering of the stiff chains or sequences
(iii) Deforming the swollen network uniaxially or biaxially to induce segmental orientation
(iv) Removing the solvent, at constant length or at constant force, in the hope of causing a first-order transition to a *single-phase*, homogeneous, and highly-ordered material.[1-15]

Films prepared using this technique have been characterized by polarizing microscopy, birefringence, and stress-strain measurements in uniaxial and biaxial extension to rupture. Some of the polymers that are processible by these techniques have the additional advantage of being biodegradable. The utility of these techniques is illustrated using recent

Polymers and Other Advanced Materials: Emerging Technologies and Business Opportunities. Edited by P. N. Prasad *et al.*, Plenum Press, New York, 1995

results obtained on cellulose esters,[8,9] cellulose ethers,[8,9,11] polyisocyanates,[12] and gelatin.[13-15]

THEORY

The theoretical aspects of these methods have been described in detail in a number of original papers and reviews,[1-7,10] and are therefore given in only outline form[15] in this report.

It has been known that real polymer chains might be described in terms of freely-jointed chains with a properly defined Kuhn segment length. The length-to-width ratio x for such a segment is usually around three, but would be considerably larger for the semi-flexible chains of interest in this work. A successful method for the calculation of the combinatory ("steric") factor and the orientational factor of the configuration partition function for a system of freely-jointed rods has been proposed by Flory and coworkers. This lattice theory has been used to enumerate the combinatory factor, although the orientation of rods are allowed to be continuous. This is done by considering a rod oriented in the k-th direction (at angles q_k, f_k). According to theory, it can be represented by y_k sequences of x/y_k segments, with x being the axial ratio already mentioned, and X the direction of stretching.

This basic theory by Flory has now been extended to such systems in a state of deformation. The result of greatest importance centers on the orientational order parameter

$$S = \langle P_2(\cos\theta)\rangle = \frac{3}{2}\langle\cos^2\theta\rangle - \frac{1}{2} \qquad (1)$$

where θ is the angle between the rod and the X axis. For uniaxial deformation, this parameter is given by

$$S \cong \frac{1}{5m}(\lambda^2 - 1/\lambda v_2)\left(1 + \frac{5xv_2}{64[1 - (\frac{x}{x_c})v_2]}\right) \qquad (2)$$

where m is the number of Kuhn segments, λ the elongation, and v_2 the volume fraction of polymer in the swollen, cross-linked network. The quantity $x_c = 3/[4/\pi - 1] \cong 10.98$ is the critical or "bifurcation" value of x in the Flory-Ronca model described above. When the ratio x/x_c becomes significantly large, then the factor in square brackets becomes important and phase separation occurs.

The theory has also been applied to thermotropic systems by elaborating the mean-field approach proposed by Flory and Ronca. The attractive dispersion forces related to the anisotropy of the polarizability tensors for chain segments contribute to the mean field, which gives an additional contribution to the alignment of polymer segments in the direction of stretch. The result is now

$$S = \frac{1}{5m}(\lambda^2 - \frac{1}{\lambda v_2})(1 + \frac{5 x v_2}{64[1 - (\frac{x}{x_c})v_2]})\{1 - \frac{1}{5}\tilde{T}^{-1}[1 - \frac{5 x v_2}{64(1 - \frac{x}{x_c}v_2)}]\}^{-1} \qquad (3)$$

where $\tilde{T}$ is the reduced temperature. The quantity in the curly brackets represents the mean-field correction for the orientation function S. With decreasing temperature, S increases in comparison with the athermal case (for which $\tilde{T}^{-1} = 0$). Similarly, knowledge of the free energy enables calculation of the stress and chemical potentials.

The above analysis yields plots of segmental orientation S against axial ratio x, composition v_2, and extent of deformation λ. The information thus obtained provides guidance on how to control the phase separation so that the entire sample becomes anisotropic, with all of the chains in a given layer very nearly aligned in the same direction.

The phase separations exhibited by undeformed semi-flexible and rigid-rod polymers are now well understood in terms of the thermodynamics and statistical mechanical theories of the liquid-crystalline state, as is documented in a recent review.[5] These extensions of the basic theories to polymeric systems in the *deformed* state is also showing considerable promise.

EXPERIMENTAL RESULTS

Cellulose Acetate and Hydroxypropylcellulose

The cellulose ester cellulose acetate[8,9] and the cellulose ether hydroxypropyl cellulose[8,9,11] have been studied using this novel orientation technique. Well-ordered liquid-crystalline states have been observed under a variety of conditions for these cellulosics. Some typical results obtained on a film sample of hydroxypropyl cellulose (HPC) having a molecular weight between cross links M_c of 20.3 x 10^3 g mol^{-1} are shown in Figure 1. Drying the sample while under uniaxial elongation is seen to greatly increase its modulus and ultimate strength in the direction parallel (||) to the extension direction during drying. The system is of course anisotropic, and this is evidenced by the *decrease* in some of the mechanical properties in the perpendicular ($\perp$) direction.

Measurements were also carried out on the relationship between the tensile modulus and the birefringence of the oriented films. It was found to be non-linear, with the modulus increasing more rapidly with increase in birefringence at higher degrees of orientation.[9]

Poly(isocyanates)

This polymer is also known to form liquid-crystalline phases, and films prepared from it have been treated by these orientation techniques.[12] In this case, the polymer was obtained by copolymerizing n-hexyl isocyante with allyl isocyanate, to obtain samples of poly(hexylallyl isocyanate) (PHAIC). This was done to provide unsaturated sites useful for

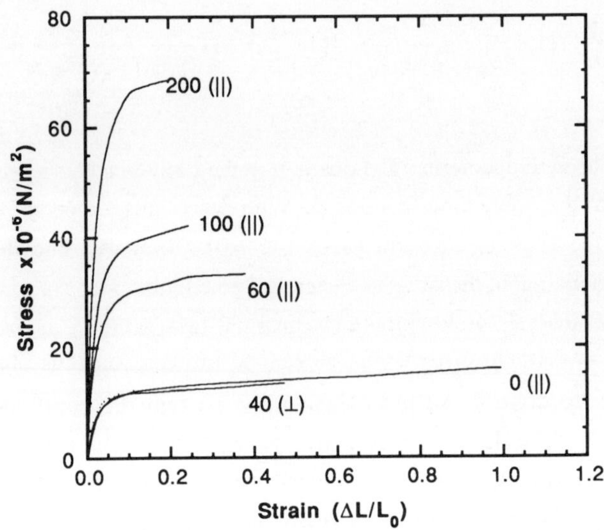

Figure 1. Stress-strain behavior of oriented films prepared from HPC.[9] Labels on the curves are the values of the extension ratio maintained during drying under uniaxial elongation. The mechanical testing was carried out either parallel (||) or perpendicular (⊥) to the orientation axis of the sample.

introducing the required cross links for the desired network structure. The cross linking was carried out using either free-radically by thermolysis of an organic peroxide, or by direct addition to the double bonds in a hydrosilation reaction.[12]

Some of the increases in modulus obtained by drying the samples in the uniaxially-extended state are shown in Figure 2.[12] They are seen to be substantial.

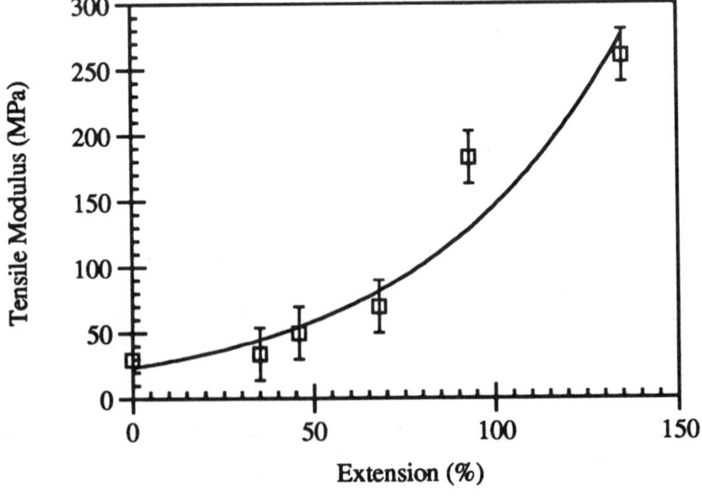

Figure 2. Tensile modulus as a function of the extension ratio during drying, for oriented films of PHAIC.[12]

Gelatin

This material is composed of the protein collagen, a naturally-occurring material, and thus has the considerable advantage of biodegradability. The samples employed were designated type A (from porcine skin, approximately 300 bloom, where "bloom values" characterize gel-forming capabilities) and type B (from bovine skin, approximately 225 bloom).[14] Although gelatin is soluble in water, a search for alternative solvents was carried out. 2,2,2-Trifluoroethanol (CF_3CH_2OH) became of particular interest since gelatin displays high solubility in it without forming the collagen folds, at any concentration at room temperature. In addition, this solvent's miscibility with some diisocyanates enables preparation of the networks with urea linkages, which should have no adverse impact on the environment upon disposal.

Cross-linked films of these two gelatin samples, swollen with either water or CF_3CH_2OH to a volume fraction v_2 of polymer, were given the drying-under-orientation treatment, both in uniaxial extension and in equi-biaxial extension.[14] (The biaxially-oriented samples were prepared by deforming the swollen type A gelatin gels in two perpendicular direction at the same time, to equal extensions). As was the case for uniaxial extension,[14] the mechanical properties were found to increase with increase in extension ratio during drying (from 0 to 115 %). A four-fold increase in tensile strength, a 150 % increase in modulus, and a fourteen-fold increase in toughness were observed for films for an extension ratio during drying of only 115 %! Typical mechanical property results are shown in Figure 3, and a summary of results are presented in Table 1.[14] In brief, the improvements in mechanical properties increased monotonically with increase in elongation during drying, for both uniaxial and equi-biaxial strains. Equi-biaxial orientation, however, gave almost twice the improvements in terms of strength and modulus obtained in the simpler uniaxial orientations (at essentially the same extension rato).

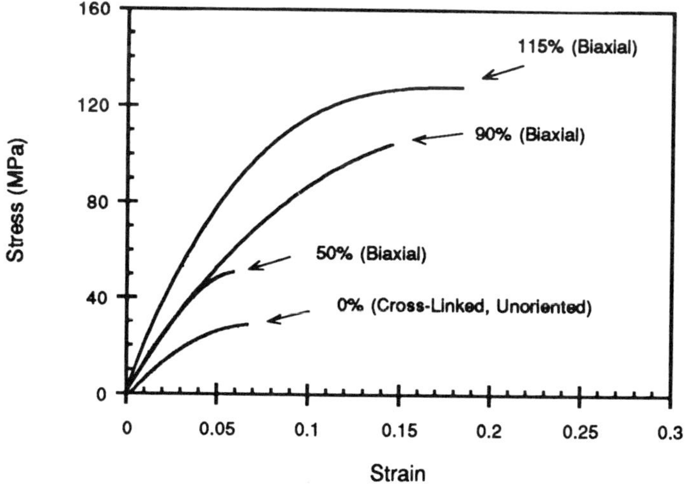

Figure 3. Stress-strain behavior of biaxially-oriented type A gelatin. Swelling solvent: water; v_2 at stretching: 0.2 - 0.25.[14]

Table 1. Maximum Values of Mechanical Properties Obtained after Orientation[14]

	Tensile Strength[a] (MPa)	Tensile Modulus[a] (MPa)	Toughness[a] (MPa)	Elongation at Break[a] (%)
Type B in water (Uniaxial)	84.3 (25.6)	1300 (470)	4.10 (1.24)	8.24 (7.65)
Type B in CF_3CH_2OH	65.3 (25.6)	900 (470)	8.74 (1.24)	16.57 (7.65)
Type A in water (Uniaxial)	140.0 (28.4)	1690 (630)	23.68 (1.20)	22.54 (6.67)
Type A in water (Biaxial)	128.4 (28.4)	1570 (630)	17.45 (1.20)	18.39 (6.67)

[a] The data within the parentheses are reference values obtained before orientation.

The birefringence of some of the films prepared from gelatin (type A, swelling solvent: water) after uniaxial orientation is shown in Figure 4.[14] There is a monotonic and substantial increase in birefringence with increase in extension ratio during drying. A dramatic fifty-fold increase in birefringence in the oriented direction was observed at an extension ratio of only 165 %. The increases in birefringence and the corresponding degrees of orientation are consistent with the observed improvements in mechanical properties.

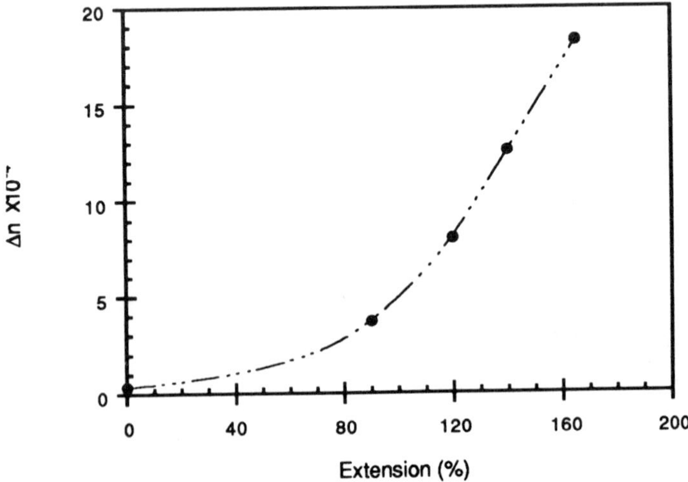

Figure 4. Birefringence as a function of extension ratio during drying for uniaxially-oriented type A gelatin.[14]

ACKNOWLEDGEMENTS

It is a pleasure to acknowledge the financial support provided by the Procter & Gamble Company through its University Exploratory Research Program and by the National Science Foundation through Grant DMR-942223 (Polymers Program, Division of Materials Research), and the fellowship support provided by the Raychem Corporation.

REFERENCES

1. B. Erman, I. Bahar, A. Kloczkowski, and J. E. Mark, A lattice model for segmental orientation in deformed polymer networks. 1. Contribution of intermolecular correlations, *Macromolecules* 23:5335 (1990).
2. I. Bahar, B. Erman, A. Kloczkowski, and J. E. Mark, A lattice model for segmental orientation in deformed polymer networks. 2. Effect of chain stiffness and thermotropic interactions, *Macromolecules* 23:5341 (1990).
3. B. Erman, T. Haliloglu, I. Bahar, and J. E. Mark, Segmental orientation in uniaxially deformed networks: A higher-order approximation for finite chains and large deformations, *Macromolecules* 24:901 (1991).
4. B. Erman, I. Bahar, A. Kloczkowski, and J. E. Mark, Stress, strain, and segmental orientation in networks with semi-rigid chains, *in*: "Elastomeric Polymer Networks", J. E. Mark and B. Erman, eds., Prentice Hall, Englewood Cliffs, NJ (1991).
5. B. Erman, I. Bahar, Y. Yang, A. Kloczkowski, and J. E. Mark, Novel orientation techniques for the preparation of high-performance materials from semi-flexible polymers such as the cellulosics, *in*: "Polymer Solutions, Blends, and Interfaces", I. Noda and D. N. Rubingh, eds., Elsevier, Amsterdam (1992).
6. A. Kloczkowski, J. E. Mark, B. Erman, and I. Bahar, Isotropic-nematic phase transitions in polymer networks deformed in the dry and swollen states, *in*: "Polymer Solutions, Blends, and Interfaces", I. Noda and D. N. Rubingh, eds., Elsevier, Amsterdam (1992).
7. B. Erman, I. Bahar, A. Kloczkowski, and J. E. Mark, Networks with semi-flexible chains, in "Synthesis, Characterization, and Theory of Polymeric Networks and Gels", S. M. Aharoni, ed., Plenum, New York (1992).
8. Y. Yang, A. Kloczkowski, J. E. Mark, B. Erman, and I. Bahar, A novel orientation technique for semi-rigid polymers. 1. Preparation of cross-linked cellulose acetate and hydroxypropylcellulose films having the required permanent anisotropy in the swollen state, *Coll. Polym. Sci.* 272:284 (1994).
9. J. E. Mark, Y. Yang, A. Kloczkowski, B. Erman, and I. Bahar, A novel orientation technique for semi-rigid polymers. 2. Mechanical properties of the cellulose acetate and hydroxypropylcellulose films, Coll. Polym. Sci. 272:393 (1994).

10. Y. Yang, A. Kloczkowski, J. E. Mark, B. Erman, and I. Bahar, Theoretical analysis of the elastomeric and optical properties of networks of semi-rigid chains in the swollen state, *Macromolecules* 28:000 (1995).
11. Y. Yang, A. Kloczkowski, J. E. Mark, B. Erman, and I. Bahar, Experimental studies of elastomeric and optical properties, and strain-induced crystallization, in deformed hydroxypropylcellulose networks in the swollen state, *Macromolecules* 28:000 (1995).
12. W. Zhao, H. Zimmer, and J. E. Mark, A novel orientation technique for networks of semi-flexible polyisocyanate chains, *Preprints, Div. Polym. Mats. Sci. Eng.* 71:399 (1994).
13. W. Zhao, A. Kloczkowski, J. E. Mark, B. Erman, and I. Bahar, Preparation of high-performance materials from biodegradable gelatin using a novel orientation technique, *Preprints, Div. Polym. Mats. Sci. Eng.* 72:86 (1995).
14. W. Zhao, A. Kloczkowski, J. E. Mark, B. Erman, and I. Bahar, Oriented gelatin – A new source for high-performance materials, submitted to *J. Macromol Sci., A-Pure Appl. Chem.*
15. J. E. Mark, Novel orientation techniques for the preparation of high-performance materials from semi-flexible polymers, submitted to special Hanoi issue of *Macromol. Chem. Phys.*

RECENT DEVELOPMENTS AND PERSPECTIVES IN ABS RESIN

Hyungsu Kim and Sang S. Woo

R&D Center, Chemicals Division
Cheil Industries Inc.
332-2 Gochun-Dong, Euiwang-Shi
Kyoungki-Do, Korea

INTRODUCTION

ABS is a family of thermoplastics that contains three monomeric units: acrylonitrile, butadiene and styrene. The versatility of ABS resins in many applications is widely recognized and the world-wide consumption has been gradually increased to nearly 6 billion lbs in 1993. Along with the expected increase in demand, a number of ABS makers are expanding production capacity and furthermore, numerous efforts have been made to achieve higher levels of performance, which lead to better mechanical properties and environmental compatibility. Among the various aspects of such activities, it may be of particular interest for most producers to improve the chemical resistance for refrigerator use and minimize or replace the hazardous components involved in flame retardant grade. In addition, developments of ABS based alloys and specialty grades are frequently practiced to meet the diversified needs required in the markets. Technical backgrounds relating to the above issues are briefly reviewed and up-to-date progresses are discussed.

FACTORS AFFECTING PERFORMANCE OF ABS

Impact strength

In ABS, a styrene/acrylonitrile copolymer(SAN) matrix contains discrete butadiene based elastomer particles for toughening. Elastomer particles are grafted with SAN to achieve necessary interaction with the matrix polymer. There are numbers of variables which control

Polymers and Other Advanced Materials: Emerging Technologies and Business Opportunities. Edited by P. N. Prasad *et al.*, Plenum Press, New York, 1995

the mechanical behavior of ABS resin. These may include the size, morphology and graft ratio of the elastomeric phase, rubber phase volume fraction, molecular weight of matrix SAN, and overall chemical composition, to mention only a few. In recent studies[1,2], several well characterized SAN grafted rubbers of various acrylonitrile(AN) compositions were blended with a series of SAN copolymers having wide range of AN contents. The major focus was to relate the effect of graft and matrix AN content on mechanical properties of ABS. It was found that the maximum impact strength was consistently obtained at a mismatched compositions of the matrix and graft rather than when the two AN levels were equal which presumably would maximize the possibility of interpenetration of graft and matrix SAN chains. The observed behavior was successfully explained by the interplay of three major factors: (a) interfacial adhesion, (b) state of rubber particle dispersion, and (c) inherent ductility of matrix SAN. The first two depend on the difference between the AN levels in the matrix SAN and the graft SAN while the latter is well characterized by measuring the mode I critical stress intensity factor, K_{IC} which linearly increases with increasing AN content. Because of the increase in toughenability of SAN copolymers as the AN content increases, ABS materials should generally become tougher as the matrix AN content increases provided the adhesion and dispersion issues remain within acceptable limits. The proposed model emphasizes that the inherent ductility of the matrix polymer provides a rational basis for understanding the properties of toughened SAN copolymers by SAN emulsion grafted rubbers.

Solvent Resistance

Improving solvent resistance of ABS material is of considerable importance in the application of refrigerator liner which contacts with foamed polyurethane layer. Due to the adverse effect of CFC-11, a widely used foaming agent, on the ozone layer, it is being replaced by environmentally more safe material like HCFC-141b. In the resin side, however, conventional ABS materials are susceptible to stress-cracking induced by such a new foaming agent. Accordingly, a new type of ABS resin with enhanced solvent resistance is inevitable. In line with the observed trend of inherent ductility of SAN copolymers, it is generally recognized that the chemical resistance of ABS is improved by adding more AN in SAN copolymer. Thus, as an effective way to overcome the drawback of conventional ABS, it may be desirable to separately prepare a suitable matrix SAN copolymer containing higher AN content and then melt-blend it with an existing rubber concentrate. Of course, a critical AN concentration required for sufficient solvent resistance should be determined by well defined evaluation for stress-cracking and some modifications on the structure of rubber concentrate may be also necessary for successful application. Provided the production technology needed for the control of AN composition is fully developed, the previously described model should offer a useful guideline for designing an optimum ABS material having well balanced mechanical properties and solvent resistance.

Heat Resistance

Incorporation of α-methyl styrene in SAN has been a prevalent way to raise practical use temperature of ABS resin, especially in automotive applications. While this type of grade allows rather confined level of heat resistance, significant enhancement with higher ranges of heat distortion temperature(HDT) is readily available by blending ABS with styrene-maleic anhydride (SMA) copolymers and/or imidized SMA containing N-phenyl maleimide. Although these copolymers are very effective in elevating the HDT of ABS resin, it should be noted that the toughness of related blends is deteriorated due to the insufficient inherent ductility of such highly heat resistant styrene copolymers[3,4]. As demonstrated in previous papers[1,4], the effect of MA content in SMA should be also considered here in light of changes in component miscibility and inherent ductility of mixed matrix of SAN/SMA blends. The limited toughenability of SMA copolymers is well understood by their lower entanglement density than other brittle polymers which prefer to deform by crazing[5] and it might be beneficial to apply the larger rubber particles up to a few microns than those in ABS for the partial retrival of loss in blend toughness.

Flame Retardance

Flammability of ABS resin has been dominantly reduced by incorporation of halogen based compounds. While halogen systems, especially the bromine compounds, are acknowledged as the most cost-effective flame retardants for engineering plastics and styrenics, they are facing an unfriendly regulatory climate due to the possible hazards caused by combustion byproducts. Despite studies by independent laboratories shows little evidence on the health risks from combustion or overheating, the poly brominated diphenyl ethers(PBDE's) are alledged to be convertible to dioxins and furans during fire. In addition, it has been reported that chlorine and bromine relased during fire combine with moisture to form acidic deposits, which may be detrimental to electric/electronic equipments. Although the specific compounds and their maximum allowables and phase-in deadlines are still controversial, some restrictions on the brominated flame retardants are likely to be adopted and there is little doubt that the big users will be inquiring more reliable compounds. Therefore, it would be advantageous for the resin makers to produce PBDE free grades for most applications and it may be carefully expected that the non-halogen based flame retartdant grades proliferate.

ALLOYS WITH OTHER ENGINEERING POLYMERS

Multiphase polymer alloys provide a route to combining the desirable features of different materials into one product and sometimes synergistic performance characteristics can be achieved. There are numbers of useful polymer alloys involving various types of ABS, in which the latter mainly improves toughness and processability of the other component. A commercially important alloy is formed by blending polycarbonate, PC, (a strong, heat resistant material whose toughness is very dependent on notch geometry and part thickness) with rubber toughened ABS materials. PC/ABS alloys can show synergisms in certain properties and exhibit toughness in thick parts and with sharp cracks. These materials are fabricated into parts for automobiles and appliances. Unlike most other recently developed polymer alloys, the current generation of products do not contain any compatibilizer(block or graft copolymer) bacause the thermodynamic affinity of PC and SAN is good enough to lead to adequate adhesion[6]. However, due to the intrinsic incompatibility between the components[7], it is difficult to accomplish a stable morphology and consistent performance. Unless properly compatibilized, even though a good dispersion is achieved during compounding, phase coalescence may occur during low stress conditions or molding of large parts which will cause deterioration of mechanical properties[8,9]. Thus, in order to improve the performance and market opportunities, it is critical to somehow compatibilize the mixture. The preferred solution may include the in situ copolymer formation via reactive compounding process, which can activate the introduction of the second generation PC/ABS alloy.

SPECIALTY GRADES

To accomodate the specific requirements of various applications, special characteristics have been imparted to the base ABS resin. Currently available features mostly comprise transparency, high gloss, low gloss, anti-static and anti-fungal capabilities, etc.. The last two functions are typically exerted by utilizing appropriate additives of desired purpose, whereas the surface appearances relating to gloss and optical clarity are mainly controlled by changes in the size and composition of the elastomeric phase. An apparent problem associated with the use of antistatic agents consisting of low molecular weight species is the fact that the presence of sufficient moisture level is needed for satisfactory performance and furthermore, the migrated species from the bulk are prone to be romoved by common solvents or frequent contacts with other parts. Consequently, consistent and prolonged antistatic property may not be guaranteed by additive type anti-static ABS. In order to overcome such shortcomings, one remarkable progress has been recently made on the development of permanently anti-static ABS, which is essentially an alloy of ABS with a highly hydrophillic polymer. Comparing with the previous grades, the novel product offers quite powerful and stable function with comparable mechanical properties and processability. Continuous efforts to extend proprietary features would no doubt add high value to ABS resins.

REFERENCES

1. H. Kim, H. Keskkula, and D.R. Paul, *Polymer* 31: 869 (1990).
2. H. Kim, H. Keskkula, and D.R. Paul, *Polymer* 32: 1447 (1991).
3. H. Kim, H. Keskkula, and D.R. Paul, *Polymer* 32:2372 (1991).
4. J.H. Kim, H. Keskkula, and D.R. Paul, *J. Appl. Polym. Sci.* 40: 183 (1990).
5. S. Wu, *Polymer International* 29: 229 (1992).
6. J.D. Keitz, J.W. Barlow, and D.R. Paul, *J. Appl. Polym. Sci.* 29: 3131 (1984).
7. T.A. Callaghan, K. Takakuwa, D.R. Paul, and A.R. Pawda, *Polymer* 34: 3796 (1993).
8. D. Quintens, G. Groeninckx, M. Guest, and L. Aerts, *Polym. Eng. Sci.* 30: 1474 (1990).
9. T.W. Cheng, H. Keskkula, and D.R. Paul, *J. Appl. Polym. Sci.* 45:1245 (1992).

RECENT ADVANCES IN POLYPROPYLENE-BASED MATERIALS, PROPERTIES AND APPLICATIONS

T. Simonazzi - HIMONT ITALIA - Ferrara (Italy)
J. C. Haylock - HIMONT Inc. - Wilmington (U.S.A.)

INTRODUCTION

Polypropylene (PP) has had a remarkable growth rate during the last 30 years, and today it continues to be the fastest growing of the thermoplastic polyolefins. This has resulted from both progress in Ziegler-Natta catalytic systems and production technologies, together with an ease of post-treatments that can modify its structure and broaden its field of application. Today, we are presenting some recent innovations that are based on a methodology of advanced, 4th generation catalysts (Reactor Granule Technology), on retailoring post-treatments (Hivalloy, foamable PP), and on polymer/processing machinery developments (PP stretch-blow molding).

REACTOR GRANULE TECHNOLOGY

In the fourth generation Ziegler-Natta catalyst, the physical and chemical structure of the active sites provide a three-dimensional spherical architecture that can be duplicated by the growing polymer particle. Control of the morphology of this polymer particle allows it to become a polymerisation reactor itself, and a change from one monomer to another can produce a second, different polymer within the granule, each polymer chain intimately dispersed throughout the preformed, porous, solid structure of the host polymer matrix. The process can be repeated to generate multipolymers, multiphase alloys and blends directly in the rector. The Catalloy® process is a new, highly sophisticated, modular technology, that has been designed to maximise the polymer properties possible from these fourth generation Ziegler-Natta catalysts.

New polypropylene-based materials obtained from this process are able to satisfy the increasingly sophisticated performance requirements of the plastics marketplace. For example, recent developments in automobile bumper designs require innovative and improved characteristics such as reduced wall thickness, improved molding productivity, functional yet aesthetically appealing covers, as well as a much wider range of physical and mechanical properties. This can vary from one automobile producer to another. One manufacturer may require a tough, low modulus resin for a flexible, high energy absorbing system, while a second manufacturer's stiff, impact resistant design requires a higher modulus, less flexible resin.

Polymers and Other Advanced Materials: Emerging Technologies and Business Opportunities. Edited by P. N. Prasad *et al.*, Plenum Press, New York, 1995

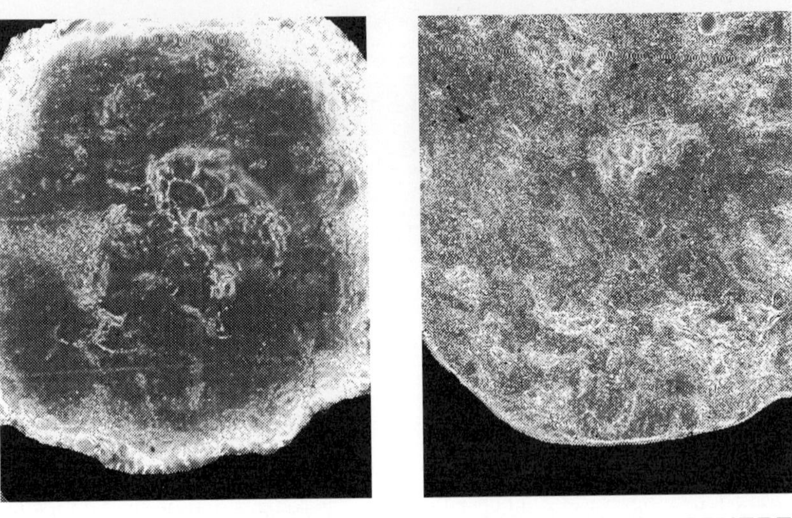

Figure 1. Comparison of polymer spheres: Particle Cross Section.

• MEDIUM FLOWABILITY MFR "L"	2 ÷ 8	dg/min
• FLEXURAL MODULUS	900 ÷ 1000	Mpa
• IZOD IMPACT STRENGTH (-30 °C)	90	J/m
• D/B TRANSITION TEMPERATURE	<-50	°C
• ELONGATION AT BREAK	>500	%

Figure 2. Polypropylene properties for standard bumper design.

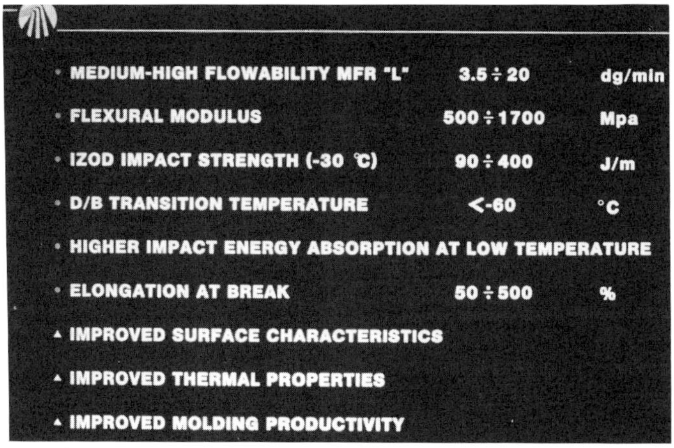

- MEDIUM-HIGH FLOWABILITY MFR "L" 3.5 ÷ 20 dg/min
- FLEXURAL MODULUS 500 ÷ 1700 Mpa
- IZOD IMPACT STRENGTH (-30 °C) 90 ÷ 400 J/m
- D/B TRANSITION TEMPERATURE <-60 °C
- HIGHER IMPACT ENERGY ABSORPTION AT LOW TEMPERATURE
- ELONGATION AT BREAK 50 ÷ 500 %
- ▲ IMPROVED SURFACE CHARACTERISTICS
- ▲ IMPROVED THERMAL PROPERTIES
- ▲ IMPROVED MOLDING PRODUCTIVITY

Figure 3. Polypropylene properties for innovative bumper design.

To meet such a wide range of targets, it is necessary to be able to tailor-make new types of polypropylene resins. This requires new structural morphologies of the polymers and is accomplished by optimising the tacticity of the matrix polymer, its molecular weight and molecular weight distribution, the content and composition of rubber copolymer components, and the fillers.

Only the polypropylene alloy approach to material development has the flexibility and cost/performance characteristics necessary for a polyolefin to meet and certify to such a wide spectrum of quality requirements. As a consequence, HIMONT's polypropylene-based alloy product line for automobile bumpers has expanded from five resins with a narrow range of moduli (1000-1100 MPa), to fifteen resins covering a much wider range of 550-1800 MPa.

HIVALLOY TECHNOLOGY

The Reactor Granule Technology approach to the production of propylene based olefinic alloys has been extended further to allow the grafting and polymerisation of non-olefinic monomers within the solid, porous polyolefin particle. This new, proprietary technology, which HIMONT calls Hivalloy®, enables the introduction, grafting and polymerisation of functionalised monomers previously thought incompatible with Ziegler-Natta catalysed polymerisation systems.

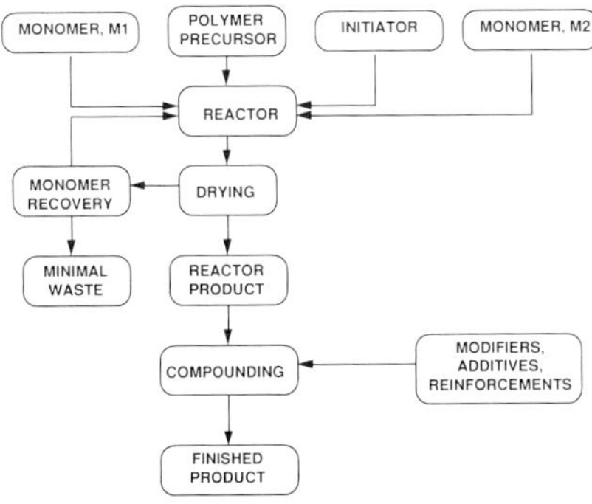

Figure 4. Hivalloy process flow diagram.

The Hivalloy® process incorporates three primary unit operations, including a downstream compounding line for the introduction of additives, polymer modifiers, and reinforcing agents. Polypropylene alloys with such functionalised polymers provide higher modulus properties at elevated temperatures, added hardness and creep resistance to the inherent chemical resistance, low density and easy processability of the polyolefin.

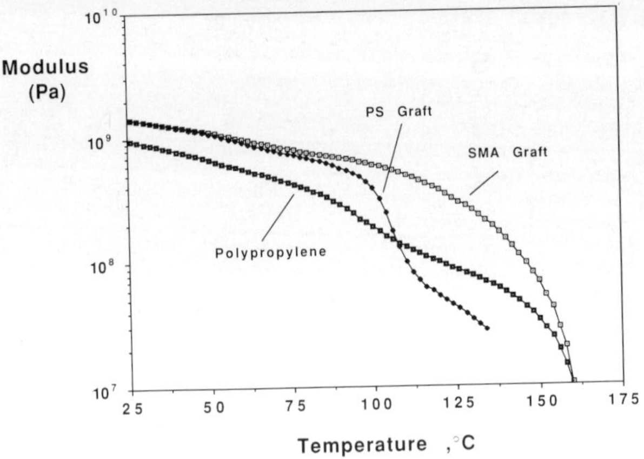

Figure 5. Modulus retention of Hivalloy®.

1) Influence of Porosity and Morphology: The structural integrity and the high internal surface nature of the Reactor Granule polyolefin matrix is essential to the Hivalloy® technology. One of the first polymer blends studied using this technology was a series of

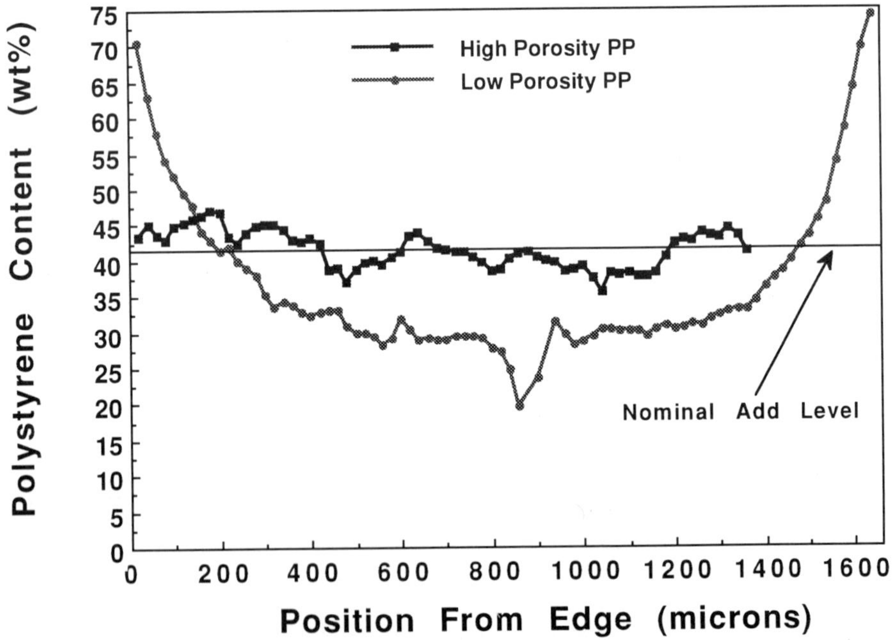

Figure 6. Composition distribution in reactor spheres.

In contrast, grafting and polymerisation on the high surface area of a porous, solid polypropylene "Reactor Granule" particle occurs primarily within the particle. This allows the production of graft copolymer alloys with much higher levels of comonomer and a wider range of processible monomer compositions.

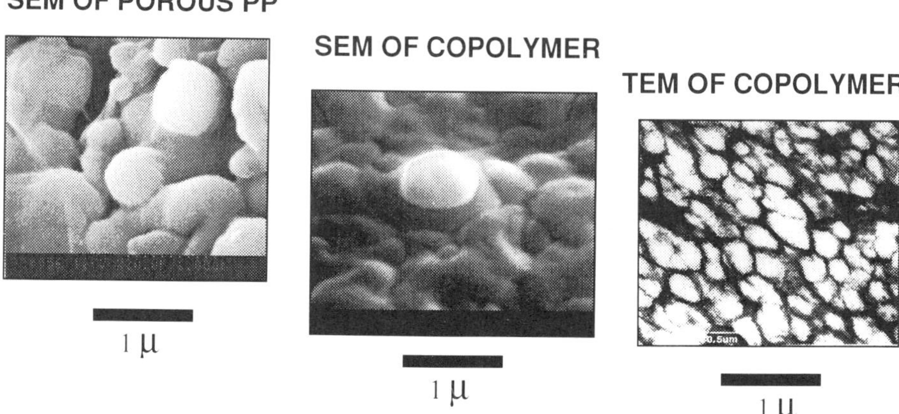

Figure 7. Morphological features of porous particle after grafting.

polypropylene-polystyrene alloys, made directly in the polymerisation reactor. If a dense polymer matrix, produced by conventional Ziegler-Natta catalyst technology for polymerising propylene, is used as the Reactor Granule, a non-olefinic monomer such as styrene does not diffuse far into the polyolefin matrix and the polymerisation reaction produces an obvious core-shell structure.

Figure 7 demonstrates how the grafting and polymerisation of styrene takes place on the internal surfaces of the porous polypropylene particle, giving an alloy with a co-continuous morphology, the dark areas representing the polystyrene rich areas. During melt extrusion of the reactor spheres, the co-continuous structure breaks down into a discontinuous structure with the polystyrene finely dispersed within the polypropylene matrix.

2) Mechanical Properties of Grafted Reactor Alloy: Such a unique alloying technology, with the incorporation of a high Tg amorphous polymer such as styrene in polypropylene, significantly increases its HDT and thus increases the modulus temperature resistance properties. The material's response to Ethylene-Propylene Rubber (EPR) modification, a conventional means of impact modifying polypropylene, is limited to the olefinic phase. On the other hand, styrene block copolymers, and more specifically, styrene butadiene-styrene triblock structures, are more effective modifiers due to interactions between this rubber and the polystyrene dispersed phase. Combinations of "olefinic-like" and "styrene-like" impact modifiers do in fact, provide a synergistic enhancement of impact resistance while at the same time producing a modulus intermediate between the levels for the two impact modifiers alone. As can be seen in Figure 8, the extent of this synergistic impact response can be quite dramatic and is connected with a much more timely dispersed polystyrene phase.

Like their parent polypropylene resins, the Hivalloy® technology alloys with polystyrene are very responsive to mineral and/or glass reinforcement. By combining optimal impact modifiers with the glass fiber reinforcement of the alloy, excellent rigidity together with exceptional toughness, low warpage and dimensional stability (due to low moisture uptake) are possible, making a "High Performance Polyolefin" alloy competitive in properties with more expensive engineering resins such as modified PPE, PBT and Nylon.

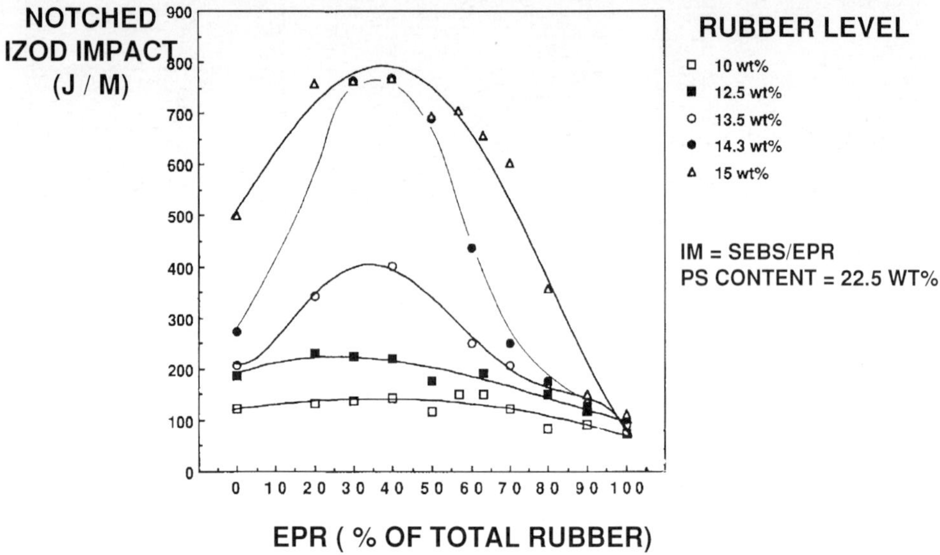

Figure 8. Effect of rubber composition on impact resistance.

POLYPROPYLENE FOAM FOR PASSIVE SAFETY

Mechanical energy absorbers for passive safety in cars (in both bumpers and interiors), is a growing application for polymeric foams. Up till now, polyurethane has been the material of choice in this application. Polypropylene has not been able to compete because its low extensional viscosity (or poor melt strength), has prevented its widespread use in low density foams. The processing window for polypropylene, unlike many amorphous polymers such as polystyrene, is relatively narrow (only about 4°C). The lower boundary of the foaming window is controlled by the melting point of the polymer, and the upper limit is simply the temperature at which the cell walls begin to collapse because the viscosity is too low. Now a new polypropylene has been developed by HIMONT that has a high melt strength (HMS), which makes it possible to produce low density foams by melt extrusion. This increase of melt strength is obtained by a proprietary electron beam radiation process that treats polypropylene under specific conditions that generate radicals along the polymer chain which, with a subsequent rearrangement, produces a long chain branched structure. This long chain branched polymeric structure results in an increase in polypropylene's extensional viscosity, thus making the production of low density, closed cell foams on conventional extrusion equipment possible.

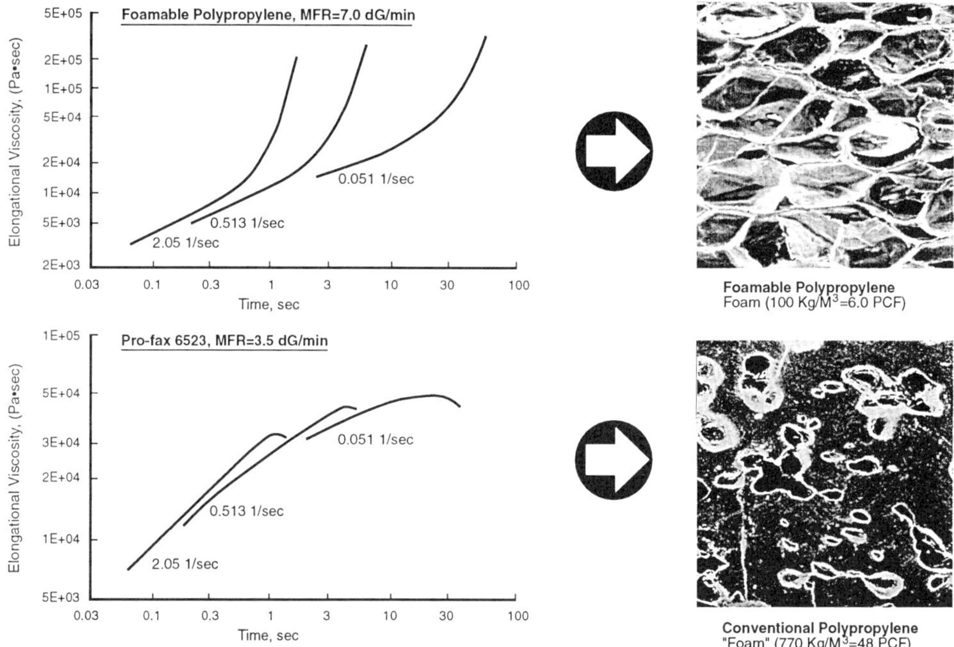

Figure 9. HMS Foamable Polypropylene: extensional viscosity and foam structure.

The rheological behavior of long chain branched polypropylene satisfies the requirements for a uniform bubble growth and allows control of the foam's final density. Such results are attainable using conventional extruders equipped with properly designed heads and with hydrocarbons instead of fluorocarbons as the blowing agents. The product from the extruder can be in the form of a bead, and articles can then be manufactured via steam sintering of the beads loaded into the mold, which is placed between the plates of a press.

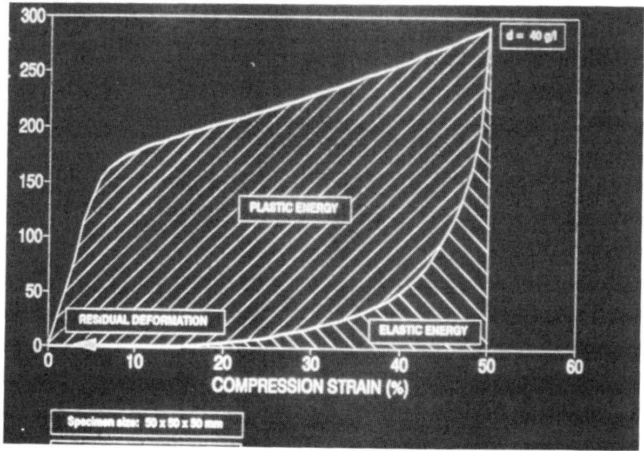

Figure 10. PP foam molded beads; compression hysteresis at 21°C.

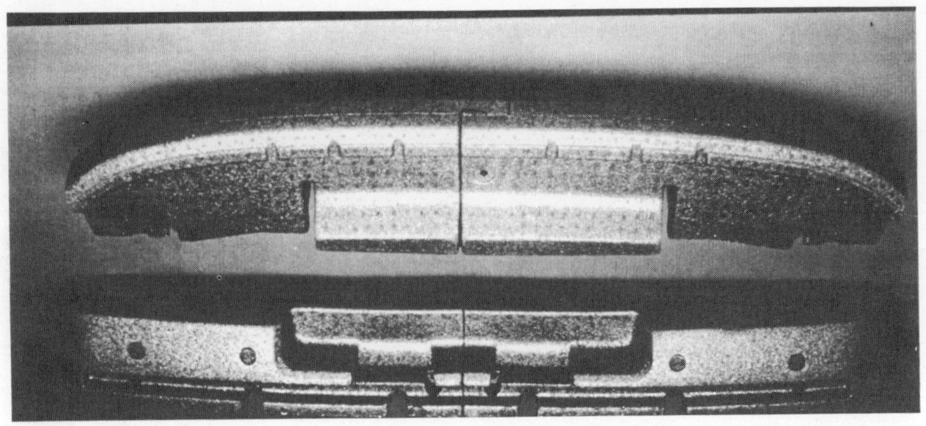

Figure 11. PP foam bumper structures.

1) Mechanical Properties of the Foamed Polypropylene. The role of an energy-absorbing structure (i.e., of a foam in a car bumper system) is to absorb mechanical energy and, at the same time, to lower the impact force so that the transmitted forces do not overcome a fixed threshold. In terms of compression stress-strain curve, polypropylene foams are intermediate in behavior between an ideal plastic and an elastic material.

Such a property makes, therefore, this material particularly suitable to play the role of mechanical energy absorber because it has the ability to dissipate a relatively large fraction of the impact energy during a compressive deformation. If, moreover, one considers that the polypropylene deformation is reversible at high percent compression (in contrast to rigid foams like polystyrene where deformation is irreversible at very low percent compression), it is clear why automotive designers have shown a high interest in this material.

The compressive properties of a foam are related not only to the nature of the material, but also to the open cells content and principally, if the other parameters are kept constant, to its density. This is explained by the fact that the resisting force of a foam under a dynamic loading consists of two components: the deformation of the cellular structure, and the air pressure generated either by the flow through the porous structure of the open cells, or by the pressure build-up within the closed cells. Thus, the energy is dissipated not only by the inherent hysteresis of the matrix material, but also by the physical process of forcing air to be compressed or to flow in and out of the foam. Figure (10) illustrates polypropylene foams made by this unique technology and recently used by Fiat in the bumpers of a new, high level coupe car.

Further expansion of this technology is predicted in the area of passive safety, such as side and front absorbers and new dashboard designs.

POLYPROPYLENE STRETCH BLOW MOLDING

With the commercial introduction of Polyethylene Terephthalate (PET) containers in 1977, stretch-blow molding has become a common processing technique in the blow molding industry. In the last few years, however, the European packaging industry has been pushing for a substitution for PET with less expensive materials, primarily polypropylene, in those market niches where barrier properties were not mandatory, such as natural water and detergent bottles. Pushed by end-users, machine producers have started to develop suitable machines for PP stretch-blow molding, based on one-stage or two-stage processes. The main problem observed with PP is the narrow processability window due to its semi-crystalline structure, when compared with amorphous thermoplastic polyester. High flexural rigidity, good transparency and acceptable impact properties are also necessary. The main investigation activities in collaboration with machine producers have been the developments of the correlations between the structural parameters of the polymer and the various operative conditions, in particular, thermal history and orientation phenomena of the different stretch-blow molding technologies.

1) Stretch-Blow Molding Technologies: There are basically two ways to make stretch-blow molding items (the one and two-stage processes, injection stretch-blow molding (ISBM) and the reheat stretch-blow molding (RSBM) processes), and their thermal histories are illustrated in Figure 12.

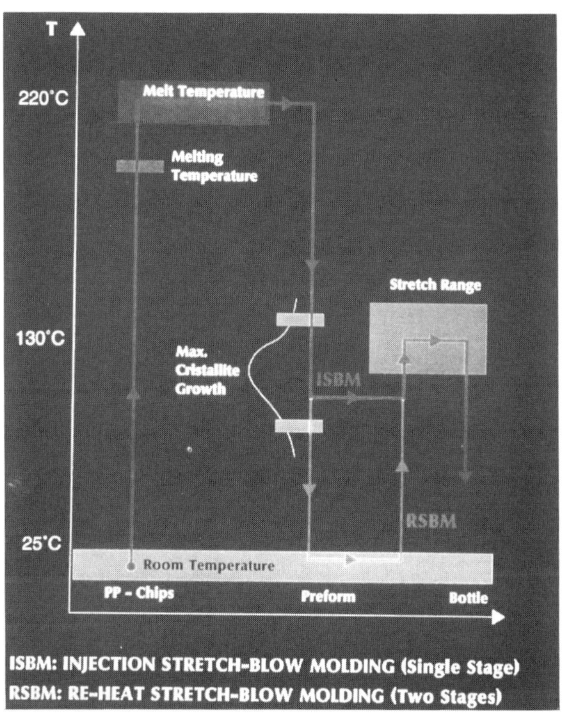

Figure 12. Thermal history of the two different stretch-blow molding techniques.

The production capacity of these machines ranges presently between 360 and 2000 one-liter bottles per hour.

Although both of these technologies involve the stretch-blowing of a preform at a uniform temperature of nearly 130°C, in the case of polypropylene, the thermal history that the polymer undergoes during the heating and equilibration of the RSBM process is completely different from the one of the ISBM process.

In the RSBM process, in fact, the preform tends to pass as quickly as possible through the zone of crystallite growth, while in the ISBM, the preform is kept for quite a long time in a temperature region of maximum spherulite growth. The heating and conditioning stages are also different for the two types of processes, and their relative advantages and disadvantages are summarized below.

INJECTION STRETCH BLOW MOLDING

Advantages:

- Mold relatively easy to change over.
- Relatively low volume production.
- All processes in one machine.
- Fewer blemishes than the two-stage.

Disadvantages:

- Produces a heavier bottle.
- Lower barrier properties.
- When the system is down, total production is down.
- Overall lower efficiency than the two-stage process.
- Requires more skilled operators.

RE-HEAT STRETCH BLOW-MOLDING

Advantages:

- Lower bottle weight.
- Best barrier properties.
- Centralized preform production feeding different blow molding sites.
- Higher flexibility in making or buying preforms.
- Possible planning for cyclical production.
- High productivity.

Disadvantages:

- High cost of investment.
- More blemishes than ISBM.
- Restriction on bottle design (e.g., no ovalized shapes).
- Only narrow neck containers or bottles.

From the economical point of view, the break-even between the two technologies stands in the range of 12 and 14 million bottles per year.

Figure 13. PP stretch-blow molded bottles.

Polypropylene Based Product Development for ISBM and RSBM

The main problem observed when trying to substitute PET with PP is the narrow processability window shown by the semi-crystalline polyolefin when compared to the amorphous thermoplastic polyester. High flexural rigidity, good transparency and acceptable impact properties are absolutely necessary, together with a wide processability window for the acceptance of a PP based polymer by the actual PET consumers. Measurements off the molecular orientation in the bottle wall by birefringence techniques are very useful for developing a better understanding of the influence of the various structural parameters on the final characteristics of the bottle.

Acknowledgments

The authors wish to acknowledge the invaluable contributions of Dr. A. DeNicola, Dr. A. Medri and Dr. D. Romanini to the preparation of this manuscript.

SURFACE MODIFICATION OF SILICONES

Michael J. Owen and Jennifer L. Fritz

Dow Corning Corporation
Midland, MI 48686-0994

INTRODUCTION

The ability to controllably modify the surface properties of a material with treatments that are sufficiently thin not to significantly affect bulk properties has become an important aspect of the development and application of polymers and advanced materials. This is the case with silicone polymers such as polydimethylsiloxane (PDMS). Although PDMS is one of the lowest surface tension polymers in common use, needs exist to lower as well as raise surface energy. Examples of the former type include a variety of low-soiling, anti-deposition coatings and sealants for substrates as diverse as textiles, masonry and solar-energy devices, as well as specialized release applications such as release liners for PDMS-based silicone pressure-sensitive adhesives. The most familiar example of the latter type of need to raise surface energy is to enhance adhesion to PDMS substrates, but there are numerous other needs including biomaterials such as tissue culture surfaces with optimum cell growth characteristics, ocular materials with good tear wettability, and low friction catheter polymers.

Controlled surface property modification is critical in a variety of research areas. A notable example is the demonstration[1] that PDMS has the bulk properties required for use in the Johnson, Kendall and Roberts direct measurement of work of adhesion of polymers[2], without the uncertainties of the usual contact angle approaches to such studies. Surface modification has enabled a variety of surfaces to be characterized in this manner[3], and the technique is proving useful in fundamental studies of wetting, adhesion, friction and adsorption. Surface modification of silicones such as PDMS can be accomplished in a variety of ways, including:

A. Plasma or corona treatment
B. Silane modification
C. Organic monomer/polymer modification
D. Ion beam implantation
E. Deposition of inorganic material
F. Chemical oxidation
G. Radiation treatment (UV, X-rays, γ-rays)
H. Flame treatment

Polymers and Other Advanced Materials: Emerging Technologies and Business Opportunities. Edited by P. N. Prasad *et al.*, Plenum Press, New York, 1995

The list is ranked in order of relevance of the approach to our interests. We first briefly review activity in the important areas of silane modification, organic monomer/polymer modification, and ion-beam implantation followed by a few comments on techniques of lesser importance, and then more fully consider the most important of these techniques, plasma or corona treatments. These two approaches to surface modification are considered together because of the similarity of the modification mechanisms involved.

SILANE MODIFICATION

Silane coupling agents of the type $RSiX_3$, where X is Cl, OMe or OEt, are widely used to promote adhesion between hydroxylated substrates and organic polymers. They can also be used to directly modify the surface of silicone polymers such as PDMS by applying in a solvent such as ethanol which allows swelling and penetration of the monomers into the polymer. Hydrolysis to silanol of the X groups and subsequent condensation will form an interpenetrating network in the surface region. As the range of R groups is wide, considerable variation in surface energy is possible. Values for the critical surface tension of wetting of polymerized layers of coupling agents range from 14 mN/m to 47 mN/m[4] so both lowering and raising PDMS surface energy is possible. Lowering PDMS surface energy requires use of fluoroalkyl-functional silanes[3,6].

A combination of surface modification techniques is often used and one of the best approaches to silane surface modification of PDMS is first to plasma-treat the surface, then to adsorb self-assembling silanes of the structure $Cl_3Si(CH_2)R$. Table 1 shows water contact angle data for several different terminal R groups, illustrating the wide range of wettabilities that can be achieved by this approach.

ORGANIC MONOMER/POLYMER MODIFICATION

Modification of silicones by organic monomers and polymers is a very wide field, not only have many different polymer treatments been investigated, but there are many

Table 1. Water contact angles on $Cl_3Si(CH_2)_nR$ functionalized PDMS

Terminal Group R	n	θ_A deg.	θ_R deg.	Ref.
$-(CF_2)_7CF_3$	2	113	105	3
$-CH_3$	11	112	110	3
-Br	11	90	85	3
$-OCH_3$	11	80	76	3
$-CO_2CH_3$	11	76	73	3
$-S(CH_2)_3(OCH_2CH_2)_3OH$	11	60	--	5
$-(OCH_2CH_2)_3CO_2CH_3$	3	54	--	5

Table 2. Water contact angles on PVP-grafted PDMS

	θ_A deg	θ_R deg
Baszkin et.al.[10]	103	64
	107	66
	90	57
Holly and Owen[11]	70	37
	71	43
This Work	40	20
	30	10

variations in the polymerization technique, e.g. chemical grafting by free-radical or ionic polymerization, radiation grafting or plasma polymerization. The graft-substrate interface is usually much thinner for the plasma polymerization approach than for chemical or radiation surface grafted techniques which can have gradients between graft and bulk extending over hundreds or thousands of Angstroms.[7] Polymer architecture can also vary from surface grafted, homogeneous (bulk) grafted or interpenetrating network. Surface modification can also be achieved through incorporation by blending of random, alternating or block copolymers.[8] Clearly PDMS has many desirable features as a substrate for hydrophilic polymers such as hydrogels judged by the wealth of articles and patents on the topic. A representative example of this interest is the study by Ratner and coworkers[9] on radiation grafted hydroxyethyl methacrylate and acrylamide hydrogels on silicone elastomer. The differences in composition that were evident at air and water interfaces dramatically illustrate the importance of polymer surface dynamics in such systems.

Table 2 shows water contact angle data for various poly(vinyl pyrrolidone) (PVP) treated PDMS samples. The Baszkin et.al.[10] materials are various levels of grafting PVP using a radiochemical procedure, the Holly and Owen[11] samples are bulk grafted commercial materials supplied by Essilor (Paris) and the new substrates were prepared by plasma treatment of the surface, swelling of the surface in monomer solvent solution, followed by a second plasma treatment to polymerize and crosslink the monomer treatment.

ION BEAM IMPLANTATION

Argon ion beams are routinely used in X-ray photoelectron spectroscopy (XPS) to obtain compositional depth profiles. Williams and Davis[12] showed that there was little sputter-induced compositional change in methylsiloxane elastomers and resins during argon ion XPS sputtering. Dwight and Beck[13] similarly showed that silicone rubber was little changed in composition and bonding, although the polymer was reduced by the ion beam, using a NASA argon ion thruster engine.

A broader range of ion beam modification of silicone rubber has been studied by Suzuki et al.[14] H^+, He^+, C^+, N^+, O^+, O_2^+, Ne^+, Na^+, Ar^+ and K^+ were examined at various dosages. Generally, the elements were implanted in a Guassian-like distribution in the silicone polymer matrix but implanted He^+, Ne^+ and Ar^+ could not be detected. Ion implantation broke both Si-O and C-H bonds forming new radicals such as OH, CH_2, SiH,

C=O and amine, depending on the species employed. Water contact angles as low as 50° were reported[15] with the decrease in contact angle immediately after ion implantation being directly proportional to the increase of the mass of the implanted ions. In common with almost all silicone surface modification techniques, these surfaces exhibited aging effects. As time elapsed, the water contact angles increased, approaching the original circa 100° value in 20-30 hours.

OTHER TECHNIQUES

There are numerous studies in the literature of chemical and plasma assisted deposition of inorganic films from such precursors as tetraalkyl orthotitanates and orthosilicates on a variety of polymer substrates that occasionally include silicones. Tetraisopropyl orthotitanate was found in internal studies to be reasonably effective, particularly when used with swelling solvents such as n-hexane.[16] Chemical oxidation can also be effective in raising the surface energy of silicones. For example, soaking in both sulfuric acid and potassium permanganate solution has been found to be a satisfactory[16] treatment of silicone elastomers. In the case of sulfuric acid, a 70% solution at 75°C for 5-30 minutes produces a surface that is fully wetted by water after prolonged soaking without impairing elasticity. Ozone treatment has also been shown to be an effective way of oxidizing the surface of PDMS.[17]

In order for light energy to initiate chemical change it must be absorbed by the molecules of the material in question. Silicones do not absorb near-UV radiation in the 300-400 nm region, the wavelength range in the atmosphere at the ground level. The far-UV and vacuum UV regions are a different matter. UV radiation at wavelengths lower than 290 nm in the presence of air induces the cleavage of methyl radicals from the silicon producing methane, carbon monoxide and carbon dioxide.[18] As only the C-H and Si-C bonds are ruptured, not the Si-O polymer backbone bonds,[19] this is a promising technique of surface modification without too drastic an effect on bulk polymer properties. In fact, UV treatment is a significant part of the effect of plasma treatment, considered more fully in the next section. Esumi and coworkers[20] have quantified the effect of UV radiation on the wettability of silicone polymers. For example, using octane in water and methylene iodide in air contact angle data, they calculate that the surface energy of a PDMS film increases from 16.5 mN/m to 39.8 mN/m after one hour of exposure to low pressure mercury lamp UV irradiation.

The effect of nuclear radiation on PDMS is much more severe than UV radiation. Gamma radiation causes it to crosslink rather than undergo chain scission but its properties are affected by lower levels of radiation than many other elastomers[21] which limits use of this type of approach to silicone surface modification. Its main application is in grafting of organic polymers to silicone substrates.

PLASMA OR CORONA TREATMENT

Plasma modification of polymers is a process of growing importance in polymers and advanced materials today for the control of surface properties. Only a few molecular layers are changed by the plasma, and it offers an environmentally acceptable process without the use of volatile solvents or highly reactive chemical treatments and produces no problematical wastestreams. For these reasons it, and related processes such as corona treatments, are receiving considerable attention academically and industrially.[22] Our primary interests in these two processes are different. We are particularly interested in plasma modification of silicones to enhance adhesion[23] whereas our interest in corona treatment derives mainly from its relevance to the performance of silicone rubber high-voltage insulation[24]. We consider them together as their effects on PDMS surfaces are quite similar.

The effect of plasma treatment on polymers surfaces is complex. Ablation, the removal of surface material including contaminants by molecular scission and evaporation, can occur but so can crosslinking. Surface activation by the substitution of surface atoms with plasma atoms (ions or activated neutral species) also occurs resulting in chemical changes such as oxidation. Plasmas also produce radiation in the visible region (the glow or colored discharge) or in the UV region. This latter radiation can cause significant surface chemical changes to polymer surfaces.

In our studies, exposure to a variety of plasmas (oxygen, nitrogen, helium, and argon) or air corona forms a wettable silica-like layer on the surface of most PDMS elastomers. This is shown by atomic composition and binding energy changes in XPS (X-ray photoelectron spectroscopy) studies. Details are given in the original papers[23,24] but, in general, before treatment the atomic composition closely reflects that of the PDMS polymer: 50% C, 25% O, 25% Si. Fillers such as silica and aluminum trihydrate (in the high-voltage insulator case) are almost never seen in the XPS sampling region, approximately 50 Angstroms thick in the usual, non-angle-resolved configuration. Nor does there seem to be any concentration of catalysts, pigments, contaminants, etc. in this surface region.

On exposure to various plasmas or corona, the oxygen content of the surfaces increases, depending on treatment conditions, up to 50%, with carbon reducing to 25%, and silicon remaining virtually unchanged. Examples of this behavior are shown in Table 3 where both the plasma and corona treatments have been taken to the point where they showed very little change in surface composition with further treatment. The Si 2p peaks broaden on treatment and can be resolved by curve fitting into a peak at 103.5 eV, consistent with a silica-like chemical shift, and another at 101.5 eV, similar to the PDMS silicon in the control (based on a 285.0 eV C 1s peak assignment).

The compositional changes, which suggest partial oxidation perhaps from Me_2SiO to Me(HO)SiO, can be reconciled with the binding energy evidence of more complete oxidation by assuming a mixture of components in the surface, for example, undamaged PDMS units coexisting with oxidized silica-like forms. That this is so, and that the heterogeneity is a function of depth within the normal circa 50 Angstrom sampling depth, is illustrated by the angle-resolved XPS data shown in Table 4. The grazing angle data, which represent the near surface region (circa top 20 Angstroms), suggest a composition corresponding to mildly oxidized PDMS. The normal angle XPS data (signal from as

Table 3. Atomic composition comparison of plasma and corona treatment of PDMS

Treatment	Atomic Composition (%)		
	O	C	Si
Untreated	27.1	50.3	22.6
Oxygen plasma (70W, 70Pa, 5s)	46.5	29.7	23.8
Untreated	26.8	49.2	24.0
Corona (14 kV, 10 min.)	42.0	32.3	25.2

Table 4. Angle-resolved XPS studies

Treatment	Angle of Resolution	Atomic Composition (%) O	N	C	Si
Oxygen Plasma (70W, 70Pa, 30s)	Grazing	32.4	nd	44.1	23.5
	Normal	46.7	nd	30.7	22.6
Argon Plasma (10W, 70Pa,30s)	Grazing	30.1	nd	42.0	27.9
	Normal	44.7	nd	34.1	21.2
Helium Plasma (10W, 70Pa, 30s)	Grazing	29.8	nd	43.4	26.8
	Normal	43.8	nd	32.6	23.6
Nitrogen Plasma (70W,50Pa,30s)	Grazing	28.6	1.2	40.8	29.4
	Normal	48.7	1.9	28.9	20.5
Corona (14kV, 10min.)	Grazing	39.5	nd	35.9	24.6
	Normal	54.6	nd	17.6	27.6

nd = not detected (<0.5%)

deep as 80 to 100 Angstroms), which are more sensitive to the bulk of the material, indicate highly oxidized PDMS. Although we are generalizing the data, there are important differences which should not be overlooked. For example, the corona data imply a deeper oxidation than the plasma treatments. Also, the nitrogen plasma incorporates a small but significant amount of nitrogen into the PDMS surface.

The XPS data can be rationalized by the existence of an outermost layer of untreated PDMS covering a thin silica-like layer, possibly with a deeper region of oxidized carbon species. Some of these states are considerably more oxidized than carbinol, for example, peroxide formation is sometimes evident. Hollahan and Carlson[25] in their original mechanistic study suggested both $SiCH_2OH$ carbinol from the attack of the methyl group by an excited oxygen atom and peroxide formation as the result of singlet oxygen reaction with methyl groups. Peroxide formation on corona-treated, medical-grade silicone sheets has been proven by reaction with 2,2-diphenyl-1-picrylhydrazyl[26].

These mechanistic speculations are consistent with much of the recent literature on the effect of plasma and corona treatments. Triolo and Andrade[27] concluded that a silica-like surface was produced with a helium plasma treatment. They offered no evidence for the presence of silanol groups but suggested oxidized carbon species such as carboxyl and aldehyde groups. A silica surface from an oxygen plasma treatment was earlier suggested by Feneberg and Krekeler[28]. Morra et.al.[29] refute the presence of aldehyde groups on O^{18} modified PDMS; they found surface silanol groups and considerable unaltered $SiCH_3$ groups. All these studies are with radio frequency (RF) plasmas as is our study. Expectedly, microwave plasmas behave rather differently. Urban and coworkers [30,31] (nitrogen and argon microwave plasmas) found carbonyl groups, low MW uncrosslinked PDMS chains and evidence for reduced silicone in the form of SiH groups. Most RF plasma studies claim increased crosslinking in the surface.

HYDROPHOBIC RECOVERY

Although refuted by some researchers[32], it is a common experience that once plasma or corona treatment, and most other surface modification techniques as well, ceases, there is a gradual loss of wettability with time. The possible mechanisms involved in hydrophobic recovery may be:

A. Reorientation of surface hydrophilic groups away from the surface ("overturn" of polar groups in the surface)[33]
B. Surface silanol condensation preventing chain reorientation[34]
C. Migration of untreated polymer (particularly low MW species) through the bulk matrix to the surface[35]
D. Migration of untreated polymer chains through microcracks to the surface[23]
E. Migration of treated polymer chains from the surface to the bulk
F. Loss of volatile oxygen-rich or other polar entities to the atmosphere
G. Changes in surface roughness
H. External contamination of the polymer surface

Several of these mechanisms are likely to be involved in the hydrophobic recovery of PDMS. Mechanism A would be favored by a flexible polymer well above its glass transition temperature. PDMS has one of the lowest glass transition temperatures of any polymer. Its high free volume, virtually free rotation about siloxane bonds, low intermolecular forces between methyl groups, open backbone bond angles and long bond lengths all contribute to a uniquely flexible polymer[36] that should reorientate readily. Mechanism B is specific to silicone polymers and works against chain flexibility. It must be involved in the development of a silica-like network.

The migration mechanisms (C, D, and E) are also plausible candidates for explaining silicone hydrophobic recovery. They are most likely to occur if low molecular weight polymer chains are present. PDMS usually contains low molecular weight cyclic and linear components of lower surface tension than the main polymer matrix[37]. Thus on surface energetic considerations such low molecular weight material should accumulate at the air/polymer surface of both treated and untreated PDMS. Migration through a crosslinked silica-like network should be more difficult than through a PDMS elastomer. However, the silica-like layer is brittle and prone to crack. Such microcracks could provide a facile pathway for migration. They occur in both corona and plasma treated materials. Figure 1 is a tilt angle scanning electron micrograph which give a better impression of the depth of these microcracks than those previously published[23]. It should also be noted that in all cases we have studied, conditions can be found where the treatment is sufficient to produce excellent hydrophilicity with advancing water contact angles close to zero without inducing microcracking.

Presumably, the cracks result from shrinkage in the treated layer as oxidation proceeds, the density of silica being considerably higher than that of PDMS. The effect has been seen in silicone applications as diverse as intraocular lenses[38], where an oxygen plasma was found to be considerably more damaging than a CO_2 plasma, and copolymer coatings for spacecraft applications in low earth orbit[39], where the atomic oxygen plasma has proved to be unexpectedly erosive. Taylor and Wolf[40] were the first to report such cracking. Sample handling can also induce cracking because of the difference in modulus of the treated surface and the underlying polymer. In particular, the XPS measurement environment (UHV, X-rays?) causes cracking. This measurement artifact makes XPS an unsuitable technique for monitoring hydrophobic recovery.

Mechanism F must also play a part in the changes occurring at the surface. Silanol condensation produces water molecules which could be lost to the atmosphere and also in the UHV XPS environment, a further problem with the use of XPS in hydrophobic

Figure 1. Scanning electron micrograph of PDMS exposed to O_2 plasma (70W, 70Pa, 10 min.) at X15,000, 55° tilt angle

recovery studies. Ozone is another volatile oxygen-rich species generated during corona treatment but this probably does not involve the polymer surface. Changes in surface roughness (Mechanism G) are evident in both corona and plasma treatment which limits the usefulness of contact angle as a tool for following hydrophobic recovery. Gubanski and Vlastos[41] have described the wide range of contact angles obtained on actual high-voltage insulator surfaces. The prevalence of microcracking also contributes to uncertainties in much of the plasma contact angle data. The widespread observations of hydrophobic recovery in a variety of controlled laboratory studies make it unlikely that Mechanism H, external contamination, plays a significant role in such experiments, although in actual use, such contamination must be considerable, particularly in high-voltage insulators operating in exposed installations for many years. Morra, Occhiello and Garbassi[42] have commented that the external contamination by ubiquitous hydrocarbons can be ruled out if the rate of aging is increased as the storage temperature is increased. Such a trend is opposite to the one that would be produced by an adsorption-driven aging from the gas phase and points clearly to a thermally activated mechanism. An example of temperature facilitated hydrophobic recovery for our plasma treated materials is shown in Figure 2.

Once cracking occurs, the difficulties associated with obtaining consistent contact angles multiply. Despite this, we have carried out sufficient studies with the same conditions repeated several times to establish certain trends[43].

Hydrophobic recovery occurs with both cracked and uncracked treated surfaces in every case. Thus the explanation of those instances reported elsewhere where hydrophobic recovery did not ensue cannot be that Mechanism C, migration through the

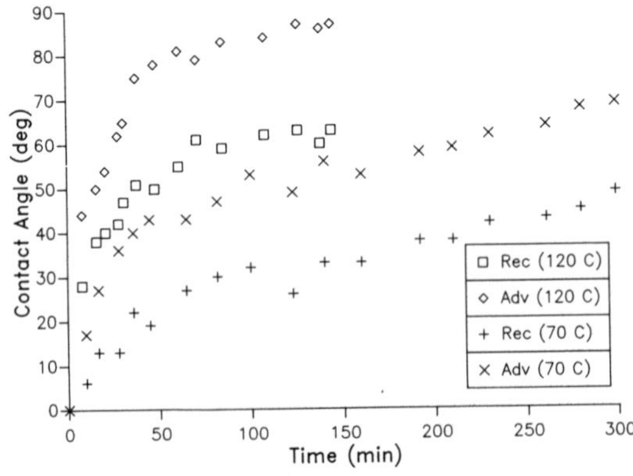

Figure 2. Hydrophobic recovery of plasma treated PDMS at various storage temperatures.

bulk matrix, does not occur. There is an initial jump in contact angle of 10 to 25 deg., depending on plasma gas used, at the onset of cracking. Thereafter the recovery of both cracked and uncracked surfaces broadly parallels each other with virtually complete recovery of original hydrophobicity within one week (water contact angles circa 100 deg.). These effects can be accounted for by rapid diffusion of low molecular weight material out of fresh cracks followed by slower bulk diffusion through the polymer matrix. Significant differences in recovery rates are also evident between different plasma gases. For example, surfaces treated in argon recover more slowly in the first 24 hours than surfaces treated in oxygen. This implies different extents of the oxidized region and the pattern of cracking in each case.

SUMMARY

There is often a need in application of high performance materials such as silicones to modify the surface properties without affecting those of the bulk material. Such needs include both raising the surface free energy to improve wetting or adhesion and lowering it to provide anti-fouling surfaces. We have studied a wide range of possible treatments of polydimethylsiloxane (PDMS) ranging from harsh acid treatment to increase surface energy to incorporation of fluorosilicone-PDMS block copolymers to reduce it. Plasma treatment with various gases is an environmentally benign, cost-effective method of increasing PDMS surface energy. It is also a useful way to prepare the surface for reaction with fluoroalkyl-containing, self-assembling silane monomers which provide surfaces of lower surface energy than PDMS. We have utilized contact angle studies, optical and scanning electron microsopy and X-ray photoelectron spectroscopy to investigate the nature of plasma and corona oxidation, the subsequent hydrophobic recovery that ensues once treatment is discontinued, and the occurrence of microcracks under more aggressive treatment conditions.

ACKNOWLEDGMENTS

Many of our colleagues contributed to the studies reviewed in this paper. We would particularly like to thank Manoj Chaudhury, John Fitzgerald, Pat Smith, Becky Durall, Peter Halm, Tom Lim and Keith Polmanteer.

REFERENCES AND NOTES

1. M.K. Chaudhury and G.M. Whitesides, Langmuir, 7:1013 (1991).
2. K.L. Johnson, K. Kendall and A.D. Roberts, Proc. R. Soc. London, Ser.A. 324:301 (1971).
3. M.K. Chaudhury, *J. Adhesion Sci. Technol.*, 7:669 (1993).
4. M.J. Owen, in "Silicon-Based Polymer Science," Eds. J.M. Zeigler and F.W.G. Fearon, Advances in Chemistry Series 224, American Chemical Society, 1990, p. 720.
5. J.H. Silver, R.W. Hergenrother, J.C. Lin, F. Lim, H.B. Lin, T. Okada, M.K. Chaudhury and S.L. Cooper, submitted to *J. Biomed. Mater. Res.*
6. M.J. Owen and D.E. Williams, J. Adhesion Sci. Technol., 5:307 (1991).

7. B.D. Ratner, in "Surface and Interfacial Aspects of Biomedical Polymers. Vol. 1. Surface Chemistry and Physics," Ed. J.D. Andrade, Plenum Press, New York, 1985, p. 373.
8. M.J. Owen in "Block Copolymers Science and Technology," Ed. D.J. Meier, MMI Press, Symposium Ser. Vol. 3, 1983, p. 129.
9. B.D. Ratner, P.K. Weathersby, A.S. Hoffman, M.A. Kelly and L.H. Scharpen, J. Appl. Polym. Sci., 22:643 (1978).
10. A. Baszkin, M.M. Boissonade, J.E. Proust, S. Tchaliovska, L. Ter-Minassian-Saraga and G. Wajs, J. Bioeng., 2:527 (1978).
11. F.J. Holly and M.J. Owen in "Physicochemical Aspects of Polymer Surfaces," Vol. 2, Ed. K.L. Mittal, Plenum Press, New York, 1983, p. 625.
12. D.E. Williams and L.E. Davis, ACS Div. Org. Coat. Plast. Chem. Preprints, 36:249 (1976).
13. D.W. Dwight and B.R. Beck, ACS Div. Org. Coat. Plast. Chem. Preprints, 40:494 (1979).
14. Y. Suzuki, M. Kusakabe, M. Iwaki and M. Suzuki, Mater. Res. Soc. Symp. Proc. (Interfaces Polym. Met. Ceram.), 153:223 (1989).
15. Y. Suzuki, C. Swapp, M. Kusakabe and M. Iwaki, Nucl. Instr. and Meth., B46:354 (1990).
16. E. Gontko, C.M. Woods, K.E. Polmanteer and W.E. Reister, unpublished Dow Corning Corporation studies.
17. J. Morvan, M. Camelot, P. Zecchini and C. Roques-Carmes, *J.Colloid Interface Sci.*, 97:149 (1984).
18. C.W. Lentz, Ind. Res. and Dev., 22(4):139 (1980).
19. A.D. Delman, M. Landy and B.B. Simms, *J. Polym. Sci.*, A1, 7:3375 (1969).
20. K. Esumi, K. Meguro, A.M. Schwartz and A.C. Zettlemoyer, Bull. Chem. Soc. Jpn., 55:3019 (1982).
21. J.K. Sieron and R. G. Spain, in "Environmental Effects on Polymeric Materials," Vol. 2., Eds. D.V. Rosato and R.T. Schwartz, John Wiley and Sons, New York, 1968, p. 1617.
22. Two recent issues of *J. Adhes. Sci. Technol.*, Vol. 7(10) (1993) and Vol. 8(4) (1994) are devoted to plasma surface modification. Also a recent conference, International Symposium on Polymer Surface Modification: Relevance to Adhesion, Las Vegas, Nevada, November 1993 reported in Vol. 8(10) (1994) and subsequent issues have numerous contributions on plasma and corona treatments.
23. M.J. Owen and P.J. Smith, *J. Adhesion Sci. Technol.*, 8:1063 (1994).
24. P.J. Smith, M. J. Owen, P.H.Holm and G.A. Toskey, Proc. IEEE CEIDP Conference, Victoria, B.C., 1992, p. 829.
25. J.R. Hollahan and G.L. Carlson, *J. Appl. Polym. Sci.*, 14:2499 (1970).
26. T. Okada and Y. Ikada, Makromol. Chem., 192:1705 (1991).
27. P.M.Triolo and J. D. Andrade, *J. Biomed. Mater. Res.*, 17:129 (1983).
28. P. Feneberg and U. Krekeler, U.S. Patent 3,959,105 (1976).
29. M. Morra, E. Occhiello, R. Marola, F. Garbassi, P. Humphrey and D. Johnson, *J. Colloid Interface Sci.*, 137:11 (1990).
30. M.T. Stewart and M.W. Urban, Polym. Mater. Sci. Eng., 59:334 (1988).
31. S. R. Gaboury and M. W. Urban, Polym. Communications, 32(13):390 (1990).
32. This was a topic of considerable debate at the conference described in Ref. 22. The paper by M.A. Hozbor and M. McPherson, "Plasma Promoted Bonding of Silicone Rubbers," is an example where hydrophilic stability was observed.
33. Y. Ikada, T. Matsunaga and M. Suzuki, Nippon Kagaku Kaishi, 6:1079 (1985).

34. C.L. Lee and G.R. Homan, "Silicone Protective Coatings for High Voltage Insulators," in: Annual Report (81CH1668-3) Conference on Electrical Insulation and Dielectric Phenomena, p. 435, IEEE (1981).
35. R.S. Gorur, E.A. Cherney, R. Hackam and T. Orbeck, "The Electrical Performance of Polymeric Insulating Materials Under Accelerated Aging in a Fog Chamber," IEEE Trans. on Power Delivery, 3(3):1157 (1988).
36. M.J. Owen, in "Siloxane Polymers," Eds. S.J. Clarson and J.A. Semlyen, Prentice-Hall, Englewood Cliffs, NJ, 1993, p. 309.
37. M.J. Owen, T.M. Gentle, T. Orbeck and D.E. Williams, In "Polymer Surface Dynamics," Ed. J.D. Andrade, Plenum Press, New York, 1988, p. 101.
38. H.J. Hettlich, F. Otterbach, Ch. Mittermayer, R. Kaufmann and D. Klee, Biomaterials, 12:521 (1991).
39. S.K. Rutledge and J.A. Mihelcic, in "Mater. Degrad. Low Earth Orbit, Proc. Symp.," Eds. V. Srinivasan and B.A. Banks, Miner. Met. Mater.Soc., Warrendale, PA, 1991, p.35.
40. G.N. Taylor and T.M. Wolf, *Polym. Eng. Sci.*, 20:1087 (1980).
41. S.M. Gubanski and A.E. Vlastos, "Wettability of Naturally Aged Silicone and EPDM Composite Insulators," IEEE Trans. on Power Delivery, 5(3):1527 (1990).
42. M.Morra, E. Occhiello and F. Garbassi, Polym-Solid Interfaces, Proc. Int. Conf. 1st, Namur, Belgium, Section IV, 407 (1992).
43. J.L. Fritz and M.J. Owen, in press, *J. Adhesion*.

DESIGN OF SIDE-CHAIN LIQUID CRYSTALLINE POLYMERS

Yusuke Kawakami

Graduate School
Japan Advanced Institute of Science and Technology
Asahidai15, Tatsunokuchi, Ishikawa 923-12 Japan

INTRODUCTION

Much attention has been recently given to side-chain liquid crystalline polymers not only because of scientific interests but also because of their potential applications for electronic devices. Extensive researches have been carried out to combine the properties of polymers and electrooptic property of the low molecular mass mesogen. Polysiloxanes and polyacrylates are usually used as main chain components[1-3]. There are only limited numbers of examples in which other main chain structures like polyolefin, poly(vinyl ether), polyphosphazene, or polyisocyanate are used. When an application for optical shutter or display material is considered, low refractive index of the polymer is important. For this purposes, polystyrene is of interest as a main chain component of a side-chain liquid crystalline polymer.

It is generally considered that the spacer groups are acting as flexible connecting groups between main chain and mesogenic groups which make it easy for the mesogenic groups to aggregate and to be organized into liquid crystalline state. This situation can be said to fit in the case of a polymethacrylate main chain, whose Tg is as high as 100°C, with more flexible oligomethylenes as spacer groups. In the cases of polysiloxane main chain polymers, the situation is a little different. Namely, the main chain itself is very flexible(Tg< -100°C), and can be said even more flexible than the spacer groups.

We have been interested in the effects of the chemical structure of main chain, the length and the structure of the spacers, and the structure of mesogenic group on the liquid crystalline phase exhibited by the side-chain liquid crystalline polymers. As can be seen that polysiloxane is used as a flexible main chain component in side-chain liquid crystalline polymers, the siloxane linkage could be used as a flexible component in the spacer by taking advantage of its low rotational energy barrier. In the reported studies, the siloxane linkage is used in the middle of the alkylene spacer group[4, 5]. This seems to be because the steric requirement of dimethyl-substituted siloxane linkage interferes the

Polymers and Other Advanced Materials: Emerging Technologies and Business Opportunities. Edited by P. N. Prasad *et al.*, Plenum Press, New York, 1995

aggregation of mesogenic groups to form mesophase if the dimethyl-substituted silyl group is introduced close to the main chain of vinyl polymers.

In this study, siloxane linkage is selected as a constituent of spacer group. Although, as already mentioned, siloxane linkage is not usually considered as a good spacer in vinyl type side-chain liquid crystalline polymers because of its bulkiness in spite of its flexibility, it should act as a good spacer if the steric hindrance in the case of the vinyl polymers could be avoided. Substituted biphenyl mesogenic groups were attached, through spacer having disiloxane linkages which are directly attached to the main chain, in every four carbons in polydiene main chain, not every other carbon as in the case of vinyl polymers. Steric hindrance can be avoided by this molecular design[6,7].

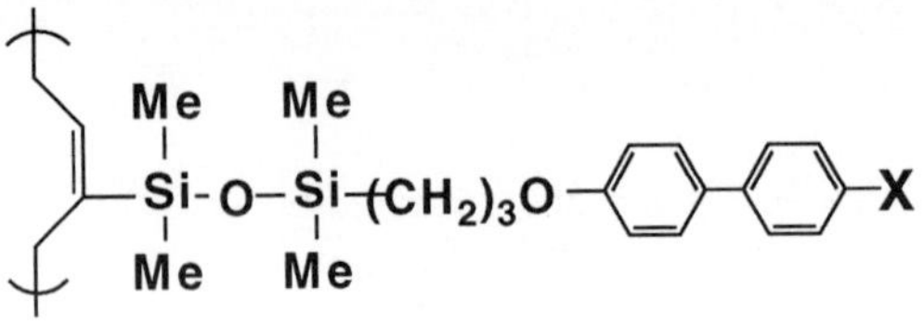

PolyB-X-S6 X = OR, CN, F

In order to study, specially, the effects of spacer group on mesophase structure, polymers having not only oligomethylene spacer containing disiloxane but also hexamethylene spacer or oligomethylene spacer containing ether linkage were synthesized, and their thermal behavior was also studied.

Polystyrene is a candidate of low refractive main chain of a side-chain liquid crystalline polymers. The siloxane segment is introduced in the spacer, not only from the point of flexibility but also from its incompatibility with other polymer segments.

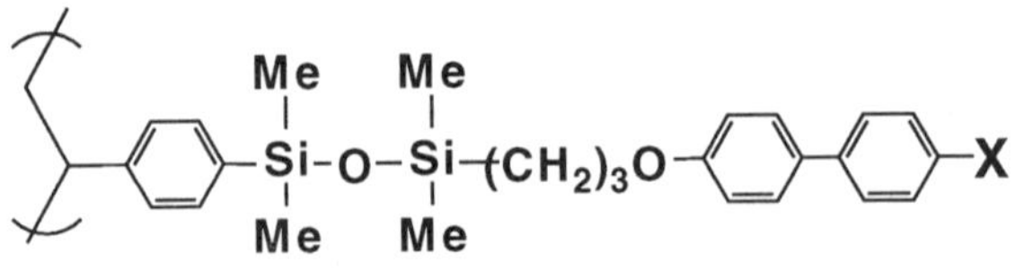

PolyS-X-S6 X = OR, CN, F

It has been recognized that polystyrenes are synthesized by various methods. Molecular weight, polydispersity, and tacticity of polystyrenes can be controlled by precise polymerization such as living and coordination polymerization. In this study, the effect of molecular weight and its distrobution of the polymer on the liquid crystalline behavior is reported.

EXPERIMENTAL

Details of the synthesis of polydiene derivatives were reported[6,7]. Those of polystyrene derivatives will be reported soon, elsewhere. Only the eaction scheme of the synthesis of monomers is given.

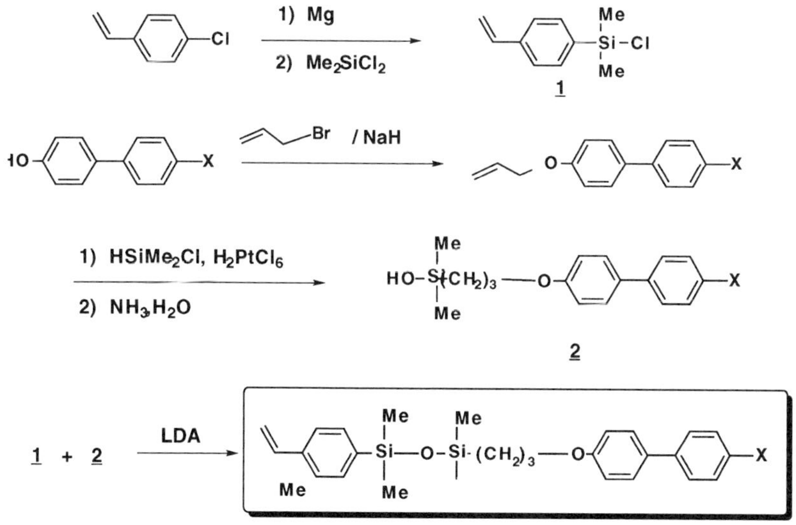

Synthetic Route to Styrene Monomers

RESULTS AND DISCUSSION

Polymer Structure

There are four structural isomers, namely, cis-1,4-, trans-1,4-, 1,2-, 3,4- structures in the polydiene main chain having a substituent at 2-position. In the case of polydiene with siloxane linkage in the spacer, cis-1,4- and trans-1,4- structure correspond to E-1,4- and Z-1,4 structure, respectively. By proton NMR it can be seen that the polymer was formed by the consumption of diene function also in the case of PB-CN-6. The newly seen signals in the region of 5.88 ppm as a singlet correspond to the olefinic proton of the main chain. The area of the singlet is one-fourth of that of methylene protons of the main chain. This fact excludes the possibility that polymers were formed through 3,4- or 1,2-addition to the diene functions. With respect to the structure of the PB-X-S6's, it can be said that only one isomeric structure is predominant. However, the mode of addition, namely, 1,4-trans or 1,4-cis was not clear at this point.

The DEPT spectrum supports that the signal at 5.88 ppm is that of CH, which is consistent with the assignment of the signal to trisubstituted olefinic proton. Interaction between methyl protons of trimethylsilyl group and the olefinic proton was observed in the NOESY spectrum the polymer, which gives a support that the olefinic proton and trimethylsilyl groups are situated at cis relation concerning the carbon-carbon double bond. This cross relation was not observed in COSY spectrum. The polymer main chain takes one predominant stereochemistry, namely, E-1,4-. This stereochemistry was also suggested by Nakahama[8]. This may be because that the bulky mesogen-containing substituent tends to be incorporated at trans position to polymer main-chain(E form) in repeating monomer unit. Similarly the main-chain structure of other PB-X-S6's is also concluded to take E-1,4. The stereochemistry is generally seen for other polydienes with siloxane linkage in the spacer.

Thermal behavior of monomers and polymers

Transition temperatures of diene monomers are shown in Table 1.

Table 1. Transition temperatures of diene monomers

Monomer	DSC(heating) Transition(°C);ΔH(mJ/mg)
B-CN-6	53.6;120.3
B-F-6	30.8;68.9, 49.5;29.3
B-1-6	84.5;138.7(broad), 118.7;1.9
B-6-6	91.5;21.6, 115.0;23.6(broad)
B-CN-S6	35.38;26.3
B-F-S6	oil
B-1-S6	72.5;30.7
B-6-S6	41.5;12.5

Diene monomers with S6 spacer except for B-F-S6 showed only one transition. These monomers did not show any liquid crystalline state. Monomers with hexamethylene spacer except for B-CN-6 showed two transition temperatures. Further accurate thermal analyses are needed.

The results of DSC analysis of polymers are shown in Table 2.

Table 2. Thermal behavior of polymers

Polymer	DSC(heating) Transition(°C)[a]	Texture
PB-CN-6	137.6	fan
PB-F-6	114.5	n.d.[b]
PB-1-6	133.0	batonnet
PB-6-6	146.5, 151.6	batonnet
PB-CN-S6	57.4	fan
PB-F-S6	51.9, 73.4	fan
PB-1-S6	116.3	n.d.[b]
PB-6-S6	48.4, 107.8, 110.3	fan

a) Determined by DSC at the starting point of transition.

b) n.d. : Phase could not be determined.

Polymers with hexamethylene and S6 spacers except for PB-6-6, PB-F-S6, and PB-6-S6 showed single transition temperature. PB-6-6, PB-F-S6, and PB-6-S6 showed two transition temperatures. Roughly speaking, the transition temperatures of polymers having the same spacer become higher in the order of fluoro, cyano, methoxy and hexyloxy group as the tail group. Although it is very difficult to see a consistent trend, it can be said that hexyloxy group is a good tail group to stabilize the mesophase to result in the higher transition temperature.

We can see the importance of the chemical structure, namely the flexibility of the main chain in determining the transition temperature(clearing point) by comparing the transition temperatures of the diene polymer having hexamethylene spacer and cyanobiphenyl as a mesogen(PB-CN-6) with those of ordinary liquid crystalline polymers having same spacer and mesogen. The glass transition temperature and clearing point of polymers of polysiloxane, polyacrylate, and polymethacrylate as main-chain are (12°C; 166°C), (35°C; 135°C), and (60°C;100°C), respectively[9]. The more flexible main chain gives the wider temperature range of liquid crystalline state. Polydiene polymer, PB-CN-6, showed the clearing point at 137.6°C. This temperature is between that of polysiloxane and polyacrylate main-chain polymers. This fact is considered to reflect that the flexibility of polydiene main-chain is between that of polysiloxane and polyacrylate. Although the glass transition temperature was not observed, PB-CN-6 is assumed to show considerably wide temperature range of liquid crystalline state.

When polymers of different spacer group with hexyloxy tail group are compared, interesting trend is noticed; namely, transition temperature drops in the order of hexamethylene, ether containing spacers, and disiloxane containing spacer. Introduction of disiloxane linkage in the spacer is very effective to bring down the transition temperature.

Only the transition of polydiene side-chain liquid crystalline polymers having the spacer with disiloxane linkage was enantiotropic. This may reflect the importance of the flexibility of the main-chain and spacer of these polymers in making the mesogens to be easily organized. The transition temperatures of the polymers having the spacer with disiloxane linkage ranges from the room temperature to around 115°C. This might be an advantage when an application is considered.

Yield and thermal transition of styrene type monomers is shown in Table 3. The transition temperatures are given at the points where transition start determined by DSC in second heating process.

Table 3. Yield and thermal transition of styrene-type monomers

X	Yield (%)	Transition temperature (°C)[a)]	Texture (°C)[b)]
OMe	49.6	44.5; 75.6	batonnet,fan (70.0)
OBu	70.0	48.0; 78.5	fan (67.5)
OHex	16.0	28.6; 49.6	fan (44.8)
CN	13.2	-49.3; -36.7	-
F	28.5	-11.3; 39.5	fan (70.0)

Phase Structureof the Polymers under Optical Polarization Micrograph

As mentioned in the preceding section, diene monomers with hexamethylene spacer except for B-CN-6 showed two transition temperatures. Judging from the enthalpy of the transition, there is a possibility for these monomers to take liquid crystalline state. However, liquid crystalline state could not be seen, so far, under optical polarization micrograph. No monomer was concluded to take liquid crystalline state by optical polarization micrograph.

Chemical structure of the main chain polymer is an important factor to determine the liquid crystalline state. The flexible polydiene main chain made it possible for these polymers to take liquid crystalline phase. Polymers with hexamethylene spacer(PB-Z-6) and those with spacer containing siloxane linkage(PB-Z-S6) took smectic liquid crystalline state. The batonnet structure taken by both PB-6-6 and PB-6-S6 grew up to a well-developed focal conic fan structure only in the case of PB-6-S6. PB-6-S6 showed two transition temperatures at 107.8°C and 110.3°C. Although it could not be elucidated by now, there is a possibility for this polymer to take two different liquid crystalline phase between these temperatures. A trend was observed that the mesophase structure of the polymers with disiloxane linkage in the spacer developed better compared to that of the polymers with hexamethylene spacer. The structure of spacer group is one of the important factors to determine the liquid crystalline state.

The structure of mosogenic group is also one of the important factor in determining the mesophase structure. Although PB-1-S6 immediately forms fine structure on cooling near the transition temperature, the phase could not be determined. PB-1-6 gave only batonnet structure, and the structure was not developed further. Apparently, methoxybiphenyl is not a good mesogenic group for polydiene to take well-organized structure. Contrary to the fact that polymers with methoxybiphenyl as a mesogen did not give a well-developed mesophase, cyanobiphenyl as a mesogen usually gives better-developed mesophase. Optical polarization micrograph of PB-CN-S6 was already reported[6]. That of PB-F-6, those of PB-Z-6 and change of those of PB-6-S6 with time. The smectic A phase of PB-6-S6 shown by annealing at the temperature slightly below the main transition temperature(108.5°C) turned into unidentified fine structure on further cooling. In accordance with the generally admitted fact that side-chain liquid crystalline polymers with polar cyanobiphenyl mesogenic group gives well organized smectic phase, so does polydiene with cyanobiphenyl as the mesogenic group(PB-CN-6, PB-CN-S6). The polar interaction between mesogenic groups makes it easy for them to organize. Polymer with fluorobiphenyl and siloxane containing spacer also showed similar smectic phase with focal conic fan structure(PB-F-S6).

The reason for the formation of the well-developed smectic structure of these polymers is considered not only because of the polar tail group, but also because of the flexibility of the polydiene main-chain, and the very mobile disiloxane in the spacer which also spaces the mosogenic groups apart from each other along the polymer main-chain, compared with ordinary vinyl polymers.

Polystyrene main chain also can give rise to liquid crystalline phase. The effects of molecular weight and its distribution are shown in Figures 1 and 2.

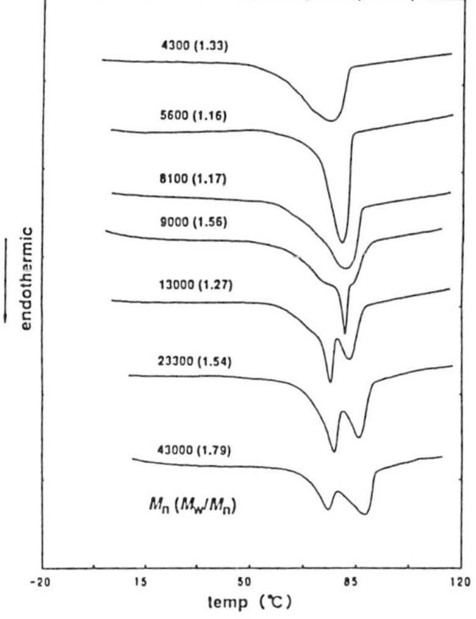

Figure 1. Influence of the degree of polymerization

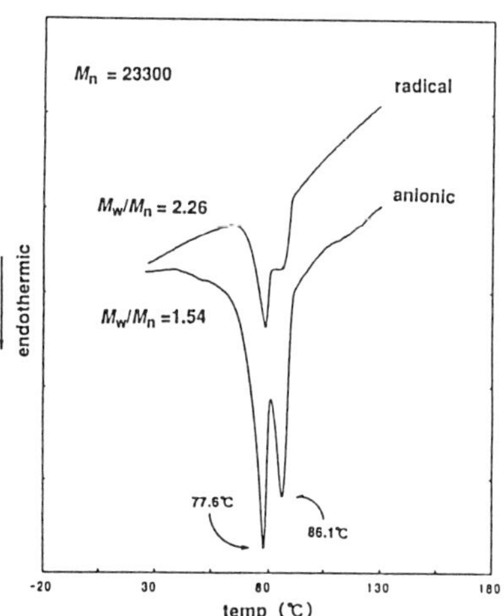

Figure 2. Influence of the moleecular weight distribution

Certain molecular weight seems to be necessary for the polymers to take liquid crystalline state. Polymers with narrower molecular weight distribution gave sharp transition temperature.

CONCLUSION

Polydiene and polystyrene were found to act as suitable main chain components for a side-chain liquid crystalline polymer with various biphenyl derivatives as the mesogenic group. Introduction of siloxane linkage into the spacer group is an effective molecular design of the chemical structure of the polymer to give better organized mesophase of lower transition temperature. Importance of the flexibility of the spacer was stressed.

ACKNOWLEDGMENT

Financial supports from a Grant-in-Aid for Scientific Research(07246219, 07651078) are gratefully acknowledged.

REFERENCES

(1) N. A. Plate, V. P. Shibaev Ed.; Comb-Shaped Polymers and Liquid Crystals : Plenum: New York(1987).

(2) C. B.McArdle Ed.; Side Chain Liquid Crystal Polymers : Chapman and Hall: New York(1989).

(3) A. A.Collyer Ed.; Liquid Crystal Polymers: From Structures to Applications: Elsevir; Barking(1992).

(4) M. Engel, B. Hisgen, R. Keller, W. Kreuder, B. Reck, H. Ringsdorf, H.-W. Schmidt, P. Tschirner, *Pure & Appl. Chem.*, 57: 1009 (1985).

(5) Y. Nagase, Y. Takamura, *Makromol. Chem.*,193, 1225 (1992).

(6) Y. Kawakami, K. Toida, Y. Ito, *Macromolecules*, 26: 1177(1993).

(7) Y. Kawakami, K. Toida, *Macromolecules*, 28: 816(1995).

(8) K. Takenaka, T. Hattori, A. Hirao, S. Nakahama, *Macromolecules*, 22:1563(1989).

POLYMERS IN SPACE ENVIRONMENTS

V.N. Krishnamurthy

Vikram Sarabhai Space Centre
Trivandrum 695022

Polymers are used in a number of applications in the space systems - space craft, space station - like thermal control coatings, adhesives, tapes, potting compounds, toughening/damping materials, seals, thin film substrates, thermal insulations in addition to composite structures. Space research demands very stringent requirements from these polymers because of the extreme environments experienced by these materials[1]. The polymers used outside the space system are the most affected ones. As we move from space exploration to space commercialization, these polymers have to last for 15-20 years of their designed life. Operational requirement needs not only these materials do their job but also have low outgassing to prevent contamination of surrounding optical systems. Hence, there is a need for understanding the effect of space environments on the polymeric materials in use and improvement/synthesis of new polymeric materials.

The space environment varies with the altitude/orbit in which the space craft is moving and is given in Table 1.

Table 1. A comparison of Geosynchronous and Low Earth Orbital environments.

Material Degrading environments	Geosynchronous (GEO)	Low Earth (LEO)
Altitude (Kms)	35,900	upto 700
Vacuum (Torr)	$>10^{-12}$	10^{-7}-10^{-9}
Atomic Oxygen	No	Yes
UV radiation	Yes	No
Micrometeoroids Debris	Minor	Major
Electron/Proton bombardment	Yes, Major	Yes
Thermal cycling (°C)	-190 TO +150	±125
Diurnal temp.cycle	24 hrs	90 minutes

The low earth orbit (LEO) which extends from earth's atmosphere to 700 kms, experiences vacuum, atomic oxygen, ultraviolet (uv) and charge particle radiations, micrometeorides and thermal cycling while the geostationary earth orbit (GEO) experiences similar environments except atomic oxygen. The general damaging effects of these environments on the polymers is summarised in Table 2. It can be seen that ultraviolet and charged particle radiations cause chain scission and cross-linking depending on

Polymers and Other Advanced Materials: Emerging Technologies and Business Opportunities. Edited by P. N. Prasad *et al.*, Plenum Press, New York, 1995

dosage while vacuum accelerates the diffusion of low molecular weights and additives. The mechanical properties degrade by thermal cycling. Atomic oxygen oxidizes the material and alters the surface properties. The combined interaction of the environments cause more severe damages to the polymers than the individual environments as is present in space.

Table 2. Some Damaging effects of space environment upon polymeric materials.

Environment	Damage
Ultraviolet radiations	——Creation of lattice defects in crystalline materials ——Chain scission of organic materials, free radical formation ——Crosslinking of organic materials
Charged particle radiation (electron/proton)	——Creation of lattice defects in crystalline materials ——Chain scission of organic materials ——Crosslinking of organic materials ——Secondary radiation damages
Vacuum	——Volatilization of low vapor pressure fractions and materials ——Diffusion ——Vacuum welding
Thermal	——Mechanical degradation, softening or embrittlement ——Chemical degradation ——Acceleration/deceleration of above environmental effects
Micrometeoroid	——Mechanical ——Fracture/Puncture
Atomic Oxygen	——Oxidation ——Surface erosion ——Cracking/crazing

Initially, the candidate materials are evaluated in the laboratory. For example, to study the effect of debris and micrometeorides impact on coatings, drop weight impact test as well as bullet test are conducted and the surface was observed for cracks and pin holes after the test by acoustic emission technique. NASA evaluates the materials after satisfactory laboratory tests in space environment by flying them in space shuttle for short durations upto 2 weeks. To evaluate long term exposure, a long duration exposure facility was built and launched abroad the shuttle challenger in April 1984. After six years of space exposure, the facility was recovered and the materials characterised for physical, chemical, mechanical and optical properties[2]. These data are being used in designing future systems to withstand long-term exposure to space environment.

VACUUM INDUCED CONTAMINATION

The atmospheric pressure decreases with increase in altitude. As seen earlier, it promotes migration of volatiles and degraded products. The impurities that are released are the catalyst, unreacted monomers, plasticizers, degraded products and materials adsorbed on the surface like finger oils, water, lubricants etc. During the service period in high vacuum environment, the volatiles evaporate, migrate through the vacuum and eventually settle and fixate in the cold regions causing malfunctioning of sensitive systems such as solar panels, sensors, optical components etc. Hence, it is crucial to have polymeric materials with low volatile content to retain reliable functioning of all sensitive components and systems.

The polymeric materials are evaluated for volatile matter as per ASTM Standard E-595-7 where in the polymer is heated at 125°C for 24 hrs under 10^{-6} torr vacuum and the volatiles collected at 25°C[3]. The total mass loss (TML) and collected volatile condensable material (CVCM) are calculated from the weight loss and weight gain respectively. The requirement for space application is 0.1% max for CVCM and 1% max for TML. The outgassing of some polymers used in space systems are given in Table 3.

Table 3. Outgassing data of some polymers.

Material	Use	%TML+	%CVCM++
Acrylic sheet	Structure	0.51	0.05
Kapton PEP film	Insulation	0.31	0.18
Kapton H 5 (Mil)	Insulation	0.77	0.02
Polyethylene	Film	0.41	0.16
Teflon	Insulation	0.02	0.00
Mylar	Insulation	0.59	0.00
Kapton shielded polyimide	Insulation	0.44	0.01
Buna-N	0-ring	14.74	6.25
Viton	0-ring	0.10	0.00
Fibre optic cable cc-100	Optical	0.27	0.07

+ %TML = Percentage total mass loss ; ++ %CVCM = Percentage collected volatile condensable material

A certain amount of correlation exists between the polymer structure and the temperature of thermal break down in vacuum. The bond strengths of the structure rank approximately as

$$\mathrm{C{-}F} > \mathrm{C{-}H} > \mathrm{C{-}Cl} > \mathrm{C{-}\overset{|}{\underset{|}{C}}{-}C} > \mathrm{C{-}\overset{|}{\underset{\underset{C}{|}}{C}}{-}C}$$

Hydrogen attached to tertiary and/or quaternary carbon atoms are less stable compared to those on secondary or primary carbons. Fluorine substitution for hydrogen tends to stabilize the chain but hydrogen and fluorine on adjacent carbon atoms are likely to go off as HF reducing the stability.

Loss of constituents from polymers in vacuum also causes degradation in physical, electrical and/or mechanical properties especially if the loss is more than 2%. Another concern is the toxicological effect of the volatile by-products (which contribute to the cabin atmosphere) on the occupants of the cabin in manned space missions. It is therefore essential to know in all applications the long term effect of vacuum by conducting accelerated tests. For example, a material which lost 0.1% of its weight in 2 weeks at 130°C could be specified to lose no more than this amount in one year at 100°C [4]. The vacuum is synergistic with material degradation by atomic oxygen, ultra violet and ionizing radiations, in the sense that the degraded products formed can migrate under vacuum easily thereby hastening the process.

ULTRAVIOLET (uv) RADIATIONS

Next to atomic oxygen, uv radiation is the most damaging space environment. The uv dosage is the same for LEO as well as GEO and is more in space than on ground due to the absence of ozone layer in space, which absorbs these radiations. In space, the polymers receive radiation doses of 10^{12}-10^{15} erg/g/year from the short wave length of sunlight which is well above the damage threshold of most polymers ie. 10^8-10^{10} erg/g/year[5]. The majority of the damages is on the surface of the polymer which primarily affects the optical properties of the polymer, that is, an increase in solar absorptivity (α). The change in optical properties of thermal control coatings and films can cause the spacecraft to heat up and fail.

Besides affecting optical properties, uv radiations also causes discoloration and degradation in mechanical, physical and electrical properties of the polymeric materials due to chain scission or crosslinking[6]. Films of commercial phenyl methyl silicone, PETF, PVC, PMMA etc underwent crosslinking on exposure to space sun light in few days[7]. Teflon films flown in LDEF showed 30% reduction in mechanical properties[8]. Addition of stabilisers improved the time to damage in vacuum by a factor of 3 to 10 times but their choice is limited due to vacuum effect. Addition of pigments to paints retard the damages.

THERMAL CYCLING

In the ultimate sense, the thermal balance of the space system is determined by the exterior surface ie. the surface decides whether the heat is to be removed or added so that the temperature remains constant and the electronic components inside function normally. The key ratio α/ϵ of the polymeric coating decides the choice. Thermal cycling causes crazing, cracking and delaminations of structural components. It can also produce stresses in materials resulting in mechanical defects or failures. It is observed that thermoplastic composites are better suited than thermosets probably because of better interaction between fibre/matrix[9]. Special thermal control coating have been developed to reduce the temperature variations[10].

IONIZING RADIATIONS

High energy protons and electrons are trapped by the geomagnetic field of the earth in the inner and outer VAN ALLEN radiation belts. LEO spacecrafts are below the inner belt but GEO space crafts are at the fringe of the outer belt. During solar flares, solar winds push the trapped electrons and protons farther out thereby affecting polymers in these orbits. The energy of these electrons ranges from Kev (10^3ev) to several Mev (10^6EV) and that of the protons varies approximately with the square root of atomic number of the material. These high energy radiations affect the surface as well as the bulk properties of the polymeric materials. They promote ionization of molecules which promote either chain scission or cross linking similar to the effects of uv radiation. The radiation effects on the mechanical, electrical, optical and other properties is very similar to that caused by neutrons and γ-radiations in nuclear reactor studies[11].

Empirical correlation between polymer structure and radiative stability are known in literature[12]. Presence of aromatic group in the side chain of polyethylene increases the stability while the presence of phenylene or ether or sulfide linkages in the main chain decreases the stability of the polymer. Radiative stability increases with increasing heat of polymerization, rigidity of the polymer chain and increased reinforcement content. Anti rads have been developed to improve the radiative stability. The stability of rubber is improved upto 10 times by the addition of anti rads. All polymers are not affected equally by radiation. Besides, permanent radiation damages, temporary electrical conductivity may also be induced in polymers on exposure to radiation[13]. Electrons because of its smaller mass and increased mobility strike the spacecraft surface more than protons resulting in spacecraft surface developing negative charges. In some cases, the negative charge build up (can be more than 10 Kev) until there is discharge as the space craft is covered with insulative polymeric materials having different dielectric constants[14].

Space craft charging is also suspected of increasing the contaminants rate. Satellite data[15] indicate that outgassed molecules ionized by the sun are attracted back to the negatively charged space craft. Before these experiments, it was thought that contaminants could only deposit on the surface that were in the line of sight of contaminate.

ATOMIC OXYGEN (AO)

Atomic oxygen with an energy level of about 5 ev are produced by the photo dissociation of molecular oxygen, predominates in the atmosphere upto 700 kms [16]. The atomic oxygen flux is 10^{13}-10^{16} atoms/cm^2 sec in LEO, Though all organic polymers react at about the same rate 2-4 cm^3/atom of atomic oxygen, fluoro polymers and silicones are less reactive as seen from Table 4. In case of teflon , the slow reactivity is attributed to the higher bond strength of C-F bond and the endothermic nature of the oxidation reaction. Silicone forms a protective coating of glassy silica layer on reaction with atomic oxygen. Atomic oxygen causes surface erosion of polymers thereby changing the surface morphology, optical, thermal and mechanical properties in addition to undergoing mass loss. Kapton thermal blankets were completely eroded away by atomic oxygen and Teflon blankets became cloudy due to surface erosion[17]. These polymers are protected by coatings which can be organic or inorganic. Inorganic oxide coatings are brittle and crack in addition to difficulties in its application. Organic coatings are easy to apply and flexible. Different coatings have been tried with varied success. The coatings tried include silicones, siloxane imides and epoxies [18], polycarboranes[19], polycarborane siloxanes[20], decaborane[21] based preceramic polymers and boron based polymers. Decaborane based polymers on oxidation shows little change in volume unlike silicones coatings due to change in volume.

Table 4. Reactive efficiency of some polymers with Atomic Oxygen

Material	Reactive Efficiency cm^3/atoms x 10^{24}
Polyimide	3.0
Polyester (Mylar)	3.4
Polyethylene	3.7
Polysulfone	2.4
Polystyrene	1.7
25% polysiloxane/75% polyimide	1.3
Teflon	<0.05
Silicone	<0.05

The major draw back of the proposed coatings is the degradation caused by the synergistic interactions of the other space environments. For example Teflon embrittle and discolor on exposure to uv radiation, crack and craze during thermal cycling can lose its protectiveness for atomic oxygen. Atomic oxygen problem is the critical one for space station materials and is yet to be fully solved. Coatings may to be periodically refurbished in orbit to compensate for the space environments.

MACROMETEOROIDS AND SPACE DEBRIS

Collision with micrometeoride or space debris is another hazard of the space environment. Even small particles can damage the space craft because of high impact velocities of about 16 km/sec[22]. In many cases, the protective coatings were penetrated by micrometeorides or debris, exposing the substrate material to the attack of atomic oxygen and ultraviolet radiations. The probability of being struck depends on the duration of the mission, exposed area and particle size of the debris.

COMBINED EFFECTS

The combined effect of all the environments tended to accelerate the degradation of many of the materials as seen in the long duration exposure facility (LDEF) experiments. Although many of the materials on LDEF suffered some level of damage from space environment, several materials survived the six year exposure with little or no degradation. Fibre reinforced aluminium composites, materials protected by preceramic coatings and several types of thermal control paints displayed excellent resistance to the effects of space environment. The data generated by LDEF will be useful in designing future systems to withstand long-term exposure to the space environment.

The choice of materials for a prescribed mission is based on laboratory simulation studies, previous and present spacecraft data and theoretical considerations of inherent material properties. The duration and position in space must also be taken into account for design. Well directed materials research and development is the need of the hour.

REFERENCES

1. Ray A. Cull, "Elastomers", p.23 (1989)
2. R.C.Tennyson, "Composite materials in space: Results from LDEF satellite " Proceedings of Canadian Space agency(6), Canada, (1990).
3. M.France Rose. "IEEE Tansaction on Electrical Translations" Vol. E1-22, p 555 (1987)
4. J.A.Thorne and C.C.Whipple, "11th National Symposium on the effect of space environments on materials" p.243, St.Louis (1967)
5. J.Mc Keller and N.S.Allen, "Photochemistry of man made Polymers", Applied Science Publications, London (1979)
6. L.D. Jaffe, "Chemical Engineering Progress Symposium Science",Vol59, p.81 (1963)
7. J.E. Cousling, A. Alexander, F.Noonam, R.Kagarise and S.Stokes. "First Symposium on Surface effects on spacecraft materials", p.364, New York (1960)
8. R.R. Kamenetzky, "AIAA Space Programs and technologies conference and exhibit Hunts ville,HL", AIAA-93-4103, Sept 21-23, (1993).
9. E.Bowles and J.W. Connel, "International Astronautical Federation Congress". IAF-92-0321, Washinton D.C. USA. August 28-Sept 5, (1992)
10. Dr Tenny, S.S.Trompkins and G.F.Sykes, "Large Space Antenna Systems Technology Conference", NASA Langley Research Centre (1984)
11. F.A.Borey. "Effects of Ionizing Radiation on Natural and Synthetic High Polymers", Interscience Publishers, New York (1958)
12. P.Y.Feng and J.W.Kennedy. "Journal Amer.Chem.Society" Vol 77. P.847 (1958)
13. T.Strganac,"AIAA Structures Structural dynamics and Materials " , AIAA/ASME/ASCE/AHS/ASC, AIAA-93-1577, Conference on (34) Lajolla, CA USA. April 19-22 (1993).
14. R.L.Hameel. "AIAA Annual Structures and Materials Conference" Palm Springs, California (1964)
15. HKA Kan, "Space Environment effects on Space craft Surface Materials" Aerospace Corporation Report. (1980)
16. C.K.Purvis, "NASA/SDIO Space Environmental Effects on Materials Workshop",Part I, pp.5, Conference Publication 3035, NASA (1988).
17. D.R.Coulter, "Effects of Energetic Proton bombardment on Polymeric materials: Experimental studies and degradation models" NASA-CR-177161, (1986)
18. E.A.Brown. "Proceedings of NASA Space Station Workshop" p.44, Nashville, Tennesse (1987).
19. D.Schwam, J.Kulig and M.H.Litt, "Chem.Materials 3, 616" (1991).
20. Joseph Kulig, Virginia Jefferis, David Schwamm and Morton Litt. "Inorganic and Metal Containing Polymeric Materials" Ed.by J.Sheats etal, Plenum Press, New York, p.225 (1990).
21. S.Packirisamy, D.Schwam and M.H. Litt, "J.Materials Science" 30, 308 (1995).
22. A.S.Levine, "69 Months in space: A history of the first LDEF", NASA . 149 (1990).

ADVANCED TURBULENT DRAG REDUCING AND FLOCCULATING MATERIALS BASED ON POLYSACCHARIDES

Ram Prakash Singh

Materials Science Centre
Indian Institute of Technology
Kharagpur-721302, India

ABSTRACT

Synthetic polyacrylamide based polymers and polysaccharides such as guar gum, xanthan gum carboxymethylcellulose and starch have been used for long. The polyacrylamides are easily amenable to shear degradation although they are very efficient drag reducing and flocculating agents even at low ppm concentrations. On the other hand polysaccharides are fairly shear stable but are not very efficient drag reducers and flocculants. Their aqueous solution are also subjected to rapid biodegradation. It was contemplated that in the case where polyacrylamide chains are grafted on polysaccharide backbones, efficient shear and biodegradation drag reducing and flocculating agents may be developed.

In the author's laboratory, detailed investigation on synthesis and characterisation of graft copolymers of polysaccharides and polyacrylamide was undertaken. It has been convincingly shown that these graft copolymers offer efficient, shear and biodegradation resistant drag reducing and flocculating systems. The laboratory scale synthesis has also been scaled up to batch reactor level.

The present paper outlines the details of synthesis and characterisation, structure-property relationship in regards to drag reduction, flocculation, shear stability and biodegradation resistance as well as the applications of these graft copolymers in systems of industrial importance. The graft copolymers are found to be highly solvated. The shear stability is correlated with the solvation number of graft copolymers. This indicates their applicability as super-absorbent materials. They can also be blended with polyolefines in order to develop biodegradable films. The various characteristics of these systems have been discussed in terms of current molecular theories of drag reduction and flocculation. The paper also covers some salient features of author's work on application of drag reducing polymers in reducing the energy requirement of the sprinkler irrigation and reduction of percolation losses of soil as well as development of slow release urea based on blending with polysaccharides.

INTRODUCTION

Turbulent Drag Reducing Polymers

The turbulent drag reduction is the phenomenon of drastic reduction of drag by a factor of two or more below that of the solvent by addition of a small quantity of certain materials such as high polymers, fibres, soaps, surfactants and their mixtures. Mysels[1] and Tom[2] observed the phenomenon independently. The phenomenon has been one of most fascinating subjects in fluid mechanics and polymer science. Despite extensive research ever since its discovery, the mechanism for drag reduction is still eluding satisfactory explanation. The developments in the field have been recently reviewed by Singh[3], Morgan and McCormic[4] and Kulicke et al.[5] As the phenomenon is of great potential benefit to several industrial processes and operations, the search for efficient and shear stable drag-reducing agents is being sustained. However soluble polymers are the most potential drag reducing agents mainly because the drag reducion upto 80% can be obtained by addition of a few tens of the polymer in a particular solvent. But the drag reducing polymers suffer from mechanical and biodegradation.

Several attempts have been made to enhance drag reduction and shear stability of polymeric drag reducers. It has been observed that the drag reducing polymers when used in combination with other drag reducing polymers or fibres can offer combinations having synergistic effect in drag reduction and shear stability[6]. Similarly by grafting the chains of flexible highly effective drag reducing polymers on rigid and shear stable but poorly drag reducing polysaccharide polymers, it is possible to develop highly drag-reduction effective and shear stable drag reducing agents[7-13]. Interacting and associating polymer systems have also shown higher drag reduction efficiency and shear stabilities[14,15]. A detailed investigation has been carried out in author's laboratory on the development of highly drag reduction and shear stable polymeric systems. The present paper summarises the results of this extensive investigation.

Polymeric Flocculants

Polymeric flocculants are extensively used for treatment of industrial effluents and mineral processing mainly because of their characteristic attributes[16,17]. Polymers are convenient to use and do not affect the pH of the medium. They are used in very small quantities (≅ 1-5 ppm) and the flocs formed during flocculation are bigger and stronger. Anionic, cationic and nonionic synthetic polymers and natural polymers are in vogue. Among the natural polymers, polysacchasrides such as guar gum and particularly starch have been used extensively. Polyacrylamide and polyacrylamide based synthetic polymers are very efficient flocculants and are mainly used for treatment of industrial effluents and mineral processing. Polysaccharides are easily biodegradable but otherwise are cheap, fairly shear stable and reproductive materials from agriculture. Hence it was envisioned that by grafting flexible polyacrylamide chains on polysaccharides such as guar gum, xanthan gum and starch, it is possible to develop efficient, shear stable and biodegradable flocculants for the treatment of industrial effluents and mineral processing. In these flocculants, the flexible chains of polyacrylamide are grafted on the rigid backbone of polysaccharides, hence the approachability of polyacrylamides chains for metallic and nonmetallic contaminants increases significantly (Figure 1). Thus they offer highly efficient flocculating attributes.

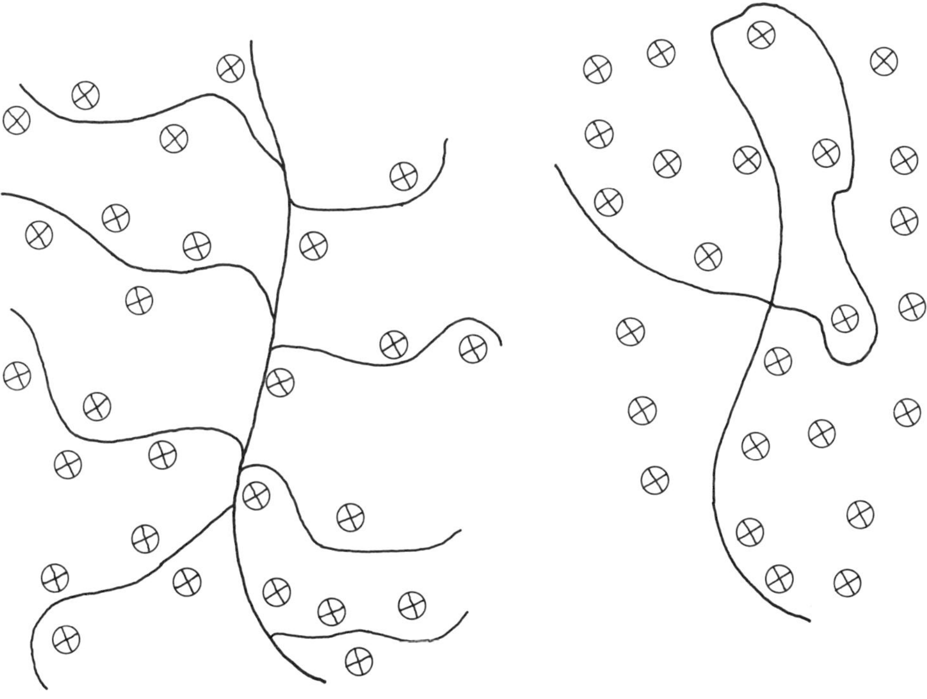

Figure 1. The approachability of linear and grafted polymers to contaminant particles.

A comprehensive programme for development of guar gum/xanthan gum/starch grafted polyacrylamides[18-21] was initiated at Materials Science Centre in late eighties mainly in pursuit of efficient and shear stable drag reducing agents. Several graft copolymers have been developed and their flocculating efficiency for industrial effluent treatment and mineral processing has been evaluated. The laboratory scale synthesis has been scaled up to batch reactor level. It has been found that starch grafted polyacrylamides are most efficient ones. The present paper summarises our investigation for several years.

Applications

The drag reducing polymers have been successfully applied for potential benefits in various industrial processes and operations[22,23] such as long distance transportation of liquid, oil well operations, sewage and flood water disposal, fire fighting, transport of suspensions and slurries, irrigation, water heating and cooling circuits, jet cutting, marine and biomedical systems.

The drag reducing polymers reduce the drag in turbulent flow while increasing the drag in laminar flow due to an increase in the shear viscosity. This feature of drag reducing polymers has been utilised in reducing the energy requirement of sprinkler irrigation system and in enhancing water throughput as well as ara of coverage of sprinkler irrigation system[24,25]. The water containing drag reducing polymers percolates slowly in the soil, thus reducing the percoation losses of water. Utilising this, a slow release urea has been developed by blending drag reducing guar gum with urea. The present paper highlights the investigation in this field carried out at IIT Khrargpur.

The graft copolymers of guar gum/xanthan gum/starch and polyacrylamide have been applied for treatment of industrial effluents containing Pb, Cd, Zn, Cu, Iron and Coal. It has been convincingly shown that these graft copolymers are efficient flocculating agents. A comparative study indicates that the starch based graftcopolymers are the most efficient ones[26,27]. In several situations of pH, shear rates, dilution and flocculant dose, these graftcopolymers show a better performance than the commercial ones.

In present investigation, the materials science and engineering approach encompassing all its elements i.e., synthesis, characterisation, structure property relationships, processing and performation has been adopted. The present paper outlines the essentials of our investigations covering almost two decades.

SYNTHESIS OF GRAFTCOPOLYMERS

Materials

The commercial guar and xanthan gums were purified to remove protein, fat and other two molecular weight impurities by soxhlet extraction with 95% ethanol for 5 days as suggested by Whitcomb et al.[30] CMC was purified by washing with methanol containing dil. HNO_3 to remove inorganic impurities. Soluble potato starch, amylose and poly(vinyl alcohol) (Mw=125000) were used as received. Acrylamide was recrystallised from acetone prior to use. Ceric amonium nitrate and nitric acid, analar grade, were used as received. Doubly distilled water was used during synthesis and solution preparation.

Synthesis

Several methods are available for synthesis of graftcopolymers. However for polyol type of substrates, ceric ion initiated solution polymerisation technique is most widely used because of its simplicity and the fact that the free radicals are generated exclusively on the substrate molecule, thereby minimising the formation of homopolymers. Thus in the present case, graftcopolymers have been synthesised by grafting acrylamide onto the substrate by using ceric ion redox initiator. The details of the synthesis have been described elsewhere[7-13,18-21]. In all reactions, amount of substrates was 1 gram and 300 ml of distilled water was used as solvent. Reaction temperature was 30±1°C and reaction time was kept 24 hrs mostly (4 hrs in case PVA graft copolymers). The graft polymerisation reaction is very sensitive to presence of traces of O_2 and metallic impurities. Hence extreme care is taken in use of inert atmosphere and chemicals free of metallic contaminants.

The graftcopolymers of guar gum, xanthan gum, CMC, starch and PVA denoted by GAm, XAm, CAm, SAm and PAm respectively. Several series of graftcopolymers have been synthesised by varying ceric ion and acrylamide concentrations. In case, acrylamide concentration is constant, the ceric ion concentration is increased, this will result graftcopolymers with increased number of grafts per backbone molecules but with shorter chain as one goes down the series. In case ceric ion concentration is constant but acrylamide concentration is increased. This will result constant number of grafts but the graft length increasing with increase in acrylamide concentration. The synthetic details are given in Tables I and II.

Scaling up to 40-L-Reactor

The grafting reaction in case of GAm, XAm and SAm has been scaled up in a 40-L reactor[1,18-21]. It consists of cylindrical vessel (d=360 mm, h=460 mm) made of brass which has copper heat exchanger. A glass beaker of 10 L was placed inside the copper spiral in the cylindrical vessel. The vessel can be closed by brass cover with holes for stirrer, gas inlet and outlet and injector of catalyst respectively. The lid has a grove to fit the 'O' ring placed on the flange of cylinder vassel. This gives vacuum fitting to the reactor. The reaction is carried out as in 1-L Ehrenmeyer flask keeping the ratio of reactants the same. The yield of graftcopolymers obtained in scaled up synthesis is higher than that of laboratory scale due to better mixing, control of temperature and N_2 atmosphere.

Characterisation of Graftcopolymers

The molecular weights of graftcopolymers have been evaluated from intrinsic viscosity. Grafting has been confirmed using IR, thermal analysis, rheological and X-ray diffraction techniques[7-13, 18-21].

Table I The details of synthesis, characterisation and drag reduction characateristics of grafted polysaccharides.

Sample No.	Moles in reaction mixture		Yield g	$[\eta]$ mlg	Max. DR (conc. for max DR)	DR before and after degradation			
						100 ppm		400 ppm	
	Acryamide	Ce x10				Initial DR	DR after 100 asses	Initial DR	DR after 100 passes
GAm3	0.14	0.05	8.835	834	70.0(50)	65.05	43.0	60.83	57.54
GAm4	0.14	0.10	11.106	688	68.2(75)	68.68	41.76	64.28	53.67
GAm5	0.14	0.20	11.264	478	65.5(500)	52.29	28.74	63.59	50.54
GAm6	0.14	0.25	9.702	319	61.4(1000*)	43.37	23.4	58.5	42.15
GAm7	0.14	0.30	9.621	280	61.8(1000*)	38.82	20.77	58.23	38.65
GAm8	0.14	0.05	3.523	703	65.2(200)	61.95	34.78	65.76	56.52
GAm9	0.14	0.05	10.908	938	67.0(50)	64.83	48.90	65.76	56.52
GAm2	0.14	0.10	5.75	1884	68.2(75)	67.04	44.32	61.36	52.21
XAm3	0.14	0.20	9.65	1026	67.1(100)	67.03	42.86	64.83	54.94
XAm4	0.14	0.30	10.24	812	64.7(200)	61.76	35.29	62.92	32.94
XAm5	0.14	0.40	10.88	528	62.4(1000*)	40.33	21.17	56.42	42.31
CAm1	0.14	0.05	3.96	1150	66.875)	65.38	31.64	61.54	48.43
CAm2	0.14	0.10	9.17	734	58.6(1000*)	37.35	15.61	54.65	31.38
CAm3	0.14	0.20	9.18	605	57.0(1000*)	27.48	3.30	50.0	15.50
CAm4	0.14	0.30	10.0	541	56.1(1000*)	21.51	1.30	43.0	4.75
CAm5	0.21	0.10	13.83	850	66.86(75)	65.08	31.69	62.13	50.94
CAm6	0.28	0.10	15.48	900	67.9(75)	67.4	33.72	62.34	53.92
SAm1	0.14	0.03	8.51	512	68.8(50)	68.23	51.17	63.3	53.93
SAm2	0.14	0.03	9.23	397	65.7(600)	58.33	39.02	64.88	45.47
SAm3	0.14	0.10	10.04	305	62.2(100)	46.66	28.88	57.9	37.90
SAm6	0.21	0.05	12.21	410	65.2(100)	65.17	43.59	63.0	51.09
SAm7	0.28	0.05	16.45	509	67.6(50)	66.47	49.85	63.0	54.18
SAm8	0.35	0.05	16.94	546	67.9(50)	66.66	46.66	60.6	55.26
PAm1	0.14	0.063	6.45	476	64.77(1000*)	56.6	35.0	58.3	40.81
PAm2	0.14	0.125	8.9	400	62.22(1000*)	47.5	23.7	55.8	35.15
PAm3	0.14	0.25	9.1	310	55.95(1000*)	33.33	9.66	42.0	24.78
PAm4	0.14	0.375	10.9	220	20.93(1000*)	5.25	1.01	7.5	1.21
PAm5	0.14	0.438	10.7	100	17.44(1000*)	4.75	0.95	6.45	1.10

The figures in the bracket in the sixth column indicate conc. inppm at which Max. DR was observed. The * indicates that maxima in DR was not achieved even at that concentration. In case of graft copolymers of PVA, shear stability were carried out upto 40 pass member at 100 ppm and 300 ppm.

Table II The details of synthesis and characterisation of PAM grafted poly(vinyl alcohol)

(1)	(2)	(3)	(4)	(5)	(6)	(7)	(8)	(9)
Gr. I	Ia	0.008	0.14	0.0625	7.81	2.24	54.55	470
	Ib	0.008	0.14	0.125	15.63	1.12	79.09	400
	Ic	0.008	0.14	0.25	31.25	0.56	81.82	310
	Id	0.008	0.14	0.375	46.88	0.37	89.55	220
	Ie	0.008	0.14	0.438	54.75	0.32	87.27	100
	IIa	0.008	0.035	0.0625	7.81	0.56	61.79	115
	IIb	0.008	0.070	0.125	15.63	0.56	75.83	300
Gr. II	IIc	0.008	0.14	0.25	31.25	0.56	81.82	310
	IId	0.008	0.21	0.375	46.88	0.56	95.94	330
	IIIa	0.008	0.035	0.25	31.25	0.14	87.50	90
	IIIb	0.008	0.070	0.25	31.25	0.28	81.67	140
Gr. III	IIIc	0.008	0.14	0.25	31.25	0.56	81.82	310
	IIId	0.008	0.21	0.25	31.25	0.84	78.13	350

Column (1) indicates - Group
Column (2) indicates - Polymer
Column (3) indicates - Poly vinyl alcohol (milimole)
Column (4) indicates - Acrylamide (mole)
Column (5) indicates - [Ce(iv)] (millimole)
Column (6) indicates - [Ce(iv)] [Poly(vinyl alcohol)]
Column (7) indicates - [Acrylamide] / [Ce)iv)] ($x10^3$)
Column (8) indicates - Percent yield
Column (9) indicates - [η]

RESULTS AND DISCUSSION

Turbulent Drag Reduction and Shear Stability of Grafted Polysaccharides

The drag reduction percentage (DR) is calculated from the formula ;

$$DR = \frac{\Delta P_w - \Delta P_s}{\Delta P_w} \times 100$$

where ΔP_w = pressure drop in water and ΔP_s = pressure drop in test solution

The flexible polymers such as poly(ethylene oxide) and polyacrylamide cause maximum drag reduction upto 80% in turbulent pipe flows at the conentration of few tens of ppm but are easily shear degradable[27]. On the other hand, rigid polymers particularly the polysaccharides such as guar gum, xanthan gum and carboxymethylcellulose are fairly shear stable but cause maximum drag reduction upto 60% at the concentration of few hundred ppm[28]. However, their aqueous solutions are easily biodegradable. Thus there exists a scope to develop new drag reducing agents having reasonable shear stability, high drag reduction effectiveness and low cost.

It has been established in author's laboratory that by grafting polysacrylamide chains on guar gum, xanthan gum and carboxymethylcellulose, it is possible to develop graftcopolymers which are more drag reducing than the backbone polymers and have shear stability of the order of the backbone polymers. The maximum in drag reduction is also found at the concentrations ranging from 50 to 100 ppm. The grafting also alters the regularity of molecular structure inhibiting the enzyme attack[29] and thus enhancing the biodegradation resistance of the graft copolymers. Starch and poly(vinylalchol) are also water soluble though with lower molecular weight are not drag reducing. In the author's laboratory, a large number of graft copolymers of guar gum, xanthan gum/carboxymethylcellulose/starch and polyacrylamide have been developed. The synthetic parameters such as concentrations of polysaccharide, acrylamide and ceric ion initiator were varied as to synthesise a large number of graftcopolymers with varying number and length of grafted chains. The synthesis details are given in Tables I & II.

The following characteristics of these drag reducers have been observed.

- It has been obsrved that purification of guar gum and xanthan gum by the method of Whitcomb et al.[30] enhances the drag reduction of these polymers as removal of low molecular impurities like fats and proteins increases the effective concentration of high molecular weight polymers hence the improve the drag reducing ability (Figure 2).
- The drag reduction performance of graftcopolymers depends on number and length of grafted chains. The graft copolymers with fewer and longer chains cause more drag reduction and have higher shear stability. Their molecular behaviour is akin to flexible molecules (Figure 3).
- The maximum in drag reduction in drag reduction vs. concentration profile is obtained at concentrations of 50-100 ppm in case of graftcopolymers of guar and xanthan gum.
- Solutions of graftcopolymers in water are biodegradation resistant (Figure 4).
- Drag reduction performance of graftcopolymers is not affected by salts present in sea water (Figure 5).
- Graftcopolymers of starch and poly(vinyl alcohol) (Mw=1,25,000) exhibit drag reduction effectiveness even though the backbone polymers being low molecular weight ones, are not drag reducing (Figure 6).
- The drag reducing charactristics of graft copolymers give credence to Brostow Polymer Flow Model[31,32] (Figure 7).
- The graftcopolymers having higher solvation have higher

shear stability[33].

o It has been observed that graft copolymers having fewer number of longer grafts showed more shear stability than the graft copolymers with more number of shorter grafts[33]. The shear stability of xanthan gum graft copolymers is comparable with that of xanthan gum[11]. The guar gum and CMC were found to be more shear stable than their graft copolymers. This may be due to the fact that during grafting reaction, the backbone polymer is scissioned. Thus the grafts have smaller molecular dimensions than those of the original backbone polymers. The graftcopolymers of xanthan gum are more shear stable than the mixtures of xanthan gum and polyacrylamide having the same composition as in graftcopolymers[11].

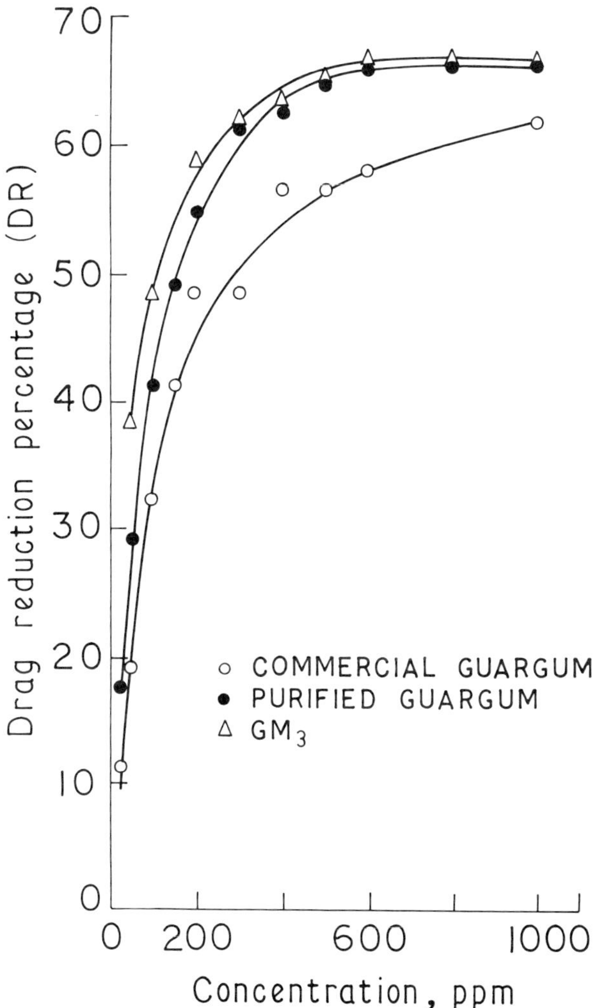

Figure 2. The DR versus concentration in commercial and purified guargum at Reynolds number (Re) = 14,000.

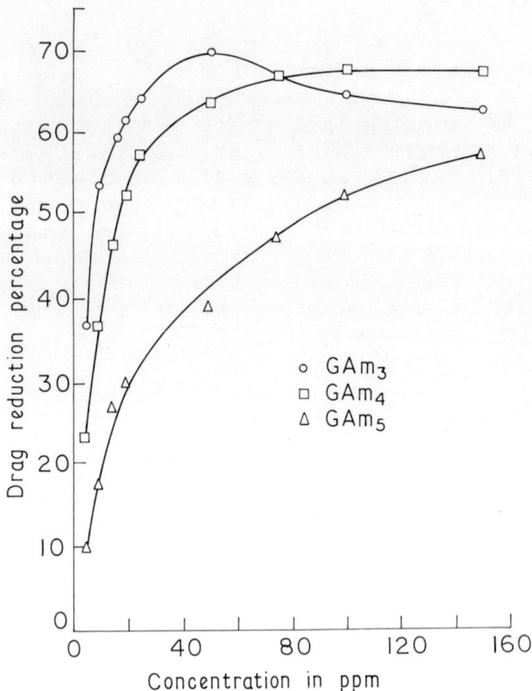

Figure 3. The DR versus concentration profiles for grafted guargum with varying number and length of grafted chains at Re = 14,000.

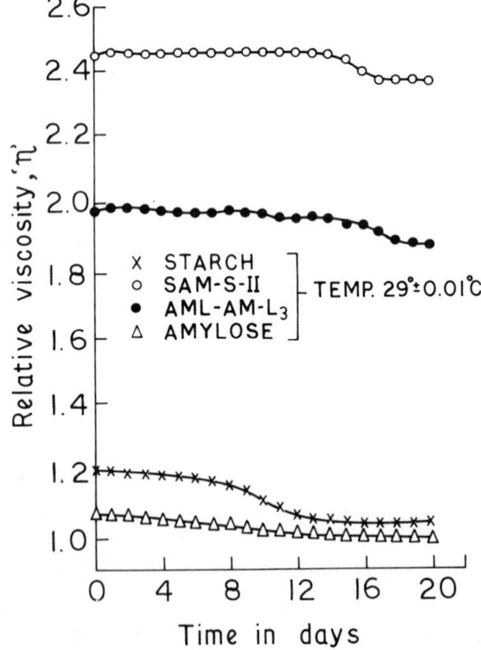

Figure 4. The variation of viscosity with time in aqueous solutions of starch & amylose and grafted starch & amylose.

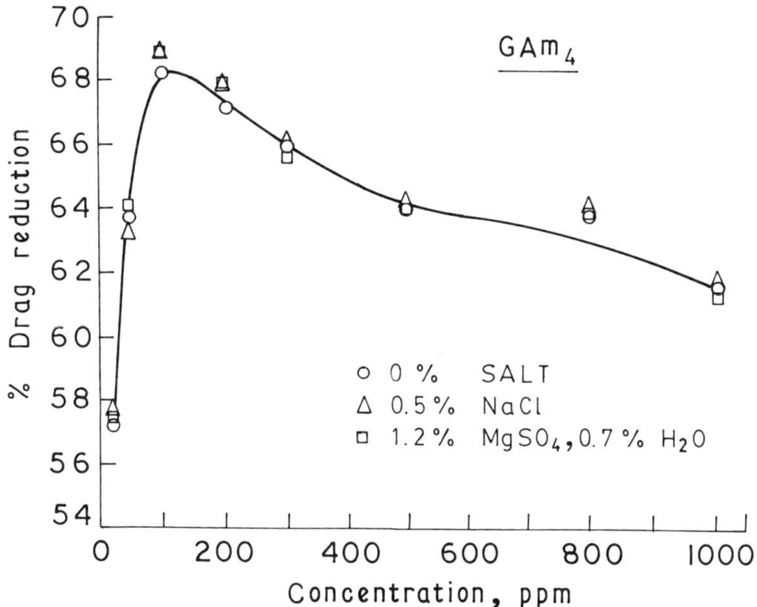

Figure 5. The effect of salts present in sea water on DR versus concentration profile of grafted guargum at Re = 14,000.

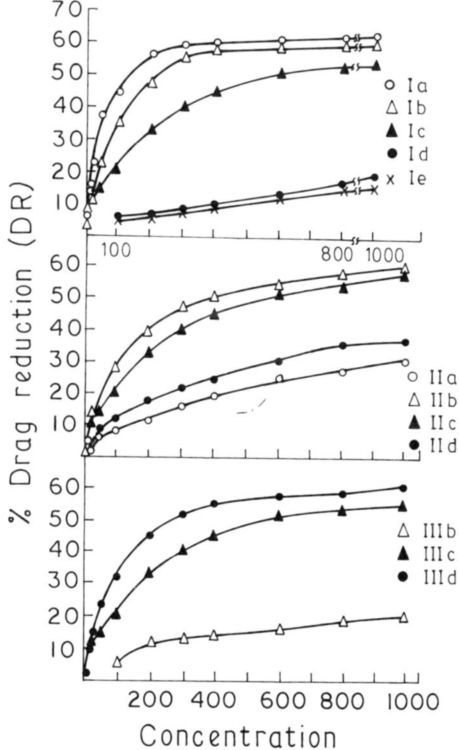

Figure 6. The DR versus concentration in grafted poly(vinyl alcohol) at Re = 14,000.

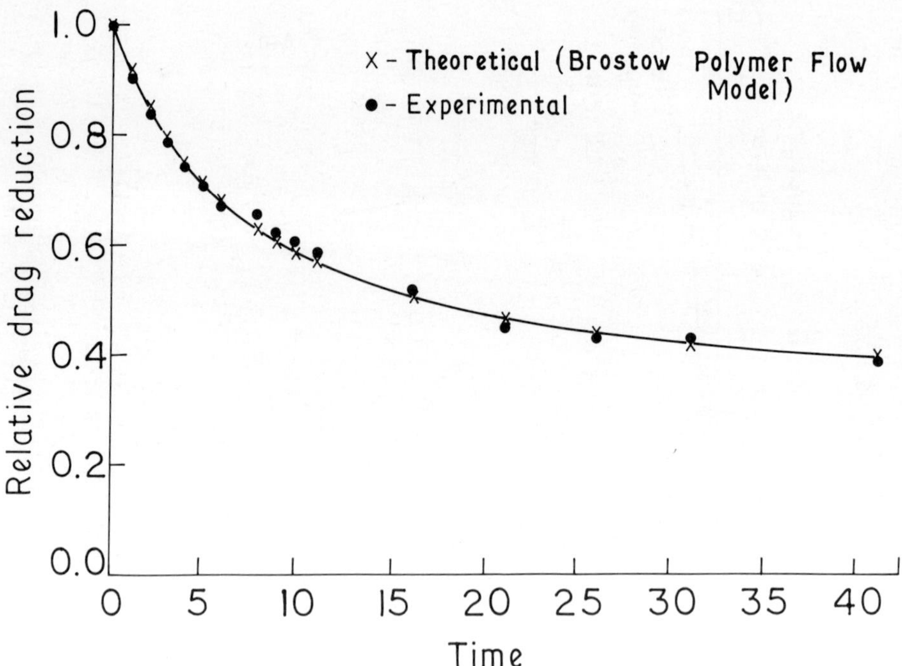

Figure 7. The variation of relative drag reduction with time for the mixture of pure guargum (20 ppm) and polyacrylamide (20 ppm) in lead effluent.

The recent experiments point out that in turbulent flows, the mid chain scission of the extended chains takes place due to axial components of turbulent stresses[34,35]. Thus giving credence to original Beuche theory[36]. In the branched polymers, mid chain point may fall at the branches hence even after degradation in turbulent flow, the polymers still remain effective drag reducers. Hence shear stability variation with length and number of branching is consistent with drag reduction effectivenes behaviour of the graft copolymers in accordance to existing molecular approach to drag reduction and shear stability.

The Flocculation Characteristics of Grafted Polysaccharides

It has been pointed out in section 1.2 that due to their molecular structural characteristics, the grafted polysaccharides are more effective and shear stable flocculants in comparison with polyacrylamide based flexible flocculants[37]. An extensive investigation was carried out to synthesise, characterise and to study the flocculation characteristics of grafted polysaccharides. The three grafted polysaccharides i.e., the guar gum[18,19]/xanthan gum[20]/starch[21] and polyacrylamide based graftcopolymers have been investigated extensively. Among them only those graftcopolymers which gave the maximum drag reduction effectiveness and shear stability i.e., having fewer but longer chains were selected. Their synthesis was scaled up. First their effectiveness was studied in synthetic effluents. Later on the industrial effluents were treated with these graftcopolymers. The following is the gist of this investigation.

o The effect of molecular parameters of various graft copolymers of guar gum and polyacrylamide was tested in the synthetic effluent of lead[18]. The maximum efficiency was shown by GM_3, followed by GAm_4 and GAm_5. It appears the presence of lead ions in the effluent caused straightening effect on the polymer chains. Due to this effect, the polymer chain has not assumed globular form thus making available all hydrogen bond forming sites to participate in the flocculation. GAm_5 with more number of shorter grafts showed poor performance but still its effect was much higher than that of commercial polyacrylamide based flocculant (Figure 8). The synthesis of GM_3 was scaled up and tested for various synthetic and industrial effluent. It was observed [19] that GM_3 performs better than guar gum and the commercial flocculant based on polyacrylamide.

o Based on above observations, the synthesis of XAM_3 was scaled up[20]. XAM_3 is the most effective drag reducer having longer and fewer grafted polyacrylamide chains on xanthan gum backbone. Its efficacy was tested in synthetic effluent containing lead and paper mill effluents. In lead effluent, its performance is better than commercial purified xanthan gum and polyacrylamide. However GM_3 performs better than XAm_3. The molecular basis of this observation has been discussed in detail elsewhere[37]. In paper mill waste treatment, the polymer acts as flocculant aids only. It is used alongwith alum.

o Starch is used extensively for mineral treatment[38] and hence it was contemplated that Grafted starch may have better performance as flocculant. Again on above considerations, SAm_1 was chosen for scale up and testing for synthetic as well as industrial effluents. A comparative study[21,26] was conducted for synthetic lead slurry. It has been found that it provides the best flocculating performance. At alkaline pH and higher shear rates, this flocculant performs better than the flocculants available commercially (Figure 9).

Starch consists of amylose and amylopectin. The amylose is linear polymer and amylopectin is branched one.The grafted amylose performs inferior to grafted starch. It is surmised that grafted amylopectin may be better flocculant than grafted starch which is in line with our model propounded in section 1.2. The work is in progress to prove this hypothesis.

o In most of cases, the flocculation was found to have taken place by bridge mechanism[38].

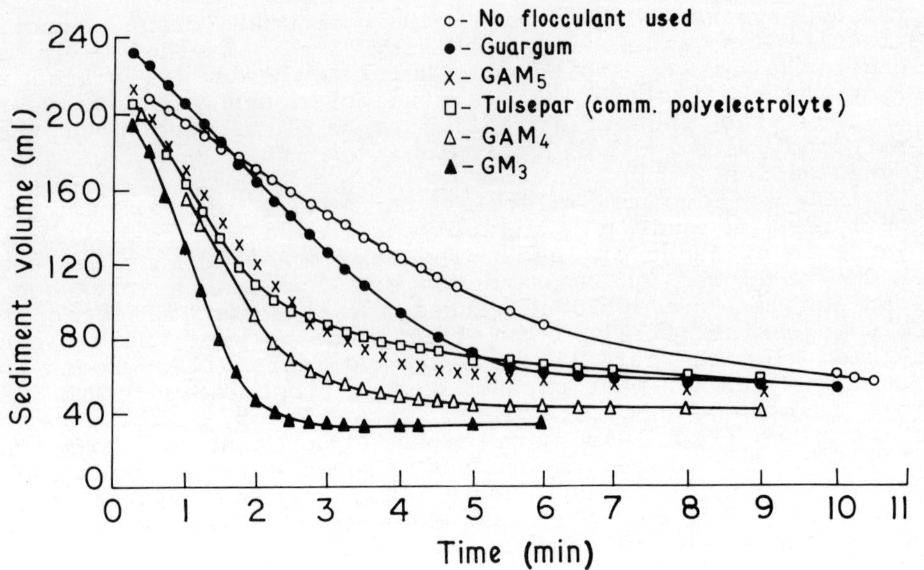

Figure 8. The effect of graft copolymers on flocculation of lead synthetic effluent.

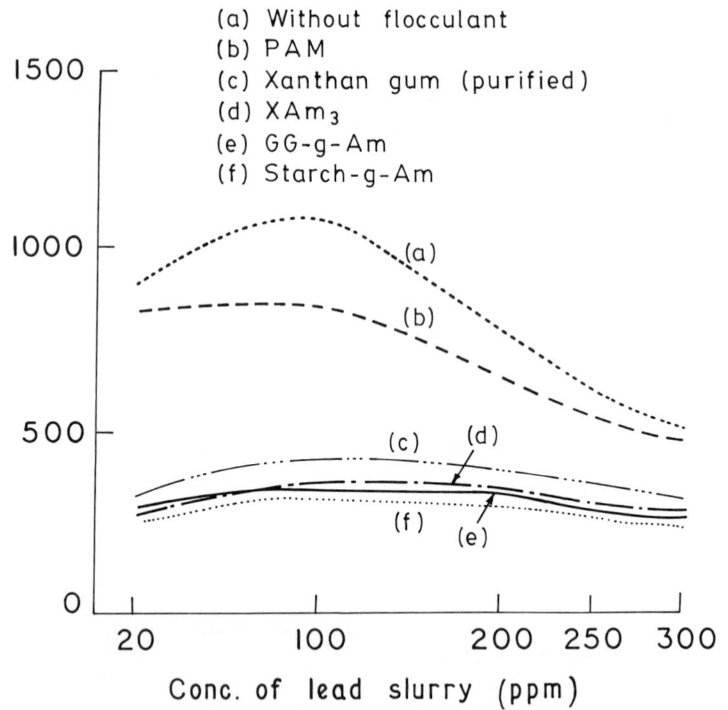

Figure 9. The variation of sedimentation time with concentration of lead slurry by addition of 2 ppm of various flocculants.

APPLICATIONS

The drag reducing polymers reduce the drag in turbulent flow while increase the drag in laminar flow due to increase in shear viscosity. The water containing drag reducing polymers percolates slowly in soil. These features of drag reducing polymers have been successfully applied in reducing the energy requirements of sprinkler irrigation system and in reducing the percolation losses through soil as well as in development of guar gum blended slow release urea.

Sprinkler Irrigation

The pervasive impact of commercial irrigation, promising incentive returns has motivated the farming community worldwide to adopt sprinkler irrigation system. In tank and canal systems, water flows through unlined earth channels. Sprinkler irrigation can be used to prevent seepage losses. It is proven fact that sprinkler irrigation has higher application efficiency as compared to surface irrigation system and can be recommended to undulating areas having high infiltration rates and limited availability of water. About 30-40% of water can be saved by application of sprinkler irrigation[39]

A major fator preventing the extensive use of sprinkler irrigation is the high energy requirement. The sprinkler irrigation system is operated in turbulent flow condition both in distribution pipes, rises and ejected water jet. Hence application of turbulent drag reduction phenomenon can be useful in reducing the energy requirement and thereby saving power and fuel requirement. The first demonstrative experiment was conducted by Unior Carbide Company, USA in 1966. They observed increase in radius of coverage and reduction in energy requirment. Significant research was also carried out in Czechoslovakia to explore feasibility of efficient application of drag reducing polymers in sprinkler irrigation in eighties[40].

In India, research in this field began in 1978 at Indian Institute of Technology, Kharagur. By extensive investigation[41-46] varying the kind of drag reducing polymers, operating pressures and number of bends in sprinkler irrigation systems, it has been established that 46.0% reduction in energy requirement is possible using 30 ppm poly (ethylene oxide) (polyox). The radius of coverage can be extended to about 40% with application of 100 ppm xanthan gum. Even indigeneously available guar gum can reduce the energy requirement upto 28% and increase radius of coverage upto37% at the doses of 100-250 ppm in sprinkler irrigation system having one bend[46] (Figure 10-12). Most of the polysaccharide drag reducers are biodegradabe and do not affect the crop yield. The synthetic drag reducers also do not affect crop yield.

The water added with drag reducing polymer percolates slowly in the soil thus reducing percolation losses of water which is an added advantage in water management through sprinkler irrigation system, particularly in those areas having high infiltration losses[45]. This latter aspect has also given rise to development of slow release urea by blending urea with guar gum[24,25,47].

The extent of reduction of hydraulic conductivity is exponential. The relation between percentage reduction in hydraulic conductivity (k'') and the concentration (C) of purified guar gum is given by following eauation[25].

$$K'' = 100 - (103.4\ e^{-0.00303C})$$ where, k'' is given by

$$K'' = [K_w - K_s] / K_w$$

K_w and K_s represent the hydraulic conductivity of water and polymer added water respectively.

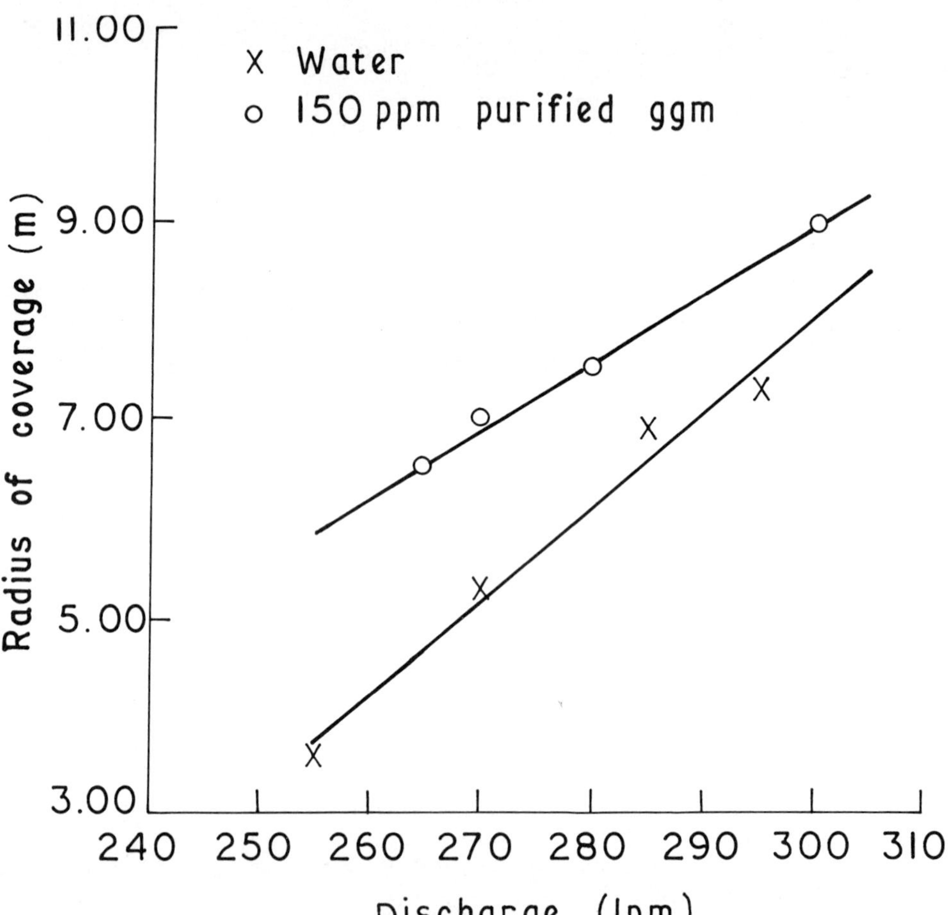

Figure 10. The variation of radius of coverage of sprinkler irrigation system with flow rate for water and polymer added water.

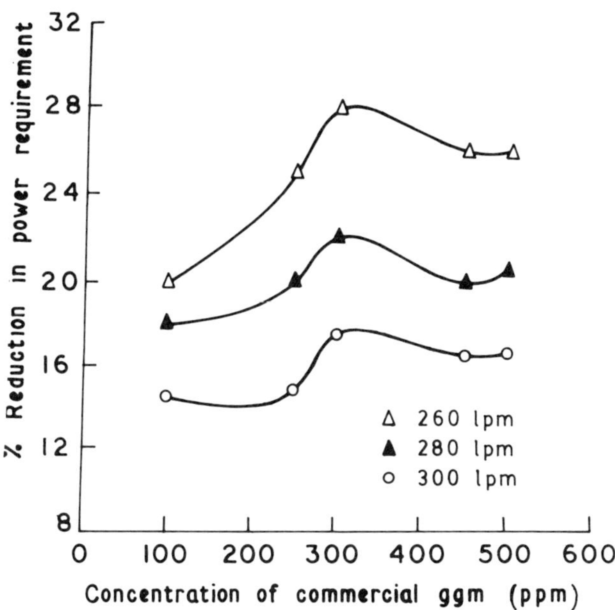

Figure 11. The reduction in energy requirement of sprinkler irrigation system with the concentration for commercial guargum at various flow rates.

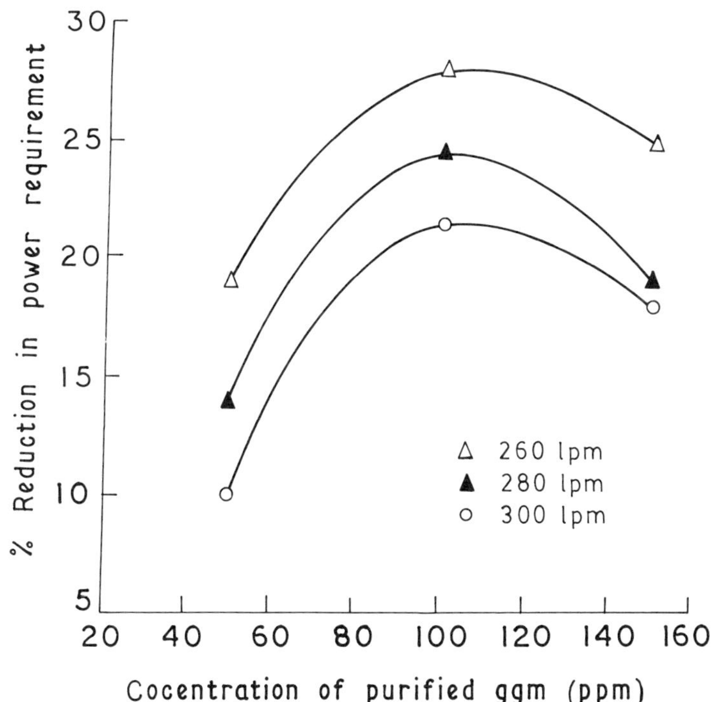

Figure 12. The reduction in energy requirement of sprinkler irrigation system with the concentration for purified guargum at various flow rates.

Slow Release Urea

Nitrogen is an important fertilizer for plant growth. The performance of urea as nitrogeneous fertilizer is due to its relatively low cost and high content of nitrogen. Urea being quickly soluble, large percentage of losses occur due to leaching and run off. This has motivated the development of slow release urea fertilizer. The various processes which contribute to nitrogen losses in agriculture field, particularly paddy fields are leaching, ammonia volatilization, nitrification, denitrification and run off. The other transformations are biological immobilization and ammonia fixation. Any process which interferes with these transformations will increase nitrogen use efficiency. Slow release fertilizers are those which have low rate of dissolution.

In general, urea super granule, sulphur, glass, lac and neem extract coated ureas have been used as slow release ureas. Sulphur coated urea has been developed in USA. It did not beccome popular due to its high cost and limited availability. It also induces acidity in soils. Lac and neem extract coated ureas are still in experimental stage and their cost is also high. It has been recently pointed out that if urea is coated with brittle coatings such as glass or sulphur, as soon as the coating fractures the fertilizer is immediately released and percolates through soil quickly. Hence the blending is a better process of slow release of fertilizers[48]. Polysaccharides such as guar gum, xanthan gum, carboxy methylcellulose and starch are known to be superabsorbent materials[49]. It was thought that coatings with urea in hydrated form would act as diffusion retardant. Performance of guar gum coating was found to be the best.

Various experiments for evaluation of infiltration, hydraulic conductivity and release characteristics in soil columns having sandwitched urea under laboratory conditions were conducted[24,25]. To study the dissolution rate, a separate expriment was carried out in which pellets of urea and guar gum blended urea were suspended in water and dissolution was studied (Table III). It is observed that the blending imparts resistance to dissolution of urea. Both melt blended and physically blend ureas were tried. The performance of melt blended urea is better but melt blending process is costlier. It involves melting of urea with guar gum and then granulation. The other process is cheaper as it involves only physical blending in tumbling mill and does not add to the cost of urea to more than 10%.

The preliminary laboratory studies were then extended to field trials. In this study physical blended urea was chosen for the treatment. The efficiency of this blend was compared with other commercially available slow release urea preparations (Table IV). However, other slow release ureas are much costiler than guar gum blended urea. Recently a comparative study involving urea supergranule, lac and neem coated ureas, and guar gum blended urea (supplied by us) has been undertaken by Kotur[47] for cauliflower and brinjal crops. Guar gum treated urea gave best results in case of brinjal and comparable results in case of cauliflower crop. In both cases there is net gain. The other treatments being much costlier, guar gum coated urea shows much better future perspective in application as a slow release urea[47,48]

Table - III Time taken by different urea/urea-guarguam blends to get dissolved completely

Sl.No.	Weight (g)	Urea/Urea blends		Average time taken (min)
1	12.6	urea		2.67
2	12.6	0.2%	O.B.	3.67
3	12.6	0.4%	O.B.	5.50
4	12.6	0.6%	O.B.	10.17
5	12.6	0.8%	O.B.	12.83
6	12.6	1.0%	O.B.	16.50
7	12.6	1.2%	O.B.	19.67
8	12.6	0.2%	O.B.	7.0
9	12.6	0.4%	M.B.	8.0
10	12.6	0.6%	M.B.	16.33
11	12.6	0.8%	M.B.	18.67
12	12.6	1.0%	M.B.	28.00
13	12.6	1.2%	M.B.	38.33

Table - IV Effect of Nitrogen Source and Nitrogen Level on Grain Yield of Rice (Kg/ha)

	Levels of Nitrogen, Kg/ha				
	0	40	80	120	Mean
Nitrogen source control	22.89				
Urea (split treatment) (50:25:25)	-	35.98	43.37	48.89	42.79
Rock phosphate coated urea	-	33.85	41.46	45.48	40.10
Gypsum coated urea	-	34.07	41.25	44.61	39.98
Nitrohumic acid coated urea	-	32.97	40.06	42.88	38.64
Guargum coated urea	-	35.81	42.07	46.46	41.45
Urea super granules (1.0 g size)	-	36.68	45.05	50.20	43.92
Mean	22.89	34.81	42.21	46.42	
CD 5% Two levels including control			0.81		
two sources			1.11		
Interaction, two sources means for same level			1.925		
Two means at same sources			1.932		

Application of Grafted Polysaccharides in Treatment of Industrial Effluents

The various industrial effluents have been treated by grafted guar gum and grafted starch. The results may be summarised as follows.

- For the industrial effluents containing nickel, copper and lead ions, the performance of grafted guar gum (GM_3) is better than that of guar gum and Tulespar (a commercial flocculant)[19].
- For industrial effluents from a copper industry (HCL, Ghatsila, India) the grafted starch (SAM-S-II) works better than Magnafloc-1011 (Allied Colloids, U.K.) at higher rpm of flocculators blades[21]
- For effluents from Steel industry (TISCO, Jamshedpur, India), the efficiency of graftes starch (SAM-S-II) is found to be better than Magnafloc-1011 at the lower concentration of the flocculants[21].
- For effluent of paper industry, the coagulation is first achieved by using alum and then grafted starch is found to be effective as flocculant[21].
- In flocculating hematite slimes, the grafted starch is found to be an efficient flocculant with a small dose at par with Magnafloc-1011 in the normal and acidic pH range. In alkaline pH, grafted starch works even better than Magnafloc-1011.

REFERENCES

1. K.J.Misels, Flow of Thickened Fluids, U.S.Patent 24 92173, Dec. 27 (1949).
2. B.A.Toms, Some Observations on the Flow of Linear Polymer Solutions Through Straight Tubes at Large Reynolds Numbers, Proc. 1st. Rheology Congress Part 2, North Holland, 135 (1949).
3. R.P.Singh, Drag Reduction and Shear Stability Mechanism, in Encyclopedia of Fluid Mechanics, Polymer Flow Engineering,I.P.Cheremisinoff, Ed., Gulf Publishing, Houston, USA, Ch.14, Vol.9, 425 (1990).
4. S.E.Morgan and C.L.McCormick, Water Soluble Polymers in Enhanced Oil Recovery, ***Prog. Polym. Sci.*** 15, 103 (1990).
5. W.M.Kulicke, M.Kotter and H.Grager, Drag Reduction Phenomenon with Special Emphasis on Homogeneous Solutions, ***Adv. Polym. Sci***., 89, 1 (1989).
6. W.K.Lee, R.C.Vaselski and A.B.Metzner, Turbulent Drag Reduction in Polymeric Solutions Containng Suspended Fibres, ***AIChE Journal*** 20, 128 (1975).
7. S.R.Deshmukh, P.N.Chaturvedi and R.P.Singh, ***J.Appl.Polym.Sci.*** 30, 4013 (1985).
8. S.R.Deshmukh and R.P.Singh, Drag Reduction Effectiveness, Shear Stability and Biodegradation Resistance of Guar Gum-Based Graft Copolymers, ***J.Appl. Polym. Sci.*** 33, 1963 (1987).
9. S.R.Deshmukh and R.P.Singh, Drag Reduction Characteristics of Graft Copolymers of Xanthan Gum and Polyacrylamide, S.R.Deshmukh an R.P.Singh, ***J.Appl.Polym.Sci.***, 32, 6163 (1986).
10. R.P.Singh, S.R.Deshmukh and S.K.Majumdar, Studies on Turbuent Drag Reduction by Graft Copolymers, Proc. Xth. International Congress on Rheology, Sydney, Plenum Press, New York, Vol.2, 278 (1988).

11. S.Ungeheur, H.W.Bewersdorff and R.P.Singh, Turbulent Drag Effectiveness and Shear Stability of Xanthan-Gum-Based Graft Copolymers, *J.Appl. Polym. Sci.*, 37, 2933 (1989).
12. S.R.Deshmukh, K.Majumdar and R.P.Singh, Drag Reduction Efficiency Shear Stability and Biodegradation Resistance of Carboxymethyl Cellulose-Based and Starch-Based Graft Copolymers, *J.Appl.Polym.Sci.*, 43, 1091 (1991).
13. S.Majumdar and R.P.Singh, Turbulent Drag Reduction by Graftcopolymers of Poly(vinyl Alcohol) and Polyacrylamide, in Macromolecules, Current Trends, S.Venkatachalam, R.Ramaswamy, V.C.Joseph and V.N.Krishnamurthy Eds., Allied Publishers Ltd., New Delhi, Vol. II, 1076 (1995).
14. R.M.Kawalik, I.Duvdelani, D.F.Pfiffer, R.D.Lundenberg, K.Kitano and D.N.Shultz, Enhanced Drag Reduction Via Interpolymer Association, *J.NonNewtonian Fluid Mech.* 24, 1 (1987).
15. S.Malik, S.N.Shintre and R.A.Mashelkar, Enhancing the Shear Stability in Drag-Reducing Polymers Through Molecular Associations, *Macromolecules* 26, 55 (1993).
16. J.Bratby, Coagulation and Flocculation, Uplands Press Ltd., Croydon, England (1980).
17. J.Gregory, Flocculastion by Polymers and Polyelectrolytes, In solid/Liquid Dispersions, Th.F.Tadros Ed., Academic Press Inc., London, 163 (1987).
18. K.Kannan, Guar Gum Based Graft Copolymers : Flocculastion and Rheological Behaviour, M.Tech. Thesis, Materials Science Centre, I.I.T., Kharagpur, India (1988)..
19. S.K.Jain, Flocculating Agents Based on Guargum-g-Acrylamide for Industrial Effluents Treatment, M.Tech. Thesis, Materials Science Centre, I.I.T., Kharagpur, India (1989).
20. N.T.Lan,Flocculant Based on Xanthan-g-Acrylamide, M.Tech. Thesis, Materials Science Centre, I.I.T., Kharagpur, India (1990).
21. G.P.Karmakar, Flocculation and Rheological Properties of Grafted Polysaccharides, Ph.D.Thesis, Materials Science Centre, I.I.T., Kharagpur, India (1994).
22. R.H.J.Sellin, J.W.Hoyt, J.Pollert and O.Scrivener, The Effect of Drag Reducing Additives on Fluid Flows and Their Industrial Applications : Part II, Present Appliations and Future Proposals, *J.Hydraulic Res.*, 20, 235 (1982).
23. R.H.J.Sellin, Industrial Applications for Drag REducing Polymer Additives : A Review Proc. Third International Conf. on Drag Reduction, Bristol, U.K., R.H.J.Sellin and M.T.Moses Eds. I 3-1 (1984).
24. R.P.Singh, J.Singh, D.Kumar and A.Kumar, Novel Applications of Drag Reducing Polymers in Agriculture, in Drag Reduction in Fluid Flows : Techniques for Friction Control; R.H.J.Sellin and M.T.Moses, Eds. Ellis Horward, Chichester, U.K., D.4-1 (1990).
25. R.P.Singh, J.Singh, S.R.Deshmukh, D.Kumar and A.Kumar, Application of Drag-Reducing Polymers in Agriculture, *Current Science*, 68, 631 (1995).

26. R.P.Singh, Novel Flocculasnts Based on Grafted Polysaccharides, ***Popular Plastics and Packaging,*** 39(5), 69 (1994); R.P.Singh, G.P.Karmakar, N.Lan and S.K.Jain, Development and Investigation on Bidegradable Flocculants for the Treatment of Industrial Effluents. Souvenir Masterials Research Society of India, 6th. Annual General Meeting, I.I.T., Kharagpur, India, PM9 (Feb.8-10, 1995).
27. J.W.Hoyt, Drag Reduction, in Encyclopedia of Polymer Science and Engineering, Vol. 5, Second Edition, John Wiley and Sons INc., 129 (1986).
28. J.W.Hoyt, Drag Reduction in Polysaccharide Solutions, ***Trends in Biotechnology***, 3, 1 (1985).
29. J.K.Seaman 'Guar Gum' In Handbook of Water Soluble Gums and Resins, R.L.Davidson Ed., McGraw Hill Inc., New York 6 (1980).
30. P.J.Whitcomb, J.Gatowski and W.M.Howland, Rheology of Guar Gum Solutions, ***J.Appl.Polym.Sci.*** 25, 2815 (1980).
31. W.Brostow, Drag Reduction and Mechanical Degradation in ***Polymer Solutions in Flow, Polymer***, 24, 631 (1983).
32. W.Brostow, H.Ertepinar and R.P.Singh, Flow of Dilute Polymer Solutions : Chain Conformations and Degradation of Drag Reducers., ***Macromolecules*** 23, 5109 (1990).
33. R.P.Singh, Turbulent Drag Reduction by Polymer Based Mixtures and Graftcopolymers, Current Science, 68, 736 (1995); S.Majumdar and R.P.Singh, ***J. Acous. Soc. India,*** 21, 229 (1993).
34. A.F.Horn and E.W.Merrill, Extensional Flow Induced chain Scission of Macromolecules in Dilute Solution. Effectof Solvent Power in a Mixed Solvent System, ***Polymer Communcations***, 28, 172 (1987).
35. A.Keller and J.A.Odell, The Extensibility of Macromolecules in Solution : A New Focus for Macromolecular Science, ***Colloid and Polym. Sci.*** 263, 181 (1985).
36. F.Beuche, Mechanical Degradation of HIgh Polymers, ***J.Appl.Polym. Sci***. 4, 101 (1960).
37. R.P.Singh, S.K.Jain and N.Lan, Drag Reduction, Flocculation and Rheological Characteristics of Grafted Polysacchasrides. In Polymer Science Contemporary Themes, S.Srivaram Ed., Tata McGraw Hill Publishing Company Ltd., New Delhi, 716 (1991).
38. G.M.Moody, The Use of Polyacrylamides In Mineral Processing, ***Minerals Enginering*** 5, 479 (1992).
39. R.K.Sivanappam, Thrust on Irrigation Harvesting, ***The Hindu Survey of Indian Agriculture, Madras, India,*** 155 (1994).
40. V.Elias and J.Vocel, Long Term Field Expriments with Polymer Additives for Sprinkler Irrigation, ***Communications of the Institute of Hydrodynamics, Prague***, No. 10 (1980).
41. S.H.Holey, R.P.Singh and J.Singh, Application of Drag Reducing Energy Requirement of Sprinkler Irrigation System, ***J.Agri. Engg.***, India 16, 53 (1979).
42. G.Sankar, R.P.Singh and J. Singh, Applications of Drag Reducing Polymers in Reducing Energy Requirement of

Sprinkler Irrigation System II, *J.Agri. Engg.*, India 19, 9 (1982).

43. M.B.Nagdeve, K.N.Tiwari and R.P.Singh, The Reduction in Energy Requirement of Sprinkler Irrigation System by Purified Guar Gum. *J.Agri. Engg.*, India 23, 310 (1986).
44. M.B.Nagdeve, K.N.Tiwari and R.P.Singh, The Effect of Drag Reducing Guar Gum on Hydraulic Conductivity of Soil, *J.Agri. Engg.*, India 23, 217 (1986).
45. A.Kumar, R.P.Singh and J.Singh, Application of the Guar Gum Polymer for Controlling Pecolation Losses under Ponded Conditions in Paddy Fields, *J.Indian, Water Resources Soc.* 8, 53 (1988).
46. S.Phukan, Application of Drag Reducing Polymers in Reducing. The Energy Requirement of Sprinkler Irrigation System, M.Tech. Thesis, Agriculture Engg. Deptt., I.I.T., Kharagpur, India (1993).
47. S.C.Kotur, Relative Efficacy of Different Forms of Urea on Rainy-season Cauliflower and Brinjal on Alfisol of Chhotanagar, *Ind. J. Agri. Sci.* 63, 436 (1993).
48. C.Hepburn and R.Arizal, A Controlled Release Urea Fertilizer Part I : The Encapsulation of Urea Fertilizer by Rubber : Processing and Vulcanisation Procedures, *Plastics and Rubber Proc. Appl.* 12, 129 (1989).
49. S.Majumdar, S.H.Holay and R.P.Singh, Adiabatic Compressibility and Solvation of Drag Reducing Polymers in Aqueous Solutions, *Eu. Polym. J.* 16, 120 (1982).
50. G.P.Karmakar, R.P.Bhagat, D.M.Chakraborty and R.P.Singh, Flocculation Studies on Hematite Mineral Using Starch-g-Polyacrylamide, In Polymer Science, Recent Advances, Vol. 2, I.S.Bhardwaj Ed., Allied Publishers Ltd., New Delhi 990 (1994).

ACKNOWLEDGEMENT

Various results discussed in drag reduction studies were obtained by my scholars Drs. S.R.Deshmukh and S.Majumdar. The flocculation results were similarly obtained by my scholars, Ms. K.Kannan, S.K.Jain, Mrs. N.T.Lan and Dr. G.P.Karmakar. The flocculation works of grafted guar gum was undertaken in collaboration with scientists, Dr. N.Bhattacharya, Dr. K.D.Banerjee and Mr. H.Banerjee of Chloride Industries Ltd., Calcutta. The application of drag reducing polymer in Agriculture was studied by my scholars, Ms. S.H.Holay, G.Sankar, V.K.Jain, A.Kumar, D.Kumar, M.B.Nagdeve, S.Phukan and Miss. U.Kiran. Most of the applications in agriculture were conducted in collaboration of my colleagues Professor J.Singh and Dr. K.N.Tiwari of Agricultural Engineering Department of IIT, Kharagpur. The contributions of these scholars and colleagues are earnestly acknowledged.

ORIGIN OF CRYSTALLINITY IN BLOCK AND RANDOM TEREPHTHALATE-SEBACATE COPOLYMERS AS STUDIED USING MONTE CARLO METHODS.

Tarek M. Madkour

Department of Chemistry
Helwan University
Cairo, Egypt

ABSTRACT

Random and block terephthalate-sebacate copolymers are known to have different thermal transitions while maintaining the same chemical structures. A considerable amount of research has been done in order to search for the origin of this behavior. While it is widely believed that sequence distribution in these copolymers is the primary cause, no theory was able to predict the characteristics of the thermal transition of the copolymers. Following up the Windle approach in generating copolymeric chains using Monte Carlo methods, one hundred chains have been simulated in order to allow for a search of crystallinity in these copolymers. According to the amount of crystallinity found in these copolymers at various feed compositions, the melting points of the different samples have been predicted. Other physical properties such as the interfacial free energy, the standard free energy of fusion and Young's modulus at small extensions were also predicted. The work is also capable of predicting the size of crystals and the minimum sequence length required for crystallization.

INTRODUCTION

Random and block copolymers made of one or more crystallizable components and having similar degrees of crystallinity might posses different thermodynamic and mechanical properties. On a qualitative basis, it was shown[1] that this difference in properties is probably due to the dependence of the melting temperature of a copolymer on the nature of its sequence distribution and not directly on its composition. For a random type copolymer, the sequence propagation probability, the probability that a crystallizable unit is succeeded by another such

Polymers and Other Advanced Materials: Emerging Technologies and Business Opportunities. Edited by P. N. Prasad *et al.*, Plenum Press, New York, 1995

unit, will equal the mole fraction of the crystallizable component. However, for an ordered block copolymer, this probability approaches unity and the depression of the melting temperature from that of the corresponding homopolymer will be only of a slight quantity. On the other hand, for an alternating type copolymer, the probability will be much less than the mole fraction and a drastic reduction in the melting temperature is expected. Of a particular interest here, is the crystallization process in poly(ethylene terephthalate-co-ethylene sebacate) copolymers. The interest lies within the ability of these copolymers to segregate and form materials of a crystalline nature both of random and block types. Moreover, the considerable difference in the melting points, due to the structural factors, between the two monomers makes a proper study of the crystallization process of one monomer possible.[2]

Hachiboshi and coworkers[3] could indeed show by etching of copolyesters and chemical analysis of the residual crystals and removed amorphous molecular portions that sebacate repeating units enter the ethylene terephthalate crystal only to a minor degree. The repeating units in these copolymers are long enough so that even in the mid-range of concentration some crystallinity remains. Kenney[4] also showed, in agreement with theoretical predictions, that while there is a monotonic decrease in the melting temperature with increasing concentration of the non-crystallizing sebacate units in random copolymers, the melting point remains constant for a large sebacate unit content in block copolymers. The above conclusions could indeed be explained qualitatively on the basis of the early work of Flory[5] and Wunderlich[6] and quantitatively by using computer simulations.[7-9]

Monte Carlo simulations were thus used to investigate the crystallization process in poly(ethylene terephthalate-co-ethylene sebacate) copolymers on a microscopic level. Several thermodynamic and mechanical properties of the copolymer under study both of random and block nature were evaluated. The study is also extended to predict the various properties of semi-blocky type copolymers, for which no experimental data is readily available.

SIMULATION METHODS AND RELATED THEORY

Simulation of the Copolymeric Chains

Copolyesters of ethylene terephthalate and ethylene sebacate units were simulated as to resemble real-time polymerization process. Edgar and Ellery[2] have indicated the inconvenience of preparing the copolymer by the direct condensation of the mixed acids with an equimolar quantity of the ethylene glycol. Consequently, they have carried out the preparations indirectly by heating together in vacuo the appropriate quantities of the esters. The low molecular weight esters referred to as "monomers" were thus used in the copolymerization process. Each monomeric unit will in turn consist of a number of repeat units that could be controlled during the preparation of these low molecular weight esters. During the copolymerization process, a variety of these esters could thus be used to produce different types of the copolymers. For instance, if the monomeric units consist of ten repeat units each, then a block copolymer is expected. On the other hand, if each monomeric unit consists of only one repeat unit of the comonomer, then a random type copolymer is expected.

Accordingly, simulation of the copolymeric chains was carried out by considering monomeric units of different lengths. One hundred chains, each containing a hundred units, were simulated according to the theory of copolymerization.[7,8] Reactivity ratios used in the calculations were taken as unity, and only "terminal" chain effects is considered (only the prior unit on the growing chain influences the type of the unit being added). This eventually will lead to the formation of a random type copolymer with a sequence length that equals at least the number of repeat units per the monomeric unit. For random, semi-blocky, and block copolymers, referred to throughout the figures as R, S, B, the numbers of such repeat units are one, three, and ten, respectively.

Search for Crystallinity

The search for crystallinity was carried out according to the Windle approach[10,11] and took place at the hypothetical temperature of 100°C so that ethylene sebacate sequences (T_m = 77°C) would remain amorphous regardless of their lengths. Ethylene terephthalate sequences having sufficient sequence length and number of matches with neighboring sequences would be expected to crystallize.[10] Crystallites formed during slow crystallization would grow longitudinally until meeting a unit of the other type, and laterally so long as sequences of the crystallizable sequences of a minimum sequence length were available from the amorphous regions.[5]

Melting Points

The extent to which the melting point of the crystallizable units in a polymer is depressed by the presence of the other units is given by[8,13]

$$\frac{1}{T_m} - \frac{1}{T^0_m} = -\frac{R}{\Delta H_u} [\ln(L)] \tag{1}$$

where T^0_m and T_m are the melting points of the pure crystallizable polymer and the polymer under study, respectively, ΔH_u is the enthalpy of fusion per mole of crystallizable units, R is the gas constant, and L is the degree of crystallinity. The depressed melting points observed experimentally[2] can be compared with those obtained from the simulations. The sequence length and the number of exact sequences giving the best fit of the experimental data can be taken as the minimum value required for crystallization of the terephthalate units.

Chain Matching

Sliding the chains past one another longitudinally to search for the largest possible matching densities was done to model annealed samples.[8,10,11] The longitudinal movement of the chains relative to one another, out of register, approximately models the lateral sorting out of sequences in polymeric chains during slow crystallization.

MODELING RESULTS

Identification of Crystalline Regions

It was estimated from the comparison results with the experimental melting point depressions[2] that three ethylene terephthalate units is the minimum number for a sequence to participate in the formation of a crystallite. Nevertheless, this was the same number predicted by Wunderlich[6] on applying the theory of 'cold' crystallization. It was also determined from the simulations that three matching sequences on three neighboring chains were also required for the formation of a crystallite. The low value of the minimum sequence length of the terephthalate comonomer when compared to values determined previously for other systems[8] is probably due to the greater length of the terephthalate unit. Some graphical representations of the results obtained are shown in Figures 1 through 3. All pertain to poly(ethylene terephthalate-co-ethylene sebacate) copolymers having a composition of 60 mol%

(a)

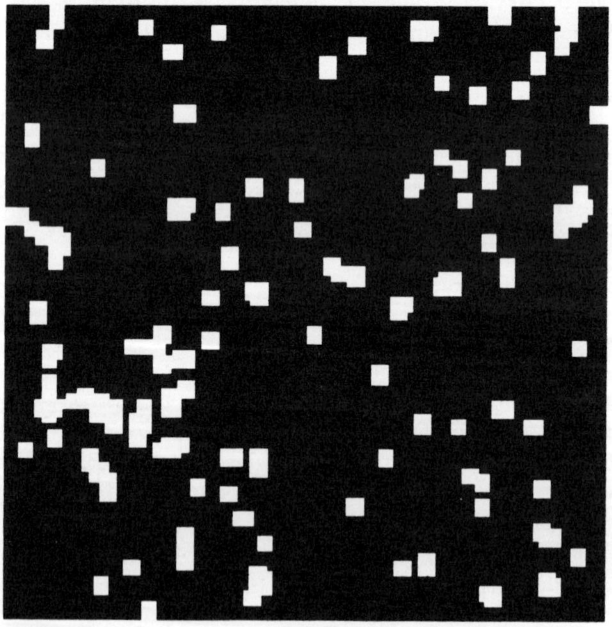

(b)

Figure 1. Simulated copolymeric structures (a) for the random copolymer, and the corresponding crystalline regions (b) after annealing. See text.

Figure 2. Simulated copolymeric structures (a) for the semi-blocky copolymer, and the corresponding crystalline regions (b) after annealing. See text.

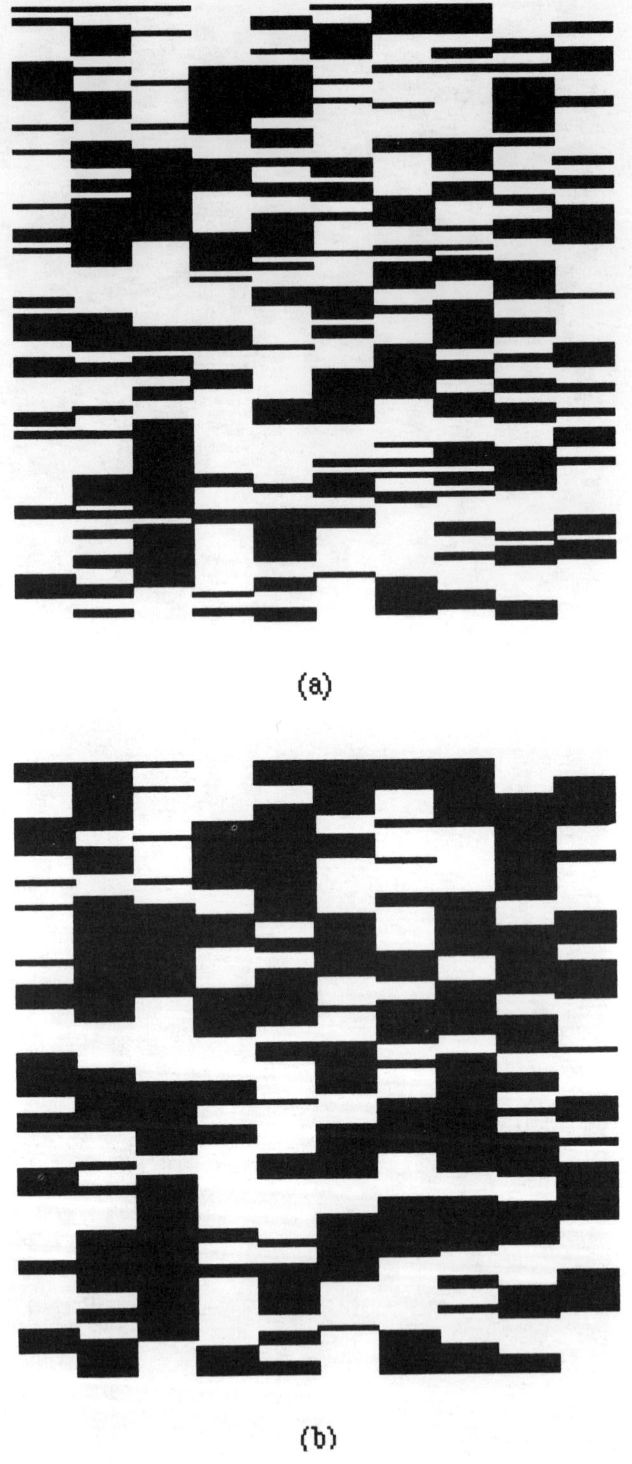

Figure 3. Simulated copolymeric structures (a) for the block copolymer, and the corresponding crystalline regions (b) after annealing. See text.

terephthalate units. The figures correspond to random, semi-blocky and block copolymers, respectively. In (a) sections of the figures, the black squares represent sebacate units whereas the white squares represent terephthalate units. Like squares neighboring each other in the same row coalesce into blocks the lengths of which are under scrutiny. The (b) sections show the extent of the crystallized regions in the different samples as represented by the white blocks, black regions in this case representing units of the non-crystallizable component, or sequences of the crystallizable component that were not long enough to participate in the crystallization. It is obvious from the figures that semi-blocky and block copolymers posses more crystallinity than random copolymers do. This increased crystallinity is responsible for the high melting points of block copolymers over a wide range content of the terephthalate component. Moreover, these crystallites presumably act as crosslinking sites and reinforcing domains, providing the additional toughness block copolymers are known to have over their random counterparts.[14]

Degree of Crystallinity

Values of the degree of crystallinity L of the various simulated samples were determined by counting the units involved in the matching sequences with respect to the total number of units (10,000) of the samples.

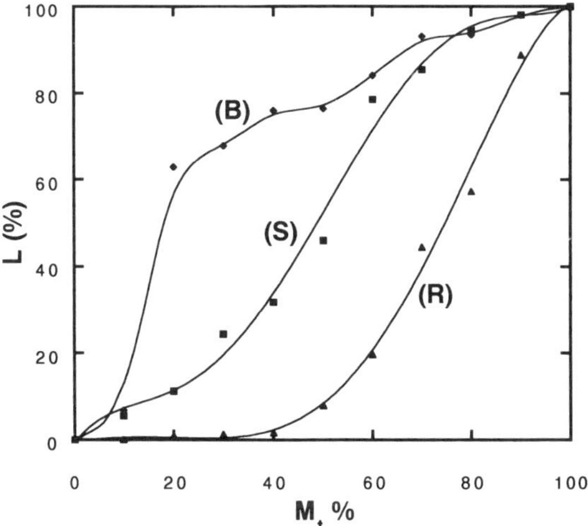

Figure 4. Degree of crystallinity of the copolymers as a function of the amount of terephthalate in the feed composition. Here, (R), (S), and (B) denote random, semi-blocky and block copolymers, respectively.

Results are presented in Figure 4 as a function of the terephthalate monomer feed composition $M_t\%$. They suggest that the degree of crystallinity did change greatly with the change in composition, as expected. Specifically, the degree of crystallinity ranged from 0% at 100% sebacate units (which agrees with the lower melting point of the homopolymer)[12], to an ideal value of 100% at 100% terephthalate content. Block copolymers showed more crystallinity content than random copolymers since it contains sequences of at least ten-unit long with increased sequence propagation probability. Such crystallinity percentages of block copolymers which rise sharply at the range of 10-20 $M_t\%$ would explain the corresponding sharp increase in the melting points of the same samples. Semi-blocky copolymers with sequences of at least three-unit long will lead to crystallinity values that fall right inbetween those of the random and block copolymers.

Prediction of The Melting Points

All the simulated melting points were obtained for the annealed case, with the chains permitted to slide past one another for a full length to locate the best matching density nearest to the starting points. The chain sections were relocated from one end to the other so as to keep the number of comparison pairs constant.

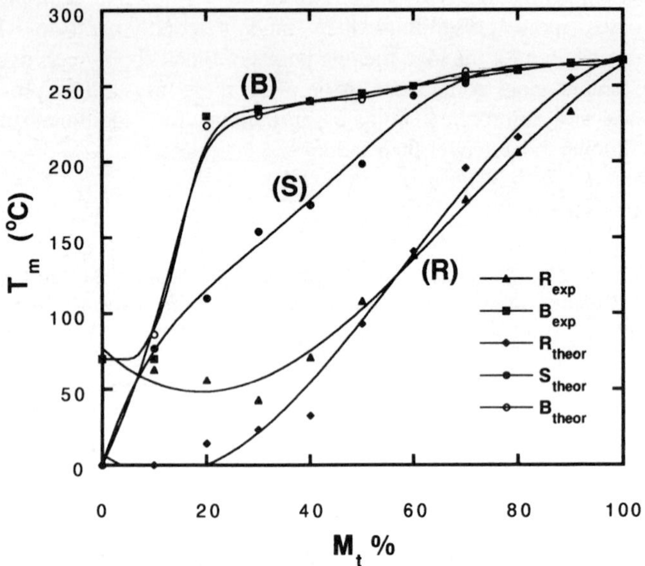

Figure 5. Melting Points of the copolymers as a function of the amount of terephthalate in the feed composition. Here, (R), (S), and (B) denote random, semi-blocky and block copolymers, respectively.

Figure 5 represents a comparison between the experimentally determined and the theoretically predicted melting points of various random (R) and block (B) copolymers. The predicted melting points were obtained by using the simulated values of the percent crystallinity in equation (1), the melting point depression relationship. It should be recalled that it was the comparison of experimental values and simulated values for the polymer that was used to estimate the minimum sequence length for crystallization. The figure clearly shows the excellent agreement between both mentioned melting points for random and block copolymers. The agreement thus indicates the validity of the used simulation techniques in the evaluation of origin of crystallinity in copolymers. It should be mentioned, however, that the discrepancy in the comparison at the lower M_t% values is due to the fact that contribution of the sebacate units to the crystallinity of the copolymers was ignored during the simulation. Those melting points that fall below the hypothetical 100°C should not be compared to experimental values. Nonetheless, the sharp increase in the melting points in case of block copolymers at 20 M_t% is probably due to the high sequence propagation probability at this low M_t% value. Moreover, the melting points of systems such as semi-blocky ones for which no experimental data is available were also predicted and shown in the figure.

Thermodynamics of Crystallization

The Flory theory[13] also provides equations for several physical quantities associated with crystallization. The arguments are based on the assumption that matching sequences of a minimum sequence length ζ of neighboring chains would coalesce to form a crystallite. It is also suggested that addition of a sequence of crystallizable units of length equal to or greater than ζ to a growing crystallite depends on: (i) the availability of neighboring sequences in the amorphous region, and (ii) the decrease in standard free energy gained by incorporation of amorphous polymer units into the crystalline state.

The probability P_ζ of finding a unit residing adjacent to a terminal unit of the crystallite and belonging to a sequence of ζ units will be related to the standard free energy of fusion ΔF_ζ of a sequence of ζ units from a crystallite of length ζ. It is given by[13]

$$P_\zeta = P^e_\zeta = \exp(-\Delta F_\zeta / RT) \qquad (2)$$

Here, ΔF_ζ is expressed in molar units and P^e_ζ is measured under equilibrium conditions (as indicated by the superscript). This probability was calculated from the simulated results by counting the units adjacent to a crystallite, with potential for crystallization, as a function of $M_t\%$.

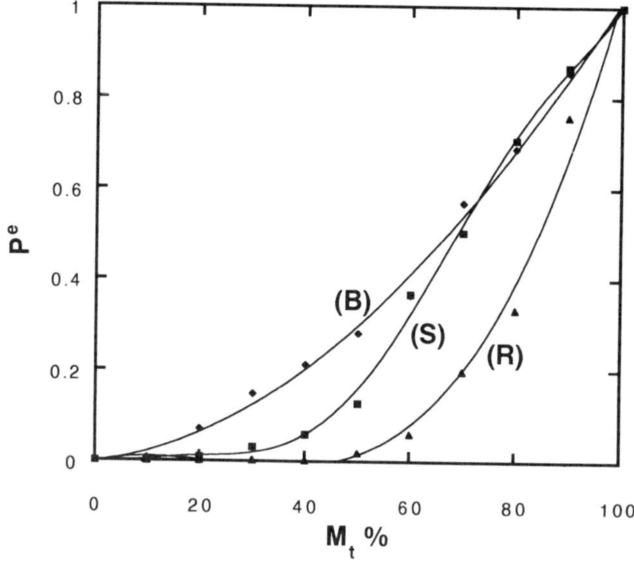

Figure 6. Probability of additional crystallizable units in the copolymers as a function of feed composition. Here, (R), (S), and (B) denote random, semi-blocky and block copolymers, respectively.

The values of this probability are shown in Figure 6. As expected, the values increase with the increase in the amount of the terephthalate component and are much larger for the semi-blocky samples and even greater for the block ones. The values of ΔF_ζ calculated from equation (2) are similarly shown in Figure 7. The results indicate a decrease in the standard free energy of fusion for a sequence of ζ units upon increase in the number of the crystallizable units. This is probably related to changes in the crystallite size.

Interfacial Free Energies

Another important crystallization variable is σ_e, the interfacial free energy per unit, which represents the excess free energy per mole associated with a unit at the end of a crystallite. Values of this quantity were also calculated from the simulation results.

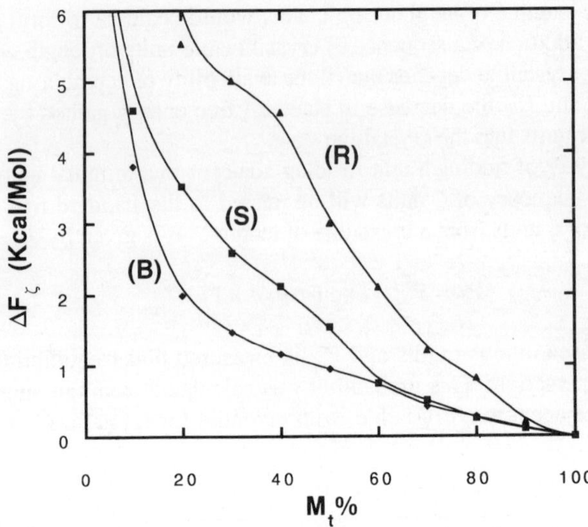

Figure 7. Standard free energy of fusion of a sequence of ζ units as a function of feed composition. Here, (R), (S), and (B) denote random, semi-blocky and block copolymers, respectively.

Specifically, the standard free energy of fusion per mole of units ΔF_u was calculated from

$$\Delta F_u = \Delta H_u - T\Delta S_u \tag{3}$$

and σ_e from

$$\Delta F_\zeta = \zeta \Delta F_u - 2\sigma_e \tag{4}$$

Estimates of the required values of ΔH_u and ΔS_u were obtained from the literature.[6,12] The values of σ_e thus obtained are shown in Figure 8. This interfacial free energy is an important variable since it takes into account the decreased stability of the shorter crystallites. It can be used as an indicator of the status of the predicted crystallinity in the copolymers. For example, it suggests that at all cases the shorter crystallites would be unstable and be reincorporated into the melt. The sharp downturn in block copolymers in the 10-20 M_t% range that is consistent with values of the melting points and degrees of crystallinity at this region indicates the greater instability of short crystallites at this feed composition range as a consequence of the formation of the larger crystals.

Young's Modulus

The theory of rubberlike elasticity[8,13,15,16] relates the modulus of elasticity to the number of crystalline sequences acting as cross links, with the sequences in the amorphous state contributing as elastomeric chains. If the number of crystalline sequences per unit volume is represented by υ_c, then Young's modulus at infinitesimal deformations is given by

$$Y_o = 3RT\,(\upsilon_c / N_1) \tag{5}$$

where N_1 is the number of the units actually participated in the formation of the crystallites. Values of Young's modulus predicted from the simulated values of υ_c are shown in Figure 9.

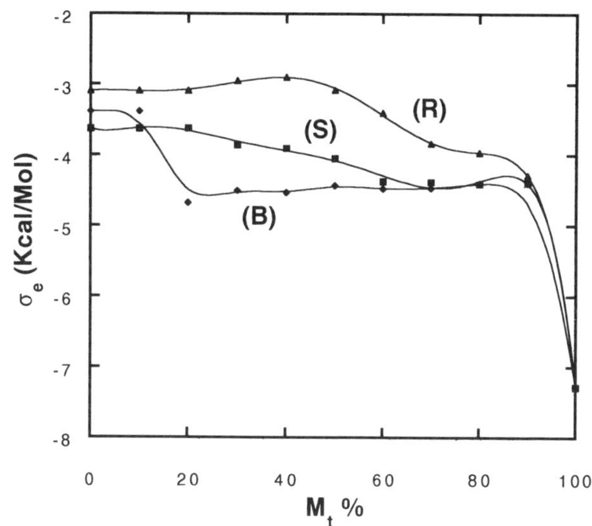

Figure 8. Interfacial free energy per mole of units as a function of feed composition. Here, (R), (S), and (B) denote random, semi-blocky and block copolymers, respectively.

Of a major interest here, is the predicted Young's moduli of the semi-blocky samples which show greater values than those of the random and block ones. The highest predicted value is being at 60 M_t%. The reason behind the high values of the semi-blocky copolymers is the dependence of the modulus of elasticity on the number of crystalline sequences acting as cross links, with the sequences in the amorphous state contributing as elastomeric chains.

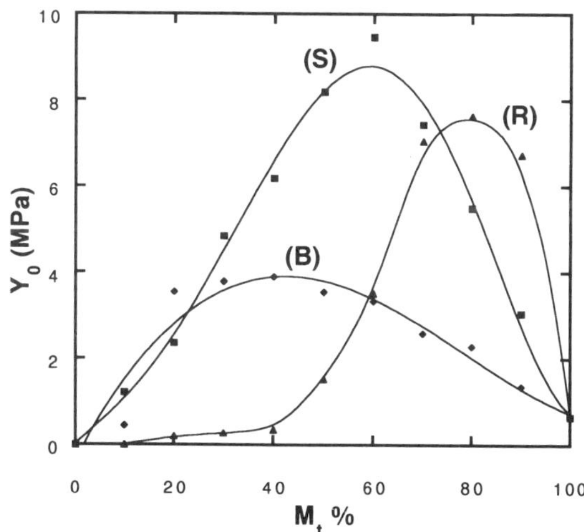

Figure 9. Young's modulus at infinitesimal extension as a function of feed composition. Here, (R), (S), and (B) denote random, semi-blocky and block copolymers, respectively.

Even though two different semi-blocky and block samples might have similar amounts of crystallinity, the number of sequences in the amorphous state will be higher in the former.

This is because of the scattering of the small crystals all around the sample acting as reinforcing domains and contributing to the moduli of the copolymers.

REFERENCES

1. J. E. Mark, A. Eisenberg, W. W. Graessley, L. Mandelkern and J. L. Koenig. "Physical Properties of Polymers," American Chemical Society, Washington, D.C. (1984).

2. O. B. Edgar, and E. J. Ellery, Chem. Soc. 2633 (1952).

3. M. Hachiboshi, T. Fukuda, and S. J. Kobayashi, Macromol. Sci. Phys. 3:525 (1969).

4. J. F. Kenney, Polym. Eng. Sci. 8:216 (1968).

5. P. J. Flory, Trans. Faraday Soc. 51:848 (1955).

6. B. Wunderlich, Chem. Phys. 29:1395 (1958).

7. T. M. Madkour, and J. E. Mark, Comput. Polym. Sci. 4:79 (1994).

8. T. M. Madkour, and J. E. Mark, Comput. Polym. Sci. 4:87 (1994).

9. T. M. Madkour, A. Kloczkowski, and J. E. Mark, Comput. Polym. Sci. 4:95 (1994).

10. S. Hanna, and H. Windle, Polymer 29:207 (1988).

11. S. Hanna, B. L. Hurrell and A. H. Windle, in: "Crystallization of Polymers," Dosiere, ed., Kluwer Academic Publishers, Dordrecht (1993).

12. J. Brandrup and E. Immergut. "Handbook of Polymer Science," Wiley, New York (1975).

13. P. J. Flory. "Principles of Polymer Chemistry," Cornell University Press, Ithaca, New York (1953).

14. B. Wunderlich. "Macromolecular Physics," Academic Press, New York (1980).

15. L. R. G. Treloar. "The Physics of Rubber Elasticity, Third Edition," Oxford University Press, Clarendon (1975).

16. J. E. Mark and B. Erman. "Rubberlike Elasticity. A Molecular Primer," Wiley-Interscience, New York (1988).

STRUCTURAL DESIGN SYNTHESIS OF MARINE CRAFTS USING FRP MATERIALS IN SANDWICH CONSTRUCTION

Abdul Razak Bin Mohd Sidek,[1] and Mohd Ramzan Bin Mainal[2]

[1]Lecturer
[2]Senior Lecturer
Faculty of Mechanical Engineering
Universiti Teknologi Malaysia
Locked Bag 791, 80990 Johor Bahru
Malaysia

INTRODUCTION

The potential use of composite materials in marine structures is been realized with its various applications. It ranges from hull structure of displacement vessels, fishing vessels, small crafts, high performance craft (surface effect ships and hovercrafts) to other applications such as deck houses to submersibles. The advantages of *Fibre Reinforced Plastic (FRP)* are well documented[1]. It includes in particular relatively low density and good resistance to fatigue and to the harsh marine environment. In addition, the low stiffness property of the material compared to metals may be an advantage in some applications. By varying the lay-ups and the constituent matrix and fibers, many specific properties are obtained against non structural requirements such as reflectivity and acoustic damping.

The development of new vehicle forms other than displacement types potentially to attain higher speeds require weight reductions as one of its performance driven parameters in its design philosophy. As these high performance vehicles require design from first principles, they offer a great chance for advanced composite materials to be used in the pursuit for the lightest marine vehicles. The lighter structural weight enhanced its hydrodynamic performance as well as leading to higher payload in the final design.

Majority of contemporary marine structures in *Glass Reinforced Plastics* (*GRP*) are very conservatively designed on the basis of available *Classification Society Rules & Regulations*. These *Rules* are based on E-glass and polyester coupled with the assumption of isotropy of composite materials in its formulations and assumptions. The safety factors[2] considered range from a value of 2 to 10 for the various loads as shown in Table 1.

Polymers and Other Advanced Materials: Emerging Technologies and Business Opportunities. Edited by P. N. Prasad *et al.*, Plenum Press, New York, 1995

Table 1. Safety Factors for Types of Load

Load	Safety factor
Static (short term)	2
Static (long term)	4
Variable	4
Repetitive	6
Fatigue	6
Impact	10

These *Rules* are generally adequate for most marine applications where weight considerations rarely the important issue concerned. In cases of high performance large vessels and special purpose naval ships, the application of these *Rules* is not possible mainly due to the hull dimensions are out of range of its scantling tables. Folowing the *Rules* despite the earlier facts' results in the weight achieved to be unacceptably large. So in cases involving high performance marine vehicles where weight is the primary concern in its performance driven limits, a more rationale structural design procedure needs to be followed as opposed to design by *Rules*.

The paper aims at presenting the structural synthesis concept as a framework within which the quantifiable portions of the marine design are formulated. This process is rational optimum design of structural systems. Within this framework, particular emphasis will be on formulations and the associated path and solutions that have been proven and implemented successfully by various researchers within the marine and composite world.

MARINE SANDWICH STRUCTURES

Sandwich Structural Concepts

The increase usage of higher performance composite materials as compared to the common composite materials of glass *FRP* has lead to the reduction of skin thickness in hull and decks. The situation arises where a closely framed structure is required due to the relatively thin and delicate skin. To overcome the less attractive options of having a closely framed structures various concepts are used in the design. One of the structural concepts that used in the ship and boat industry is the *FRP* sandwich. The characteristics of the sandwich construction mean that large panels are possible without the needs for additional stiffening thus giving smooth surfaces on both sides. The extra advantange of increased stiffness is also achieved without the expense of weight. Additional advantage obtained is built-in thermal insulation provided by the core materials used. In many cases on high speed vessels, longitudinal stiffening is virtually eliminated.

Skin Laminates

By far the most commonly used combination in marine applications has been glass reinforced polyester comprising of E-glass arranged as woven fabrics or more recently in a form of multi axial reinforcements. *Aramid* fibres such as *Kevlar* has been used extensively in performance yatchs and racing craft. The properties of aramid fibres with good tensile and stiffness properties but poor in compression have made it rather unattractive as skin laminates in sandwich constructions. Nevertheless the combinations of *Aramid* fibres with glass and carbon to form a hybrid laminate taking the distinctive advantage of hybridized effect is being sought in design where weight is very critical and high costs can be justified. With increasing use of these advanced composite materials in other fields their price is now reducing. More general use of these materials can be contemplated in marine field in order to take the advantage of its distinct properties as shown in Table 2.

Table 2. Representative of some reinforcing fibres.

Fibre Type	Density (gm/cm^3)	Young's Modulus (GPa)	Poisson ratio	Tensile strength (GPa)	Failure strain (%)	Rrelative cost
E-Glass	2.55	72	0.20	2.40	3.00	1.00
S2-R-Glass	2.50	8	0.20	3.40	3.50	8.00
High Strength Carbon	1.74 - 1.81	248 - 345	-	3.10 - 4.50	0.90 - 1.80	45 - 50
High Modulus Carbon	2.00 - 2.18	520 - 826	-	2.10 - 2.20	0.30 - 0.40	250 - 2700
Aramid (Kevlar 49)	1.45	124	-	2.80	2.50	15.0

The most frequently used resin in boat building is orthophthalic polyester. Isophthalic polyester has better properties particularly with resistance to water absorption. This type of polyester is generally used in naval vessels and higher quality commercial vessels. Epoxy resin that is generally better than polyester resin is currently restricted to smaller constructions. This is mainly due to the need of high temperature cure. The advancements of epoxy resin formulation leading to reduced toxicity level with lower temperature cure of 60° -70° C with a prolonged working time currently to 2 to 3 hours, enable larger working hours. The possibility of using a vacuum bag production method by curing in a simple enclosure for several hours at 60° - 70° C is feasible thus opening up its usage in the marine structures. Vinyl esters with high fibre volume and is also used in cases where cost is less significant but toughness is more important.

Sandwich Cores

Crosslinked *Polyvinyl Chloride* (*PVC*) foam is the most common sandwich core material, typically in the range of 80 kg/m^3 for superstructures and 130-200 kg/m^3 for bottom structures. Linear *PVC* foams are also used but with inferior static properties to the crosslinked foams. However it has a higher ductility with high degree of damage tolerance. Other types of cores used are end grain balsa wood and polymethacrylic imide (*PMI*). Due to brittle nature of *PMI* its usage is generally confined to superstructures. For end grain balsa wood a very good bond between the care and the skin is generally difficult to obtain thus susceptible to water absorption.

Bonding between core and skin laminates by direct method require priming of the core with resin. A layer of chopped strand mat is laid before the main reinforcement fabrics. It is also possible to glue the core by glue with the aids of vacuum-bag system enabling skin laminates to be made separately.

PRELIMINARY STRUCTURAL DESIGN

Sandwich Design Scheme

The basic approach for sandwich design of vessels is to develop a schematic steps to reduce the associated complex structure to a manageable units and analyzing it as a plates or beams. A widely used design scheme for composite structures are as follows:

1. Define panels
2. Description of possible loads, including combinations of loads
3. Evaluation of material(s) to be used
4. Design and analyse laminates
5. Optimize

Defination of Panels. The complexity of structural system is initially reduce to a manageable unit comprising its basic components representing principle support system to the hull and deck. Such unit ranges from longitudinal and transverse stiffeners, bulkheads, engine beds and frames. The panels that set the structural standard is then isolated with reference to standard guidelines to *Classification Society Rules and Regulations.* The isolated panels normally comprises of the midship hull bottom panels, forward hull bottom panel, foredeck and deckhouse front panel. Judgment is made as to the other relevant panels that need to be isolated with respect to the design requirements. The end conditions are then defined for every panels defined accordingly based on the values suggested and practiced in the normal design of steel vessels as proposed by *Classification Society*.

Loads Defination. The first step normally taken in the structural design of conventional ship is the determination of shear force and bending moment distributions either in hogging and sagging condition along the ship length. The ship is treated as a free-box girder. The curves obtained are treated as a major loading condition from which the basis of calculations is performed. Any local loading is treated as secondary, mostly used for secondary detailing and reinforcement. For *FRP* crafts that is limited in length, the predominant loads are the local loads which determined the required scantlings. Nevertheless it is advisable to consider the effect of longitudinal loads as well.

The actual loads are governed by the nature of the ship whether it is a displacement, semi planing and planing mode. For displacement craft the major load is the hydrostatic loads while semi-planing and planing craft, the most significant loads is the slamming loads on the bottom panels. Various theoretical formulas have been proposed and the best combination approach is to calculate the maximum acceleration of the vessels using the Savitsky's formula[3] and the slamming pressure using Allen and Jones's formulation[4]. Other major loads that might be of importance is the shock loading from underwater explosions in cases of naval crafts. Various research in this matter has been conducted[5].

Material Characterization. The availability of various alternatives materials apparently lead to a difficulties of selecting the material systems best suited. In some cases the choice of materials system is governed by previous experience, the availability of the production techniques or the availability of materials. Nevertheless a paramatric studies involving a predetermined objectives such as cost, weight or both will dictate the probable optimum solution.

The problem is compounded further by the values of mechanical properties of the materials to be considered. These values are influenced by many factors such as fibre content, workmanship and surrounding environment. The actual properties could only be obtained by specimen testing at the real production environment. In preliminary design, mechanical properties can be estimated from the approximate formula or nominal values obtained from the manufacturers of the materials.

Laminate Design. The laminate design basically involves the process of estimating the response of sandwich panels to the associated dimensioned loads. In sandwich construction, plate shear and thickness direction stresses and deformations as well as bending and membrane effect may be important. Local and global buckling phenomena may also play a role.

Stress, deformation and buckling analysis of sandwich panels need to take into account correctly of the material anisotropy. Stress-strain relationship assuming orthrothopic elasticity at a point is the related by the expression[6]:

$$\begin{Bmatrix} \sigma_1 \\ \sigma_2 \\ \sigma_3 \\ \sigma_4 \\ \sigma_5 \\ \sigma_6 \end{Bmatrix} = \begin{bmatrix} Q_{11} & Q_{12} & Q_{13} & 0 & 0 & 0 \\ Q_{12} & Q_{22} & Q_{23} & 0 & 0 & 0 \\ Q_{13} & Q_{23} & Q_{33} & 0 & 0 & 0 \\ 0 & 0 & 0 & 2Q_{44} & 0 & 0 \\ 0 & 0 & 0 & 0 & 2Q_{55} & 0 \\ 0 & 0 & 0 & 0 & 0 & 2Q_{66} \end{bmatrix} \begin{Bmatrix} \varepsilon_1 \\ \varepsilon_2 \\ \varepsilon_3 \\ \varepsilon_4 \\ \varepsilon_5 \\ \varepsilon_6 \end{Bmatrix} \tag{1}$$

where

$$\begin{aligned}
Q_{11} &= E_{11}(1-\nu_{23}\nu_{32})/\Delta \\
Q_{22} &= E_{22}(1-\nu_{31}\nu_{13})/\Delta \\
Q_{33} &= E_{33}(1-\nu_{12}\nu_{21})/\Delta \\
Q_{44} &= G_{23}, Q_{55} = G_{13}, Q_{66} = G_{12} \\
Q_{12} &= (\nu_{21}+\nu_{31}\nu_{23})E_{11}/\Delta = (\nu_{12}+\nu_{32}\nu_{13})E_{22}/\Delta \\
Q_{13} &= (\nu_{31}+\nu_{21}\nu_{32})E_{11}/\Delta = (\nu_{13}+\nu_{12}\nu_{23})E_{22}/\Delta \\
Q_{23} &= (\nu_{32}+\nu_{12}\nu_{31})E_{22}/\Delta = (\nu_{23}+\nu_{21}\nu_{13})E_{33}/\Delta \\
\Delta &= 1-\nu_{12}\nu_{21}-\nu_{23}\nu_{32}-\nu_{31}\nu_{13}-2\nu_{21}\nu_{32}\nu_{13}
\end{aligned}$$

Relating these relationships to the x, y and z coordinate systems

$$\begin{Bmatrix} \sigma_1 \\ \sigma_2 \\ \sigma_3 \\ \sigma_4 \\ \sigma_5 \\ \sigma_6 \end{Bmatrix} = \begin{bmatrix} \overline{Q}_{11} & \overline{Q}_{12} & \overline{Q}_{13} & 0 & 0 & 0 \\ \overline{Q}_{12} & \overline{Q}_{22} & \overline{Q}_{23} & 0 & 0 & 0 \\ \overline{Q}_{13} & \overline{Q}_{23} & \overline{Q}_{33} & 0 & 0 & 0 \\ 0 & 0 & 0 & 2\overline{Q}_{44} & 2\overline{Q}_{45} & 0 \\ 0 & 0 & 0 & 2\overline{Q}_{45} & 2\overline{Q}_{55} & 0 \\ 0 & 0 & 0 & 0 & 0 & 2\overline{Q}_{66} \end{bmatrix} \begin{Bmatrix} \varepsilon_1 \\ \varepsilon_2 \\ \varepsilon_3 \\ \varepsilon_4 \\ \varepsilon_5 \\ \varepsilon_6 \end{Bmatrix} \tag{2}$$

where

$$\left[\overline{Q}\right] = [T]^{-1}[Q][T] \tag{3}$$

$$[T] = \begin{bmatrix} m^2 & n^2 & 0 & 0 & 0 & 2mn \\ n^2 & m^2 & 0 & 0 & 0 & -2mn \\ 0 & 0 & 1 & 0 & 0 & 0 \\ 0 & 0 & 0 & m & -n & 0 \\ 0 & 0 & 0 & n & m & 0 \\ -mn & mn & 0 & 0 & 0 & (m^2 - n^2) \end{bmatrix} \tag{4}$$

where $m = \cos\theta$, $n = \sin\theta$

The integrated constitutive equation can be written as follows:

$$\begin{bmatrix} N_x \\ N_y \\ N_{xy} \\ M_x \\ M_y \\ M_{xy} \end{bmatrix} = \begin{bmatrix} A_{11} & A_{12} & 2A_{16} & B_{11} & B_{12} & 2B_{16} \\ A_{12} & A_{22} & 2A_{26} & B_{12} & B_{22} & 2B_{26} \\ A_{16} & A_{26} & 2A_{66} & B_{16} & B_{26} & 2B_{66} \\ B_{11} & B_{12} & 2B_{16} & D_{11} & D_{12} & 2D_{16} \\ B_{12} & B_{22} & 2B_{26} & D_{12} & D_{22} & 2D_{26} \\ B_{16} & B_{26} & 2B_{66} & D_{16} & D_{26} & 2D_{66} \end{bmatrix} \begin{bmatrix} \varepsilon_{x0} \\ \varepsilon_{y0} \\ \varepsilon_{xy0} \\ \kappa_x \\ \kappa_y \\ \kappa_{xy} \end{bmatrix} \tag{5}$$

where (x, y) represents an arbitrary coordinate system

$[\varepsilon_{x0} \quad \varepsilon_{y0} \quad \varepsilon_{xy0}]$ = vector of the midplane strains

$[\kappa_x \quad \kappa_y \quad \kappa_{xy}]$ = vector of the curvatures

$[N_x \quad N_y \quad N_{xy}]$ = vector of the stress resultant

$[M_x \quad M_y \quad M_{xy}]$ = vector of the moment resultant

The equation above can be solved for the midplane strains and curvatures as follows:

$$\begin{bmatrix} \varepsilon_0 \\ \kappa \end{bmatrix} = \begin{bmatrix} A & B \\ B & D \end{bmatrix}^{-1} \begin{bmatrix} N \\ M \end{bmatrix} \tag{6}$$

where A, B and D are the extensional, coupling and stiffness matrix defined as

$$\left[A_{ij},\ B_{ij},\ D_{ij}\right]=\sum_{k=1}^{n}\left(\overline{Q}_{ij}\right)_k\left(z_k - z_{k-1},\frac{z_k^2 - z_{k-1}^2}{2},\frac{z_k^3 - z_{k-1}^3}{3}\right)$$

Solutions can then be obtained with the above integrated constitutive equations together with equilibrium equations and strain displacement relations for the beam and plate structures to be analyzed. By assuming that a sandwich construction is consists of three ply consisting of core and its two upper and lower faces, numerous solutions available for composite plate can be used in solving sandwich problems.

For determination of shear stress ie σ_{xz} and $\sigma_{yz} \neq 0$ it is assumed that the transverse shear stresses are distributed parabolically across the laminate thickness. A continuos function *f(z)* is used as a weighting functions which is consistent with Reissner:

$$f(z)=\frac{5}{4}\left[1-\left(\frac{z}{h/2}\right)^2\right] \tag{7}$$

From equation (2)

$$\begin{aligned}\sigma_{xzk} &= 2\overline{Q}_{55}\varepsilon_{xzk} + 2\overline{Q}_{45k}\varepsilon_{yz}\\ \sigma_{yzk} &= 2\overline{Q}_{45k}\varepsilon_{xz} + 2\overline{Q}_{44k}\varepsilon_{yz}\end{aligned} \tag{8}$$

hence

$$\begin{aligned}Q_x &= 2\left(A_{55}\varepsilon_{xz} + A_{45}\varepsilon_{yz}\right)\\ Q_y &= 2\left(A_{45}\varepsilon_{xz} + A_{44}\varepsilon_{yz}\right)\end{aligned} \qquad \text{where } A_{ij}=\frac{5}{4}\sum_{k=1}^{n}\left(\overline{Q}_{ij}\right)_k\left[h_k - k_{k-1} - \frac{4}{3}\left(h_k^3 - h_{k-1}^3\right)\frac{1}{h^3}\right]$$

Other solution of inclusion of shear stress effect is also available with improved Mindlin theory for the determination of improved transverse shear stiffness for anisotropic sandwich plate[7]. Limited number of solutions are available to this plate problem with the inclusion of transverse effect. Elasticity solutions are also available in solving anisotropic sandwich plate as shown by Pagano,[8,9,10] but quite limited in its available solutions with respect to plates. In preliminary design classical laminate theory is adequate without the effect of shear stress taken into account.

Considerations will now be focused on the various failure modes to provide the necessary design criteria. Generally a foam cored panel can fail by facing failure in tension and compression, core failure in shear and in compression, general and local buckling plus bond line failure.

Facing stress has been shown earlicr to be the sum of moments applied to the panel and possibly an in-plane load. If the stresses exceed the critical strength of the materials, the facings will fail either in compression or tension. As proposed by the Classification Society a Tsai-Wu failure criterion is favoured.

The foam core will also fail if the shear stress exceeds its shear strength. Under transverse loading, the core is predominantly in a state of stress. In DNV *Rules*, the limit is currently set at 35 or 40 % of ultimate shear strength. Attempts have been made to model

PVC core materials nonlinearly[11] in a finite element analysis. Under pure shear, the anisotropic yield criterion reduces to the same simple condition as the von Mises criterion under pure shear ie $\sigma_y = |\tau|$ where σ_y is the yield stress and τ is the shear stress in cartesian coordinates.

General buckling of sandwich panels under in-plane loading is a function of end fixity, aspect ratios and also mechanical properties of materials. Local buckling or general known as face wrinkling is as a results of the incapability of the core to sustain the facings laterally. The face wrinkling phenomena can be written as:

$$\sigma_{critical} = \phi\left(E_f E_c G_c\right)^{\frac{1}{3}} \text{ where G is the shear modulus}$$

$$\text{and } \phi = \left(\frac{9\left(1-\nu_c^2\right)}{2\left(1+\nu_c\right)\left(3-4\nu_c\right)^2\left(1-\nu_f^2\right)}\right)^{\frac{1}{3}} \tag{9}$$

Inadequate bonds between facing and cores are responsible for more sandwich failures than failures due to facings or cores. The bond line must be continuos and of consistent thickness.

Optimization. The process of optimization of the respective units with its dimensioned loads should be considered as an essential step in weight critical structures. The respective failure modes can be considered as the governing constraints that could be associated and model in the optimization process. Generally in composite various optimization have been conducted ranging from the optimal design of composite based on classical laminate theory without the effect of transverse shear stress[12,13,14,15] to the process of optimization using finite element modeling which take into account of the shear stress effect[16]. The optimization routines differs in range from continuos non-linear functions to discrete algorithm. The respective objective is to find the best ply ratios and orientations possible with the acting loads either in-plane or lateral with the weight as the governing factors.

CONCLUSION

The rationally based design of sandwich composite structures is essential as design from *Rules* by *Classification Society* does not cover advanced composite materials. These Rule are based on contemporary composite materials.

The design of any units of structure need to be addressed independently. A set of predefined parameters as well as design constraints and objectives need to be formulated. The optimum solution for the resulting non-linear formulation can be solved by multiple programming techniques or the optimal criteria. Finite element formulations using suitable elements for composite structures could also be the tools in solving the problems.

The rationally based design of sandwich construction has the advantage of:

a. arriving at absolute minimum weight for a given type of sandwich, loading and materi system.
b. having the ability to select the best face and core material to minimize weight structure.
c. the best stacking sequence for faces composed of laminated composite structures

REFERENCES

1. Smith C. S., 'Design of Marine Structures in Composite Materials', Elsevier Applied Science, London, England, 1990.
2. Gibbs and Cox Inc., 'Marine Design Manual for Fibre Reinfirced Plastics', McGraw-Hill, New York, 1960.
3. Savitsky D., Brown W., 'Procedures for Hydrodynamic Evaluation of Planing Hulls in Smooth and Rough Waters", Marine Technology, Vol. 13, No. 4, pp 381-400.
4. Allen R. G., Jones R. R., 'A Simplified Method for Determining Structural Design-Limit Pressures on High Performance Marine Vehicles", AIAA/SNAME Advanced Marine Vehicles Conference, San Diego, Paper No 78-754.
5. Moyer Jnr. E. T., Amir G. G., Olsson K. A., Helbratt S. E., 'Response of GRP Sandwich Structures Subjected to Shock Loading', Proc. 2nd International Conference on Sandwich Construction 2, Gainsville, Florida, USA, 1992, EMAS, Warley, UK , 1992.
6. Vinson J. R., Sierakowski R. L., 'The Behaviour of Structures Composed of Composite Materials', Martinus Nijhoff Publs., Dodrecht, 1986.
7. Rohwer K., 'Improved Transverse Shear Stiffness for Anisotropic Sandwich Plates and Shells', Proc. 2nd International Conference on Sandwich Construction, Gainsville, Florida, USA, 1992, EMAS, Warley, UK, 1993.
8. Pagano N. J., 'Exact Solutions for Composite Laminates in Cylindrical Bending', J. Comp. Materials, 3(3), pp. 398-411, 1969.
9. Pagano N. J., 'Exact Solutions for Rectangular Bidirectional Composites and Sandwich Plates', J. Comp. Materials, 4, pp. 20-34, 1970.
10. Pagano N. J., Wang S. D., 'Further Study of Composite Laminates under Cylindrical Bending', J. Comp. Materials, 5, pp. 521-528, 1971.
11. Rothschild Y., Echtermeyer A. T., 'Simulation of Yield and Plastic Flow of Three Dimensional Sandwich Foam Core', Proc. 2nd International Conference on Sandwich Construction, Gainsville, Florida, USA, 1992, EMAS, Warley UK, 1993.
12. Park W. J., 'An Optimal Design of Simple Symmetric Laminates under First Ply Failure Criterion', J.Comp. Materials, 16, pp. 341-345, 1982.
13. Hirano Y., 'Optimum Design of Laminated Plates under Shear', J. Comp. Materials, 13, pp.329-334, 1979.
14. Zacharapoulos A., Willmer K. D., ' An Optimality Criterion Method for Structures with Stress, Displacement and Frequency Constraints', Computers and Structures, No. 4, pp. 621-629, 1984.
15. Flanagan G., Palzotto N., 'Composite Laminate Optimization Program Suitable for Microcomputers', Computers and Structures, Vol 22, No. 6, pp. 995-1009, 1986.
16. Kam T. Y., Lai M. D., 'Multilevel Optimal Design of Laminated Composite Plate Structures', Computers and Structures, Vol. 31, No. 2, pp.197-202, 1989.

THE EFFECT OF FLUORINE GAS MIXTURES ON THE SURFACE PROPERTIES OF POLYOLEFINS

Francois J. du Toit[1], Ronald D. Sanderson[1], Jakobus B. Wagener[2]

[1] Institute for Polymer Science
University of Stellenbosch
Stellenbosch
South Africa
[2] Atomic Energy Corporation of South Africa
Pretoria
South Africa

INTRODUCTION

The durability of fluoropolymers that result from the strength of the carbon-fluorine bond, have long interested polymer scientists, but high costs and difficulties in processing have limited rapid expansion of the fluoropolymer industry. Early work on surface fluorination of polymers involved attempts to make materials with properties that resembled the chemical inertness of polytetrafluoroethylene from relatively inexpensive or easy processable materials, by a post-treatment fluorination technique.

Controlled surface fluorination of polymers was done as early as the mid-1930s when attempts were made to passivate rubber gaskets with fluorine gas [1]. These gaskets were intended for use in atmospheres of fluorine during the production of fluorine. Edward Kropa reported in 1950 on the fluorination of partially fluorinated polymers such as polychlorotrifluoroethylene, which could be used as materials for corrosive and heat resistant seals [2].

Although fluorine gas is perhaps the oldest fluorinating reagent, the use of elemental fluorine was avoided for long periods by companies and researchers because of difficulties in controlling the reactions of elemental fluorine and the safe handling of fluorine gas. The major problem in the direct fluorination of hydrocarbons is the generation of excessive quantities of heat during the very exothermic reaction. Ideally, fluorination should result only in substitution of hydrogen atoms with fluorine atoms, but because of the reactivity of fluorine, side reactions such as chain scission or crosslinking can be expected. Lagow and Margrave made an important contribution in changing the negative attitude towards surface fluorination of polymers by developing the LaMar process [3]. They stressed that in order to control the reaction between a hydrocarbon polymer and fluorine gas, the fluorine gas should be diluted with an inert gas such as nitrogen. As the reaction propagated, the fluorine concentration could be increased without breaking the carbon-carbon backbone, or burning or even destroying the material.

Today, fluorine is an industrial chemical and it can be handled much more readily than during the past. Since Alfred Rudge [4] exposed polyethylene to fluorine gas in 1954, the development of surface-active analytical techniques has ensured that today, the chemistry of fluorinated surfaces is much better understood. As a result there is a renewed interest in surface fluorination and in the surface properties that can be achieved by exposing a hydrocarbon polymer to fluorine gas mixtures. This paper will show that in addition to the polytetrafluoroethylene-like insoluble or non-stick surface properties that can be achieved today, radically different properties can be obtained such as fluorinated surfaces with high susceptibility for printing inks or conventional adhesives.

Today the driving force behind fluorination is no longer the need to make "Teflon" by a cheaper process, but rather to improve the properties of existing hydrocarbon polymers. Characteristics such as low cost, good mechanical properties and processibility have established polyolefins such as polyethylene and polypropylene as the largest tonnage plastic materials. However, certain disadvantages limit the use of these materials in many applications. Among the industries in which plastics are consumed, the packaging industry is by far the largest. Unlike traditional packaging materials such as glass or metal, polymers are permeable. Polyolefins are characterized by a very high permeation rate of non-polar liquids through them.

The first breakthrough of commercial importance in surface fluorination followed a report by Stephen Joffre in 1957 in which he stated that surface fluorination of polyethylene containers resulted in a decrease of the permeation of liquids such as allyl caproate through the polymer [5].

Air Products commercialised surface fluorination by developing the AIROPAK process, in which the standard blow-moulding process was used to impart barrier properties to high density polyethylene containers by replacing the plant air with a dilute mixture of fluorine gas in nitrogen [6]. Today, different processes are used by different companies. In addition to in-line fluorination, some companies prefer post-fluorination, in which containers are exposed to dilute fluorine mixtures. However, the result is the same. The chemical modification of polyethylene containers using fluorine gas results in the formation, on the polymer surface, of a very thin permanent fluorocarbon layer, highly resistant to mechanical abrasion and chemical attack. Without exception, these different processes are used for high density polyethylene containers, and the literature contains no information on the barrier properties of other polyolefins. In this study, other polyolefins were evaluated as possible materials for petroleum containers.

A further disadvantage of polyolefins is the difficulties in adhesion to these surfaces. Polyolefins require some form of pretreatment in order to create a surface with an acceptable degree of adhesion for printing inks, paints or conventional adhesives. Recently, it was shown that large increases in adhesion to polyolefins can be obtained by treating the surfaces of these polymers with fluorine gas [7]. It will be shown in this study that surface fluorination, although it has been a process in use for many years, now offers an interesting new alternative in modifying polymer surfaces to improve adhesion.

Fluorination, if performed by well-trained staff, can be performed safely and with no threat to health or the environment. Fluorination does not result in harmful residues that could contaminate the contents of fluorinated containers. It can result in polyolefins with much improved barrier properties, without affecting the reprocessibility of the polymers, which is not the case with coatings or laminates. In fact, studies in our laboratories indicated a slight improvement in some mechanical properties upon reprocessing of fluorinated containers.

THE EFFECT OF FLUORINE ON HYDROCARBON POLYMER SURFACES

Fluorine, one of the most reactive elements, reacts with hydrocarbon polymers in a highly exothermic reaction by a free-radical chain reaction mechanism to form a partially fluorinated surface layer [8]. The thickness of this layer is controlled by fluorine gas diffusion through the sample. A fluorine depth profile through the thickness of a polyethylene film, determined with Rutherford backscattering, is shown in Figure 1. This points clearly to a diffusion-controlled reaction, confirming earlier reports on kinetic studies [8]. The nature and thickness of the layer depend on reaction variables such as fluorine concentration, reaction temperature and reaction time. In spite of the reactivity of fluorine gas, it seems that fluorination of hydrocarbon polymers is selective and a function of fluorine substitution, because of the deactivating effect of substituted fluorine. When a partially fluorinated polymer such as polyvinylidene fluoride was exposed to diluted fluorine gas mixtures (10% fluorine in nitrogen), no reaction occurred [8]. This, together with possible side reactions such as crosslinking or oxidation make perfluorination of the outer surface layers unlikely.

Lagow and Margrave reported on the importance of diluting the fluorine gas with an oxygen-free inert gas to prevent fragmentation of carbon-carbon bonds [3]. Nitrogen, helium and argon proved to be successful for this purpose. It is however still believed by many researchers that oxidation always accompanies fluorination, even in atmospheres substantially free of oxygen [9,10], since commercial fluorine is known to contain trace amounts of oxygen as an impurity. In the presence of substantial amounts of oxygen, additional functionalization can be achieved by the process known as oxyfluorination.

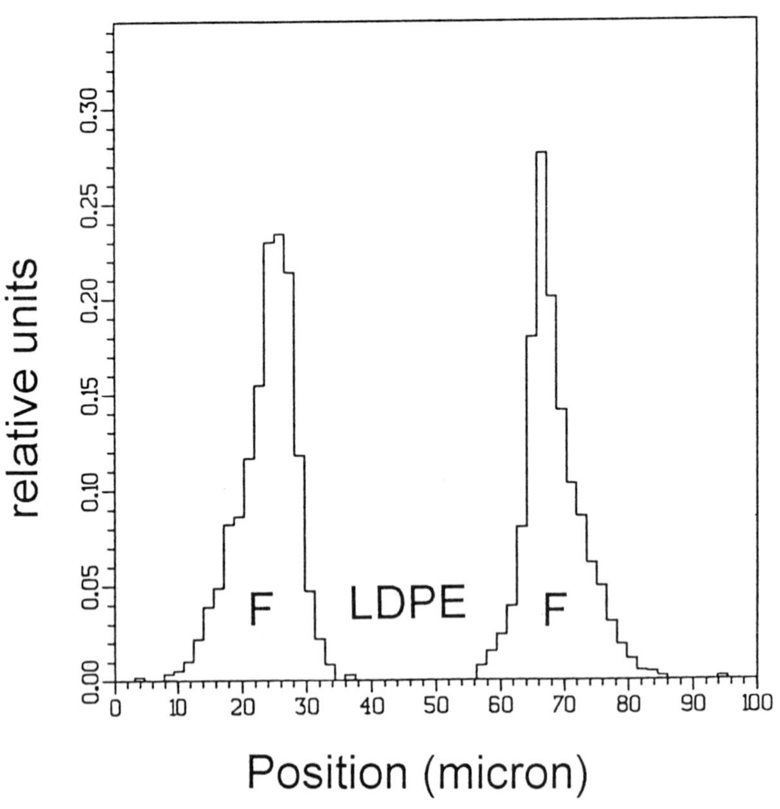

Figure 1. Fluorine depth profile through the thickness of a polyethylene film

Fluorination

Photoacoustic infrared spectra of untreated polypropylene and polypropylene, fluorinated with a 10% fluorine-in-nitrogen gas mixture, are presented in Figure 2. Fluorination resulted in the formation of a broad band at about 1200 cm^{-1} in the infrared spectrum of polypropylene. This band resulted from the stretching vibration of the C-F bond. No indication of oxygen-containing functionalities could be found upon fluorination. XPS results reported for fluorinated polyolefins showed that the hydrocarbon repeating units were replaced by a complex structure of partially fluorinated units [11].

Oxyfluorination

It is clear from Figure 2 that in addition to hydrogen substitution by fluorine atoms, oxygen is incorporated in the surface of polypropylene during oxyfluorination. This is evident in the band at 1850 cm^{-1}, which represents acid fluoride groups. Upon reaction with water or atmospheric moisture, the acid groups were hydrolysed to the corresponding acid groups.

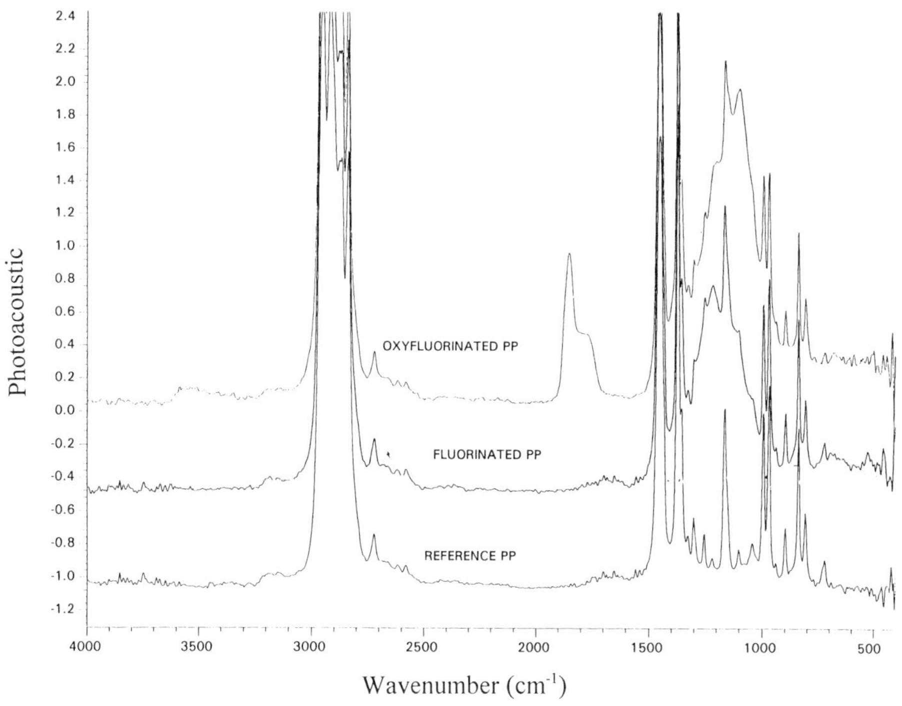

Figure 2. Infrared spectra of untreated, fluorinated and oxyfluorinated polypropylene

REACTION RATES OF FLUORINATION AND OXYFLUORINATION REACTIONS

Fluorine gas reacts with hydrocarbons as follows:

$$RH + F_2 \rightarrow R\cdot + HF + F\cdot$$

If no other reactive species are present, chain propagation occurs in which each reaction site consumes a reactive particle, but again generates another radical:

$$R\cdot + F_2 \rightarrow RF + F\cdot$$

Reaction rates of polypropylene film are presented in Figure 3. In addition to the functionalization of polypropylene by the formation of acid groups, the reaction rates for oxyfluorination were much lower than that of fluorination, and the surface layers that were formed during oxyfluorination were much thinner than those formed during fluorination. This was to be expected since oxygen, which is very reactive towards radicals, inhibits the reaction by reacting with the radicals that are formed during fluorination to form peroxy radicals which are much less reactive than the R· radical [12]:

$$R\cdot + O_2 \rightarrow ROO\cdot$$

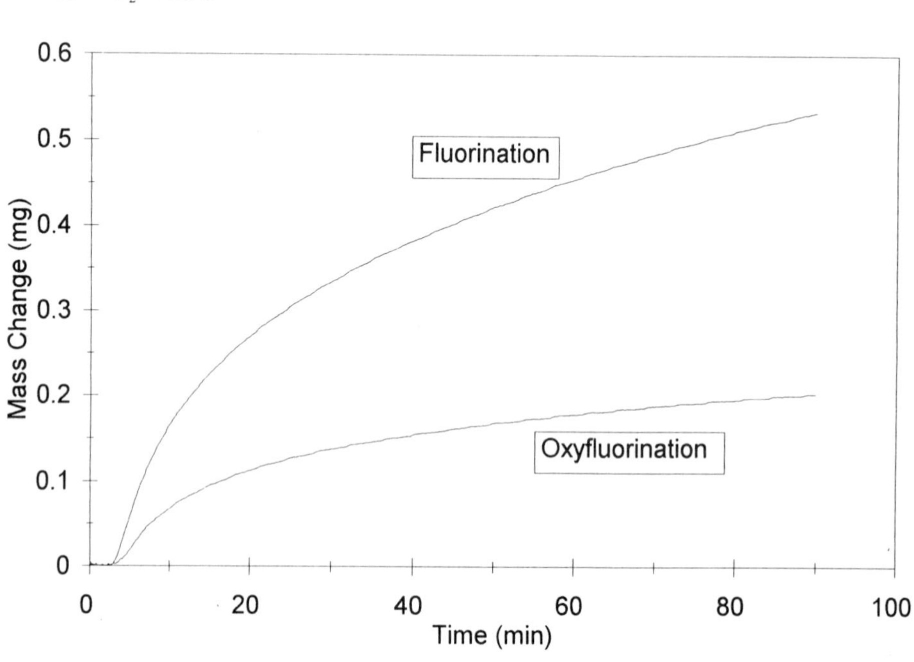

Figure 3. Fluorination and oxyfluorination rates of polypropylene

THE EFFECT OF FLUORINATION AND OXYFLUORINATION ON THE SURFACE PROPERTIES OF POLYPROPYLENE

Contact angle studies provide a very effective method to detect changes that occur at the outer surface layers since, compared with other surface analytical techniques, contact angle studies refer to relatively thin surface layers. Contact angle studies give information about the outer 5 - 10 Å of a solid [13], whereas techniques such as ATR-infrared spectroscopy samples a much thicker layer (>1 μm). The surface tension of fluorinated polypropylene, calculated from contact angle data, as a function of fluorination time is presented in Figure 4. Surface tensions were determined with Wu's Harmonic-mean approach using contact angle data of water and methylene iodide [14]. Contact angles were determined with a CAHN Dynamic Contact Angle Analyser which uses the Wilhelmy plate technique. Although untreated polypropylene is non-polar, the total surface tension of the polymer included a small polar component, which probably resulted from surface oxidation during processing. During fluorination, the polar component of the surface tension increased initially to almost 14 dynes cm^{-1}, but decreased again as the fluorination time was increased and higher levels of fluorination were achieved. The increase in the polar component of the surface tension resulted from dipole moments of the partially fluorinated layer rather than from oxidation accompanying fluorination. At the same time, the dispersive component of the surface tension was lowered to values below 20 dynes cm^{-1}. The

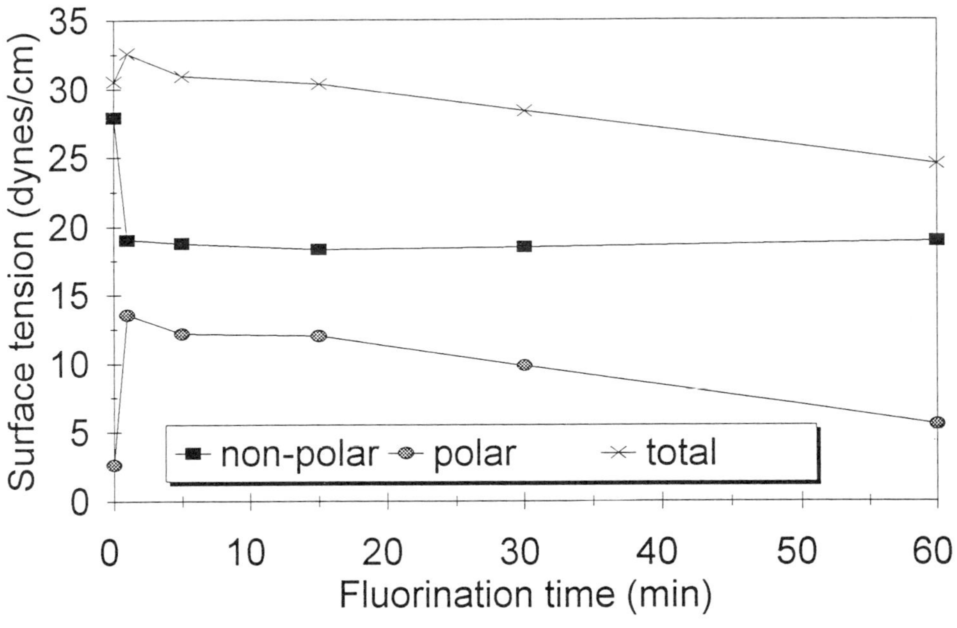

Figure 4. Surface tension of fluorinated polypropylene

net effect was that the total surface tension of polypropylene could be lowered considerably to match the surface energies of highly fluorinated polymers such as polytrifluoroethylene, by exposing it to fluorine gas diluted with nitrogen. Not only did the fluorine penetrate the sample deeper with increasing fluorination times, but the structure of the outer 5 - 10 Å of the fluorinated layer, which is the portion of the layer that is measured by contact angle studies, changed progressively with fluorination time, approaching perfluorination of the surface.

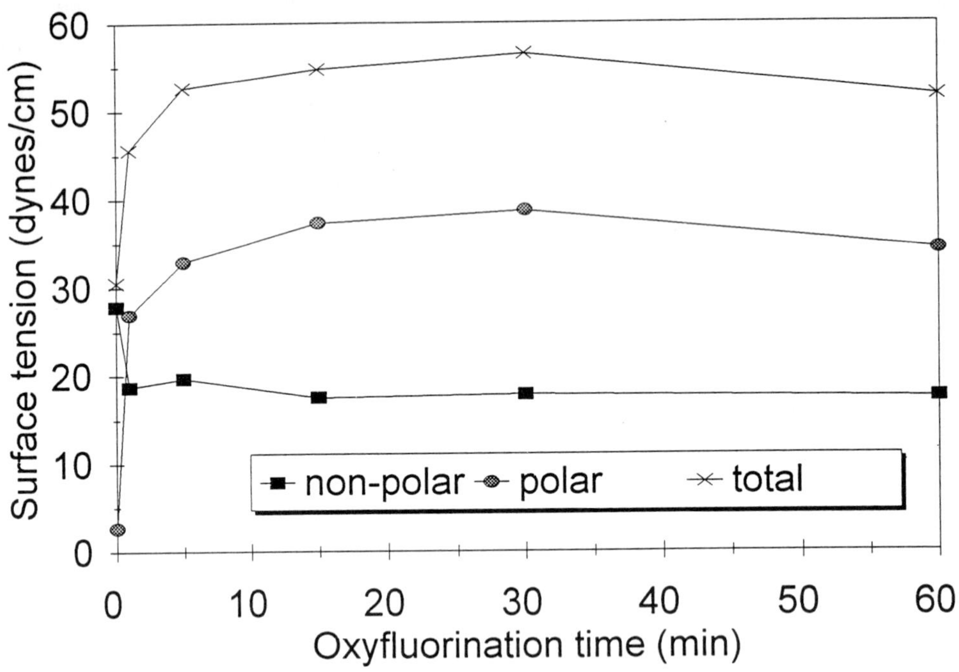

Figure 5. Surface tension of oxyfluorinated polypropylene

Oxyfluorination of polypropylene resulted in large increases in the total surface tension of the polymer (see Figure 5). Surface tension values of 55 dynes cm^{-1} were obtained which were much higher than that reported for polymers such as nylons. This included a very large increase in the polar component of the surface tension, which resulted from the introduction of oxygen-containing functional groups during oxyfluorination, together with a similar decrease in the dispersive component of the surface tension that was observed during fluorination. Significant to possible commercialization of the process, was the fact that wetting of the polypropylene surface was much improved even after only a few seconds in the oxyfluorination reactor.

INVERSION OF FLUORINE-TREATED SURFACES

It has been reported recently that oxyfluorinated polyethylene surfaces reverted to the original polyethylene surface upon heat treatment [15]. This effect was observed only upon heat treatment at 100 °C, whereas no effect was observed at 30 °C and 50 °C. Although the surface tension of the heated sample resembled that of the untreated polymer, the oxygen functionalities were still present and could be detected with attenuated total reflectance infrared spectroscopy. That suggested that the apparent disappearance of the polar groups from the surface resulted from migration of the polar groups into the bulk of the material rather than from chemical changes, the driving force behind the migration being the minimization of the surface free energy. No reversion was obtained with fluorinated surfaces. The relative stability of oxyfluorinated surfaces at room temperature compared with that obtained by other pretreatment techniques such as plasma treatment, offers an advantage in that printing, bonding or painting, need not be done immediately after treatment. However, for optimum adhesion, treated materials should be left in air to hydrolyse completely before adhesive bonds are prepared.

FLUORINE TREATMENT AS A PRETREATMENT TECHNIQUE FOR ADHESION IMPROVEMENT

The reasons for the poor adhesion of polyolefins and the effects of the various modification processes have been the subject of controversy for many years. In general, the various pretreatment techniques used today alter the surfaces of polyolefins in one or more of the following ways:

- removal of the weak boundary layer;
- change in surface topography;
- change in chemical nature of the surface; and
- modification of the physical structure.

The introduction of oxygen in the surface of polypropylene during oxyfluorination as acid fluoride groups, which upon reaction with moisture were hydrolysed to acid groups, affected the surface of polypropylene as follows:

- increased wettability;
- increased polarity significantly;
- increased the total surface tension markedly; and
- roughened the surface.

Lap-shear results for oxyfluorinated polypropylene using an epoxy vinyl ester and a polyester adhesive are presented in Figure 6. Oxyfluorination of polypropylene resulted in large improvements in adhesion using both adhesives, with a highest lap-shear value of 12,8 MPa obtained by using the polyester adhesive. In both cases, large improvements were obtain within seconds of fluorine treatment and the effect levelled off with longer treatment times.

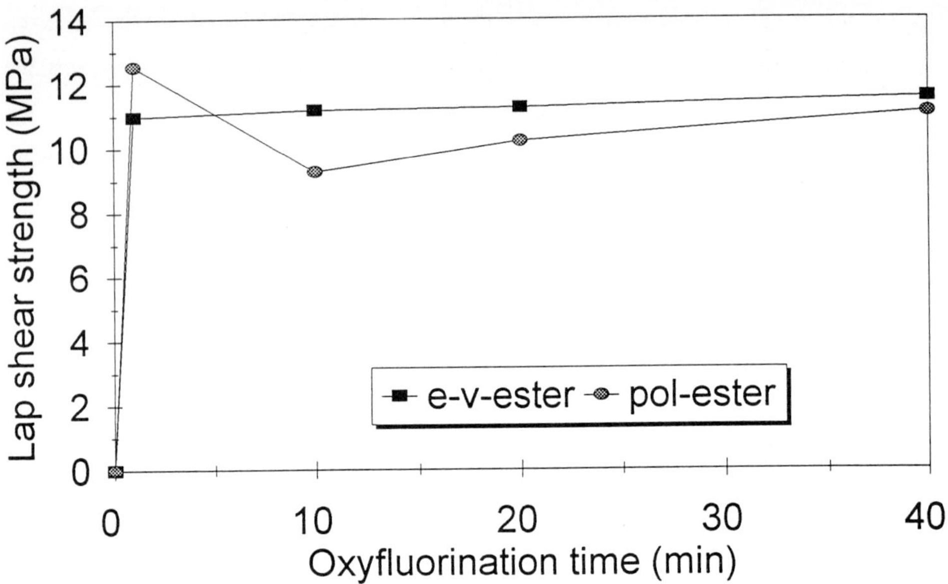

Figure 6. Lap-shear values of an oxyfluorinated polypropylene surface

Fluorination of polypropylene resulted in the formation of a surface layer consisting of partially fluorinated units of propylene and its effects on the surface properties of polypropylene included the following:

- initial slight increase in wettability, but decrease after longer treatment times;
- increase in surface polarity after short fluorination times, but decrease in polarity as perfluorination was approached at long fluorination times; and
- surface roughening resulting from the exothermic nature of the reaction.

Lap-shear results for fluorinated and oxyfluorinated surfaces using a polyester adhesive are presented in Figure 7. Although when compared with that of untreated polypropylene, the wettability of fluorinated polypropylene was not significantly improved, and was even lower than that of the untreated polymer after long treatments, lap shear results were surprisingly high. No evidence of oxygen functionalities was observed during fluorination and the increase in surface polarity could be attributed to dipole moments of the partially fluorinated surface. However, this could not have been the only factor that accounted for the improved adhesion, since partially fluorinated polymers such as polyvinylidene fluoride have poor adhesion properties even though they are significantly polar.

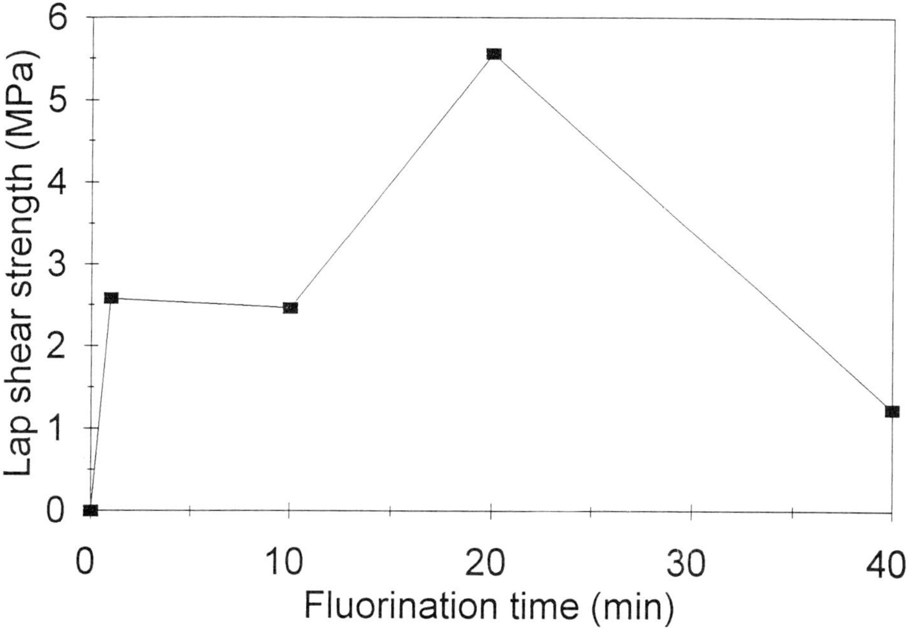

Figure 7. Lap-shear values of a fluorinated surface

In addition to changes in the advancing contact angles, fluorine treatment resulted in changes in the receding contact angles. That was also evident in the large increases in contact angle hysteresis of the polypropylene surfaces upon fluorination and oxyfluorination. Contact angle hysteresis of surfaces can result from surface roughness, surface heterogeneity or restructuring of functionalities at the surface. An increase in the surface area of the fluorinated surface could have an effect on adhesion performance, since significant surface roughening upon fluorination has been reported [16]. Advancing contact angles are sensitive to the less wettable part of a surface, while receding contact angles are very sensitive to small fractions of high surface energy materials [17]. Although no evidence of oxygen-containing functionalities could be detected with infrared spectroscopy or advancing contact angles, it is known that commercial fluorine contains oxygen as an impurity and several authors are convinced that oxidation always accompanies fluorination [9,10]. Since it is the receding contact angle and not the advancing angle that is sensitive to small amounts of high energy material, the hysteresis and improved adhesion of fluorinated surfaces may be justified in terms of a slight oxidation of the polypropylene surfaces; this could be the cause of the reduction in the receding angles, although the surface tensions of the fluorinated samples (which were determined from advancing data) were considerably lower than that of the untreated polymer.

FLUORINATION REDUCES LIQUID PERMEATION

Because of their low cost, corrosion resistance and good processability, polyolefins such as polypropylene and polyethylene are strongly established in the container industry. Expansion in this industry, especially in the containment of petroleum-based products, is limited by the high permeation rate of most hydrocarbons through these polymers; even thick-walled containers are adversely affected. In addition, swelling leads to a deterioration of physical properties of the containers, resulting in secondary effects such as panelling [18].

Since Joffre reported that fluorination could cause a decrease in permeation through polyethylene containers [5], surface fluorination has been employed by different companies in in-line processes as well as in post-treatment processes. Without exception, these processes are used for high density polyethylene containers. The automobile industry in small countries such as South Africa is not large enough to justify the cost and capital layout of blow-moulded petrol tanks. Rotational moulding is, at this stage, a more economically viable process, and it was therefore important to evaluate materials other than high density polyethylene as candidates for the containment of non-polar liquids such as petroleum.

In this study, all the permeation tests were done according to the regulations from the Department of Transport of the United States of America [19]. These regulations limit the amount of allowable permeation of hazardous materials to 2% by weight during an accelerated test comprising a 28 day period at 50 °C.

Xylene weight losses through untreated and fluorinated polypropylene 200 ml bottles are shown in Figure 8. As was reported for high density polyethylene, fluorination of polypropylene bottles resulted in large decreases in xylene permeation rates. The barrier properties of surface-treated polypropylene were functions of the thickness of the fluorinated layer. The weight loss of the fluorinated container with layer thickness 20 μg cm^{-2} was 98%, whereas it was only 2% for a layer ten times thicker. It has been reported that good barrier properties for high density polyethylene are associated with fluorinated layers thicker than 45 μg cm^{-2} [20]. Fluorination can therefore effectively decrease the permeation of non-polar liquids such as xylene through polypropylene containers, but in order to create a satisfactory barrier, the fluorine content of the surface layers must be considerably higher for polypropylene than for high density polyethylene.

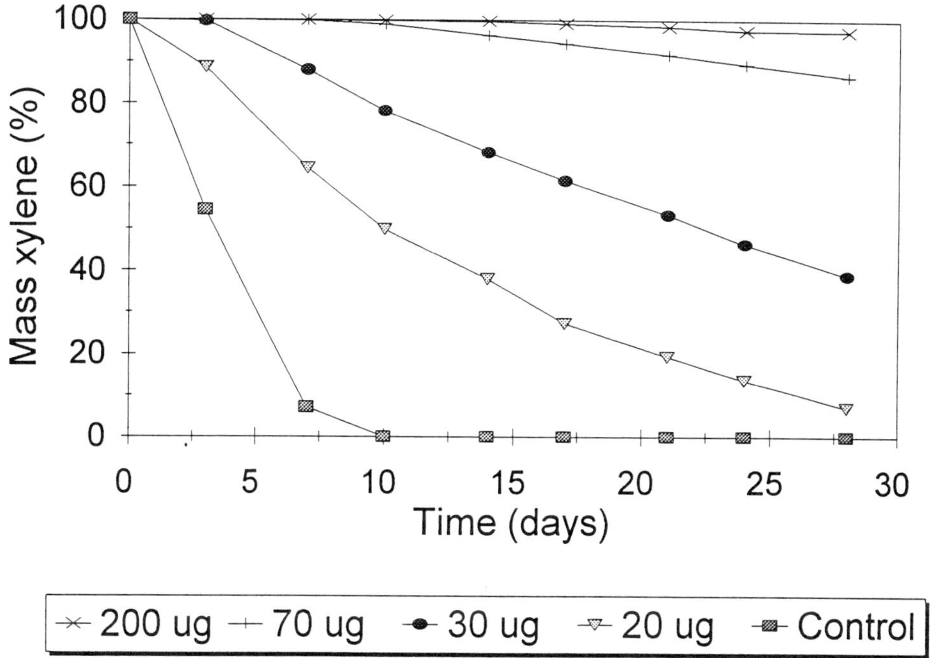

Figure 8. Xylene weight loss through polypropylene containers

Polypropylene bottles were oxyfluorinated and fluorinated to exactly the same fluorine content (30 μg cm^{-2} as determined by neutron activation analysis). Comparative permeation results for the two different treatments are presented in Figure 9. Although the fluorine contents of both surfaces were the same, the presence of oxygen atoms, which were incorporated in the oxyfluorinated surfaces, had an adverse effect on the barrier properties of fluorine-treated polypropylene. These results stress the importance of excluding oxygen from fluorination reactions when fluorination is used to decrease permeation of non-polar liquids.

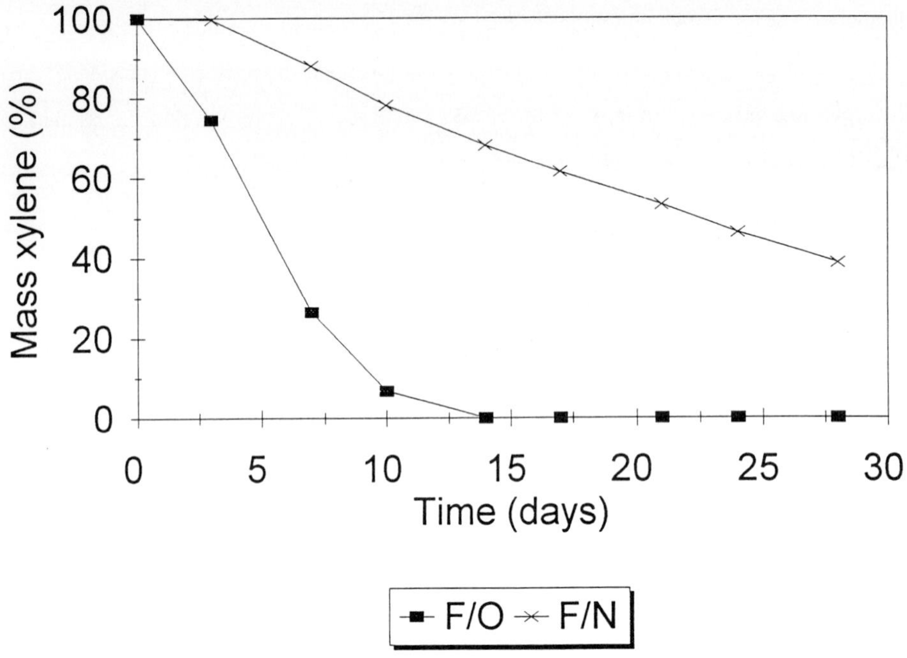

Figure 9. Xylene permeation through fluorinated and oxyfluorinated polypropylene containers

Polyolefin sheets were separately treated under exactly the same conditions with equal amounts of fluorine diluted with nitrogen. Barrier properties of untreated and fluorinated polyolefin sheets were evaluated using stainless steel permeation cells. Total petroleum weight losses during the 28-day test period for the various untreated polymers are presented in Figure 10. The permeation problem of untreated polyolefin containers is clearly evident, with large percentages of the contents of all containers being lost. Permeation results for the various polyolefin containers, prepared and fluorinated under exactly the same conditions are compared in Figure 10. Although significant improvements were obtained for all materials upon fluorine treatment, when the barrier properties of the various polymers were compared, only high density polyethylene gave satisfactory results at the specific fluorine layer thickness, with the improvement of linear low density polyethylene being surprisingly low. The random copolymer as well as the polypropylene homopolymer containers maintained more than 90% of their contents during the 28-day testing period.

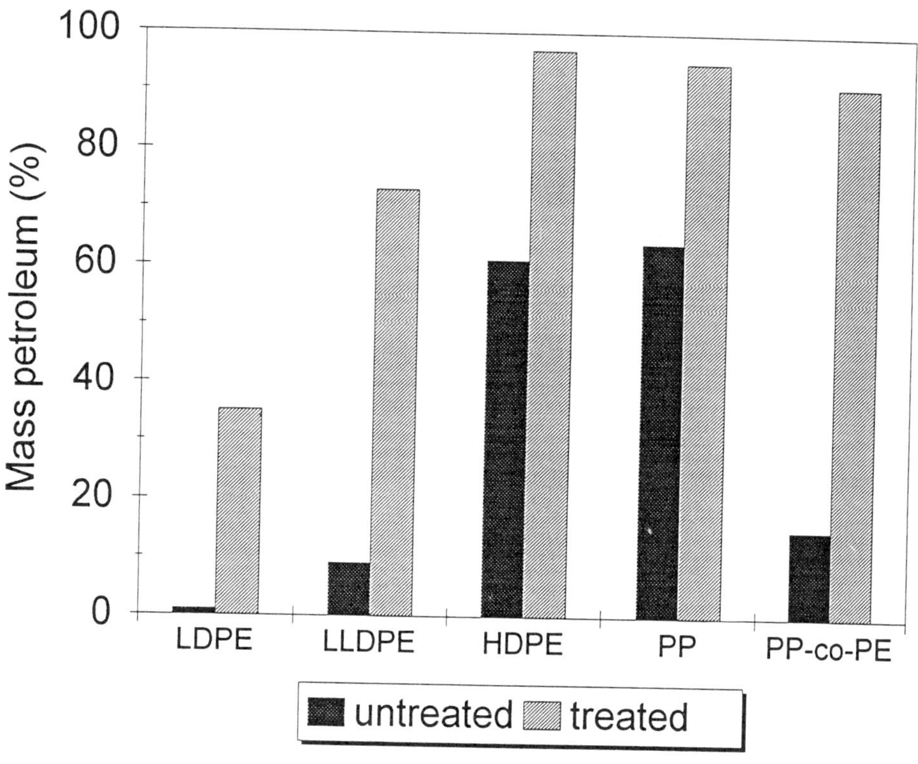

Figure 10. Petroleum permeation results of various polyolefins

STABILITY OF FLUORINATED SURFACES

The packaging industry places a strong premium on chemical inertness of packaging materials. Plastic containers must not change their contents by chemical interaction, physical extraction or migration. It has already been mentioned that, unlike oxyfluorinated surfaces, fluorinated surfaces showed no inversion. In addition, the effect of possible contamination of products was studied by filling fluorinated polyethylene bottles with water immediately after fluorination. The quantity of fluoride that leached from the bottles was determined over a period of 6 weeks by ion chromatography. It is clear from Table 1 that the quantity of fluoride that leached from the bottles levelled off at less than 3 ppm, which was well below the maximum allowable level [21]. It has been recently reported that foodstuffs are effectively contained in fluorinated containers, with fluorination providing an effective method to avoid flavour loss. The fluorinated as well as the oxyfluorinated layers change the surface properties of the treated polymers completely, but are thin enough not to have a negative effect on the reprocessability or recycling of the polymers.

Table 1. Fluoride leaching from fluorinated bottles

Time (days)	Fluoride concentration (ppm)
0	0
5	1.2
16	1.8
22	2.7
25	2.8
38	2.8
45	2.8

CONCLUSIONS

Fluorine treatment offers a relatively new, inexpensive and effective method for pretreatment of polymer surfaces to improve their adhesion properties. Unlike traditional methods, such as plasma treatment, treated surfaces do not revert to the untreated surfaces quickly and adhesive bonds do not have to be prepared immediately after treatment. Fluorination as a means to create barriers towards non-polar liquids can be extended to polyolefins other than high density polyethylene. Although in comparitive tests, high density polyethylene outperformed other polyolefins as barrier materials, with increasing fluorinated layer thickness, barrier properties were also obtained for polymers such as polypropylene and its copolymers, thereby broadening the range of materials that can be used widely by, for example, the petroleum industry. While oxyfluorination proved to be superior to fluorination in improving adhesion properties,the presence of oxygen in the reaction mixture should be avoided when barrier properties towards non-polar liquids are required.

REFERENCES

1. J. Soll, United States Patent, 2,129,289 (1938).
2. E. L. Kropa, United States Patent, 2,497,046 (1950).
3. R. J. Lagow and J. L. Margrave, *Prog. Inorg. Chem.* 26:161 (1979).
4. A. J. Rudge, British Patent, 710,523 (1954).

5. S. P. Joffre, United States Patent, 2,811,468 (1957).

6. D. D. Dixon *et al.*,United States Patent, 3,862,284 (1975).

7. I. Brass *et al., Int. J. Adhes. Adhes.* 11:150 (1991).

8. R. D. Sanderson *et al., J. Therm. Anal.* 41:563 (1994).

9. L. J. Hayes, *J. Fluorine Chem.* 8:69 (1976).

10. J. D. Le Roux *et al., J. Membrane Sci.* 90:37 (1994).

11. D. T. Clarke *et al., Polym. Sci. Technol.* 5:373 (1974).

12. T. Volkmann and H. Widdecke, *Kunststoffe* 8:743 (1989).

13. G. S. Ferguson and G. M. Whitesides, chapter 6, in: Modern Approaches to Wettability, M. E. Schrader and G. I. Loeb, ed., Plenum Press, New York (1992).

14. S. Wu, chapter 3, in: Polymer Interface and Adhesion, S. Wu, ed., Marcel Dekker, INC., New York (1982).

15. F. J. du Toit *et al., J. Fluorine Chem.* (in print).

16. G. Kranz *et al., Int. J. Adhes Adhes.*14:243 (1994).

17. M. Morra *et al., Adv. Colloid Interfac.* 32:79 (1990).

18. R. K. Metha *et al.*, Air Products Technical Paper presented at the Petroleum Packaging Committee Technical Meeting, Memphis (1987).

19. P. D. Marsh *et al.*, Air Products Technical Paper presented at the Annual Technical Conference Society of Plastic Engineers, Boston (1986).

20. H Sorek, *J. Radioanal. Nucl Chem., Letters.* 93:97 (1985).

21. United States Department of Health and Human Services, Federal Register, 48 (168) (1983).

SYNTHESIS AND BIOLOGICAL ACTIVITIES OF ENDO-3,6-EPOXY-1,2,3,6-TETRAHYDROPHTHALIMIDE AND ITS POLYMERS

Gi Cheol Kim[1], Jin Gon Jeong[1], Neung Ju Lee[2], Chang Sik Ha[1] and Won Jei Cho[1]

[1]Department of Polymer Science and Engineering. Pusan National University, Pusan 609-735, Korea
[2]Department of Premedical Sciences, Kosin University, Pusan 602-735, Korea

INTRODUCTION

The high molecular weight antitumor compounds(HMAC) are attracting much interest from the standpoints of polymers synthesis as well as drug development, because of their high specificity of action, low toxic side effects and longer duration of drug action as compared with low molecular weight antitumor compounds(LMAC).[1-9] HMAC may be combined with large molecules in the living body and be slowly released into the biological fluids, giving effective concentrations over long periods.

It has been known that many polyanions exhibit antitumor activities. Among the polyanionic polymers, the 1:2 regularly alternating copolymer(DIVEMA) of divinyl ether(DVE) and maleic anhydride(MAH), first reported by Butler, has been extensively studied for its structure and its antitumor activity. DIVEMA has been shown to possess antitumor, antiviral, antibacterial, and antifungal activities as well as interferon inducing ability. However, it has also toxicity such as enlarged liver and spleen. Afterwards, many attempts have been made to obtain DIVEMA-like HMAC to reduce side effects as well as to enhance the antitumor activity. We have synthesized and characterized homopolymers and copolymers of exo-3,6-epoxy-1,2,3,6-tetrahydrophthalic anhydride(ETA)[10], exo-3,6-epoxy-1,2,3,6-tetrahydro phthalic glycinyl maleimide(ETGI)[11], N-glycinyl maleimide(GMI)[12,13].

In this study, monomeric ETPI was synthesized by reacting maleimide and furan in diethylether. Poly(ETPI) and poly(ETPI-co-acrylic acid) were prepared by the photopolymerization. Synthesized monomer and polymers were identified by IR and ^{1}H-NMR spectroscopies. The composition of the copolymer containing ETPI was determined by elemental analysis. *In vitro* cytotoxicities of prepared ETPI and

Polymers and Other Advanced Materials: Emerging Technologies and Business Opportunities. Edited by P. N. Prasad *et al.*, Plenum Press, New York, 1995

polymers containing ETPI moiety were evaluated against Fibrolast cell and K-562 human leukemia cell as target cell. *In vivo* antitumor activities of prepared ETPI and polymers containing ETPI moiety were investigated using tumor bearing mice(Sarcoma 180).

EXPERIMENTAL

Materials and instruments

Maleimide(MAI, Aldrich Co.), furan(Aldrich Co.), vinylacetate(VAc, Junsei chem.), acrylic acid(AA, Junsei chem.), maleic anhydride(MAH, Junsei chem.) and dimethoxy benzoin(DMB) were purified by conventional methods. 5-fluorouracil (5-FU, Aldrich Co.) was used without further purification. Fibroblast cell, K562 human leukemia cell, BALB/c mice and Sarcoma 180 as tumor cell were purchased from the Center of Genetic Engineering, Korea Institute of Science and Technology(KIST).

IR spectra were taken on a Jasco FT/IR-5300 spectrophotometer using KBr disk. ^{1}H-NMR spectra were recorded on a FT-300 MHz Bruker A-3000 spectrophotometer. Elemental analysis was made on a Carlo Erba Instruments Model EA 1108 elemental analyzer.

Synthesis of ETPI and polymers containing ETPI

Furan and MAI were dissolved in diethylether and stirred at room temperature for 2 days. The white precipitate was filtered and washed with diethyl ether. The melting point of ETPI was 122℃.

ETPI and dimethoxy benzoin(DMB) were dissolved in the 2:1(v/v) mixture of 2-butanone and acetone and introduced into a dry pyrex polymerization tube and placed in photochemical chamber reactor using λ_{max}=313nm lamps at 25 ± 0.1℃ for 12 hours. The polymer solution was precipitated in diethylether. Precipitated poly(ETPI) was collected by filtration, washed with diethylether and dried under vacuum until it reached to a constant weight.

Poly(ETPI-co-AA) was prepared by the copolymerization of ETPI and AA. The copolymerizations was carried out by the same procedure as that of poly(ETPI).

In vitro cytotoxicity test

Fibroblast and K562 human leukemia cell line were used as target cell for *in vitro* cytotoxicity test. Cells were continuously cultured in RPMI1640 medium. 3×10^5 cells of target cell line and distilled water solutions of compounds were mixed, then placed in 96-well flat-bottomed plates and cultured for 60 hours in a CO_2 incubator at 37℃. After adding 0.1ml of trypan blue (0.4%), cell numbers were counted using a hemocytometer.

In vivo cytotoxicity test

To evaluate the antitumor activities of ETPI and polymers, mice bearing Sarcoma 180 tumor cells are used. BALB/c mice were first intraperitonearly(i.p.) implanted with Sarcoma 180 cells(2×10^5). The animals were then treated with the saline of sample on days 1-4. Three different dosages were tested : 0.8, 80 and

800 mg/kg.

For comparison, the antitumor activity of free 5-FU was also tested by the same method. A control group was divided into two groups : one group was treated with Sarcoma 180 cells along with the same volume of saline and the other was treated with only Sarcoma 180 cells. Each group was consisted of 10 animals.

RESULTS AND DISCUSSION

Characterization

The structure of ETPI was identified by IR and ^{1}H-NMR spectroscopies. The IR spectrum of ETPI shows characteristic absorption bands at 3210cm^{-1}(NH), 1188cm^{-1}(C-O-C), 1716cm^{-1}(C=O), 2790 cm^{-1}(CH) and 1564cm^{-1}(CH=CH). In the ^{1}H-NMR spectrum of ETPI, the absorptions due to two methine protons of imide ring, two olefinic protons and two methine protons in cyclic ether group and imide hydrogen were shown at 3.4 ppm, 6.5 ppm, 5.2 ppm and 9.6 ppm, respectively.

^{1}H-NMR spectrum of poly(ETPI) shows methine protons of imide ring at 3.4 ppm, methine protons in polymer backbone at 2.9 ppm and methine protons of cyclic ether at 5.2 ppm and the absorption assingnable to the olefinic protons of ETPI monomer is not observed at 6.5 ppm.

The structure of poly(ETPI-co-AA) were also identified by its ^{1}H-NMR spectrum.

Solubility of ETPI and polymers containing ETPI moiety

Solubilities of ETPI, poly(ETPI) and poly(ETPI-co-AA) were listed in Table 1. ETPI and polymers containing ETPI moiety were soluble in polar solvents such as water, DMF, DMSO and methanol but insoluble in diethylether and chloroform.

Table 1. Solubility of ETPI and polymers containing ETPI moiety[1]

Solvent	ETPI	Poly(ETPI)	Poly(ETPI-co-AA)
Water	○	○	○
Dimethyl sulfoxide	○	○	○
N,N-Dimerhylformamide	○	○	○
Methanol	○	△	○
Acetone	○	×	○
2-Butanone	△	×	×
Tetrahydrofuran	○	×	×
Ethylacetate	○	×	×
Chloroform	×	×	×
Diethyl ether	×	×	×
Benzene	×	×	×
n-Hexane	×	×	×

[1]○ : Soluble, △ : partially soluble, × : insoluble

Composition of polymers containing ETPI

The content of ETPI unit in poly(ETPI-co-AA) was determined 35% by elemental analysis. Composition of polymers containing ETPI was listed in Table 2.

Table 2. Composition of polymers containing ETPI.

Sample	Composition (%)			ETPI fraction (%)
	C	N	H	
ETPI	58.8	8.6	4.4	
Poly(ETPI)	57.6	6.9	5.7	
Poly(ETPI-co-AA)	51.0	4.4	5.4	35.4

In vitro cytotoxicity test

Cytotoxicity of ETPI and polymers containing ETPI moiety against Fibroblast cell(Normal Cell) and K562 human leukemia cell were shown in Figure 1 and 2. The cytotoxicity against Fibroblast cell(Normal Cell) and K562 human leukemia cells increased with the increase of the sample concentration. Monomer and polymers show considerable cytotoxicities at higher concentrations(5.0, 1.0, 0.1 mg/ml). The order of cytotoxicity against Fibroblast cell(Normal Cell) at 0.1 mg/ml is as follows : poly(ETPI-co-AA) > ETPI > poly(ETPI). The order of cytotoxicity against K562 human leukemia cell at 0.1mg/ml is as follows : ETPI >> poly(ETPI-co-AA) > poly(ETPI).

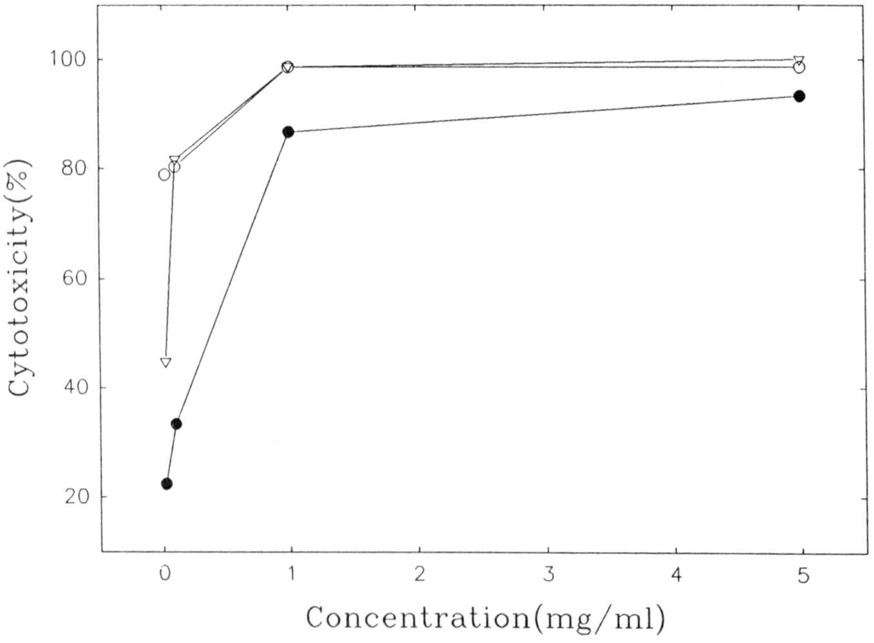

Figure 1. Cytotoxicity of ETPI and polymers containing ETPI moiety against Fibroblast cell(Normal Cell) ; ETPI[○], poly(ETPI)[●], poly(ETPI-co-AA)[▽].

In vivo cytotoxicity test

The results of antitumor activities of ETPI, polymers containing ETPI moiety, 5-FU and control group against Sarcoma 180 cells were showed in Figure 3. The life span of mice treated with poly(ETPI) at the concentrations of 800 and 80 mg/kg was increased by 229% and 195% as compared with the control group but 5-FU was increased by 50% and 140% in the same condition. The order of antitumor activity at a dosage of 80mg/kg is as follows : ETPI > poly(ETPI) > poly(ETPI-co-AA) > 5-FU > Control group.

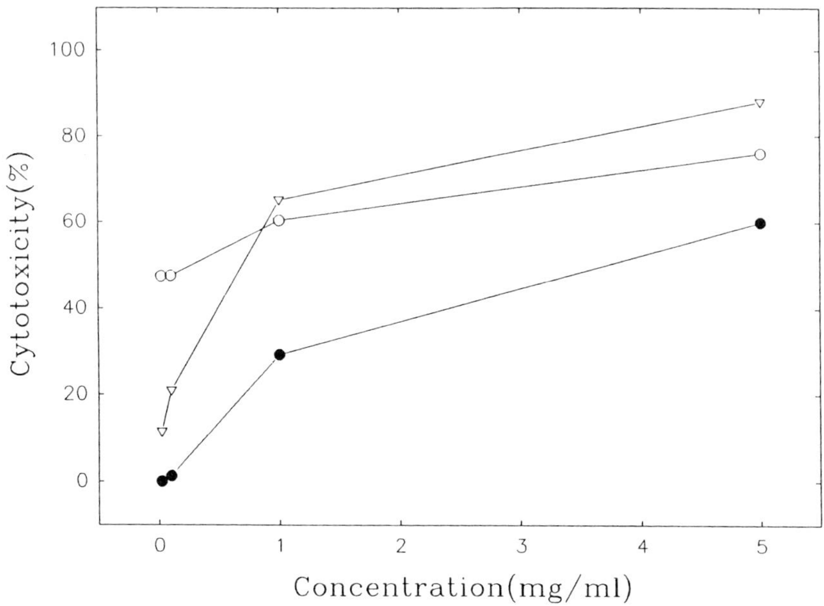

Figure 2. Cytotoxicity of ETPI and polymers containing ETPI moiety against K562 human leukemia cell line ; ETPI[○], poly(ETPI)[●], poly(ETPI-co-AA)[▽].

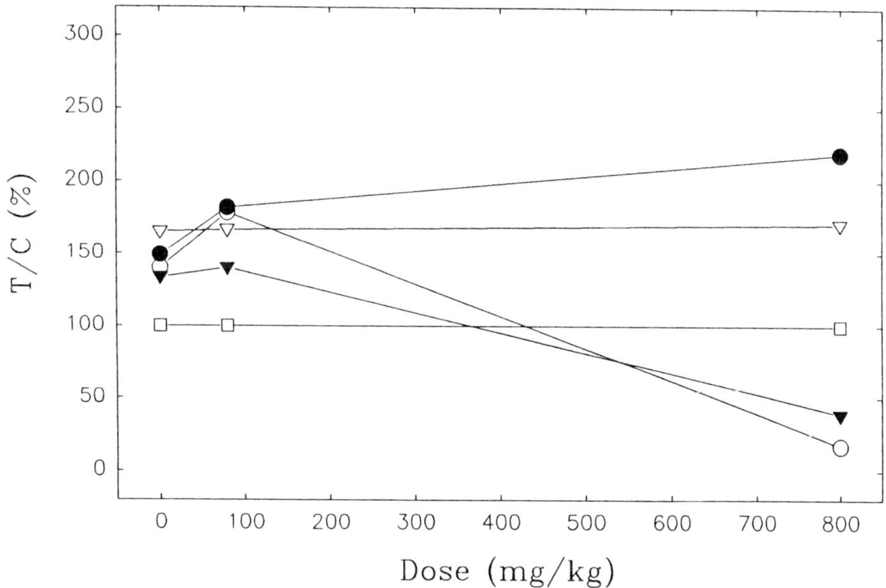

Figure 3. Antitumor activity of ETPI, polymers containing ETPI moiety, 5-FU and control against Sarcoma 180 ; ETPI[○], poly(ETPI)[●], poly(ETPI-co-AA)[▽], 5-FU[▼] and control group[□].

CONCLUSIONS

ETPI was synthesized by Diels-Alder reaction of furan and maleimide in diethylether at room temperature. Poly(ETPI) and poly(ETPI-co-AA) were prepared by photopolymerization at room temperature and their structures were identified by IR and ^{1}H-NMR spectroscopies. ETPI fraction in poly(ETPI-co-AA) was 35.4%. Solubilities of ETPI, poly(ETPI) and poly(ETPI-co-AA) are soluble in water, DMF, DMSO and methanol. The order of cytotoxicity against Fibroblast cell(Normal Cell) at a dosage of 0.1 mg/ml is as follows : poly(ETPI-co-AA) > ETPI > poly(ETPI). The order of cytotoxicity against K562 human leukemia cell at a dosage of 0.1mg/ml is as follows : ETPI >> poly(ETPI-co-AA) > poly(ETPI). *In vivo* antitumor activity of synthesized polymers was greater than that of 5-FU at a dosage of 80 mg/kg. The order of antitumor activity at a dosage of 80mg/kg is as follows : ETPI > poly(ETPI) > poly(ETPI-co-AA) > 5-FU > Control group.

REFERENCES

1. S.Ozaki,Y.Ike,H.Mizuno,K.Ishikawa and H.Mori,*Bull.Chem.Soc.Japan*,50,2406 (1977).
2. P.P.Umrigar,S.Ohashi and G.B.Butler,*J.Polym.Sci.,Polym.Chem.Ed.*,17,351(1979).
3. M.Akashi,K.BEPPU,I.Kikuchi.andN. Mizauchi,*J.Mcromol.Sci-Chem.ED.*,17,351 (1986).
4. R.M.Ottenbrite,*J.Mcromol.Sci-Chem.*,A22(57),819(1985).
5. T.Ouchi,H.Juzei,Y.Sakamoto,H.Chikachita and O Vogl,*J.Polym.Sci.,Polym.Chem.* Vol.23,2059(1986).
6. R.M.Ottenbrite,J.Sunamoto,T.Sato,K.Kojima,K.Sahara,K.Hara,M.Oka,J.Bioact.& Compati.*Polym.*3,184,(1988).
7. G.Butler,*J.Macromol.Sci-Chem.*,A13(3),351(1979).
8. N.J.Lee,C.S.Ha, and W.J.Cho,*J.Macromol.Sci-Chem.*,29,2,161(1992).
9. N.J.Lee,C.S.Ha, and W.J.Cho,Bioact.& Compati.*Polym.*7,39(1992).
10. J.G.Jeong,N.J.Lee,C.S.Ha and W.J.Cho,*Proc. 3rd Pacific Polymer Conf,Gold Coast, Australia*, Dec.13-17,pp.75-76(1993).
11. D.Y.Lee,N.J.Lee,J.G.Jeong,C.S.Ha and W.J.Cho,*The Society of Polymer Science, Japan*, 797page10/12-10/14.(1994).
12. G.T.Gam,J.G.Jeong,N.J.Lee,C.S.Ha and W.J.Cho,*J.Appl.Polym.Sci.* in print(1995).
13. G.T.Gam,J.G.Jeong,N.J.Lee,C.S.Ha and W.J.Cho, *IUPAC International Symposium on Functional and High Performance Polymers, Taipei, Taiwan*, Nov. 14-16,p.(1994).

ACKNOWLEDGEMENT

The authors are grateful for the financial support of the Korea Science and Engineering Foundation(Grant No. 931-0300-014-2).

OPTIC CHARACTERIZATION OF OLIGO-THIOPHENE MULTILAYERS

M. Muccini[1], R.F. Mahrt[1], R. Hennig[1], U. Lemmer[1], H. Bässler[1], F. Biscarini[2], R. Zamboni[2], and C. Taliani[2]

[1]Fachbereiche Physikalische Chemie und Physik, WZMW, Philipps-Universität, D-35032 Marburg,, Germany.

[2]Istituto di Spettroscopia Molecolare, CNR via Gobetti 101, 40129-Bologna, Italy.

INTRODUCTION

Polythiophene is an interesting system for its remarkable charge transport and optical properties as well as for the investigation of the electronic excitations and their dynamics in one-dimensional structures. However, conformation and configuration defects in the real material result in physical properties depending on sample preparation and processing[1-3]. On the contrary, thiophene oligomers are well-defined chemical systems with a precise structural definition of the condensed phase and provide considerable insight into the physical properties of polythiophene[4-6]. Heteromultilayer organic structures represent one further step for the understanding of basic theoretical aspects such as the electronic level ordering and the dynamics of optical excitations. Considerable theoretical attention has been recently devoted to the linear and non-linear optical properties of organic multilayer systems[7-10]. Besides the electronic properties of a single molecule the disorder present in the thin films leads to strong implications. It is, in fact, well kown that the optical and electrical properties of organic heteromultilayer structures are highly sensitive to the degree of structural order. Recently, it has been shown that highly oriented oligo-thiophene polycrystalline thin films can be grown by vacuum sublimation[11].

Polymers and Other Advanced Materials: Emerging Technologies and Business Opportunities. Edited by P. N. Prasad *et al.*, Plenum Press, New York, 1995

By applying a series of optical low-temperature measurements on an oligo-thiophene based multilayer structure (tetrathiophene (T4)/esathiophene (T6)/tetrathiophene (T4)) we show in this paper that (i) interface excitons are formed between T4 and T6 adjacent layers and that (ii) energy transfer from excited states of T4 to lower energy states (interface excitons and T6) occurs.

EXPERIMENTAL

Tetrathiophene and esathiophene were thermosublimated in high vacuum (10^{-7} mbar) from two different quartz cells to grow thin films onto quartz substrates which were kept at T=300K during the entire sample preparation. The deposition rate was 0.5-1 Å/s. Fig. 1 shows the three layer sample consisting of a T4/T6/T4 structure on a quartz substrate.

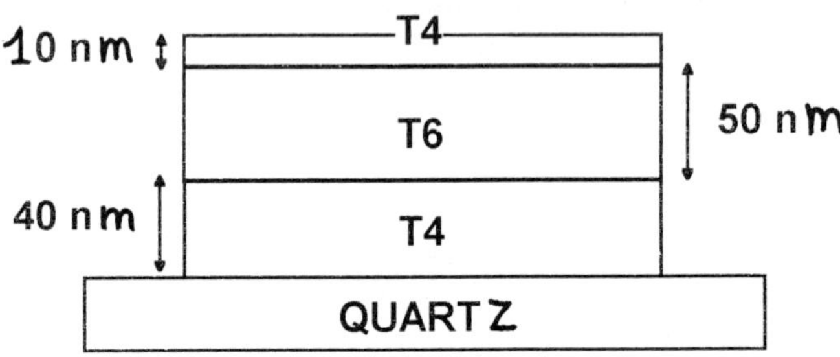

Figure 1 Three layer structure of tetrathiophene (T4) and esathiophene (T6) grown by vacuum evaporation on a quartz substrate.

Site-selective photoluminescence spectra were taken using a tunable pulsed dye laser with spectral width of less than 1 cm^{-1} as excitation source. The emission was detected by a cooled GaAs photomultiplier after dispersion in a double monochromator. A box-car averager with a gate width of 5 ns was used for processing the signal. The photoluminescence excitation spectra were measured by scanning the excitation energy and monitoring an emission band.

The time resolved photoluminescence spectra were measured with the streak camera technique. The sample was excited with 120 fs pulses with 3.12 eV photon energy. The luminescence was dispersed by a grating monochromator and detected by a streak camera with a time resolution of 30 ps.

RESULTS AND DISCUSSION

Fig. 2 shows low temperature photoluminescence (PL) spectra of the multilayer structure taken using different excitation energies. Though the main emission peaks of the multilayer system (indicated by dashed lines in Fig. 2) are at the same energetic positions like the peaks of the pure T4 and T6 PL spectra, the relative intensity of the peaks cannot be explained by a simple superposition of the T4 and T6 emission spectra. Different excitation geometries were used to check if the relative intensity of the PL bands is determined by absorption and re-absorption of each layer, but no significant difference was found in the optical spectra.

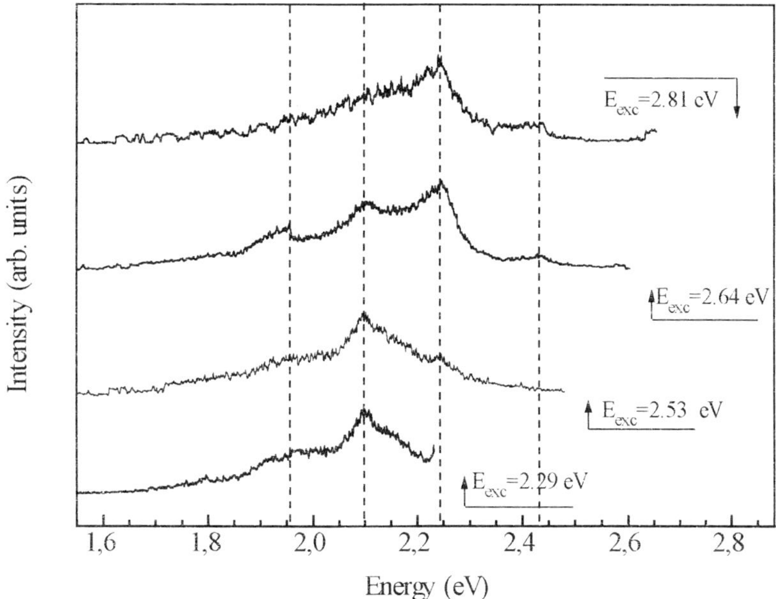

Figure 2 Photoluminescence spectra of the multilayer system at T= 10K. The excitation energy for each spectrum is indicated by arrows.

The multilayer structure emission spectra can be explained by assuming that new excitonic states are formed at the interfaces between T4 and T6 adjacent layers. New electronic excitations at the interfaces between different layers in organic multilayer structures have been predicted[10]. The related phenomenon of surface excitons blue-shift in organic crystals (for a review see Ref. 12) is well explained by the absence of neighbours for the molecules in the surface monolayer. Similarly, the interface excitons formed by the interaction of T4 and T6 adjacent layers are likely to be energetically located between the T4 and the T6 states.

We attribute the enhanced intensity of the emission peaks at 2.24 eV and 2.10 eV observed in Fig. 2 to a superposition of the emission spectrum of the interface excitons with, respectively, the 0-0 electronic transition of T6 and the vibronic progression of T4. For excitation energies higher than 2.6 eV the absorption in the multilayer structure is dominated by T4 molecules. The excited T4 states can either decay radiatively or transfer energy to interface exciton states or to T6 states. Energy transfer to interface excitons is likely to be the most efficient process since the emission at 2.24 eV dominates the spectra. When exciting with energy lower than 2.6 eV the absorption of T4 decreases while the absorption of interface excitons and of T6 increases. The relative intensity of the peaks at 2.10 eV and 1.95 eV indicates that no efficient energy transfer occurs between interface excitons and T6 molecules. The emission peaks in T5 are close in energy to the interface excitons peaks. However, since these bands are not observed in the simultaneously grown single layer reference samples of T4 and T6 we rule out a significant contribution of T5 impurities.

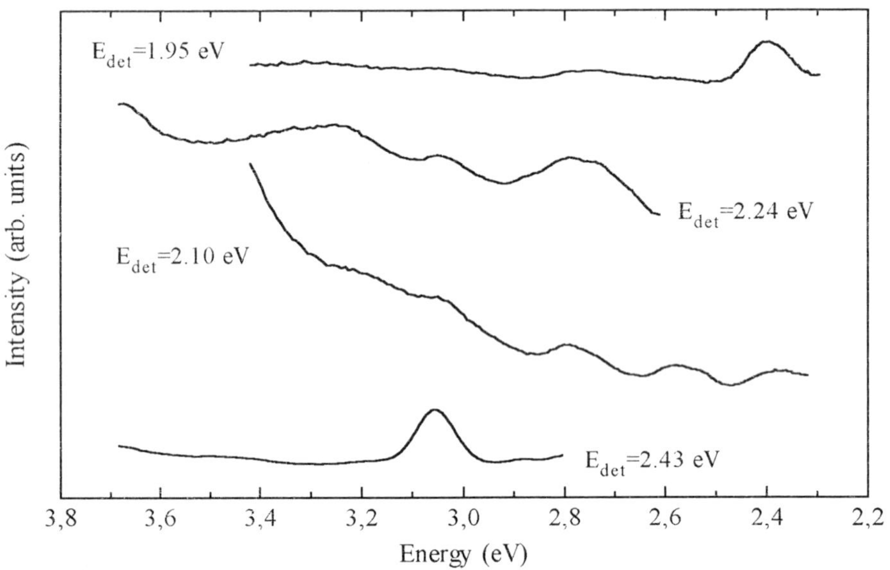

Figure 3 Photoluminescence excitation spectra taken at T=10K by monitoring the T4/T6/T4 emission bands at the spectral positions indicated by the dashed lines in Fig. 2.

Photoluminescence excitation (PLE) spectroscopy of the multilayer structure confirms the above interpretation. Fig. 3 depicts four PLE spectra of the multilayer system measured by monitoring the four emission peaks indicated by the dashed lines in Fig. 2. The excitation spectra clearly show that new species other than T4 and T6 contribute to the emission bands at 2.24 eV and 2.10 eV. In fact, the PLE spectra of the pure T4 and T6 luminescence peaks (2.43 eV and 1.95 eV) are considerably different from the PLE spectra monitored at the interface excitons emission bands.

In order to study the dynamics of the photoexcitations we performed time-resolved photoluminescence spectroscopy. In Fig. 4 are compared the time resolved spectra of the T4 and T6 luminescence peaks in pure films and in the multilayer structure.

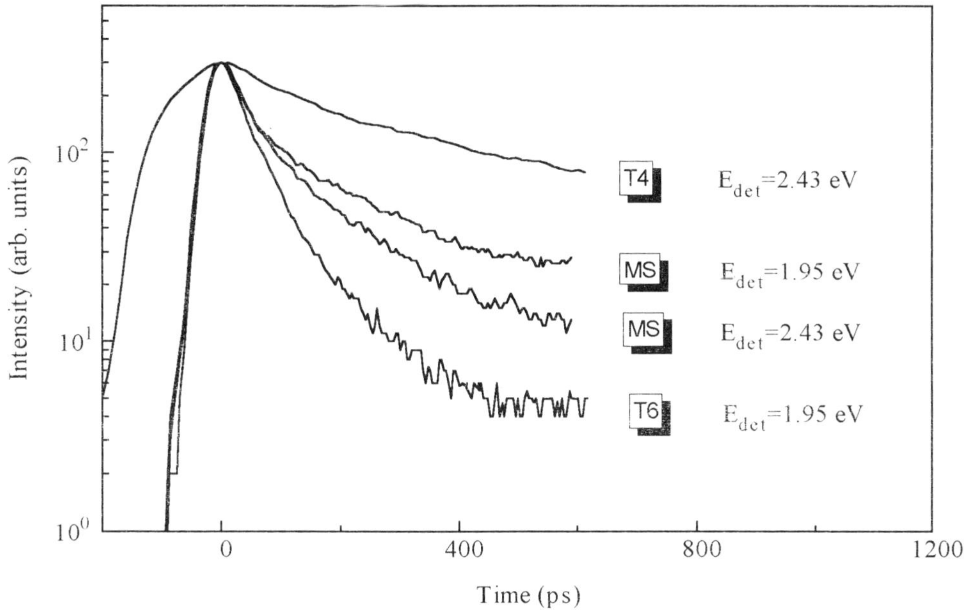

Figure 4 Time resolved PL spectra of the same emission peaks in the multilayer structure (MS) and in pure T4 and T6 films. The detection energy for each spectrum is reported nearby. The excitation energy was E_{exc}=3.1 eV and the temperature T= 10K.

The T4 photoluminescence decay at 2.43 eV is faster in the multilayer structure than in the pure sample. On the contrary, the T6 emission at 1.95 eV becomes slower in the multilayer structure due to refilling processes from higher energy states. These findings give strong indication of energy transfer from excited states of T4 to T6 states.

In conclusion, by applying optical spectroscopic techniques we showed evidence of interface excitons formation at the interface between adjacent T4 and T6 layers in a multilayer structure. We also showed that efficient energy transfer processes from excited states of T4 to lower lying states occur.

REFERENCES

1. C. Taliani, R. Danieli, R. Zamboni, P. Ostoja, and W. Porzio, Synth. Met., **18** (1987) 117.
2. M. Kobayashi, J. Chen, T.C. Chung, F. Moraes, A.J. Heeger, and F. Wudl, Phys. Rev. **B 30** (1984) 702.
3. Z. Vardeny, E. Ehrenfreund, J. Shinar, and F. Wudl, Phys. Rev. **B 35** (1987) 2498.
4. N. Periasamy, R. Danieli, G. Ruani, R. Zamboni, and C. Taliani, Phys. Rev. Lett. **68** (1992) 919.
5. C. Taliani, R. Danieli, R. Lazzaroni, N. Periasamy, G. Ruani, and R. Zamboni, Mol. Cryst. Liq. Cryst., **217** (1992) 101.
6. R. Zamboni, N. Periasamy, G. Ruani, and C. Taliani, Synth. Met. **54** (1993) 57.
7. V.M. Agranovich, Mol. Cryst. Liq. Cryst. **230** (1993) 13.
8. V.M. Agranovich, R.D. Atanasov, and G.F. Bassani, Chem. Phys. Lett. **199** (1992) 621.
9. V.M. Agranovich, Physica Scripta **T49** (1993) 699.
10. V. M. Agranovich, and O.A. Dubovsky, Chem. Phys. Lett. **210** (1993) 458.
11. R. Marks et al., submitted for publication.
12. J.M. Turlet, Ph. Kottis, and M.R. Philpott, Advan. Chem. Phys.**54** (1983) 303.

ACKNOWLEDGMENT

Financial suppport by the European Commission Human Capital and Mobility SELMAT project and the EC basic research action 8013 - LEDFOS is greatfully acknowledged.

THE PRESENT STATE OF, AND FUTURE PROSPECTS FOR, LARGE-AREA FULL-COLOR DC PDPS

Junro Koike

NHK(Japan Broadcasting Corporation)
Science and Technical Research Laboratories
Tokyo 157, Japan

INTRODUCTION

Since 1991, Hi-Vision(HDTV) programs have been transmitted via a broadcast satellite in Japan. A metre-wide display of high quality is indispensable to fully enjoy the sensation of reality provided by a Hi-Vision picture. Several types of Hi-Vision receivers are now on the market. Conventional receivers with CRTs and projection displays are, however, too heavy and bulky. Plasma display panels(PDPs) are considered the most promising for a hang-on-the-wall Hi-Vision large-screen display because of their easy fabrication and other features such as wide viewing angle, fast response and good color reproduction characteristics.

We first developed a pulse memory drive scheme to radically increase the brightness of DC PDPs.[1] Afterward, we proposed a planar pulse memory panel with a simple structure.[2] In 1992, we developed a 40-inch color PDP with an aspect ratio of 16:9, and high quality Hi-Vision pictures were reproduced on it.[3] Several problems remain to be solved before commercial production of this panel can be attempted, including improving the luminous efficiency, prolonging life, and achieving a finer cell pitch. These problems are all closely related to increasing the panel size. This paper describes the present state of, and future prospects for, large-area full-color DC PDPs.

FLAT DISPLAYS FOR HI-VISION

Table 1 shows an example of a Hi-Vision flat display specification, based on CCIR documents.[4] A display of at least 55-inch diagonal is required to produce a strong subjective sensation of reality from Hi-Vision pictures. The number of display cells and the pixel size in the table originate from the Hi-Vision system. The values for luminance and contrast ratio etc. are the minimum required levels.

Table 2 compares various possible Hi-Vision flat panel display technologies. All flat panel displays except for the PDP entail some serious problems for Hi-Vision. With regard

Polymers and Other Advanced Materials: Emerging Technologies and Business Opportunities. Edited by P. N. Prasad *et al.*, Plenum Press, New York, 1995

Table 1. Target specifications of flat Hi-Vision display

Screen size	0.8 m² min. (0.7 m x 1.2 m, 55-in (dia.))
Number of display cells	1035 x 1920 x 3
Pixel pitch	0.65 mm max.
Viewing distance	3H (H:Screen height)
Horizontal viewing angle	30°
Luminance	150 cd/m² min.
Contrast ratio	50 : 1
Power consumption	200 W max.
Luminous efficiency	1.0 lm/W min.
Thickness	100 mm max.

Table 2. Comparison of possible Hi-Vision flat displays

	PDP	LCD	Flat CRT	VFD	EL	LED
Chromaticity	○	○	○	△	▲	▲
Response	○	△~○	○	○	△	○
Dot pitch	△	○	○	○	○	○
Screen size	○	▲	■	■	▲	▲
Lifetime	△~○	○	○	○	○	○
Uniformity	○	△	○~▲	△	△	▲
contrast	○	○	○	△	▲	○
Luminance	○	○	○	▲	▲	▲
Efficacy	△	△	○	▲	▲	▲

○ possible △ may be possible ▲ difficult ■ virtually impossible

to feasibility, we reached the conclusion that PDPs, especially DC PDPs, would be the most suitable for Hi-Vision displays on account of the ease of fabrication of large panels and their good color reproduction characteristics.

THE PRINCIPLE OF THE COLOR DC PDP

Figure 1 is a schematic illustration of the operational principle of color DC PDPs. Vacuum ultra-violet(VUV) light is generated by a gas-discharge ignited between the anode and cathode electrodes, both of which are exposed in a cell filled with discharge gas. The VUV emission excites the phosphor layer deposited on the inner surface of the front glass plate. Photo-luminescence converts VUV light into visible light, the spectrum of which depends on the phosphor material.[5]

The major problem with DC PDPs in the early stages was the lack of any memory function. This resulted in lower luminance as the screen became larger. We developed the pulse memory drive scheme[1] for TV displays based on an earlier method.[6] Figure 2 shows the principle of the pulse memory scheme. Memory operation is obtained by using two phenomena peculiar to gas-discharge: (1) the quantity of metastable particles generated by gas-discharge decreases slowly with time even after the discharge pulse is turned off, (2) the breakdown voltage that ignites gas-discharge can be lowered with metastable particles. Sustain pulses are constantly applied to every cells at a sufficient level to maintain gas-discharge once it has been ignited by the write pulse. The cell can hold either of two states for the sustain pulse; the ON state where repetitive pulse discharges are maintained, and the

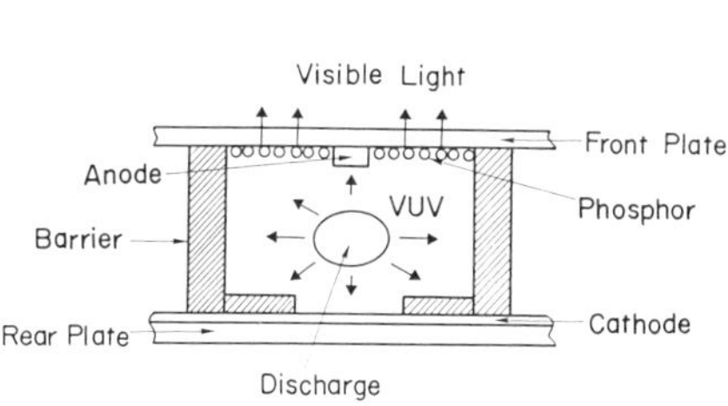

Figure 1. Principle of the color DC PDP

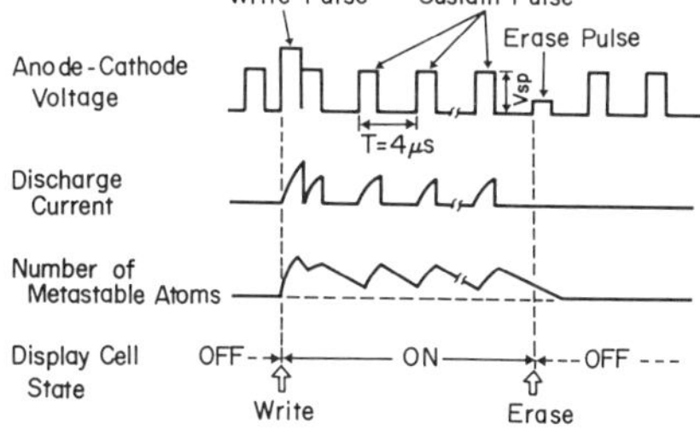

Figure 2. Principle of the pulse memory drive scheme

OFF state where no discharge occurs. It has also been established that auxiliary cells are essential in order to obtain sufficient operational speed for the write and erase operations, as described below.

A DESCRIPTION OF THE 40-IN. DC PDP

Panel Structure

Figure 3 shows the structure of the 40-in. full-color plasma display panel. Auxiliary cells are arranged between the display cells in order to lower the display cell's breakdown voltage and accelerate the write access operation. They provide a priming effect to the display cells through the priming space.

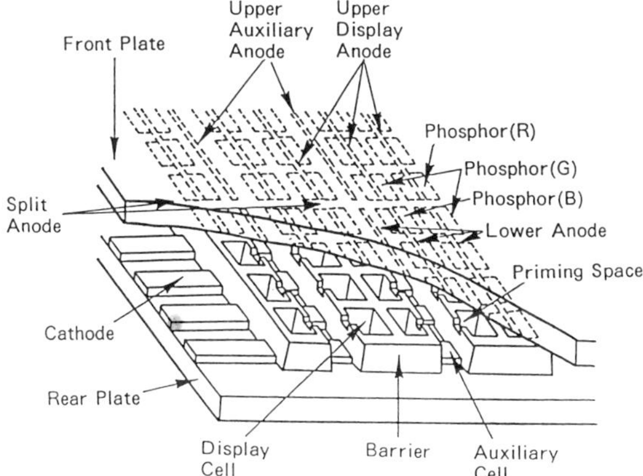

Figure 3. Structure of 40-in. panel

The panel consists simply of two glass plates, with no parts that are difficult to make.[2] All components except the phosphor layers were screen printed as thick films. The phosphors are deposited on the inner surface of the front plate by photo-lithography. It seems comparatively easy to manufacture a large panel with a small cell pitch at low cost.

Table 3 shows the specifications of the 40-in. full-color panel. The panel is 6 mm thick, weighs 8 kg, and has the Hi-Vision aspect ratio of 16:9. The display cell pitch of 0.65 mm is the same as the pixel pitch required for a future 55-in. panel. It is still insufficient for Hi-Vision demonstrations. The resolution of the display is enhanced by the diagonal

Table 3. Specifications and performance of 40-in. panel

Screen Size (mm)	520× 874
Number of Display Cells	800×1,344(1,075,200)
Average Display Cell Pitch (mm)	0.65
Cell Arrangement	Auxiliary Cell, Display Cell; G R / BG GR / BG GR; 0.65mm; 1.3mm
Panel Thickness (mm)	6
Panel Weight (kg)	8
Peak White Luminance (cd/m²)	93(27 fL)
Luminous Efficacy (White, lm/W)	0.15
Number of Gray Levels	256
Display Unit Thickness (cm)	8

arrangement of the green cells, since green carries 70 % of the detail in a picture.

Panel Fabrication Process

We have developed various fabrication processes for making large-area DC PDPs. Figure 4 shows the flow chart for the 40-in. panel fabrication process. On the rear glass plate, the cathode terminals, cathodes and barriers are formed in that order by screen printing. Similarly, the anodes are formed on the front plate, then three phosphors are deposited and photoetched to form display cell dots. The cross sectional views of both glass plates are shown in Figure 5. The front and rear glass plates so prepared are assembled and sealed with frit glass and then evacuated. Finally, a He-Xe gas mixture is introduced to the panel, and Hg is diffused by heat treatment to protect the cathode surface against sputtering.

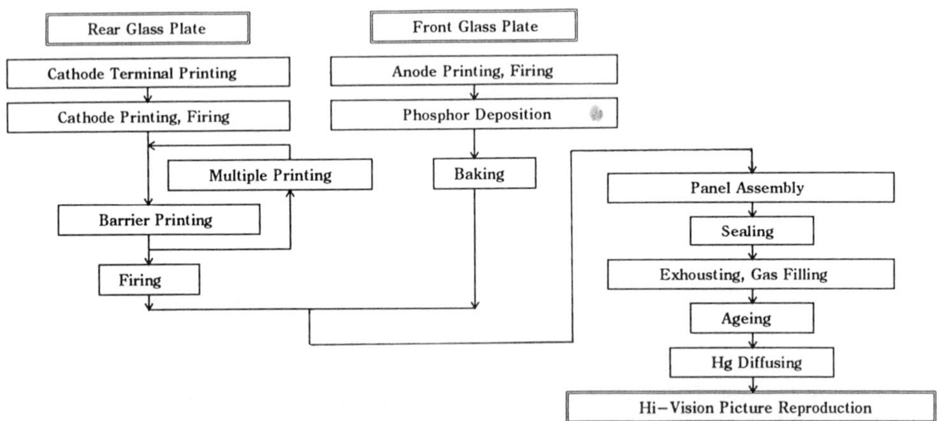

Figure 4. Panel fabrication process

The barriers are formed by multiple printing with single screen to a height of about 200 μm, as shown in Figure 5. The operational condition of PDPs depends greatly on the distance between the anode and the cathode, as determined by the barrier height. We developed fine thick film printing technologies, including devices to optimize the printing

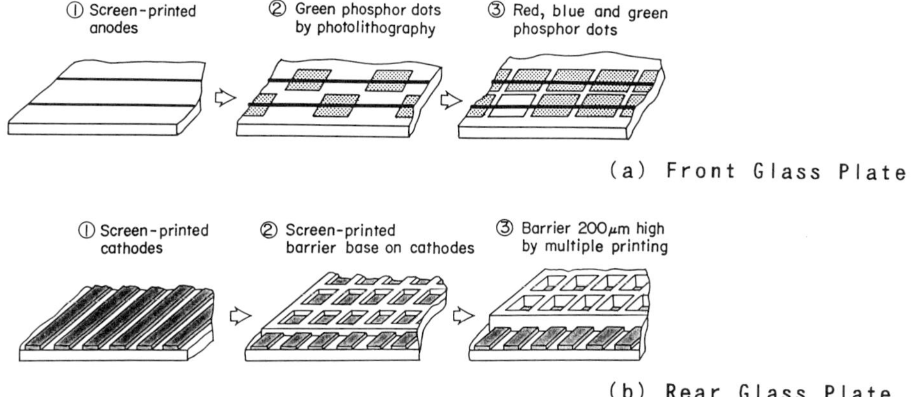

Figure 5. Schematic views of front and rear glass plates after each process

quality and a new strengthened screen. These developments produce stable memory operation for the whole active area of the panel, as described below.

HI-VISION DISPLAY PERFORMANCE OF THE 40-IN. PDP

Display System

Figure 6 shows the overall design of the system to display Hi-Vision pictures on the 40-in. panel.[3] The display system consists of a signal processing unit and a display unit. The former supplies display data and control signals derived from the Hi-Vision signals to the display unit, which is an assembly of a display panel and drive circuitry driving the panel directly. Anode lines are grouped with the upper and the lower halves divided at almost the center of the panel, as shown in Figure 6. Accordingly, the driving circuits are composed of the upper and lower anode drivers and the cathode driver.

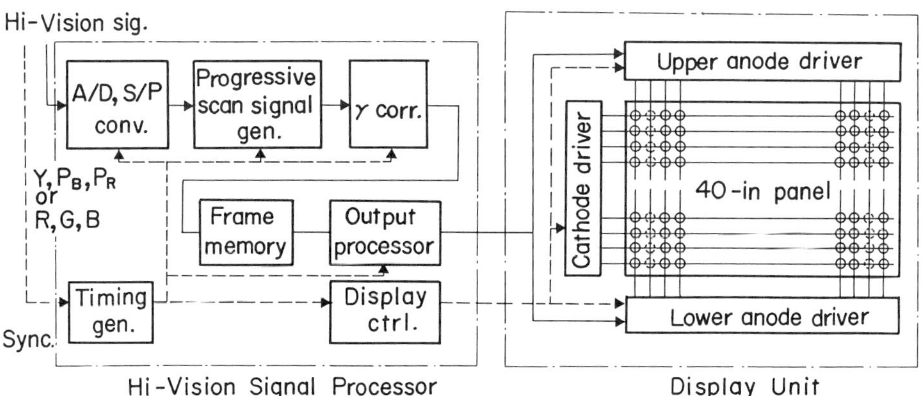

Figure 6. Hi-Vision display system for 40-in. panel

The sampling frequency is 74.25 MHz(1920 samples/line,8 bits/sample). Since write access in the discharge panel is performed line-sequentially, interlaced scanning signals are converted to progressive scanning signals. After these signals are rearranged into the pulse memory drive format in the frame memory, they are fed to the anode drivers.

Display Performance

The performance of the 40-in. display is shown in Table 3. A Hi-Vision picture is displayed in color with 256 gray levels. The luminance is about 1.5 times brighter than that of the earlier 33-in. panel with the same structure[7] due to the increase in aperture ratio of the display cell. As shown in Figure 7, a Hi-Vision picture with high luminance has been reproduced on the 40-in. wall-hanging plasma display.

Figure 7. 40-in. Hi-Vision PDP

FUTURE PROSPECTS FOR DC PDPs

Figure 8 compares current pulse memory DC PDPs with those anticipated in the near future. The solid lines show the characteristics of the 40-in. PDP. It seems that there are several problems such as luminous efficiency, contrast ratio, lifetime and cell pitch etc. The future estimates are based on fundamental studies of small experimental panels and the 26-in. DC PDP[8,9] etc. described below. The figure shows that the characteristics of the latest PDPs are greatly improved to practical levels, compared with the earlier 40-in. PDP.

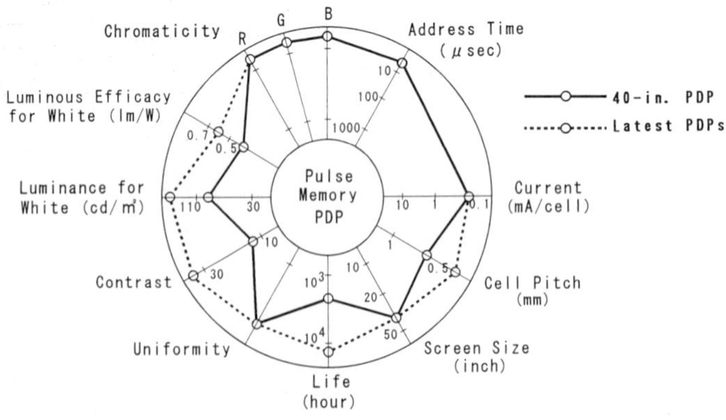

Figure 8. Comparison of present and future pulse memory PDPs

The major reason for the shortening of the lifetime of the DC PDP, which is determined as the time until luminance is reduced to half, is sputtering of the cathode material. As shown in Figure 9, we found that the lifetime is drastically increased when using a higher gas pressure and a lower discharge current.[10] Thus, a panel with relatively high gas pressure and a current limiting resistor in each cell can achieve a lifetime of over 10,000 hours which is long enough for practical use.[11]

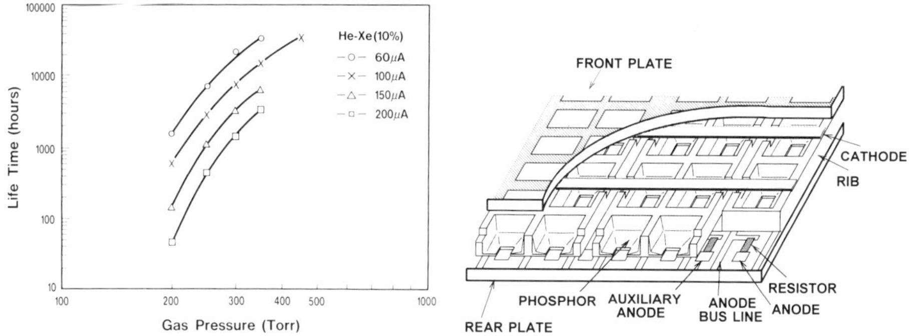

Figure 9. Relationship between lifetime and gas pressure

Figure 10. Structure of 26-in. panel

Figure 10 shows the structure of the 26-in. DC PDP(896x512 cells,16:9) recently developed at Matsushita Electronics Corporation in association with NHK. The gas pressure in the panel was increased to 1.5 times that of the 40-in. panel. The discharge current for each cell is controlled by the resistor-in-cell structure. In order to increase the luminance of the panel, phosphors were deposited on the inner surface of each cell of the panel. As a result, a peak white luminance of 150 cd/m^2, an efficiency of 0.4 lm/w and a life time of over 10,000 hours were achieved.

To improve the contrast ratio, we have developed color filter technologies which can provide a much larger contrast ratio. These technologies were used to fabricate a 25-in. high contrast pulse memory PDP[12] with a contrast ratio of 50:1 under 200 lx illumination.

To reduce the power consumption, a new pulse memory drive scheme(the CPM method)[13] has been devised which reduces the energy loss due to the charging and discharging of the panel capacitance. The power consumption can be reduced to one fifth of that of the conventional driving scheme. Since the circuits for the CPM method are simple, it is comparatively easy to produce ICs for the driver circuit. The anode and cathode driver ICs are now under development. The power consumption of the whole receiver will be reduced by developing signal processing LSIs.

The cell pitch will be reduced to less than 0.5 mm and the panel size will be increased to 55-in. by improvements in the panel fabrication equipment, materials and techniques. The performance of this display is expected to surpass that of other displays in the near future.

CONCLUSION

The present state of, and future prospects for, large-area full-color DC PDPs have been described. Although further improvements in luminous efficiency, luminance, lifetime, and cell pitch are necessary to produce a practical Hi-Vision display, we have already identified several approaches to address these problems.[14-19]

The start of practical Hi-Vision broadcasting by the successor to the BS-3 broadcasting satellite in 1997 and the Nagano Winter Olympic Games in 1998 are thought to offer the best opportunity for wide-scale domestic acceptance of Hi-Vision.

In Oct. 1994, a Hi-Vision PDP Joint Development Consortium was organized to develop a practical Hi-Vision 40-in. PDP receiver in time for the Nagano Winter Olympic Games and to develop a prototype 55-in. Hi-Vision PDP receiver by 1999, with the membership of NHK and leading manufacturers in all related technical areas. We all expect to enjoy beautiful and impressive Hi-Vision pictures of the next Winter Olympic Games on a 40-in. PDP receiver.

ACKNOWLEDGMENT

The author wishes to express his gratitude to T. Izumi, S. Yoshikawa, T. Yoshino, H. Murakami, and his colleagues in the PDP research group for their helpful assistance and advice.

REFERENCES

1. H. Murakami et al.,"A Pulse Discharge Panel Display for Producing a Color TV Picture with High Luminance and Luminous Efficacy",IEEE Trans,Electron,Devices,29(6),pp988-994(1982).
2. H. Murakami et al.,"Planar Pulse Discharge Panel for a TV Display",IEEE Letters,6(3),pp132-134(1985).
3. T. Yamamoto et al.,"A 40-in.-Diagonal HDTV Plasma Display",SID'93 Digest,pp165-168(1993).
4. CCIR Documents 11/1007(Rev.1)-E(1990) and 11/1008-E(1990).
5. J. Koike et al.,"New Tricolor Phosphors for Gas Discharge Display",J.Electrochem.Soc.,126(6),pp1008-1010(1979).
6. G.E. Holtz,"Pulsed Discharge Display with Memory",'72 SID Int.Symp.,Dig.Tech.Papers,pp70-71(1972).
7. H. Murakami et al.,"A 33-in.-Diagonal HDTV Display Using Gas Discharge Pulse Memory Technology", SID'91 Digest,pp713-716(1991).
8. K. Takahashi et al.,"A Long-Life 26-in. DC Pulse-Memory Color PDP with Resistor-in Cell Structure", SID'94 Digest,pp715-718(1994).
9. H. Kono et al.,"A Long-Life DC Color PDP with Concave Cathode",SID'95 Digest, pp823-826(1995).
10. T. Sakai et al.,"A Method for Extending the Life of a DC Gas-Discharge Color Memory Panel", Proc.Eurodisplay'93,pp289-292(1993).
11. M. Ushirozawa et al.,"Color DC PDP with an Improved Resistor Design in Each Cell", SID'94 Digest,pp719-722,(1994).
12. Y. Takano et al.,"A 25-in. High-Contrast Color DC-PDP",SID'92 Digest,pp720-723(1992).
13. Y. Takano et al.,"A 40-in. DC-PDP with New Pulse-Memory Drive Scheme",SID'94 Digest,pp731-734(1994).
14. M. Seki et al.,"An 8-in. Pulse Memory Color DC-PDP without Auxiliary Cell",Proc.12th IDRC,pp617-620(1992).
15. Y. Murakami et al.,"Computer Simulation of Pulsed Discharge for a Color DC Plasma Display Panel ", Proc.Eurodisplay'93,pp555-558(1993).
16. Y. Motoyama et al.,"Long-Life Color DC PDP with Transparent Cathode",Proc.IDRC D'94,pp460-463,(1994).
17. Y. Terao et al.,"Fabrication of Fine Barrier Ribs for Color Plasma Display Panels by Sandblasting", SID'92 Digest,pp724-727(1992).
18. H. Fujii et al.,"A Sandblasting Process for Fabrication of Color PDP Phosphor Screens", SID'92 Digest,pp728-731(1992).
19. J. J. Felton et al.,"Thick-Film Materials for Plasma Displays", SID'95 Digest,pp488-490(1995)

DIELECTRIC AND DIPOLE MOMENT STUDY OF CHARGE TRANSFER COMPLEXES OF POLY(N-VINYLCARBAZOLE) AND 2,4,7-TRINITRO-9-FLUORENONE

Mohammed Fiaz*

Speciality Polymer Group, Chemical Engineering and Applied Chemistry, University of Aston In Birmingham, Aston Triangle, Birmingham, B4 7ET, United Kingdom

ABSTRACT

Poly(N-vinylcarbazole), PVK, is one of the most important photoconductive polymer. The stereostructure of PVK is now understood to influence significantly its electro-active properties. PVK was prepared by azobisisobutyronitrile (sample labeled F1) and borontrifluoride etherate (sample labeled F2) showed isotacticities of 25% and 55%, respectively as determined by 1H and 13C N.M.R. Spectroscopy. Dipole moments of these two stereostructural samples of PVK complexed with 2,4,7-trinitro-9-fluorenone were determined in solution in 1,4-dioxane. The results are explained in terms of polymer conformation and type of charge transfer complex. The influence of temperature on dipole moments is also discussed.

INTRODUCTION

In the earlier publication(1) the temperature coefficients of dipole moments of the poly(N-vinylcarbazole) samples which is strongly related to the stereostructures and tacticity of the PVK polymers was discuused. Sample F1 with the least isotactic content (prepared with a free radical catalyst), had the smallest dipole moment per repeat unit but the largest positive temperature coefficient. Contrasting behaviour was shown by sample F2 (synthesised by cationic catalyst), which showed the largest dipole moment but the smallest temperature coefficient. The dipole moment of the model compound, N-ethylcarbazole had no measureable temperature coefficient. The trends of the results were explained in terms of polymer stereostructures, the conformations of the polymers and the intramolecular cancelling of dipole moments. The various stereoregular forms of PVK, therefore provide an interesting opportunity to further extend this work by including PVK complexes. Useful information, concerning the structures of the donor / acceptor complex can be derived by measuring the dipole moment of the complex moiety. In principle, it should be possible to calculate, using measured dielectric permittivities, the molecular dipole moment of the complex formed between the carbazole side group and TNF. However, there are a number of difficulties

* Present address where all correspondence should be made: Institute of Polymer Technology & Materials Engineering, Loughborough University of Technology, Loughborough, Leicestershire, LE11 3TU, United Kingdom.

associated with this exercise. Firstly, it is important to have an accurate estimate of the concentration of complex formed in solution. This is because it is essential that the dielectric increment is accurately apportioned to free (uncomplexed) carbazole, free TNF molecules and the amount of PVK and TNF that are complexed. Secondly, the experimentally determined dielectric increment must be very accurate. In practice this is a difficult constraint to satisfy mainly due to the rather poor solubility of the PVK : TNF complex. However, despite these drawbacks concerning the interpretation of the solution dielectric permittivities it is concluded that they are sufficiently accurate to be used in the determination of the dipole moments of the complexes.

In this paper, calculations are presented of the dipole moments of PVK:TNF complexes. Also, a study is described of the influence of the temperature on the dipole moments of these complexes. A method by Weiser and Seki[2] was employed to obtain the values of the equilibrium constants which were then used to calculate the concentration of the PVK:TNF complexes in solution in 1,4-dioxane. The equilibrium constant provides an indicator of the degree of charge transfer interaction. Perhaps, more importantly it enables the concentration of the complex to be accurately determined. This is essential if specific properties, for example, dipole moment and molecular Kerr constant, of the complex are to be studied. By relating these interactions to the molecular structures of PVK and TNF, specific characteristics of PVK:TNF complex, such as temperature coefficients of the dipole moment and molecular Kerr constant can be investigated.

DETERMINATION OF THE CONCENTRATION OF TNF : PVK COMPLEXES IN 1,4-DIOXANE USING VISIBLE SPECTROSCOPY AND CALCULATION OF DIPOLE MOMENTS

For measurements of equilibrium constants, solutions of the charge transfer complexes were prepared by dissolving the polymer samples in 20ml of 1,4-dioxane in a 25ml volumetric flask and then adding the TNF to the solution and diluting to volume. The polymers were left to dissolve completely for a week before the TNF was added. Deep red solutions were immediately formed when TNF was added to the polymer solutions. Using quartz cells (2mm path length) the samples were analysed on ultra-violet and visible spectrophotometer (wavelength 450 -700nm). The concentration of the polymer was maintained at 2% (w/v) but the TNF concentration was varied from 0.25 to 2% (w/v). Since the main aim of this part of the work was to study the temperature coefficients of dipole moment of PVK:TNF complexes via the analysis of uv-visible, spectra were obtained at temperatures over the range of 298 - 338K. The optical density at 500nm was read from the spectra and used to calculate the equilibrium constants at the various temperatures as described by Weiser and Seki[2]. These concentations were then used to calculate the appropriate dielectric permittivity of charge transfer complexes in solution ($\varepsilon_{complex} = \varepsilon_{solution} - \varepsilon_{freePVK} - \varepsilon_{free\ TNF}$). The dielectric data for solutions of the various PVK: TNF complexes are presented in Tables 2. Dielectric data for solution of PVK in 1,4-dioxane and for solutions of TNF in 1,4-dioxane are also listed Table 2.

Mean square dipole moments $<\mu^2>$ per repeat unit of polymer chain were calculated using an appropriate from the Guggenheim-Debye equation[3]. Viz

$$<\mu^2> = \frac{27kT}{4\pi N_L\ (\varepsilon_1 + 2)\ (n_1{}^2 + 2)} (\Delta/C)_{c->0} \quad (1)$$

where ε_1 is the dielectric permittivity of solvent, ε_{12} is the dielectric permittivity of polymer dissolved in solvent, n_1 is the refractive index of solvent, k is the Boltzman constant, N_L is the Avagadro's number, $\Delta = (\varepsilon_{12} - n_2{}^{12}) - (\varepsilon_1 - n^2{}_1)$, and $(\Delta/C)_{c->0}$ is the limiting gradient of the plot of Δ versus concentration of the complexes C. By subsituting values of (Δ/C) into the Guggenheim-Debye equation (equation 1) the dipole moment of PVK : TNF complexes were calculated. The temperature coefficient of the dipole moments is conventionally defined by $d(\ln<\mu^2>)/dT$. Plots of $\ln <\mu^2>$ versus temperature are shown in Figure 1.

Table 1 Concentration of complexes formed in solution in 1,4-dioxane at various temperature.

Temperature (K)	Concentration of TNF (mol cm^{-3})	Concentration of Polymer (mol cm^{-3})	Derived concentration of F1 : TNF (mol cm^{-3})	Derived concentration of F2 : TNF (mol cm^{-3})
298	1.65 x10^{-5}	1.04 x 10^{-4}	1.86 x10^{-6}	3.46 x10-6
	3.20 x10^{-5}	1.04 x 10^{-4}	6.26 x10^{-6}	6.35 x10^{-6}
	4.80 x10^{-5}	1.04 x 10^{-4}	9.30 x10^{-6}	8.97 x10^{-6}
	6.35 x10^{-5}	1.04 x 10^{-4}	2.15 x10^{-5}	1.12 x10^{-5}
308	1.65 x10^{-5}	1.04 x 10^{-4}	1.85 x10^{-6}	3.38 x10^{-6}
	3.20 x10^{-5}	1.04 x 10^{-4}	6.13 x10^{-6}	6.38 x10^{-6}
	4.80 x10^{-5}	1.04 x 10^{-4}	9.11 x10^{-6}	8.78 x10^{-6}
	6.35 x10^{-5}	1.04 x 10^{-4}	2.07 x10^{-5}	1.10 x10^{-5}
318	1.65 x10^{-5}	1.04 x 10^{-4}	1.84 x10-^{6}	3.30 x10-^{6}
	3.20 x10^{-5}	1.04 x 10^{-4}	5.99 x10^{-6}	6.24 x10^{-6}
	4.80 x10^{-5}	1.04 x 10^{-4}	8.92 x10^{-6}	8.61 x10^{-6}
	6.35 x10^{-5}	1.04 x 10^{-4}	2.01 x10^{-5}	1.08 x10^{-5}
328	1.65 x10^{-5}	1.04 x 10^{-4}	1.83 x10^{-6}	3.23 x10^{-6}
	3.20 x10^{-5}	1.04 x 10^{-4}	5.87 x10^{-6}	6.11 x10^{-6}
	4.80 x10^{-5}	1.04 x 10^{-4}	8.72 x10^{-6}	8.43 x10^{-6}
	6.35 x10^{-5}	1.04 x 10^{-4}	1.95 x10^{-5}	1.06 x10^{-5}
338	1.65 x10^{-5}	1.04 x 10^{-4}	1.81 x10^{-6}	3.16 x10^{-6}
	3.20 x10^{-5}	1.04 x 10^{-4}	5.85 x10^{-6}	5.99 x10^{-6}
	4.80 x10^{-5}	1.04 x 10^{-4}	8.58 x10^{-6}	8.29 x10^{-6}
	6.35 x10^{-5}	1.04 x 10^{-4}	1.89 x10^{-5}	1.04 x10^{-5}

Table 2 Dielectric permittivity of TNF, F1, F2 and their complexes in solution in 1,4-dioxane at various concentrations at temperatures range 298 - 338K.

Temperature (K)	ε_{TNF}	ε_{F1}	ε_{F2}	$\varepsilon_{F1:TNF}$	$\varepsilon_{F2:TNF}$
298	2.2682	2.2523	2.2742	2.1826	2.1776
	2.2899	2.2557	2.3164	2.1892	2.4586
	2.2974	2.2651	2.3505	2.1900	2.6164
	2.3123	2.2796	2.3573	2.2076	2.8628
308	2.2560	2.2432	2.2653	2.1636	2.1722
	2.2756	2.2473	2.3075	2.1766	2.4309
	2.2902	2.2573	2.3382	2.1830	2.5558
	2.3029	2.2711	2.3481	2.1985	2.7467
318	2.2150	2.2123	2.2302	2.1497	2.1503
	2.2387	2.2155	2.2692	2.1556	2.3931
	2.2474	2.2257	2.3001	2.1576	2.5379
	2.2614	2.2371	2.3105	2.1713	2.6767
328	2.1962	2.1932	2.2103	2.1126	2.1334
	2.2153	2.1925	2.2492	2.1229	2.3839
	2.2264	2.2032	2.2783	2.1306	2.5113
	2.2384	2.2186	2.2883	2.1532	2.6275
338	2.1685	2.1733	2.1884	2.0912	2.1228
	2.1891	2.1734	2.2261	2.1045	2.3824
	2.2009	2.1842	2.2531	2.1183	2.4977
	2.2101	2.1962	2.2625	2.1358	2.5934

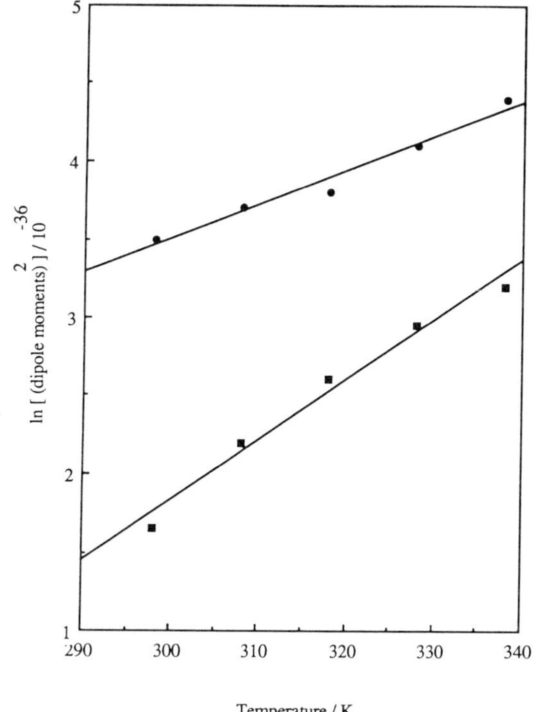

From Figure 1 the following observations can be made:

(i) The dipole moments of complexes formed between TNF and F1 (lowest isotacticity) have the lowest dipole moment of the various types of PVK but they exhibit the largest temperature coefficient ($0.039K^{-1}$).

(ii) The complexes formed between TNF and polymer F2 (largest isotacticity) have the highest dipole moment of the various types of PVK but they exhibit the smallest temperature coefficient ($0.022K^{-1}$).

DISCUSSION

These observations are discussed in detail below:

The proton and carbon-13 NMR indicated that polymer F1 synthesised with AZBN catalyst possessed 30% isotacticity whereas for sample F2 prepared with cationic catalyst (boron trifluoride etherate) the isotacticity was estimated to be 55%. From the X-ray studies(4,5), it is evident that the polymer of type F1 would be expected to possess mainly syndiotactic stereostructures with the possibility of forming 2/1 helixes that are not as sterically hindered as the 3/1 helixes believed to be formed in isotactic rich polymer F2. The degree of complex formation and the strength of the charge transfer differs considerably and depends on the conformation and tacticity of the polymer chains. The distances between the adjacent pendant carbazole groups in a helix in the 2/1 helix differs considerably when compared to the same distance in helices composed largely of 3/1 helices. These distances were estimated from desk top molecular modelling program (by Polyhedron Software Ltd.) and in the case of 2/1 helical chains the pitch distance is 4.33Å whereas in 3/1 type structures the distance is 6.10Å. The PVK and TNF form sandwich type charge transfer complexes(6) where the π orbital of the carbazole group would interact with the electron deficient TNF. However, for the 2/1 helical conformation (predominant in polymer F1) there is not sufficient room for the TNF molecule to adopt this conformation. However, it is important to emphasise that the charge transfer complexes are formed with all types of PVK polymers, it is only the amount of complexation and relative stability that differs. The higher dipole moments and virtually zero temperature coefficient observed for F2 : TNF complex provides evidence that the charge transfer complexes are relatively stronger then of F1:TNF. The strength of the F2:TNF complexes seem to be little affected by an increase of temperature. As a result the dipole moment remains virtually constant. Positive temperature coefficients were obtained for F1: TNF complexes, suggesting that these complexes were thermally less stable compared to F2 : TNF. At low temperatures, the chains are very much restricted in movement, and also tend to occupy the more energetically favourable helical conformations. The adoption of helical conformations results in the cancellation of dipole moments. As the temperature increases, there is more thermal energy and hence a greater degree of freedom of the pendant groups and as a consequence this results in the formation of non helical structures. This change in the proportion of helical structures leads to decrease in the cancellation of the dipole moments of the repeat units.

CONCLUSIONS

A modified method of Weiser and Seki(2) was used to estimate the equilibrium constants of charge transfer complexes of PVK and TNF. These values were then used to calculate the concentrations of the charge transfer complexes in solutions of 1,4-dioxane. The knowledge of the concentration levels of complexes facilitated calculation of the dielectric increments of charge transfer complexes. It was found that all PVK samples formed complexes with 2,4,7-trinitro-9-fluorenone. However, the amount of complexation and the thermal stability of the complexes differed significantly. The complex of polymer type F2 (highest isotacticity) and TNF was the most stable having the highest dipole moments and almost zero temperature coefficient for the dipole moments. The complexes formed between polymer sample F1 (lowest isotacticity) and TNF were, on the other hand, found to have lowest dipole moments but the largest dipole moment temperature coefficient .

ACKNOWLEDGMENT

The financial support for this work was provided by Science and Engineering Research Council.

REFERENCES

1. M. Fiaz and M.S. Beevers, "Advanced Materials-93" eds. A.ul Haq, F.Habiby, A.Q.Khan, Pub. Dr A.Q. Khan Res. Lab., Rawalpindi, Pakistan (1994).
2. G. Weiser and H. Seki, Charge Transfer Complex Formation of Trinitrofluorenone With Poly(N-vinylcarbazole and Ethylcarbazole, *IBM Journal of Research and Development,* 16(6):598 (1972).
3. E.A. Guggenheim, A Proposed Simplification in the Procedure For Computing Electric Dipole Moments, *Trans. Far.Soc.,* 45:714 (1949).
4. H. Mikawa K. Kasai, S. Kusahoyashi, H. Hirata and A.Kimura, Crystallinity of Poly(N-vinylcarbazole), *J. polym. Sci.*, 8:613 (1970).
5. C.H. Griffiths, Folded Chains Crystallisation of Poly(N-vinylcarbazole), *J. polym. Sci., Phys. Edn.*, 13:1167 (1975).
6. S. Tazuke and Y.Matsuyama, Donor Acceptor Interaction in Polymeric Systems, *Makromol. Chem.,* 176:3167 (1975).

COMPARATIVE STUDY OF THE RAMAN SPECTRA OF DIFFERENT FORMS OF POLYANILINES

G. Louarn, S. Quillard and S. Lefrant

Laboratoire de Physique Cristalline, Institut des Matériaux *
Université de Nantes, 2 rue de la Houssinière
44072 Nantes Cedex 03, France.

INTRODUCTION

The polymerization of aniline, and subsequently the properties of oligoanilines, polyanilines and substituted derivatives, have led to a large amount of studies in the last decade. This is due to the conducting properties of protonated or doped polyanilines which primarily offer real potential applications such as electronic devices, antistatic thin films, etc.[1,2]

It is now generally accepted that the name "polyaniline" represents a family of compounds of the following formula :

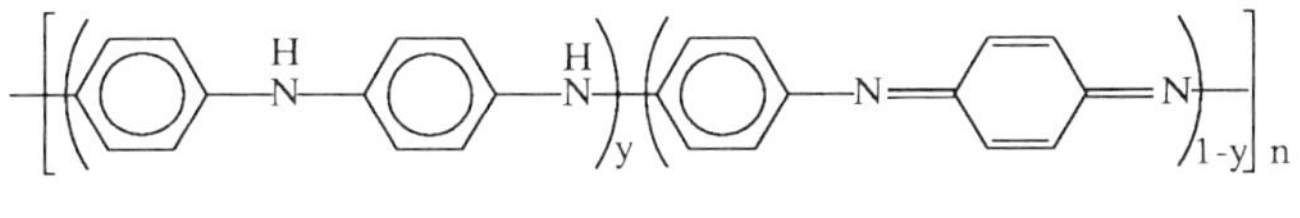

where y : represents the fraction of reduced units
1-y : represents the fraction of oxidized units

In principle, y can change continuously within $0 \rightarrow 1$. However two extreme and one intermediate forms are usually distinguished in the literature,[3] namely : the totally reduced form : leucoemeraldine (y = 1) ; the totally oxidized form : pernigraniline (y = 0) and the half-oxidized form : emeraldine (y = 0.5).

The above formula shows only the base forms of polyaniline. However due to the presence of basic sites (amine and imine nitrogens), polyaniline can be protonated with a sufficiently strong acid and protonated emeraldine is the only conductive form of polyaniline.

In order to contribute to the understanding of the fundamental molecular properties of polyaniline, we present in this paper vibrational studies of polyemeraldine base and protonated polyemeraldine. In fact, protonation process inevitably causes significant changes in the vibrational and electronic properties of the polymer. Thus, both Raman and uv-vis-nir

Polymers and Other Advanced Materials: Emerging Technologies and Business Opportunities. Edited by P. N. Prasad *et al.*, Plenum Press, New York, 1995

spectroscopic methods are extremely sensitive tools for studying the acid base chemistry of polyaniline.[4-8] Also, we used deuterated compounds (*d*-EB and *d*-ES) which have proved to be significantly important as model compounds in vibrational spectroscopic investigations. As in polyparaphenylene[9] (PPP) or polythiophene[10] (PTh) for which deuterated compounds are available, a clear relationship between the isotopic downshift of vibrational frequencies and the assignment of these modes can be exploited in order to have a better knowledge of the chemical structure and of the electronic properties of the related polymer.

Finally, spectroelectrochemical studies of polyanilines were carried out in 1M HCl aqueous solution using uv-vis-nir and Raman spectroscopy. Spectra of the samples polarized from -150 mV to +650 mV vs SCE have been recorded with λ_{exc} = 514.5 nm. We show that resonant conditions, which are constantly changed during electro-chemical cycling, strongly influence the resulting Raman spectra.

EXPERIMENTAL

Chemical synthesis of polyaniline powder : polyaniline was prepared chemically by the synthesis of pristine emeraldine hydrochloride (~ 42%) protonated by the oxidative polymerization of aniline by ammonium peroxydisulfate as reported previously.[11] The pristine emeraldine hydrochloride was immediately converted to the emeraldine base (EB, y ~ 0,5) by treatment with 0,1 M NH_4OH. The fully deuterated emeraldine base was synthesized similarly using deuterated aniline, heavy water and DCl.

Electrochemical synthesis of polyaniline films : Polyaniline was synthesized either on a platinum electrode (for cyclic voltammetry and Raman studies) or on a conducting glass electrode (ITO) (for uv-vis-nir-studies). A classical three electrodes configuration with a platinum counter-electrode and SCE reference was used. The polymerization aqueous medium consisted of 0.25 M solution of aniline in 1M HCl or deuterated aniline in 1M DCl.

Typical polymerizations were carried out by potential scanning between -150 and 700 mV vs SCE with the rate of 50 mV/s using an EG&G PAR 273 galvanostat-potentiostat.

Uv-vis-nir "in situ" spectroelectrochemical studies : Optical absorption investigations in the uv-vis-nir region were performed on a Varian Cary 2300 spectrometer using a specially designed spectroelectrochemical cell depicted in Fig. 1. The measurements were strictly correlated with cyclic voltammetry i.e. spectra were recorded for the potentials in which characteristic features of the cyclic voltammogram are observed[12] (onset and maximum of the peak, plateau between two peaks etc...).

Resonance Raman Scattering : Resonance Raman Scattering (RRS) spectra were recorded with a standard equipment in a 90° scattering geometry using either argon or krypton laser lines. A T64000 (Jobin-Yvon) triple monochromator in a triple subtractive configuration was used in all experiments. The signal was detected with a cooled CCD detector. In order to avoid local degradation of the compounds, the laser beam power was limited to 20 mW and the incident beam was defocused (~ 3 mm^2). All experiments were performed at room temperature. FT-Raman spectra were measured on a RFS 100 (Brucker) Fourier transform spectrophotometer. A Nd:YAG laser was operated at 1064 nm for excitation. In this case, the Raman scattering radiation was collected in a back scattering configuration and the spectral resolution was 4 cm^{-1}.

"In situ" Raman spectroelectrochemical studies were carried out in the same spectroelectrochemical cell used for uv-vis-nir studies (Fig. 1). The scattering signal was collected at 90° except for FT Raman spectra where a back scattering configuration was used. Although exact determination of the power of the laser beam is very difficult in our case, we can reasonably estimate that it did not exceed 100 mW.

Similarly as uv-vis-nir, Raman spectroscopic studies were also correlated with cyclic voltammetry.

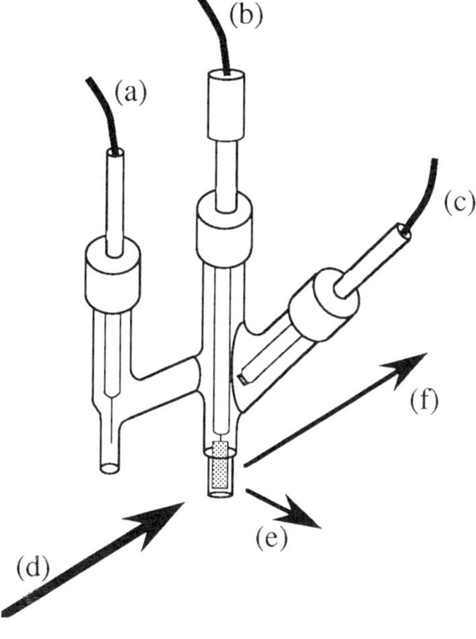

Figure 1. Electrochemical cell for "in situ" spectroelectrochemical studies using uv-vis-nir and Raman spectroscopies : a) counter electrode, b) working electrode, c) reference electrode, d) incident beam, e) scattered radiation (Raman), f) optical absorption.

RESULTS

Uv-vis-nir "in situ" spectroelectrochemical studies

Resonance Raman scattering requires a good knowledge of the optical transitions of the compounds. The optical absorption spectrum of emeraldine base (EB) presents two large bands,[13] one at 3.8 eV (330 nm) corresponding to the $\pi \rightarrow \pi*$ transition and a second centered at 1.94 eV (640 nm) associated to the exciton band. More details about these different absorption features are discussed by Brédas.[14]

Spectra of protonated emeraldine are presented Figure 2. These data were recorded for different potentials on polyaniline films deposited electrochemically on ITO. The results are in good agreement with those reported in the literature.[4,5]

Spectral changes imposed by increasing potential can be briefly described as follows :

At E = -150 mV, the spectrum is characteristic of the leucoemeraldine base as expected. However, a small shoulder on the low energy side of the $\pi \rightarrow \pi*$ transition peak in leucoemeraldine seems to indicate that a small fraction of oxidized units exists in the polymer. The oxidation at 200 mV, i.e. at the potential of the maximum of the first oxidation peak, results in a drastic change of the uv-vis-nir spectrum. Three absorptions now can be seen at 3.8 eV, 2.9 eV and 1.0 eV. These features are characteristic of polyaniline in its conducting form, i.e. protonated emeraldine.[15]

Further increase of the potential to 450 and 650 mV leads to gradual decrease of the 430 nm peak with simultaneous increase of the low energy peak, which is, in addition, slightly blue shifted.

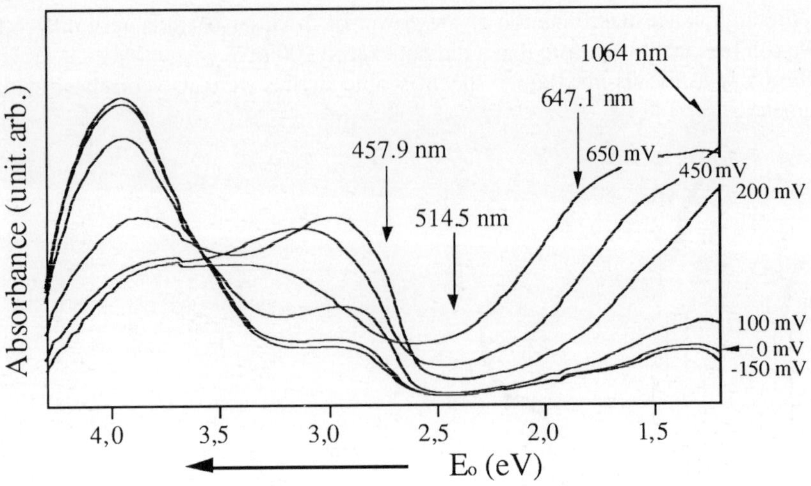

Figure 2. Uv-vis-nir spectra of polyaniline recorded at different potentials. The arrows indicate the positions of the excitation lines used in Raman with respect to the given spectrum of polyaniline

Raman Spectroscopy

The complete vibrational analysis of emeraldine base has been achieved from comparison with the leucoemeraldine and the pernigraniline bases. Previous assignments have been made comparing the Raman spectra of emeraldine base (EB) with those of leucoemeraldine (LB) at 457.9 nm and of pernigraniline (PNB) at 676.4 nm.[16] These studies on the half-oxidized form of polyaniline are here extended to the deuterated polymer, labelled *d*-EB. Raman spectra of EB and *d*-EB are shown in figure 3. Three different excitation wavelengths have been chosen (457.9, 514.5 and 676.4 nm) as we know one can get informations from the resonance conditions, since the uv-vis-nir spectra have revealed several characteristic features.

Comparing those spectra at each excitation line, one can notice that major differences are observed in the 1150-1250 cm^{-1} frequency range.

Above 1300 cm^{-1}, the Raman spectra are very similar. From our previous studies on base forms,[16] we assign the three bands around 1550-1620 cm^{-1} to the carbon-carbon stretching modes of the rings. In this region, these bands appear to be only little affected by the deuteration. Increasing the excitation wavelength up to 676.4 nm is known to enhance the vibrational modes of the oxidized units. From this remark, the CC ring stretching of the quinoid ring is peaked at 1587 cm^{-1} in emeraldine base while it appears clearly at 1553 cm^{-1} in *d*-EB. This small shift is consistent with the nature of the vibration which implies mainly C-C bonds of the ring. The same behavior is observed in deuterated PPP.[9]

The chemical nature of conjugated polymers leads also to another resonance originating spectral feature i.e. the dependence of the frequency position of C=C stretching modes on the energy of the excitation line. This effect observed in polyacetylene,[17] poly(n-alkythiophene)[18] but also pernigraniline[19] is usually attributed to the distribution of conjugation lengths which inevitably must exist in macromolecular systems such as those mentioned above. The intense Raman band at 1480 cm^{-1} in EB is characteristic of C=N stretching vibration. This band shows a slight effect of frequency dispersion (1492 cm^{-1} with 457.9 nm to 1474 cm^{-1} using 676.4 nm), and that suggests a significant sensitivity toward the conjugation length. A similar phenomenon is present in the deuterated compound (1480 cm^{-1} with 457.9 nm to 1462 cm^{-1} using 676.4 nm), in the same manner.

For λ_{exc} = 457.9 nm the Raman spectrum of EB exhibits an intense band at 1160 cm^{-1}, still present with the other excitation lines. This vibrational band is clearly assigned to the C-H in plane bending deformation in quinoid-type ring, coming from the 9a vibration (Wilson notation) of the benzene molecule. As a consequence, this mode is clearly very sensitive to the deuteration and it has totally disappeared on the *d*-EB spectra. Theoretically, from the Teller and Redlich relation,[20] it should be shifted in a ratio around 1.4 and the similar mode in the deuterated polymer is then expected at 870 cm^{-1}. That is not quite visible on the spectra of *d*-EB.

In the same spectral range, the C-N stretching vibration, which appears at 1219 and 1218 cm^{-1} for the leucoemeraldine and pernigraniline respectively, is clearly present in EB at 1219 cm^{-1}. However, the similar mode in *d*-EB is mainly observed using the 676.4 nm excitation at 1210 cm^{-1}.

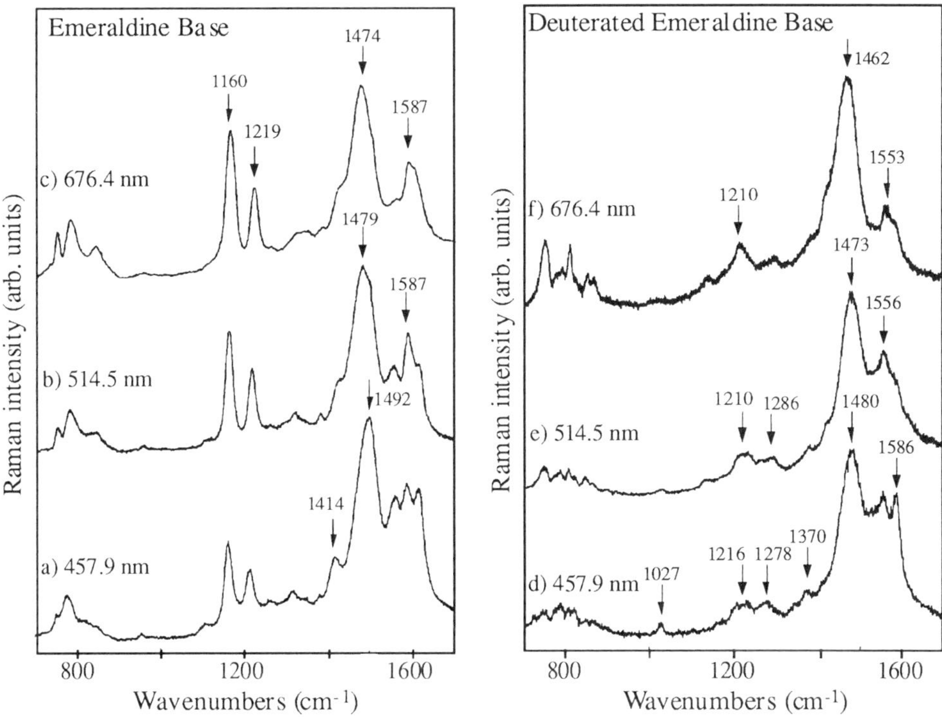

Figure 3. Raman spectra of polyemeraldine (EB) and deuterated polyemeraldine (*d*-EB) base ($y \approx 0.5$) using different excitation lines

The other "in plane" vibrations are assigned to bands appearing in selected regions. Their assignments are not always unambiguous and dynamical calculations (Potential Energy Distribution and Cartesian Displacements) are necessary for define assignments. However, this information is too voluminous to deserve a detailed description in this paper. However, in Table I we have collected all the Raman frequencies in the 700-1700 cm^{-1} frequency range and we propose the different assignments and filiations for LB, PNB, EB and *d*-EB.

Table I. Experimental frequencies of leucoemeraldine (LB), pernigraniline (PNB), emeraldine and deuterated emeraldine base Raman modes. (B) benzenoid ring, (Q) quinoid ring.

LB	PB	Emeraldine Base		*d*-Emeraldine Base		DESCRIPTION OF VIBRATIONS
457.9 nm	647.1 nm	457.9 nm	676.4 nm	457.9 nm	676.4 nm	
1618	1612	1620	1618	1586	1586	C-C stretching (B)
1597	1553	1551	1553	----	----	C-C stretching (B)
	1579	1587	1587	1556	1553	C=C stretching (Q)
	1480	1492	1474	1480	1462	C=N stretching
	1418	1414	shoulder	1370	shoulder	C-C stretching (Q)
		(1330)				residual protonated struct.
1219	1215	≈ 1220	1219	1216	1210	C-N stretching
1181	----	----		----	865	C-H bending (B)
	1157	1162	1160	----	848	C-H bending (Q)
868	----	880	870	786		ring deformation (B)
820		828	826	----	805	amine def. (X-sens.)
	788		788		745	ring deformation (Q)
	749	720	750	----	----	imine deformation

Figure 4 shows the Raman spectra of the emeraldine salt form obtained using three different excitation wavelengths : 457.9, 676.4 and 1064 nm. In the case of protonated emeraldine and considering the results of uv-vis-nir absorption, it is interesting to perform Raman experiments with the n.i.r. excitation line. Indeed, the conducting form of polyaniline shows a free carrier tail absorption extended to the nir region.

For $\lambda_{exc} = 457.9$ nm the Raman spectrum of ES exhibits an intense band at 1192 cm^{-1}. This vibrational band is clearly assigned to the C-H in-plane bending deformation in benzoid-type ring. This band is not observed in *d*-ES, being downshifted to 876 cm^{-1}. For the red (676.4 nm) and infrared (1064 nm) excitations, the maximum of this C-H in-plane bending peak is observed in ES at $\approx$ 1170 cm^{-1} i.e. at a value characteristic of this type of deformation in a semi-quinoid type ring (for comparison the quinoid mode is peaked at 1160 cm^{-1} in emeraldine whereas the benzenoid one at 1185 and 1181 cm^{-1} for emeraldine and leucoemeraldine respectively).

In the spectra of *d*-ES, the similar vibration is also downshifted in the same ratio of 1.4, and is peaked at 860 cm^{-1}. However, one can notice that probably, some non-deuterated ring still exist in *d*-ES, as evidenced by a Raman band at 1160 cm^{-1}.

In the frequency range of 1550-1650 cm^{-1} and for $\lambda_{exc} = 457.9$ nm, single sharp lines at 1626 cm^{-1} in ES and 1596 cm^{-1} in *d*-ES are present, which means that the dominating units are benzenoid-type. For longer excitation wavelength, these lines broaden and a clear shoulder appears on the low energy side of these Raman peaks. A similar behavior in the base forms is noticed. This characteristic region of CC ring stretching shows two overlapping Raman bands which can be assigned to vibrations centered on the quinoid-type ring (1578 cm^{-1} in ES ; 1553 cm^{-1} in *d*-ES) and on the benzenoid-type ring (1620 cm^{-1} in ES ; 1580 cm^{-1} in *d*-ES). Thus, the red excitation line (676.4 nm) reveals the co-existence of benzenoid and quinoid units, the signal due to the presence of the latter being strongly resonantly enhanced. Infrared excitation (1064 nm) gives rise to one broad band centered at 1595 and 1565 cm^{-1} for ES and *d*-ES respectively.

On the ES spectrum at 676.4 nm, two Raman bands are observed in 1450-1550 region, one of them associated to a C-C stretching + C-H bending deformation (9a in the Wilson notation) on the semi-quinoid ring (1492 cm^{-1}) and the second is associated to a N-H bending deformation (1520 cm^{-1}).

The latter mode is no more seen in *d*-ES and furthermore, the deuteration induces a new Raman band at 1084 cm^{-1}, assigned to the N-D bending deformation. The former mode is also downshifted in a lower ratio, at 1428 cm^{-1}. In agreement with resonance conditions, no Raman lines are observed if λ_{exc} = 457.9 nm is used . Infrared excitation (1064 nm) gives rise to one broader peak at ca 1493 cm^{-1} in ES and at 1439 cm^{-1} in *d*-ES.

The most interesting features observed in protonated emeraldine occur in the frequency range of 1300-1350 cm^{-1}. It should be pointed out that none of the basic forms of polyaniline (i.e. leucoemeraldine, emeraldine and pernigraniline) gives rise to vibrational modes in this spectral region. The existence of peaks in the vicinity of 1300-1350 cm^{-1} is therefore inherently associated with the protonation process via a poly(semiquinone) radical formation mechanism. The internal redox process associated with this type of protonation results in the lowering of the bond order between C and N which now becomes intermediate between the amine (=C-N-) and imine (>C=N-) bonds. Since the peaks characteristic of >C=N- stretching are located in the vicinity of 1470 cm^{-1} and those due to >C-N- stretching around 1220 cm^{-1}, an intermediate semiquinone radical type of bond should give an Raman peak around 1300-1350 cm^{-1}.

For the red (676.4 nm) and infrared (1064 nm) two bands are observed both being resonantly enhanced. In the case of ES, the two components of this splitted band are observed at 1317, 1338 cm^{-1}, while in the deuterated polymer, they are observed at 1308, 1340 cm^{-1}. The small downshift of this mode is consistent with its proposed assignment i.e. a >C-N- stretching deformation of the semiquinone radical.

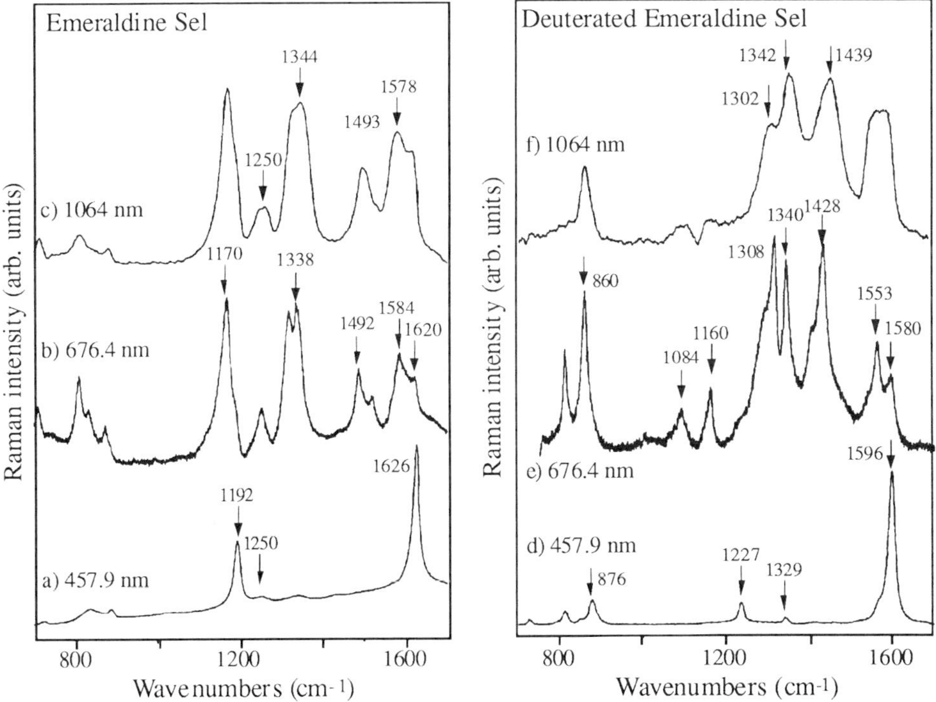

Figure 4. Raman spectra of polyemeraldine salt (ES) and fully deuterated polyemeraldine salt (*d*-ES) using different excitation lines

Raman "in situ" spectroelectrochemical studies

Raman spectra recorded "in situ" at different potentials are already published for $\lambda_{exc} = 488.0$ nm,, $\lambda_{exc} = 676.4$ nm and $\lambda_{exc} = 1064$ nm respectively.[12] These spectra clearly confirm that, in the case of polyaniline, the excitation wavelength plays a very important role.

The use of an excitation line from the blue region (457.9 nm) leads to Raman spectra which are only slightly dependant on the potential of the electrode. The application of nir and red excitation lines (1064 and 676.4 nm respectively) gives rise to different resonance conditions for which an enhancement of the lines originating from oxidized (quinone type) units is evident. Similarly as in the case of the 457.9 nm excitation line, the Raman spectra are only a little dependent on the potential.

One other excitation line (514.5 nm) is situated between the main absorptions characteristic of uv-vis-nir spectrum at polyaniline (Fig. 2). Intuitively, one may expect that spectra obtained with this excitation line should be more sensitive to the changes of the oxidation state of polyaniline since in this case, neither the absorptions due to the reduced units not those characteristic of oxidized units should exhibit strong (and selective) resonance enhancement. Indeed this is the case, as shown in Fig 5. It is seen that for $\lambda_{exc} = 514.5$ nm Raman spectra change drastically with the applied potential.

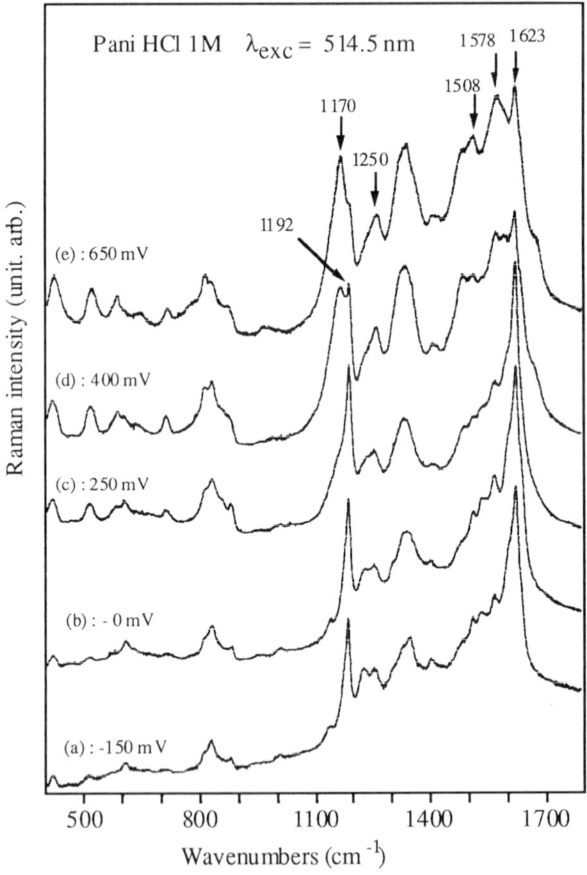

Figure 5 : Raman spectra of polyaniline recorded at different polarization potentials ($\lambda_{exc} = 514.5$ nm)

Thus, from a spectroelectrochemical point of view, the co-existence of several species and their mutual interconversion must be considered. Additionally, if a semiquinone model of protonation is valid, the oxidation of reduced polyaniline should lead to the same product as the protonation of partially oxidized polymer. In the latter case, protonation will lower the CN bond order which will become intermediate between C=N and C-N and this process should be put in evidence in the Raman spectra by the appearance of "semiquinone-radical" bands.

Comparing the spectra obtained at different polarization potentials, we can roughly separate them into two groups. One group would contain the spectra performed for the potentials from -150 mV/SCE up to 250 mV/SCE and data show very similar features. In particular, two intense Raman bands are present at 1192 cm^{-1} and at 1623 cm^{-1}, characteristic of benzenoid-type rings. According to the uv-vis-nir absorption spectra, in these cases, we are in resonance with the absorption situated at 2.8 eV. This band increases with the increasing potential values, until V = 200 mV /SCE, i.e. until the first redox wave is reached. Then, it decreases with the potential and is shifted toward higher energies.

This behavior is related to the creation of polarons when the oxidation degree increases. In a first step, the oxidation process leads to the appearance of polarons along the chain. The second step occurs when the number of these polarons has reached a value of 2 per 4 rings. Then, this step consists into an interconversion of the polaron lattice into a semiquinoid structure, according the following scheme :

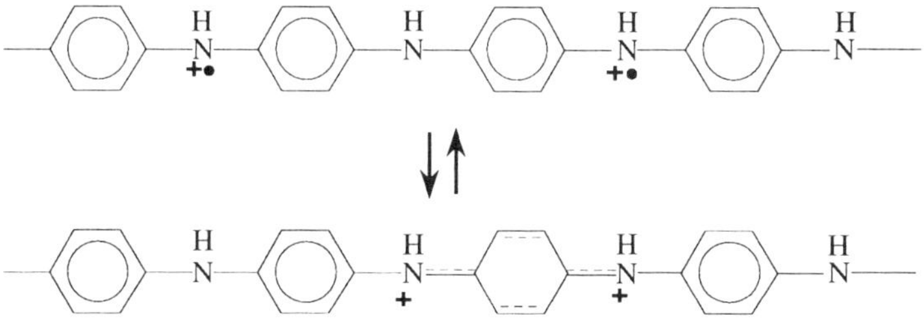

Using 514.5 nm as excitation line, for the potentials higher than 250 mV, we are in resonance with the second absorption band, around 1.5 eV. Thus, the spectra obtained for V = 400 mV and V = 650 mV are also very similar. Both show the spectroscopic signatures of a semiquinoid structure with a broad band centered at 1330 cm^{-1}, and with the appearance of quinoid-type rings, mainly revealed by the Raman bands at 1578 cm^{-1} and 1170 cm^{-1}.

CONCLUSIONS

In this paper, we have presented a Raman analysis of emeraldine and deuterated emeraldine in their base or salt forms. Previous assignments of the vibrational Raman modes of emeraldine base have been confirmed by the behavior of these bands after deuteration. Similar shifts of these characteristic peaks have also been observed in the salt compounds. In addition, we have studied the spectroelectrochemical behavior of polyaniline, using the green excitation line, which appeared to be the most appropriate wavelength to put in evidence the formation of semiquinoid structure during the metal/insulator transition.

Generally speaking, gradual oxidation of polyaniline in an acidic medium can be elegantly revealed by the use of intermediate excitation lines (488.0 and 514.5 nm). In this case, the bands characteristic of the semi-quinoid units gradually increase in intensity with increasing potential. The use of these lines allows to confirm the strong protonation of the oxidized units via semiquinone radical formation mechanism.

ACKNOWLEDGMENTS

We would like to thanks Drs A. Pron and M. Lapkowski for their active participation in the different studies presented in this paper.

* The Institut des Matériaux is 'Unité Mixte de Recherche' CNRS - Université de Nantes.

REFERENCES

1. Y. Yang and A. J. Heeger, *Nature*, 372:344 (1994).
2. J. Joo and A. J. Heeger, *Appl. Phys. lett.* 65:2278 (1994).
3. W. S. Huang, B. D. Humphre and A. G. MacDiarmid, *J. Chem. Soc. Faraday Trans.* 82:2385 (1986).
4. Y. Furukawa, F. Ueda, Y. Hyodo, I. Harada, T. Nakajima and T. Kawagoe, *Macromolecules*, 21:1927 (1988).
5. H. Kuzmany, and N. S. Sariciftci, *Synth. Met.* 18:353 (1987).
6. M. C. Bernard, S. Cordoba-Torresi and A. Hugot-Le Goff, *Solar Energy Materials and Solar Cells*, 25:225 (1992).
7. A. Hugot-Le Goff and M. C. Bernard, *Synth. Met.*, 60:115 (1993).
8. A. P. Monkman, in Conjugated Polymeric Materials, NATO ASI Series, 182:273 (1990).
9. J. P. Buisson, J. Y. Mévellec, S. Zeraoui and S. Lefrant, in "Lower-Dimensional Systems and Molecular Electronics", Edited by R.M. Metzger et al.,Plenum Press,New York, (1991).
10. Y. Furukawa, M. Akimoto and I. Harada, *Synth. Met.*, 18:151 (1987)
11. A. G. MacDiarmid, J. C. Chiang, A. F. Richter, N. L. D. Somasiri and A. J. Epstein , in "Conducting Polymers", edited by L. Alcacer, Reidel Dordrecht, (1987).
12. S. Quillard, K. Berrada, G. Louarn, S. Lefrant, M. Lapkowski and A. Pron, *New J. Chem.*, 19 (1995) in press.
13. R. P. McCall, J. M. Ginder, J. M. Leng, H. J. Ye, S. K. Manohar, J. G. Masters, G. E. Asturias, A. G. MacDiarmid and A. J. Epstein, *Phys. Rev. B*, 41:5202 (1990).
14. J. L. Brédas, in "Proceeding of the 1991 Nobel Symposium on Conjugated Polymers and Related Materials", edited by W. R. Salaneck, Oxford University, New York, (1993).
15. S. Stafström, J. L. Brédas, A. J. Epstein, H. Woo, D. B. Tanner, W. S. Huang and A. G. MacDiarmid, *Phys. Rev. Lett.*, 59:1464 (1987).
16. S. Quillard, G. Louarn, S. Lefrant and A. G. MacDiarmid, *Phys. Rev. B*, 50:12496 (1994),.
17. G. A. Arbuckle, A. G. MacDiarmid, S. Lefrant, T. Verdon, E. Mulazzi, G. P. Brivio, X. Q. Yang, H. S. Woo and D. B. Tanner, *J. Electroanal. Chem.*, 220:67 (1987).
18. G. Louarn, J. Y. Mévellec, J. P. Buisson and S. Lefrant, *J. Chim. Phys.*, 89:987 (1992).
19. S. Quillard, G. Louarn, J.P. Buisson, S. Lefrant, J. Masters and A. G. MacDiarmid, *Synth. Met.*, 55-57:475 (1993).
20. O. Redlich , Z. Physik Chem., B28:371 (1935).

THE LOWEST EXCITED STATES IN OLIGO(PHENYLENEVINYLENE)S: A THEORETICAL INVESTIGATION OF OPTICAL ABSORPTION SPECTRA AND GEOMETRY RELAXATION PHENOMENA

J. Cornil[1], D. Beljonne[1], Z. Shuai[1], D.D.C. Bradley[2], and J.L. Brédas[1]

[1] Service de Chimie des Matériaux Nouveaux
Centre de Recherche en Electronique et Photonique Moléculaires
Université de Mons-Hainaut
Place du Parc 20, B-7000 Mons, Belgium

[2] Department of Physics and Centre for Molecular Materials
University of Sheffield
Hicks Building, Hounsfield Road, Sheffield S37RH, U.K.

INTRODUCTION

Conjugated polymers have attracted a growing interest as active layers in light-emitting diodes; since the initial discovery by Burroughes *et al.* that poly(p-phenylenevinylene), PPV, could emit light in an efficient way[1], many efforts have been devoted to the design of novel devices with enhanced characteristics[2]. Due to their ease of synthesis and processing under the form of stable thin films[3] and to their remarkable luminescence properties[4], PPV and its substituted derivatives are among the most thoroughly investigated systems to be exploited in LED's. With time, several strategies have been exploited to significantly improve the performances of these electro-optic devices; as a matter of fact, the internal quantum efficiency (defined as the number of photons emitted over the number of electrons injected) has evolved from an initial value around 0.01%[1] to values up to 10%[5].

In this context, quantum-chemistry calculations can prove useful by providing guidelines to chemical synthesis as well as a basic understanding of the phenomena taking place in the polymer layers of the LED's. It is interesting to note that the study of conjugated oligomers has emerged as an appropriate tool to shed light into the complex behavior of the corresponding polymers; this approach is justified by the fact that actual samples are often better described in terms of disordered molecular materials rather than as infinite polymer chains[6]. Moreover, the study of oligomers presents the advantage that model molecules are better suited to most experimental spectroscopic techniques and that sophisticated theoretical treatments preferentially deal with finite-size systems.

In this contribution, we theoretically investigate the nature of the lowest-lying one-photon-allowed singlet excited state (^1B_u state) from which originates the photoluminescence or electroluminescence processes following radiative decays of the polaron-excitons. A

Polymers and Other Advanced Materials: Emerging Technologies and Business Opportunities. Edited by P. N. Prasad *et al.*, Plenum Press, New York, 1995

particular attention is also paid to the lowest triplet excited state that can be reached when carriers of opposite charge but same spin recombine or when intersystem crossings between the singlet and triplet manifolds occur. First, on the basis of well-resolved experimental optical absorption spectra of oligo(phenylenevinylene)s (measured at 80K in a poly(methylmethacrylate) (PMMA) matrix, as previously reported[7]), we set up a theoretical methodology that allows us to fit the vibronic progression. We then take advantage of these information to better characterize the ^1B_u state, especially the extent of the relaxation process occurring upon photoexcitation. In a second part, the trends derived from the analysis of the experimental spectra are compared to the results of calculations that provide direct estimates of lattice relaxations in both the lowest singlet (denoted hereafter as S_1) and lowest triplet (T_1) excited states.

THEORETICAL METHODOLOGY

In this work, we have investigated unsubstituted PPV oligomers containing from 2 to 5 phenylene rings; we sketch in Figure 1 the chemical structure of the four-ring oligomer and the bond numbering adopted when introducing the geometric data.

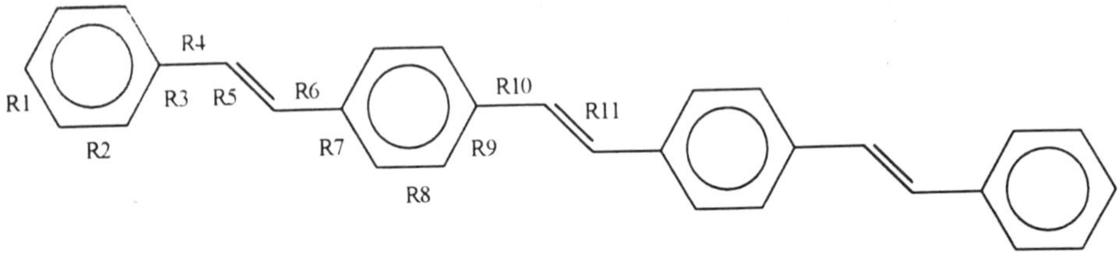

Figure 1. Chemical structure of the four-ring PPV oligomer.

The geometries of the oligomers have been optimized by means of the semiempirical Hartree-Fock Austin Model (AM1) method[8], assuming fully planar conformations. On that basis, we make use of the semiempirical Hartree-Fock Intermediate Neglect of Differential Overlap (INDO) Hamiltonian coupled to a Single Configuration Interaction (SCI) technique[9] to determine the transition moments and transition energies to be plugged into the Sum-Over-States (SOS) expression of the polarizability[10]. The CI development is built with the help of singly excited configurations obtained by promoting an electron from one of the 20 upper occupied levels to one of the 20 lower unoccupied ones. The optical absorption spectra are then simulated by taking the imaginary part of the polarizability and multiplying it by the incident frequency.

In order to fit the well-resolved experimental spectra[7], we have incorporated in the SOS expression the vibronic contributions that are coupled to the electronic transitions. The vibronic contributions are treated within the Franck-Condon formalism, as shown in Equation 1:

$$\alpha_{ij}(-\omega;\omega) = \frac{1}{\hbar} P(i,j;-\omega,\omega) \sum_{m} \sum_{v^x} \frac{\langle g|\mu_i|m\rangle \langle m|\mu_j|g\rangle \prod_{x=1}^{3N-6} \langle 0|v^x\rangle^2}{(\omega_{mg} + v^x E_{ph}^x - \omega - i\,\Gamma)} \tag{1}$$

where the overlap between the vibrational wavefunctions is given by:

$$<0|v^{x}>^2 = \frac{e^{-S_x}\, S_x^{v}}{v!} \tag{2}$$

We provide the best fit of the experimental data by modulating: (i) the value of the Huang-Rhys factor (S) that controls the relative intensities of the vibronic peaks and (ii) the amplitude of the damping factor (Γ) that defines the width of the absorption features. In the present work, we explicitly take into account the two most dominant effective modes at 0.16 eV and 0.21 eV in a 1/2 ratio, on the basis of site selective fluorescence spectra[11].

Estimates of the lattice relaxations are also obtained through direct optimization of the geometry of the lowest singlet and triplet excited states by means of the AM1 method. To do so, we perform calculations on the basis of a level occupancy where one electron is promoted from the HOMO level to the LUMO level, one spin being flipped for the triplet. This way of proceeding provides geometries in the excited states that are consistent with those obtained by imposing bipolaron-type relaxation or by making use of empirical bond-order/bond-length (BOBL) relationships[12]. In order to assess in more depth the geometric deformations that characterize the S_1 and T_1 states, the AM1-optimized geometries are fitted to Equation 3 which represents the changes in bond lengths following the creation of a confined soliton-antisoliton pair upon photoexcitation, as derived from the Su-Schrieffer-Heeger (SSH) theory[13]:

$$r_n^e = r_n^g + (-1)^n \times \phi \times \tanh\frac{2R}{\zeta} \times \left[\tanh\frac{n-N_0-R}{\zeta} + \tanh\frac{n+1-N_0-R}{\zeta} - \tanh\frac{n-N_0+R}{\zeta} - \tanh\frac{n+1-N_0+R}{\zeta}\right] \tag{3}$$

Here, r_n^g (r_n^e) is the nth bond length in the AM1-optimized geometry of the ground (excited) state, N_0 is the center of the chain, ζ is the coherence length of the defect, ϕ (ϕ_R for the phenyl rings) is the amplitude of bond-length deformation, and 2R is the distance between the "soliton" centers.

RESULTS AND DISCUSSION

We present in Figure 2 the experimental[7] and INDO/SCI-simulated optical absorption spectra of the PPV oligomers. We stress the very good agreement obtained between theory and experiment when looking at the absolute values of the excitation energies between the ground state and the ^1B_u excited state, which mainly originates from an electron transition between the HOMO and LUMO levels; both the experimental and calculated data show the existence of a linear relationship between the transition energies and the inverse number of carbon atoms on the shortest pathway between the two ends of the chain. These results primarily point to the fact that no drastic conformational changes occur when the size of the system is increased.

The existence of several vibronic features grafted on the electron transitions indicates that there is a significant displacement of the equilibrium position of the potential energy curves when going from the ground state to the ^1B_u state; in other words, such a lineshape clearly reflects the electron-phonon coupling characteristic of conjugated systems. We report in Table 1 the fitted parameters used when plotting the theoretical spectra. We note that we have to consider a weaker value for the damping factor (Γ=0.09 eV) with stilbene while a similar broadening (Γ=0.12 eV) is adopted for the other members of the series. This trend is rationalized by the fact that long oligomers possess a larger number of degrees of

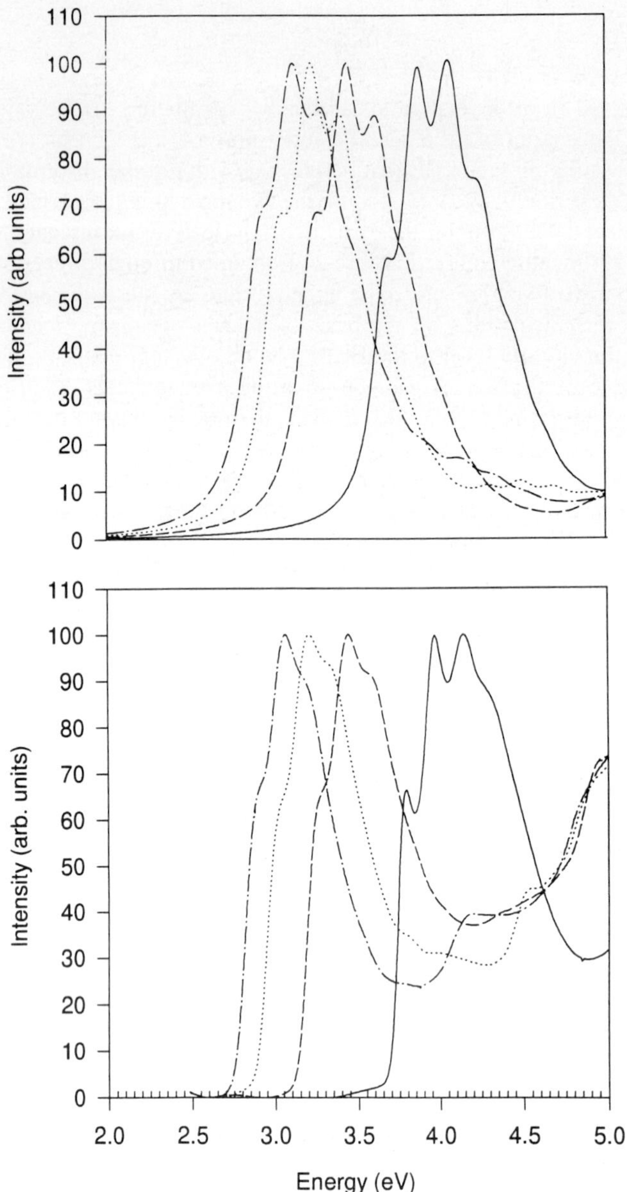

Figure 2. Experimental (lower part) and INDO/SCI-calculated (upper part) absorption spectra of the two- (solid line), three- (dashed line), four- (dotted line) and five-ring (dash-dotted line) PPV oligomers.

freedom that can give rise to a broader statistical distribution of conformations.

On the other hand, the fitted values of the Huang-Rhys factor are found to diminish with increasing chain length, as predicted by earlier theoretical calculations[14]. This evolution is consistent with the experimental absorption spectrum of improved PPV, where the dominant feature is found to be the 0-0 vibronic transition[15], in contrast to what is obtained with the oligomers investigated in this work.

Table 1. Parameters obtained by fitting the theoretical model to the experimental optical absorption spectra of the two- (PPV2), three- (PPV3), four- (PPV4), and five- (PPV5) ring oligomers of poly(p-phenylene vinylene). Γ (in eV) is the damping factor and S, the Huang-Rhys factor.

	PPV2	PPV3	PPV4	PPV5
Γ	0.09	0.12	0.12	0.12
S	1.95	1.60	1.60	1.60

The knowledge of the fitted values of the Huang-Rhys factor allows us to estimate the strength of the lattice relaxation taking place in the ^1B_u state. To do so, we express the relaxation energy in the excited state as a sum of separate contributions that are each described by the energy of a vibronic mode weighted by the amplitude of the Huang-Rhys factor associated to this mode. In the present case, where we exclusively deal with two dominant modes at 0.16 eV and 0.21 eV in a 1/2 ratio, the development gets simplified to give:

$$E_{rel} = S_1\, 0.16 + S_2\, 0.21 \qquad with \quad S_{tot} = S_1 + 2\, S_1 \tag{4}$$

Applying this expression to the results of our fitting procedure gives a relaxation energy in the ^1B_u excited state on the order of 0.39 eV for stilbene while 0.34 eV is obtained for the other members of the series.

We report in Table 2 the AM1-optimized geometries and the corresponding relaxation energies with respect to a vertical excitation for the lowest singlet (S_1) and triplet (T_1) states of PPV oligomers containing from 2 to 4 rings. Analysis of these results shows that the S1 state in stilbene presents relatively weak geometric deformations with respect to the ground state; in the phenyl rings, small bond-lengths deformations (on the order of 0.01 Å) slightly increase the quinoid character already observed in the ground state while, within the vinylene moieties, a more significant decrease in the bond-length alternation is observed (going from 0.11 Å in S_0 to 0.02 Å in S_1) due to the elongation of the double bonds and shortening of the single bonds. These results are in excellent agreement with those reported by Soos and co-workers on the basis of exact Pariser-Parr-Pople (PPP) calculations[16]. In contrast, the triplet polaron exciton is characterized by a more pronounced geometric relaxation; indeed, the phenyl rings acquire a semiquinoid character and the single- and double-bond character in the vinylene linkage is reversed. As a result, the relaxation energy in the T_1 state is found to be almost twice as big as in the S_1 state (0.60 eV instead of 0.28 eV).

In the lowest singlet state of the longer oligomers, weak bond-length deformations take place at the center of the molecule while the geometry of the external part is almost unchanged with respect to the ground state. As a result, weak and similar relaxation energies are calculated for the three- and four-ring compounds; this behavior is in full agreement with the trends derived from the fit of the experimental spectra. In contrast, the geometry of the T_1 state displays strong and localized deformations; these data suggest that a sole phenylene ring is needed for a proper accommodation of the triplet defect. This assumption is consistent with the spatial extent of a triplet exciton that is deduced from optically-detected magnetic resonance (ODMR) measurements[17].

We have analyzed in more detail the geometric deformations in the lowest singlet and triplet states by fitting Equation 3 to the AM1-optimized geometry of the four-ring oligomer (Table 3). As expected, the parameters R, representing the degree of confinement of the electron-hole pair, and ζ, denoting the coherence length of the photogenerated species, have

Table 2. Carbon-carbon bond lengths (in Å) for the singlet (S_1) and triplet (T_1) polaron-excitons in the two- (PPV2), three- (PPV3), and four- (PPV4) ring oligomers, as optimized at the AM1 level. The relaxation energies, E_{rel} (in eV), with respect to vertical excitations are also given. The bond length labels are those from Figure 1.

	PPV2		PPV3		PPV4	
	S_1	T_1	S_1	T_1	S_1	T_1
R1	1.390	1.399	1.396	1.394	1.394	1.393
R2	1.387	1.381	1.389	1.392	1.392	1.393
R3	1.421	1.432	1.413	1.408	1.407	1.403
R4	1.416	1.382	1.433	1.435	1.447	1.456
R5	1.419	1.446	1.364	1.389	1.351	1.344
R6			1.426	1.382	1.440	1.446
R7			1.421	1.452	1.412	1.413
R8			1.376	1.349	1.381	1.372
R9					1.421	1.439
R10					1.418	1.374
R11					1.383	1.437
E_{rel}	0.28	0.60	0.22	0.40	0.23	0.35

smaller values for the triplet than for the singlet; this is also the case for the amplitude ϕ and ϕ_R since the triplet is characterized by much more significant bond modifications. We show in Figure 3 the adiabatic potential related to the S_1 and T_1 states of the four-ring compound, as calculated within the AM1 formalism. This plot indicates that the energy of the singlet polaron exciton slowly evolves with R, due to the small amplitude of the geometric deformations in the S_1 state; on the other hand, the localized character of the triplet exciton is illustrated by the lineshape of its adiabatic potential versus R since the triplet state gets quickly destabilized when R is made to increase. For large values of R, we note a crossing between the two curves; the vibration modes associated to the R parameter should thus be included in the nonradiative decay from S_1 to T_1 following photoexcitation.

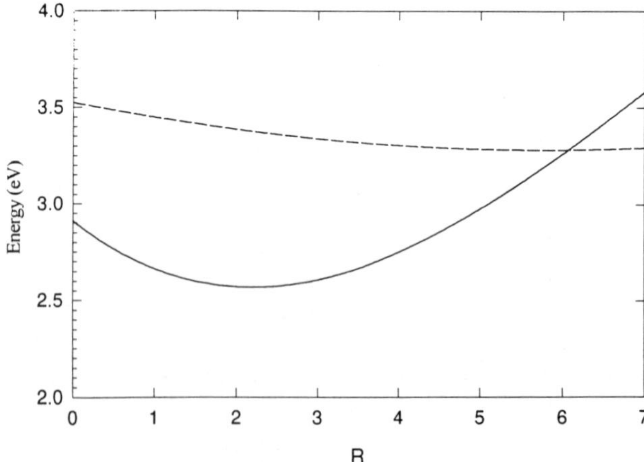

Figure 3. Adiabatic potentials for the triplet (solid line) and the singlet (dashed line) polaron-excitons with respect to parameter R, in site number, from Eq. (3).

Table 4. Parameters in Eq. (3) obtained by fitting to the AM1-optimized geometries of the singlet (S_1) and triplet (T_1) polaron-excitons in the four-ring PPV oligomer.

	ϕ	ϕ_R	ξ	R
S_1	0.020	0.008	2.713	3.844
T_1	0.057	0.035	1.795	2.090

SYNOPSIS

We have simulated well-resolved optical absorption spectra of PPV oligomers by explicitly taking into account the vibronic contributions. Considering the two most dominant effective modes at 0.16 eV and 0.21 eV, we have modulated the values of the damping factor (Γ, related to the broadening of the absorption features) and of the Huang-Rhys factor (S, governing the relative intensities in the vibronic progression) in order to the best agreement between theory and experiment. The results point to the fact that; (i) there is a linear relationship between the lowest transition energy and the inverse number of carbon atoms along the conjugated path; and (ii) the amplitude of the Huang-Rhys factor, and thus the relaxation energy in the ^1B_u state, is found to decrease as the chain grows.

These trends are consistent with direct estimates of the strength of the lattice relaxations that are calculated within the AM1 formalism. However, we note that the values derived from the fitting procedure are systematically overestimated when compared to the AM1 results. This discrepancy could be attributed to the fact that the experimental absorption spectra do not originate from fully planar conformations, as assumed in our theoretical approach. We have also optimized the geometries of the PPV oligomers in both the S_1 and T_1 states with the help of the AM1 method. The results show that the triplet exciton is characterized by strong and localized bond-length modifications. In contrast, the singlet polaron exciton presents weaker structural deformations extending over a larger part of the system. The strong confinement of the triplet makes that its relaxation energy is almost twice as big as for the singlet. This theoretical characterization of the nature of the lowest-lying excited states in PPV oligomers significantly contributes to a better understanding and control of the intrinsic properties of the corresponding polymer; information on the relative locations of the lowest singlet and triplet excited states are of prime importance to try setting up new strategies that can prevent at best intersystem crossings in the polymer layer of LED's, and hence lead to the achievement of improved quantum efficiencies.

REFERENCES

1. J.H. Burroughes, D.D.C. Bradley, A.R. Brown, R.N. Marks, K. Mackay, R.H. Friend, P.L. Burn, and A.B. Holmes, Nature 347:539 (1990).
2. D. Braun, A. Brown, E.G.J. Staring, and E.W. Meijer, Synth. Met. 65:85 (1994).
3. D.D.C. Bradley, R.H. Friend, T. Hartmann, E.A. Marseglia, M.M. Sokolowski, and P.D. Townsend, Synth. Met. 17:473 (1987).
4. I.D.W. Samuel, B. Crystall, G. Rumbles, P.L. Burn, A.B. Holmes, and R.H. Friend, Chem. Phys. Lett. 213:472 (1993).
5. D.R. Baigent, N.C. Greenham, J. Grüner, R.N. Marks, R.H. Friend, S.C. Moratti, and A.B Holmes, Synth. Met. 67:3 (1994).
6. R. Kersting, U. Lemmer, R.F. Mahrt, K. Leo, H. Kurz, H. Bässler, and E.O. Göbel, Phys. Rev. Lett. 70:3820 (1993).
7. J. Cornil, D. Beljonne, Z. Shuai, T. Hagler, I. Campbell, C.W. Spangler, K. Müllen,

D.D.C. Bradley, and J.L. Brédas, Chem. Phys. Lett., submitted for publication.
8. M.J.S. Dewar, E.G. Zoebisch, E.F. Healy, and J.J.P. Stewart, J. Am. Chem. Soc. 107:3902 (1985).
9. M.C. Zerner, G.H. Loew, R. Kichner, and U.T. Mueller-Westerhoff, J. Am. Chem. Soc. 102:589 (1980).
10. B.J. Orr and J.F. Ward, Mol. Phys. 20:513 (1971).
11. S. Heun, R.F. Mahrt, A. Greiner, U. Lemmer, H. Bässler, D.A. Halliday, D.D.C. Bradley, P.L. Burn, and A.B. Holmes, J. Phys.: Condens. Matter 5:247 (1993).
12. D. Beljonne, Z. Shuai, R.H. Friend, and J.L. Brédas, J. Chem. Phys. 102:2042 (1995).
13. J.P. Sethna and S. Kivelson, Phys. Rev. B 26:3613 (1982).
14. Z. Shuai, J.L. Brédas, and W.P. Su, Chem. Phys. Lett. 228:301 (1994).
15. K. Pichler, D.A. Halliday, D.D.C. Bradley, P.L. Burn, R.H. Friend, and A.B. Holmes, J. Phys.: Condens. Matter 5:7155 (1993).
16. Z.G. Soos, S. Ramasesha, D.S. Galvao, and S. Etemad, Phys. Rev. B 47:1742 (1993).
17. L.S. Swanson, J. Shinar, A.R. Brown, D.D.C. Bradley, R.H. Friend, P.L. Burn, A. Kraft, and A.B. Holmes, Phys. Rev. B 46:15072 (1992).

ACKNOWLEDGMENTS

The Mons-Sheffield collaboration is performed in the framework of the Human Capital and Mobility Network "Novel Third Order NLO Molecular Materials". The work is partly supported by the Commission of the European Union ESPRIT project LEDFOS-8013; the Belgian Federal Government SSTC ("Programme d'Impulsion en Technologie de l'Information" and "Pôle d'Attraction Interuniversitaire en Chimie Supramoléculaire et Catalyse"); FNRS/FRFC; and an IBM Academic Joint Study. One of us (JC) is Aspirant of the Belgian National Fund for Scientific Research (FNRS). The work in Sheffield is also supported by Toshiba Corporation, Japan. DDCB wishes to thank the University of Sheffield for its financial support in the form of start-up funds.

CHEMICAL AND ELECTRONIC STRUCTURE OF INTERFACES WITH CONJUGATED POLYMERS: SYSTEMS OF INTEREST IN MOLECULAR ELECTRONICS APPLICATIONS

G. Iucci[1], K. Xing[1], C. W. Spangler[2], M. Lögdlund[1],
A. Holmes[3], and W. R. Salaneck[1]

[1]Department of Physics, Linköping University,
S-58183 Linköping, Sweden.
[2]Department of Chemistry, Northern Illinois University,
DeKalb, IL, 60115, USA.
[3]Department of Chemistry, University of Cambridge,
Cambridge CB3 OHE, UK.

INTRODUCTION

The doping of *conjugated polymers* by electronic oxidation or reduction leads to the high electrical conductivity of *conducting polymers*. In the doping process, electronic charge is transferred either to (reduction) or from (oxidation) the conjugated polymer. The electrons transferred are stored in new electronic states, the energies of which fall with in the otherwise forbidden electron energy gap of the pristine (undoped) polymer. In polymers with a degenerate ground state, such as polyacetylene, spinless charged solitons are formed, while for polymers having a non-degenerate ground state, such as poly(*para*-phenylenevinylene), the charges are accommodated in singly-charged polarons or doubly-charged spinless bipolarons [1, 2]. In this work, n-type doping is studied, since the addition of electrons can be studied by photoelectron spectroscopy, where added holes (the

Polymers and Other Advanced Materials: Emerging Technologies and Business Opportunities. Edited by P. N. Prasad *et al.*, Plenum Press, New York, 1995

absence of electrons) is more difficult. In the doping of non-degenerate ground state polymers, the very first electrons go into polaron states. As the number of polarons increases, with increased doping, bipolarons may be formed, depending upon which form of charge storage species has the lowest energy in the particular polymer system, and a polaron-bipolaron transition takes place.

Recently, poly(*p*-phenylenevinylene), or PPV, and related polymers have received considerable interest, because of advantageous electroluminescence properties which may find applications in polymer-based light emitting diodes (LED's) [3]. Light emission in conjugated polymer-based LED's derives from an electron-hole recombination process in the polymer. The use of a low work function metal as the cathode increases the efficiency of the LED's by increasing the electron injection efficiency at the cathode. For this reason it is important to understand the nature of the interaction between the surfaces of conjugated polymers and metals. Many studies have been reported, where in the interface between clean PPV (or substituted PPV's) and metals such as, for example, calcium and aluminium have been investigated [4-8].

Ultraviolet photoelectron spectroscopy (UPS) may be used to observe directly energies of soliton and bipolaron states induced in n-type doping of conjugated polymers and model conjugated molecular systems. UPS provides a direct measure of the binding energy of electrons in the band of valence level (occupied) molecular orbitals, including, for n-doped polymers, any new states created, by the doping, in the previously forbidden electron energy band gap. Also, experimental results are directly compared with the results of quantum chemical calculations of the band structure of polymers, enabling a definitive interpretation of the UPS spectra. As an important additional issue is the interpretation of spectra of polymers from studies of model conjugated molecules; in the condensed thin film form, conjugated molecules may be studied, for which high resolution UPS spectra are obtained and high level quantum chemical calculations may be performed [9].

For n-doping of PPV in UHV using sodium, studied by UPS, the formation of bipolaron bands was observed [10]. Similar results were obtained, but with better resolution in the UPS spectra, by studying the interaction of sodium with *trans,trans*-1,4-distyrylbenzene (3PV), a phenyl-capped dimer of PPV consisting of three phenyl units and two double bonds [11]. In addition, the *trans*-diphenylpolyenes, and especially (α,ω-diphenyltetradecaheptaene), denoted as DP7, since it has seven C=C double bonds in the polyene chain, have been studied as model molecules for both *trans*-polyacetylene, because of the localisation in energy of the frontier molecular states, and PPV, because of the presence of the benzene rings [12]. In DP7, for example, the two highest energy occupied molecular orbitals are mainly localised on the polyene part of the molecule, while the next three are mainly localised on the benzene ring part of the molecule. When sodium atoms are deposited on DP7 in UHV, the sodium atoms become Na^+ ions, the DP7 is doped with sodium, and the charge transferred is stored in the form two confined solitons, because of the finite length of the polyene chain [13]. The confinement of the solitons on the finite length of the polyene portion of the molecule leads to the splitting of the otherwise degenerate energy levels. On the other hand, when calcium is deposited on DP7, the calcium atoms become Ca^{++} as DP7 becomes doped, with the formation of more bipolaron-like states, indicating the strong influence of the counterion on the structure of the doped polyene [14]. Aluminium atoms, however, form covalent bonds with the polyene part of the molecule, strongly modifying the structure of the π-system in the near surface region of the molecular solid film [14].

In this contribution, some results of the study of the interaction of atoms of a metal with a very low work function, rubidium, deposited on the surfaces of two different conjugated polymers and a model molecular solid, using photoelectron spectroscopy (both UPS and XPS), are reviewed. In particular, the investigation of the interaction of rubidium with two polymers, poly(*para*-phenylenevinylene), or PPV, and poly(2-methoxy,5(2'-ethyl)hexoxy-1,4-phenylenevinylene), or MEHPPV, and with the model molecule, α,ω-diphenyltetradecaheptaene (or DP7), are reviewed. The chemical structures of these molecular systems are shown in Fig. 1. UPS data have been interpreted with the help of the results of quantum chemical calculations carried out using the Austin Model 1 (AM1) and valence effective Hamiltonian (VEH) methods on undoped, singly charged and doubly charged DP7, and on undoped and fully-doped PPV, and by comparison with the previously published results obtained on sodium-doping of the same materials.

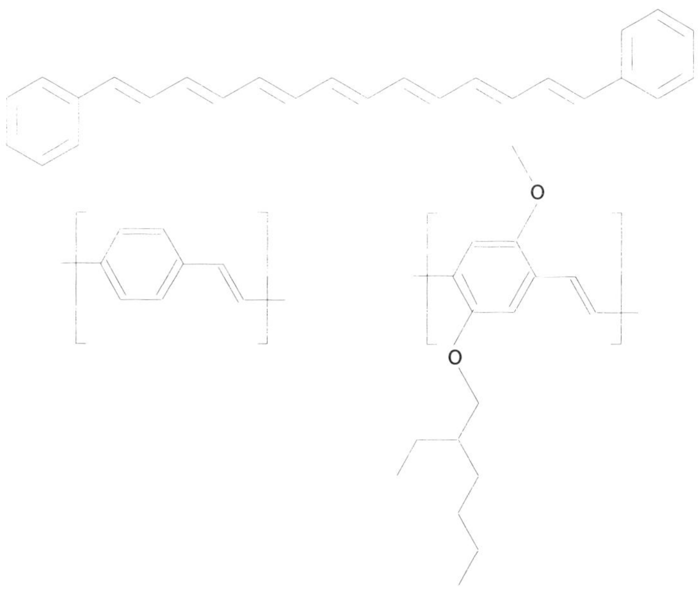

Figure 1. The molecular structures of the model molecule for PPV, α,ω-diphenyltetradecaheptaene (or DP7), and the two polymers, poly(*para*-phenylenevinylene), or PPV, and poly(2-methoxy,5(2'-ethyl)-hexoxy-1,4-phenylenevinylene), or MEHPPV, are shown.

Rubidium, like sodium, is a first group metal. Both rubidium and sodium have workfunctions considerably less than that of calcium, such that both would be attractive candidates for electron-injecting contacts on polymer-based LED's. But, like sodium, rubidium might be expected to give up electrons to the polymer, leading to n-type doping. On the other hand, the larger ionic radius of the Rb^+ ion compared with the Na^+ ion might be expected to inhibit the bulk diffusion of the metal, such that a low workfunction contact might be achieved despite charge transfer. Finally, because of the larger size of the rubidium cation, a lower stabilisation energy is expected for the coulomb interaction between the negatively-charged chain defects and the counterion, which also might tend to inhibit diffusion of cations.

EXPERIMENTAL

Photoelectron spectroscopy (XPS and UPS) measurements were performed using an instrument of our own design and construction. The ultra-high vacuum (UHV) system consists of separate sample preparation and analysis chambers, pumped with a special combination of ion and cryogenic pumps, so that the ultimate base pressure in the system is below 10^{-10} torr. XPS measurements were carried out with a non-monochromatised $MgK_{\alpha 1\alpha 2}$ X-ray source ($h\nu$=1253.6 eV) and with an electron energy analyser resolution such that the FWHM (full width at half maximum) of the $Au(4f_{7/2})$ was 0.9 eV. UPS was carried out using monochromatised HeI ($h\nu$=21.2 eV) photons from a He-resonance lamp.

DP7 condensed molecular solid films on clean gold substrates were prepared *in situ*, in UHV by physical vapour deposition (PVD), using a filament-heated borosilicate glass crucible, essentially a simple Knudsen cell, through a cryogenic shield held at 20 K onto sputter-cleaned gold substrate. The deposition was carried out with the gold substrate at room temperature. The films obtained were free from oxygen, as determinated from XPS; their thickness, calculated from the attenuation of the $Au(4f_{7/2})$ XPS line, was about 100 Å.

PPV samples were prepared from tetrahydrothiophene leaving-group precursor polymer solutions in methanol, spin-coated onto optically flat Si(110) substrates; the films obtained were less than 50 nm thick and were converted to PPV by warming at 240 ^{O}C for about 15 hours inside the UHV spectrometer. The purity was monitored by XPS; the maximum contamination detected was approximately one oxygen atom every seven monomer units.

MEHPPV films were prepared by spin-coating directly from chloroform solution onto Si (natural oxide) or Au substrates; measured film thicknesses were about 100 Å; the C/O ratio, measured from XPS spectra, is in good agreement with the theoretical value (C/O=8.5).

Deposition of rubidium atoms was carried out stepwise *in situ* in UHV using an SAES getter source, by exposing the samples to 10^{-9} torr pressure of rubidium in carefully timed steps. The temperature of the sample during metal deposition was varied. At first it was thought that rubidium would form a metallic electrode contact; diffusion of the atoms into the bulk was not anticipated. Subsequently, however, following the observation that rubidium atoms dope the polymer surfaces and eventually diffuse into the films, the substrate temperature was adjusted to be low enough that the sticking coefficient would be large enough that the Rb atoms had a reasonable probability of sticking, and high enough that the eventual diffusion effects took place on a fast enough time scale that steady state measurements could be made without waiting too long. The final sample temperatures used were -50 ^{o}C for DP7 and room temperature for MEHPPV. The resultant doping level was calculated from the ratio of the areas of the C(1s) and Rb($3d_{5/2,3/2}$) XPS signals, normalised to the photoionization cross-sections. In sodium-doping experiments reported previously, the saturation doping levels were found to be one sodium atom per monomer unit for PPV [10] and two sodium atoms per molecule for the model molecules [13]. The maximum doping level reached in rubidium-doping of PPV and DP7 corresponds to the saturation level found in sodium-doping of the same materials. For MEHPPV, it was necessary to perform higher doping in order to study the variations in the XPS O(1s) signal, due to the interaction of the rubidium atoms with the oxygen atoms of the side chains in the monomer unit. For the two polymers, indicated doping levels correspond to the ratio rubidium atoms/monomer units, given as percentage.

THEORETICAL CALCULATIONS

The semiempirical Austin Model 1 (AM1) method [15] has been used for calculating the ground-state geometries of neutral and doped molecules, in order to determine the changes in the molecular structure upon doping. Calculations on singly and doubly charged DP7 molecule have been performed by adding one or two electrons respectively to the molecule. The geometry parameters obtained served as an input for the electronic structure calculations, that were performed using the valence effective Hamiltonian (VEH) method [16, 17].

The VEH method is a non-empirical pseudopotential method based on the use of an effective Fock Hamiltonian. It has been applied to many conjugated systems, and provides an accurate estimate of the location of the gap states in doped conjugated materials [18]. The density of valence states (DOVS) curves are calculated by taking the inverse derivative of the band energies with respect to the momentum. Before making a direct comparison with solid state UPS data, the DOVS curves are treated in a standard way, that is: (1) convoluted by a Gaussian, having a full width at half maximum of 0.7 eV, in order to match the experimental resolution in the UPS data; (2) contracted along the energy scale by a factor of 1/1.3, since the Hartree-Fock method is know to overestimate the width of the valence bands, due to the neglect of coulomb correlation; and (3) shifted rigidly to lower binding energy in order to align the spectra with the experimental binding energies axis, because VEH calculations are performed on isolated chains and do not take into account intermolecular relaxation effects in the solid state.

AM1 and VEH calculations carried out on undoped and fully-doped PPV and CN-substituted PPV have been described in detail by Fahlman [10, 19], while results of calculations carried out on undoped and doubly charged DP7 have been previously reported by Lögdlund [13].

The deposition of metals onto the surface of conjugated polymers and model molecules induces changes in their electronic structure that usually shift the position of the Fermi level; for this reason, both valence band and core level spectra have been referred to the vacuum level, in order to give a correct interpretation of the changes in the UPS and XPS spectra upon doping. The work function of the doped and undoped samples has been measured from the cut-off of the secondary electron background in the UPS spectra.

RESULTS AND DISCUSSION

Both for the two polymers as well as the model molecular solid, the interaction with rubidium atoms results in n-doping of the conjugated system. From the C(1s) and Rb(3d) XPS spectra, the nature of the bond between the doped material and rubidium is found to be ionic. The binding energy of the Rb($3d_{5/2}$) signal, as referred to the vacuum level, is -114.2 to -114.4 eV in all the doped systems, which can be compared to the value measured by Rousseau and Estrade-Sczwarckopf for ionic rubidium-graphite intercalation compounds [20]. The XPS C(1s) signal shifts to lower binding energy relative to the vacuum level upon doping (about 1 eV for the polymers and about 0.6 eV for DP7, from the undoped to the fully-doped system), because the electron-transfer reaction from the metal to the conjugated material produces an increase of negative charge in the π-system. An increase in the peak width as the doping level increases is also found for all the studied systems, as expected in case of n-doping. Moreover, a decrease in the work function of all the materials investigated was always observed (about 2 eV for PPV and MEHPPV, but 1.2 eV for DP7, from the undoped to the fully doped system), due to appearance, in the otherwise empty energy band gap, of new electronic states which shift the position of the Fermi level to higher values.

In Fig. 2 is shown the UPS spectra of the low binding energy portion of the valence band of DP7 during step wise doping of the condensed molecular solid film using sodium atoms in UHV. The progression from the pristine state (upper curve) to the fully doped case (lowest curve) clearly shows the development of two new states in the otherwise forbidden energy gap. The sodium atoms are in the Na^+ ionic state, according to the XPS spectra. Note that the LUMO level in the pristine molecule moves up into the gap and becomes the lower bipolaron (confined soliton) band in the completely doped system.

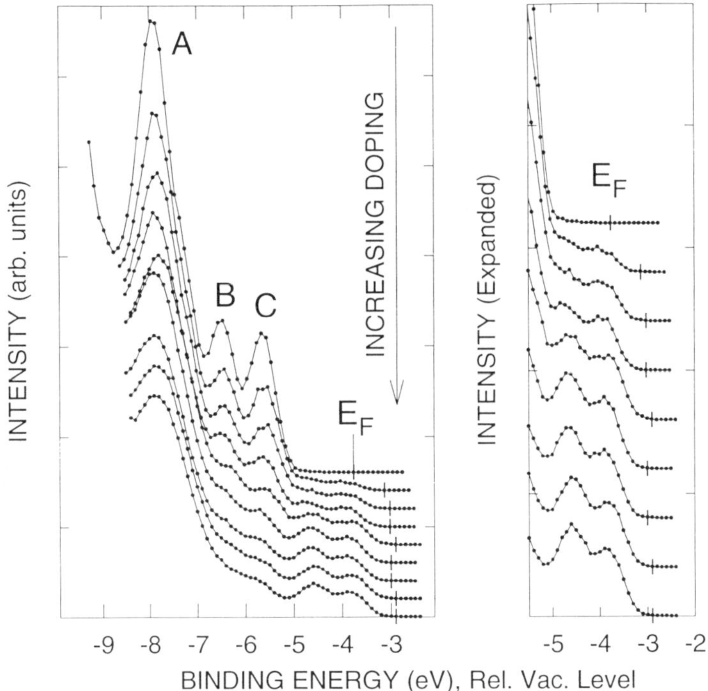

Figure 2. The low binding energy portion of the UPS spectra of sodium on DP7 are shown.

Maximum doping corresponds to two Na^+ ions per DP7 molecule. Further addition of sodium to the system results in diffusion of sodium atoms through the molecular film and accumulation of metallic sodium at the gold substrate under the DP7 film.

In Fig. 3 is shown the UPS spectra of the low binding energy portion of the valence band of DP7 during step wise deposition of calcium atoms on to the surface of the condensed molecular solid film in UHV. The progression from the pristine state (upper curve) to the saturation case (lower curve) clearly shows the development of two new states in the otherwise forbidden energy gap, just as in the case of sodium on DP7. The calcium appears to dope the near surface region of the DP7 film, according to the binding energy of the Ca(2p) line in the XPS spectra. Saturation in the case of calcium, however, means that metallic calcium begins to accumulate on the surface of the sample, in contrast to the case of sodium, where the excess sodium diffuse through the film.

In Fig. 4 is shown the UPS spectra of the low binding energy portion of the valence band of DP7 during step wise deposition of rubidium atoms on the surface of the condensed molecular solid films in UHV. As in the case of sodium, the difference between the pristine state (lowest curve) and the fully doped case (upper curve) clearly shows the existence of two new states in the otherwise forbidden energy gap.

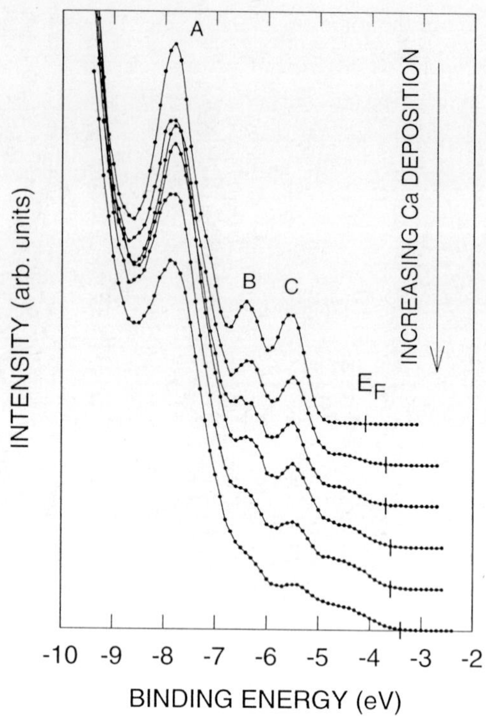

Figure 3. The low binding energy portion of the UPS spectra of calcium on DP7 are shown.

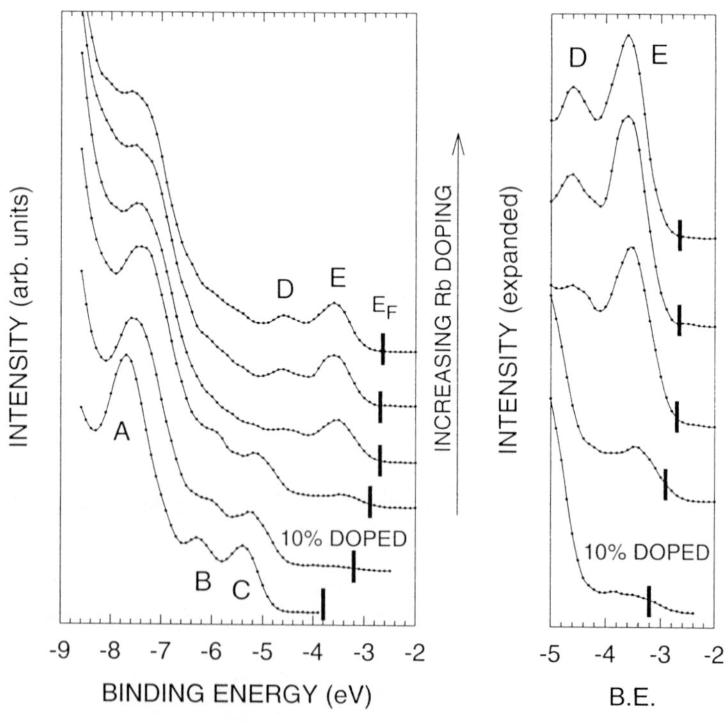

Figure 4. The low binding energy portion of the UPS spectra of rubidium on DP7 are shown.

The rubidium atoms are in the ionic state, according to the XPS spectra. As in the case of sodium on DP7, the LUMO level in the pristine molecule moves up into the gap and becomes the lower bipolaron (confined soliton) band in the completely doped system. Maximum doping corresponds to two rubidium ions per DP7 molecule (defined as 100% doping). A significant difference exists in the series of Fig. 4 (rubidium on DP7) compared with that of Fig. 2 (sodium on DP7), however, as seen in the curves corresponding to low doping with rubidium. The first step of rubidium deposition shown (for about 10% doping; or one rubidium ion per five DP7 molecules) corresponds to the formation of polaron states, both according to the theoretical modelling and the fact that there is a finite density-of-states observed at the Fermi energy. The apparent overflow of the density-of-states above the Fermi energy seen in the 10% doping curve of Fig. 4 arises from the finite energy resolution allowed by molecular solids in UPS studies [21]. Thus the series of doping curves in Fig. 4 reveals the presence of a polaron to bipolaron transition upon continuous doping of DP7 with rubidium in UHV. And thus rubidium is seen to behave more as an n-type dopant for DP7 as in the case of sodium, rather than an electrode material as for calcium.

In Fig. 5 is shown the UPS spectra of the low binding energy portion of the valence band of PPV during step wise deposition of rubidium atoms on the surface of the polymer films in UHV.

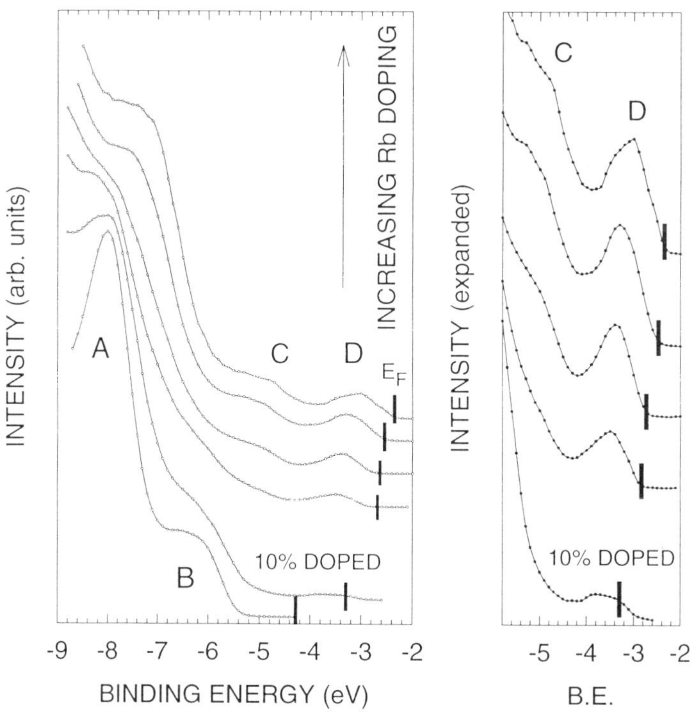

Figure 5. The low binding energy portion of the UPS spectra of rubidium on PPV are shown.

As in the case of rubidium on DP7, the difference between the pristine state (lowest curve) and the fully doped case (upper curve) clearly shows the existence of two new states in the otherwise forbidden energy gap. The rubidium atoms are in the ionic state, according to the XPS spectra. As in the case of sodium on PPV, the LUMO level in the pristine molecule moves up into the gap and becomes the lower bipolaron (confined soliton) band in the completely doped system. Maximum doping corresponds to two rubidium ions per PPV repeat unit. The first step of rubidium deposition shown (for 10% doping; or one rubidium ion per ten PPV repeat units) corresponds to the formation of polaron states, both according to the theoretical modelling of the 10% doping case and the fact that there is a finite density-of-states observed at the Fermi energy. The apparent overflow of the density-of-states above the Fermi energy seen in the 10% doping curve of Fig. 5 arises from the finite energy resolution allowed by molecular solids in UPS studies [21]. As in the case of DP7, the series of doping curves of PPV reveals the presence of a polaron to bipolaron transition upon continuous doping of PPV with rubidium in UHV. The results for rubidium on MEHPPV are essentially identical to those for rubidium on PPV, and will not be reproduced here.

SUMMARY

The exposure to thin films of poly(*para*-phenylenevinylene), or poly(2-methoxy,5(2'-ethyl)hexoxy-1,4-phenylenevinylene), or condensed molecular films of the model molecule, α,ω-diphenyltetradecaheptaene (DP7), results in charge transfer reactions similar to the case of the doping of PPV, MEHPPV and DP7 by sodium atoms. The time for diffusion of the rubidium atoms into the bulk of the thin films varies with temperature, and, presumably because of the larger ionic size of the rubidium cation compared with the sodium cation, is generally slower than for sodium doping. In contrast to the case of sodium doping, the presence of polaron states at low doping levels is observed for all three of the systems studied, with a transition to bipolaron states as the doping increases. Just why the polaron states, which are expected, have not been observed in the sodium cases, is not understood at this time.

ACKNOWLEDGEMENTS

Research on conjugated polymers and molecules in Linköping is supported by grants from the Swedish Natural Science Research Council (NFR), the Swedish Research Council for Engineering Sciences (TFR), the Swedish National Board for Industrial and Technical Development (NUTEK), the European Commission programme SCIENCE (project 0661 POLYSURF), the ESPRIT Network of Excellence, NEOME, and the ESPRIT Basic Research Action LEDFOS 8013.

REFERENCES

1. "Handbook of Conducting Polymers", T. A. Skotheim, ed., Markel Dekker, New York (1986).
2. "Conjugated Polymers: The Novel Science and Technology of Highly Conducting and Nonlinear Optically Active Materials", J. L. Brédas and R. Silbey, ed., Kluwer, Dordrecht (1991).
3. J. H. Burroughes, D. D. C. Bradley, A. R. Brown, R. N. Marks, K. Mackay, R. H. Friend, P. L. Burns and A. B. Holmes, Light-emitting diodes based on conjugated polymers, *Nature* 347:539 (1990).
4. P. Dannetun, M. Lögdlund, W. R. Salaneck, C. Fredriksson, S. Stafström, A. B. Holmes, A. Brown, S. Graham, R. H. Friend and O. Lhost, New results on metal-polymer interfaces, *Mol. Cryst. Liq. Cryst.* 228:43 (1993).
5. T. P. Nguyen, V. Massardier, V. H. Tran and A. Guyot, Studies of the polymer-metal interface in metal-ppv-metal devices, *Synth. Met.* 55-57:235 (1993).
6. E. Ettedgui, H. Razafitrimo, K. T. Park, Y. Gao and B. R. Hsieh, An x-ray photoemission spectroscopy study of the role of sample preparation on band bending at the interface of Al with poly(*p*-phenylenevinylene), *J. Appl. Phys.* 75:7526 (1994).
7. Y. Gao, K. T. Park and B. R. Hsieh, X-ray photoemission investigations of the interface formation of Ca and poly(*p*-phenylenevinylene), *J. Chem. Phys.* 97:6991 (1992).
8. P. Dannetun, M. Fahlman, C. Fauquet, K. Kaerijama, Y. Sonoda, R. Lazzaroni, J. L. Brédas and W. R. Salaneck, Interface formation between poly(2,5-diheptyl-p-phenylenevinylene) and calcium: Implications for light emitting diodes, *Synth. Met.* 67:113 (1994).
9. W. R. Salaneck, S. Stafström and J. L. Brédas. "Conjugated Polymer Surfaces and Interfaces", Cambridge University Press, Cambridge (1996).
10. M. Fahlman, D. Beljonne, M. Lögdlund, R. H. Friend, A. B. Holmes, J. L. Brédas and W. R. Salaneck, Experimental and theoretical studies of the electronic structure of na-doped poly(p-phenylenevinylene), *Chem. Phys. Lett.* 214:327 (1993).
11. P. Dannetun, M. Lögdlund, M. Fahlman, C. Fauquet, D. Beljonne, J. L. Brédas, H. Bässler and W. R. Salaneck, The evolution of charge-induced gap states in degenerate and non degenerate conjugated molecules and polymers as studied by photoelectron spectroscopy, *Synth. Met.* 67:81 (1994).
12. P. Dannetun, M. Lögdlund, J. L. Brédas, C. W. Spangler and W. R. Salaneck, Evolution of charge-induced gap states in short diphenylpolyenes as studied by photoelectron spectroscopy, *J. Phys. Chem.* 98:2853 (1994).

13. M. Lögdlund, P. Dannetun, S. Stafström, W. R. Salaneck, M. G. Ramsey, C. W. Spangler, C. Fredriksson and J. L. Brédas, Soliton pair charge storage in doped polyene molecules: evidence from photoelectron spectroscopy studies, *Phys. Rev. Lett.* 70:970 (1993).

14. P. Dannetun, M. Lögdlund, R. Lazzaroni, C. Fauquet, C. Fredriksson, S. Stafström, C. W. Spangler, J. L. Brédas and W. R. Salaneck, Reactions of low workfunction metals, na, al, and ca, on α,ω-diphenyltetradecaheptaene: Implications for metal/polymer interfaces, *J. Chem. Phys.* 100:6765 (1994).

15. M. J. S. Dewar, E. G. Zoebisch, E. F. Healy and J. J. P. Stewart, AM1: A new general purpose quantum mechanical molecular model., *J. Am. Chem. Soc.* 107:3902 (1985).

16. J. L. Brédas, R. R. Chance, R. Silbey, G. Nicolas and P. Durand, A nonempirical effective Hamiltonian technique for polymers: Application to polyacetylene and polydiacetylene, *J. Chem. Phys.* 75:255 (1981).

17. J. M. André, J. Delhalle and J. L. Brédas, "Quantum Chemistry Aided Design of Organic Polymers", World Scientific, Singapore (1991).

18. J. L. Brédas, B. Thémans, J. G. Fripiat, J. M. André and R. R. Chance, Highly conducting polyparaphenylene, polypyrrole and polythiophene chains: An ab initio study of the geometry and electronic-structure modifications upon doping, *Phys. Rev. B* 29:6761 (1984).

19. M. Fahlman, P. Bröms, D. A. dos Santos, S. C. Moratti, N. Johansson, K. Xing, R. H. Friend, A. B. Holmes, J. L. Brédas and W. R. Salaneck, Electronic structure of pristine and sodium-doped cyano-substituted poly(2,5-dihexyloxy-p-phenylenevinylene): A combined experimental and theoretical study, *J. Chem. Phys.*, in press.

20. B. Rousseau and H. Estrade-Szwarckopf, Photoelectron core level spectroscopy study of K- and Rb-graphite intercalation compounds-II. Clean surface studies, *Solid State Commun.* 85:793 (1993).

21. C. B. Duke, W. R. Salaneck, T. J. Fabish, J. J. Ritsko, H. R. Thomas and A. Paton, The electronic structure of pendant-group polymers: molecular ion states and dielectric properties of poly (2-vinyl pyridine), *Phys. Rev. B,* 18:5717 (1978).

SYNTHESIS OF POLY(2-METHOXY-5-METHYL-1,4-PHENYLENE VINYLENE) ; SUBSTITUENT EFFECTS ON THE ELECTRICAL AND NONLINEAR OPTICAL PROPERTIES

Jeong-Ik Lee,[1] Hong-Ku Shim,*[1] Geon Joon Lee[2] and Dongho Kim[2]

[1]Department of Chemistry, Korea Advanced Institute of Science and Technology, Taejon 305-701, Korea
[2]Spectroscopy Lab., Korea Research Institute of Standards and Science, Taejon 305-340, Korea

INTRODUCTION

Conjugated polymers are currently considerable interest as materials for a wide variety of applications in last afew decades.[1,2] Among them, the interest in poly(1,4-phenylenevinylene) (PPV) and its derivatives has been increased because they have high environmental and thermal stability, good processibility and high film quality, etc. They can be readily prepared in high molecular weight through the water-soluble precursor route.[3] This route has allowed for preparation of a number of derivatives, including alkoxy, alkyl and halogen substituted derivatives.[4-9] The substitution on the phenylene ring of the monomer with different type of substituent shows significant difference in electronic structure of the corresponding polymer,[4-14] and thus affects the nonlinear optical and electrical properties of the polymer.

PPV films can be easily doped with strong oxidizing agents such as AsF_5 and $FeCl_3$ to give high electrical conductivity but they cannot be effectively doped with weaker oxidizing dopants such as I_2. The presence of strong electron-donating substituents on the phenylene ring leads to the reduction in band gap and ionization potential of the conjugated polymers, which facilitates oxidation with weak oxidants such as I_2. For example, poly(2,5-dimethoxy-1,4-phenylenevinylene) (PDMPV) can be doped with I_2 to give electrical conductivity of 10^1 S/cm whereas the precursor polymer of PDMPV cannot be drawn to any extent during or prior to the elimination reaction.[15] The drawability of the precursor polymer is also an important factor in electrical conductivity because the drawn films exhibit about 10 times higher conductivity along the drawn direction than undrawn ones.

Generally, asymmetrically substituted PPV derivatives can be easily stretched. For example, poly(2-methoxy-1,4-phenylenevinylene) (PMPV) films can be readily stretched to show the electrical conductivity of 10^0 S/cm after I_2 doping.[16] Despite of stretching the PMPV film, its electrical conductivity is not higher than that of PDMPV. It can be explained that PMPV having only one methoxy substituent on each phenylene ring has higher oxidation potential to oxidative doping than PDMPV which contains two methoxy groups on each phenylene ring, and this results in lower susceptibility of PMPV toward I_2.

Therefore, we have been interested in the synthesis of PPV derivatives which can be easily stretched and have low oxidation potential to oxidative doping. Asymmetrically disubstituted poly(2-methoxy-5-methyl-1,4-phenylene-vinylene), PMMPV is one of the candidates. In this article, we report the synthesis, electrical conductivity and 3rd order nonlinear optical properties of PMMPV and investigate the asymmetric substituent effects on the electrical conductivity. The synthetic routes and polymer structures are shown in synthetic scheme.

Polymers and Other Advanced Materials: Emerging Technologies and Business Opportunities. Edited by P. N. Prasad *et al.*, Plenum Press, New York, 1995

Synthetic scheme

EXPERIMENTAL

Instrumentation. The synthesized compounds were identified by ^{1}H-NMR spectra that were obtained using Bruker AM 200 Spectrometer. Melting points were determined using a Electrothermal Model 1307 digital analyzer. FT-IR spectra were measured by Bomem Michelson Series FT-IR spectrophotometer and UV-visible spectra were obtained with a Shimadzu UV-3100S. Thermogravimetric analysis (TGA) and Differential Scanning Calorimeter (DSC) of precursor polymers were performed under nitrogen atmosphere at a heating rate of 10 °C/min with Dupont 9900 analyzer. Elemental analyses were performed by the Analytical Department of the Korea Research Institute of Chemical Technology. Electrical conductivities of doped polymer films were measured by the four-in-line probe method at room temperature.

Third order nonlinear optical coefficient of the final polymer was measured by Third Harmonic Generation (THG) technique with Nd:YAG laser system. The thicknesses of the spin-coated polymers were measured using the Alpha step 200 profilometer. The photoluminescence spectrum of PMMPV film was obtained using Perkin-Elmer LS-50 luminescence spectrometer.

1-Bromo-4-methoxy-2,5-dimethylbenzene (1). 30.0 g (220 mmol) of 2,5-dimethylanisole was dissolved in 150 mL of carbon tetrachloride and 40.5 g (250 mmol) of Br_2 was added dropwise to the solution at 0 ℃. When dropping was finished, the mixture was stirred at room temperature for 2 h. The reaction was quenched by aqueous NaOH solution, then organic layer was extracted by methylenechloride. The extract was dried over anhydrous magnesium sulfate and the solvent was removed by rotary evaporator. The product was obtained by vacuum distillation and the yield was 45.0 g (95.1 %): ^{1}H-NMR ($CDCl_3$) d 7.3(s, 1H), 6.7(s, 1H), 3.8(s, 3H), 2.4(s, 3H), 2.2(s, 3H). Anal. Calcd for $C_9H_{11}OBr$: C, 46.38, H, 4.76. Found : C, 46.19, H, 4.65 ; bp 85 °C (ca. 0.5 Torr).

1-Bromo-2,5-bis(bromomethyl)-4-methoxybenzene (2). Compound **(2)** was prepared by reacting 20.0 g (93.0 mmol) of **(1)** with 36.4 g (200 mmol) of N-bromosuccinimide (NBS) in 150 mL of carbontetrachloride. A small amount of benzoyl peroxide was added as an initiator. The reaction mixture was refluxed for 3 h under a nitrogen atmosphere. The completion of the reaction was indicated by the appearance of succinimide and consumption of the NBS. Succinimide was filtered out and crude solution was concentrated under reduced pressure. The white solid product was obtained by precipitation in n-hexane and the yield was 24.3 g (70.1 %): ^{1}H-NMR ($CDCl_3$) d 7.5(s, 1H), 6.9(s, 1H), 4.5(s, 2H), 4.4(s, 2H), 3.9(s, 3H). Anal. Calcd for $C_9H_9OBr_3$: C, 25.32, H, 2.12. Found : C, 25.48, H, 2.30 ; mp 137 °C.

1-Bromo-2,5-bis(hydroxymethyl)-4-methoxybenzene (3). A mixture of 16.0 g (42.9 mmol) of **(2)** and excess of $CaCO_3$ in 140 mL of dioxane and 100 mL of water was refluxed for 15 h. When the reaction was finished, the extra $CaCO_3$ was quenched by HCl solution. The white solid product was obtained by precipitation in water and dried in vacuum oven. The product yield was 8.5 g (80%): ^{1}H-NMR (DMSO-d_6) d 7.5(s, 1H), 7.1(s, 1H), 5.5(t, 1H), 5.2(t, 1H), 4.5(m, 4H), 3.8(s, 3H). Anal. Calcd for $C_9H_{11}O_3Br$: C, 40.78, H, 4.18. Found : C, 40.62, H, 4.09 ; mp 151 - 152 °C.

1-Bromo-2,5-bis(tetrahydropyran-2-yloxymethyl)-4-methoxybenzene (4). 1.6 g of pyridinium-p-toluenesulfonate (PPTS) and 8.2 g (96 mmol) of 3,4-dihydropyran (DHP) were added to the solution of 8.0 g (32 mmol) of **(3)** in 100 mL of methylenechloride and then the mixture was refluxed for 4 h. The mixture was washed with water and the organic layer was dried with magnesium sulfate. The volatiles were removed under reduced pressure and the product was separated by the column chromatography. The white solid was obtained and the yield was 11.3 g (85 %): ^{1}H-NMR ($CDCl_3$) d 7.5(s, 1H), 7.0(s, 1H), 4.7(m 4H), 4.5(m, 2H), 3.9(m, 2H), 3.8(s, 3H), 3.5(m, 2H), 1.5 - 1.8 (m, 12 H). Anal. Calcd for $C_{19}H_{27}O_5Br$: C, 52.67, H, 6.28. Found : C, 52.49, H, 6.16 ; mp 84 - 86 ℃.

1-Methoxy-2,5-bis(tetrahydropyran-2-yloxymethyl)-4-methylbenzene (5). 9.2 g (22 mmol) of **(4)** was dissolved in 150 mL of dried ether, and 26.0 mL (66.0 mmol) of n-BuLi was added to this solution at -77 ℃. After 1 h of reaction, 7.0 mL (66 mmol) of dimethylsulfate was added to the reaction mixture and the reaction was carried out at -77 ℃ for 2 h. The reaction was quenched by water and the organic layer was dried with magnesium sulfate. The white solid product was obtained by column chromatography and the yield was 4.6 g (60%): ^{1}H-NMR ($CDCl_3$) d 7.2(s, 1H), 6.9(s, 1H), 4.7(s, 4H), 4.5(m, 2H), 3.9(m, 2H), 3.8(s, 3H), 3.5(m, 2H), 2.3(s, 3H), 1.8 - 1.5 (m, 12H). Anal. Calcd for $C_{20}H_{30}O_5$: C, 68.55, H, 8.63. Found : C, 68.47, H, 8.71 ; mp 66 - 67 °C.

1,4-Bis(hydroxymethyl)-2-methoxy-5-methylbenzene (6). A mixture of 4.5 g (13 mmol) of **(5)** and 0.3 g (1 mmol) of PPTS in 100 ml of ethanol was reacted at 55 °C for 12 h. The volatiles were removed under reduced pressure and then white solid was obtained. This solid was purified by washing with methylenechloride. The yield was 1.8 g (79 %): ^{1}H-NMR (DMSO) d 7.1(s, 1H), 6.9(s, 1H), 5.0(t, 1H), 4.8(t, 1H), 4.4(m, 4H), 3.7(s, 3H), 2.1(s, 3H). Anal. Calcd for $C_{10}H_{14}O_3$: C, 65.92, H, 7.74. Found : C, 65.85, H, 7.76 ; mp 80 - 81 °C.

1,4-Bis(chloromethyl)-2-methoxy-5-methylbenzene (7). The reaction of 1.6 g (9.0 mmol) of **(6)**, 3.2 g (27 mmol) of N-chlorosuccinimide and 5.7 g (22 mmol) of triphenylphosphine dissolved in 100 mL of THF was carried out at room temperature for 12 h. The product was separated by column chromatography and then white solid was obtained.

The yield was 1.6 g (82 %): ^{1}H-NMR ($CDCl_3$) d 7.2(s, 1H), 6.8(s, 1H), 4.6(s, 2H), 4.5(s, 2H), 3.9(s, 3H), 2.3(s, 3H). Anal. Calcd for $C_{10}H_{12}O_1Cl_2$: C, 54.82, H, 5.52. Found : C, 54.75, H, 5.59 ; mp 109 - 110 °C.

1-Methoxy-4-methyl-2,5-phenylenedimethylenebis(tetrahydrothiophenium chloride) (8). 1.5 g (7.0 mmol) of compound **(7)** and excess tetrahydrothiophene were reacted at 50 °C for 24 h in 20 mL of methanol. As the reaction proceeded, this suspension changed to the clear solution. The final salt monomer **(8)** was obtained by concentration of the reaction solution, precipitation in cold acetone, filtration, and vacuum drying. The product yield was 1.9 g (70 %): ^{1}H-NMR (D_2O) d 7.2(s, 1H), 6.9(s, 1H), 4.4(s, 2H), 4.3(s, 2H), 3.7(s, 3H), 3.4-3.2(m, 8H), 2.2(s, 3H), 2.3-2.0(m, 8H). Anal. Calcd for $C_{18}H_{28}OCl_2S_2$: C, 54.69, H, 7.14. Found : C, 54.47, H, 7.09 ; mp 240 - 241 °C (dec). The salt monomer for PPV, (9), was prepared similarly from a,a'-dichloro-p-xylene and tetrahydrothiophene following the literature method.[17, 18]

Polymerization. The procedures used for the preparation of homopolymer and copolymers were the same as reported by us[6, 18] and others[17]. The monomer sulfonium salt solution was polymerized in a 1.0 N NaOH solution at 0 °C. A homogeneous and viscous solution was obtained. For the copolymers, the total moles of the two monomers, **(8)** and **(9)**, were kept constant, but their mole ratios were varied. The polymerization reaction was quenched by neutralization with 0.5 N HCl and the precursor polymer solutions were dialyzed against deionized water for 5 days to remove unreacted monomers and low molecular weight oligomers using a dialysis tube with a molecular weight cutoff at 12,000. The films cast or spin-coated from these aqueous precursor polymer solutions were subjected to thermal elimination in vacuo (10^{-2} Torr) at 230 °C for 12 h to transform them into the final polyconjugated polymer films. If desired, the precursor polymer films were uniaxially drawn at 120 °C using a zone-heating apparatus.

RESULTS AND DISCUSSION

For the first time, the salt monomer of PMMPV, **(8)** was synthesized from 2,5-dimethylanisole through the somewhat long synthetic steps as shown in Synthetic scheme. It was necessary to build the functional group on the benzylic position and then to protect this functional group for the introduction of methyl group to phenylene ring. Tetrahydropyrane was used as a protection group because this group could be easily attached and dettached and the protected compound could be readily handled as a solid form at room temperature. The deprotection was carried out and then the final salt monomer was synthesized in moderate yield.

The thermal analyses of the precursor polymers were performed to find the temperature at which the elimination reaction occurred. Figure 1 shows the DSC and the TGA thermograms of the precursor polymer of PMMPV. Two endotherm peaks are appeared. The low temperature endotherm peak at 100 °C corresponds to the loss of absorbed water during film casting and the higher temperature one between 120 °C and 170 °C to the elimination reaction of HCl and tetrahydrothiophene. The TGA thermograms are well-consistent with the DSC results. The weight loss of absorbed water was appeared below 100 °C. The elimination reaction was found between 100 °C ~ 200 °C. The polymers were stable up to 400 °C and above that temperature, decomposition occurred. The analyses of DSC of the fully eliminated films show that no melt transition occurred. This is due to the high rigid nature of the polymer chains. From these thermal analyses, the elimination reactions of the polymers were finished before 200 °C. So elimination reaction of the precursor polymers was performed at 230 °C.

UV-visible spectra of the PMMPV, PDMPV and PMPV are presented in figure 2. The broad, long wavelength absorptions around 350 ~ 560 nm are due to π-π^* transitions of the polyconjugated systems. The maximum absorption peak for π-π^* transition and the band edge were found at 460 nm and 560 nm for PMMPV. The maximum absorption peak value and the band edge for PMMPV are placed between those of PMPV and PDMPV, and it is closer to those of PMPV than those of PDMPV. This suggests that electron donating effect of methyl group is not strong as we expected. The interests in electroluminescence of PPV derivatives have been spawned in recent years. Therefore, the photoluminescence spectrum was obtained for characterizing the polymer as electroluminescence devices (Figure 2d). Excitation wavelength is 400 nm, and emission maximum is 570 nm which corresponds to

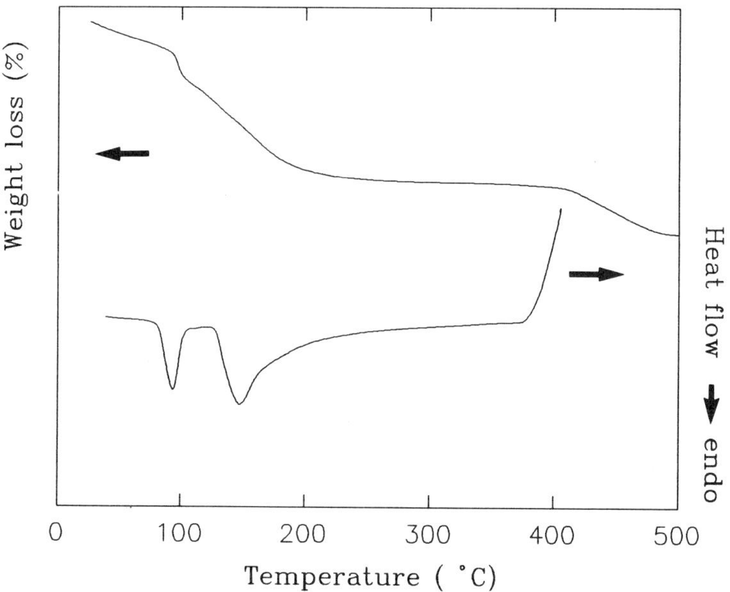

Figure 1. DSC and TGA thermograms of precursor polymer of PMMPV

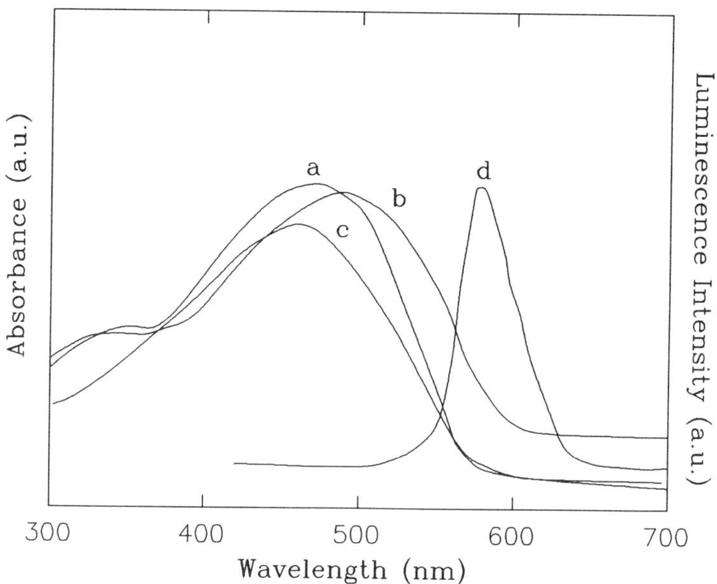

Figure 2. UV-visible spectra of a) PMMPV, b) PDMPV, c) PMPV and d) photoluminescence spectrum of PMMPV

orange color whereas the emission maxima of PDMPV and PMPV are 600 nm and 560 nm, respectively.

Electrical conductivity of the polymers. As a part of the synthesis of well dopable and stretchable PPV derivatives, PMMPV was synthesized through somewhat long synthetic steps. PMMPV film which was stretched 5 times, showed the electrical conductivities of 27

S/cm and 7.4 S/cm, repectively when doped with $FeCl_3$ and I_2. Table 1 shows the draw ratios and the maximum conductivities of $FeCl_3$ and I_2 doped PMMPV and other comparable PPV derivatives. The precursor film of PMMPV could be easily drawn up to draw ratio of 5, whereas that of PDMPV could not be drawn. This result is explained that there are good polymer chain packing in PDMPV due to the symmetry of the two methoxy substituents, while the chain packing abilities between the polymer chains are significantly reduced in asymmetrically disubstitued PMMPV. And the electrical conductivity value of PDMPV film is higher than that of PMMPV film because the two strong electron-donating methoxy groups in PDMPV lead to the reduction in band gap and ionization energy of the conjugated polymer chains, to enhance the electrical conductivity than the case of PMMPV. Especially the electrical conductivity of 5 times stretched PMMPV film is lower than that of unstretched PDMPV film. This means that the electrical conductivity is more strongly affected by electron-donating power of the methoxy groups and the packing ability between polymer chains than the polymer chain drawability. PMPV, PBPV and PDPV are another asymmetrically substituted PPV derivatives which can be stretched. Especially, as the side chain length is increased, the drawability is enhanced though the electrical conductivities of these polymers are lower than that of PMMPV. This fact also supports that the electrical conductivity is more strongly affected by electronic property of substituent and the packing ability. And also, we can conclude that the drawability of the PPV derivatives strongly depends upon the structural symmetry of the substituents at 2,5-positions of the phenylene ring of the polymer main chain.

Table 1. Maximum conductivities of $FeCl_3$- and I_2-doped PMMPV and copolymers

Polymers	Electrical Conductivity (S/cm)			
	$FeCl_3$ doping		I_2 doping	
	Undrawn	Drawn (Draw ratio)	Undrawn	Drawn (Draw ratio)
PMMPV	3.6	27 (5)	0.52	7.4 (5)
PDMPV	~ 10^2	--	~ 10^2	--
PMPV	2.810^{-2}	4.0 (6)	1.3	3.2 (6)
[1]PBPV	3.2×10^{-2}	1.8 (6)	1.0	6.6 (6)
[2]PDPV	8.6×10^{-3}	1.2×10^{-2} (10)	2.7×10^{-4}	2.2×10^{-3} (10)

[1]PBPV stands for poly(2-n-butoxy-1,4-phenylenevinylene) [6]

[2]PDPV stands for poly(2-n-dodecyloxy-1,4-phenylenevinylene) [6]

Third-order nonlinear optical susceptibility. To determine the third order nonlinear optical susceptibility of PMMPV film, the third harmonic generation (THG) intensity of polymer was measured in function of the angle of incidence. THG intensity from the BK-7 glass substrate was also measured as a reference. Figure 3 shows Maker-fringe patterns of the polymer and BK-7 glass substrate. The $\chi^{(3)}$ value of the PMMPV was calculated by comparing the measured THG peak intensity of the film and that of the substrate using the equation reported elsewhere, assuming the film thickness to be much thinner than the coherence length of the film.[19, 20] We also assumed that the effect of the refractive index difference between the PMMPV and the substrate for $\chi^{(3)}$ calculation is negligible. Measured $\chi^{(3)}$ value was 7.5×10^{-12} esu at 1.91 μm. This $\chi^{(3)}$ value is slightly lower than that of PDMPV, 5.4×10^{-11} esu at 1.85 μm reported by Kaino et al. and slightly higher than that of PPV, 3.0×10^{-12}

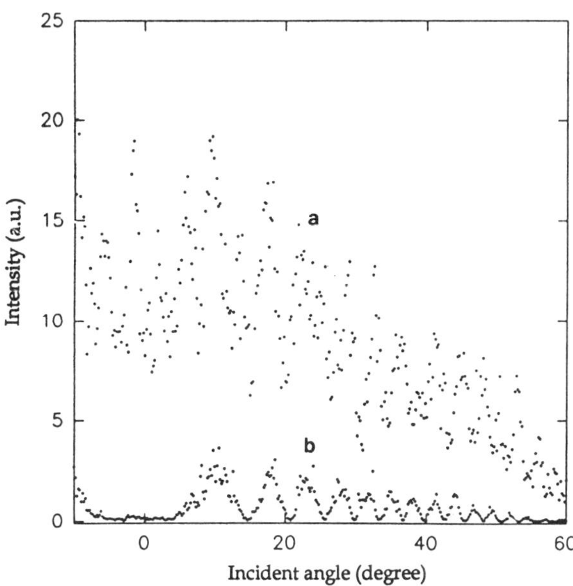

Figure 3. Maker-fringe patterns of a) PMMPV and b) BK-7 glass

esu at 1.91 μm reported by us.[21, 27] This result is explained by the band gap energy difference of the polymers. It is well known that $\chi^{(3)}$ value is in inversely proportional to 6th power of band gap energy. The positions of absoption edges are 610 nm for PDMPV, 550 nm for PMMPV and 530 nm for PPV, respectively. So, the order of band gap energies is PDMPV < PMMPV < PPV. This order of band gap energies is well consistent with that of $\chi^{(3)}$ values. The trend of increasing third order nonlinearities shows important contribution derived from the electron-donating substituents at the 2,5-positions of the phenylene ring of PPV.

REFERENCES

1. J. L. Bredas; R. Silbey, "Conjugated polymers" Kluwer Academic Publishers, 1991
2. P. N. Prasad, D. J. Williams, "Introduction to nonlinear optical effects in molecules and polymers" John Wiley & Sons, Inc., 1991
3. R. A. Wessling, R. G. Zimmerman, U.S. Patent 3, 401, 152, 1968; U.S. Patent 3, 706, 667, 1972
4.. H. Hoerhold, N. Helbig, *Macromol. Chem. Macromol. Symp.* 12: 229 (1987)
5. I. Murase, T. Ohnishi, T. Noguchi, M. Hirooka, *Synth. Met.* 17: 639 (1987)
6. H. K.Shim, D. H. Hwang, J. I. Lee, K. S. Lee, *Synth. Met.* 55-57: 908 (1993)
7. I. N. Kang, G. J. Lee, D. Kim, H. K. Shim, *Polymer Bull.* 33: 89 (1994)
8. J. I. Jin, C. K. Park, H. K. Shim, *Macromolecules* 26: 1799 (1993)
9. H. K. Shim, H. G. Noh, *Mol. Cryst. Liq. Cryst.* 247: 23 (1994)
10. R. W. Lenz, C. C. Han, J. Strenger-Smith, F. E. Karasz, *J. Polym. Sci. Polym. Chem. Ed.* 26: 3241 (1988)
11. W. B. Liang, R. W. Lenz, F. E. Karasz, *J. Polym. Sci. Polym. Chem. Ed.* , 28: 2867 (1990)
12. R. K. McCoy, F. E. Karasz, A. Sarker, P. B. Lahti, *Chem. Mater.* 3: 941 (1991)
13. J. I. Jin, H. K. Shim, C. K. Park, H. J. Kang, S. H. Yu, J. C. Kim, *Synth. Met.* 41-47: 271 (1991)
14. D. R. Gagnon, J. D. Capistran, F. E. Karasz, R. W. Lenz, *Polym. Bull.* 12: 293 (1984)

15. R. Y. Jen, L. W. Shacklette, R. Elsenbaumer, *Synth. Met.* 22: 179 (1987)
16. H. K. Shim, D. H. Hwang, K. S. Lee, *Macromol. Chem.* 194: 1115 (1993)
17. D. R. Gagnon, J. P. Capistran, F. E. Karasz, R. W. Lenz, *Polymer* 28: 567 (1987)
18. J. I. Jin, C. K. Park, H. K. Shim, *J. Polym. Sci. Part A: Polym. Chem.* 29: 93 (1991)
19. T. Kaino, K. Kubodera, S. Tomaru, T. Kurihara, S. Saito, T. Tsutsui, S. Tokito, *Electronic Letter* 23: 1095 (1987)
20. K. Kubodera, *Nonlinear Optics* 1: 71 (1991)
21. T. Kaino, H. Kobayashi, K. Kubodera, S. Tokito, *Appl. Phys. Lett.* 54: 1619 (1989)
22. D. H. Hwang, J. I. Lee, M. Lee, H. K. Shim, *Bull. Korean Chem. Soc.* 15, 25 (1994)

A SOLUBLE AND ELECTRICALLY CONDUCTING POLYMER SYSTEM INCORPORATING THIOPHENE AND ACETYLENIC MOEITIES

Abhijit Sarkar,† Leong L. San, Hardy S. O. Chan and Siu C. Ng

Department of Chemistry
National University of Singapore
Kent Ridge Crescent, Singapore 0511

INTRODUCTION

For the last two and a half decades, conducting polymers have been at the forefront of research activity all over the world. The chemists as well as physicists are interested in this class of polymers due to their potential applications. Several research groups have investigated polythiophene due to its high environmental stability and its conjugated system in both neutral-insulating as well as doped-conducting state.[1,2] Researchers have been trying to engineer the monomer system, either by introducing substituents on the thiophene rings or by introduction of spacers, in order to get polymers with desired properties. It has been reported in the literature that the conductivity is largely compromised if an ethylene or an acetylenic spacer is introduced in between the two thiophene rings of the bithiophene monomer.[3] However, we have found in our laboratory that such a monomeric system gives polymers with conductivity as high as 2 S cm^{-1} in doped form apart from having low band gap energy. Moreover, by functionalising the thiophene units, we were able to get electroactive polymers soluble in organic solvents.

EXPERIMENTAL

Synthesis of Monomers

A series of monomers, 1-5, were synthesised in our laboratory.[4] In all the cases pure monomer was obtained by doing a silica gel column chromatography using hexane as the eluent. The detailed experimental details will be reported elsewhere.

2,2'-(1,2-Ethynediyl)bisthiophene, 1 The synthesis of 1 was carried out as reported in the literature.[4] M.P. 93.5 - 95°C. Elemental analysis : Calc. for $C_{10}H_6S_2$: C 63.15%, H 3.16%, S 33.68%, Found : C 62.83%, H 3.05%, S 33.98%; NMR(DMSO, TMS) ^1H : 7.57(dd,2H), 7.30(dd,2H), 7.08(dd,2H).

†Present address : Institute for Chemical Reaction Science, Tohoku University, 2-1-1 Katahira, Aoba-ku, Sendai 980, JAPAN.

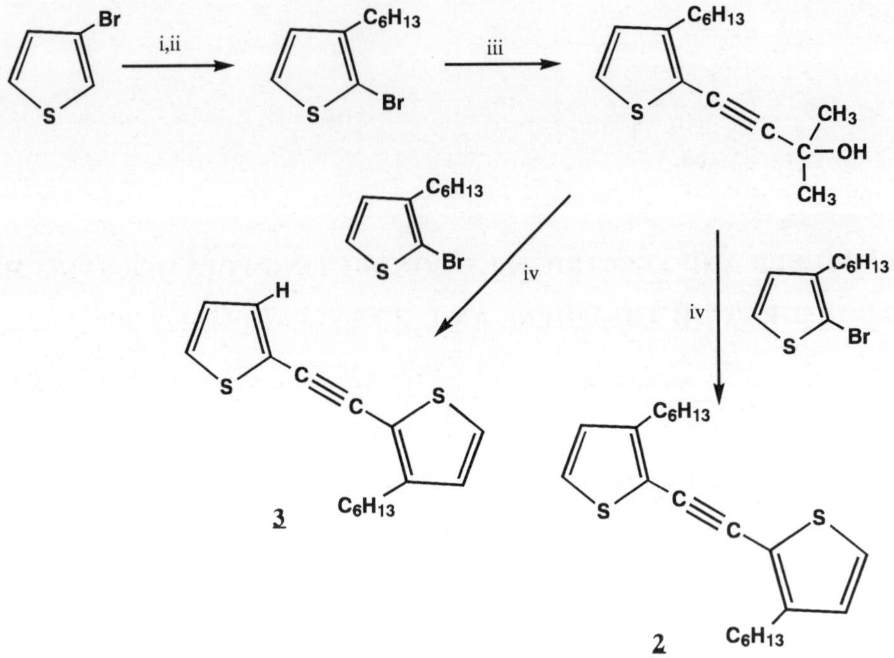

i) $C_6H_{13}Br$, Mg, Ether, Ni(dppp); ii) NBS, DMF, AcOH; iii) $H{\cdot}C{\equiv}C{-}C(CH_3)_2OH$, Pd(0), CuI, C_6H_6, N_2-atm., reflux; iv) reflux.

Scheme 1

3,3'-Dihexyl-2,2'-(1,2-ethynediyl)bisthiophene, 2 Synthesis of 2 was carried out as shown in Scheme 1. 2-Bromothiophene was converted into 3-hexyl-2-bromothiophene using Grignard's reagent and Ni(dppp) catalyst. This was then reacted with 2-methylbut-3-yn-2-ol using $(Ph_3P)_2PdCl_2$ and cuprous iodide under nitrogen atmosphere to obtain 2-methyl-4-(2'-thienyl)-but-3-yn-2-ol in situ. 3-Hexyl-2-bromothiophene was then again syringed into the reaction mixture to obtain the monomer, 2 as a pale yellow liquid. B.P. 210°C. Elemental analysis : Calc. for $C_{22}H_{30}S_2$: C 73.74%, H 8.34%, S 17.88%, Found : C 73.75%, H 8.41%, S 17.91%; NMR($CDCl_3$, TMS) ^1H : 7.21(dd,2H), 7.05(dd,2H), 2.74(t,4H), 0.80-1.71(m,22H).

3-Hexyl-2,2'-(1,2-ethynediyl)bisthiophene, 3 The monomer 3 was prepared using similar procedure as followed for 2(Scheme 1). In case of 3, instead of adding second portion of 3-hexyl-2-bromothiophene in the final step, 2-bromothiophene was added. B.P. 150°C. Elemental analysis : Calc. for $C_{16}H_{18}S_2$: C 70.07%, H 6.57%, S 23.35%, Found : C 68.99%, H 7.04%, S 23.35%; NMR($CDCl_3$, TMS) ^1H : 7.29(dd,2H), 7.27(dd,1H), 7.25(d,1H), 7.23(d,1H), 7.05(dd,1H), 6.93(d,1H), 1.32-1.66(m,8H), 0.87(t,3H).

3,3'-Dibromo-2,2'-(1,2-ethynediyl)bisthiophene, 4 Monomer 4 was prepared following similar route as for 1, starting with 2,3-dibromothiophene. M.P. 70.5-71°C. Elemental analysis : Calc. for $C_{10}H_4S_2Br_2$: C 34.48%, H 1.14%, S 18.39%, Found : C 34.99%, H 1.14%, S 17.90%; NMR($CDCl_3$, TMS) ^1H : 7.20(dd,4H).

3,3'-Dichloro-2,2'-(1,2-ethynediyl)bisthiophene, 5 The above monomer was obtained by heating 4 with CuCl in DMF for 48 hours. After work up, monomer 5 was then purified by column chromatogaphy. M.P. 69.5-70°C. Elemental analysis : Calc. for $C_{10}H_4S_2Cl_2$: C 46.33%, H 1.54%, S 24.71%, Found : C 46.34%, H 1.54%, S 24.71%; NMR($CDCl_3$, TMS) ^1H : 7.11(dd,4H).

Chemical Polymerisation

All the monomers were polymerised chemically using $FeCl_3$ as oxidant in dry chloroform at 0°C. The monomer and $FeCl_3$ were reacted in 1:4 molar ratio. The reaction was carried on for 24 hours in each case. After the reaction, the product mixture was filtered through Buchner's funnel, washed with water, methanol, chloroform and finally with acetone. The polymer, which was obtained as residue, was then subjected to soxhlet extraction, at first with methanol and then with acetone. The final polymer was then dried under suction and used for further studies.

Conductivity Studies

The polymers obtained from all the monomers by $FeCl_3$ polymerisation were pelletised for electrical conductivity measurement. Conductivity was measured using a Keethley four probe apparatus. The polymers were doped with iodine and conductivity was also measured for the doped polymers.

Electrochemical Polymerisation

The monomers **1**-**5** were polymerised by electrochemical method. The electrochemical polymerisation were performed in a single compartment cell having three electrode system under nitrogen atmosphere. The polymerisation was done using potentiodynamic mathod. Cyclic voltammetry experiments were carried out at 50mV s^{-1} using EG & G PARC model 273 potentiostat. The counter electrode used was platinum wire while the polymerisation was done vs. Ag/Ag^+ using platinum foil as the working electrode. The Bu_4NBF_4-CH_3CN electrode system was used. The first scan yielded the monomer peak anodic potential(E_{pa}). Polymer films for spectroscopic measurements were grown onto Indium Tin Oxide(ITO)-coated glass electrodes from 0.1M monomer solution in acetonitrile containing 0.2 M electrolyte. The films were rinsed with acetonitrile and were dedoped by exposing to NH_3 vapours before recording their absorption spectra.[5]

RESULTS AND DISCUSSION

Brownish polymer was obtained from **1** upon $FeCl_3$ polymerisation which was insoluble in common organic solvents. When doped with iodine, this polymer exhibited maximum conductivity of 2 S cm^{-1} while iodine intake was 50% by weight. This is in comparision to polybithiophene which shows conductivity of 3.5 S cm^{-1} in doped state.[6] These values suggest that the introduction of an acetylenic spacer group in between the two thiophene rings has not adversely affected the conductivity of the polymer. Interestingly, we have prepared another monomer 2,3'-(1.2-ethynediyl)bisthiophene, **1a**. This monomer was also polymerised using $FeCl_3$ oxidant. The black polymer formed was insoluble in common organic solvents. The doped form exhibited conductivity in the order 10^{-4} S cm^{-1}. This result was expected since the polymer from **1a** would not have conjugated backbone as was the case for the polymer from **1**.

FTIR for polymers from **1** and **1a** gives insight to the structural difference in these polymers(Fig. 1). Polymer from 1 exhibits peak at 796 cm^{-1} which corresponds to the β C-H bending vibration of the thiophene moeity whereas the peak for the α C-H bending (700 cm^{-1}) is relatively weak. Monomer **1** showed both these peaks of equal intensity which implies that the polymerisation of **1** has gone through α–α coupling of the thiophene moeities and thus resulting in a significantly stereoregular polymer. On the other hand, the polymer from **1a** shows peaks of almost equal intensity for α- and β- C-H bending vibrations indicating that the polymer is relatively less stereoregular.

The monomer **2** was prepared with an aim to obtain soluble electroactive polymer. The n-hexyl chains at 3 and 3' position of thiophene rings are expected to contribute towards the polymer's solubility. The added advantage of such a system is the regioselectivity expected in the resulting polymer which is necessary in order to achieve low band gap energy and higher conductivity.[7] Reddish brown polymer was obtained which was soluble in

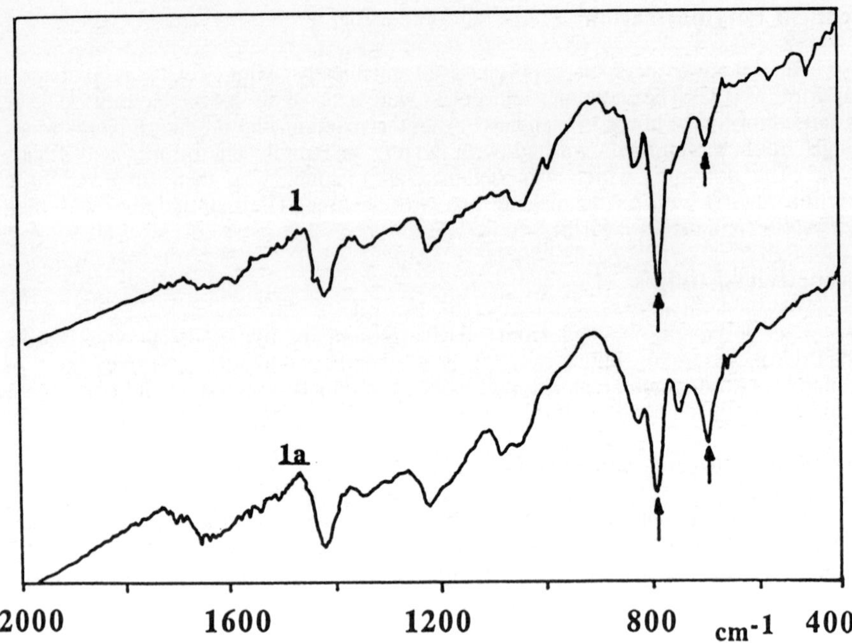

Fig. 1 : FTIR for the polymers from (a) **1** and (b) **1a**

chloroform and THF solvents. The band gap energy as obtained from the absorption band edge of the undoped polymer was 2.1 eV. The solid polymer upon doping with iodine showed a maximum conductivity of 1.4 S cm^{-1}. This value is not very much different from the polymer from **1** or for polybithiophene. Thus in this case, the conductivity of the polymer is retained over and above the advantage that the polymer is soluble.

Monomer **3** where only one of the two thiophene rings has hexyl moiety at 3-position yielded brown polymer on $FeCl_3$ polymerisation. It was slightly soluble in chloroform and THF. The pristine polymer was insulator of electricity but the I_2 doped polymer exhibited a maximum conductivity of the order 10^{-3} S cm^{-1}. The polymers from **4** and **5** were black powder and were insoluble in common organic solvents. The conductivity of the iodine doped polymers were below 10^{-7} S cm^{-1} for both these cases suggesting that the bromo and chloro substituents at the 3 and 3' position of thiophene rings have adversely affected the conjugated backbone and hence the electrical property of the polymer.

Table 1 : Oxidation potential for monomers **1**-**5** and band gap energy of their polymers

	Monomer Oxidation Potential(V)	Polymer Band Gap Energy(eV)
1, R = R' = H	1.03	1.97
2, R = R' = C_6H_{13}	0.82	2.13
3, R = H, R' = C_6H_{13}	1.00	1.82
4, R = R' = Br	1.15	2.07
5, R = R' = Cl	1.10	2.14

We were unable to get very good films of the polymers for all the monomers upon electrochemical polymerisation because the oligomer formed in each case was soluble in the electrolytic solution. A typical cyclic voltammogram(CV) obtained for the polymer of 1 is shown in Figure 2. The oxidation peak gradually decreases in intensity with each cycle although the polymer is almost fully reversible between oxidised and reduced states. The reason for this behaviour might be due to the solubility of the polymer into the electrolyte solution thereby reducing the film thickness on the electrode. The oxidation potential for all the monomers have been obtained by electrochemical method which are listed in Table 1. It can be observed that the oxidation potential for the parent monomer, 1 is significantly low Even after introduction of substituents, it has not been influenced much. In fact, monomer 3 having a single hexyl substituent has lower oxidation potential. Electrochemical polymerisation was also carried out on ITO glass electrodes. Golden coloured films were obtained for all the polymers. The band gap energies for the polymers derived from the absorption spectra are tabulated in Table 1.

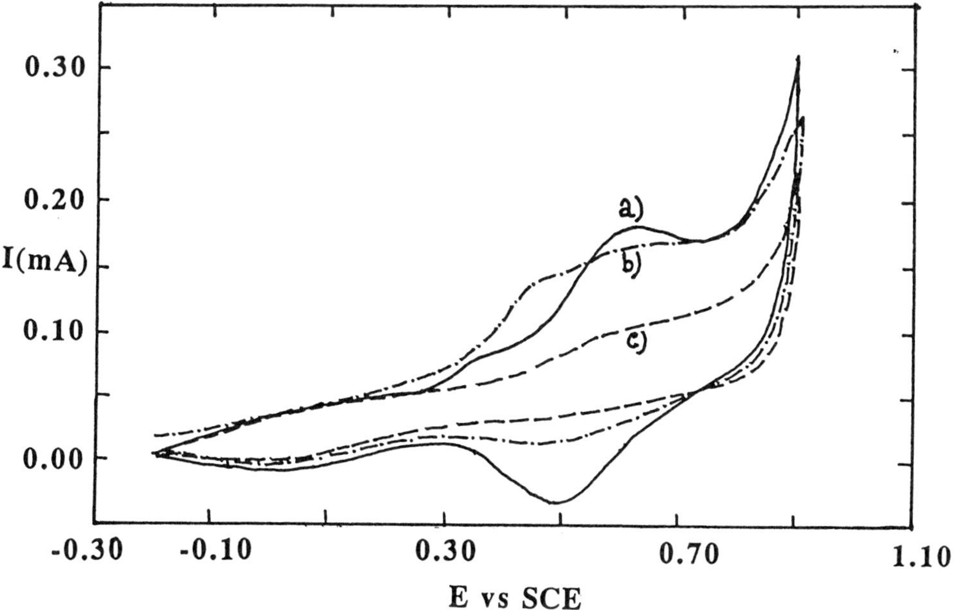

Fig 2 : CV of polymer film from 1 on Pt. (a) second scan, (b) fifth scan and (c) tenth scan.

Band gap energy(E_g) for all the polymers are within 0.82-1.15 V, the lowest value being that for polymer from 2 which have hexyl substituents in the two thiophene rings. Moreover, this polymer has the added advantage of being soluble.

CONCLUSIONS

The substituent on the thiophene rings play a vital role in determining some of the important properties of the polymer viz. solubility, band gap energy and oxidation potential. The polymer system developed by us is of significance since we have been able to invoke a combination of all the above mentioned properties into it with little compromise. Further investigations are in progress to fine tune all these properties to our advantage.

ACKNOWLEDGEMENTS

One of the author, AS, is grateful to National Science and Technology Board(NSTB), Singapore for award of a fellowship.

REFERENCES

1. B. Krische, J. Helberg and C.Lilja, J. Chem. Soc. Chem. Comunn., 1496, 1987; B. Krische and M. Zagorska, Synth. Met., 41-43, 503, 1991.
2. A. Sarkar, PhD Thesis, IIT Bombay, 1993; A. Sarkar, A. Kulkarni, A. Q. Contractor and S. S. Talwar, in "Polymer Science-Contemporary Themes", S. Shivram, ed., Tata McGraw Hill, New Delhi, pp. 267, 1991.
3. M. Onodo, T. Iwasa, T. Kawai and K. Yoshino, J. Phys. D : Appl. Phys., 24, 2076, 1991; G. Koßmehi, Macromol. Chem. Macromol. Symp., 4, 45, 1986; E. Cernia, L. D. Ilario, G. Ortaggi, M. Scarsell, R. Scialis and G. Sleiter. Gazzetta Chimica Italiana, 119, 309, 1989.
4. A. Carpita, A. Lessi and R. Rossie, Synthesis, 571, 1984.
5. J. P. Ferraris, A. Bravo,W. Kim and D. C. Hrncir, J. Chem. Soc. Chem. Commun., 991,1994.
6. H. S. O. Chan, S. C. Ng, S. H. Seow and M. J. G. Moderscheim, J. Mater. Chem., 2, 1135, 1992.
7. T. A. Chen and R. D. Rieke, Synth. Met., 60, 174, 1993.

LIMITING FACTORS IN OPERATION AND FABRICATION OF ORGANIC ELECTROLUMINESCENCE DEVICES

Jan Kalinowski

Istituto di Fotochimica e Radiazioni d'Alta Energia del C.N.R., Via P. Gobetti 101, 40 129 Bologna, Italy, and
Department of Molecular Physics, Technical University of Gdansk, Ul.G. Narutowicza 11/12, 80-952 Gdansk, Poland

INTRODUCTION

A satisfactory explanation of the interrelation between performance parameters and fundamental electronic processes in organic electroluminescence (EL) devices (of which light-emitting diodes - LEDs, display screens and light transducers are most apparent) clearly exceeds the frames of this brief presentation. Therefore, the paper will be focused on some selected important aspects of operation and fabrication of organic LEDs.

In EL electrical energy is directly transformed into light. It is thus understood that the answer to the question how well the input electrical power (Ui) is transformed into output light flux (Ghν), that is determination of the electroluminescence efficiency (η=Ghν/Ui), is one of the most important characteristics of the EL diodes. The photoluminescence yield is an obvious factor limiting this ratio. The excitation mode of EL plays an important role in this transformation process. It is directly related to the material bandwidths.

BAND WIDTH

High-field (intrinsic) EL could be expected in wide-band organic materials. In order to accelerate electrons to 2-3eV allowing impact generation of excited states, the drift velocity $\approx 10^8$ cm/s must be reached. This is possible at high fields ($\approx 10^6$ V/cm) with mobilities ≈ 100 cm^2/Vs. Mobilities of this order of magnitude have been observed in some organic single crystals such as naphthalene (holes) or perylene (electrons) at temperatures below 77K.[1] However, the vast majority of organic materials, especially at room temperature, exhibits mobilities much below this value,[2] making this material property an important limiting factor in operation mechanisms of organic LEDs. This also makes difficult to understand intrinsic EL characteristics observed in pioneering works on organic EL by Bernanose and coworkers,[3-5] and later by other authors.[6,7]

The search for wide-band organic materials is a challenge to design and to fabricate organic intrinsic EL devices. The band width also plays an important role in charge injection processes from metallic electrodes in the recombination-type of EL.

Polymers and Other Advanced Materials: Emerging Technologies and Business Opportunities. Edited by P. N. Prasad *et al.*, Plenum Press, New York, 1995

CARRIER INJECTION CONTACTS

Maximization of the EL efficiency requires recombination time (τ_{rec}) to be shorter than the charge carrier transit time (τ_T) over the thickness (d) of the cell:

$$\tau_{rec} < \tau_T = d/\mu F \,, \tag{1}$$

where μ is the carrier mobility and F is the electric field operating in the cell.

If the carrier recombination can be described in terms of the Langevin theory of a totally random process, as usually is the case with organic materials,[2,8] then for hole mobility $\mu^+ \approx \mu >> \mu^-$ (electron mobility), $\tau_{rec} \approx \varepsilon\varepsilon_o/n$ is inversely proportional to the charge concentration n, and according to (1)

$$j \geq j_{crit} = \varepsilon\varepsilon_o \mu U^2/d^3 \,. \tag{2}$$

This relation assumes the current flow ($j=en\mu F$) driven by the nominal field F=U/d, where U is the voltage applied to the EL cell, and tells us that electroluminescence to be recombination-limited must be detected under space-charge-limited current (SCLC) conditions.[9]

Often, however, $\tau_{rec} > \tau_T$, so that the current is injection-limited, we deal with injection-limited EL (IL EL). A part of injected carriers avoid recombination, lowering the EL efficiency (Figure1). Injection ability of electrodes stands thus for an important factor limiting performance of organic LEDs.

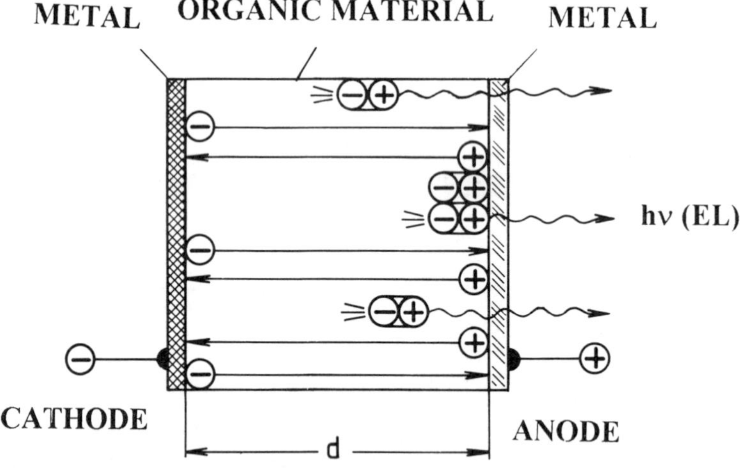

Figure 1. Injection-limited electroluminescence.

How the injection limitation of EL can be observed in experiment ? An obvious way is to mesure the current-voltage characteristics and fit them to various models of current flow in solids. However, interestingly, the brightness-current relation appears to be sensitive to injection conditions at the electrodes. The EL intensity (Φ_{EL}) is proportional to the squared concentration of charge carriers since the Langevin recombination is a second order process with a rate constant γ transforming charge carriers into singlet excitons, S(x), which decay with a total rate constant k_S and a radiative rate constant k_F,

$$\Phi_{EL} = k_F \int_0^d S(x)\, dx = (k_F/k_S)\, \gamma\, n^2 d\,. \tag{3}$$

Since under SCLC conditions (2) n is proportional to the applied voltage,[9] $n=\varepsilon\varepsilon_0 U/ed^2$,

$$\Phi_{EL} = (k_F/k_S)\, \gamma\, \varepsilon\, \varepsilon_o\, (e^2\mu)^{-1}\, j\,, \tag{4}$$

the EL intensity should be a linear function of the current flowing through the cell.

The situation changes for IL EL. The charge density is no more proportional to the applied voltage . It is now proportional to suitable (electron and hole) injection current densities (j_i^e, j_i^h). EL intensity is thus proportional to their product:

$$\Phi_{EL} = (k_F/k_S)\, \gamma\, (e^2\mu^+\mu^- F^2)^{-1}\, j_i^e j_i^h\,. \tag{5}$$

If, in particular case, both injection currents follow the one-dimensional Onsager model at high fields with the source current $j_S=j_O\exp(-\alpha x)$ and with α^{-1} standing for the penetration depth of the primary injected charge carriers, that is can be approximated by[10]

$$j_i = j_o \exp\,[-\alpha(\beta kT/eF)^{1/2}]\,, \tag{6}$$

where $\beta=e^2/16\pi\varepsilon\varepsilon_0 kT$, and $\alpha_e = (2\sqrt{2}/\hbar)(m_{eff}^e I_A)^{1/2}$ and $\alpha_h = (2\sqrt{2}/\hbar)(m_{eff}^h E_g)^{1/2}$, the LED brightness appears to be a power - type function of the current density,

$$\Phi_{EL} \propto j^n \tag{7}$$

with the power

$$n = 1 + \left(\frac{m_{eff}^e I_A}{m_{eff}^h E_g}\right)^{1/2} \tag{8}$$

being dependent on basic material parameters such as width of conduction and valence bands (through the effective masses of electrons , m_{eff}^e, and holes, m_{eff}^h, respectively) and positions of energy levels through the energy gap (E_g) and electron affinity (I_A). Due to a broad spectrum of these parameters for organic materials one would expect a vast range of values of the power n. Indeed, n from below 1 to 6 has been experimentally observed up to date. Superlinear, power-type increase of EL intensity with current can be seen starting from single-layer (SL) EL cells of anthracene and tetracene-doped anthracene films (Figure 2a), through PPV derivatives-based EL cells (Figure 2b) up to double-layer EL structures exemplified here by a series of phtaloperinone derivatives (Figure 2c). It is interesting to note that addition of an electron-transport layer to the double heterostructure HTL-EML with derivative E transforms the strongly non-linear plot in a linear dependence.[14] n close to six has been found by us for an EL double structure with Alq_3 as the emitter (EML) and quinacridone as a hole transporting layer (HTL).[15]

In light of these considerations it is apparent that the requirement of good injecting electrodes and, especially, electron-injecting contacts brings considerable limitation of the operation mode of organic EL diodes. Moreover, it brings serious problems to their

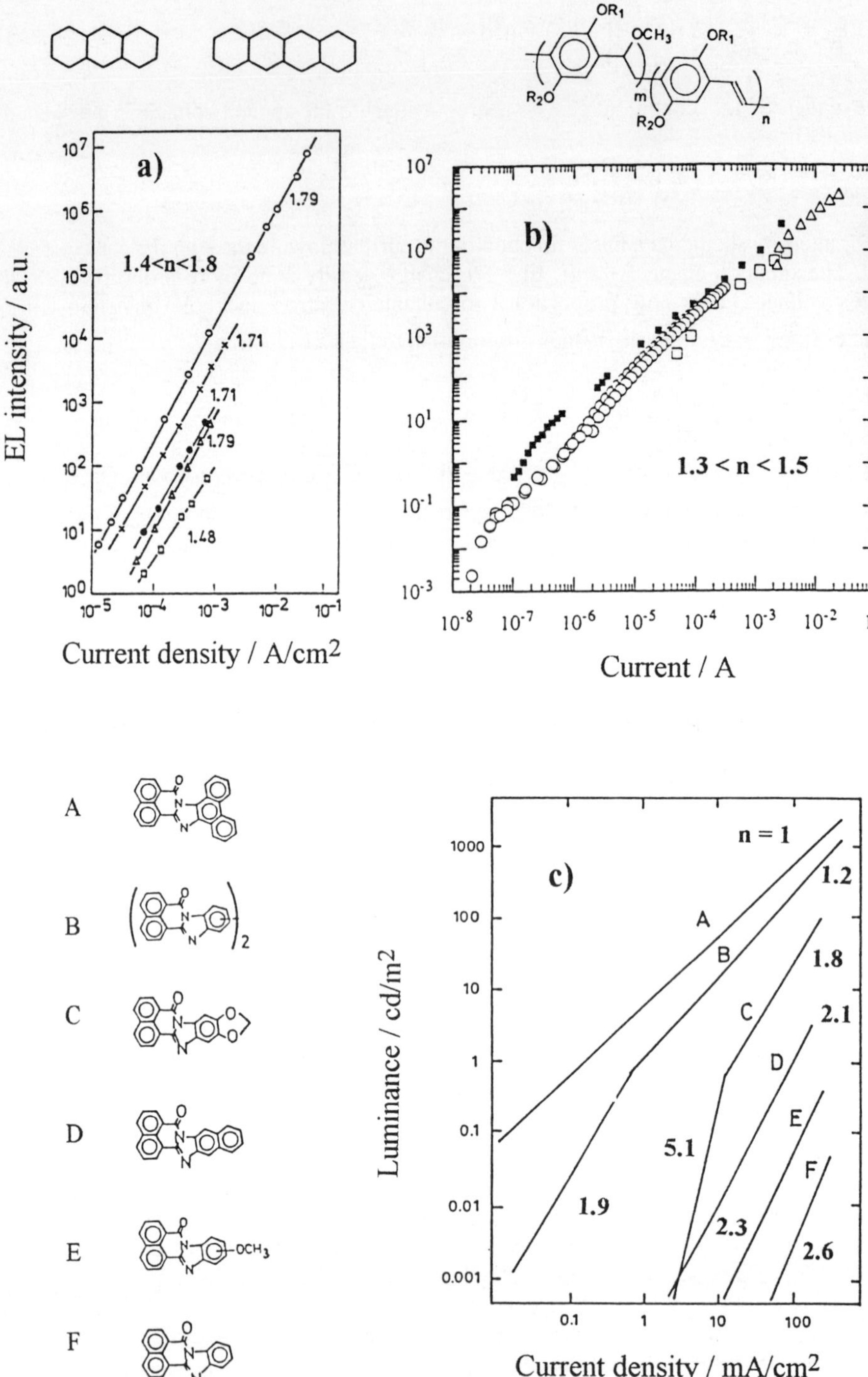

Figure 2. EL intensity versus current in various EL diodes: (a) single layer (SL) LEDs based on anthracene and tetracene-doped films (slopes are given near by the plots; for details see Figure 5 of ref. 11) - figure obtained from ref. 11; (b) SL LEDs based on poly(dialkoxy-*p*-phenylenevinylene) - figure obtained from ref. 12; (c) double-layer EL heterostructures based on six phthaloperinone derivatives as EML - figure obtained from ref.13.

fabrication since electron-injecting contacts are usually made of low-work function metals difficult for handling and protection against chemical reactions.[16, 17]

The above picture of interrelation between recombination and transport processes can, however, be much complicated by the presence of charge carrier traps and excitonic interactions.

CARRIER TRAPPING & EXCITONIC INTERACTIONS

Due to spin statistics three times more triplet than singlet states are produced from electron-hole recombination in a trap-free material. This ratio can be much higher in the presence of deep carrier traps (Figure 3). For instance, in order to create a singlet exciton from a recombination event of a free hole with such a deeply trapped electron an energy barrier, E_S-E_{eh}, has to be overcome and the probability to yield a singlet exciton

$$\alpha_1 = (1/4) \exp [-(E_S-E_{eh})/kT] \tag{9}$$

is lower than 1/4. For $E_S >> E_{eh}$, $\alpha_1 \rightarrow 0$. Only triplet excited states are created. Consequently, only phosphorescence or delayed fluorescence due to triplet-triplet annihilation can be detected in the EL signal. Since in the latter case for each emitting singlet two triplets are required, a quadratic EL intensity-current relationship should be obtained. A mixed prompt-delayed EL signal will show up in the power n between 1 and 2. This, however, appears at high triplet exciton concentrations, that is at high currents, when their quenching by charge carriers becomes of importance.

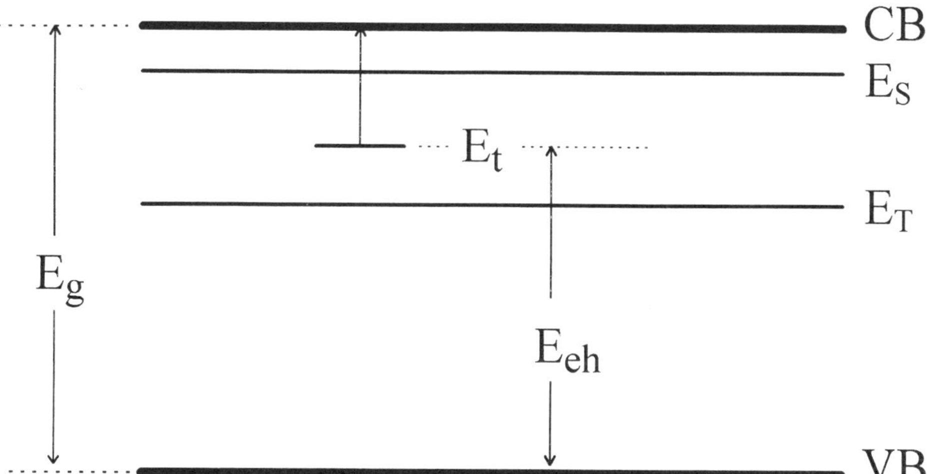

Figure 3. Recombination between a free hole and a deeply trapped electron.

Both singlet and triplet excitons can interact with free and/or trapped charge carriers. This can be observed as photo-enhanced current[18-20] and/or luminescence quenching.[21-23] Since the quenching rate

$$k_{quench} = K\, n \tag{10}$$

is proportional to the charge density, n, (K is the second order rate constant for exciton quenching), an inhomogeneity in the spatial charge distribution can be reflected in the spatial distribution of the EL intensity. Figure 4 shows a striking example of singlet and triplet exciton quenching by inhomogeneously trapped holes in the anode region of an EL cell based on a tetracene single crystal. The near-contact emission zones at the lower voltage change their shape at a higher voltage, showing a dark narrow subzone close to the anode as due to singlet exciton quenching, and a splitting up at a larger distance due to triplet exciton quenching. The latter reflects the fact that the deeper located EL subzone originates from singlet excitons created by triplet-triplet annihilation. The triplet-to-singlet exciton ratio increases towards the bulk because the concentration of trapped holes is a decreasing function of distance from the injecting contact. This leads to decreasing occupancy of shallow traps ($E_t < E_g - E_S$); relatively more recombination events takes place on deeply trapped holes producing triplet states. The effect can be observed with a strong gradient of the trapped charge only. Otherwise, the difference between coordinate variation of directly (e+h) and indirectly (T+T) generated singlets is not sufficient to create well separated prompt and delayed EL components. The lack of the cathode zone splitting may be associated with a weak-gradient spatial distribution of trapped electrons and/or shallow electron trapping.[24]

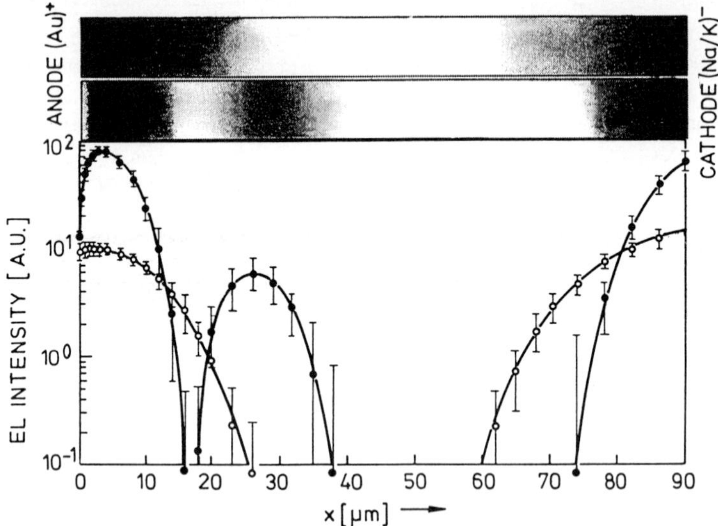

Figure 4. Spatial distribution of EL intensity in a 90 μm thick tetracene crystal at two different voltages. U=750V (o), U=950V (•). Figure obtained from ref. 24.

The factors limiting increase of the EL efficiency are the second order rate constant for exciton quenching, K, and the lifetime, τ_S, of the emitting singlet exciton. Increasing n above

$$n_c = (K\tau_S)^{-1} \tag{11}$$

makes the concentration of singlet excitons to drop down, so that the efficiency should decrease or, at least, slow down its increase with current. Taking for K the value 10^{-7} cm^3/s (singlet-free charge carrier interaction rate constant) as found for single organic crystals[22,25,26] and making an extrapolation to other organic materials we see that the critical value of charge density falls far above the high-current free charge carrier densities in single crystals and degenerate ground state (DGS) conjugated polymers, but it can be easily met in high-brigthness EL cells based on polycrystalline layers, non-degenerate ground state (NDGS) conjugated polymers and molecularly-doped non-conjugated polymers (Table 1).

Table 1. Critical and free charge carrier densities at high current conditions for different types of organic materials.

EL cell based on	Single crystals	DGS conjugated polymers	Polycrystalline layers	NDGS conjugated polymers	Molec.-doped non-conjugated polymers
K [cm^3/s]	10^{-7}	10^{-7}	10^{-9} (♦)	10^{-9} (♦)	10^{-9} (♦)
n_c [cm^{-3}] with τ_S = 10 ns	10^{15}	10^{15}	10^{17}	10^{17}	10^{17}
μ [cm^2/Vs]	≈ 1	≈ 1	10^{-5} - 10^{-3}	≈ 10^{-4}	10^{-7} - 10^{-5}
$n = j/e\mu F$ with $j = 10^2$ mA/cm² and $F = 10^6$ V/cm	10^{12}	10^{12}	10^{17} - 10^{15}	10^{16}	10^{19} - 10^{17}

(♦) Due to low mobility of charge carriers the exciton-trapped charge carrier interaction has been assumed for exciton quenching.[22,25,26]

In order to preserve fast increase of the EL efficiency with charge density, materials with short exciton lifetime and high carrier mobilities are needed.

CONCLUDING REMARKS

The following factors have been shown to limit operation and fabrication of organic EL diodes:

- Width of conduction and valence bands (limiting the excitation mode of EL materials and influencing the brightness-current relationship),
- Energy level positions (limiting the injection efficiency of the electrodes and influencing the brightness-current relationship),
- Injection ability of electrodes (determining the operation mode of the EL cells: recombination- or injection-limited EL; limiting EL cell lifetime due to chemical reactions of low-work function metals),
- Charge carrier trapping and excitonic interactions (limiting the EL efficiency and time resolved EL characteristics, influencing the brightness-current relationship).

ACKNOWLEDGMENT

This work has been done in the framework of the Program “Novel Materials” (Progetto Strategico “Materiali Innovativi”) of the National Research Council (C.N.R.) of Italy.

REFERENCES

1. N. Karl, Getting beyond impurity-limited transport in organic photoconductors, *in*: “Defect Control in Semiconductors”, vol.II, K. Sumino, ed., North Holland, Amsterdam (1990).
2. M. Pope and C.E. Swenberg. “Electronic Processes in Organic Crystals”, Clarendon Press, New York (1982).

3. A. Bernanose, M. Compte and P. Vouaux, On a new mode of luminescence from some organic compounds, *J. Chem. Physique* 50: 64 (1953) (in French).
4. A. Bernanose and P. Vouaux, Organic electroluminescence: a study of the emission mode, *J. Chem. Physique* 50: 261 (1953) (in French).
5. A. Bernanose, Electroluminescence of organic compounds, *Brit. J. Appl. Phys., Suppl. No.4*: 554 (1955).
6. A. G. Goldman, M.W. Kurik, Ya.I. Vertzimakha and B.N. Korolko, Electroluminescence of carbazole, *Zhurn.Prikl.Opt.* 14: 235 (1971) (in Russian).
7. N.P. Sinha, Y. Misra, L. Tripathi and M. Misra, Electro-optical properties of doped anthracene films, *Solid State Commun.* 39: 89 (1981).
8. R. Morris and M. Silver, Direct electron-hole recombination in anthracene, *J. Chem. Phys.*50: 2969 (1969).
9. W. Helfrich, Space-charge-limited and volume-controlled currents in organic solids, *in*: "Physics and Chemistry of the Organic Solid State", D. Fox, M.M. Labes and A. Weissberger, eds., Wiley, New York (1967).
10. J. Godlewski and J. Kalinowski, Injection-limited currents in insulators, *Japan. J. Appl. Phys.*28: 24 (1989).
11. J. Kalinowski, J. Godlewski and Z. Dreger, High-field recombination electroluminescence in vacuum-deposited anthracene and doped anthracene films, *Appl. Phys. A*37: 179 (1985).
12. D. Braun, E.G.J. Staring, R.C.J.E. Demandt, G.L.J. Rikken, Y.A.R.R. Kessener and A.H.J. Venhuizen, Photo- and electroluminescence efficiency in poly (dialkoxy-*p*-phenylenevinylene), *Synth. Metals* 66: 75 (1994).
13. T. Tsutsui and S. Saito, Organic thin films for electroluminescence displays, *in*: "Polymers for Microelectronics - Science and Technology" (47), Y. Tabata, ed., Kodansha Press, Tokyo (1990).
14. C. Adachi, S. Tokito, T. Tsutsui and S. Saito, Organic electroluminescence device with a three layer structure, *Japan. J. Appl. Phys.*27: L713 (1988).
15. P. Di Marco, J. Kalinowski, N. Camaioni, V. Fattori and G. Giro, A thermally stable organic light-emitting diode, *in*: "Advanced New Materials and Emerging New Technologies", P.N. Prasad, E. Mark and J.F. Fung, eds., Plenum Press, New York (1995).
16. L. M. Do, E. M. Han, Y. Niidome, M. Fujihara, T. Kanno, S. Yoshida, A. Maeda and A. J. Ikushima, Observation of degradation processes of Al electrodes in organic electroluminescence devices by electroluminescence microscopy, atomic force microscopy, scanning electron microscopy, and Auger electron spectroscopy, *J. Appl. Phys.* 76: 5118 (1994).
17. W.R. Salaneck, Chemical and electronic structure of interfaces with conjugated polymer: system of interest in molecular electronics applications, *in*: "Advanced New Materials and Emerging Technologies", P.N. Prasad, E. Mark and J.F. Fung, eds., Plenum Press, New York (1995).
18. J. Godlewski and J. Kalinowski, Photo-enhanced currents in organic insulators, *Phys. Stat. Sol. (a)* 53: 161 (1979).
19. J. Kalinowski and R. Signerski, Exciton-enhanced double injection currents in tetracene crystals, *Phys. Stat. Sol. (b)* 118: K147 (1983).
20. J. Godlewski and J. Kalinowski, Photoenhanced conductivity in organic crystals, *Mol. Cryst. Liq. Cryst.*228: 61 (1993).
21. J. Kalinowski and J. Godlewski, Luminescence modulation in organic crystals by exciton-charge carrier interaction, *Acta Phys. Polonica A* 46: 523 (1974).
22. J. Kalinowski, J. Godlewski and R.R. Chance, Evidence for trapped-exciton fluorescence in anthracene crystals at room temperature, *J. Chem. Phys.* 64: 2389 (1976).
23. J. Kalinowski and J. Godlewski, Multi-charge carrier trapping in organic crystals: quantized internal macrotrap energy levels, *Mol. Cryst. Liq. Cryst.* 205: 101 (1991).
24. J. Kalinowski, Space-resolved recombination electroluminescence in organic crystals, *Synth. Metals* 64: 123 (1994).
25. J. Kalinowski and J. Godlewski, Singlet exciton-charge carrier interaction in anthracene crystals, *Phys. Stat. Sol.(b)* 35: 789 (1974).
26. J. Kalinowski and J. Godlewski, Spatial behaviour of the charge created at the illuminated interface molecular crystal/electrolyte, *Chem. Phys.* 32: 201 (1978).

ELECTRICAL PROPERTIES OF IMPURITY DOPED ION CONDUCTING POLYMERS

S.Radhakrishna and S.Anandan

Institute of Advanced Studies
University Malaya
59100 Kuala Lumpur

ABSTRACT

Generally polymers have been considered to be insulators, but they can now be "tailored" as conducting polymers and can be synthesized for applications, especially for the semiconductor and electronic industries. Polymers can be made to conduct either electronically or ionically. Polymer materials that conduct ions can be described as ion conducting polymer electrolytes. While as prepared polyethylene oxide (PEO) samples are insulators with very low conductivity (10^{-10} Scm^{-1}), impurity ion doped PEO showed conductivity values in the range of 10^{-5} Scm^{-1}. The conductivity of electron conducting polymers can reach values as high as 10^{5} Scm^{-1}. Doping of polymers such as PEO can change its electrical properties significantly and infact make it a semiconducting material.[1] This property has opened up new areas of applications such as solid state batteries, electrochromic windows and fuel cells.

PEO doped with various divalent cations from first transition metal group have been investigated. These materials are prepared as thin films from solution casting technique and are further characterized for their suitable applications. Structural studies are carried out spectroscopically and electrical properties are determined by ac and dc conductivity measurements. Other techniques such as DTA and XRD have been used to get a better understanding of these materials.

INTRODUCTION

Polymers are long chain molecules of very high molecular weight, often in hundreds of thousands. They are basically organic materials bound by covalent bonds. Polymers were traditionally considered to be insulators or bad conductors. This is because usually carbon atoms form strong covalent bonds. For example, in polymers like polyethylene, each carbon atom shares one of its four electrons with each of the carbons next and one each with the

Polymers and Other Advanced Materials: Emerging Technologies and Business Opportunities. Edited by P. N. Prasad *et al.*, Plenum Press, New York, 1995

two hydrogens attached to it. The electrons that form these covalent bonds are free to move. These localised electron makes the material a very poor conductor.

However it is now possible to change this and in some cases we have conducting carbon chains. For example, if we remove one of the hydrogen atoms from each carbon chain of polyethylene, there will be an extra electron available for conduction. The extra electron will form a second bond with the neighbouring carbon atom, thus giving rise to a double bond. Finally, this will result in an alternating sequence of single and double bond carbon chains along the polymer backbone. The extra electron in this conjugated configuration can be induced to move along the chain and this will make the polymer have relatively higher conductivity.

Conducting polymers can generally be divided into two categories, depending on their charge carrier types. The two types are :

- electron conducting polymers
- ion conducting polymers

ELECTRON CONDUCTING POLYMERS

Electron conducting polymers are made by providing electrons which are not fixed in their positions. Usually this is found in a conjugated polymer such as polyacethylene.[2] As described earlier , the extra electron along the backbone chain can be induced to move, forming electric current. However in this paper we present results of polymers which have good ionic conductivity.

ION CONDUCTING POLYMERS

Ion conducting polymers are polymer electrolytes which contain mobile anion or cation species. These polymers are made by 'dissolving' ionic salts in an appropriate polymer. A balanced ion-polymer interaction must be achieved to conduct ionically. The interaction must be strong yet weak enough to permit the ions to move under the electric field. The conductivity value depends on the number of mobile ions, their charge, mobility and the host polymer's phase constitution.[3]

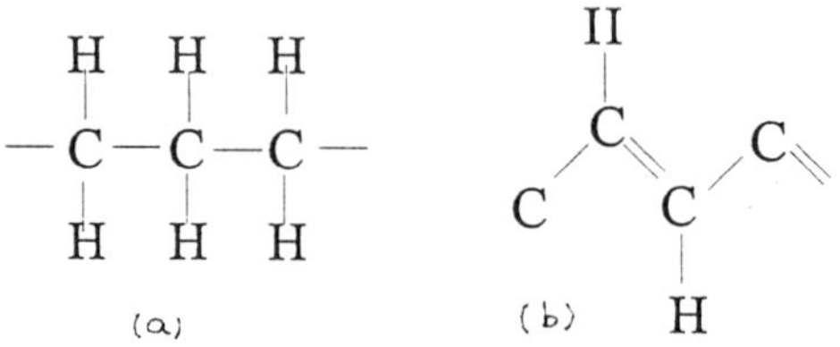

(a) (b)

Figure 1. Structures of non conducting polyethylene (a) and of conducting polyacethylene (b)

Some examples of electron conducting polymers are :

- poly pyrrol (PP)
- poly p-phenylene (PPP)
- poly p phenylenevinylene (PPV)
- poly phenylene sulfide (PPS)
- poly aniline
- poly thiophene

Figure 2 shows some of the electron conducting polymers.

Polymer	Chemical structure	Conductivity (S/cm)
PA		10^3 - 10^5
PPP	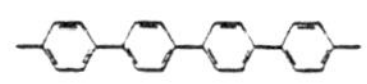	10^2 - 10^3
PPS	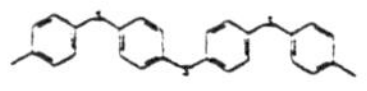	100 - 500
PPV	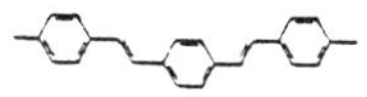	3 - 5 x 10^3
PAni	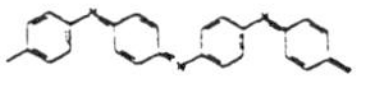	30 - 200
Ppy	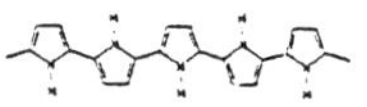	102 - 7.5 x 10^3
PT	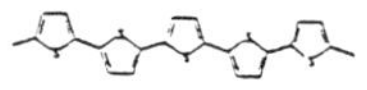	10 - 10^3

Figure 2. Some electron conducting polymers [4]

Below are some examples of ion conducting polymers.

- poly ethylene oxide (PEO)
- poly pyrrol oxide (PPO)
- poly ethylene glicol (PEG)
- poly oxymethylene oligo oxyethylene (POMOE)

Figure 3 shows the structure of one of the better known ion conducting polymer PEO.

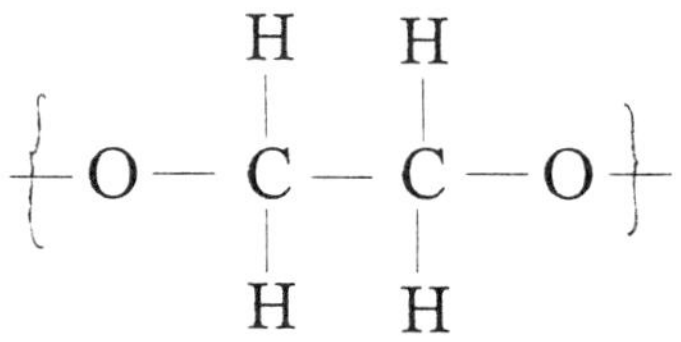

$$\left[O - \underset{H}{\overset{H}{C}} - \underset{H}{\overset{H}{C}} - O \right]$$

Figure 3. Structure of PEO

Electron conducting polymers have found their way into many applications because of their conductivity. These materials are being investigated as substitutes to the conventional materials such as silicon and galium arsenide in the semiconductor industry. Some of the applications of electron conducting polymers are:

1. rechargeble batteries
2. anti static device
3. light emitting diode (LED)
4. metal-insulator-semiconductor (MIS) device

Their main advantages over non polymeric organic semiconductors is the possibility of processing the polymers to form useful and also robust structures. These materials also show nonlinear response to electronic excitation, suggesting the possibility of using them as electroluminescent materials. For example , light emitting diodes based on conjugated polymers have been fabricated.[5]

Current work was carried out to investigate the electrical and optical properties of PEO-MX system. PEO or poly ehtylene oxide was chosen as the host material due to its polymer ion interactions with many types of ionic salts. M is divalent cations from the first row of transition metal group. X is an anion of the halide group or thiocynate ions. Some of the polymer electrolyte systems that are currently being investigated are as below:

PEO-$NiCl_2$ system
PEO-$NiBr_2$ system
PEO-$Co(SCN)_2$ system
PEO-$CoCl_2$ system

Results from PEO-$Co(SCN)_2$ are discussed. Besides the conductivity, results from other experimental techniques are also shown to further strengthen and to correlate the results.

EXPERIMENTAL METHOD

Polyethylene oxide from Aldrich chemicals of average molecular weight 900,000 was used. $Co(SCN)_2$ was obtained from Sigma chenicals. Samples were prepared by solution casting technique. Appropriate amounts of PEO and $Co(SCN)_2$ were added and stirred in 50 ml of acetonitrile to ensure complete dissolution. The solution was cast in glass rings and allowed to evaporate for 24 hours. The resulting films were further dried. The samples that were prepared contained the ether oxygen to cation ratios of 128:1; 64:1; 32:1 and 16:1.

The following experimental techniques were used to characterize the samples prepared. UV-Vis spectroscopy experiments were done using a Shimadzu UV-Vis 3010 model spectrophotometer. The thermal analysis data were obtained from Perkin-Elmer DTA 7 series differential thermal analyser. A Hioki model 3025 was used to obtain AC impedance data and the DC conductivity values were obtained from a Keithley High Voltage source unit model 267. The X ray diffractograms were plotted out from the Phillips PW1180 XRD model.

RESULTS AND DISCUSSION

UV-Visible

The dark bluish samples of PEO doped with Co ions give spectra as shown in figure 4. This spectra shows clearly that as the $Co(SCN)_2$ salt concentration in the polymer matrix is increased, the broad peak appearing between 550 and 650 nm increases in intensity. This is an evidence of the existance of Co^{2+} in the polymer. The width of the spectral band is supportive of the fact that Co^{2+} can form complexes with PEO.

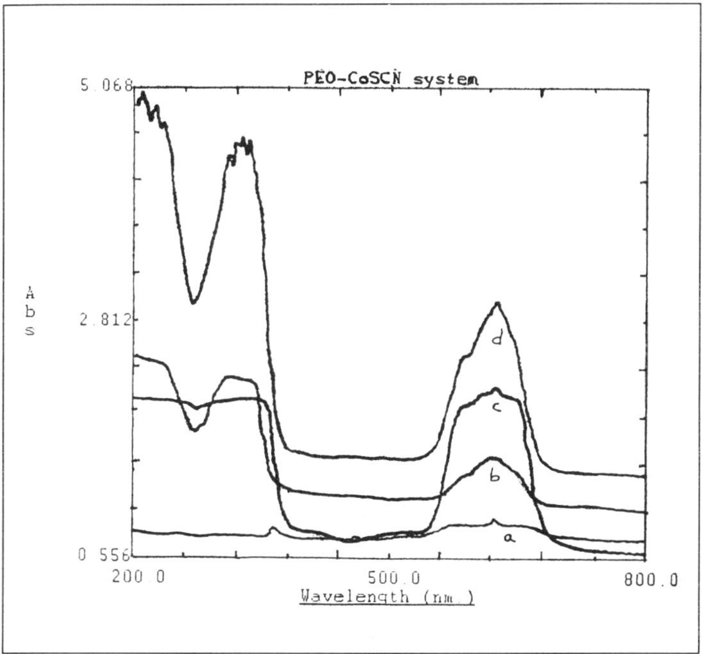

Figure 4. UV-Vis spectra of the various PEO doped Co(SCN)2 samples (a) 128:1 (b) 64:1 (c) 32:1 (d) 16:1

Differential Thermal Analysis

The DTA curves for all the concentrations are shown in figure 5. Generally they indicate the melting point of crystalline PEO phase at about 60 degrees C. A step in the curves before the melting point shows the existence of glass transition temperature Tg at or below the room temperature. The large width of the DTA peaks could be attributed to the PEO-$Co(SCN)_2$ salt complexation.

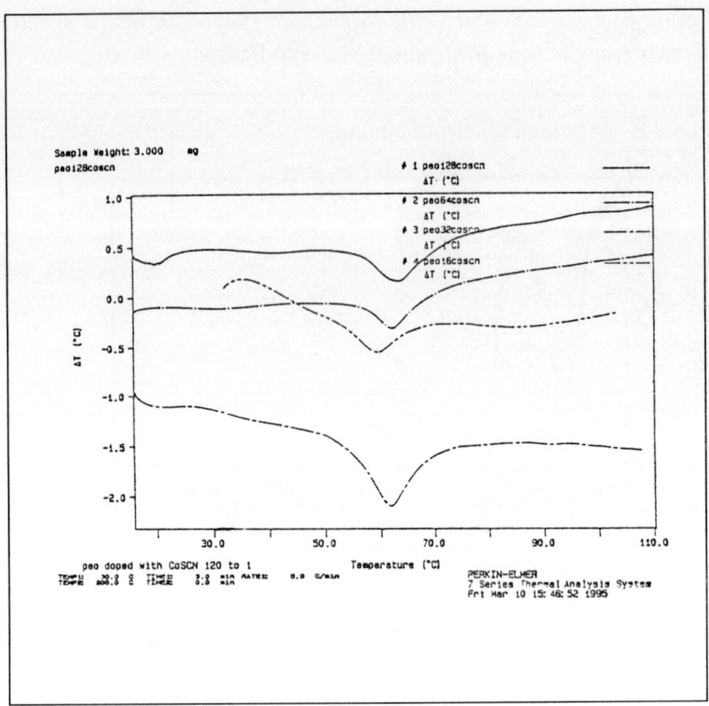

Figure 5. DTA curves of PEO-Co(SCN)2 system (a) 128:1 (b) 64:1 (c) 32:1 (d) 16:1

XRD Analysis

The diffractograms for the various concentration are shown in figure 6. Compared to the pure PEO, these diffractograms do not show formation of any other phases except the two phases of PEO-Co$(SCN)_2$ salt complex phase and the amorphous PEO phase. It also shows that generally addition of salt into the system decreases the crystallinity of the samples.

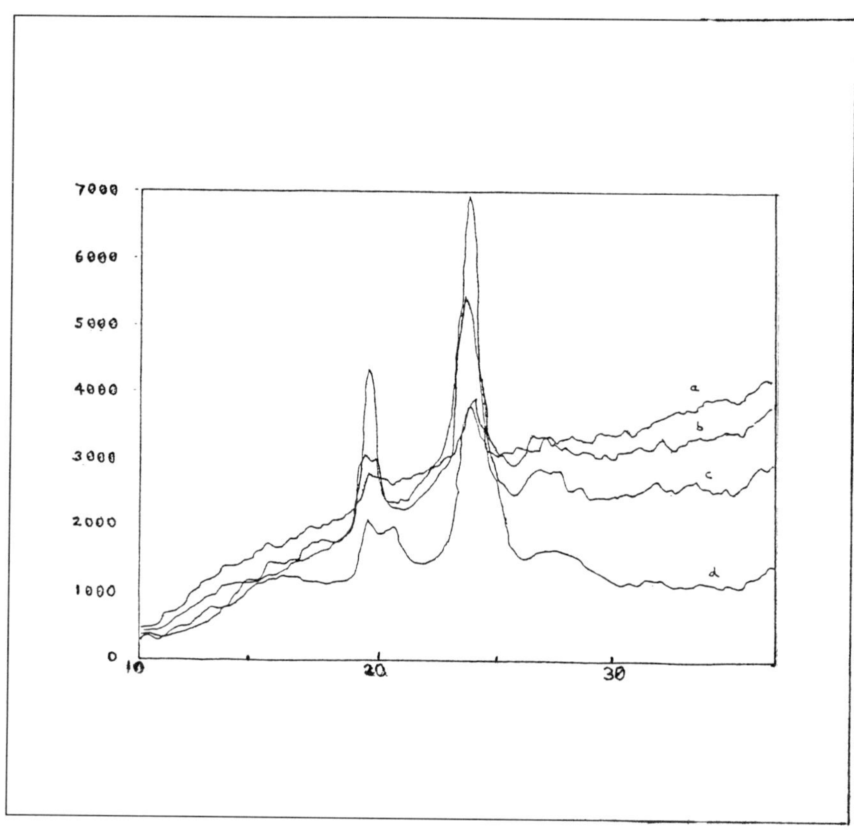

Figure 6. Diffractograms of PEO-Co$(SCN)_2$ system (a) 128:1 (b) 64:1 (c) 32:1 (d) 16:1

AC Impedance

The cole-cole plots are shown in figure 7. All the plots show the usual trend of polymer electrolytes consisting of a skewed semicircle with a tilted tail at the end. From these plots, the bulk resistance was estimated and the conductivity at room temperature was calculated. All samples showed conductivity values in the range of ~ 10^{-6} Scm-1. The highest among these samples was for the sample with ether oxygen to cation ratio of 64:1, giving σ value of 3×10^{-6} Scm^{-1}.

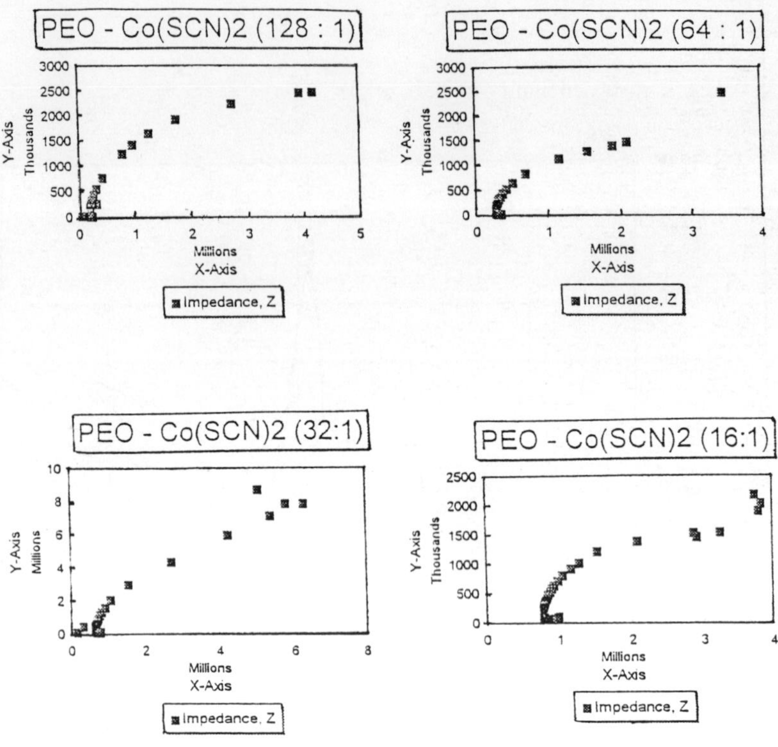

Figure 7. Cole-cole plots of PEO-$Co(SCN)_2$ system

DC Impedance

The Arhennius plots of conductivity of conductivity versus 1000/T are shown in figure 8. All plots show an increase in conductivity with temperature with a sharp increase at about 60 degrees C. This is due to the melting of the crystalline PEO phase. The highest conductivity reached at about 100° C is shown by the sample with ratio of 16:1.

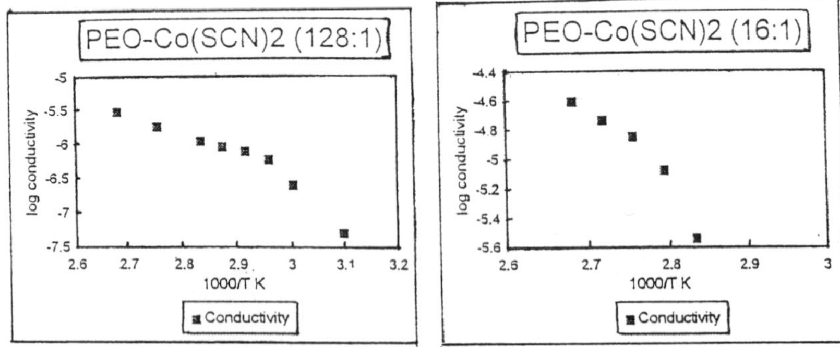

Figure 8. Arhennius plots of the PEO-$Co(SCN)_2$ system

BATTERY PERFORMANCE

For the PEO-$Co(SCN)_2$ system that were studied , the sample that gives the best conductivity data is identified and from this sample a battery is made.
A battery is prepared using the materials and configuration as shown below.

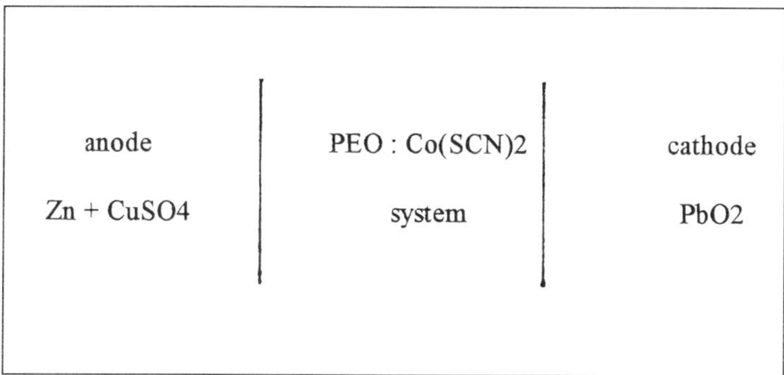

Figure 9. Battery configuration

Equal proportions of anode and cathode materials are prepared and then brushed on both sides of the polymer film as anode and cathode. After drying , it is placed in an air tight chamber and discharged at 3 nA. Readings of voltage vs time are taken. A typical discharge curve for the PEO-$Co(SCN)_2$ system studied is given in figure 10 below. The curve shows a battery which lasts for a about two hours. The performance could be improved and further work is being carried out with different cathode and intercalation materials.

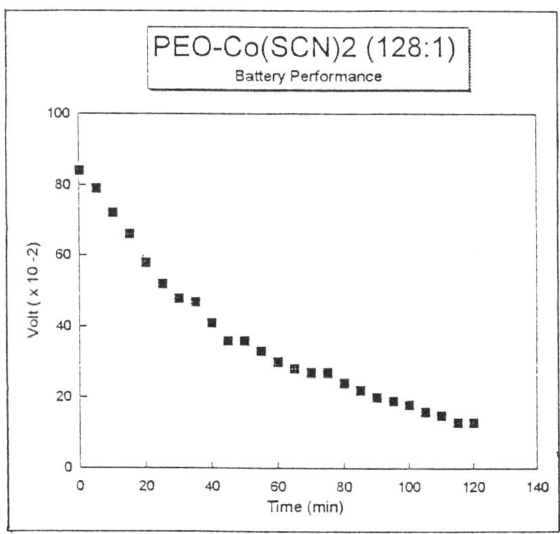

Figure 10. Discharge curve of PEO-Co(SCN)2 (128:1)

CONCLUSION

PEO-$Co(SCN)_2$ system does shows to be a comparable polymer electrolyte material, giving conductivity value in the order of 10^{-6} Scm^{-1}. It was also further shown that this particular polymer electrolyte material can be made to be a battery, even though the battery performance is still far from satisfactory for commercial application.

REFERENCES

1. M.B.Armand , *Solid State Ionics* , 69 , 320-335, (1994)
2. W.R.Salaneck, *Contemporary Physics*, 30, 403-416, (1989)
3. J.R.Owen , Fast ion conducting polymers, in: “Superionics Solids and Solid Electrolytes”, Academic Press, (1989)
4. Peter Bauerle , *Adv. Mater.* , 12 , 879-885, (1993)
5. J.H Burroughes et al , *Nature* , 347, 539, (1990)

THE OPTICAL PROPERTIES AND THE AC. CONDUCTIVITY OF THE GAMMA-IRRADIATED ABS MATERIAL

M. E. Kassem And M. R. Issa

Department of Physics , Faculty of Science ,
Qatar University , P.O.Box 2713 , Doha,Qatar

ABSTRACT

The analysis of the optical absorption determined by Shimadzu Spectrophotometer in the wavelength range 200-1100 nm is presented for non-radiated and gamma-irradiated Acrylate-Butadiene-Styrene(ABS) material. In addition the a.c. conductivity in the frequency range 0.1 - 100kHz is studied for the same material . Detailed calculations of the optical energy and the type of transition of energy absorptions are presented . The AC. conductivity for the material follows the power-law formula :

$\sigma(\omega) \sim \omega^{S}$, with (s) varying between 0.94 to 0.98 .

INTRODUCTION

Acrylate- butadiene -styrene (ABS) is of special importance because of its wide practical applications in thermoelastic fields and electronics. This material is a good candidate for pyroelectric applications [1]. Dandurand et al. [2] observed that the addition of anti-static and processing were responsible for the modification of depolarization. In ABS samples, the low temperature mode is characteristic of the structure of acrylic domains. The major use for ABS is in pipe fitting, particularly in draining waste and in installations in buildings. It is also used in automotive parts and various appliances, including refrigerator door liners, computer and telephone housings, electrical conduits, and for shielding electromagnetic radio frequencies.

The structure of ABS is not that of a random terpolymer. It can be considered as a blend of glassy copolymer (styrene-arcylate) and rubbery domains (primarily a butadiene polymer). The best impact strength of ABS is obtained when styrene-acrylate copolymer matrix is grafted to rubber domains to produce a two phase structure [3].

Radiation effect influences the structure of the polymers and hence their physical properties. Studies of the effect of radiation on the electrical conductivity of polymers indicate an enhancement of conductivity, owing to changes brought about in the polymer

Polymers and Other Advanced Materials: Emerging Technologies and Business Opportunities. Edited by P. N. Prasad *et al.*, Plenum Press, New York, 1995

structure as a result of the effect of radiation [4,5,6]. Research into the effect of irradiation on the optical properties and A.C. conductivity are rare. This paper seeks to enhance understanding of the inner structure of ABS material.

EXPERIMENTAL

Materials and Sample Preparations

The material used in this study is a Novadur ABS-1PV product of Bayer AG Company, Germany. The samples were compressed and moulded into disc-like specimens of 2.5 centimetres diametre. The discs were of varying thicknesses, ranging from 0.02 to 0.2 cm. They were prepared in a compression moulding machine at 30 Mpa and at 165°C for 20 minutes. The disc-like specimens were γ-irradiated at room temperature in air, using a ^{60}Co gamma cell 220 manufactured by the Atomic Energy Agency of Canada. The exposure dose rate was 0.9 Mrad and a set of samples, covering integrated doses ranging from 1 to 200 Mrad, was prepared.

RESULTS AND DISCUSSION

The Optical Properties

The UV-160 Shimadzu spectrophotometer in the wavelength range 200-1100 nm was used for the absorption measurements. The instrument provides direct measurements of the absorbance (optical density).

The Optical Absorption and the Energy Gap

Exponential Absorption

At photon energies below E_g (energy gap), the absorption, in almost all materials, varies exponentially with energy as follows[7]:

$$\alpha(\hbar\omega) = \text{const.} \exp((E_e - \hbar\omega)/\Delta E) \qquad (1)$$

where E_e is the exciton energy of the order of E_g. ΔE is the tail of the densities of states. It is also linked to the average electric field between the centres. The lack of long range order inherent in the amorphous solids results in conduction and valence bands' tails. The variation in the width of the tail depends on positional and compositional disorder [8].

Figures (1.a,b) show the spectrum of the absorption coefficient at different radiation doses. The increase in radiation doses reduces the absorption up to 5 Mrad dose, whereafter the material is saturated and irradiation ceases to affect the absorption coefficient.

Figures (2.a,b) present exponential behaviour. This absorption behaviour in the lower energy region can be explained in terms of the broadening of the width of the tail as a result of strong electron-phonon interaction [9]. The least squares fit for the linear part gives the tail's width of the densities of states. These are presented in Table 1. The γ-interaction with ABS material is reflected in the changes of the absorption edge and the energy tail. The breaking of the bonds is most effective at the lower dose of irradiation (1 Mrad), while the strengthening of the bonds takes place in doses of 2 Mrad and above.

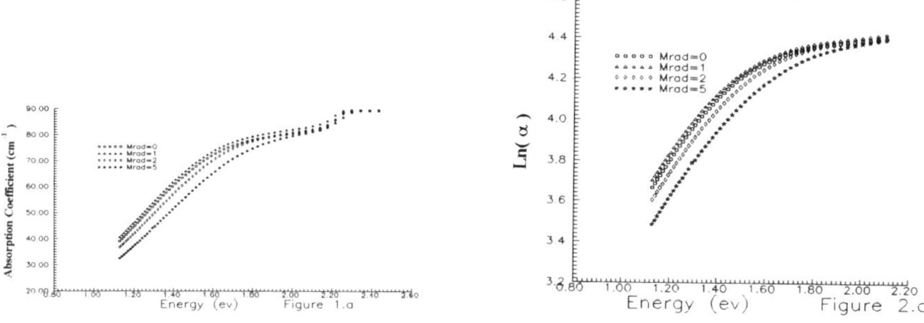

Figure 1 The absorption coefficient versus energy in ev at different radiation doses: (a) at (0, 1, 2, 5 Mrad); (b) at (10, 20, 100, 200 Mrad)

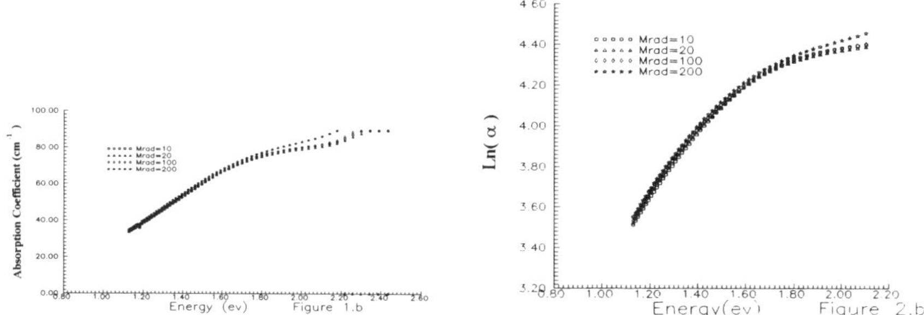

Figure 2 ln (α) versus energy in ev. The dose and captions are as in Figure 1

Table 1 : The Energy Gap and the Width of the Tail at Different γ-Doses

Dose (Mrad)	ΔE (ev)	E_g (ev)	E_{edge} (λ nm)
0	0.534	1.195	545
1	0.507	1.195	555
2	0.563	1.23	545
5	0.630	1.285	535
10	0.608	1.277	535
20	0.602	1.265	535
100	0.598	1.273	545
200	0.585	1.268	570

The Power Law Absorption Region

The densities of states just beyond the mobility edges are represented by power laws [7] as:

$$g_i(-E) \sim E^p \qquad (2.a)$$

$$g_f(E) \sim (E - E_g)^q \qquad (2.b)$$

where energies are measured from the valence and mobility edge. The conduction band mobility edge is at (E_g). The absorption coefficient takes the form References 7,8,9 :

$$\alpha \hbar \omega \quad = \text{const.} \, (\hbar \omega - E_g)^{p+q+1} \quad = B \, (\hbar \omega - E_g)^n \qquad (3)$$

The index (n) depends on the type of transition. The absorption coefficient in equation (3) was tested for several values of (n). The best fit was taken when (n = 1/2), suggesting that the transition is direct and allowed [8,10] The optical energy gap E_g calculated from Figure (3a,b) for different irradiation doses are listed in Table 1. The irradiation widened the energy gap relatively slowly.

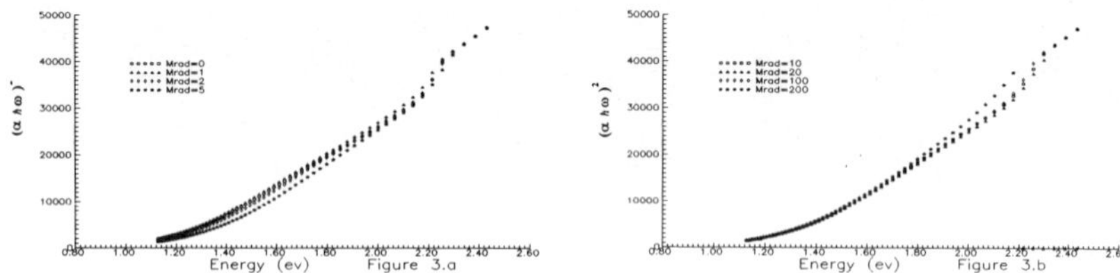

Figure 3 $(\alpha\hbar\omega)^2$ versus energy in ev. The dose and captions are as in Figure 1

The A.C. Conductivity

For electrical measurements cylindrical discs 32 mm in diameter were used. The measurements were performed in the frequency range 0.1 - 100 kHz ,using Hoiki- 3520 LCR tester at room temperature. The impedance $|Z|$, the phase angle (ϕ), the capacitance C, and the loss factor tan δ data were read directly from the impedance meter.

The electrical conductivity was calculated. The results are presented in Figures (4a,b) at different γ-irradiation doses. The shape of the conductivity variations with frequency conforms to the phonon-assisted hopping between localized states mechanism. In this mechanism the conductivity [9, 11] follows the form:

$$\sigma(\omega) \sim \omega^s \qquad (4)$$

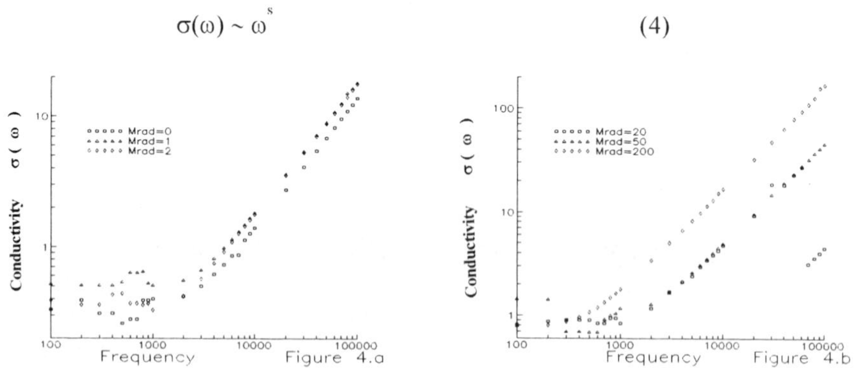

Figure 4 A.C. conductivity versus frequency at different radiation doses: (a) at (0, 1, 2 Mrad); (b) at (20, 50, 200 Mrad)

The log-log scale demonstrates the above relation and the index (s) at different doses is presented in Table 2. The variations of the index (s) at different γ-doses are slight. This is in agreement with the variations of the width of the energy tail (ΔE), which is related to the index (s) as follows [9]:

$$1 - s = 6kT/\Delta E \tag{5}$$

where k is the Boltzman constant and T is the absolute temperature. In general the effect of radiation tends to increase the conductivity of the ABS material over a range of decades of frequencies.

Table 2: The Variation of index (s) with Dose

Dose	0	1	2	20	50	200
Index (s)	0.97	0.98	0.99	0.91	0.94	0.99

CONCLUSION

The optical absorption of ABS follows the behaviour of multi-component materials. The average energy gap is (1.2 ev) the width of the energy tail $\Delta E \sim 0.58$ ev. The conduction mechanism conforms to the phonon - assisted hopping model with an average index of (0.95). The radiation enhances the conductivity of ABS material over the frequency range considered.

REFERENCES

1. A. Burns, D. Chatain and C. Lacabanne, 3rd ICPADM, Tokyo, Japan, July 8 - 12th, (1991).
2. J. Dandurand, C. Lacabanne and J.P. Baulet, Proceedings of the 3rd. International Conference on Properties and Applications of Dielectric Materials 1087, 8 - 12 July (1991), Tokyo, Japan.
3. M. Matsuo, Polymer Engineering Science, 9, 206, (1969).
4. M.E. Kassem, A.Y. Kandeil and L. Al-Houty, Journal of Polymer Materials, 9, 113, (1992).
5. M.E. Kassem, A.A. Higazy, R.R. Zahran and A.Y. Kandeil, Materials Letters, 16, 236, (1993).
6. A.Y. Kandeil and M.E. Kassem, Thermochemica Acta, 191, 333, (1991).
7. G.A. Connel in 'Optical Properties of Amorphous Semi-conductors', edited by M.H. Brodsky, pub. Springer-Verlag, (1988), p. 73.
8. A. Madan and M.P. Shaw in 'The Physics and Applications of Amorphous Semi-conductors', pub. Academic Press Inc., (1990), p. 36
9. N.F. Mott and E.A. Davis in 'Electronic Processes in Non-crystalline Materials', pub. Clarendon Press, (1978), p.198.
10. M.A. Angadi and V. Thanigaimani, Phys. State Sol. (a), 143, K 119, (1994).

PROCESSIBLE AROMATIC POLYAMIDES DERIVED FROM 2,5-BIS(4-AMINOPHENYL)-3,4-DIPHENYL THIOPHENE AND AROMATIC DIACID CHLORIDES

Won-Kyu Lee,[1] Kwang-Sup Lee[1], Hyun Hoon Song[1] and Soo-Min Lee[2]

[1]Department of Macromolecular Science
[2]Department of Chemistry
Han Nam University
Taejon 300-791, Korea

INTRODUCTION

Rigid polymers have been of great interest in recent years because of their outstanding mechanical properties and potential applications for the high performance materials. The rigid polymers, however, are almost nonprocessible and structural modifications to improve processibility have been the subject of investigation thereafter. Attempts were made by introducing the flexible linkages or nonlinear aromatic moieties in the polymer main chain[1-7] and by coupling of flexible side chains.[8-13] Attachment of flexible alkoxy side chains to the several rigid polymers were proven to be successful to lower the transition temperatures and to enhance the solubility of polymers. The flexible side chains apparently act as a bound solvent, also yielding a layered mesophase of the polymer chains. The layered mesophase and systematic changes with side chain length also have been of interest in these rigid polymers. In this work we prepared polyamides having bulky rigid moiety in the main chain and alkoxy side chains. We report the synthesis and properties as well as molecular packing behavior associated with the side chains.

MONOMER AND POLYMER SYNTHESIS

Monomer synthesis: 2,5-Bis(4-aminophenyl)-3,4-diphenylthiophene (BAPT) were synthesized by nitration and reduction of 2,3,4,5,-tetraphenylthiophene as reported in literatures.[15,16] 2,5-Dialkoxyterephthaloyl chlorides (DATC) were also prepared by saponification and chlorination after Claisen condensation reaction of diethyl 2,5-dihydroxyterephthalate with alkyl bromide with different carbon numbers.[9]

Polymer synthesis: All polymers were synthesized by the low temperature interfacial polycondensation process. A typical polymerization method of PA series of

BAPT and DATC is as following. In a reactor equipped with mechanical stirrer were placed 4.104g (0.01mol) of BAPT and 1.06g of sodium carbonate in 75ml of H_2O at 0℃. To this was slowly added 0.01mol of DATC in 120ml of methylene chloride at the same temperature, and the mixture was stirred for 30min. The contents were poured into 500ml of methanol. The precipitated polymers were collected, washed thoroughly with water, methanol, and dried under vacuum at 100℃ for 2 days. The yields of polymer PA series were between 79% and 94%.

MEASUREMENT

A Shimadzu IR-435IR spectrometer was employed for IR analysis. NMR spectra were obtained with a Bruker AM 300 (300MHz) NMR spectrometer. Thermo-gravimatric analyzer (Model 9900) of Du pont was used for the study of thermal stability of polyamides. X-ray scattering intensities were collected using a Picker diffractometer and a Rigaku rotating anode X-ray generator (RU200) (CuKα) operating at 45kV and 70mA. The X-ray beam was collimated by the two pin holes and monochromated by a graphite monochromator. All polymer samples were in powder form and the intensities were measured by the transmission technique at room temperature.

POLYMER CHRACTERIZATION AND PROPERTIES

A series of polyamides containing tetraphenyl thiophene groups and alkoxy side chains were synthesized by the low temperature interfacial polycondensation of BAPT and DATC

H2N S NH2 + OCnH2n+1 ClOC COCl OCnH2n+1

Na2CO3

CH2Cl2 / H2O

S N H N H C O OCnH2n+1 C O OCnH2n+1

n= 8(PA-8), 10(PA-10), 12(PA-12), and 14(PA-14)

The polymer structure was identified by the IR and NMR spectroscopy. Relatively strong absorption peak of secondary amine groups resulting from polymerization appeared at 3300cm^{-1}, CH-streching vibration of alkoxy side chain at 2850-2950cm^{-1}and a carbonyl group of amide linkage at 1660cm^{-1}. The peaks at 0.9-2.3ppm and 4.3ppm in the ^{1}H-NMR spectra were attributed to the alkoxy side chains. Aromatic proton substituted to the thiophene ring exhibited signals at 7.0-7.5ppm and amine proton showed signal at 9.8ppm.

As summarized Table 1, satisfactory C/H/N ratios by elemental analysis were obtained for all polymers. The yields of PA series were 85-94% after purification. The inherent viscosities of the polymers ranged 0.62-0.86dl/g. Solubility of the polymers is increased with the increase of alkoxy side chain lengths. This phenomenon is due to bound solvent effect of side chains. All of the PA series were soluble in phenol/tetrachloroethane (6:4 wt. %) and chloroform/trifluroacetic acid (20:1 vol. %) mixture. Polymers with longer side chain lengths, PA-12 and PA-14, were dissolved in common organic solvents such as DMAc, DMSO and m-cresol.

Table 1. Elemental analysis data, yields and inherent viscosities(η_{inh}) of polyamids with various side chain lengths

Polymers	C(%)		H(%)		N(%)		Yield	η_{inh}[a)]
	Cal.	Found	Cal.	Found	Cal.	Found	(%)	(dℓ/g)
PA-8	77.58	76.90	7.01	7.37	3.48	3.65	85	0.75
PA-10	78.10	77.91	7.49	7.64	3.25	3.68	88	0.86
PA-12	78.56	78.02	7.91	8.06	3.05	3.50	94	0.62
PA-14	78.97	79.24	8.28	8.63	2.88	3.37	90	0.79

a) Measured in chloroform/trifluroacetic acid (20:1 vol.%) at 25℃(0.5g/dℓ)

Figure 1 shows TGA curves of the polymer samples. Initial decomposition appeared in the range of 297-335℃ in nitrogen atmosphere. A significant decrease of the char yield at 600℃ can be noted with the increase of the alkoxy side chain length attached to the polymer backbone.

MOLECULAR PACKING

Wide angle X-ray patterns of the polymers with 4 different side chain lengths (n=8, 10, 12 and 14) are compared in Figure 2. All four patterns are very similar to each other, showing a strong small angle reflection near $2\theta=5^\circ$ and several weak but discernible peaks as well as a broad halo centered near $2\theta=20^\circ$ (d=4.5Å). The strong small angle reflection changes in the peak position as the side chain length varies, indicative of the layered structure where the interdigited side chains determine the layer spacing.[9-13] D-spacings calculated are listed in Table 2 and plotted against the side chain length (number of CH_2 units) in Figure 3. Extrapolating the linear fit

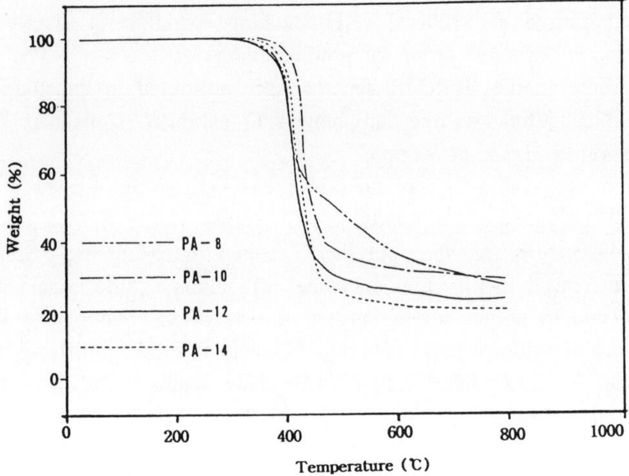

Figure 1. TGA curves of several rigid polyamides with flexible side chains

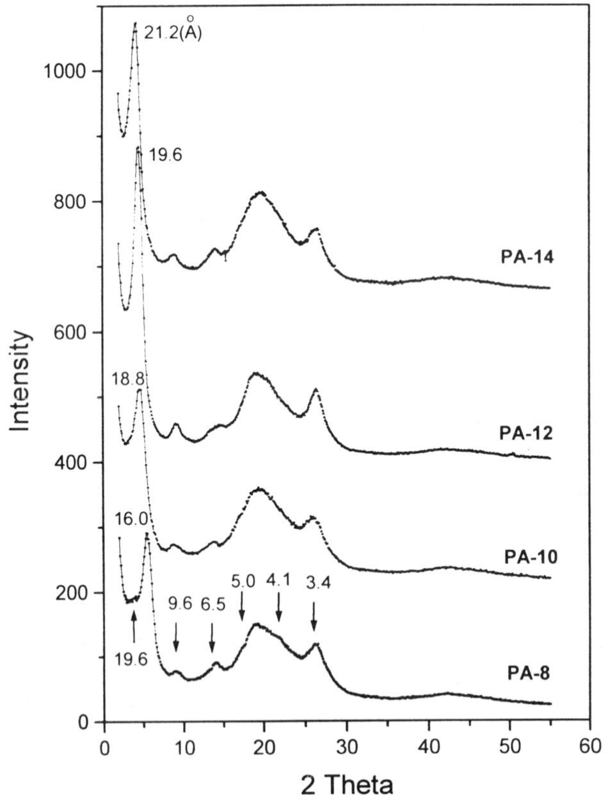

Figure 2. Wide angle X-ray patterns of polyamides at room temperature

Table 2. D-spacings of the peaks in the X-ray diffraction pattern. The first column denotes layer spacings.

Polymers	D-spacing (Å)					
PA-8	16.0	9.6	6.5	5.0	4.1	3.4
PA-10	18.8	9.7	6.4	5.0	4.1	3.4
PA-12	19.6	9.7	6.5	5.1	4.1	3.4
PA-14	21.2	9.7	6.4	5.0	4.1	3.4

of the plot to n=0 results in 9.86Å, representing the mean diameter of the polymer backbone. The slope of the curve, i.e. layer spacing increment per CH_2 unit, is 0.82 Å. The value is somewhat smaller than the length of CH_2 (1.27Å), suggesting the tilted layer structure, as depicted in Figure 4, and/or the alkyl chains are not fully extended.

D-spacings of other relatively weak reflections, listed in the Table 2, remain unchanged regardless of the side chain length. The results, therefore, eliminate the possibility that the peaks are higher order reflections of the layer spacing. Sequence of the peak positions, however, implies that the peaks could be the multiple orders of a single diffraction unit, more than 10Å apart. The structural origin then could be found from the intra-chain order along the stiff polymer backbone which is likely to be in an extended conformation. A cylindrically averaged scattering function is calculated from the fully extended single chain of 5 repeating units of trans conformation. The intensity profile along the chain axis with the X-ray beam perpendicular to the chain axis is shown in Figure 5. The first two peaks appear to be very intense but the intensity is in fact very low, when compared to the intensities from the inter-chain correlation. Agreement with the X-ray data in the peak positions can be noted with an exception at the first peak. The first peak of d=20.0Å apparently associated with the repeat distance of the backbone, is not seen in our X-ray pattern. This is probably due to the fact that the strong reflection of layer spacing locates nearly at the same diffraction angle. The polymer PA-8 having the shortest layer spacing (16Å), however, shows the peak as a weak hump near d=20.0 Å as indicated by an arrow in Figure 2.

It is questionable that the peak at d=3.4Å belongs to one of those diffraction peaks of the single chain. The peak is well defined and rather strong in intensity compared to other peaks. Similar peak (d=3.4-3.5Å), in fact, has often been observed in polyamides with long side chains.[17,18] Such a short spacing could be identified with the correlation of main chains by the hydrogen bonding within the layers, as suggested by others.[17,18] The bulky non-planar structure (note torsion angle (~80°) of the two pendent phenyl groups in Figure 6), also strongly suggests the tilted layers. The tilt angle estimated from the layer spacing increment of 0.82Å per CH_2 unit (assuming the extended side chain) is ~40°, yielding distance of ~4.5Å between the main chains within layer.

The scattered intensity from the intra-chain order along the backbone is usually too weak to be observed. Observing the multiple orders of (00l) peaks in the X-ray

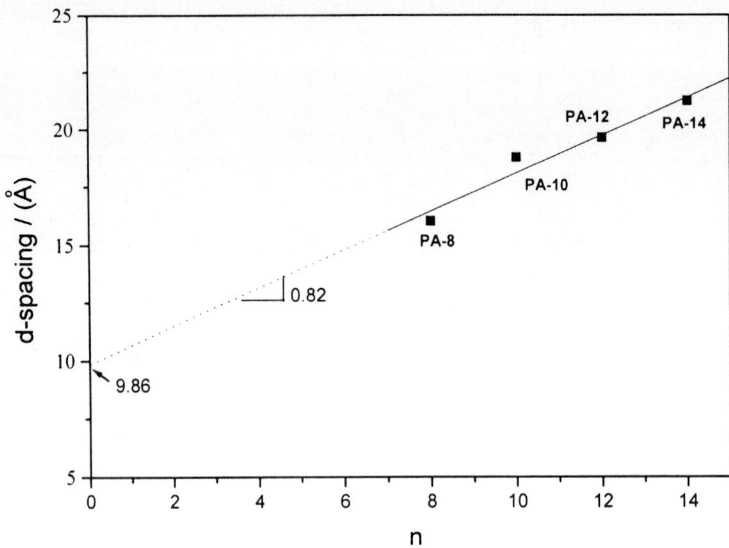

Figure 3. Plot of layer spacings vs. number of carbon atoms in the alkoxy side chain.

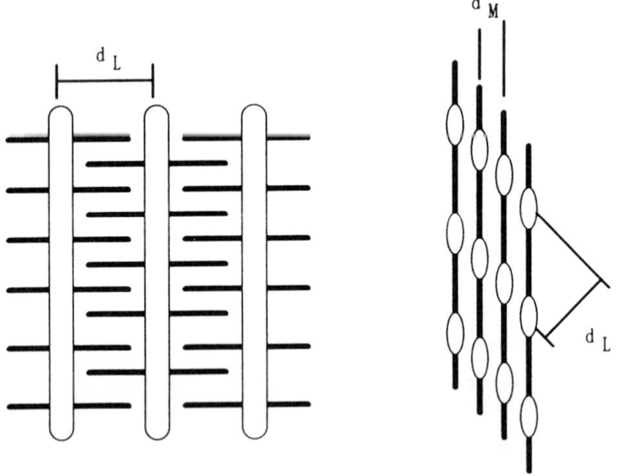

Figure 4. Schematics of layered structure

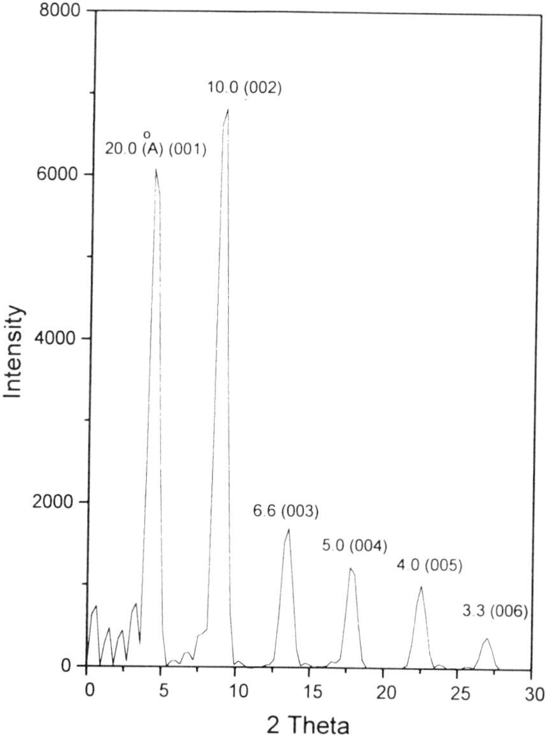

Figure 5. Intensity profile along the chain direction, obtained from the cylindrically averaged scattering function of an extended chain 5 repeating units of PA-8.

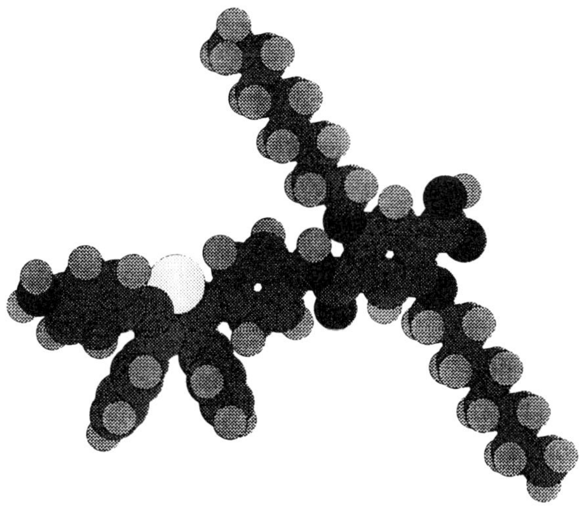

Figure 6. Energy minimized chemical structure of polyamide monomer unit for PA-8.

diffraction patterns, therefore, demonatrates a long coherence length and thus well extended chain conformation. The sharp and intense reflection from the layer spacing demonstrates a good packing order between the layers as well. The origin of the amorphous halo then could be found from the structural units within the layers. Non-crystalline packing of the alkoxy side chains within the layers is probably the main cause for the amorphous halo centered near 20° (d=4.5Å), recalling that the average distance between the main chains is 4.5Å. Low packing order between the non-planar backbones within the layers as well as correlation between the phenyl group and/or hetero-cycles could also make contributions to the intensity of the amorphous halo.

In summary, X-ray diffraction patterns of the polyamides synthesized in our laboratory demonstrate the polymers form the layer structure, confirming that the layer structure is a general phenomenon common to stiff polymers with long flexible side chains. The layer spacing increment per CH_2 unit (0.82Å) and the short packing distance (0.34Å) between the main chains within the layer suggest the tilted layer structure. The several weak reflections are from the structural order along the polymer backbone, suggesting that the chains are well extended. The alkoxy side chains also appear to be extended, but exhibit non-crystalline packing order within the layers, causing the amorphous halo.

ACKNOWLEDGEMENTS

The authors express thanks to the Basic Science Institute, Han Nam University for the finacial support.

REFERENCES

1. K.B. Seymour and G.S. Kirshenbaum, ed., "High Performance Polymers: Their Origin and Development", Elsevier, NY, 1986.
2. B.P. Griffin and M.K. Cox, *Brit. Polym. J.* 12:147(1980).
3. S.M. Lee, K.-S. Kim, K.-S. Lee, and S.-K. Lee, *Polymer(Korea)* 13:888(1989).
4. J.M. Rhee, K.-S. Lee, and K.-Y. Choi, *Polymer(Korea)* 14:448(1990).
5. M.G. Dobb and J.E. McIntyre, *Adv. Polym. Sci.* 60:63(1984).
6. H.-J. Jeong, A. Kobayashi, M. Kakimoto, and Y. Imai, *Polymer J.* 26:373(1994).
7. K.-S. Kim, S.M. Lee, K.-C. Ryu, and K.-S. Lee, *Polym. Bull.* in press.
8. J. Majnusz, J.M. Catala, and R.W. Lenz, *Eur. Polym. J.* 19:1043(1983).
9. M. Ballauff, *Makromol. Chem., Rapid Commun.* 7:407(1986).
10. M. Ballauff, *Angew. Chem., Int. Ed. Engl.* 28:253(1989).
11. R. Stern, M. Ballauff, and G. Wegner, *Makromol. Chem., Makromol. Symp.* 23:373(1989).
12. K.-S. Lee, J.C. Won, and J.C. Jung, *Makromol. Chem.* 190:1547(1989).
13. K.-S. Lee, H.M. Kim, J.M. Lee, and S.M. Lee, *Makromol. Chem.* 192:1033(1991).
14. K.-S. Lee, W.K. Lee, S.M. Lee, H.O. Kim, and B.W. Lee, *Mol. Cryst. Liq. Cryst.* 254:37(1994).
15. M.G. Voronkov, V.E. Udel, *Khim. Getewtsikl. Soedin. Akad. Nauk. Latu.S.S.R.*, 1965, 683; Chem. Abstr. 64:11148h(1966)..
16. W. Dilthey and E. Graef, *J. Prakt. Chem.* 151:275(1938).
17. M. Ballauff and G.F. Schmidt, *Makromol. Chem., Rapid Commun.* 8:93(1987).
18. A. Adam and H.W. Spiess, *Makromol. Chem., Rapid Cummun.* 11:249(1990).

POLYMER DISPERSED LIQUID CRYSTALS FOR ELECTRONIC APPLICATIONS

J.C. Dubois, L. Bouteiller, P. Le Barny, P. Robin

LCR, Thomson-CSF,
Domaine de Corbeville,
91404 Orsay Cedex, France

The electrooptic characteristics of liquid crystal-polymer composites (PDLC, PNLC, NCAP...) are strongly dependent on surface interactions between polymer and liquid crystal molecules. In order to clarify this dependence, we have investigated the chemical modification of the surface of the polymer, without affecting its bulk properties. We show that it is possible to greatly modify the electrooptic performances of composite cells, and thus to optimise them for an application in visualisation.

INTRODUCTION

Liquid crystal-polymer composites (called PDLC, PNLC or NCAP according to their structure or their method of preparation), form a class of materials with many potential applications[1], and are under development for projection television[2] and certain types of direct view displays[3]. Their advantages over existing technologies are :

- simple manufacturing process,
- no alignment layers,
- no polarizers,
- large viewing angle.

However, the behavior of these composites is very complex, and the influence of all parameters is not completely understood. In particular, the chemical nature of the polymer strongly influences the electrooptic characteristics of liquid crystal-polymer composites. Indeed, if the same liquid crystal mixture is used, the characteristics of the composite depend on :

- optical properties[4] of the polymer (refractive index, or indices in the case of liquid crystalline polymer),
- dielectric properties[5] of the polymer (resistivity, dielectric constant),
- thermodynamical properties[6] of the polymer (size and shape of liquid cristal domains in the composite is governed by the phase separation process),
- surface interactions[7] between the polymer and liquid crystal.

In particular, the effect of the last point is difficult to assess, because it is hard to make two identical composite cells which differ only by the nature of the surface of the polymer. On the other hand, theoretical studies have been made on models (alumina membranes[8]), but these models are too different from the liquid crystal-polymer composite structure to give a direct relationship between the electrooptical properties and the nature of the surface.

Polymers and Other Advanced Materials: Emerging Technologies and Business Opportunities. Edited by P. N. Prasad *et al.*, Plenum Press, New York, 1995

In this paper, we describe a method to selectively modify the surface of the polymer after the composite has been formed, and we present the preliminary results showing the strong influence of the chemical nature of the polymer surface on electrooptic properties.

The principle of our method (shown on scheme 1) is to remove the liquid crystal from a liquid crystal-polymer composite cell, and

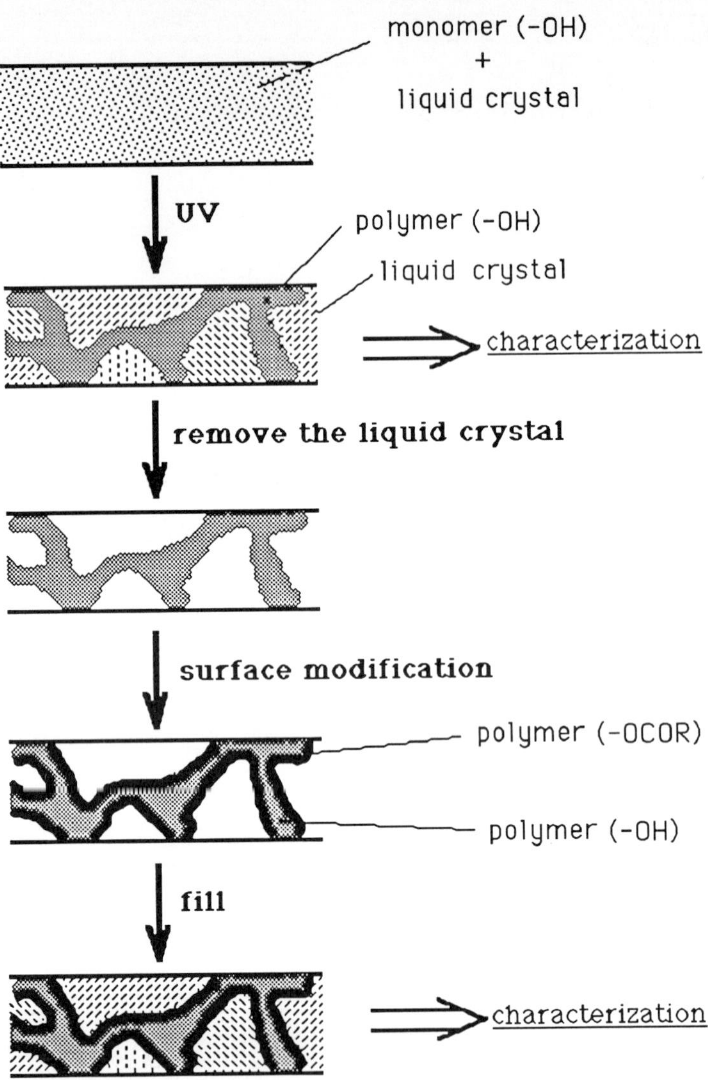

Scheme 1. Principle of our method.

then to chemically modify the surface of the exposed polymer network, without altering the bulk of the polymer. The differences between the electrooptic properties of the cell before and after the modification show what the influence of surface interactions really is. This would not be possible if we simply compared two cells made from two different polymers, because their phase separation processes would be different, their resistivity and refractive indices would not be the same...

EXPERIMENTAL

Realisation of a liquid crystal-polymer composite cell

The cell is made from two ITO coated glass plates, covered by a 0,6 μm thick layer of polyvinyl alcohol, to prevent gazeous HCl (which is a side product of the chemical modification) from reacting on ITO. These two plates are separated by mylar spacers and fixed together, so that the cell thickness is 25 μm.

The monomer mixture (KHEA) is composed of 20% HydroxyEthyl Acrylate (Aldrich) and 80% acrylate mixture Kayarad HX620 (Nippon Kayaku), to which 2,5% of photoinitiator Darocure 1173 (Merck) is added.

A 20% mixture of this monomer in the liquid crystal BL012 (Merck BDH) is introduced in a cell, and photopolymerized (Hg lamp, $9 mW/cm^2$ at 365 nm, for 17 minutes, at 40,0°C) so that a porous polymer network in a continuous liquid cristal phase is obtained (PNLC).

RESULTS AND DISCUSSION

Description of the method

Step 1 : polymerization. The composite cells are made by photochemical polymerization induced phase separation (UV-PIPS), and their preparation has been described previously[9].

The formation of the composite is a very complex phenomenon, even if it is technically a simple step, because each component of the prepolymer mixture influences the phase separation process through its solubility and its polymerization kinetics. Indeed, our monomer mixture, is composed of 20% HydroxyEthyl Acrylate (HEA) and 80% Kayarad HX620 (acrylate mixture from Nippon Kayaku), but both the infrared spectrum and the elementary analysis (performed on the polymer recovered from a composite cell) show that the composition of the copolymer is in fact 25% HEA and 75% Kayarad. Of course, the unpolymerized Kayarad remains in the liquid crystal.

Step 2 : removing the liquid crystal. The liquid crystal is removed from the composite by immersing the cell in a solvent. The slow diffusion of the solvent inside the cell, followed by the dissolution of the liquid crystal yields a cell containing only the polymer network and the solvent.

The solvent must meet three requirements : first it must be volatile, so that later, it may be easily removed. Secondly, it must be a good solvent of the liquid crystal : if not the process is very slow. (In the case of the liquid crystal BL012 (Merck), ethanol and (trichlorotrifluoro)ethane are poor solvents, and it takes a week to wash a 1 cm^2 cell instead of a night with better solvents). Thirdly, the solvent should not swell the polymer network too much, or else the polymer can be damaged. (In the case of BL012, if we use dichloromethane or tetrahydrofuran we can see that lumps of polymer are torn away.)

We selected 6 solvents meeting these conditions, and we observed their effect on the electrooptical properties of composite cells. To do this, we simply evaporated the solvent and then filled the cell with fresh liquid crystal.

Every solvent yields a composite whose electrooptical curve is shifted toward higher voltages. Moreover, we can see on figure 1 that if we reproduce the same cycle on the same cell, the electrooptical curve remains essentially the same as it was after the first cycle.

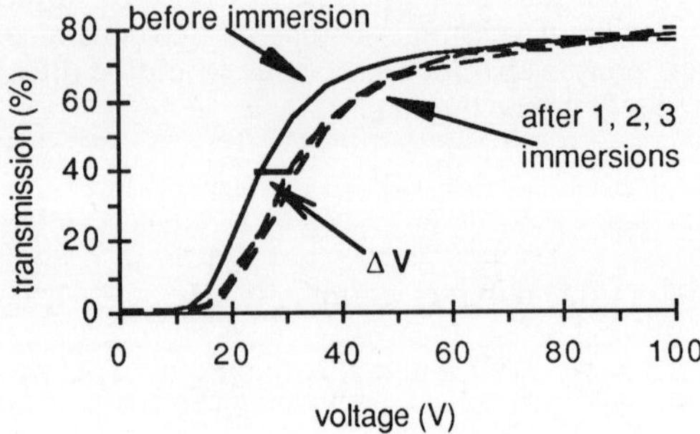

Figure 1 . Electrooptical curves of a cell before and after 1, 2, or 3 immersions in ether.

The value ΔV measuring the shift of the curves depends on the nature of the solvent : figure 2 shows this value as a function of the swelling ratio[10] of the polymer in the corresponding solvent. We can see that the more the polymer is swollen by the solvent, the higher is the shift.

This shift toward higher voltages may be due to the dissolution of short oligomers, which may modify the roughness of the surface of the polymer.

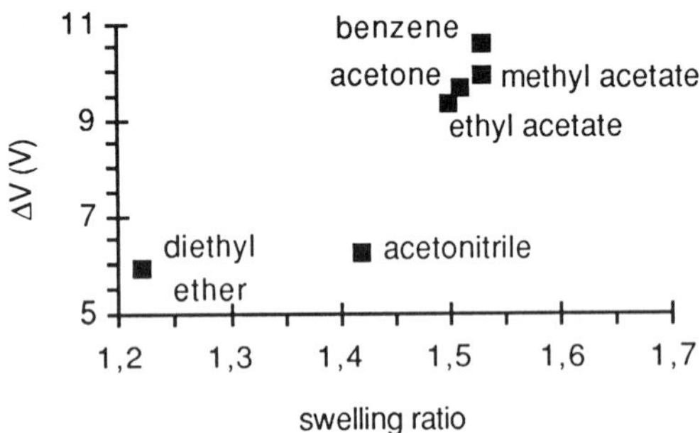

Figure 2 . Shift of the electrooptical curve due to the immersion in a solvent, versus the swelling ratio of the polymer in the same solvent.

Diethyl ether was chosen for the rest of this study as it gives the smallest shift and has the lowest boiling point.

Step 3 : drying the polymer. We reported in a previous paper[9] a technique which consisted in simply evaporating the solvent, and then filling the cell with fresh liquid crystal. This technique does work, but since then, we have realized that during evaporation of the solvent, the polymer network crumbles, and that it is swollen back by the fresh liquid crystal. This is shown by the following experiment.

Figure 3. SEM photograph showing the contrast between the polymer of the composite and the adhesive.

Six composite cells were immersed in ether, dried, and then filled with a UV-curable adhesive (Norland NOA 60). Each cell was stored in the dark for a different time, and then irradiated so as to cure the adhesive. The cells were then broken, and observed with a scanning electron microscope (SEM). Figure 3 shows a photograph of such a cell where we can see the polymer of the composite, because when the cell is broken, the hard adhesive yields a flat surface, whereas the soft polymer yields an irregular surface.

Thus, we can measure the width of the layer occupied by the polymer. Figure 4 shows the ratio of the width of this layer to the cell gap. We can see that it increases from 13% to 100% : the polymer is completely swollen by the adhesive, and we believe the same phenomenon occurs when we fill the cell with liquid crystal.

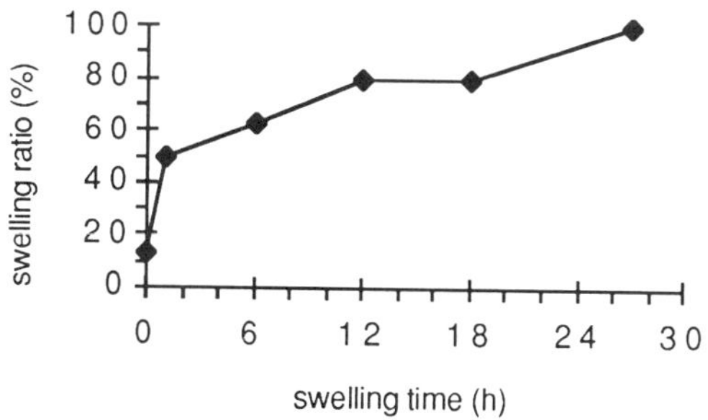

Figure 4 . Width of the layer occupied by the polymer (in percent of the cell gap), versus time of storage before curing.

Indeed, figure 5 shows the transmission of a composite cell that has just been filled with fresh liquid crystal as a function of time. At first the cell is nearly transparent, because it is roughly composed of a layer of polymer and a layer of liquid crystal. But if we wait a few hours, the cell scatters again, because the polymer network has recovered its initial shape.

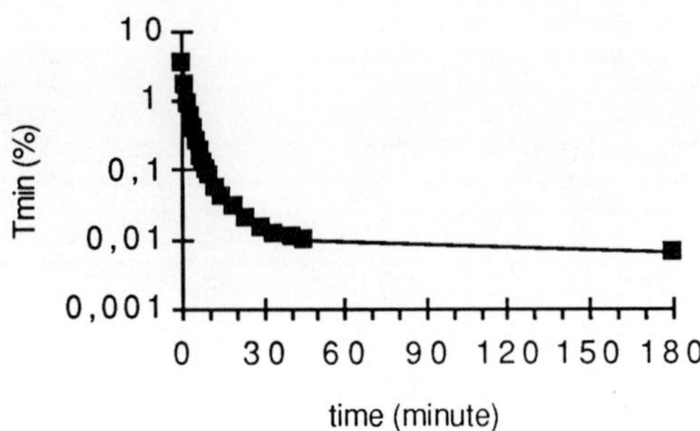

Figure 5 . Transmission of a cell, versus time after filling with liquid crystal.

Thus the polymer recovers its initial shape, but we cannot be sure it will do so after modification of its surface[11]. That is why we have modified our process, so that the polymer does not crumble.

It has been shown in other fields that surface tension of a liquid can be responsible for the collapse of porous material[12], and that this collapse can be avoided by drying the solvent by a supercritical process. Indeed, if the pressure and temperature of the liquid solvent in the porous material are increased above its critical point, and then if the pressure is reduced, all the physical properties of the solvent are continuously changed from those of a liquid to those of a gas. There is no formation of interface, so that no surface tension is applied on the porous polymer.

We have applied this process to our case in the following manner. After the composite cell has been immersed in ether, we put it in a 300ml autoclave together with solid carbon dioxide (150g), and we closed the autoclave. The set up was kept at room temperature a few hours, so that ether was replaced by liquid CO_2, then the temperature was increased up to 40°C, and the pressure (fixed by the amount of CO_2 introduced) went up to 100bar (the critical parameters of CO_2 are Tc=33°C, Pc=73bar). At this point, carbon dioxide was slowly vented out (at 40°C), and the autoclave was opened. Figure 6 proves that we obtain a porous polymer network which is not crumbled.

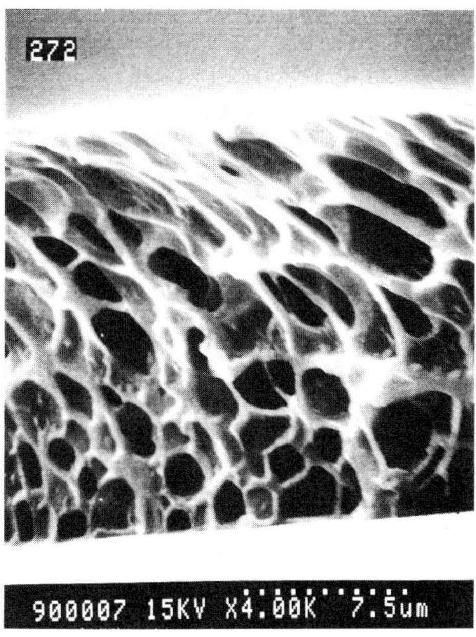

Figure 6. SEM photograph of the polymer network afer supercritical drying.

Step 4 : chemical modification of the polymer surface

Chemical modification of a plain polymer film. The free standing porous polymer network can now be modified. Choosing the nature of the chemical reaction is not so straightforward, because it must fulfill a few criteria : this heterogeneous reaction must give a good yield, in mild conditions (temperature...). It should not "pollute" the polymer with side products (such as ions...). If a solvent is used, it shoud wet but not swell the polymer, because the reaction should be, as much as possible, selective of the surface. Finally, the reaction must not be specific of a particular reagent, so that several kinds of reagents may be used, and may be compared.

After several attempts, we decided to esterify alcohol functions attached to the polymer with a vaporized acid chloride reagent, according to Ogino et al[13]. The reagent we used to prove the feasability of the process was *para*butyl benzoic acid chloride, because its Infra Red spectrum is easily identifiable, thus allowing a kinetic study of the esterification.

$$\text{polymer—OH} + \text{Cl—C(=O)—}C_6H_4\text{—}C_4H_9 \longrightarrow \text{polymer—O—C(=O)—}C_6H_4\text{—}C_4H_9 + HCl\uparrow$$

The experimental conditions of the chemical reaction were tested on a 10µm thick polymer film : it was exposed to the reagent vapor at 60°C, under a pressure of 9mmHg.

The chemical modification of the surface was monitored by measuring the contact angle of a drop of glycerol on the film, after different reaction times. On the same sample, the chemical modification of the bulk was monitored by measuring the intensity of Infra Red absorption through the film at 1604cm^{-1} (FTIR).

The results, shown on figure 7, prove that the reaction proceeds a lot faster on the surface than in the bulk, so that it is possible to modify alcohol functions on the surface before affecting those which are deeper in the film. This can be done by selecting a reaction time short enough.

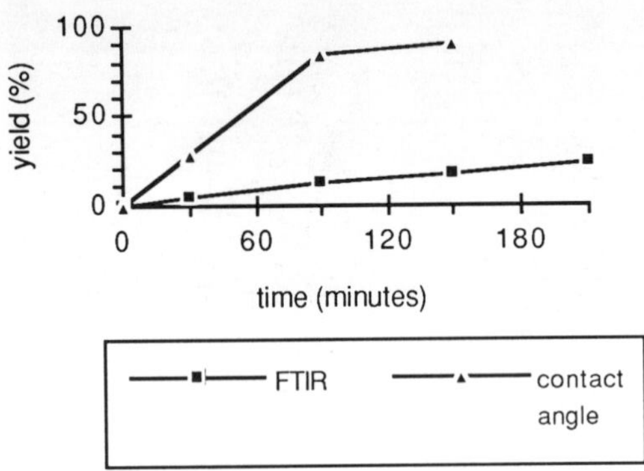

Figure 7. Surface and bulk modification versus reaction time.

Chemical modification of the polymer surface in the composite. The same procedure was applied to composite cells where liquid crystal had been removed by dissolution in ether and evaporation of the solvent.
Figure 8 shows the transmission/voltage curves of 5 different cells, reacted respectively for 1, 5, 5, 7, and 10 minutes.

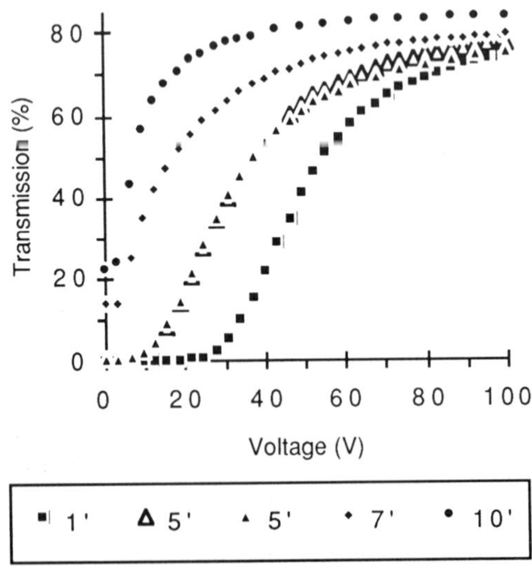

Figure 8 . Electrooptic curves function of reaction time.

First, it can be noticed that the two cells reacted for 5 minutes have very similar electrooptic curves, showing that a good reproducibility is achieved. Secondly, a trend is clear on this figure : the longer the reaction time, the lower the driving voltage (this will be discussed later), and the more transparent the cell at 0 volt. This last effect shows that the polymer network (which had collapsed during evaporation of the solvent) does not swell back for long reaction times.

It proves that simple evaporation of the solvent is not satisfactory. Thus we have used the improved method involving the supercritical drying procedure.

Results of the chemical modification

Modification with an alkyl moiety. Figure 9 shows the electrooptical curves of 3 cells modified for 0, 45, or 90 minutes with heptanoyl chloride ($C_6H_{13}COCl$). We can see that they are shifted toward lower voltages.

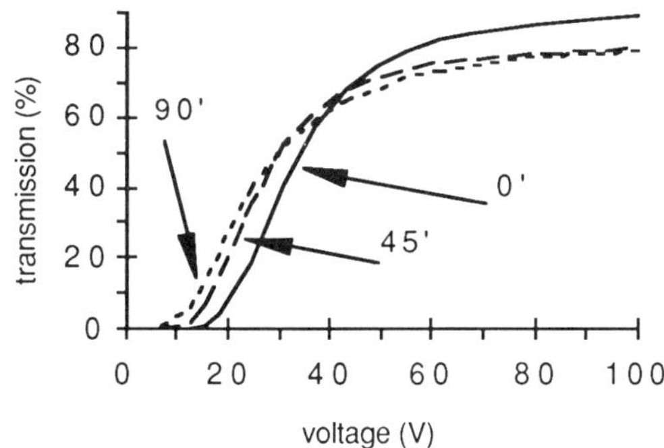

Figure 9. Electrooptical curves of cells modified for 0, 45, or 90 minutes with C6H13COCl (90' corresponds to a modification of the surface of 100%).

Figure 10 shows the effect of the modification on response times : rise time (t_{ON}) is decreased by 2 orders of magnitude, and decay time (t_{OFF}) is slightly increased.

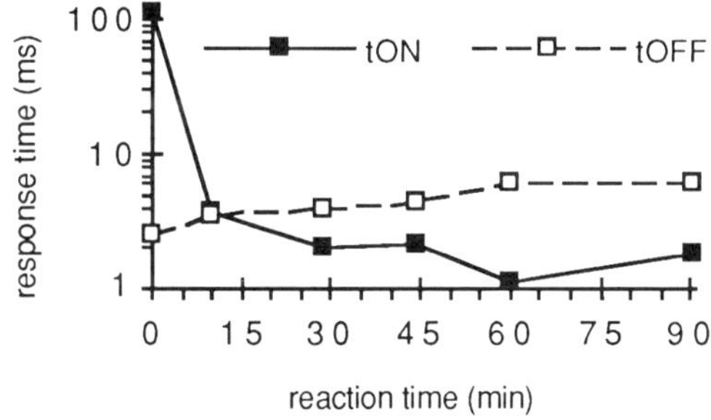

Figure 10 . Response times versus reaction time.

The effect on response times and driving voltage (V_{90}) is consistent with a diminution of polar interactions at the interface. Indeed, it reduces the anchoring energy, so that it is easier to align the liquid crystal parallel to the electric field (V_{90} and t_{ON} are lowered), and it takes longer for the liquid crystal to relax (t_{OFF} is slightly increased).

This reduction of anchoring energy also affects the contrast. Figure 11 shows that the off-state transmission measured before applying an electric field ($Tmin_0$) is independent of the surface modification. On the other hand, we can see that the off-state transmission measured after an electric field has been applied (Tmin) increases when the anchoring energy is reduced. The average director in each droplet does not recover its initial orientation, because the torque exerted by the surface is not as strong as before.

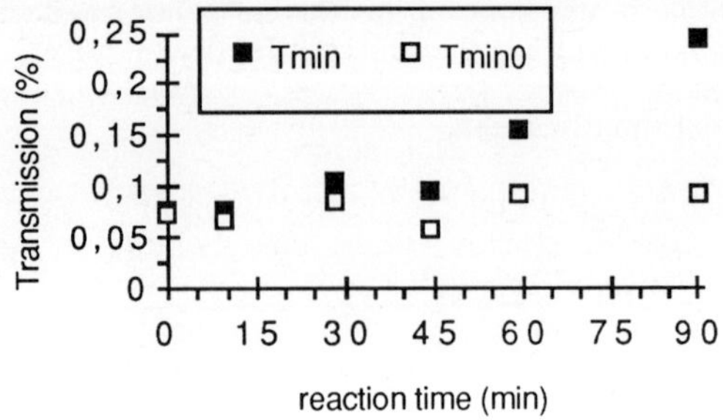

Figure 11. Evolution of the off-state transmission (Tmin) and the off-state transmission measured before applying an electric field ($Tmin_0$), versus reaction time.

Modification with fluoroalkyl moieties. In order to amplify the effects noticed with an alkyl moiety, we modified other cells with the corresponding semi- and perfluorinated alkyl moieties ($C_2F_5C_4H_8COCl$ and $C_6F_{13}COCl$). By reducing the surface energy of the polymer, we expected to further reduce interactions at the interface.

In fact, the modification of plain films (monitored by contact angle goniometry) shows that the reaction works, that there is no subsequent hydrolysis of the ester group, and that there is no rearrangement of the fluorinated chains at the surface. Nevertheless, we do not observe any significant variation of the electrooptical characteristics of composite cells modified by the fluorinated moieties.

A possible explanation for this behavior may be that the strong decrease of surface energy induced by the fluorinated moieties is responsible for a dramatic modification of the director configuration (for instance from bipolar to radial), so that the effect on the electrooptical curves is more complicated. If the reduction of surface interactions is accompanied with a change of director configuration, the two effects might cancel each other.

CONCLUSION

We have presented a method which makes it possible to study the effects of surface interactions on electrooptical characteristics of liquid crystal-polymer composites. With this method, we have shown that a small reduction of the surface energy (alkyl moiety) of the polymer reduces the driving voltage and the rise time. On the other hand, a strong reduction of this surface energy (fluorinated moiety) induces a more complicated phenomenon. This phenomenon is under investigation.

Acknowledgements

This work has been supported by the French government under the SERICS contract.

REFERENCES

1. J.W. Doane, Liquid Crystals Applications and Uses, World Scientific, 1990, ch 14, p361.
2. E. Ginter, E. Lueder, T. Kallfass, S. Huttelmaier, M. Dobler, V. Hochholzer, D. Coates, M. Tillin, P. Nollan, "Optimized PDLCs for active-matrix addressed light valves in projection systems", Proc. 13th IDRC, 1993, p105.
3. H. Yoshida, K. Nakamura, H. Tsuda, M. Ohashi, I. Tomita, M. Okabe, "A direct view polymer-dispersed LCD with crossed Nicols and a uniaxial film", Proc. 13th IDRC, 1993, p121.
4. N.A. Vaz, G.P. Montgomery, "Refractive indices of polymer-dispersed liquid crystal film materials: epoxy based systems", J.Appl.Phys., vol 62(8), p3161, 1987.
5. D. Seekola, J. Kelly, "A comparative study of the dielectric and optical response of PDLC films", SPIE, vol 1455, p19, 1991.
6. Y. Hirai, S. Niiyama, H. Kumai, T. Gunjima, "Phase diagram and phase separation in LC/prepolymer mixture", SPIE, vol 1257, p2, 1990.
7. H. Nomura, S. Suzuki, Y. Atarashi, "Interfacial interaction between nematic liquid crystal and polymer in composite film...", Jap.J. Appl.Phys., vol 30(2), p327, 1991.
8. G.P. Crawford, R.J. Ondris-Crawford, S.Zumer, S. Keast, M. Neubert, J.W. Doane, "Alignment and ordering mechanisms at a liquid crystal-solid interface", Liquid Crystals, vol 14(5), p1573, 1993.
9. L. Bouteiller, P. Le Barny, F. Massie, P. Robin, "Influence of the chemical nature of polymer surface on electrooptic properties of PNLC", Proc. 13th IDRC, 1993, p325.
10. The swelling ratio measures the increase in volume of a piece of polymer immersed in a solvent. It is measured by weighing.
11. In ref 9, the polymer in cells modified for more than 7' do not recover their initial shape. This explains the dramatic increase of Tmin.
12. J. Fricke, A. Emmerling, "Aerogels", J.Am.Ceram. Soc., vol 75(8), p2027, 1992.
13. K. Ogino, T. Otsuka, Kobunshi Ronbunshu, 42(12), 857, 1985.

OPTICAL DEVICES UTILIZING NONLINEAR ORGANIC MATERIALS

J.R. Davy, S.R. Rees, J. Staromlynska, J.A. Hermann,
M.P. Gillyon, T.J. McKay and P.B. Chapple

DSTO Land, Space and Optoelectronics Division
Dept of Defence, PO Box 1500, Salisbury, SA 5108, Australia

INTRODUCTION

Research into novel optical devices at DSTO has featured the use of materials which exhibit large optical nonlinearities. This research activity can be loosely subdivided into (a) the development of both single and multiple stage tunable electro-optic filters for applications such as laser bathymetry (measurement of ocean depth), laser warners and laser anti-dazzle filters, and (b) the investigation of passive optical power limiters designed to protect viewing devices from the various possible harmful effects of laser irradiation.

Various Lyot-type filters incorporating liquid crystal (LC) retarders have been designed and tested. The liquid crystal retarders allow various important functions, such as adequate wavelength tunability and corrections for inaccuracies, nonuniformities and drifts, to be performed. The specific device applications include a multiple stage LADS (laser airborne depth sounding) filter, a broad-band anti-dazzle discriminator, and a single-stage double-element laser warner device.

Factors affecting the design of passive optical power and energy limiters have also been investigated in a program involving both experimental work and theoretical modelling. The active materials employed in the optical limiter experiments were a soluble platinum ethynyl compound and some substituted phthalocyanines. Nanosecond laser pulses were used at the wavelength 532 nm and picosecond pulses were used at 1064 nm. Modelling studies involving thick nonlinear samples have addressed the issues of optimal sample thickness and optimal axial position.

TUNABLE ELECTRO-OPTIC FILTERS

Optical devices of this type, which actively discriminate between different parts of the spectrum, have a wide variety of applications which include wavelength multiplexing/de-multiplexing for telecommunications, spectral filtering for remote sensing and rapid-scan

Polymers and Other Advanced Materials: Emerging Technologies and Business Opportunities. Edited by P. N. Prasad *et al.*, Plenum Press, New York, 1995

spectroscopy, laser bathymetry, laser light detection and protection, and laser stabilization and tuning. The desirable properties common to these applications include good spectral resolution and contrast ratio, wide field of view, good transmission, and fast tuning over adequate ranges.

Anti-dazzle filter and laser warner

The phenomena of dazzle and veiling glare are commonly experienced when a sensor is irradiated by a laser (as well as by exposure to other bright light sources). Dazzle occurs at optical power levels lower by an order of magnitude or more than the optical power required for minimal optical damage. For this reason, dazzle protection tends to rely upon active devices such as the electro-optic tunable filter concept recently developed[1-3] at DSTO, in which laser light is blocked without unduly affecting the ambient light.

A simple anti-dazzle filter consists essentially of a single birefringent plate of a material such as lithium niobate in conjunction with a liquid crystal retarder (with voltage-controlled birefringence), both suitably positioned between a pair of crossed polarizers. This arrangement is depicted in figure 1. All of the optical components need to be anti-reflection coated. In the presence of laser irradiation the optical transmission exhibits periodic behaviour with respect to wavelength, as depicted in figure 2. The period (free spectral range) is much smaller than the spectral width of visible light and much larger than the spectral width of normal laser lines. Thus by setting the voltage to provide the minimum transmission, the incoming laser light can be selectively blocked.

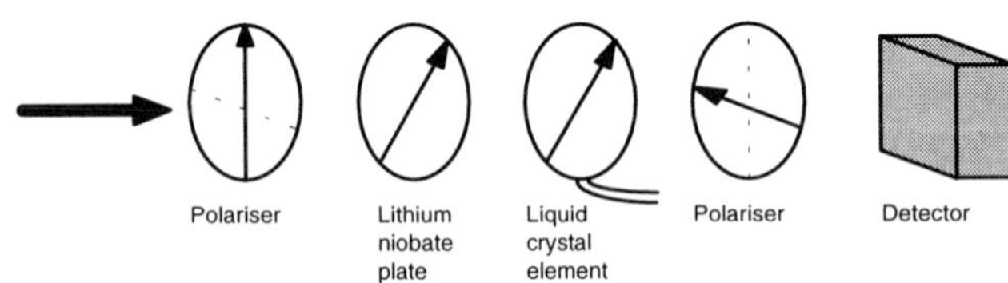

Figure 1. Tunable filter device

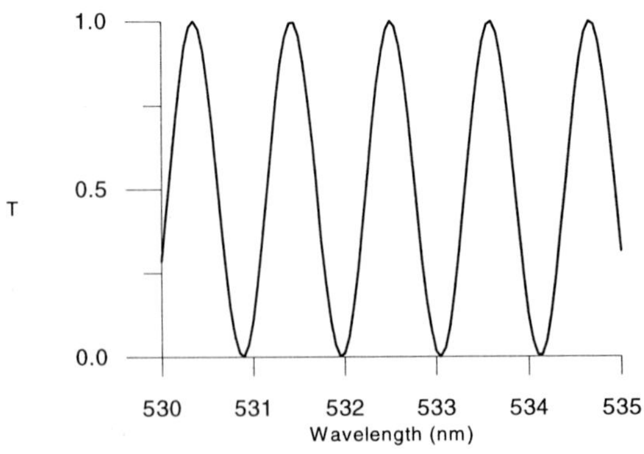

Figure 2. Transmittance of the filter as a function of wavelength

The concept demonstrators developed at DSTO have employed the nematic liquid crystals E44 (cell thickness 3.35 cm) and 2140 (cell thickness 5.19 cm). As the latter material was found to possess better spatial uniformity than the former, it was chosen for analyzing the anti-dazzle device. The fixed lithium niobate plate was 2.9 mm thick. The optical arrangement was also operated as a laser warner device, which is used to detect the presence of a laser beam. In this mode, the voltage applied to the LC cell may be ramped repetitively, at a frequency of about 5 Hz. This operation does not greatly affect the transmitted white light signal, but strongly modulates the transmission of laser light (see figure 3). Suitable control electronics can detect this modulation, thereby recognizing the presence of a laser. In our experimental work this entire operation was accomplished within about 0.5 sec, using an incident laser irradiance of around 0.1 $\mu W/cm^2$ together with a white light irradiance of 30 $\mu W/cm^2$ in order to emulate ambient light.

In the anti-dazzle mode, the servo electronics allow the LC voltage to move towards the minimum transmission voltage (this operation usually takes less than one second) and can hold it at this level indefinitely. The recorded image of a printed page exposed to dazzle from a 1 W argon laser beam is shown in figure 4. With the voltage held at the transmission maximum it is impossible to detect print on the page, however at the transmission minimum the page can be viewed clearly. Color photography has verified that good color balance can be retained.

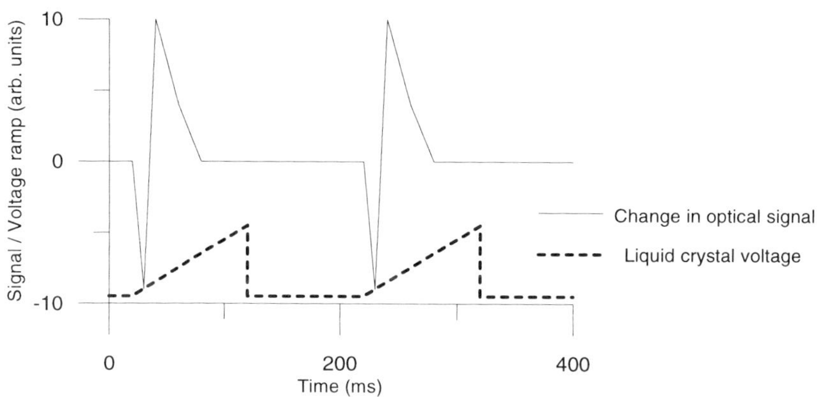

Figure 3. Typical trace showing the variation in the optical signal as the voltage is scanned.

It is useful to distinguish between the extreme cases of poor look past (PLP) where both the color balance and transmission are poor, and good look past (GLP) where at least the color balance is good. For example, a GLP device might correspond to 1 nm spacing of the transmission peaks while PLP might correspond to 200 nm spacing. A GLP device may be unsuitable for some applications, owing to restrictions to the field of view which are associated with the birefringent plate. In practice a compromise will need to be effected, involving a trade-off between the field of view and the color balance. This feature represents a major problem area for this type of device. Other problem areas include (a) excessive overall thickness and weight owing to a requirement for the LC substrates to provide sufficient rigidity, (b) the relative slowness of many LC materials (eg, milliseconds), (c) various factors which can lead to a degradation in the contrast ratio, also known as the extinction ratio (the ratio of the maximum to minimum transmissions);

Figure 4(a). Scene obscured by dazzle from a 1 W argon ion laser beam.

Figure 4(b). Scene visible, with the LC voltage at a transmission minimum.

these include random signal fluctuations associated with the atmospheric propagation, and also the impact of residual multiple reflections. Investigations of these effects have been reported recently by Rees and Staromlynska [3].

In addition to the factors mentioned above, it has been found that a strong reduction in the contrast ratio also occurs with nematic liquid crystal retarders for certain optical parameter regimes, notably for certain wavelengths and for certain conjunctions of the angles of arrival within the field of view. Some theoretical modelling results pertaining to

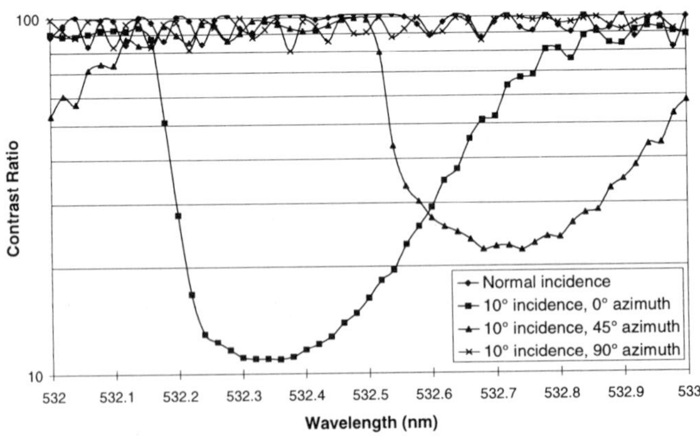

Figure 5. Wavelength dependence of the contrast ratio for different directions of the angles of arrival.

the wavelength dependence of the contrast ratio is presented in figure 5. These results reveal that a displacement of the elevation (or incidence) angle from the normal direction by only 10 degrees (with a wide choice of azimuthal angles) can have the effect of allowing the contrast ratio to range over two orders of magnitude (between 10 and 1000) within certain wavelength domains.

The dependence of the contrast ratio upon the azimuthal and elevation angles for a specific wavelength may also be represented graphically, and some of the data from this modelling work is presented in figure 6. The features of particular concern are a number of deep "holes" appearing at certain angular configurations. These are thought to be related to the manner in which the nematic liquid crystal molecules respond to the applied electric field. For this reason the use of highly birefringent organic materials other than the nematics, such as ferroelectric liquid crystals, is currently being investigated.

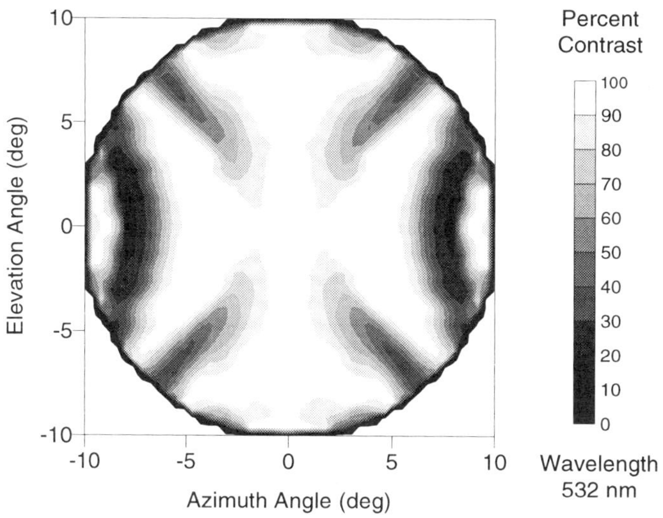

Figure 6. Dependence of the contrast ratio upon the azimuthal angle (horizontal axis) and the elevation angle (vertical axis) for the fixed wavelength 532 nm.

The LADS filter

The Laser Airborne Depth Sounder (LADS) system[4] utilizes a frequency-doubled Nd:YAG laser (pulse length 5 nsec, pulse energy 5 mJ, repetition frequency 168 Hz) to scan ocean depths by laser ranging techniques. The maximum measurable ocean depth is constrained by the signal to noise ratio, which is in turn determined by the ambient light level associated with the reflected laser light. To enable operation of the system during daytime requires the use of a six-stage LC tuned Lyot filter operating at a narrow passband (approximately 0.1 nm) with a free spectral range of around 7 nm, an overall field of view of 17 deg, approximately 25 per cent transmittance (unpolarized light), a clear aperture of 50 mm, an overall length less than 200 mm, and relative insensitivity to variations in the ambient temperature.

The essential design of a Lyot filter may be carried out using a simplified description of the birefringent optics such as that described by Yeh[5]. The individual stages (which are separated by polarizers aligned parallel to one another) each consist of a birefringent plate whose optic axis is in the plane of the plate and oriented at 45 deg (azimuth) to the pass plane of the polarizers. If d is the thickness of the thinnest plate, then the successive thicknesses are given by $2^{N-1}d$ (N being the number of stages) and the transmittance at each stage is given by $I(\lambda) = I_0 \cos^2\{\frac{1}{2}\Gamma(\lambda)\}$ where I_0 is the incident irradiance, while $\Gamma(\lambda)$ is the retardance at N=1. The precise dependence of Γ on the wavelength λ, and on the angles of arrival, are given in reference 4. The maxima in $I(\lambda)$ occur when $\Gamma(\lambda)=2\pi M$ (M is an integer). Further equations relate the bandwidth, free spectral range and field of view to the birefringence, dispersion of birefringence, number of stages N, and thickness d. A secondary dielectric filter has also been used to suppress any spectral transmittance which is not close to 532 nm.

The optical materials selected for this filter were (a) lithium niobate for the nontunable birefringent elements, and (b) the nematic liquid crystal 2140 for the variable retarder. The voltage requirements for this liquid crystal are conveniently low, and the necessarily high level of retardance uniformity can be achieved by maintaining a very high degree of parallelism between the LC cell walls. Unfortunately the latter requirement creates some undesirable and unavoidable etaloning effects. Achievement of the 0.1 nm bandwidth and 7 nm free spectral range requires at least six stages, the thickest stage being 9.6 mm. Field of view calculations indicate a transmittance of 83 per cent at the thickest stage for an overall 17 deg FOV. The optimization of the filter design involves the transmittance of unpolarized light, in addition to the figure of merit (FOM) defined as the ratio of the total integrated transmittance within the laser bandwidth to that outside the bandwidth, and also the angular distribution of illumination over the collecting aperture. Details of the modelling procedures undertaken to achieve this optimization, and of tuning procedures, are also provided in reference 4.

OPTICAL LIMITERS

Recent research on the protection of sensors from laser-induced optical damage has been mostly concerned with constructing optical switches and limiters. The limiters differ in a fundamental manner from switches in that the transmission is suppressed, rather than terminated, above the operating power (or energy) threshold. In order that a limiter may achieve adequate protection for the unaided human eye, it will desirably possess the following features: (a) a limiting threshold energy of about 1 μJ, (b) a spectrally broad response within the visible and near infrared regions, (c) low insertion and absorption losses under low-power operating conditions, (d) a field of view compatible with that of the imaging system, (e) a dynamic range spanning several orders of magnitude, (f) a sufficiently short nonlinear response time (Q-switched pulses in the 5-100 nsec range are the most important regime to be accommodated), and (g) a very high damage threshold. More generally, optical damage may occur with a wide variety of laser sources, ranging in wavelength from the ultraviolet to the far infrared. These sources can also range in pulse width from sub-picosecond laser pulses to continuous wave lasers. The diversity of these sources demands a variety of solutions to prevent damage. If the wavelength of a laser is known in advance then a fixed-line filter would normally be expected to provide adequate protection.

The techniques for optical power limiting involve nonlinear optics. Most optical limiter schemes require that the light be focused in order to increase the laser intensity at the nonlinear medium, since the optical nonlinearities of available materials are not otherwise large enough to provide the necessary level of protection. Therefore, inserting a power limiter into an optical device will normally be easier if the optics already contains a focus at some accessible position. It is unlikely that the incident beam will be Gaussian; rather, it is probably approximately top-hat-shaped at the entrance aperture with a width matching the aperture diameter. Near the beam waist, the intensity profile is much more complicated, because of diffraction effects. The rate at which the beam converges inside the system can be expressed by the f/# (f number), which is the ratio of the focal length of the optics to the beam diameter. When comparing different optical limiting schemes, it is important to take account of the different experimental conditions. The beam radius and Rayleigh length usually have a critical bearing on the degree of limiting. Equally important are the laser wavelength and pulse length, which determine which nonlinear optical mechanisms can take place in the particular medium. The length of the nonlinear medium is also of vital importance, as is the concentration (where the nonlinear optical material is in solution), along with the linear absorption at the relevant wavelength.

Optical Limiting Mechanisms

Nonlinear refraction, which embraces self-focusing, self-defocusing and self-deflection, has been studied in the context of optical limiting applications for many years[6-10]. Figure 7 shows a suitable set-up for an optical limiter based on a thin self-focusing element. A lens focuses the incoming light as shown. Self-focusing obliges the beam to focus to a smaller

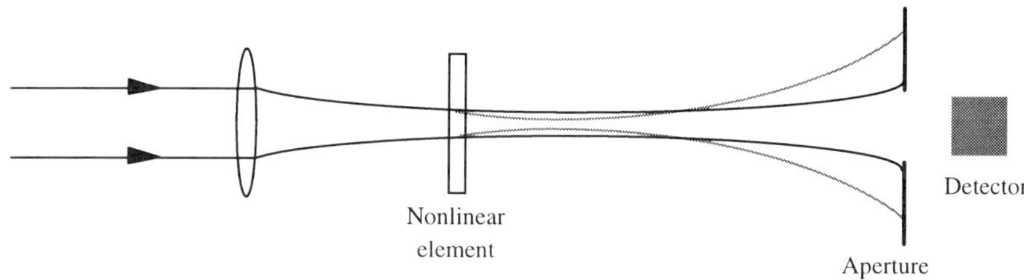

Figure 7. Self-focusing optical power limiter.

waist at higher beam powers and to spread more rapidly subsequently, so that more of the light is rejected by the aperture and less light is intercepted by the detector. Studies of the Z-scan technique for a Kerr nonlinearity have indicated where the nonlinear element should be placed in relation to the beam waist. For a thin nonlinear sample with a Gaussian beam profile and a small aperture, the most efficient limiting occurs when the sample is placed at a distance somewhat less than one Rayleigh length before the waist (the optimal distance depends upon the peak optical power).[11]

Another determinant of optical limiting behaviour is the aperture size, conveniently characterised by its linear transmittance S. If the aperture radius is r_a and the Gaussian beam radius at the aperture is w_a, then $S = 1 - e^{-2r_a^2/w_a^2}$. The most efficient optical limiting is obtained with small S (r_a much less than w_a), but this severely limits both the transmittance of the device and its angular field of view. There is therefore a trade-off between efficient limiting and performance under low-light conditions. The limiting behaviour of such a system may be calculated using the first-order thin Kerr medium theory of Hermann and Wilson,[12] which shows that for S =0.7 the optimal distance of the thin medium from the waist is 0.72 Rayleigh lengths. Strong limiting utilizing self-defocusing was demonstrated experimentally by Swartzlander *et al.*[13] recently, using 6 nsec and 30 nsec pulses at wavelength 532 nm, and exploiting the thermal nonlinearity produced by the dye nigrosin dissolved in CS_2.

Because the performance of refractive limiters depends sensitively upon the profile of the incoming laser beam as well as the exit aperture size, such devices are unlikely to match the overall versatility of the nonlinear absorptive limiters. Nonlinear absorbers are materials possessing intensity-dependent absorption coefficients. The most important absorptive mechanisms for limiting are[4] (i) excited-state absorption (ESA), which is a fluence-dependent process arising when molecules absorb light to form highly absorbing excited states. When ESA is stronger than the ground-state absorption and long-lived the effect is otherwise known as reverse saturable absorption (RSA), and is often very effective for a broad range of laser pulses. A related phenomenon is free-carrier absorption in semiconductors. (ii) Two-photon absorption (TPA), which is an intensity dependent process of importance within some organic and semiconductor media. It can be effective for picosecond pulses, which have high peak powers at low energies, but its usefulness diminishes for longer pulses. (iii) Thermally-enhanced absorption, associated with heating of the medium, and useful for longer pulses.

Another limiting mechanism is nonlinear scattering, which may occur in either solid media, solutions, or liquid suspensions. The light is scattered and possibly absorbed by the existing particles or by particles formed by the presence of the optical beam. A considerable amount of experimental work has been done[15-17] on the nonlinear scattering of carbon-black suspension in various liquids. The nonlinear scattering in this case is essentially energy or fluence-dependent, and can limit longer pulses at low powers. Because carbon is highly absorbing across the visible spectrum, and the limiting threshold is relatively low, this type of limiter is one of the most effective available for broadband visible protection from nsec to μsec pulses. A serious disadvantage is that the suspension becomes damaged for a repetitively pulsed laser, owing to the vaporization of carbon atoms, necessitating some form of circulation within the suspension.

Useful Molecules for Optical Limiting

Many organic molecules possess potentially useful optical nonlinearities, particularly where a high degree of electron delocalisation occurs, and some molecules have broadband nonlinearities extending across a significant part of the visible or infrared spectrum.[18] Although the hope has been to find molecules with large nonresonant nonlinearities, the most useful nonlinearities for optical limiting purposes to date have been those associated with absorption resonances. Molecules which have been investigated for optical limiting include the phthalocyanines and naphthalocyanines, transition metal poly-ynes and ethynyl molecules, porphyrins, indanthrone molecules, fullerene, King's complex, and many others.

Some of these molecules are dyes and many are organometallics with a large degree of electron delocalization.

A common feature of many of the molecules exhibiting RSA is the role of singlet and triplet excited states, as depicted in figure 8. Normally only transitions between the S (spin zero) states are possible, however in the presence of spin-orbit coupling (as may occur when there is a heavy metal atom in the molecule) transitions between the singlet and triplet states are possible. Such transitions can also occur due to molecular collisions. In these ways, the T_1 state may be populated from S_1 with a typical time constant of several nanoseconds. Laser pulses of suitable wavelengths can excite molecules from S_0 to S_1. If the pulses are short (of the order of psec), there is normally insufficient time for a transition from S_1 to T_1 before the pulse passes. For longer pulses (of the order of nsec), transitions to T_1 are often important, and absorption may increase because of the $T_1 \rightarrow T_2$ transition. The T_1 state has a relatively long lifetime (μsec to msec) compared with the S_1 state (nsec), so $T_1 \rightarrow T_2$ transitions become increasingly important as the pulse length increases.

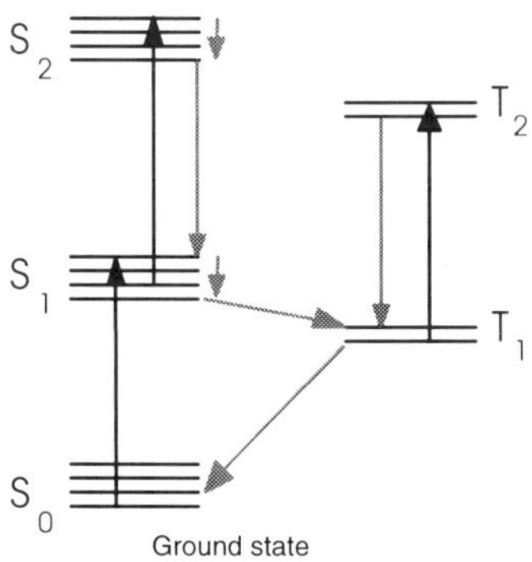

Figure 8. Typical states and excitation pathways for RSA molecules.

One of the most promising families of optical limiting molecules is the phthalocyanine and naphthalocyanine dyes.[19-27] These molecules generally have unusually high thermal and chemical stability, caused by the high degree of planar electron delocalization in the molecules. Their overall tendency to extreme insolubility can be overcome by attaching solublizing groups to either the central metal or the aromatic rings. Coulter *et al.*[19] studied

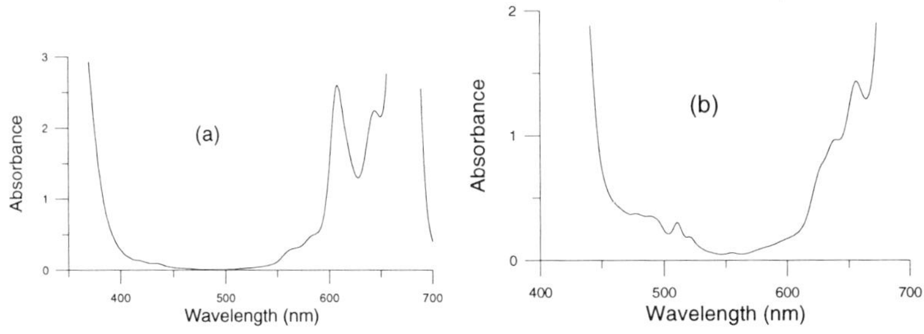

Figure 9. Linear absorption spectra measured in our laboratories.
(a) CAP in ethanol. (b) SINC in tetrahydrofuran.

optical limiting in chloro-aluminium phthalocyanine solution (CAP), using nanosecond and picosecond pulses at 532 nm. Subsequent studies[20] have also investigated the nonlinear behaviour of a silicon naphthalocyanine, SiNc(OSi(n-hexyl)$_3$)$_2$ (abbreviated SINC). Linear absorption spectra of solutions of these molecules are depicted in Figure 9.

The results for our optical limiting experiments with 11 nsec pulses at a wavelength of 532 nm are shown in Figure 10. The use of an aperture with S=0.8 produces additional limiting, owing to the presence of nonlinear refractive processes. Mansour *et al.*[24-26] have discovered improved limiting behaviour for nanosecond pulses when heavier metal atoms such as tin and lead are used in the phthalocyanine molecule. The presence of these atoms causes fast transitions from the singlet to the triplet states. Experiments have been done with phthalocyanines in solution form and doped in host materials such as polymethyl methacrylate and sol-gels. Results for tin phthalocyanine are particularly impressive[25].

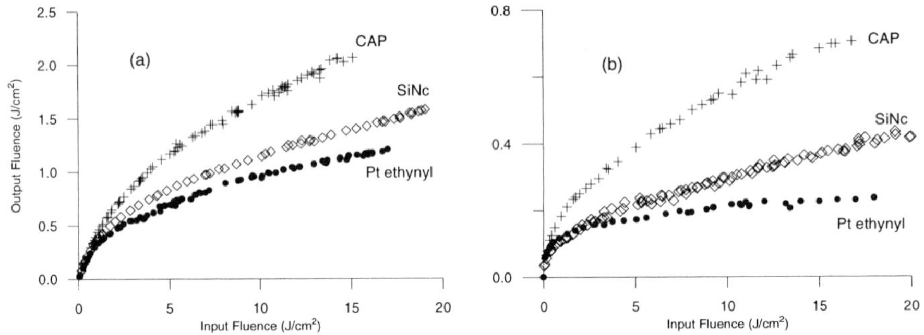

Figure 10. Optical limiting observed by our group for 11nsec pulses at 532 nm, using solutions of CAP, SINC and platinum ethynyl (see figure 11), each material possessing a linear transmission of 70 percent. (a) No aperture. (b) Aperture: $S = 0.8$.

Another family of organic molecules showing promise for optical limiting applications is the transition-metal containing poly-ynes. The transition metal atom, often platinum or palladium, is bonded to acetylenic groups, and is stabilized by tributyl phosphine ligands. Many polymers and monomers in this family exhibiting large nonlinearities have been investigated by Guha *et al.*[27-29] While at first it was thought[28] that two-photon absorption was responsible for the nonlinear behaviour, it was subsequently realised[29] that excited state absorption plays an important part. The molecules have a strong S_0 to S_1 absorption (see figure 8) in the ultraviolet region, at around 370 nm. Wittmann *et al.*[30] identified also a weak absorption in the green, due to a normally forbidden S_0 to T_1 transition. It appears that this transition plays an important part in populating the T_1 state, for excited state absorption at 532 nm.

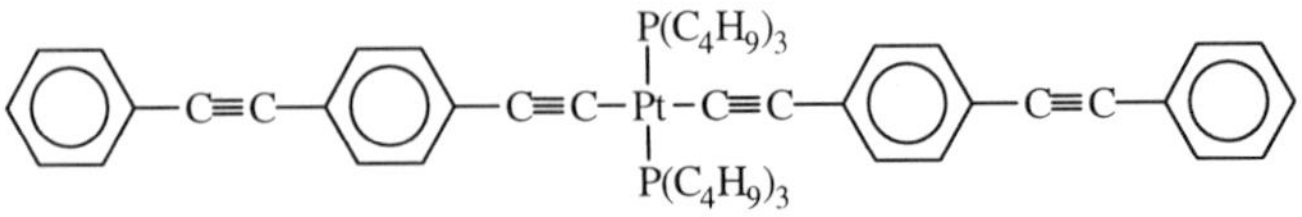

$P(C_4H_9)_3$

—C≡C— —C≡C–Pt–C≡C— —C≡C—

$P(C_4H_9)_3$

Figure 11. The platinum-ethynyl molecule studied in our laboratory: bis((4′-ethynyl(1,1′-biphenyl)-4-yl) ethynyl)-bis(tributylphosphine)-platinum.

We have investigated the platinum-ethynyl compound represented in figure 11 using 6.5 nsec and 11 nsec pulses at 532 nm and 20 psec pulses at 1064 nm.[31] The compound was dissolved in tetrahydrofuran. Figure 12 depicts the measured absorption spectrum. The optical limiting results at 532 nm are shown in figure 10, together with the results for SINC and CAP. It is clear that the platinum-ethynyl compound limits most effectively at 532 nm. At higher fluences, above 6 J/cm^2, significant nonlinear scattering was observed.

Nonlinear scattering was also observed for the 20 psec Nd:YAG pulses at 1064 nm, and this was again accompanied by optical limiting. The intensity of the scattered light varied considerably, but was most evident with fluences in excess of 0.75 J/cm^2. Transmission studies with a cw helium neon laser have demonstrated that the Nd:YAG pulses produce an absorptive effect within the solution, and that this effect persists for many milliseconds[31].

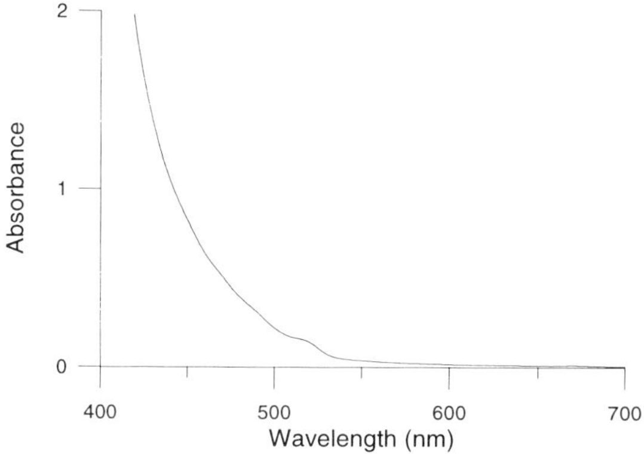

Figure 12. Absorption spectrum of platinum-ethynyl in tetrahydrofuran, showing the structure at 520 nm caused by singlet to triplet absorption.

This phenomenon suggests that the material could be effective for limiting laser pulses with durations ranging from picoseconds to milliseconds. The platinum-ethynyl molecule is not as stable as the CAP and SINC molecules, and further investigations are required to ascertain whether its stability is adequate for practical limiter requirements.

Optical Limiter Geometry

For many years it was widely believed that the unfinished task of understanding and enhancing the nonlinear optical effects within organic materials was a primary reason for the overall failure to develop devices offering protection against laser irradiation for sensitive detectors and sensors. Particular emphasis was placed upon developing techniques for increasing the magnitudes of the nonlinear refractive and absorptive coefficients. Despite this frenzy of activity the long-awaited devices, satisfying all of the essential technical requirements, have still to be realized. We believe that a major, and largely unrecognized, reason for this situation lies with the relative neglect of the design requirements for optimizing limiting devices. These demand, amongst other things, a thorough and detailed understanding of the macroscopic device physics, and in particular the propagation of optical beams of arbitrary initial shape within thick nonlinear media. The latter requirement has been alluded to in recent papers by Miles.[32-34] Only limited progress has been made in this direction, a consequence being that the existing device analyses are still very crude.

In this context an exact analysis of propagation by an initially Gaussian beam, within a thick intensity-dependent medium displaying both refractive and absorptive behaviour, at *low* optical powers (that is, an exact analytic description to the lowest order of nonlinearity) has been published recently, and has also been successfully applied to a description of optical scanning, including the Z-scan.[35] In this technique the nonlinear sample is simply translated in the propagation direction through the waist region while the transmission is monitored at the far field. The theoretical predictions are in excellent agreement with results obtained from carefully-controlled Z-scan experiments.[36] One of the consequences of the analysis is that the nonlinear "interaction length" (the length of the nonlinear medium required to achieve 90 per cent of the maximum possible change in transmission at the far field) corresponds, in the case of a refractive nonlinearity, to a sample thickness of around six Rayleigh lengths. Interestingly, the same experiments have demonstrated that this characteristic length is also correct at much higher optical powers, and is therefore highly relevant to the performance of refractive limiters. In the case of a purely absorptive nonlinearity, the corresponding interaction length (defined by reference to the normalized transmission of optical power) is predicted to be around twelve Rayleigh lengths.

The nonlinear first-order theory predicts that the normalized transmittance T of a thick nonlinear layer, evaluated on the propagation axis at the far field, will be given by[35]

$$T = 1 - \tfrac{1}{8}.\beta.\ln\left(\frac{(9+\upsilon_b^2).(1+\upsilon_a^2)}{(9+\upsilon_a^2).(1+\upsilon_b^2)}\right) - \tfrac{1}{4}.\alpha.\left(\tan^{-1}\left[\frac{4\upsilon_a}{3-\upsilon_a^2}\right] - \tan^{-1}\left[\frac{4\upsilon_b}{3-\upsilon_b^2}\right] + \varepsilon\right)$$

where α and β are the characteristic nonlinear absorptive and refractive parameters as defined in reference 35, $\varepsilon=(0,\pm\pi)$ is a shift term required in order to ensure continuity beyond the principal ranges, υ_a is the distance of the front face of the nonlinear layer (relative to the incoming beam) from the waist position, and υ_b is the distance of the rear face from the waist position (both distances being scaled by the Rayleigh length). Using this expression one may obtain exact expressions, in the extreme cases $\alpha=0$ and $\beta=0$, for the separation of the maximum and minimum, ΔT. The above equation for T makes possible an assessment of the values of α and β for a given material possessing a simple (eg, Kerr-type) nonlinearity, by performing Z-scan measurements with and without a collecting aperture. Note that a useful analytic description of the propagating electric field, exact to all nonlinear orders in thick media, has not been discovered to date (even for the case of thin media, the known double-summation solution to all orders[11,12] only converges for certain parameter domains).

Optical limiting may be optimized by choosing a suitably thick nonlinear medium. Experimental studies were carried out via the Z-scan technique, using CS_2 samples longer than a Rayleigh length.[36] A set of Z-scan traces at different power levels, exemplified by the high power result shown in figure 13, were produced. The first-order theory gives a reasonably good fit to this high-power scan, and a numerical solution gives an even better fit. At low optical powers the experimental scans are in excellent agreement with the first-order analysis. For sample lengths l longer than about 2.5 times the Rayleigh length in the medium, the minimum occurs when the beam waist coincides with the back surface of the cell (see figure 14). This is the position to place the nonlinear layer when designing a self-focusing optical power limiter. The transmittance difference ΔT for different sample lengths is shown in figure 15, where the experimental and calculated results are compared.

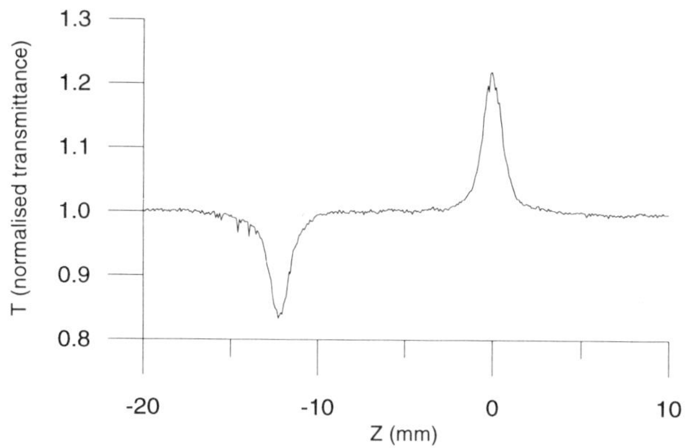

Figure 13. *Z*-scan trace for a 20 cm cell containing CS_2 (peak intensity 1 GW/cm^2)

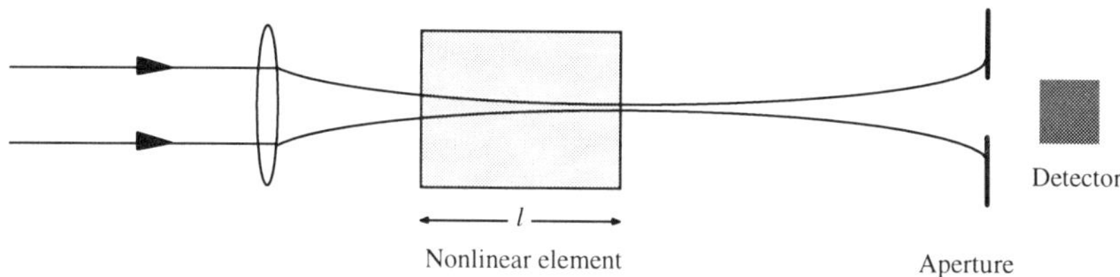

Figure 14. Diagram of a thick refractive limiter.

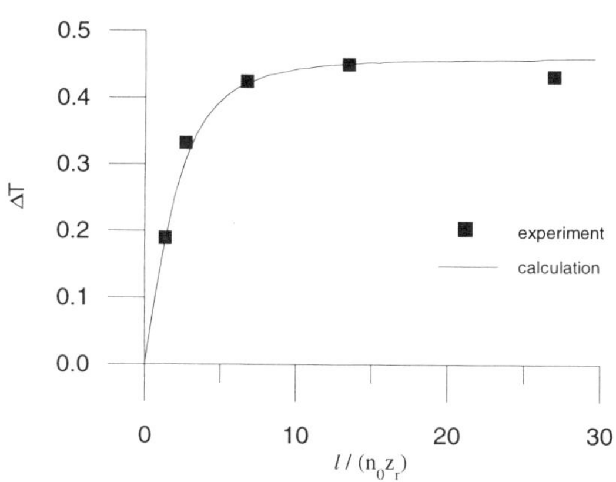

Figure 15. ΔT versus sample length (scaled to the Rayleigh length within the medium)

The results in figure 15 clearly demonstrate the saturation of ΔT with sample length. Some of the constraints usually imposed on an optical limiter intended to protect sensitive detectors or sensors relate to the available space, weight and transmission under low light conditions. Obviously these will be compromized if the active medium is excessively thick, linear transmission being a relevant consideration.

CONCLUSIONS

Studies of novel optical devices which feature nonlinear organic materials have been carried out at the Defence Science and Technology Organisation (DSTO), Australia. Tunable filters using liquid crystal technology have been developed for the Laser Airborne Depth Sounder, which makes possible the daytime scanning of ocean depths by means of laser ranging. Tunable filters have also been used as concept demonstrators for laser warners and laser anti-dazzle devices. Organic materials possessing strong optical nonlinearities have been investigated in the context of designing passive optical power limiters. Various approaches to optimizing the performance of these devices have been discussed.

REFERENCES

1. J. Staromlynska, "A double-element broad-band liquid crystal tunable filter - factors affecting contrast ratio", *IEEE J. Quant. Electron.*, **28**, 501-506 (1992).
2. J. Staromlynska and R.S. Seymour, "Field of view calculations on tunable filters designed for the selective filtering of laser radiation", *Opt. Eng.*, **32**, 793-798 (1993).
3. S. Rees and J. Staromlynska: "Automatic laser light detection and filtering using a liquid crystal Lyot filter," *Int. J. Nonlin. Opt. Phys.*, **2** (4), 661-676 (1993).
4. R.S. Seymour, S.M. Rees, J. Staromlynska, J. Richards and P. Wilson, "Design considerations for a liquid crystal tuned Lyot filter for laser bathymetry", *Opt. Eng.*, 33, 915-925 (1994).
5. P. Yeh, "Dispersive birefringent filters", *Opt. Commun.*, **37** (3), 153-158 (1981).
6. J.A. Hermann, "Beam propagation and optical limiting with nonlinear media", *J. Opt. Soc. Am. B*, **1**, 729-736 (1994).
7. M. Sheik-Bahae, A.A. Said, D.J. Hagan, M.J. Soileau and E.W. Van Stryland, "Simple analysis and geometric optimization of a passive optical limiter based on internal self-action", *SPIE Proc.*, **1105**, 146-158 (1989).
8. J.A. Hermann, "External self-focusing, self-bending and optical limiting with thin nonlinear media", *Opt.* and *Quant. Electron.*, **19**, 169-178 (1987).
9. M. Sheik-Bahae, A.A. Said, D.J. Hagan, M.J. Soileau and E.W. Van Stryland, "Nonlinear refraction and optical limiting in thick media", *Opt. Eng.*, **30**, 1228-1236 (1991).
10. E.W. Van Stryland, Y.Y. Wu, D.J. Hagan, M.J. Soileau and K. Mansour, "Optical limiting with semiconductors", *J. Opt. Soc. Am. B*, **5**, 1980-1990 (1988).
11. J.A. Hermann, P.B. Chapple, J. Staromlynska and P.J. Wilson, "Design criteria for optical power limiters", *SPIE Proc.*, **2229**, 167-178 (1994).
12. J.A. Hermann and P.J. Wilson, "Factors affecting optical limiting and scanning with thin nonlinear samples," *Int. J. Nonlinear Opt. Phys.*, **2** (4), 613-629 (1993).
13. G.A. Swartzlander, B.L. Justus, A.L. Huston, A.J. Campillo and C.T. Law, "Characteristics of a low f-number broadband visible thermal optical limiter", *Int. J. Nonlin. Opt. Phys.*, **2**, 577-611 (1993).
14. J.A. Hermann and J. Staromlynska, "Trends in optical switches, limiters and discriminators," *Int. J. Nonlin. Opt. Phys.*, **2** (3), 271-337 (1993), and references therein.
15. K. Mansour, M.J. Soileau and E.W. Van Stryland, "Nonlinear optical properties of carbon-black suspensions," *J. Opt. Soc. Am. B*, **9** (7), 1100-1109 (1992).
16. R.R. Michael, C.M. Lawson and G.W. Euliss, "Nanosecond switching in carbon microparticle suspensions," *Proc. SPIE* , **1626**, 205-216 (1992).
17. C.M. Lawson and R.R. Michael, "Nonlinear reflection characteristics at a dielectric-carbon suspension interface," *Proc. SPIE* , **2229**, 112-120 (1994).

18. J.W. Wu, J R. Heflin, R.A. Norwood, K.Y. Wong, O. Zamani-Khamiri, A.F. Garito, P. Kalyanaraman and J. Sounik, "Nonlinear-optical processes in lower-dimensional conjugated structures," *J. Opt. Soc. Am B*, **6** (4), 707-720 (1989).

19. D.R. Coulter, V.M. Miskowski, J.W. Perry, T-H. Wei, E.W. Van Stryland and D.J. Hagan, "Optical limiting in solutions of metallo-phthalocyanines and napthalocyanines," *Proc. SPIE*, **1105**, 42-51 (1989).

20. T.H. Wei, D.J. Hagan, M.J. Sence, E.W. Van Stryland, J.W. Perry and D.R. Coulter, "Direct measurements of nonlinear absorption and refraction in solutions of phthalocyanines," *Appl. Phys. B*, **54** 46-51 (1992).

21. N.Q. Wang, Y.M. Cai, J.R. Heflin, J.W. Wu, D.C. Rodenberger and A.F. Garito, "Third order optical properties of quasi-two dimensional conjugated discs: silicon naphthalocyanine," *Proc. SPIE*, **1328**, 100-107 (1990).

22. J.S. Shirk, J.R. Lindle, F.J. Bartoli, Z.H. Kafafi, A.W. Snow and M.E. Boyle, "Third-order nonlinear optical properties of metallo-pthalocyanines," *Int. J. Nonlin. Opt. Phys.*, **1** (4), 699-726 (1992).

23. S.R. Flom, J.S. Shirk, R.G.S. Pong, J.R. Lindle, F.J. Bartoli, M.E. Boyle, J.D. Adkins and A.W. Snow, "Excited state absorption and dynamics in a Pb-phthalocyanine copolymer," *Proc. SPIE*, **2143**, 229-238 (1994).

24. K. Mansour, D. Alvarez Jr., K.J. Perry, I. Choong, S.R. Marder and J.W. Perry, "Dynamics of optical limiting in heavy-atom substituted phthalocyanines," *Proc. SPIE*, **1853**, 132-141 (1993).

25. K. Mansour, P. Fuqua, S.R. Marder, B. Dunn and J.W. Perry, "Solid state optical limiting materials based on phthalocyanine containing polymers and organically-modified sol-gels," *Proc. SPIE*, **2143**, 239-250 (1994).

26. J.W. Perry, K. Mansour, S.R. Marder, K.J. Perry, D. Alvarez Jr. and I Choong, "Enhanced reverse saturable absorption and optical limiting in heavy-atom-substituted phthalocyanines," *Opt. Lett.*, **19** (9), 625-627 (1994).

27. P.L. Porter, S. Guha, K. Kang and C.C. Frazier, "Effects of structural variations of platinum and palladium poly-ynes on third-order non-linearity," *Polymer*, **32** (10) 1756-1760 (1991).

28. S. Guha, K. Kang and P.L. Porter, "Two-photon absorption-induced thermal effects in platinum poly-ynes," *Appl. Phys. B*, **53**, 308-313 (1991).

29. S. Guha, K. Kang, P.L. Porter, J.F. Roach, D.E. Remy, F.J. Aranda and D. Rao, "Third-order optical nonlinearities of metallotetrabenzo-porphyrins and a platinum poly-yne," *Opt. Lett.*, **17** (4) 264-266 (1992).

30. H.F. Whitmann, K. Fuhrmann, R.H. Friend, M.S. Khan and J. Lewis, "Optical excitations in transition metal containing poly-ynes," *Synthetic Metals*, **55** (1) 56-61 (1993).

31. J. Staromlynska, P.B. Chapple, J.R. Davy and T.J. McKay, "A platinum ethynyl compound for optical limiting," *Proc. SPIE*, **2229**, 59-66 (1994).

32. P.A. Miles, "DARPA sensor protection program support", Final Report under Contract F33615-89-C-0532, Logicon RDA, Los Angeles, CA, Sept. 1991.

33. P. A. Miles, "Bottleneck optical limiters - the optimal use of excited state absorbers," *Appl. Opt.*, **33** (30) 6965-6979 (1994).

34. P. A. Miles, "Material figures of merit for saturated excited state absorptive limiters," *Proc. SPIE*, **2143**, 251-262 (1994).

35. J.A. Hermann and R.G. McDuff, "Analysis of spatial scanning with thick optically nonlinear media", *J. Opt. Soc. Am. B*, **10**, 2056-2064 (1993).

36. P.B. Chapple, J. Staromlynska and R.G. McDuff, "Z-scan studies in the thin- and thick-sample limits", *J. Opt. Soc. Am B*, **11**, 975-982 (1994).

NONLINEAR ELECTRO-OPTICAL AND HOLOGRAPHIC STORAGE EFFECT IN FULLERENE-DOPED NEMATIC LIQUID CRYSTAL FILM

I. C. Khoo

Department of Electrical Engineering
The Pennsylvania State University
University Park, PA 16802

INTRODUCTION

There have been intense research interests in materials capable of recording holographic planar or volume gratings. These include inorganic [1] and polymeric [2] photorefractive crystals, liquid crystals [3-5], bacteriorhodopsin [6], semiconductors [7] and glasses [8]. There are various advantages and limitations associated with these storage materials. In photorefractive materials, for example, fixing the optically written holograms requires the use of high fields and/or time consuming large temperature cyclings. The liquid crystalline systems [3-5] that have been investigated in the context of persistent grating also involve large laser induced temperature and phase changes. More recent work [6-9] on other materials have shown promises in overcoming some of these limitations. We have previously reported dc field assisted transient grating effects in untreated and dye-doped nematic liquid crystal film [10]. In this paper, we describe the observation of permanent grating formation in Fullerene C_{60}-doped nematic liquid crystal film. In very lightly doped and practically totally transparent thin nematic films, permanent gratings can be generated with unfocused milliwatt power laser with the aid of dc voltages around 1 volt or so. Diffraction efficiency of over 10%, and grating constant $\lesssim$ 10 μm can be realized in these materials. These holographic gratings can be electronically turned off and on at millisecond speed.

BASIC MECHANISM

Liquid crystals possess many unique physical, electro-optical and nonlinear optical properties. Among the various mesophases of liquid crystals, nematics are perhaps the most widely used materials for electro-optical application.

It is well-known from numerous previous studies of dc-field induced current and material flows in nematic liquid crystals that space charge fields can be set up owing to the conductivity anisotropy of nematics [11]. The creation of a spatially varying director-axis reorientation profile by an incident optical intensity of a nematic film is illustrated in

Polymers and Other Advanced Materials: Emerging Technologies and Business Opportunities. Edited by P. N. Prasad *et al.*, Plenum Press, New York, 1995

figure 1a-d. The incident sinusoidal optical intensity pattern (fig. 1a) generates a corresponding ionic charge distribution and flow (caused by the applied dc field $\vec{E}_a$ in $\hat{z}$-direction) in the fullerene doped nematic film, c.f., fig. 1b. These space charges ρ give rise to a space charge field E_{sc} (fig. 1c) given by the Maxwell equation $\vec{\nabla} \cdot \vec{E} = -\rho(\vec{r})/\epsilon$. Thus the space charge field spatially distribution is phase-shifted with respect to the incident optical intensity. This is analogous to the process of photoionization and space charge field formation in photorefractive crystal [1]. The phase shift is responsible for two-beam coupling effect observed in reference [10].

There are two mechanisms that will cause director axis reorientation. One is the combined applied dc and space-charged fields ($\vec{E}_a + \vec{E}_{sc}$) that creates a dielectric torque on the nematic axis, as discussed in some detail in reference [10]. The other is the reorientational effect caused by the shear stress $\sigma \sim -\frac{\partial v_z}{\partial \xi}$. Note that the velocity gradient is spatially phase shifted from the flow velocity distribution v_z and is "in-phase" with the space-charge distribution. The net effect of these two torques is to cause a spatially phase-shifted (from the incident optical intensity function) director-axis reorientational profile given in figure 1d. The quantitative aspects of these space charges, fields and flow processes caused by <u>dc applied</u> fields may be seen in reference 11; details pertaining to the present case where the ionic charges are <u>photo-induced</u> is quite complicated and is relegated to a longer article elsewhere. It suffices to say that in fullerene C_{60}-doped films, the enhanced photo-conduction and current flow [12] resulted in permanent distortion of the orientation of the liquid crystal director on the alignment surface layers. Some photochemical processes occurring in photo-excited C_{60}-doped film under dc applied fields, hitherto undetermined, are likely to also play a role in the formation of stable permanent reorientation of the director axis.

EXPERIMENTAL RESULTS

The liquid crystal used in our studies is 5CB (Pentyl-Cyano-Biphenyl) purchased from EM Chemicals. Various dopants have been used, including: dichroic dyes D2 (from EM Chemicals), laser dye Rhodamine R6G (from Kodak), Phathalocyanine-based laser dyes (from Aldridge Chemicals), and Fullerene C_{60} (from Kodak). In terms of induced refractive index change Δn versus incident laser fluence F (in Joules/cm^2), C_{60}-doped samples perform the best. This is attributed to the greater photocurrent production ability of C_{60}-doped film compared to the other dopants, as indicated in Table 1 below.

In untreated or dye-doped samples, it was observed [10] that long lived, though not permanent, gratings can be created under similar experimental conditions reported here; these gratings decay in tens of minutes. On the other hand, in C_{60}-doped nematic films, the electro-optically written grating persist for months (since the effect was observed and the writing of this report) without degradation in diffraction efficiency.

These optical storage effects have been observed in homeotropically (director axis perpendicular to cell walls) and planarly (director axis parallel to the cell wall) aligned nematic films. Figure 2 shows the orientation of the liquid crystal axis and the incident optical fields in a homeotropic sample. The initial alignment of the liquid crystal is obtained by treating the transparent ITO-coated window with a surfactant HTAB. The

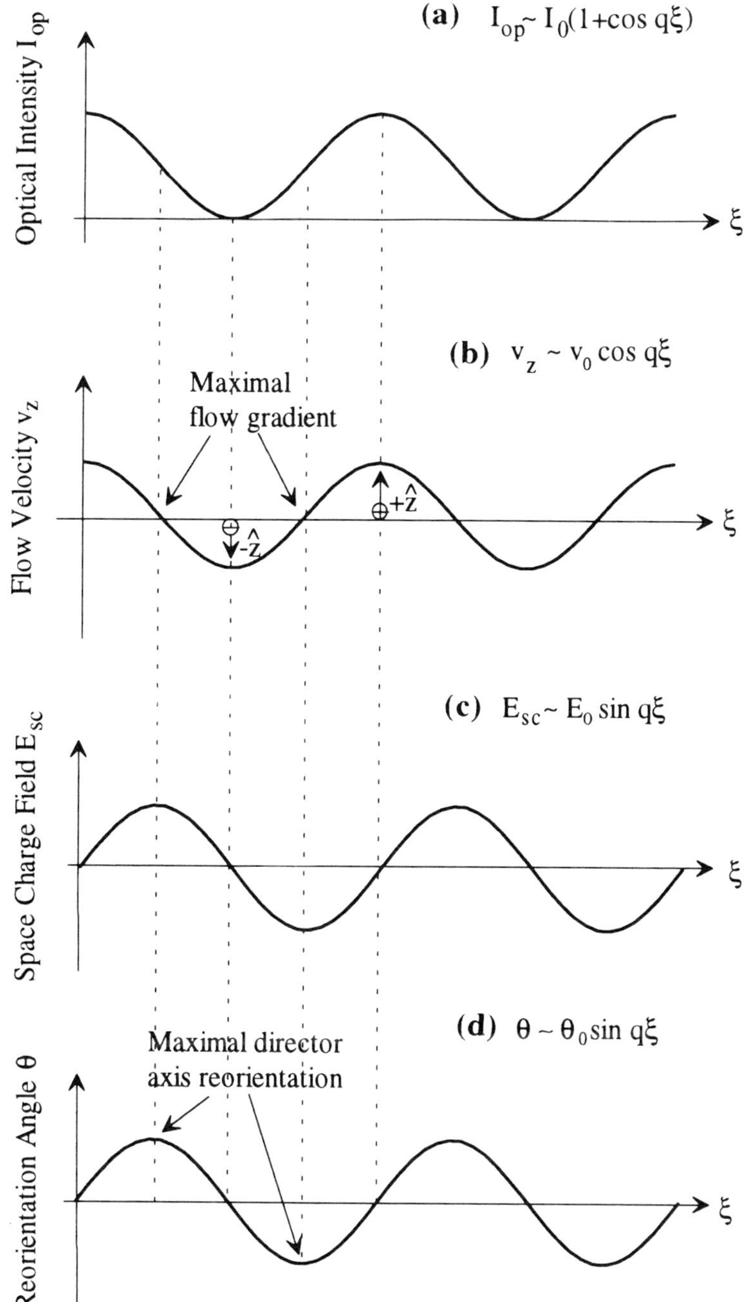

Figure 1 (a) Incident sinusoidal optical intensity spatial distribution.
(b) Flow of charges and their spatial distribution.
(c) Space-charge field distribution.
(d) Director-axis reorientation profile.

Table 1

Photoinduced current in 100 μm thick homeotropically aligned nematic films doped with several dyes. Laser power used $\approx$100 mWatt, applied dc voltage of 1 Volts. Sample area under illumination is 4 mm^2. The experimental geometry used is that of figure 2 with $\beta = 0$ and a single incident beam.

Dopants	D2	R6G	C_{60}	Undoped
I_{dark} (Amp)	3.5×10^{-8}	2.9×10^{-8}	6.0×10^{-8}	0.5×10^{-8}
$I_{illuminated}$ (Amp)	4.2×10^{-8}	1.3×10^{-7}	1.3×10^{-7}	0.5×10^{-8}
ΔI (Amp)	0.7×10^{-8}	10.1×10^{-8}	7.0×10^{-8}	0
α (cm^{-1})	373.2	101.7	30.2	0
$\frac{\Delta I}{\alpha}$ (Amp cm)	1.88×10^{-11}	9.93×10^{-10}	2.32×10^{-9}	NA

Maximum solubility dopant concentrations by weight: C_{60} (0.05%); R6G (0.2%).

two coherent incident laser beams are obtained by splitting an Argon-laser operating at the 5145Å line. They overlap at a wave mixing angle α on the sample, with the direction of propagation of the +1 beam making an angle β with the director axis. The thickness of the sample ranges from 25 μm to 100 μm. The concentration of C_{60} and 5CB, corresponding to maximum solubility is 0.05% by weight. Sample made with such dopant concentration yield an absorption constant of about 30 cm^{-1} [measurement from a 100 μm thick sample for $\beta = 45°$, where the scattering loss contribute a further 11 cm^{-1}]. These absorption and scattering losses become practically negligible in 25 μm-thick films. A cw polarized He-Ne (not shown in figure 2) is used to probe the polarization and time dependence of the induced gratings.

The photo in figure 2 shows a typical observed transmitted +1 and -1 beams and their side diffractions. In such thin phase gratings, several orders of diffraction are generated, c.f, photo in figure 2. Up to four orders are easily observable by eye under the experimental condition: wave mixing angle $\alpha = 0.02$ rad, $\beta = 22.5°$, +1 beam power = 10.3 mWatt, -1 beam power = 7.6 mWatt, beam diameter = 3mm, dc voltage = 1.5 Volt, film thickness = 25 μm.

We have observed that, in general, the induced gratings consist of a "fast" build-up/decay but transient component, and a much slower but persistent component. The decay and build-up times of the transient component (corresponding to transient electro-optically induced director axis flow-reorientation effect) are on the order of 1 sec., as schematically depicted in figure 3a. On the other hand, the build-up of permanent grating component (irreversible distortion of the director axis alignment); which follows the fast initial transient shown in figure 3a, is a very much slower process. With a

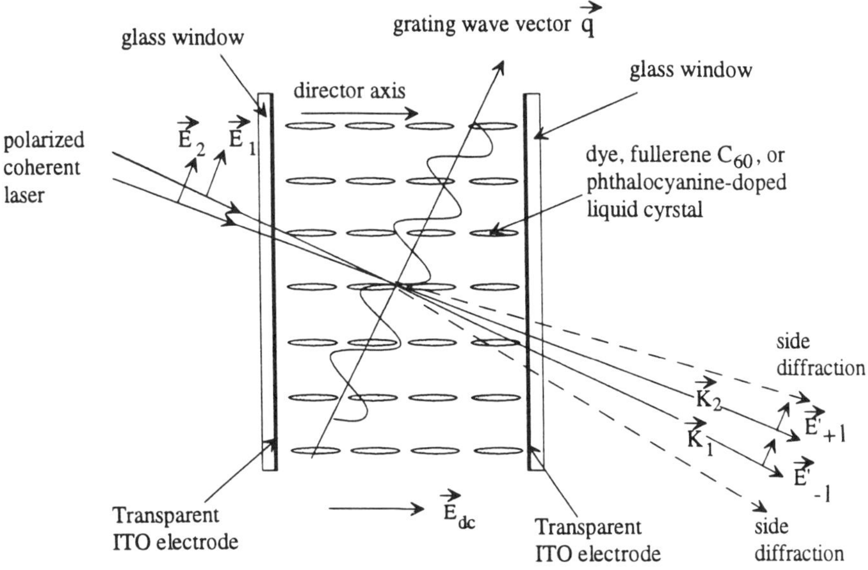

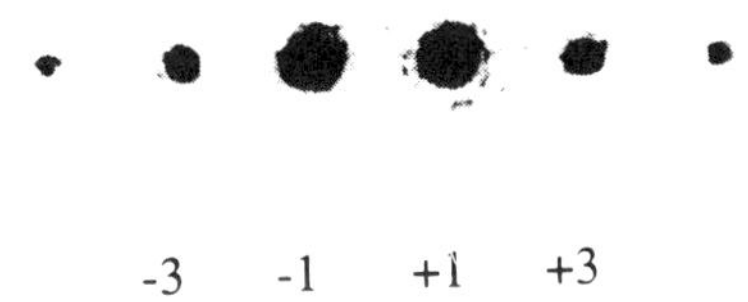

Figure 2 Schematic depiction of the geometry of interacting fields in a nematic liquid crystal film. Insert is a photograph of the transmitted input beams and their self-diffractions from a 100 μm thick C_{60}-doped film (applied voltage is 1.2 volt; wave mixing angle is ~2 mrad; input propagation wave vector makes an angle of 0.5 rad with the liquid crystal director axis).

writing beam power on the order on the order of 30 mWatt, the build-up time is about 10 minutes. If the writing beams are on for only a few seconds, as shown in figure 3b, practically no permanent grating is generated.

The time required for inducing observable permanent grating (director axis distortion) can be considerably shortened if higher-power writing beams and/or higher dc voltage are used. We have found that under writing beam power on the order of 500 mWatt (beam diameter = 3mm), and a dc voltage of 10 V, permanent gratings of a few % in diffraction efficiency can be generated in about 2 sec. High optical power/dc applied voltage, however, usually result in grating with high degree of depolarization. This is due to the higher degree of turbulence and randomness in the director axis reorientation caused by the large current and nematic flows under high optical and dc fields. More quantitatively, consider the ratio of the first order diffraction efficiencies R $= \eta_{\|}/\eta_{\perp}$ ($\eta_{\|}$ and $\eta_{\perp}$: efficiencies for He-Ne polarization parallel and perpendicular to the Argon writing beam respectively). For low power (a few mWatts) writing beams and low dc voltage (1-2 volt) R $\approx$ 20 whereas R $\approx$ 1 for writing beams power around 10^2 mWatts and/or dc voltage of several volts. Therefore, the use of high dc voltages should generally be avoided, as they tend to degrade the sample and also cause dynamic scattering effect [11].

Because of the intermolecular torque between molecules situated in the optical intensity maxima with those in the minima, and the spatial dependence of space charge fields accumulation, the diffraction efficiency of the permanent grating is strongly dependent on the grating constant. Such dependence is characteristics of similar optical [12] and conductivity-induced torques and electrodynamical effects on nematic liquid crystals under dc field [11]. Figure 4 shows the first order diffraction efficiency from permanent gratings with grating constants ranging from 5-150 μm. It is evident that there is an optimal grating constant ($\Lambda_{opt} \approx$ 30 μm) where the highest diffraction efficiency is achieved. Note that Λ_{opt} is on the order of the film thickness. These observations on the growth time for the grating, and the dependence of the grating efficiency on the grating constant (with $\Lambda_{opt} \approx$ film thickness) support the interpretation that the permanent grating formation is initiated by orientation--flows (both current and nematic material) instabilities [11] caused by the applied optical and dc field.

By placing an object such as a star-shaped aperture or a lens on one of the input beams, we can record a hologram of these objects. Holograms so constructed can be read by the reference beam; the first order diffraction (along the direction of the input object beam) from these holograms will give a reconstruction of these objects. Figure 5a shows a reconstructed holographic image of a input star-shaped aperture (figure 5b). We have also demonstrated the focusing action of a hologram of an input converging lens.

An interesting and useful feature of these holographic gratings is that they can be electronically turned on and off, as the director axis of the liquid crystal response to an applied ac field through the usual electro-optical effect [11]. We found that a 200 Hz, 94 Volt ac field will completely realign the whole sample, i.e., turn off the grating with a rather fast response time of 0.2 millisecond; when the ac field is removed, the grating will recover with a response time of 20 milliseconds.

Because of their broadband birefrigence and transparency [13] throughout the visible and near- and mid-IR spectral range (0.4 - 4 μm), and low cost, liquid crystal films have been widely used in various image processing, storage and display device applications. In our preliminary study here as one may note from figure 3, η = 10%.

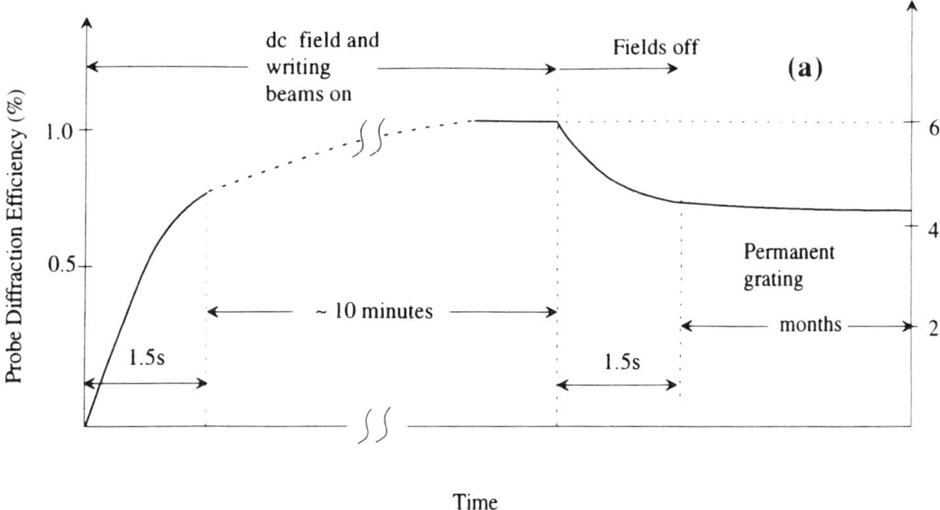

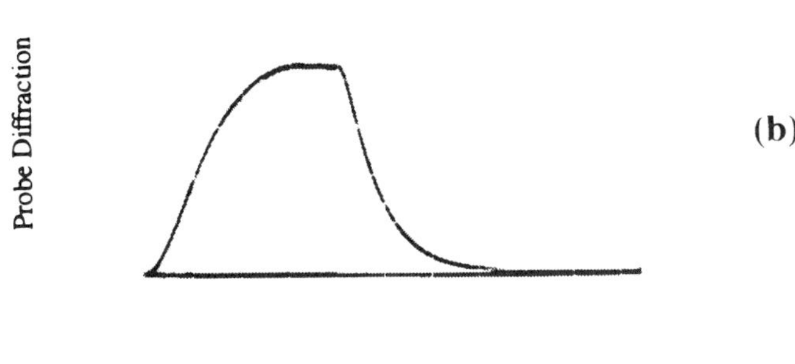

Figure 3 (a) Schematic depiction of the dynamics of transient and permanent grating formation in a C_{60}-doped nematic liquid crystal film. Experimentally observed value quoted here are for the case: film thickness = 25 μm; α_{air} = 0.024 rad; β = 22.5°; +1 beam power = 32 mWatt; -1 beam power = 21 mWatt; V_{dc} = 1.5 Volt; C_{60} dopant concentration = 0.05% by weight.

(b) Oscilloscope trace of transient grating diffraction from a 25 μm thick C_{60}-doped nematic film illuminated by 2-second long square pump laser pulses.

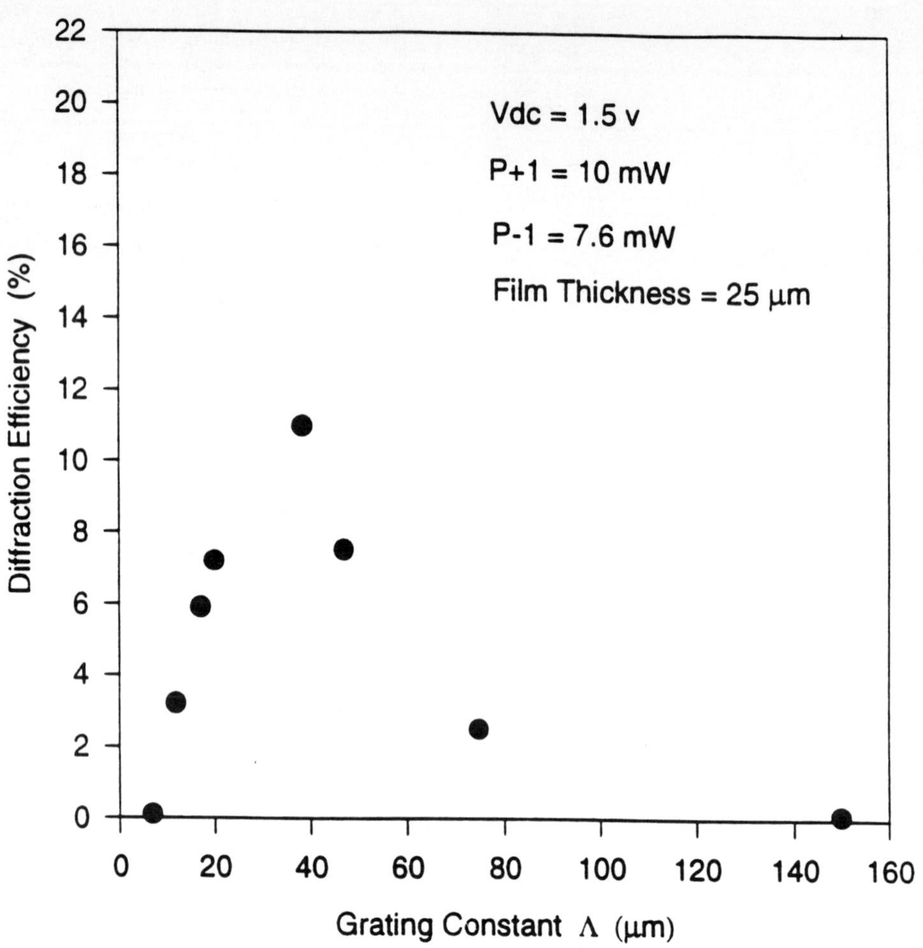

Figure 4 First order diffraction efficiency of permanent gratings of various grating constant generated in a 25 μm thick C_{60}-doped film. Writing time: 10 minutes.

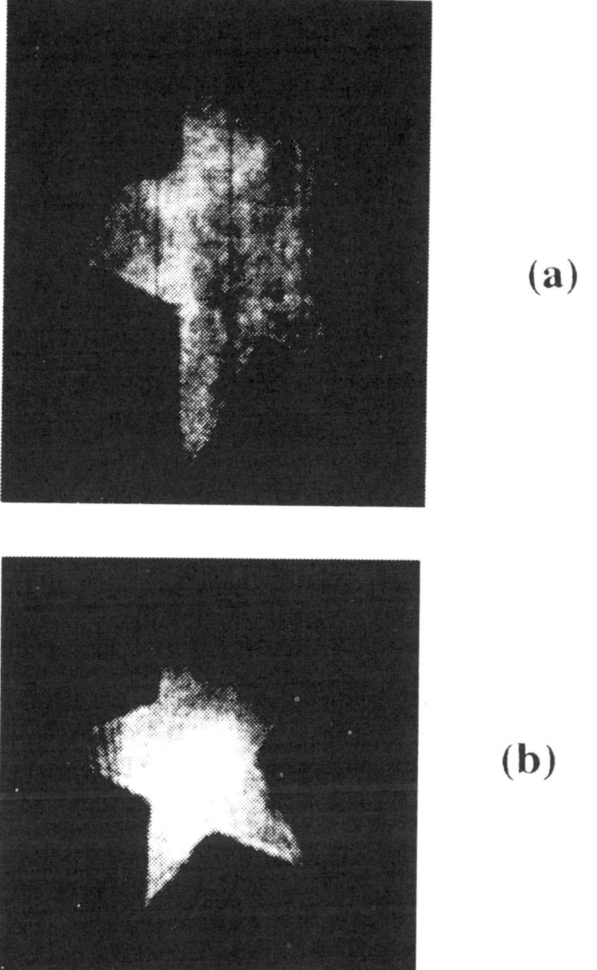

Figure 5 Permanent holograms of star-shaped object written in a 25 μm thick C_{60}-doped nematic film (a) Reconstructed image (b) Input image.

Since $\eta \sim (\Delta n\pi d/\lambda)^2$ for thin grating, this corresponds to $\Delta n \sim 2 \times 10^{-3}$. On the other hand, F ~ (10 mWatt) x 10 mins./$(3mm)^2$ = $60J/cm^2$). Therefore, the nonlinear sensitivity as defined in [8] $S = \Delta n/F$ obtained is on the order of 3×10^{-5} cm^2/J This value of S is already compared to those obtainable from photorefractive material and hydrogen treated germano-silicate glass [8]. Clearly, by using nematic liquid crystals in which higher concentration of C_{60} could be dissolved, and or different writing beam wavelength, the nonlinear coefficient can be improved. Studies along these lines are currently underway.

ACKNOWLEDGEMENT

This research is supported in part by the Air Force Phillips Laboratory and the Army Research Office.

REFERENCES

1. A. Kewitsch, M. Segev, A. Yariv and R. Neurgaonkar, Optics Lett. 18, 1262 (1993) and references therein. See also, G. Gunter and J. P. Huignard, ed., "Photorefractive Materials and Their Applications," Vol. I and II (Springer Verlag, Berlin, 1989).

2. S. Ducharme, J. C. Scott, R. J. Tweig and W. E. Moerner, Phys. Rev. Letts. 66, 1846 (1991); S. M. Silence, C. A. Walsh, J. C. Scott, T. J. Matray, R. J. Tweig, F. Hache, G. C. Bjorklund and W. E. Moerner, Opt. Lett. 17, 1107 (1992). Y. Zhang, Y. Cui and P. N. Prasad, Phys. Rev. B46, 9900 (1992).

3. W. M. Gibbons, P. J. Shannon, S. T. Sun and B. J. Swetlin, Nature 351, 49 (1991); Kunihiro Ichimura, Yasuzou Suzuki, Takahiro Seki, Yuji Kawanishi, Takashi Tamaki and Koso Aoki, "Reversible Alignment Change of Liquid Crystals Induced by Photochromic Molecular Films," Japanes J. Appl. Phys. 28, p. 289-292 (1989). Alan G.-S. Chen and David J. Brady, Optics Letts. 17, 1231 (1992).

4. I. C. Khoo and R. Normandin, Appl. Phys. Letts. 47, 350 (1985).

5. M. Eich and J. Wendorff, J. Opt. Soc. Am. B7, 1428 (1990).

6. See, for example, O. Werner, B. Fisher and A. Lewis, Opt. Lett. 17, 241 (1992).

7. R. A. Linke, T. Thio, J. D. Chadi and G. E. Devlin, Appl. Phys, Lett. 64 (1994).

8. A. Partovi, T. Erdogan, V. Mizrahi, P. J. Lemaire, A. M. Glass and J. W. Heming, Appl. Phys. Lett. 64, 821 (1994) and references therein.

9. K. Meerholz, B. L. Volodin, Sandalphon, B. Kippelen and N. Peyghambarian, Nature, 371, 497 (1994).

10. I. C. Khoo, H. Li and Y. Liang, Opt. Lett. 19, 1723 (1994).

11. W. Helfrich, J. Chem. Phys. 51, 4092-4105 (1969). See also L. M. Blinov, "Electro-Optical and Magneto-Optical Properties of Liquid Crystals" (Wiley Interscience, 1983) Chapter 5.

12. For photoconduction in fullerene, see for example, J. Mort, M. Machonon, R. Ziolo, D. R. Huffman and M. I. Ferguson, Appl. Phys. Lett. 60, 1735 (1992); Y. Wang, Nature 356, p. 585 (1992).

13. I. C. Khoo, "Liquid Crystals: Physical Properties and Nonlinear Optical Phenomena" (Wiley Interscience, NY, 1995). See also I. C. Khoo and S. T. Wu, "Optics and Nonlinear Optics of Liquid Crystals" (World Scientific, Singapore, 1993).

PATTERNING OF POLYDIACETYLENE WAVEGUIDES BY MEANS OF ARGON LASER PHOTOBLEACHING

L. Palchetti, D. Grando [1], E. Giorgetti, S. Sottini, Qu Li [2]

Istituto di Ricerca sulle Onde Elettromagnetiche - CNR,
Via Panciatichi 64, Firenze, Italy
[1] Politecnico di Bari, Via Orabona 4, Bari, Italy
[2] Shanghai Jiao Tong University,
1945 Huan Shan Road, 200030 Shanghai, People's Republic of China

INTRODUCTION

There is currently a noticeable interest in the investigation of the nonlinear optical properties of conjugated polymers. Even if the materials we could test so far, in particular poly-3BCMU, have a third order susceptivity $\chi^{(3)}$ which is not sufficient for practical purposes, however great hopes still lie in new organic materials - based on recent theoretical achievements - that can fulfil all the requirements of future nonlinear devices[1].

In particular, tests on optical waveguides, consisting of polymer films are attractive because waveguides allow longer interaction lenghts and maximize the energy density of the light pulses. Therefore, the realization of good film guides and, in particular, their patterning deserve primary attention. Here some experimental studies are presented on photoinduced bleaching of poly-3BCMU. It will be shown that bleaching is a viable technique to produce different kind of waveguide components.

PHOTOBLEACHING OF POLY-3BCMU BY AN ARGON LASER BEAM

Spin coated poly-3BCMU films were used with a thickness ranging between 0.35 and 3 μm; the bleaching was carried out with the 488 nm line from a CW Ar laser. Although the process is not yet fully understood, the following two equations are sufficient to describe the bleaching [2]. Let us suppose, indeed, that a monochromatic beam impinges normal to the surface of a film of thickness d and refractive index n. Some molecules inside the film absorb photons and become bleached, so increasing the transparency of the material: the decay rate of the absorbing molecules of density N, is given by:

$$\text{(1)} \qquad \frac{\partial N(x,t)}{\partial t} = -\eta \cdot I(x,t) \cdot N(x,t)$$

where I is the beam intensity at time t and depth x inside the film and η is the bleaching quantum efficiency. Eq. 1 is valid only for a single photon process.

The decrease of I as the beam propagates inside the film is described, as usual, by Beer's law and is given by:

Polymers and Other Advanced Materials: Emerging Technologies and Business Opportunities. Edited by P. N. Prasad *et al.*, Plenum Press, New York, 1995

$$\frac{\partial I(x,t)}{\partial x} = -I(x,t)\cdot\left[\alpha \cdot N(x,t)\right] \quad (2)$$

$\alpha \cdot N(x,t)$ is the local absorption coefficient due to unbleached material: the bleached material gives a contribution to the absorption, which can be neglected.

Eqs. 1,2 should be integrated numerically, with appropriate boundary conditions, in order to have a full description of the process. However, an approximate solution can be obtained for highly absorbing materials. In this case, although the bleaching is an indiffusion process, the effects of the impinging light are localized into a thin surface layer: the light propagates inside the film only after the complete bleaching of such layer [2]. Let us define the total transmission of the polymer film as $T(t) = I(d,t)/I_0$, where $I(d,t)$ is the transmitted intensity at time t and I_0 the impinging intentity; if the reflections at the interfaces can be neglected, T(t) is given by:

$$T(t)=\exp\{-\int_0^d \alpha \cdot N(x,t)\, dx\} \quad (3)$$

By deriving eq. 3 with respect to t and substituting from eqs. 1,2, we have:

$$\frac{dT}{dt}=T(t)\cdot \alpha \cdot \eta \cdot \int_0^d \frac{1}{\alpha}\cdot\frac{\partial I}{\partial x}dx=T(t)\cdot \eta \cdot [I_0-I(d,t)] \quad (4)$$

During the bleaching, a part of the film becomes transparent and is abruptly separated from the unbleached absorbing material; due to the high absorption, that is $T(t) << 1$, $I(d,t)$ can be neglected with respect to I_0 and the rate of change of the transmission turns out to be:

$$\frac{dT}{dt}=\eta \cdot T \cdot I_0 \quad (5)$$

A subsequent integration of eq. 5 gives:

$$T(t)= \exp\left[-A\cdot(d-vt)\right] \quad (6)$$

The constant $v = \eta \cdot I_0/A$ represents the propagation velocity of the bleaching inside the film and A is the total absorption coefficient of the unbleached film. For low bleaching intensity and $t << 1/\eta \bullet I_0$ eq.6 can be further simplified:

$$T(t)= (1+A\cdot v\cdot t)\exp(-A\cdot d) \quad (7)$$

Several measurements of T(t) have been performed. Only the reflections due to the glass substrate-air and film-air interfaces had to be taken into account, being the other reflections negligible with respect to the variations of T(t) due to the bleaching. Fig.1 shows T versus t in the case of a 0.89 μm thick film and $I_0 = 260$ W/cm^2. The straight line in the figure represents the theoretical behaviour, as predicted by eq. 6, and its slope gives the bleaching rate v: the agreement is good up to a value of $T = 0.1$, corresponding to a residual unbleached layer of only 0.19 μm. In Fig.2, T is plotted versus time for different values of I_0: such curves allow the evaluation of the bleaching rate v as a function of I_0. A plot of v

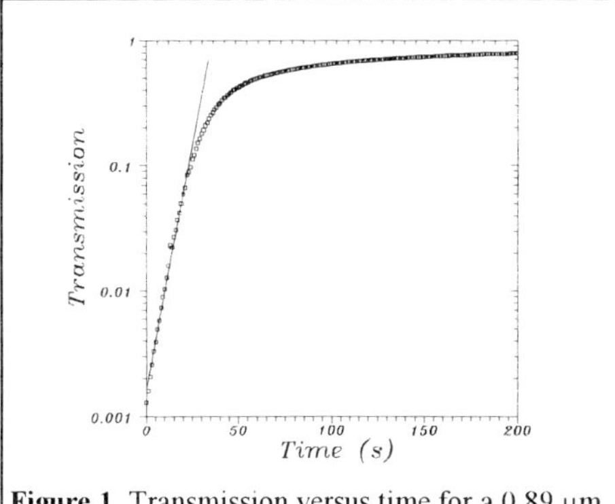

Figure 1. Transmission versus time for a 0.89 μm thick film with 260 W/cm^2 power density.

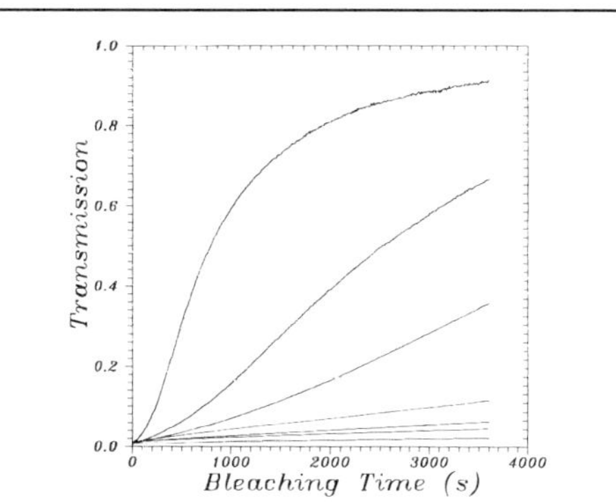

Figure 2. Transmission versus time at bleaching intensities from 1.24 to 28.9 W/cm^2 for a 1.4 μm thick film.

versus power density is shown in Fig. 3. A linear behaviour is found for $I_0 < 8$ W/cm^2; for larger values of I_0 the curve is no longer a straight line, indicating that different processes, such as multiphoton absorption or ablation occurr.

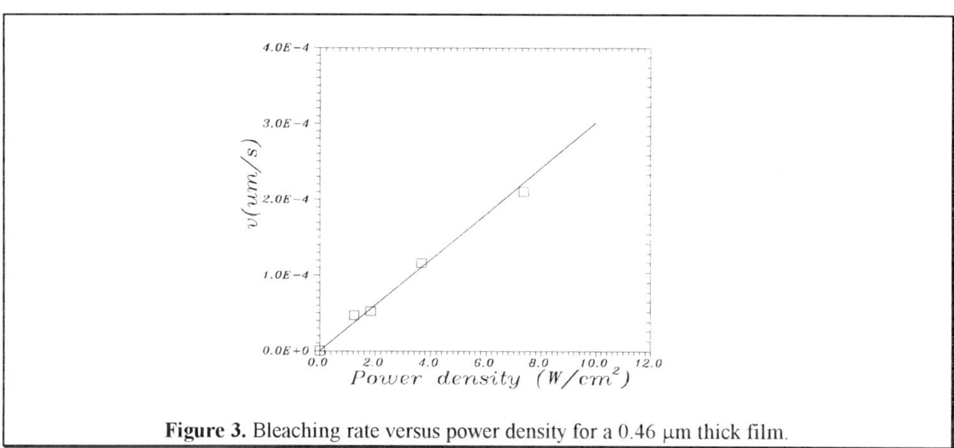

Figure 3. Bleaching rate versus power density for a 0.46 μm thick film.

The increase in transmission of bleached films is caused by the change in the absorption spectrum. As already reported in ref. 3, we have measured the absorption spectra before and after the film was bleached. The absorption peaks in the visible region become weaker as the bleaching time increases. Meanwhile, small peaks appear in the UV region indicating that the backbones are broken up into shorter ones. Indeed, analogous peaks are present also in the monomer spectrum. At the end of the bleaching process the UV peaks disappear completely indicating that also the conjugated diacetylene bonds of the monomer are broken. This allows the so called "development" of the bleached film, that can be completely removed by a proper solvent, while the unbleached film remains unaffected. For example, for a 0.35 μm thick poly-3BCMU film total exposure of about 10 KJ/cm^2 at 488 nm is needed in order to completely remove the bleached film with acetone.

The bleaching also changes the refractive index of poly-3BCMU. For example, an unbleached poly-3BCMU film on quartz has a refractive index of 1.68 for 849 nm light with electric field in the plane of the film, while the bleached one has an index of 1.51.

The fine control of v and of the refractive index difference Δn between the bleached and unbleached material can be exploited to fabricate different kinds of guided wave structures such as channel waveguides, gratings, tapered guides, waveguide lenses [4,5]. However, a controlled patterning of polymer films requires a further study of the physics of the bleaching process, in order to estabilish the optimum working conditions and the possibility of using standard photolithographic techniques.

It is well known that oxygen is necessary in photoinduced bleaching of polydiacetylene. Is the number of oxygen molecules already diffused in the poly-3BCMU film at the beginning of the bleaching process enough for completely bleaching the film? To answer this question we carried out an experiment in which a two-month-old poly-3BCMU film (1 μm thick) was bleached before and immediately after treatment in a vacuum chamber (10^{-6} Torr for 6 h). Fig. 4 shows the result. No difference can be seen between these two processes, indicating that at least the major part of the oxygen, if not all, necessary for photoinduced bleaching is not present in the film when the bleaching process begins. The oxygen molecules continuously diffuse into the film during the process.

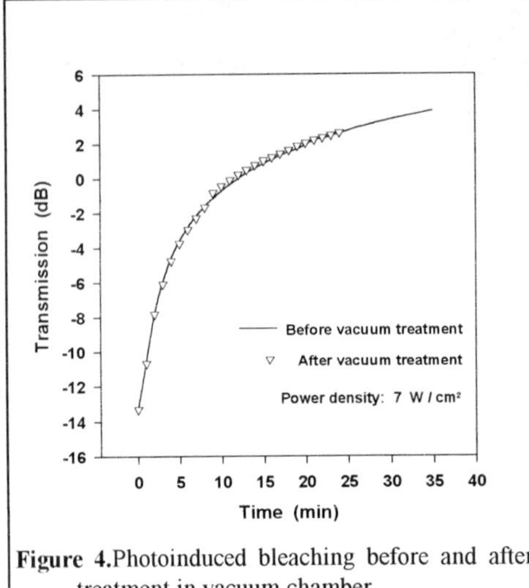

Figure 4. Photoinduced bleaching before and after treatment in vacuum chamber.

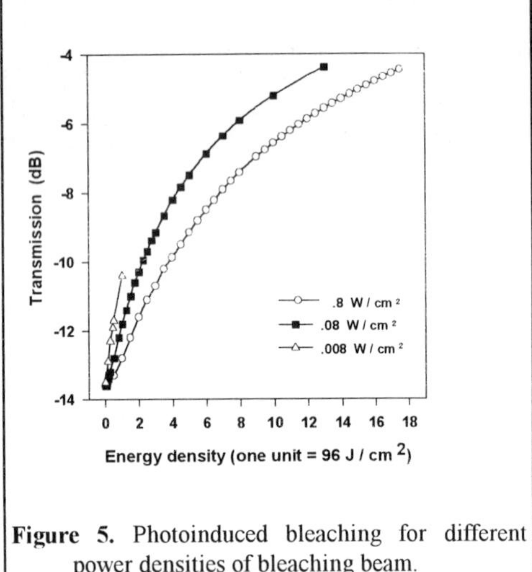

Figure 5. Photoinduced bleaching for different power densities of bleaching beam.

Since the diffusion rate of oxygen into poly-3BCMU films is limited, the degree of bleaching is expected to be determined not only by the energy density of the bleaching beam but also by the power density of it. Fig. 5 shows the bleaching processes of a 1 μm thick film with different power densities. The scale of abscissa is normalized to energy density with one unit corresponding to 96 J/cm^2. It can be seen that, as expected, the process is more efficient for lower power density, in the sense that for the same bleaching energy density the lower power density beam causes greater change in transmission. In another experiment we compared the bleaching processes with power densities being 2.8 W/cm^2, 28 W/cm^2 and 280 W/cm^2. The bleaching process at 28 W/cm^2 was less efficient than at 2.8 W/cm^2, as expected. The bleaching process with 280 W/cm^2 power density was more efficient than that with 28 W/cm^2, implying that at this high power density level some process other than bleaching, such as ablation, took place.

Some problems can arise when one tries to produce patterns using photolithographic techniques. When the photoresist layer on top of the poly-3BCMU film is exposed, the poly-3BCMU might be bleached resulting in unwanted changes in its linear and nonlinear properties. An experiment was carried out to see how significant this unwanted bleaching can be. A poly-3BCMU film (1 μm thick) was bleached before and after it was coated with a 0.4 μm S1400-17 photoresist layer. The bleaching beam was incident onto the film from the glass substrate side with power density of 7 W/cm^2. The result is shown in Fig. 6. One finds out that the photoresist coating produces a substantial reduction of the bleaching rate. A similar result was observed when the bleaching beam was incident from the air side. Since the change in transmission of the photoresist layer before and after exposure is much less than that of the poly-3BCMU film, the great difference between the bleaching rates before and after coating with photoresist was due to a change in the photoinduced bleaching process of the poly-3BCMU film. Lack of oxygen is responsible for this change. The photoresist layer prevents the oxygen from sufficiently diffusing into the poly-3BCMU film during the bleaching process, so that the process is slowed down. Curve c in Fig. 6 is based on the same data as curve a, except that the time scale is compressed by a factor of 10. Comparison between curve b and curve c shows that, during the first few minutes, the bleaching rate of the photoresist coated film was slowed down by a factor of about 10. Since the time sufficient for properly exposing a 0.4 μm S1400-17 photoresist layer with the power density used in this experiment is only a few seconds, we conclude that the production of patterns with lithographic techniques would not cause significant bleaching of the underneath poly-3BCMU film.

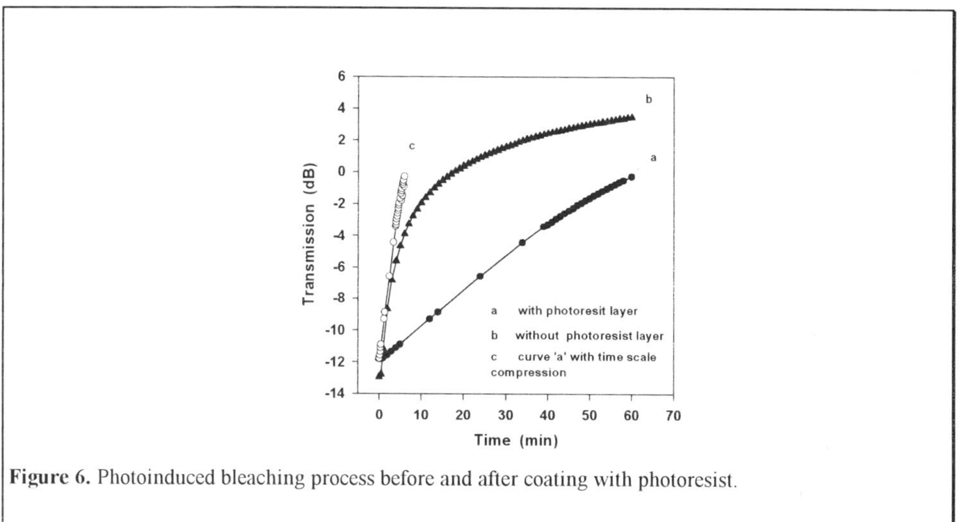

Figure 6. Photoinduced bleaching process before and after coating with photoresist.

WAVEGUIDE TAPERS, REFRACTIVE LENSES AND GRATINGS

The bleaching process has been tested by realizing different surface components as tapered waveguides, gratings, refractive waveguide lenses etc.

Tapered poly-3BCMU guides have allowed to couple efficiently light from a glass ion exchange guide, having a thickness compatible with fibre cores (4-10 μm), and a thin polymer guide, optimized for nonlinear tests [6]. The tapered coupler is shown in Fig. 7. It consists of a poly-3BCMU film, which is laid on top of an ion-exchanged glass waveguide (GIW) with non costant depth: the GIW is typically single mode in the input and output regions, while it disappears completely or is below cutoff, in the central region. These regions are connected by a smooth taper, which is designed to maximize the transmission efficiency; the organic film also has tapered edges. The structure includes a superlayer, with refractive index smaller than that of the polymer and is single mode everywhere at $\lambda = 1.321$ μm.

The poly-3BCMU film (0.85 μm thick) was tapered with the beam at 488 nm from a CW Ar-Kr laser. The beam was focused so that a power density of 33 W/cm^2 was obtained; the sample was then moved continuously, but with non costant velocity, in order to achieve a linear taper and the process was monitored in real time to evaluate the film transparency, which is related to the thickness of the unbleached film. At the end of the process, the bleached material was not removed, but, instead, it was kept as a superlayer in order to increase the symmetry of the waveguide and, consequently, the taper transmission [6].

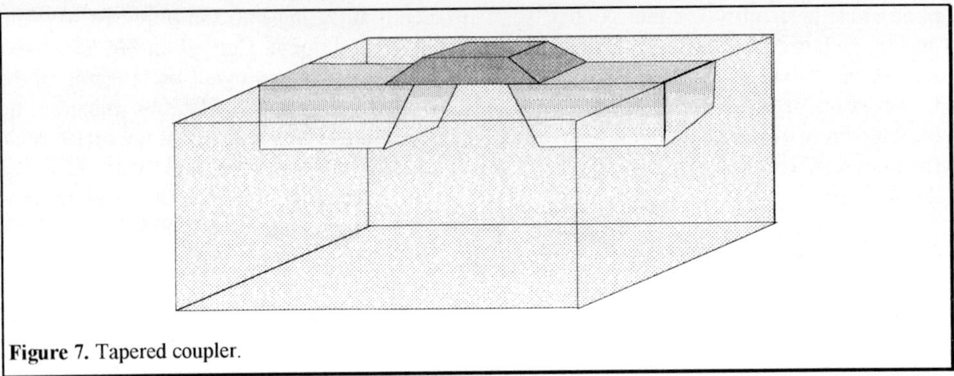

Figure 7. Tapered coupler.

Several samples of tapered coupler were fabricated and tested: the experimental insertion losses were below 3 dB at $\lambda = 1.321$ μm, in good agreement with the theoretical modelling of the structure. The influence of the polymer taper on the coupler losses is negligible, while a considerable reduction of such losses is expected by improving the fabrication conditions of the glass waveguide taper.

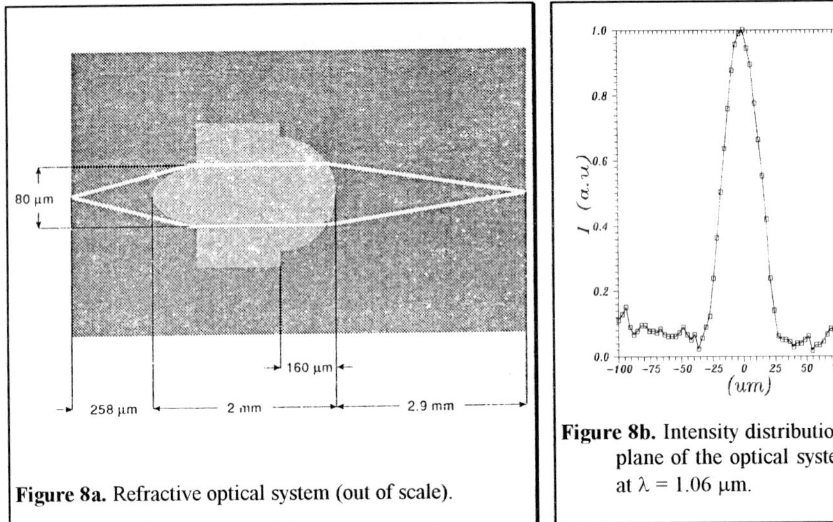

Figure 8a. Refractive optical system (out of scale).

Figure 8b. Intensity distribution in the focal plane of the optical system, measured at $\lambda = 1.06$ μm.

An optical system (Fig. 8a) based on refractive microlenses has been fabricated on a poly-3BCMU film of 3 μm thickness. The dark and light regions in Fig.8a correspond to partial and no bleaching, respectively. The system consists of an input hyperbolic interface and an output circular interface enclosing the unbleached nonlinear material. The input lens is able to transform the output of a single mode fiber with NA = 2.2 into a 80 μm wide collimated beam, while the output lens can refocus on-axis and off-axis collimated beams on the waveguide output plane. The system was fabricated by partial bleaching through a mask: the degree of bleaching was chosen so that the difference Δn between the effective indices of the guided modes in the bleached and unbleached regions satisfied the theoretical requirements of the optical system.The focusing characteristics of the system were measured at λ = 1.06 μm with a CW Nd-YAG laser and are shown in Fig. 8b. They are in good agreement with the theoretical predictions: in practice, it means that Δn could be controlled with a precision better than 3 %.

Gratings are attractive elements for nonlinear testing. Therefore, we have realized some holographic diffraction gratings, both on photoresist films laid on top of poly-3BCMU guides and directly on poly-3BCMU film guides. In the following, we will limit to the latter case. The gratings were recorded with an argon source at 488 nm, by following the procedure described in ref. 7; the depth of the bleached film was evaluated from eq. 7.

Our results can be summarized referring to a set of four gratings of period Λ = 0.94 μm. They were recorded using a power density of 0.45 W/cm^2; their diffraction efficiency η was measured at λ = 0.849 μm and for normal incidence. In this condition only the first diffracted order exists, therefore $\eta = I_1 / I_i$, where I_1 is the diffracted intensity and I_i is the impinging one. The measured diffracted efficiency was compared with the calculated one, obtained by following the theory of thin gratings (Fig.9). As the refractive index of the polymer and the index difference between the bleached and the unbleached film are known independently, the theoretical curve representing the behaviour of η versus grating thickness can be easily found (— in Fig.9). Finally, the experimental points corresponding to different exposure times (□ in Fig.9) can be positioned on the theoretical curve by taking advantage of the previously estabilished relation between depth and bleaching time (Fig.3).

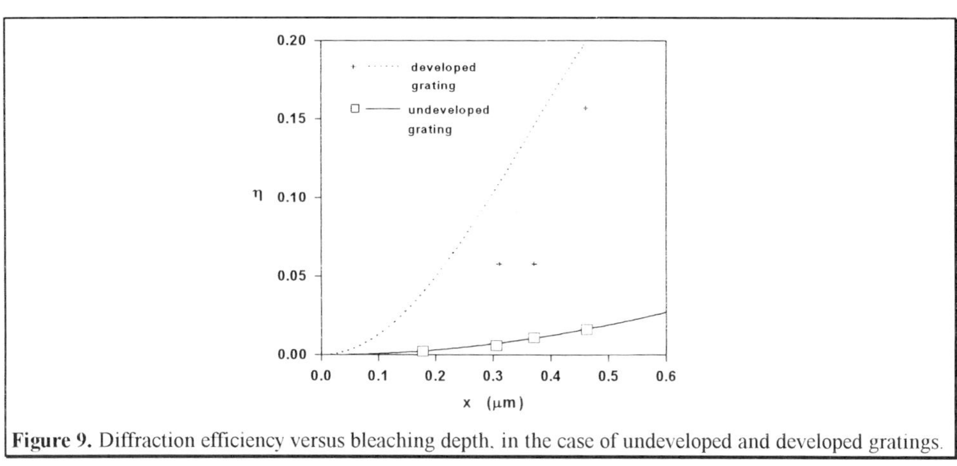

Figure 9. Diffraction efficiency versus bleaching depth, in the case of undeveloped and developed gratings.

Successively, the gratings were developed. One sample was damaged during this process; the efficiency of the remaining samples (+) was also compared with the theory (···). In this case the agreement is not good; the main reason could be related to the development procedure in acetone, which might have dissolved some unbleached polymer at the boundary, reducing the grating depth.

CONCLUSIONS

Photobleaching of spin coated poly-3BCMU films has been studied in order to confirm the interesting potentiality of this technique for polymer based integrated optics. In particular, the control of the bleaching velocity inside the films has allowed the fabrication of waveguide tapers and refractive lenses: their performances were in good agreement with the theory. Diffraction gratings showed also good performances before they were developed in acetone, while, after development, their efficiency resulted to be 15-30% lower than the expected value: this is probably due to a reduction of the grating depth during the treatment with acetone.

Due to their photosensitivity, our polymer devices must be carefully preserved by visible ligth, while no significant damage has been found so far due to exposure to low intensity NIR radiation (.8-1.5 μm).

REFERENCES

1. G. Stegeman, *Optics and Photonics News*, 26, (1993)
2. R. S. Moshrefzadeh, D. K. Misemer, M. D. Radcliffe, C. V. Francis, and S. K. Mohapatra, *Appl. Phys. Lett.*, 62, 1:16 (1993)
3. S. Sottini, D. Grando, E. Giorgetti, L. Palchetti, Qu Li, G. Gabrielli, M. Nocentini, and G. Sbrana, *Mol. Cryst. Liq. Cryst.*, 235:191 (1993)
4. K.B. Rochford, R. Zanoni, Q. Gong,, and G. Stegeman, *Appl. Phys. Lett.*, 55:1161 (1989)
5. M.B.J. Diemeer, F.M.M. Suyten, E.S. Trommel, A. McDonach, J.M.Copeland, L.W. Jenneskens, and W.H.G. Horsthuis, *Electr. Lett.*, 26:379 (1990)
6. S. Sottini, D. Grando, L. Palchetti, and E. Giorgetti, *IEEE Journal of Quantum Electronics,* 31 (in press) (1995)
7. M. Xu, R. Moshrefzadeh, U.J. Gibson, G. Stegeman, and C.T. Seaton, *Appl. Opt.*, 24:3155, (1985)

ACKNOWLEDGMENTS

This work was carried out under the Italian Projects on Telecommunications and MSTA. The authors gratefully acknowledge Dr. M. Sparpaglione, ENIChem s.p.A., Italy, for the synthesis of the 3BCMU and Dr. V. Skarda and Dr. D. Westland, ERA Technology Ltd., England, for its polymerization.

POLYMERS AS MULTI-ROLE MATERIALS FOR PHOTONICS TECHNOLOGY

Paras N. Prasad

Photonics Research Laboratory
Department of Chemistry
State University of New York at Buffalo
Buffalo, NY 14260-3000

ABSTRACT

Photonics has been labeled by many as the leading technology for the 21st century. Polymers have emerged as an important class of materials which will play multiroles such as laser light source, passive optical interconnects, optical signal processing and optical data storage devices for photonics. In this paper a brief background will be presented on polymer's multirole in photonics with a special emphasis on nonlinear optical processes, the latter providing many of the operational functions for the photonic technology. New applications such as polymer lasers, polymer amplifiers and optical circuitry are discussed. Applications of polymers and composites in three dimensional holographic data storage are also covered. Particularly interesting new media for optical data storage are photorefractive polymers. The photorefractive polymeric systems utilize the combined action of photoconductivity and nonlinear optical effect. The development of photorefractive polymers in our laboratory is discussed.

INTRODUCTION

Polymers are emerging as an important class of optical materials for photonics technology.[1,2] The most attractive feature of the polymeric materials for photonics is the flexibility of tailoring their properties at both molecular and bulk levels to introduce multifunctional optical response. At the molecular level, one can use design and chemical synthesis to introduce optically responsive groups in the main chain or in the side chain. Many optical properties are dependent on the delocalization length of π-conjugated segments. To judiciously control these properties one can prepare a well defined oligomeric unit or a block co-polymer containing a segmented conjugation unit.

At the bulk level the multifunctionality can be introduced with microstructure control by processing and field or stretch induced orientation, or by making a homogenous polymer blend. Another important concept involves a composite where the phase separation is only on

Polymers and Other Advanced Materials: Emerging Technologies and Business Opportunities. Edited by P. N. Prasad *et al.*, Plenum Press, New York, 1995

nanometer scale, much smaller than the wavelength of light, so that the material is optically transparent. In this regard an important development in our laboratory is that of a multiphasic nanostructured composites whereby the multifunctionality is introduced by optimizing each phase for one functionality but controlling the phase separation on the nanometer range. Again the result is that the composite is optically transparent even though it consists of multiphases. This class of material will be discussed later in this article.

The four principal components of the photonics technology are: (i) optical source, (ii) optical transmission system, (iii) optical processing unit, and (iv) optical data storage unit. Polymeric composite materials can play an important role for the fabrication of all these four types of components. In this article, for each of these applications, ongoing research activities in our laboratory are presented along with a discussion of the current status.

Optical Source

Optical source provides the optical signal for processing, transmission and storage. Inorganic semiconductor diode lasers are currently being used for this purpose.[3] Although, they are not broadly tunable, diode lasers do offer a wide range of choice from 1.55 μ to 1.3 μ for optical telecommunication, and at much shorter wavelength (<830 nm) for local area network. In long distance transmission such as fiber-optics telecommunication, one needs optical amplifiers to compensate for transmission loss. One promising amplifier is an erbium amplifier which works at 1.55 μ.

Recent development in the design and processing of polymeric materials has shown promise that polymeric materials may be used as an optical source and an optical amplifier. Polymeric light emitting diodes are a subject of considerable research activities.[4-9] The diodes are easy to fabricate. They have shown considerably improved efficiency, although still inferior to an inorganic LED. Another advantage of the polymeric structure is the tunability of the emission wavelength. Polymeric LEDs covering from the red to the blue spectral range have been demonstrated.[4-9] The important issue to address is the operational lifetime of the polymeric LED. Although LEDs can be used for applications in photonics, a more desirable optical source is a coherent source such as laser diodes. Our research program focuses on exploring the feasibility of producing polymeric laser diodes. In our program molecular designs of suitable structures include chemical units which are highly efficient emitters as well as units which permit charge injection and transport. A side-chain copolymer synthesized and currently being investigated in our laboratory is shown in Figure 1. It consists of unit A in the side-chain which is highly fluorescent and unit B, also in the side chain, which permits charge injection and hole transport. This polymer when blended (1%) with poly-n-vinyl carbazole has been used to achieve lasing when optically pumped by the 532 nm second harmonic of an Nd:Yag laser. The same film also shows good electroluminescence. However, the gain achieved by electroluminescence is not sufficient to achieve population inversion or lasing in this material. It will be a formidable challenge to optimize structures where electrical pumping is highly efficient to create population inversion. Our current effort is centered on addressing this issue.

Polymeric structures also offer prospect of amplification. Their flexibility to be processed in forms of channel waveguides or fibers is one of the attractive features. However, a much broader and important advantage offered by polymeric materials is that, by the proper choice of the chromophore, amplification can be achieved for a broad range of wavelengths. Recent advances have shown that organic structures can emit even at wavelengths as long as 1.3 μ.[10] In our laboratory, we have used sol-gel processed inorganic oxide (such as SiO_2) fibers containing dyes to investigate optical amplification. The sol-gel processing, being a low temperature (<200°C) chemical processing of glass, has permitted us to prepare dye doped fibers which are ~25 μ wide. In such fibers we have successfully coupled light and obtained

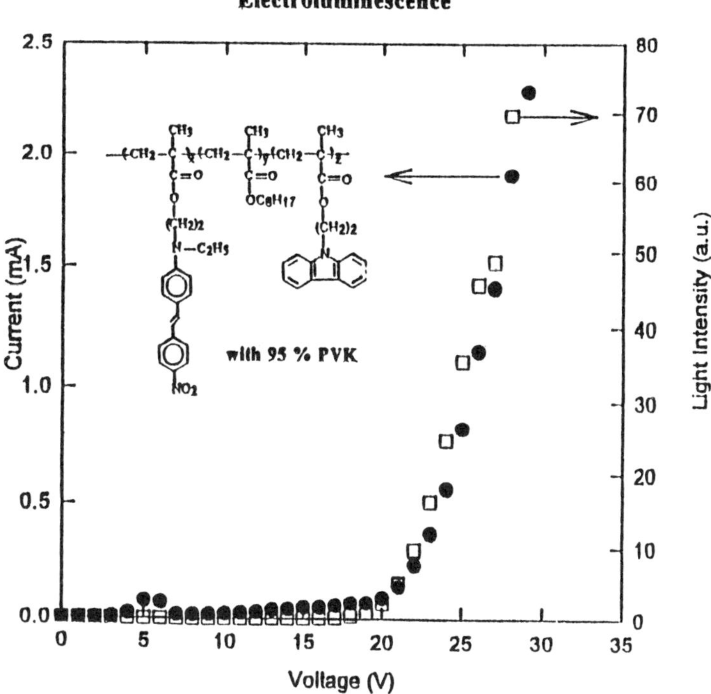

Figure 1. The electroluminescence behavior of the copolymer blended with 95% wt PVK.

superradiance which demonstrates that it acts as a gain medium. Our current effort is focusing on designing and synthesizing dyes which efficiently emit in the near IR region to be used as amplifiers at 830 nm and 1.3 μ.

Optical Transmission

The optical transmission may require a passive or an active optical circuit. The passive circuit or transmission network simply provides the function of transmission line where as an active optical circuit also performs optical signal processing discussed in the next section. In the latter case, the medium is an optically nonlinear medium.

For passive long distance transmission such as in optical telecommunication, very high quality glass fibers have been developed which exhibit no group velocity dispersion (at 1.3 μ) that leads to broadening of optical signal pulses.[3] It is difficult to imagine at this stage, how polymeric materials can replace the glass fibers for long distance transmission.

For short distance propagation, such as in optical circuit local area network, polymeric materials offer ease of fabrication. Channel waveguides can very conveniently be fabricated using a variety of processing techniques. However, channel waveguides fabricated using pure polymers still have somewhat higher optical losses; they do not offer a wide choice of refractive indices and do not exhibit the necessary thermal and environmental stability. Furthermore, the end facets of polymeric films are not sufficiently flat and polishable to efficiently use end-fire coupling to couple light into the channel waveguide. For these reasons, our laboratory has used sol-gel processed polymer:inorganic glass composites for the fabrication of optical waveguides. We have used polyvinylpyrrolidone polymer (PVP) processed with SiO_2 and TiO_2 to prepare planar and channel waveguides.[12] In our processing, the PVP polymer is mixed with the SiO_2 precursor (tetraethoxysilane) and TiO_2 precursor (titanium ethyl hexoxide) in the solution phase, their relative ratios varied to optimize the thermal stability and desired refractive index. The alkoxides are hydrolyzed under carefully controlled conditions and the sol is allowed to reach a certain viscosity before a film is cast. The film is then thermally treated above 200° C to induce cross linking in PVP. We have evaluated the performance of these films in terms of their planar waveguide optical losses. The optimum composition appears to be 50% PVP. The relative ratio of SiO_2 and TiO_2 are varied to achieve the desired refractive index. Figure 2 shows the variation of the refractive index vs. the TiO_2 relative composition. The

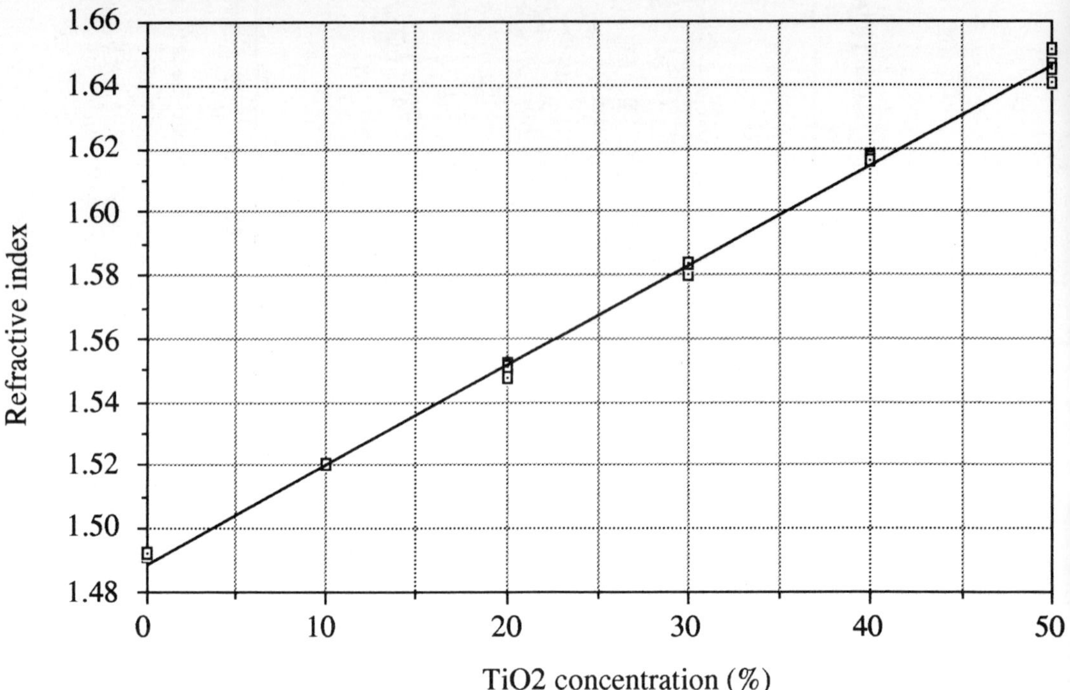

Figure 2. Refractive index change due to TiO2 concentration in PVP/SiO_2/TiO_2. PVP is fixed to be 50%.

films can be heated up to 200°C for densification before any deterioration of PVP occurs as evidenced by observed discoloration and a rapid increase in the optical waveguide loss.

To fabricate channel waveguides, we have used a combination of photolithography and selective laser densification.[12] A schematic of this fabrication approach is shown in Figure 3. It involves deposition of metal lines by using a photomask. Then the film is exposed to light which can be of any wavelength at which the PVP/SiO_2/TiO_2 composite does not absorb. The metal lines absorb and transfer the heat to the underlying films whereby densification occurs to produce regions of higher refractive index in the waveguide channels. We have produced channels which are 5 micron wide and 1 micron deep. The waveguide losses in these channels are 0.8 dB/cm at 633 nm which is considerably higher than that (<0.2 dB/cm) in the planar waveguide form. We are currently improving our processing technique to explore if channel waveguide losses can be further reduced.

Optical Signal Processing

Nonlinear optical processes provide key functions for optical signal processing. Organic and polymeric structures containing π-electrons have shown enhanced second and third order

temperature. A promising approach taken by many is where the polymeric structure is poled and, during poling, cross-linked to make it rigid and thus slowing down relaxation of poled alignment.[13-16] However, under cross-linking very often optical quality of the sample deteriorates.

In the development of electro-optic polymers in our laboratory, the objective has been threefold: (i) to use chromophores which offer wider optical transparency so that they can be used at diode laser wavelengths of 830-670 nm;(ii) to produce electrical alignment which is preserved for at least 1,000 hours after heating at 100°C and for 30 minutes after heating up to 300°C; (iii) to produce materials which can be used in a waveguide geometry with a very low propagation loss. A promising approach taken in our laboratory is shown in Figure 4.[17] The

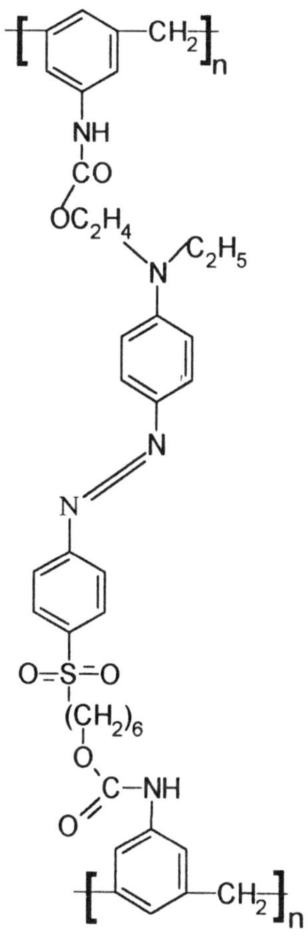

Figure 4. Cross-linked polyurethane structure containing the NLO chromophore unit.

chromophore containing -OH group is mixed with a liquid isocyanate polymer which can be thermally converted to a polyurethane cross-linked structure as shown in the figure. The film is cast after a very small degree of cross-linking and then maximum cross-linking is introduced during poling. The use of a liquid cross-linking polymer allows a more uniform mixing and cross-linking so that the optical quality does not degrade. We have achieved an r-coefficient of 20 pm/v at 633 nm. The measured planar waveguide loss at 830 nm is 1 dB/cm. By further optimizing processing conditions, we believe that this waveguide loss can be further reduced.

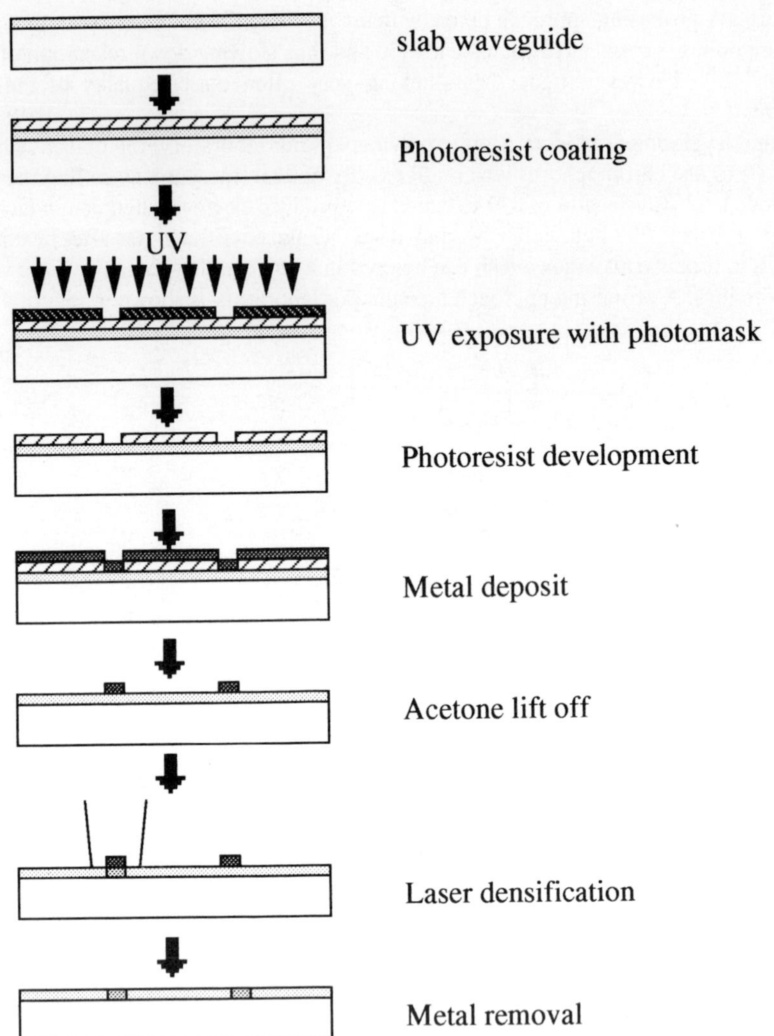

Figure 3. Fabrication technique for channel waveguides using laser densification.

nonlinear optical response even under non-resonant conditions.[1] However, it is the electro-optic material utilizing the second order nonlinear optical response which appears to be more developed and thus more likely to find applications in the near future in optical signal processing. One of the most important applications of electro-optic material is in switching devices for fiber-optics telecommunication or for a local area network. Although many polymeric systems with a large electro-optic coefficients, r_{ii}, have been reported, these electro-optic coefficients are generated by electric-field poling whereby the electro-optically active chromophores are aligned in a non-centrosymmetric arrangement. For a non-vanishing electro-optic coefficient which is related to the second-order nonlinear optical susceptiblity, $\chi^{(2)}$, symmetry restriction requires the medium to be non-centrosymmetric. A major concern in the application of the electrically poled polymeric structure is the temporal stability of the poled alignment. The poled alignment should not only be stable at room temperature but also at elevated temperature since fabrication of electro-optic devices may subject the material to high

Optical Data Storage

Polymeric media have been investigated for both permanent and erasable data storage.[2] For erasable optical data storage, a great deal of interest has been shown in photorefractive materials.[18] Photorefractivity in inorganic crystals such as $LiNbO_3$, $BaTiO_3$, $KNbO_3$, etc. have been widely investigated over an extensive period.[18] A more recent entry is by photorefractive polymers which offer advantages of ease of processing and cheaper cost over the traditional inorganic photorefractive materials.[19-22] Considering the relatively short history of photorefractive polymers, rapid advances have occurred and their photorefractive figure of merit is now comparable and even better in some regard than that of inorganic crystals.

Photorefractivity is a phenomenon which is produced by the combined action of photoconductivity and nonlinear optical properties of a material.[18] It is a result of the functions of photogeneration, transport, charge trapping, and the electro-optic effect. In a holographic experiment, two beams are crossed to produce intensity interference pattern. The action of light produces charge carriers which move to dark regions, get trapped and set up an internal space-charge modulation field. This space charge field then produces a refractive index modulation (refractive index grating) through the electro-optic effect. This is the holographic grating containing the optical information which can be read by the diffraction of a probe beam. The process of photorefractivity in polymeric systems is different from that in inorganic crystals in that the key parameters affecting the photorefractive behavior are electric field dependent for polymeric systems. The role of the applied dc electric field for the polymeric materials is crucial and multifunctional: (i) it enhances the charge-carrier generation; (ii) it stimulates the transport of the carriers via field-dependent mobility; and (iii) it forces a non-centrosymmetric alignment of $\chi^{(2)}$ chromophores. Polymeric materials for the photorefractive effect have to be optimized for charge photogeneration, transport, trapping and electro-optic effect. We have found that by using a polymeric composite structure each of these functionalities can be derived from independent components and thus separately optimized. A polymeric composite developed in our laboratory which exhibits a very high diffraction efficiency of ~50% at 633 nm consists of poly[N-vinyl carbazole] as the charge transporting matrix, fullerene C_{60} as photosensitizer, (diethylamino)nitrostyrene as the electro-optic chromophore and tricresyl phosphate as the plasticizer. The plasticizer lowers the Tg of the composite so that it can be electrically poled in situ at room temperature. Figure 5 shows the electric field dependence of the diffraction efficiency. It is apparent from this result that in order to achieve high diffraction efficiency, a very high electric field needs to be applied which is a major hurdle in practical application of these materials. The research effort in our laboratory is now directed towards creating a better understanding of the processes involved in order to design a composite which would work at lower fields. Another objective of our research on photorefractivity is to understand the kinetics of grating formation and decay. The decay process is particularly important for storage of information.

Multiphasic Nanostructured Composites

We have developed multiphase composite bulk using sol-gel processing which produces large size bulks of excellent optical quality.[23] In this approach, we prepare highly porous monolith gel (in the present case silica) and thermally process it. The pores in the silica monolith under our processing conditions are in the nanoscale region (~4 nm). This allows various molecules (such as fullerene) to be adsorbed in the pores by diffusion of a solution followed by evaporation of the solvent. Then the pores are filled with a polymerizable liquid such as methylmethacrylate (MMA) which is then polymerized in-situ. This MMA procedure first pioneered by Pope, Asami and Mackenzie[24] has been used by several groups. We have followed the procedure described by Gvishi et al.[25]

The concept of a multiphasic nanostructured composite can be used to prepare a wide variety of optical materials. We have been able to dope two (or more) different optically

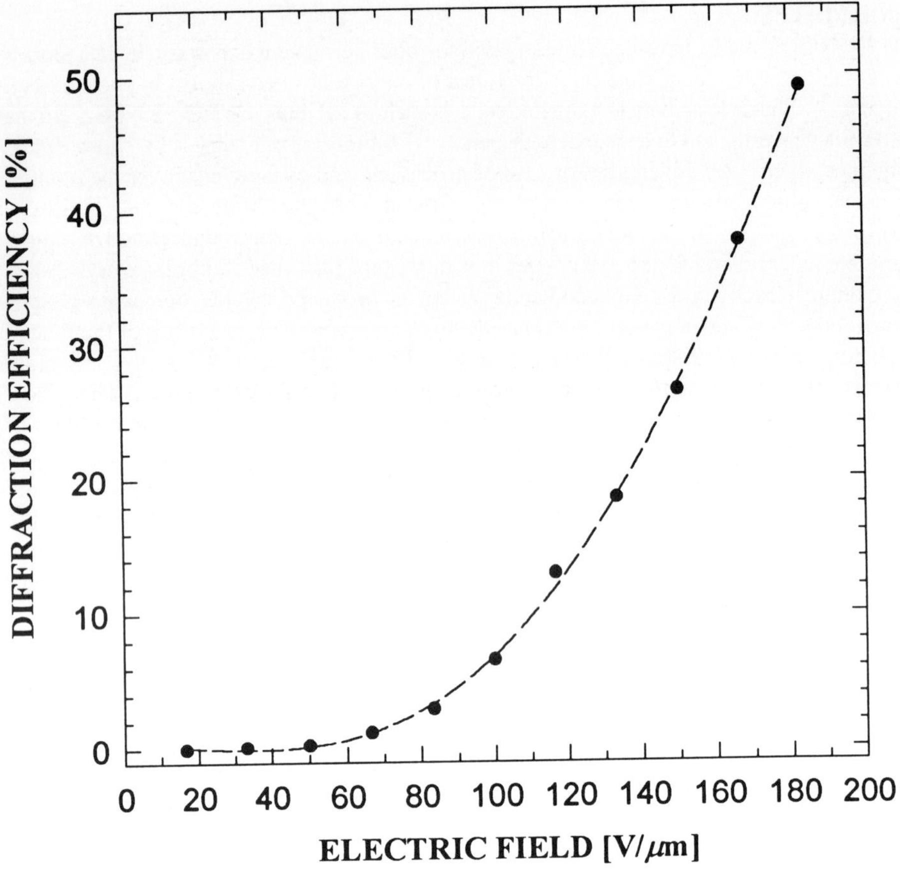

Figure 5. The electric field dependence of the diffraction efficiency in the PVK-TCP:C_{60}:DEANST polymeric composite. Data are corrected for reflection losses arising from the slanted geometry of the sample arrangement, the dashed line serves as the eye-guide.

responsive materials, each of which can be in different phases of the matrix (the silica phase, the PMMA phase and the interfacial phase), to make multifunctional bulk materials for photonics. For example, we have doped in addition to the fullerene, which is adsorbed in the interfacial phase, a fluorescent and optically nonlinear chromophore bisbenzothiazole 3,4-didecyloxy thiophene (BBTODT) in the PMMA phase.

Acknowledgments

This research was supported by the Air Force Office of Scientific Research and the Polymer Branch of Wright Laboratory, under contract F49620-93-C0017. We thank Bruce Reinhardt of the Polymer Branch of Wright Laboratory for helpful discussions and for providing us with the BBTODT sample which was synthesized in his laboratory. Thanks are also due to Mr. Makoto Yoshida for his help.

References

1. P. N. Prasad and D. J. Williams, "Introduction to Nonlinear Optical Effects in Molecules and Polymer," Wiley (1991).
2. K. L. Mittal ed., "Polymers in Information Storage Technologies", Plenum (1989).
3. B. E. A. Saleh and M. R. Teich, "Fundamentals of Photonics," Wiley, Interscience, (1992).
4. D. D. C. Bradley, Synth. Met. 56, 401 (1993).
5. D. Braun and A. J. Heeger, Appl. Phys. Lett. 58, 1982 (1991).
6. R. H. Friend, Synth. Met. 51, 357 (1992).
7. S. C. Moratti, D. D. C. Bradley, R. H. Friend, N. C. Greenham and A. B. Holmes, Proc. Mat. Res. Soc. 328, 371 (1994).
8. Z. Yang, I. Sokolik, and F. E. Karasz, Macromol. 96, 1188 (1993).
9. P. Dynekler and O. Inaganäs, Adv. Mat. 7, 43 (1995).
10. B. Wang, Z. Yao, and Z. Zhu, J. East China Institute of Chemical Technology, 19, 230 (1993).
11. L. C. Klein, ed., "Sol-Gel Optics: Processing and Applications," Kluwer Academic Publishers (1994).
12. M. Yoshida and P. N. Prasad, submitted for publication to J. Appl. Opt.
13. M. Eich, B. Reck, D. Y. Yoon, C. G. Wilson, and G. C. Bjorklund, J. Appl. Phys. 66, 3241 (1989).
14. M. Chen, L. R. Dalton, L. P. Yu, Y. Q. Shi, and W. H. Steier, Macromolecules 25, 4032 (1992).
15. R. J. Jeng, Y. M. Chen, J. I. Chen, J. Kumar, and S. Tripathy, Macromolecules 26, 2530 (1993).
16. C. Ye, N. Minami, T. J. Marks, Y. Yang, and G. K. Wong, Macromolecules 21, 2910 (1988).
17. C. K. Park, J. Zieba, C.-F. Zhao, B. Swedek, W. M. K. P. Wijekoon, and P. N. Prasad, Macromolecules (in press).
18. P. Yeh, "Introduction to Photorefractive Nonlinear Optics", Wiley-Interscience, New York (1993).
19. S. Ducharme, J. C. Scott, R. J. Twieg, and W. E. Moerner, Phys. Rev. Lett. 66, 1846 (1991).
20. Y. Zhang, Y. Cui, and P. N. Prasad, Phys. Rev. B 46, 9900 (1992).
21. B. Kippelen, K. Tamura, N. Peyghambarian, A. B. Padias, and H. K. Hall, Jr., J. Appl. Phys. 74, 3617 (1993).
22. M. E. Orczyk, B. Swedek, J. Zieba, and P. N. Prasad, J. Appl. Phys. 76, 4995 (1994).
23. R. Gvishi, G. Ruland, D. Kumar, U. Narang, and P. N. Prasad, submitted for publication to Chem. Mat.
24. E. J. A. Pope, M. Asami and J. D. Mackenzie, J. Mater. Res. 4, 1018 (1989).
25. R. Gvishi, R. Reisfeld, and Z. Biershtein, J. Sol-Gel Sci. and Tech. 4, 49 (1994).

DICHROMATED POLYVINYL ALCOHOL FOR REAL-TIME HOLOGRAM RECORDING: A STUDY ON THE EFFECT OF SOME EXTERNAL ELECTRON DONORS AND HUMIDITY

M.Barikani*, E.Simova** and M. Kavehrad**

*Polymer Research Center of Iran, 9 Etemadzadeh St., Fatemi Ave., Tehran, Iran,
**University of Ottawa, Dept. of Electr. Eng., 161 Louis Pasteur, Ottawa, Ont. KIN 6N5 Canada

INTRODUCTION

Photopolymer film have been used for hologram recording since late 1960's and over the ensuring years they have become increasingly important to the commercialization of holography. Mass production of volume phase holograms to date has primarily involved dichromated gelatin and fine grain silver halide emulsions. Both silver halide emulsion and Dichromated Gelation (DCG) require a delicte wet processing followed by thickness changes, and are environmentally sensitive. Therefore, much work is being done on self-developing photopolymers for holography that can provide good diffraction efficiency and environmental stability [1-3]. Dichromated polyvinyl alcohol (DCPVA) seems a promising photosensitive material because of its self-development, reasonable diffraction efficiency, easy preparation and low cost [4-6].

The effect of external electron donors on the photosensitive materials for hologram recording is an interesting issue. Electron donors are known to reduce photoexited dyes [7]. The effect of electron donors on the hologram efficiency in Methylene Blue sensitized DCG (MBDCG) has been studied [8-10]. An improvement of the diffraction efficiency in DCPVA due to the use of an electron donor has been reported by other authors [6]. Their effect on the photosensitivity and hologram efficiency, however, is not cleary obvious and completely understood. It has also been found that by controlling the film moisture in MBDCG at the hologram recording the sensitivity and diffraction efficiency can be increased. the effect of the moisture has been studied for transmission and reflection holograms and an optimum moisture found [8,11-12].

In this work, we present a study on the effect of four external electron donors in the formulation of DCPVA films on the photosensitivity and real-time diffraction efficiency of volume holgographic transmission gratings. The effect of the environmental humidity on the real-time diffraction efficiency in DCPVA films with and without electron donors has been

Polymers and Other Advanced Materials: Emerging Technologies and Business Opportunities. Edited by P. N. Prasad *et al.*, Plenum Press, New York, 1995

studied. Measurements of holographic characteristics are discussed. The photochemical recording mechanism in DCPVA in the presence of external electron donors is briefly discussed, as well.

EXPERIMENTAL

Preparation of the DC-PVA films

Materials: Polyvinyl alcohol (PVA) (average M.W. 13,000 -23,000, 98% hydrolyzed from Aldrich), Ammonium Dichromate (from BDH), 1,4-diazobicyclo [2,2,2] octane (DABCO), 2{[2-(Dimethylamino)ethyl] mehylamino} ethanol (DMEMA) and Triethanolamine (TEA), all from Aldrich, were of the best available grade and were used without any further purification. All solutions were prepared in distilled water.The names and structures of the amines included in the formulation of DCPVA films are given in Table 1.

Procedure: DCPVA films were prepared by the gravity settling method. The general procedure followed [10]. The required amount of PVA solution was prepared by dissolving 7 g of PVA in 100ml of preheated distilled water (70°C), The weighed amount of Ammonium Dichromate, $(NH_4)_2Cr_2O_7$ solid (0.8g) ($32x10^{-4}$mole) was added to the solution and stirred to a uniform mixture. Films wrer prepared with and without external electron donors. In the latter case, the weighed amount of 0.2, 0.5 or 0.8g of DABCO, DMEMA or TEA, respectively, was added to the solution mixture before casting it onto the glass plate. A typical composition of the photosensitive mixture is outlined in Table 2. A constant volume of 4.0 ml of the mixture was poured onto a leveled, cleaned optical grade glass substrate(size 5×7.5cm). The film was allowed to dry in a dark chamber under normal laboratory conditions (T=21°C, room humidity RH=40-60%) for 24h and a brownish/yellow transparent film was obtained.

The film thickness measured and estimated from angular selectivity measurements varied between 55-60μm. The absorption spectra of exposed and non-exposed DCPVA films are shown in [9]. The refractive index of freshly prepared films is n=1.53 and increases with the exposure energy as well as with the concentration of Dichromate [13].

Table 1. Names and structure formula of the amines used

Name	Abbreviation	Structure formula
1,1,3,3-Tetramethylguanidine	TMG	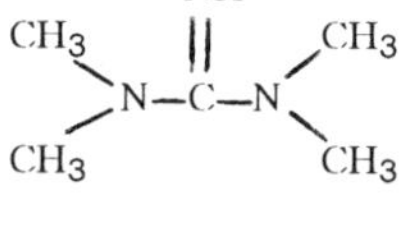 $(CH_3)_2N-C(=NH)-N(CH_3)_2$
Diazobicyclo[2,2,2] octane	DABCO	 N N
2{[2-(dimethylamino)ethyl] methylamino} ethanol	DMEMA	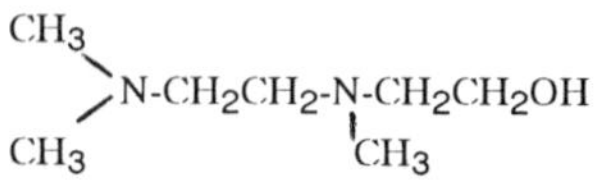 $(CH_3)_2N-CH_2CH_2-N(CH_3)-CH_2CH_2OH$
Triethanolamine	TEA	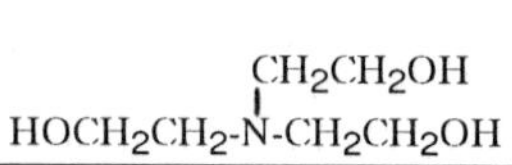 $HOCH_2CH_2-N(CH_2CH_2OH)-CH_2CH_2OH$

Table 2. Typical mixture.composition

Distilled water	100 ml
Polyvinyl alcohol	7 g
Ammonium Dichromate	0.8 g
Electron Donor	0.2-0.8 g

Storage in a constant RH environment: The DCPVA films were stored in a desiccator at constant humidity for 24 h before the exposure. The RH factors were established, as shown in Table 3.

Table 3. Material used for constant humidity

Material used	RH[%] at 20°C
Saturation sol.of KBr	84
Saturation sol.of NaBr	68
CaSO4 desicant	20

Holographic Set-up

The recording and real-time replay holographic set-up is shown in fig.1. An Ar laser at 488nm was used for the recording. The vertically polarized laser beam was split into two beams by a variable beamsplitter. Both beams were spatially Filtered, expanded, collimated and superimposed onto the photosensitive film. Transmission nonslanted holographic gratings were recorded with a period of $\Lambda=\lambda/2\sin\theta=0.58\mu m$, where λ is the recording wavelength, and 2θ is the interbeam angle of the recording beams measured in the air. The recorded gratings of this

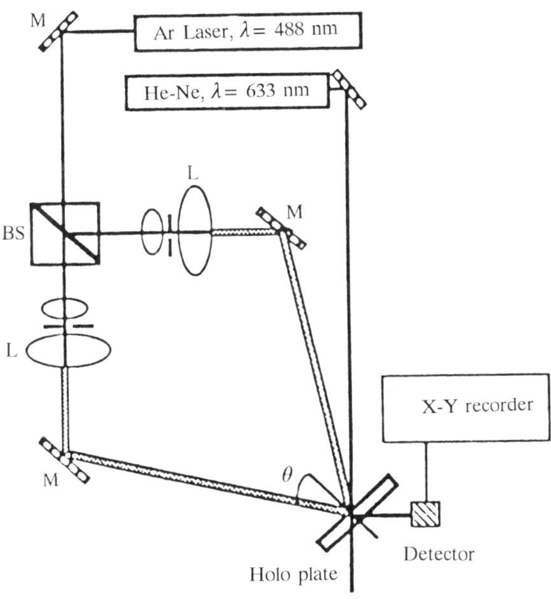

Fig. 1. Holographic recording and replay set-up

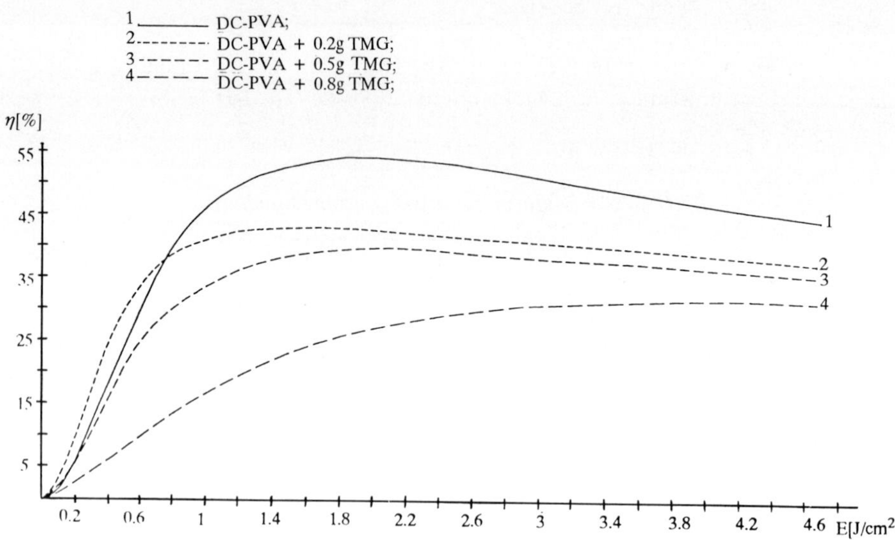

Fig. 2. Diffraction efficiency versus exposure energy for DC-PVA and varying concentration of TMG.

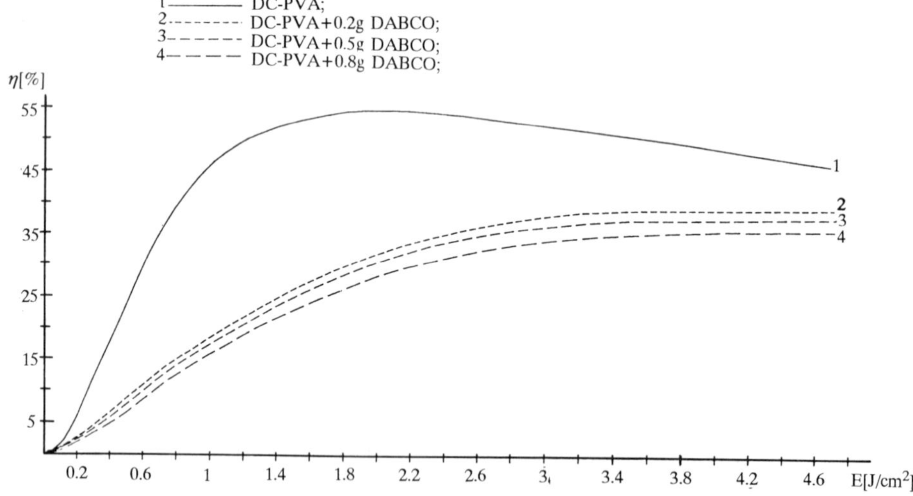

Fig. 3. Diffraction efficiency versus exposure energy and varying concentration of DABCO.

period and film thickness were well in the volume regime of diffraction and only one diffraction order was observed. The total intensity of the recording beams was 20 mW/cm^2 and the intensity beam ratio was unity. The DCPVA is not sensitive to the red spectral region, hence, the diffracted intensity can be measured with red light without disturbing the process of hologram formation at 488 nm. The real-time diffracted intensity in the first diffraction order was monitored at 632.8 nm by a He-Ne laser beam with a random light polarization. The incident angle of the monitoring beam was adjusted to the Bragg angle. The intensity was measured by a photometer calibrated for this wavelength and connected to an X-Y recorder. The recorded intensity of the diffracted beam was then normalized to the incident intensity corrected for Fresnel reflection at the surfaces as to obtain a diffraction efficiency of $\eta = I_{diffr.}/I_{inc.}[\%]$. All the experiments were carried out in the set-up shown in Fig.1 under the above described conditions. Mixed type of absorption and phase grating is recorded in the DCPVA films and no further thermal or chemical processing is required.

MECHANISM OF HOLOGRAM FORMATION

The basic recording mechanism in DCPVA is attributed to photocrosslinking process of the PVA when mixed with ammonium dichromate. The nature of crosslinking is believed to involve complexation of the polymer with the chormium ion. This type of crosslinking is suited for water soluble polymers, as well as PVA. With green-blue light the dichromate ions in the PVA matrix are photoreduced to trivalent chromium ions which provide a complex with two neighboring polymer molecules. The energy required to carry out the valence changes of chromium inos is often less than that required for most covalent bond cleavages and, hence, visible light might be sufficient to carry out these changes [7]. Therefore, The films should be kept in the dark to slow down this reaction before the exposure. On excitation by light Cr(VI) in the PVA matrix undergoes electron transfer reaction to produce Cr(V) and a polymer radical. They undergo a further reaction to generate the photoreduced Cr(III) and crosslinked polymer. Cr(V) has been confirmed as the primary species in the photoreaction, polymer radical and Cr(III) have also been confirmed by spin-trapping studies and UV-vis and ESR spectroscopy, respectively, [13]. A photochemical process may be suggested as follows:

$$Cr_2O^{2-}_7 \xrightarrow{h\nu} Cr^*(VI) \xrightarrow{\text{in polymer matrix}} Cr(V) + (PVA)^{\bullet} \xrightarrow{h\nu} Cr(III) + \text{crosslinked PVA}$$

RESULTS AND DISCUSSIONS

Effect of external electron donors.

The real - time diffraction efficiency versus the total exposure energy for TMG, DABCO, DMEMA and TEA with varying concentration from 0.2 to 0.8% are shown in Figs. 2 to 5, respectively, for a total intensity of the recording beams of 20 mV/cm^2 and room humidity of 60%. Concentrations above 1% were found to be disastrous for the efficiency which was virtually zero on the scale shown. The real - time diffraction efficiency for pure DCPVA films with a maximum value measured of η=55% is shown in all figures for comparison. TMG (Fig.2) seemed to have the least detrimental effect on the diffraction efficiency and a moderate positive effect on the photospeed in small concentrations. In Fig.3., the photospeed and the maximum diffraction efficiency are significantly reduced by adding DABCO. However, within the studied concentration in interval no significant deterioration of both characteristics occured. The detrimental effect of DMEMA can be clearly seen in Fig.4. Even small quantities of TEA proved to be disastrous for the hologram formation, as seen in Fig.5. As seen in Figures 2 to 5, the diffraction efficiency was found to decrease with increasing amine

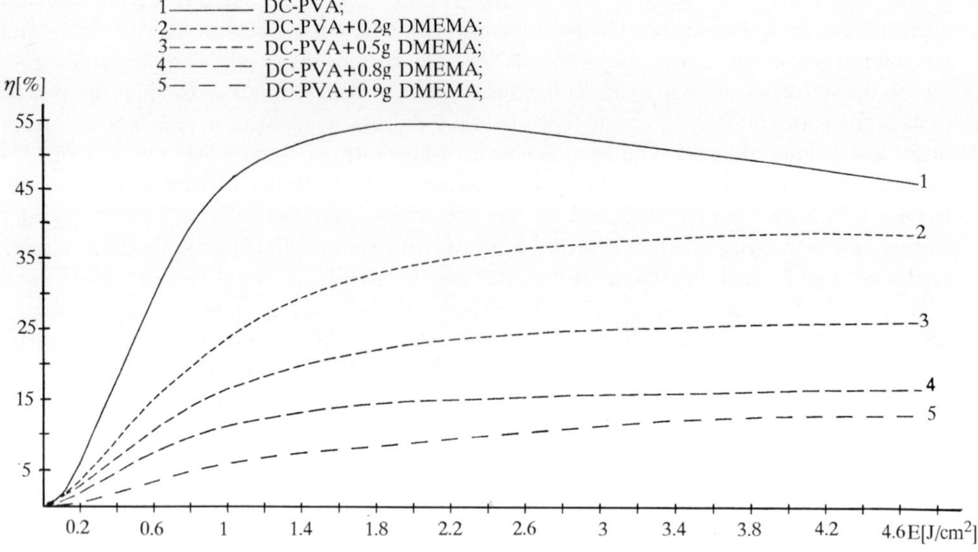

Fig. 4. Diffraction efficiency versus exposure energy for DC-PVA and varying concentration of DMEMA

concentration. The reduced diffraction effiency suggests a lower index - modulation achievable in the material maybe because of a preliminary uniform crosslinking. With the exception of TMG, the same trend can be seen in the photospeed. A lower photospeed may suggest a reduced mobility of the chromium ions possibly due to complexation[12]. Both effects are to be attributed to the amines. DMEMA caused a significant reduction of the film transparency in the nonexposed areas, and for TEA it was reduced to half of itsinitial value (Table 4). The color of the solution mixture was visibly bleached by adding amines. The order of decoloration was observed as follows:

$$\xrightarrow[\text{order of bleaching}]{\text{TEA}<\text{DMEMA}<\text{DABCO}<\text{TMG}}$$

Table 4. transmittance of the nonexposed and exposed areas in DCPVA with and without external electron donors

Sample expose	Transmittance [%]	
	nonexposed	exposed
DCPVA	82	57
DCPVA with 0.2% TMG	83	57
DCPVA with 0.5% TMG	85	59
DCPVA with 0.8% TMG	88	60
DCPVA with 0.2% DABCO	82	56
DVPVA with 0.5% DABCO	83	57
DVPVA with 0.8% DABCO	85	58
DCPVA with 0.2% DMEMA	72	44
DCPVA with 0.5% DMEMA	64	40
DCPVA with 0.8% DMEMA	60	38
DCPVA with 0.2% TEA	56	40
DCPVA with 0.5% TEA	43	38
DCPVA with 0.8% TEA	40	38

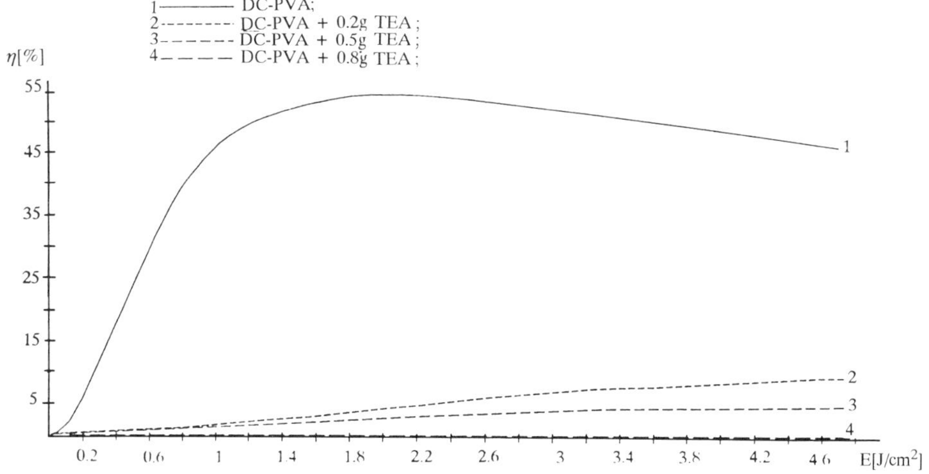

Fig. 5. Diffraction efficiency versus exposure energy for DC-PVA and varying concentration of TEA.

This might be due to the photobleaching effect of electron donors in the DCPVA matrix. This behavior was also noticed by J.Blyth[10]. A general characteristic of the amine groups that cause the fastest photobleaching effect is that they contain a nitrogen atom attached to at least two and preferably three alkyl groups and that these groups have an electron donating inductive effect on the nitrogen atom. Possibly, the resonating structure of guanidine may allow easier electron donation than with other amines, Fig.6.

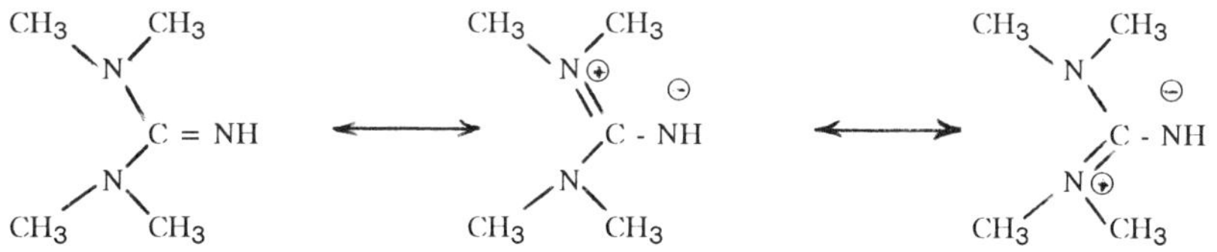

$(CH_3)_2N-C(=NH)-N(CH_3)_2 \longleftrightarrow (CH_3)_2N^{\oplus}=C(-NH^{\ominus})-N(CH_3)_2 \longleftrightarrow (CH_3)_2N-C(-NH^{\ominus})=N^{\oplus}(CH_3)_2$

Fig.6. Resonationg structure of the TMG.

Among these amines DMEMA and TEA, especially, also have the disadvantage of undergoing dark reaction in the PVA matrix.It is believed that these amines are able to form a strong chelate ring structure with chromium ions (Cr(III)) produced by the photoreduction, see Fig.7.

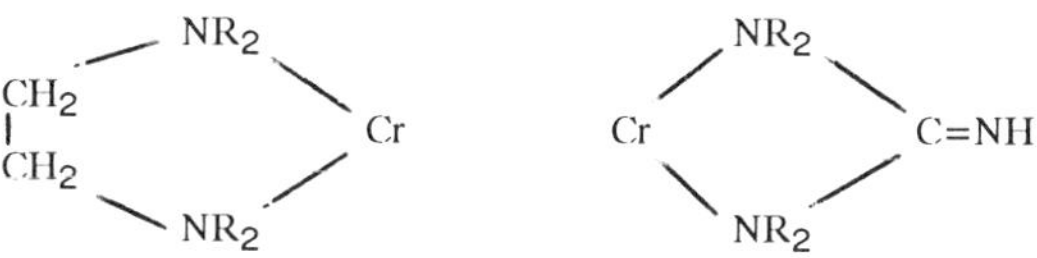

Fig.7. Amine-chromium chelate ring

So they may interfere with the crosslinking action of the trivalent chromium ion which is produced by exposing the DCPVA film. Thus, the concentration of the chelating agent is sufficient to complex the Cr(III) produced, as compared with the complexation action of the chromium ion with the polymer. PVA molecules which would normally form complexes with Cr(III) and, hence become crosslinked, are not able to complete crosslinking in the presence of amines. Crosslinking of the PVA matrix may occur either by a PVA radical or complexation of PVA with a chomium ion. In both cases, the mechanism of hologram formation includes an electron transfer reaction, and the use of amines which are good electron donors hinders the process by making chelate with Cr(III) or giving the necessary electron to the reaction Cr(VI)->Cr(III) instead of the PVA molecules. Thus, the amount of complexation or crosslinking of PVA is reduced and, correspodingly diminishes the index modulation, i.e. the difference in the refractive index of the more and less crosslinked areas in the PVA matrix. The photochemical reactions in DCPVA in the presence or absence of electron donor are considered as follows:

$$Cr(VI) \xrightarrow{h\nu} \text{excited } Cr(VI)$$
$$\text{excited } Cr(VI) + P \longrightarrow Cr(V) + P^{\circ}$$
$$Cr(V) + P^{\circ} \xrightarrow{h\nu} Cr(III) - P(\text{crosslinking, complexation})$$

Holographic characteristics:

In the second series of experiments, we recorded the real-time diffraction efficiency of transmission holograms for varying total intensity of the recording beams. The results are shown in Fig.8 for DCPVA. As seen, with increasing intensity of the recording beams the photospeed increases and the maximum of the real-time diffraction efficiency decreases. We found a constant improvement of the maximum efficiency for lower total power of the recording beams down to $4mW/cm^2$. This observation suggests that there is an optimum total power of the recording beam which enables the maximum crosslinking in the PVA to occur. Above this power, the correspondence between the intensity distribution and the more and less crosslinked areas in PVA matrix, i.e., the refractive index modulation, becomes smoothed out maybe due to higher energies of the excited Cr ions. In Fig.9, the real-time diffraction efficiency for two total intensities of the recording beams is shown for three cases: (i) both recording beams were on for the whole exposure interval; and (ii) both recording beams were blocked after the maximum efficiency had been reached; and (iii) after reaching the saturation level, the sample was uniformly exposed at 488 nm. As seen from the figure, higher intensity of the recording beams causes faster deterioration of the hologram efficiency after reaching saturation, which is to be expected. In case (ii), the hologram formation, i.e., the PVA crosslinking, continues for a few seconds after the recording process has been stopped, because of the excited Cr still available. As seen in case (iii), after reaching the saturation level, the efficiency starts to deteriorate, possibly because of uniform crosslinking by the excited chromium which tends to smooth out the reached index modulation, hence the correspondence between intensity distribution and crosslinked PVA degrades.

Effect of humidity:

The real-time diffraction efficiency in dried DCPVA films kept for 24h in different humidity environment is shown in Fig.10 versus exposure energy. As seen, the photospeed indicated by the curve slope increases significantly in higher RH environment. So does the diffraction efficiency, the maximum improves several times, degradation is faster, however. In low-humidity environment is the maximum reached was less than η=5%. In Fig.11, the real-time diffraction efficiency versus exposure energy is shown for one DCPVA film, first exposed right after completely dried at RH=60%, then after it had been kept in dark chamber for 24 and 48 h, respectively, at a room humidity RH=46-48%, and, finally, exposed after it

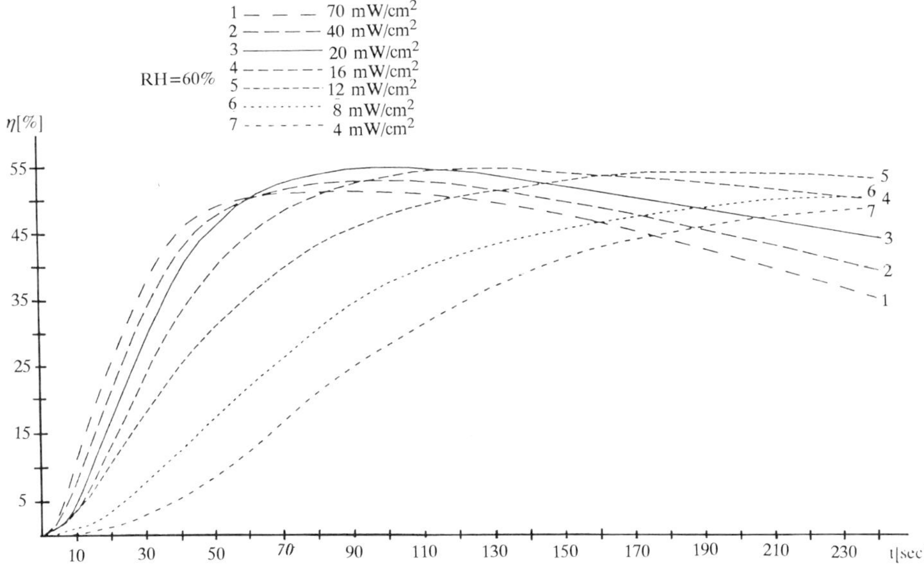

Fig. 8. Diffraction efficiency for DC-PVA and for various intensity of the recording beams.

had been kept for 2 hours at an RH=84%. As seen, the photospeed is significantly reduced and the maximum real-time efficiency drops more than three times in low humidity environment compared to the first exposure. Two hours in a high-humidity environment restored photospeed and diffraction efficiency. It is clear that the hologram formation, i.e., the chromium photoreduction with subsequent crosslinking in the PVA are very sensitive to environmental humidity. Storing the DCPVA films in a high-humidity environment significantly prolongs their shelf life as well. Films exposed in a high humidity environment (RH>60%) always showed much higher diffraction efficiency than those exposed in lower humidity, e.g., RH<46-49%. They tended to keep higher efficiency enen after being kept for few weeks in slowly varying normal humidity environment RH>46-52%. Low humidity also proved disastrous for the film causing small cracks to appear and totally destroying the film in

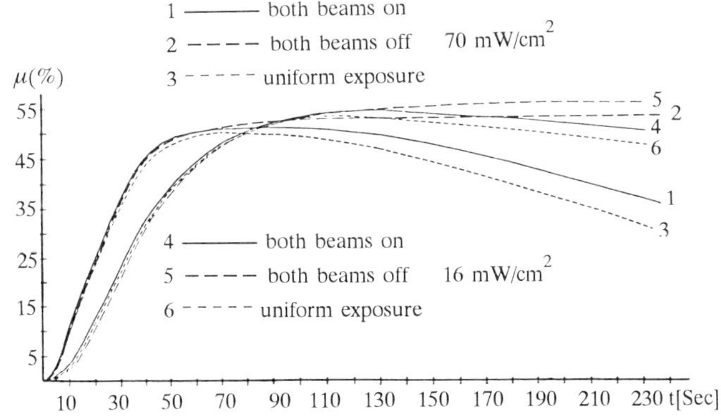

Fig. 9. Efficiency for two intensities of recording beams:
(i) both beams on for the exposure interval;
(ii) both beams blocked after reaching maximum;
(iii) uniform illumination with one beam.

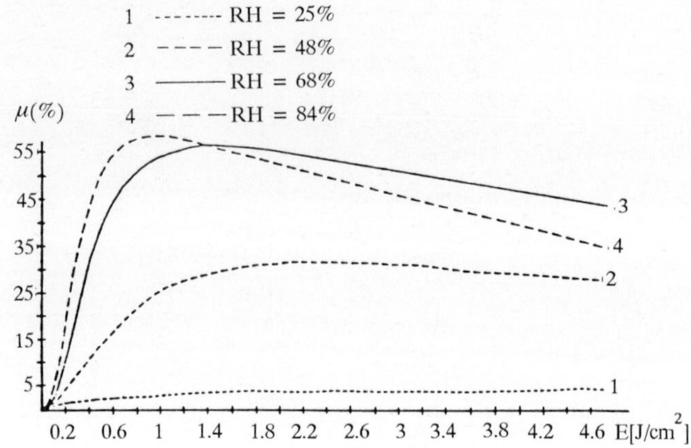

Fig. 10. Diffraction efficiency in DC-PVA exposed after 24 h at different RH environment as a function of the exposure energy.

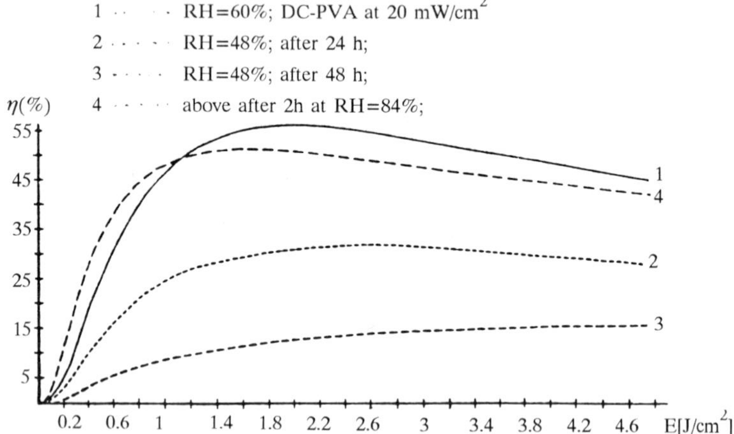

Fig. 11. Efficiency versus exposure energy in DC-PVA; (i) exposed in RH=60%; (ii) kept 24 h in RH=48%; (iii) kept 48 h RH=48%; (iv) kept 2 h in RH=48%.

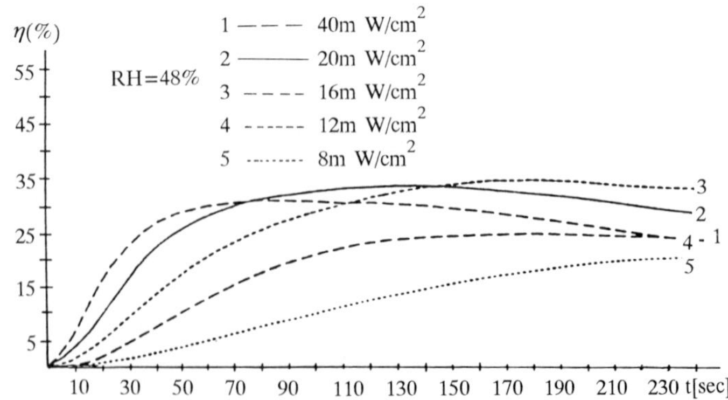

Fig. 12. Efficiency for varying intensitities of the recording beams in RH=48%.

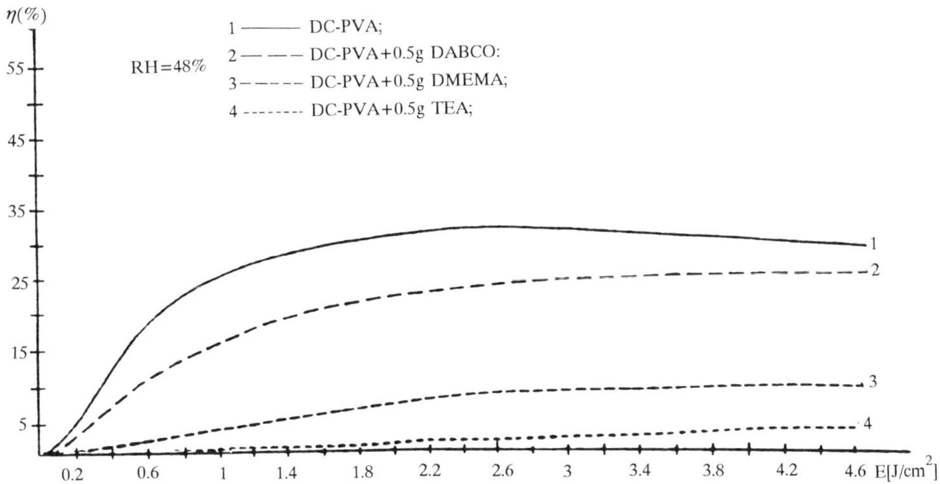

Fig. 13. Efficiency versus exposure energy in DC-PVA without and with amines, 0.5% at RH=48%.

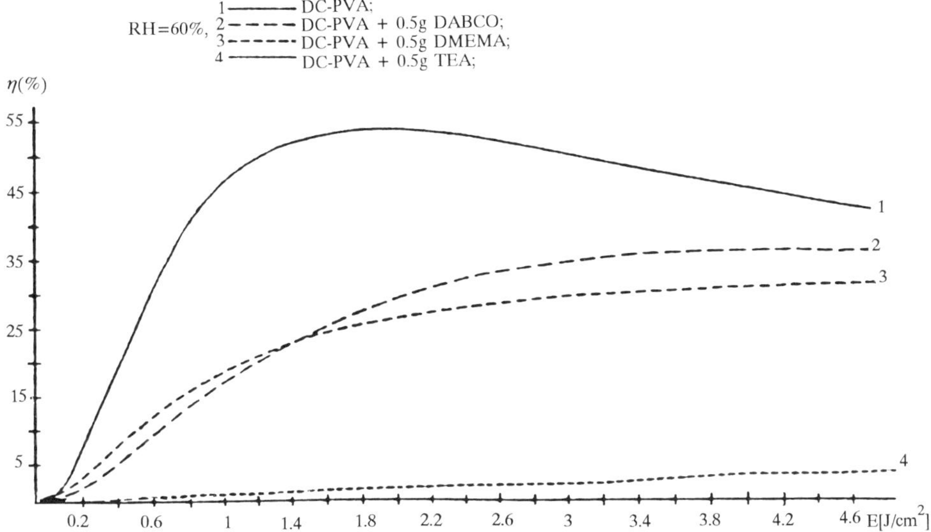

Fig. 14. Efficiency versus exposure energy in DC-PVA without and with amines, 0.5% at RH=60%.

three to four days. The positive effect of the humidity may be attributed to a plasticization effect of the water in the film which increases the degree of mobility of the polymer chain. Similar mechanisms have been suggested for DCG[14] and MBDCG[11].

The effect of humidity on the diffraction efficiency was studied for different total intensity of the recording beams. The results are shown in Fig.12 for RH=48% and Fig.8 for RH=60%. Comparing both figures, an improvement of the efficiency in higher RH for all intensities of the recording beams is seen. The photospeed tends to decrease and the efficiency tends to increase with decreasing total intensity of the recording beams. those trends remain regardless of the relative humidity. However, at lower humidity environment, an optimum total intensity of the recording beams can be noticed in Fig. 12. Below a specific intensity, e.g., 16mW/cm^2 in our case, the maximum diffracted intensity starts to decrease again. Clearly, the interval of optimum recording intensities is extended at high humidity. A comparison between the effects of three amines with a concentration of 0.5% is shown in Figs. 13 and 14 at RH=48% and RH=60%, respectively. The real-time diffraction efficiency for DCPVA without an electron donor is shown in both figures for comparison. As seen, the photospeed and the diffraction efficiency were found to decrease significantly with increasing, even small concentration for all tested amines. The observed trends remain in higher and lower humidity environment. The improvements of the diffraction efficiency in the presence of amines in small concentration at high humidity may again be attributed to the plasticization effect of the film moisture.

CONCLUSIONS:

In the present work, the effect of four external electron donors in the formulation of DCPVA films on the real-time hologram formation has been studied. The photosensitivity and real-time diffraction efficiency have been recorded for volume transmission holographic gratings. Measurements of parameters influencing the holographic performance are reported. The effect of the relative environmental humidity on the real-time diffraction efficiency of DCPVA films has also been studied. The amines were found to be detrimental to the holographic characteristics, even in small concentrations. The recording mechanism in DCPVA in the presence of external electron donors are briefly discussed. The amines may form a strong chelate ring structure with the chrromium ions and, possibly, interfere with the crosslinking action of the photoreduced trivalent chromium. The hologram formation includes an electron transfer reaction, and the use of amines which are good electron donors hinders the process by making chelate with the chromium. Our study has shown that higher relative humidity affects positively the photoreduction and, hence, significantly improves the photospeed and the real-time diffraction efficiency in DCPVA films with and without electron donors. Low-humidity environment is detrimental to its shelf life and reduces the diffraction efficiency several times.

ACKNOWLEDGEMENTS:

This work was supported by the Telecommunications Research Institute of Ontario (TRIO), Photonics Networks and Systems Thrust.

REFERENCES:

1. B.Monroe et.al., "Improved photopolymers for hologram recording I:Imaging Properties", J.Imag. Sci., 35(1), 19-25(1990).
2. B.Monroe, "Improved photopolymers for holographic recording II: Mechanism of Hologram Formation", Ibid 35(1), 25-29(1990).
3. V.Barachevsky, "Organic light-sensitive media for holographic optical elements", Proc. SPIE, Vol.1183, 143-153,1990.
4. M. Mazakova, M. Pantcheva, "Dichromated Polyvinyl Alcohol for Hologram Recording," J. of Inform. Record. Mater. 18(2), 191-196(1990).
5. S.Lelievre and J.J.A.Couture, "Dichromated Polyvinyl Alcohol Films Used as a Novel Polarization Real-Time Hologram Recording Material," Appl. Opt. 29(29), 4384-4391(1990).
6. G.Mannivanan, R.Changkakoti and R.A.Lessard, "Cr(VI)- and Fe(VI) - and Fe(IV) - Doped polymer systems as Real - Time Holographic Recording Material," Opt. Eng., 32(3), 671-676(1993).

7. G.Oster and G.Oster, "Photochemical Modification of High Polymers by Visible light," J.Polym. Sci.XLVIII, 321-327(1960).
8. C.Solano, et al., "Methylene Blue Sensitized Dichromated Gelatin as a Photo-sensitive Medium for Conventional and Polarizing Holography," Appl. Opt.26(10), 1989-97(1987).
9. R.Changkakoti, et al., "Role of External Electron donor in Methylene Blue Sensitized Dichromated Gelatin Holograms: Experimental study," Appl. Opt.27(2), 324-330(1988).
10. J.Blyth, "Methylene Blue Sensitized Dichromated Gelatin Holograms: a New Electron Donor for their Improved photosensitivity," Appl. Opt. 30(13), 1598-1602(1991).
11. T.Mizuno, et al., "Methylene Blue Sensitized Dichromated Gelatin Holograms: Influence of Moisture on Exposure and Diffraction Efficiency," Appl. Opt., 20(32), 4457-60(1990).
12. J.Blyth, "Methylene Blue Sensitized Dichromated Gelatin Holograms: a New Electron Donor for their Improved Photosensitivity," Appl. Opt. 30(13), 1598-1602(1991).
13. G.Mannivanan, R. Changkakoti, et al., "Primary Photoprocess of Cr(VI) in Real-Time Hologram Recording Material: Dichromated Polyvinyl Alcohol," J.Phys. Chem., 97(28), 7228-7233(1993).
14. M.Sasaki, K.Honda, and Kikuchi, "Studies on Photosensitive Dichromated Materials," Rep. Inst. Ind. Sci. Univ. Tokyo, 27(171), 1-58(1978).

HIGHLY EFFICIENT LIQUID CRYSTAL POLYMERS FOR QUADRATIC NONLINEAR OPTICS

D. Gonin[1], B. Guichard[1], C. Noël[1] and F. Kajzar[2]

[1]Laboratoire de Physicochimie Structurale et Macromoléculaire, ESPCI, 10, rue Vauquelin, 75231 Paris Cedex 05, France
[2]Commissariat à l'Energie Atomique, LETI/DEIN/SPE/GCO, Centre d'Etudes de Saclay, 91191 Gif sur Yvette Cedex, France

INTRODUCTION

Organic materials tailored to be transparent in the blue or near UV range are targeted as highly efficient frequency doublers for applications in the field of high density optical data storage[1]. Today, laser diodes emitting in near infrared region are commercially available at low cost. By doubling their frequency it should be possible to quadruple the information storage capability on the same surface. Other potential applications of such blue sources are in microelectronics, medicine, biology and information technology. Organic materials can be also used for the fabrication of tunable laser sources like OPO's as well as for electro-optic modulation applications. For these applications, three major device-driven material requirements have to be fulfilled:
i) broad transparency range. Typically, for blue conversion, the cutoff wavelength for linear absorption should lie below 350 nm
ii) excellent optical quality
iii) macroscopic non-centrosymmetric structure

Charge transfer biphenyls display an interesting compromise in terms of non linearity/transparency trade-off. The twist angle between two phenyl rings reduces efficiently the length of π electron conjugation, thus leading to materials which are transparent in the near UV spectral range (cutoff at 320 nm , cf. Fig. 1), but still exhibit noticeable first hyperpolarizability β values. However, the fabrication of non-centrosymmetric, good optical quality, single crystal waveguides is made difficult because of serious materials limitations. Such organic crystals are very difficult to grow with the required optical quality. Of equal but less obvious importance is the barrier which crystallization creates between molecular tailoring and material properties. The large ground state dipolar moments of many such molecules cause them to crystallize centrosymmetrically. An alternative approach is to incorporate these molecules into a transparent polymer matrix. To satisfy the requirement for non-centrosymmetry, the NLO species must be induced to point, at least statistically, in the same direction by the

Polymers and Other Advanced Materials: Emerging Technologies and Business Opportunities. Edited by P. N. Prasad *et al.*, Plenum Press, New York, 1995

application of an external electric DC field, a process known as poling. This is typically carried out by heating the polymer to above its glass transition temperature , applying a large DC voltage (corona or electrode poling), and cooling to low temperatures with the voltage still applied.

In the simplest case of solid solutions, the concentration of the active molecules is usually limited to a few percents only. If much higher concentrations are used, the NLO molecules tend to segregate before and/or after the poling procedure, forming (usually centrosymmetric) microcrystallites. These species, if made from antiparallel aligned dipolar moments, do not contribute to the nonlinear optical effect, but, in any case, may give rise to undesired light scattering. In addition, in guest-host materials a plasticizing effect is often observed which lowers the glass transition temperature implying that such systems must be kept at fairly low temperatures after poling in order to prevent relaxation of the induced polar ordering. Alternately, the active molecules can be incorporated as part of the polymer structure. Attaching the NLO groups to the polymer reduces their mobility and tendency to crystallize or phase-separate at high concentrations and leaves the glass transition temperature relatively unaffected. The poled polymeric films have a ∞mm symmetry with two non-zero components of the macroscopic second order nonlinear optical susceptibility $\chi^{(2)}_{333}$ and $\chi^{(2)}_{113}$.

Neglecting dispersion and assuming a one-dimensional NLO moiety, with enhanced first hyperpolarizability β_{zzz} tensor component one gets

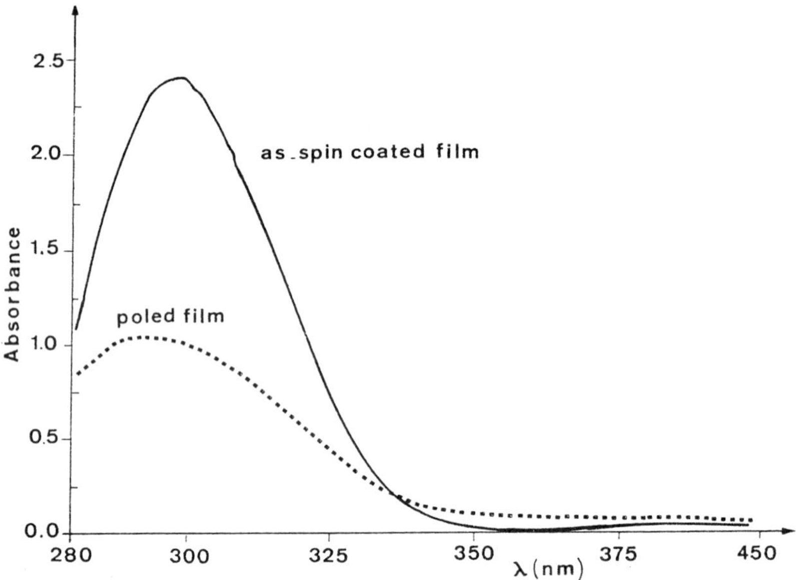

Figure 1. Optical absorption spectra of PECH 0.8 before and after poling.

$$\chi^{(2)}_{333} = N\beta_{zzz}F \langle\cos^3\Theta\rangle \quad (1)$$

and

$$\chi^{(2)}_{113} = N\beta_{zzz}F \langle\cos\Theta\sin^2\Theta\rangle/2 \quad (2)$$

where 3 is the poling direction and z is the charge transfer direction in the molecular frame. In equations (1)-(2) N is the number density of active groups, F is the local field correction factor and Θ is the angle between the charge transfer axis and the poling field direction.

The brackets <> imply an averaging over molecular orientations. The ratio of these two non zero components of $\chi^{(2)}$ can be expressed as follows

$$a = \chi^{(2)}_{333}/\chi^{(2)}_{113} = 2\langle\cos^3\Theta\rangle / \langle\cos\Theta\sin^2\Theta\rangle \tag{3}$$

It varies between 3 in systems that are isotropic before poling and ∞ for a perfect ordering (axial or polar).

In this paper we describe the synthesis and NLO properties of a number of side chain polymers functionalized with a CT biphenyl derivative which possesses mesogenic properties itself. Polyacrylate, polymethacrylate, polyepichlorohydrin and poly(3,3-bis(chloromethyl)oxetane) backbones are used. The resulting polymers exhibit either liquid crystalline or isotropic behavior depending on the chemical structure. Magnitudes of second order NLO coefficients d_{113} and d_{333} determined by optical second harmonic generation (SHG) are analyzed in the framework of the molecular statistical models developed for isotropic and ordered systems.

POLYMER SYNTHESIS AND PROCESSING

A nematic polyacrylate (number-average molecular weight $\overline{M}_n$ = 3790, weight average molecular weight $\overline{M}_w$ = 6000) and an isotropic polymethacrylate ($\overline{M}_n$ = 22800, $\overline{M}_w$ = 66850) containing 4-cyano biphenyl as the mesogen and three methylenic units in the spacer were obtained from Merck. Their chemical formulae are shown in Fig. 2. Three copolyethers based on the same mesogenic group were prepared by chemical modification of polyepichlorohydrin (PECH) ($\overline{M}_n$ = 6680, $\overline{M}_w$ = 10640) with the sodium salt of 4-cyano-4' hydroxybiphenyl as described in Ref. 2 (for synthesis route see Fig. 3).

The sodium salt was prepared by exchange reaction in dry ethanol containing 4-cyano-4'-hydroxybiphenyl (from Tokyo Kasei Kogyo Co, Ltd, Tokyo, Japan) and sodium ethoxide. Ethanol was removed and the salt was dried in *vacuo*.

5 g of PECH were dissolved in DMF and sodium 4-cyano-4' - biphenoxide (12.9 g) and tetrabutylammonium hydrogensulfate (TBAH) (2 g, Fluka) were added. The reaction mixture was stirred at 60°C under a nitrogen stream. The degree of substitution as a function of time was determined by withdrawing reaction aliquots. The latter were centrifuged in order to remove sodium chloride and then poured into methanol. The obtained polymers were purified by precipitation from $CHCl_3$ solution into methanol and soxhlet extraction. Finally, the products were dried at 60°C in *vacuo*.

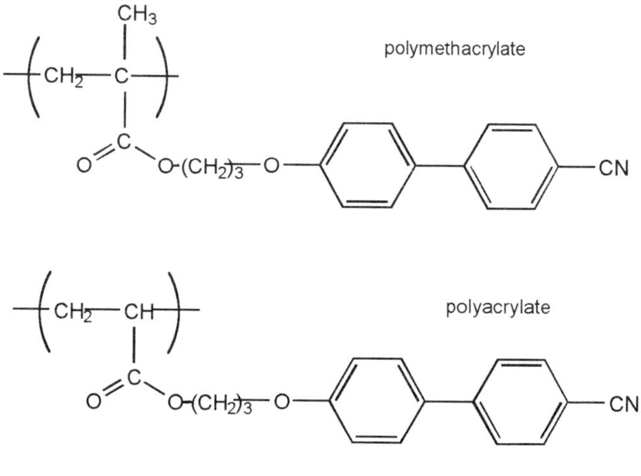

Figure 2. Chemical structures of polymethacrylate and polyacrylate.

Chemical modification of poly(3,3-bis(chloromethyl)oxetane) (PBCMO) ($\overline{M}_n$ = 1860, $\overline{M}_w$ = 2600) (SNPE) with the sodium salt of 4-cyano-4' - hydroxybiphenyl was carried out as shown in Fig. 4. An additional polyether was synthesized by cationic polymerization of 3,3 - bis {[4'-(4"-cyanophenyl)phenoxy]methyl} oxetane. This monomer was prepared by the reaction of sodium 4-cyano-4'-biphenyloxide with 3,3-bischloromethyloxetane (SNPE) in refluxing ethanol. After being stirred for 20 days, the mixture was cooled to room temperature and poured into water. The product was filtered, recrystallized from ethyl acetate and dried in *vacuo*. 3,3-bis{[4'-(4" - cyanophenyl)phenoxy] methyl}oxetane was polymerized with boron trifluoride diethyl etherate in dry methylene dichloride. The mixture was allowed to react for 3 days at room temperature and poured into methanol to precipitate the product polymer which was filtered, purified by reprecipitation using a DMF - methanol system and dried at 60°C in *vacuo*.

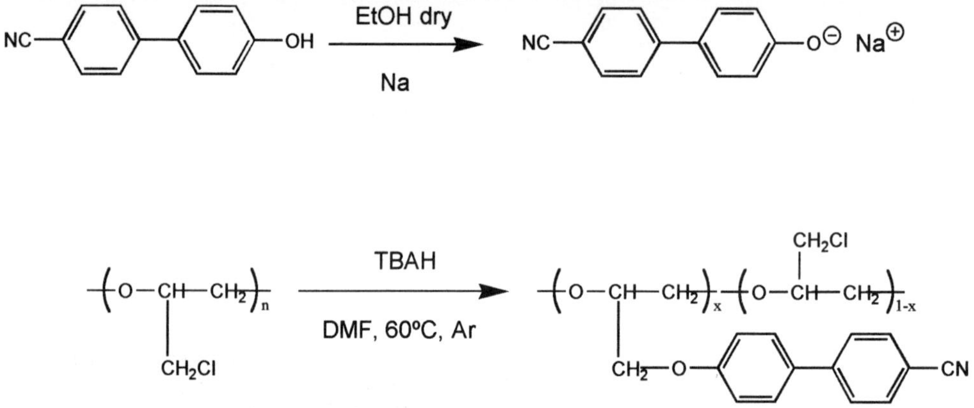

EtOH dry

Na

TBAH

DMF, 60°C, Ar

Figure 3. Synthesis route for PECH 0.3 (x = 0.3), PECH 0.55 (x = 0.55) and PECH 0.8 (x = 0.8).

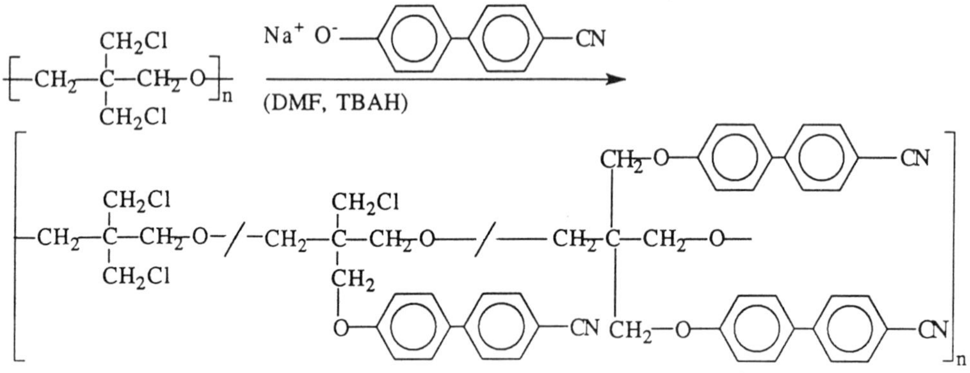

(DMF, TBAH)

(mole ratio x: y: z) x = 0.14; y = 0.43; z = 0.43

Figure 4. Synthesis route for PBCMO14/43/43.

Characterization

The structures were substantiated by elemental analysis, IR spectroscopy and NMR. IR spectra were recorded on a Nicolet FT - IR 205 spectrometer. ^1H and ^{13}C NMR spectra were recorded on a Bruker AC - 300 spectrometer. The molecular weights were estimated using a Waters gel permeation chromatograph, calibrated with polystyrene standards.

Tetrahydrofuran was used as solvent. The transition temperatures were measured using a differential thermal analyzer (Dupont 1090) operating at 20°C/min. Optical observations were made using a polarizing microscope (Olympus BHA - P) equipped with a Mettler FP 52 hot - stage and FP 5 control unit. X - ray diffraction patterns were recorded on flat films using CuK_{α} radiation. A flat graphite crystal with a pinhole collimator was used as a monochromator. The samples were kept in 1 mm Lindemann glass tubes which were mounted in an electrically heated oven.

Thin Film Fabrication and Characterization

Thin polymeric films were obtained by spin coating of filtered solutions onto glass or silica substrates. Depending on the structural characteristics of the polymer, either 1,1,2,2 - tetrachloroethane or 1,1,2 - trichloroethane were used as solvents. The solvent was allowed to evaporate at 35°C in *vacuo*, which resulted in uniform and homogeneous films of high optical quality. The final thin film thicknesses were determined using a stylus profilometer.

The UV- visible optical absorption spectra of the films were recorded on a Uvikon 810 spectrometer and the transverse electric (TE) and transverse magnetic (TM) refractive indices were measured at different wavelengths using a Metricon PC - 2000 prism coupler.

Polar orientation of chromophores was induced by the standard corona poling technique(cf. Refs. 3-4). The samples were placed on a hot plate at a distance of *ca* 20 mm from a tungsten needle to which a high voltage (5-7 kV) was applied. The poling temperature (T_p) was a few degrees higher than the polymer glass transition temperature (T_g) of polymer. The poling current was monitored during all the poling process. After a given poling time (t_p), the samples were cooled to room temperature under the applied DC field.

The thin film nonlinear optical properties were determined by the second harmonic generation (SHG) technique in transmission[5-6]. An α - quartz single crystal (d_{11}=0.5 pm/V[7]) was used as standard for the calibration of the second order nonlinear optical susceptibilities $\chi^{(2)}{}_{113}$ and, $\chi^{(2)}{}_{333}$ which were deduced from p-p and s-p fundamental - harmonic beam configuration measurements, respectively. A neodymium doped yttrium aluminium garnet pulsed (pulse duration of 13 ns) laser, operating at 10 Hz repetition rate, was used. The sample was rotated along an axis perpendicular to the beam propagation direction and the SHG intensities were collected as a function of incidence angle. First the d_{31} (= $\chi^{(2)}{}_{113}$,/2) coefficient was determined by fitting the appropriate theoretical formulas to the data obtained from the s-p polarization configuration experiment. This value was then injected into the formulas[5-6,8] describing the dependence of the SHG intensity on the incidence angle of the fundamental beam for the p-p fundamental - harmonic beam polarization configuration. Finally, the value of the d_{33} (= $\chi^{(2)}{}_{333}$,/2) coefficient was determined by fitting these data with the corresponding theoretical formulas[5-6,8]

RESULTS

Material Characterization

Optical microscopy observations indicate that the polyacrylate PA3 exhibits a nematic phase which appears as a threaded texture.The assignment of nematic phase was confirmed by X - ray diffraction. The X - ray diffraction pattern obtained with powder sample in the mesophase is characterized by a diffuse outer ring which is related to the lateral interferences between the mesogenic cores and points out the lack of lateral periodic order, and a diffuse inner ring corresponding to the half of the length of the side chains calculated for the fully extended trans conformation.The X - ray pattern also reveals a less diffuse ring at about 24 A° characteristic of S_{Ad} fluctuations. Clearly, the cyano groups induce new correlations, essentially of the dipolar nature, between molecules and favor an antiparallel arrangement of the permanent dipoles. This results in the formation of antiparallel, overlapping interdigitated dimers of length approximately 1.2 times that of the fully extended model length. The coherence length of the S_{Ad} fluctuations is roughly 100 A°. Confirmation of the existence of dimers of parallel neighbors having oppositely directed cyano groups was obtained by dielectric relaxation measurements.

By contrast, the polymethacrylate PMA3 in which the same mesogenic unit is at - tached to the polymer backbone through the same flexible spacer of three methylenic units is isotropic at rest. Clearly, the addition of a methyl group to each structural unit of the polyacrylate backbone creates a higher conformational barrier to mesogen packing. Of particular interest here is PMA3 which exhibits a virtual nematic phase. In an attempt to determine the virtual nematic/isotropic liquid transition temperature we applied the mutual miscibility method derived for low molar mass liquid crystals by preparing binary mixtures of PMA3 and PA3. These mixtures were prepared by casting thin films from solutions of the known weights of PMA3 and PA3 in chloroform. The glass transition temperatures were determined by DSC and the nematic/isotropic transitions were observed by using a polarizing microscope. The phase diagram constructed from the obtained data is shown in Fig. 5. The T_{NI} transition temperatures are situated on a curve which, by extrapolation, leads to the determination of the virtual nematic/isotropic transition temperature of PMA3. Without making any claims to great accuracy, it is nevertheless possible to locate the virtual transition in the temperature range 70 - 80 °C. Similar results were obtained by studying binary mixtures of PMA3 and PA4, a polyacrylate in which the same mesogenic unit is attached to the polymer backbone through a flexible spacer of four methylenic units (cf. Fig. 6).

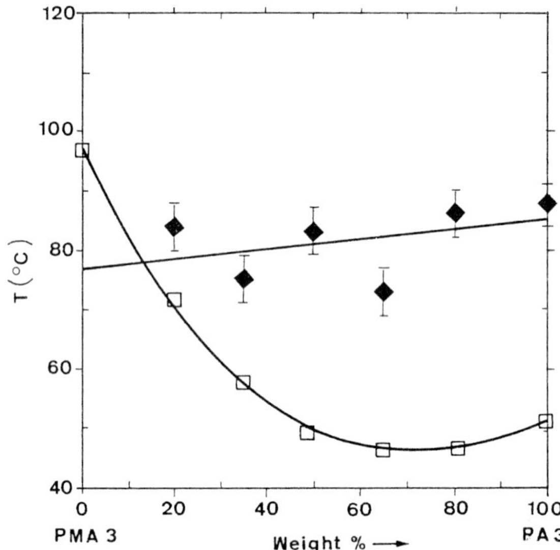

Fig. 5. Phase diagram obtained from binary mixtures of PMA3 and PA3.

Figure 7 shows that the thermal behavior of the copolyethers prepared by chemical modification of polyepichlorohydrin is strongly dependent on the degree of substitution x. There is a minimum substitution limit of approximately 65% necessary for liquid crystal formation. Below this limit, all copolyethers resemble typical amorphous polymers. There is a sharp increase in the glass transition temperature with increasing amount of substitution. It is evident that the bulky side groups cause severe hindrance to main chain motions. Once 65% substitution is reached the increase in T_g drastically decelerates. The microscopy observations indicate that at and above this limit copolyethers exhibit a nematic phase. In this work we concentrate our attention on copolyethers PECH 0.8, PECH 0.55 and PECH 0.3 which contain 80%, 55% and 30% of NLO mesogenic groups, respectively.

On heating, the nematic phase of PECH 0.8 appears as a threaded texture. On cooling, the nematic phase separates from the isotropic liquid in the form of typical droplets (cf. Fig. 8a), which after further cooling grow, coalesce and reorganize their shape until a final Schlieren texture is established (cf. Fig. 8b). As previously observed for PA3, a pronounced local order of a quasi S_{Ad} nature is developed in the nematic phase. The X - ray diffraction patterns consist of a broad, diffuse halo related to the lateral interferences between the mesogenic cores. They present an additional diffuse ring at small angles. Its position corresponds to a repeat distance along the texture axis of about 1.5 times the extended molecular length, which again can be explained by considering dimers of parallel neighbors having oppositely directed cyano groups.

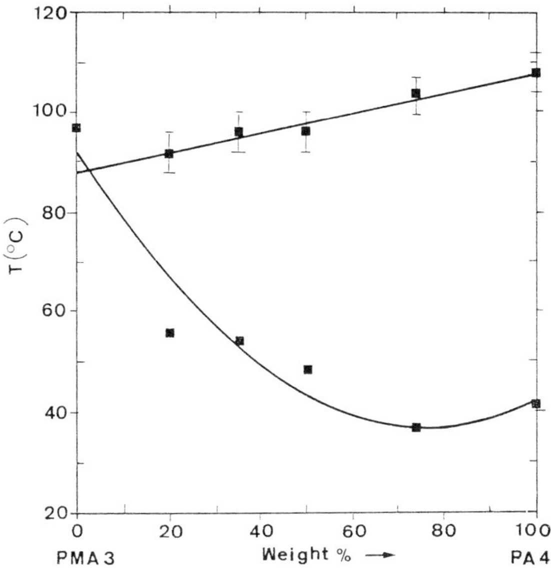

Figure 6. Phase diagram obtained from binary mixtures of PMA3 and PA4

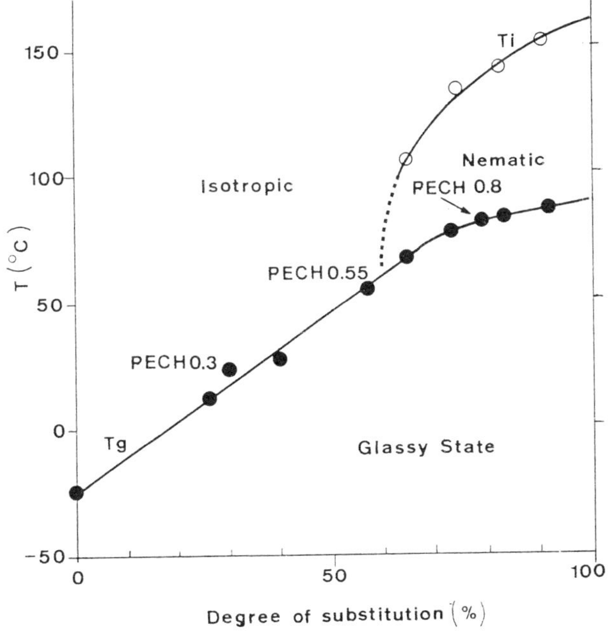

Figure 7. Thermal properties of copolyethers derived from polyepichlorohydrin.

As previously observed for PMA3, PECH 0.55 is purely isotropic at rest, but exhibits a virtual nematic mesophase. From the phase diagram established by studying binary mixtures of PECH 0.8 and PECH 0.55 it is possible to locate the virtual nematic/isotropic transition in the temperature range 56 - 40 °C. PECH 0.3 resembles typical amorphous , isotropic polymers.

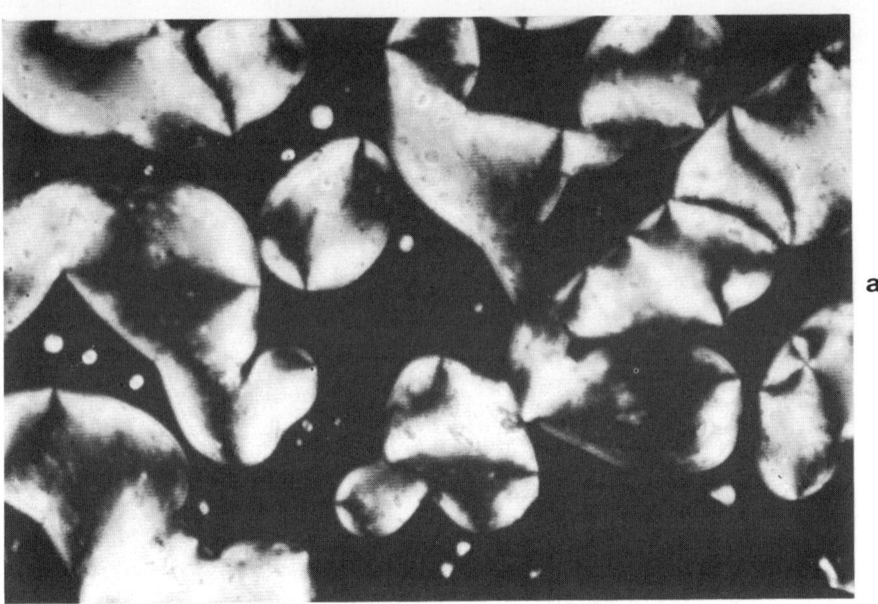

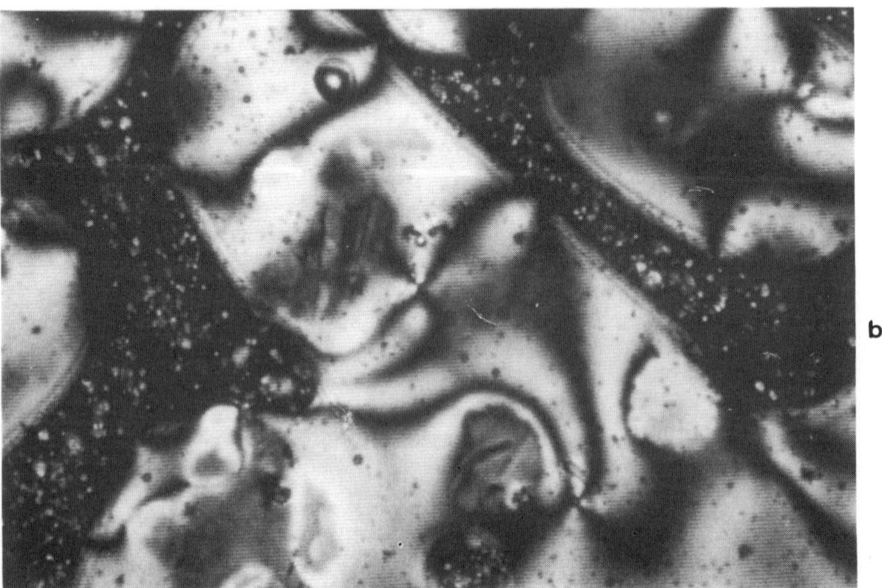

Figure 8. Droplets and Schlieren texture for PECH 0.8 (a) at 115 °C and (b) after annealing for 17 h (crossed polarizers, magnification x200).

Copolyethers PBCMO 14/43/43 and PBCMO 100 demonstrate a behavior between isotropic and liquid crystal state. Above the glass transition temperature they give a birefringent melt. However, no typical texture can be detected. This effect is observable over 20 °C. The transition temperatures of the studied polymers are listed in Table 1.

Table 1. Molecular weights and transition temperatures of studied polymers

Polymer	M_n	M_w	T_g(°C)	T_{NI}(°C)
PA3	3790	6000	51.5	90
PMA3	22800	66850	97	[70-90][a]
PECH 0.3			23	
PECH 0.55	11910	17020	56	[40-56][a]
PECH 0.8	13040	18640	79.5	146
PBCMO 14/43/43	6660	10740	106	
PBCMO 100	4280	7570	130	

[a]virtual transition temperature.

In order to control the parameters influencing the production of homogeneous thin films by the spin coating process, experiments were conducted with different solvents: 1,1,2,2 - tetrachloroethane 1,1,2 - trichloroethane, DMF, toluene using different preparation regimes (solution concentration, spin velocity, acceleration). The best results were obtained with 1,1,2,2 - tetrachloroethane or 1,1,2 - trichloroethane at a concentration range of 60-120 g/l, depending on the desired thin film thickness (0.4 - 1.25 μm). The solutions were filtered through a 0.5 μm filter and the filtrates were spun over cleaned silica (for refractive index measurements) and over glass (for SHG measurements) plates. Typical spin coating conditions were: spread cycle of 2000 rpm (acceleration 1000 - 3000 rpm/s) for 20 - 30 sec followed by a spin cycle of 4500 - 5000 rpm (acceleration 2000 - 3000 rpm/s) for 30 sec. After drying at 35°C in *vacuo* for two days the films appeared completely dark between crossed polarizers indicating that the NLO mesogenic side groups are either distributed at random (liquid - like ordering) or uniformly align with the long axes normal to the substrate (homeotropic alignment).

The thin film refractive indices for studied polymers, measured at different wavelengths are listed in Table 2.

Table 2. Polymer refractive indices

Polymer	Refractive index					
Wavelength	632.8 nm		830 nm		1314 nm	
Mode	TE	TM	TE	TM	TE	TM
PA3	1.6211	1.6192	1.6074			
PECH 0.3	1.5827	1.5832	1.5726	1.5732	1.5644	1.5631
PECH 0.55	1.6157	1.6152	1.6015			
PECH 0.8	1.6433	1.6403	1.6268			
PBCMO 14/43/43	1.6349	1.6306	1.6199	1.6144	1.6081	1.6027

The refractive index measurements of the as - spin coated films prepared from PMA3, PECH 0.3 and PECH 0.55 give the same values for the TE and TM modes in agreement with the isotropic behavior, evidenced by thermal analysis and optical microscopy measurements for these polymers. The birefringence values $\Delta n = n_e - n_o = -(19 \div 55) \times 10^{-4}$ obtained for unpoled films prepared from PA3, PECH 0.8 and PBCMO 14/43/43 are not coherent with the nematic characteristics identified by optical microscopy; they are much lower than those usually found for liquid crystalline materials. This observation leads us to conclude that the macroscopic orientation is negligible: upon spin coating from isotropic solutions, these polymers are frozen in a metastable isotropic phase, even though the negative values of Δn indicate some degree of planar alignment of the NLO mesogenic side groups. Studies of the temporal and thermal characteristics of the poling process have been carried out to understand and control the poling process with the intention of maximizing poling - induced macroscopic polar order. Figure 9 illustrates the real time *in situ* poling process observed for PA3. Corona poling field - induced growth of SHG signal is used to probe the induction of molecular orientational order at different temperatures. Under the experimental conditions, below the glass transition temperature, the poling process can be described by a single exponential with a saturation level which remains rather low. In contrast, the growth of the polarization follows a two component process at temperatures above or close to the polymer glass transition temperature T_g = 51.5°C: an initial fast response to the poling field, followed by a much slower response. A qualitative interpretation that distinguishes between the motions of the NLO mesogenic side groups decoupled from the polymer segmental movement and the motions of the side groups coupled to the polymer segmental movement can explain our results. The dynamics of side chain liquid crystalline polymers combine a number of features characteristic of low molar mass liquid crystals and polymers. In addition to the secondary relaxation processes (β, γ, ...) involving local motions of dipoles in the side chains and the typical α - relaxation behaviour associated with the segmental motions of the main chain, side chain liquid crystalline polymers exhibit a δ process, which corresponds to the reorientation of the mesogenic groups about the polymer backbone. This process is quite similar to the end-over-end reorientation of the mesogenic units about their small axis, so characteristic of low molar mass liquid crystals. The activation energy is very high and the characteristic times extremely long because of the strong coupling of the reorientational motion to the nematic surroundings. The temperature dependence of the δ frequencies clearly shows that this process has a contribution resulting from the glass transition of the material, as revealed by its non - arrhenian character: the side group rotation is controlled by free - volume effects. This peculiar feature is the starting point for the electric field induced orientation mechanism in the melt and the glassy states of the NLO polymers considered here.

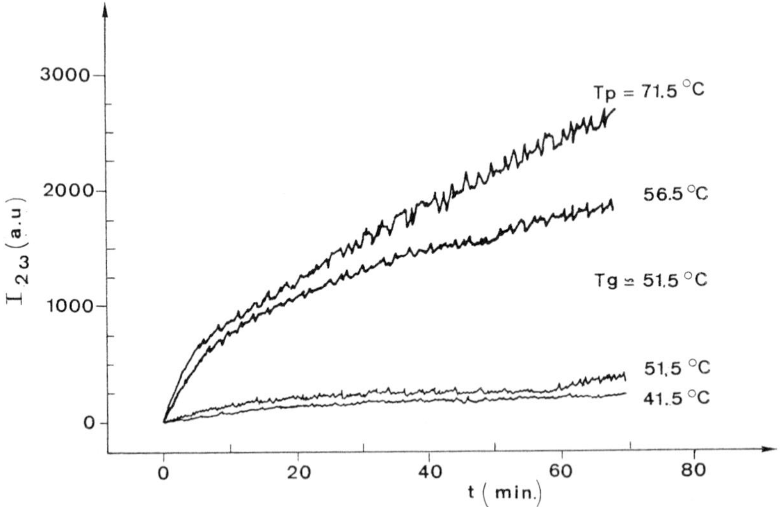

Figure 9. Real time *in situ* poling process for PA3 at different temperatures

When the poling field is first applied, an initial fast growth of the SHG signal is partly due to the fast reorientational response of the polar side chain chromophores. Therefore non-zero polarization is rapidly created. Its magnitude depends on the strength of the poling field and the local mobilities of the chromophores, which are controlled by the size of the free volume around each chromophore, the degree of decoupling between the chromophores and the polymer backbone, and the poling temperature. Since the chromophore dipoles tend to orient along the direction of the poling field, those polymer segments coupled to the chromophores are subjected to intra - and intermolecular interactions and respond by changing their conformation locally to adapt to the anisotropic orientation distribution of the poled chromophores. This creates an anisotropic local environment which enhances the orientational ordering of chromophores. Such a description leads us to associate the slow growth of the SHG signal in the poling process with the reorientational motions of both side chains and polymer backbone chain segments. Its maximum contribution depends on the coupling interaction between the chromophores and the polymer segments and the polymer segmental mobility. Different results support the proposed mechanism:

- distorsion of chain statistics away from spherical is manifested in scattering experiments: depending on the mesophase type and polymer stiffness, the chain becomes prolate (rod - like) or oblate (disc - like).
- at temperatures below the glass transition, since the polymer chains do not undergo a significant reorganization and the side chain chromophores do not orient via the δ process in the glassy state, the slow step growth is eliminated from the SHG signal.
- the time constants decrease as the flexibility of the polymer backbone increases:

$$\tau_{PBCMO\ 14/43/43} > \tau_{PA3} > \tau_{PECH\ 0.8}$$

The optical absorption spectra shown in Fig. 1 illustrate the difference between the as - spin coated films and the poled films prepared from PECH 0.8. After poling the optical absorption, at normal incidence, drastically decreases. This decrease[9] is due to the uniform alignment of the side groups induced by the electric field in the nematic phase. For polymers under interest, the dielectric anisotropy $\Delta\varepsilon$ is positive and the side groups tend to align parallel to the electric field and, as a consequence, perpendicular to the substrate. The degree of axial order can be characterized by the order parameter $\langle P_2 \rangle$ given by the following relation:

$$\langle P_2 \rangle = \frac{A_{\parallel} - A_{\perp}}{A_{\parallel} + 2A_{\perp}} \tag{4}$$

where $A_{\parallel}$ and $A_{\perp}$ are the absorbances parallel and perpendicular to the optical axis. As shown in Fig. 10, the axial ordering increases with poling time before leveling off. In the case of PECH 0.8 and PA3, a saturation limit for the order parameter $\langle P_2 \rangle$ of ~ 0.65 was observed. It was found, however, that the microscopic order parameter increased more rapidly in PECH 0.8 relative to PA3, which is likely due to the greater flexibility of the polyether backbone. The order parameters derived for PECH 0.55 and PMA3 are smaller (0.4÷0.5) but larger than those reported for other isotropic polymers. Clearly, the poling field strongly influences the state of order and induces an isotropic/nematic transformation. The actual nematic/isotropic transition temperature in the presence of a field will be larger than the temperature at which the discontinuous nematic/isotropic phase transformation would take place in the absence of an electric field. During poling, PECH 0.55 and PMA3 can be pulled through the phase transition, permitting the nematic/isotropic transition temperature to become equal to or larger than the poling temperature. The studies of the poling process for PBCMO 14/43/43 and PBCMO 100 are in progress.

Nonlinear Optical Properties.

The second order nonlinear optical susceptibilities d_{31} and d_{33} for PA3, PMA3, PECH 0.8 and PECH 0.55 polymers, determined by second harmonic generation, are listed in Table 3.

For all the polymers under investigation the values found for the ratio a (cf. Eq. (3)) are significantly larger than those reported in the literature for the amorphous, isotropic polymers (a = 3 ÷ 6). This shows that the polar order of the NLO groups can be enhanced by the axial order present in the mesophases. The nonlinear optical response is thus determined by the order imposed by the poling field as well as the order arising from intermolecular forces as quantified by the microscopic order parameter. The observed difference of behavior between the two copolyethers PECH 0.8 and PECH 0.55 and between the polyacrylate PA3 and the polymethacrylate PMA3 is likely due to the smaller

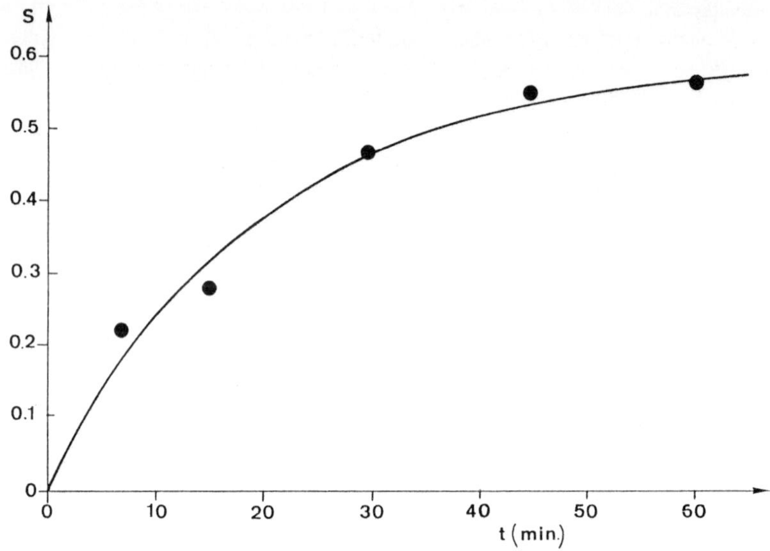

Figure 10. Variation of the microscopic order parameter <P2> as a function of poling time for PA3. Film thickness 0.23 μm, $T_p = T_g + 6°C$, V = 8 kV

axial ordering induced by the poling field in the films prepared from the isotropic copolyether (PECH 0.55) and polymethacrylate (PMA3), even though these polymers undergo an isotropic/nematic transition during poling. In the following paragraphe, these results will be analyzed in the framework of the molecular statistical models developed for isotropic and liquid crystalline polymers.

Table 3. Second order nonlinear optical susceptibilities of the studied polymers

Polymer	t_p (min)	T_p (°C)	$\langle P_2 \rangle$	λ (μm)	d_{33} (pm/V)	d_{31} (pm/V)	a= d_{33}/d_{31}
PECH 0.8	15	81	0.57	0.777	32.7	1.8	15±3
	15	81	0.57	1.0642	17.8	0.97	15±3
	60	81	0.62	1.0642	23.4	1.65	14±3
PECH 0.55	15	57	0.45	0.777	14.6	1.33	11.0±2.2
	15	57	0.45	1.0642	7.9	0.7	11±2
PA3	15	57	0.17	1.0642	9.5	1.2	8±1.5
	60	57	0.57	1.0642	35.6	2.1	17±3
PMA3	15	97	0.45	1.0642	33.5	3.3	10±2
PBCMO	30	107	0.19	1.0642	20	2.4	8±15

Orientation Models.

Several models have been proposed in order to explain the orientation mechanism of NLO chromophores during poling. For rod-like charge transfer (CT) molecules with ∞mm symmetry, the orientation distribution function G(Θ) depends only on the angle between the molecular axis (CT axis) and the preferential orientation axis (e. g. poling field direction). In the Boltzmann statistics this function is given by

$$G(\Theta) = \frac{e^{-U(\Theta)/kT}}{2\pi \int_0^{\pi} e^{-U(\Theta)/kT} \sin \Theta d\Theta} \tag{5}$$

where U(Θ) is the potential energy of the molecule in the presence of the poling field E. In the simplest isotropic model this energy is given by

$$U_E(\Theta) = -\mu_0 E \cos\Theta \tag{6}$$

where μ_0 is the fundamental state dipole moment. The orientational averaged values of $\langle\cos^3\Theta\rangle$ and $\langle\cos\Theta \sin^2\Theta\rangle$, intervening in Eqs. (1) - (3), can be obtained by expanding G(Θ) in Legendre polynomials $P_n(\cos\Theta)$ series

$$G(\Theta) = \sum_n \frac{2n+1}{2} \langle P_n(\cos\Theta)\rangle P_n(\cos\Theta) \tag{7}$$

The expansion coefficients $\langle P_n(\cos\Theta)\rangle$ are orientational averages of $P_n(\cos\Theta)$ and are defined as order parameters. Usually $\langle P_2(\cos\Theta)\rangle$ and $\langle P_4(\cos\Theta)\rangle$ are sufficient to describe the state of order in liquid crystal polymers. In the isotropic model all order parameters are equal to zero.

The orientational averaged value of any function of Θ is calculated by integrating this function over Θ using the orientational distribution function as a weight factor. The result can be expressed in terms of Langevin function series $L_n(u)$, where $u = \mu_0 E/kT$. If $u<1$, a limited development into a Taylor series gives simple expression for $\langle\cos^3\Theta\rangle$ and $\langle\cos\Theta \sin^2\Theta\rangle$ (for details see Refs. 10-11). As can be seen from Table 4 the isotropic model predicts that $\langle\cos^3\Theta\rangle$ is three times larger than $\langle\cos\Theta \sin^2\Theta\rangle/2$.

In the Ising model one assumes that the molecules are perfectly aligned. This gives two possible values for the angle Θ: 0 and π and consequently $\langle P_2(\cos\Theta)\rangle = 1$.

Singer and coworkers[12] (SKS model) modified the isotropic model by introducing an additional term $U_m(\Theta)$ into the potential energy, in order to take into account the fact that the intrinsic axial order of a liquid crystalline medium can be transferred to chromophores by mutual interactions, leading to an axial ordering of chromophores. In the presence of an external field the molecular energy U(Θ) is then given by

$$U(\Theta) = -\mu_0 E \cos\Theta + U_m(\Theta) \tag{8}$$

For $u<1$ the orientational averages of $\langle\cos^3\Theta\rangle$ and $\langle\cos\Theta \sin^2\Theta\rangle$ are given by

$$\langle\cos^3\Theta\rangle = u(\frac{1}{5} + \frac{4}{7}\langle P_2\rangle + \frac{8}{35}\langle P_4\rangle) \tag{9}$$

$$\langle \cos\Theta \sin^2\Theta \rangle / 2 = u(\frac{1}{15} + \frac{1}{21}\langle P_2 \rangle - \frac{8}{70}\langle P_4 \rangle) \quad (10)$$

The axial order parameters $<P_2>$ and $<P_4>$ are input parameters which have to be determined experimentally at zero field strength. In the case when all order parameters vanish ($<P_2> = <P_4> = ... = 0$) the SKS model reduces to the isotropic model.

Another approach was proposed by Van der Vorst and Picken[13-14] (MSVP model). To take the intrinsic mesogenic properties of the NLO molecules themselves into account, these authors extended the original model of Maier and Saupe[15-17] for axial order in field - free liquid crystals to the case of electric field poling of NLO active molecules forming a liquid crystal phase. The molecular energy has the following form

$$U(\Theta) = -\mu_0 E \cos\Theta + U_m(\Theta) + U_i(\Theta) \quad (11)$$

where $U_m(\Theta)$ describes the mesogenic properties and $U_i(\Theta)$ the influence of the electric field through the linearly induced dipole moment

$$U_i(\Theta) = -\Delta\alpha E^2 P_2(\cos\Theta) / 3 \quad (12)$$

where $\Delta\alpha = \alpha_{||} - \alpha_{\perp}$ is the anisotropy of the linear molecular polarizability.

In order to take into account the tendency toward mutual axial alignment experienced by strongly anisotropic, rod-like NLO mesogens Van der Vorst and Picken[13-14] proposed to write the $U_m(\Theta)$ energy term in the same form as that originally proposed by Maier and Saupe[15-17]

$$U_m(\Theta) = -\zeta < P_2 > P_2(\cos\Theta) \quad (13)$$

in their mean field theory for the nematic phase. The ζ parameter in Eq. (13) scales the magnitude of the interactions between mesogens and takes a constant value for a given liquid crystal. It follows from the Maier and Saupe[15-17] theory that ζ is proportional to the liquid crystal clearing temperature T_c (in the absence of external field): $kT_c \cong 0.22\zeta$.

The order parameters in MSVP model have to be calculated by numerical methods for a particular set of parameters μ_0, $\Delta\alpha$, ζ (or T_c), E and T_g. A self consistent value of $<P_2>$ is obtained by an iterative procedure which is initiated with an appropriate approximation of $<P_2>$. After substituting the $<P_2>$ value obtained in the expression for $G(\Theta)$, one can get other averages like $<\cos^3\Theta>$ and $<\cos\Theta \sin^2\Theta>$ which intervene in the expressions for quadratic NLO susceptibilities (for details see e. g; discussion in Refs. 10-11) and consequently in the ratio $a = d_{33}/d_{31}$. The corresponding quantities calculated within different theoretical models are listed in Table 4.

An important difference between the four models is whether the exact solutions are derived for the polar order parameters, either analytically or numerically. In the isotropic, Ising, and SKS models approximate analytical solutions are derived, which are linear functions of the poling field strength (cf. Table 4). The linearization is a direct consequence of the truncation of Taylor series expansion after the linear term. By using a quasi one - dimensional description of the acceptor -donor molecules the MSVP model is capable of going beyond the (usual) linear approximation, which implies that it is suitable for the strong field regime. The MSVP and SKS models give the same orientational - averaged values of $<\cos^3\Theta>$ and $<\cos\Theta \sin^2\Theta>/2$ for local field strengths smaller than approximately 100 V/μm. At higher field strengths, however, the saturation effects, which are not taken into account in the SKS model, lead to different values for order parameters. The difference between these two models increases with increasing field strength.

Comparison between experiment and theory

Experimental and theoretical values of the order parameters <P$_2$> and <P$_4$> and the ratio a = d_{33}/d_{31} are given in Table 5 for PECH0.55, PECH0.8, PA3 and PMA3. Several important features can be noted:

(i) The experimental values of <P$_2$> are smaller than those predicted by the MSVP model. This effect may be attributed to errors in the input parameters for the poling theory (e. g.: μ_0, $\Delta\alpha$, E_p, ...) and also to errors in the measured <P$_2$> themselves due to the existence of dimers absorbing at higher wavelength than the monomers. It is also worth noting that the MSVP model is not a dynamical model. It predicts only equilibrium values that are usually reached after a long annealing time in the case of polymers.

(ii) For both PECH0.8 nd PA3, which are initially nematic, the ratio a = d_{33}/d_{31} is in qualitative agreement with the SKS and MSVP models. The ratio is much higher than 3 owing to the axial order present in the mesophase. For both PECH0.55 and PMA3, which are initially isotropic, the MSVP model is the only model capable of describing the isotropic/nematic transition induced by the strong poling field.

(iii) The experimental SHG coefficient d_{33} agrees well with the SKS and MSVP models for PA3, but is found to be only 70% of the theoretical predictions in the case of PECH0.8. This discrepancy may be attributed to errors in the input parameters necessary for the conversion of the theoretical polar order parameter <cos$^3\Theta$> into the theoretical SHG coefficient d_{33} (N, F, β, μ_0, E_p). There are also errors in the measured SHG coefficient itself if the sample has not be poled long time enough to approach to an experimental steady state value. Of particular interest here is also the enhancement of the NLO performance with liquid crystalline character by a factor of 2.3 ÷ 3.3.

Dispersion

In view of the great practical importance, characterization of the wavelength dispersion of the SHG coefficients was carried out for PECH0.8. Figure 11 shows that d_{33}

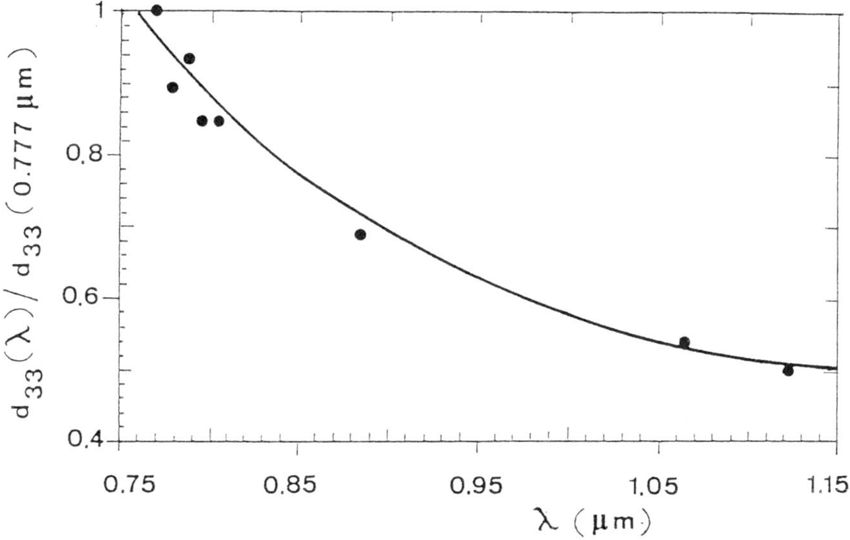

Figure 11. Dispersion of the d33 coefficient for PECH 0.8. Points show experimental values whereas the solid line is from the resonant two- level model.

decreases with increasing wavelength, as expected for an off resonance behavior. The experimental data are in good agreement with the curve predicted by the resonant two-level model[18]. The values of the resonant frequency of the first excited state (ω_0 = 34250 cm $^{-1}$)

and the damping term ($\Gamma = 4500\ cm^{-1}$) used in the calculation are close to those deduced from the optical absorption spectrum. From the position of the absorption maximum and the width of the absorption spectrum, ω_0 and Γ were found to be 34130 cm^{-1} and 5080 cm^{-1}, respectively. The existence of dimers, which absorb at higher wavelength than the monomers, but do not participate to the poling process, may explain the larger value determined experimentally for the width of the absorption band.

Relaxation

One of the important factors governing the applications of poled polymers in NLO devices is the temporal stability of the polar alignment induced by the electric field. Several studies have been done in order to describe the decay phenomena in amorphous, isotropic functionalized polymers (Ref. 19). It is clear that the molecular size, thermal history and glass transition temperature, as well as the type of bonding in polymer and between the NLO moiety and the polymer impact the stability of SHG signal.

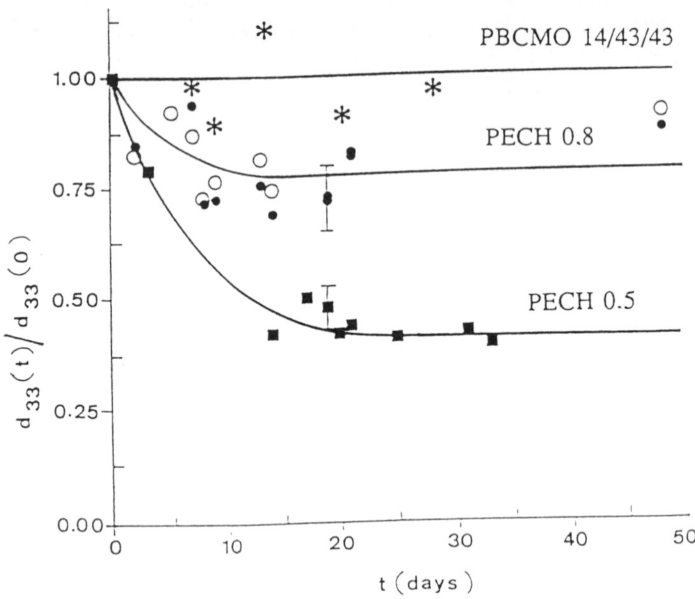

Figure 12. Temporal behavior of the d_{33} susceptibility for PECH0.8, PECH0.5 and PBCMO14/43/43 polymers. Solid lines are guides for eyes.

A number of poled films were stored at room temperature and SHG coefficients were periodically measured. The results are shown in Fig. 12 for PECH0.55, PECH0.8 and PBCMO14/43/43. Two interesting effects can be noted. First, by comparison of the data in Fig. 12 for PECH0.55 and PBCMO14/43/43 and in the literature for isotropic polymers containing NLO moieties in the side chains and characterized by glass transition temperatures in the range 80 - 110°C it is clear that initial liquid crystallinity increases the orientational stability. Second, the temperature at which the poled films are stored below T_g is important in determining the relaxation rate of SHG coefficients. It was also found in this study that the absorption spectra of the poled films do not change with time, indicating that the axial order, characterized by the microscopic order parameter $\langle P_2 \rangle$, is frozen in the glassy state. An important consequence is that the ratio $a = d_{33}/d_{31}$ remains constant, although the SHG coefficients d_{33} and d_{31} decrease as a function of time. Further studies are in progress to investigate the effect of increasing T_g as the degree of mobility of the side chains will depend on the locus of the temperature of the SHG measurements relative to the δ (reorientation of the side chains around the backbone), α (or T_g) and secondary (β or sub T_g) relaxation processes.

Table 4. Orientational averaged values of $\langle \cos^3 \Theta \rangle$ and $\langle \cos\Theta \sin^2\Theta \rangle /2$ and microscopic order parameters in four molecular - statistical models

Model	$\langle \cos^3\Theta \rangle$	$\langle \cos\Theta \sin^2\Theta \rangle/2$	$\langle P_2 \rangle = \langle 3\cos^2\Theta - 1 \rangle/2$	$a = d_{33}/d_{31}$
Ising	u	0	1	∞
Isotropic	u/5	u/15	$u^2/15$	3
SKS[1]	uA	uB		A/B
MSVP	to be calculated for a given set of parameters			

[1] $\langle P_2 \rangle$ has to be determined experimentally at zero field

$u = \mu_0 E/kT$

$$A = \frac{1}{5} + \frac{4}{7}\langle P_2 \rangle + \frac{8}{35}\langle P_4 \rangle$$

$$B = \frac{1}{15} + \frac{1}{21}\langle P_2 \rangle - \frac{8}{70}\langle P_4 \rangle$$

DISCUSSION

The present study shows that side chain liquid crystal polymers (SCLCP's) are undoubtedly an interesting class of materials for quadratic nonlinear optics. Obviously, LCP media offer competition to amorphous, isotropic functionalized polymers. The former

- have equivalent functionality
- exhibit enhanced field - induced polar ordering. The experimental values obtained for the ratio of the two polar order parameters are as high as 17±3 (cf. Table 3), which cannot be obtained with conventional isotropic functionalized polymers. Such high ratios lead to very large values of d_{33} (d_{33} = 35.6 pm/V for PA3 at 1064.2 nm), never obtained for isotropic polymers absorbing in the same frequency range.

The ease of polymer processability also makes commercial sense to try to utilize LCP's for optical waveguide fabrication. However, many problems still must be solved before NLO SCLCP's can become a routinely applied material. One of the important problems encountered with LC's is light scattering. This is especially true with nematics, whose scattering cross-section is about one order of magnitude larger than that in pure isotropic media. This arises from the orientational fluctuations of the director. However, it has been shown that in the presence of a strong electric field, the fluctuations may be strongly decreased as they become inversely proportional to E^2 (cf. Refs.10-11, 20-21.). Such field effect has been observed in low molar mass LC's by Durand and Martinaud[21]. With large enough electric fields, light scattering is expected to be similar to that observed in isotropic systems.

Often the thermal and temporal stability of polymers is also questioned. Further studies of new SCLCP's are in progress to evaluate the influence of the flexibility of the polymer backbone and the degree of orientational freedom of the side chain chromophores on the poling process with the intention of maximizing the poling induced nonlinearity and stability.

ACKNOWLEDGEMENTS

This work was supported in part by the Indo-French Centre for the Promotion of Advanced Research (IFCPAR Research Project n^0 1008-1).

Table 5. Experimental and theoretical values of the order parameters <P2> and <P4, >, the ratio a = d33/d31 and R = calculated d33 /isotropic d33 susceptibilities for PECH0.8, PECH0.55, PA3 and PMA3.

Polymer	Model	<P2>	<P4>	d33 (pm/V)	a=d33/d31	$R = \frac{d_{33}(calc))}{d_{33}(isotr.)}$
PECH0.8	Isotropic	0	0	10	3	1
	SKS	0.68[b]	0.33[b]	33.3	10.8	3.3
	MSVP	0.77[c]	0.44[c]	37.0	13	3.7
	Ising	1	1	50.1	∞	5
	Experiment[a]	0.62	-	23.4	14±3	2.34
PA3	Isotropic	0	0	10.7	3	1
	SKS	0.61[b]	0.26[b]	32.3	9.2	3.0
	MSVP	0.75[c]	0.40[c]	38.7	11.9	3.6
	Ising	1	1	53.2	∞	5
	Experiment[a]	0.57	-	35.6	17±3	3.3
PECH0.55	Isotropic	0	0		3	1
	SKS	0	0		3	1
	MSVP	0.65[d]	0.28[d]		9.3	
	Ising	1	1		∞	
	Experiment[a]	0.45	-	7.9	11±2	
PMA3	Isotropic	0	0		3	1
	SKS	0	0		3	1
	MSVP	0.61[d]	0.25[d]		8.6	
	Ising	1	1		∞	
	Experiment[a]	0.45	-	33.5	10±2	

[a]Experiments carried out at 1.0642 μm fundamental wavelength
[b]Calculated with the MSVP model at zero field strength
[c]Calculated with the MSVP model at E_p = 2.9 MV/cm (f(0) = 2)
[d]Calculated with the MSVP model at E_p = 2.07 MV/cm (f(0) = 2)

REFERENCES

1. G. Khanarian, M. Mortazavi and R. A. Norwood, Frequency doubling and parametric interactions in organic thin films, in: "Organic Thin Films for Waveguiding Nonlinear Optics: Science and Technology ", F. Kajzar and J. Swalen Eds., Gordon & Breach Sc. Publ;, in press

2. S. Piercourt, N. Lacoudre, A. Le Borgne, C. Friedrich and C. Noël, Synthesis of side - chain liquid crystalline polyethers by chemical modification of polyepichlorohydrin, *Makromol. Chem.*, **193**: 705(1992).

3. B. Comizzoli, Use of corona discharges in the semiconductor industry, *J. Electrochem. Soc.: Solid Science and Technology*, **134**: 424-29(1987).

4. G. Gadret, F. Kajzar and P. Raimond, Nonlinear optical properties of poled polymers, in: "Nonlinear Optical Properties of Organic Materials IV", K. D. Singer Ed., Proc. SPIE, vol. 1560, pp. 226-237, Bellingham 1991.

5. P. A. Chollet, G. Gadret, F. Kajzar, P. Raimond and M. Zagorska, Electrooptic organic thin films for application in waveguide devices, in: "Nonlinear Optical Properties of Organic Materials V", D. J. Williams Ed., Proc. SPIE, vol. 1775, pp. 121-132, Bellingham 1993.

6. F. Kajzar, P. A. Chollet, I. Ledoux, J. Lemoigne, A. Lorin and G. Gadret, Organic thin films for quadratic optics, in: "Organic Molecules for Nonlinear Optics and Photonics",

NATO ASI Series, Series E, vol. 194, pp. 403-432, Kluwer Academic Publ., Dordrecht 1991.
7. M. Choy and R. L. Byer, Accurate second-order susceptibility measurements of visible and infrared nonlinear crystals, *Phys. Rev. B*, **14**:1693(1976)
8. J. Swalen and F. Kajzar, in Organic Thin Films for Waveguiding Nonlinear Optics: Science and Technology , F. Kajzar and J. Swalen Eds., Gordon & Breach Sc. Publ;, in press
9. R. H . Page, M. C. Jurich, B. Reck, A. Sen, R. J. Twieg, J. D. Swalen, G. C. Bjorklund and C. G. Wilson, Electrochromic and optical waveguide studies of corona-poled electro-optic polymer films, *J. Opt. Soc. Am. B*, **7**: 1239(1990).
10. D. Gonin, Thesis, Universite Pierre et Marie Curie, Paris, 1994
11. D. Gonin, C. Noël and F. Kajzar, Liquid crystalline polymers, in: "Organic Thin Films for Waveguiding Nonlinear Optics: Science and Technology ", F. Kajzar and J. Swalen Eds., Gordon & Breach Sc. Publ;, in press
12. K. D. Singer, M. G. Kuzyk and J. E. Sohn, Second-order nonlinear-optical processes in orientationally ordered materials: relationship between molecular and macroscopic properties, *J. Opt. Soc. Am. B*, **4**: 968(1987)
13. C. P. J. M. Van der Vorst and S. J. Picken, A model for poling by an electric field of acceptor - donor molecules in a liquid crystalline phase, Proc. SPIE, vol. 866, 99(1987)
14. C. P. J. M. Van der Vorst and S. J. Picken, Electric field poling of acceptor-donor molecules *J. Opt. Soc. Am. B*, **7**: 320(1990)
15. W. Maier and A. Saupe, Einfache molekularstatistische Theorie der nematischen Phase Dispersion wechselwirkung Zustandes, *Z. Naturforsch.*, **13A:** 564(1958)
16. W. Maier and A. Saupe, Einfache molekularstatistiche Theorie der nematischen Phase, *Z. Naturforsch.*, **14A:** 882(1959)
17.W. Maier and A. Saupe, Einfache molekularstatistiche Theorie der nematischen Phase, *Z. Naturforsch.*, **15A:** 287(1960)
18. D. Gonin, C. Noël and F. Kajzar, EFISH and THG measurements on organic molecules in solution, *Nonl. Optics*, **8**, 37(1994)
19. K. Singer and M. Andrews, Quadratic nonlinear optics in poled polymer films: from physics to devices, in "Molecular Nonlinear Optics: Materials, Physics, and Devices ", J. Zyss, Ed., Academic Press Inc., Boston 1994, p. 245.
20. P. G. De Gennes, The Physics of Liquid Crystals, Clarendon Press, Oxford 1974.
21. J. L. Martinaud and G. Durand, Electric field quenching of thermal fluctuations of orientation in a nematic liquid crystal, *Solid State Commun.*, **10**: 815(1972)

NOVEL ELECTRO-OPTIC POLYIMIDE AND POLYAMIDE SIDE CHAIN AND MAIN CHAIN POLYMERS

P. Günter, P. Prêtre, P.Kaatz

Nonlinear Optics Laboratory
Swiss Federal Institute of Technology
CH-8093 Zürich

Ch. Weder, P. Neuenschwander, U.W. Suter

Polymer Institute
Swiss Federal Institute of Technology
CH-8092 Zürich

INTRODUCTION

There has been considerable interest in organic nonlinear optical (NLO) materials, because of their potential application in integrated electro-optical devices.[1] A promising approach to the development of new second-order NLO-materials is that of poled polymers. The advantages of poled polymers are large susceptibilities, fast response times, easy processability, and high physical and mechanical stability, but their NLO properties are usually not stable, due to the thermal relaxation of the chromophore orientation.[1] Different design strategies have been worked out to synthesize polymers with desirable NLO-properties. Typically the NLO chromophores have been incorporated by doping[2] (guest-host systems) or attaching them covalently[3] (side-chain systems) into amorphous or liquid crystalline polymers. To enhance the orientational stability of the chromophores, cross-linked polymers[4] or polymers which incorporate the chromophores with their dipole moments head-to-tail in the main chain[5] have been synthesized.

We have synthesized amorphous polyimide type side chain and novel polyamide main chain NLO polymers with good processability, high nonlinear susceptibilities, high glass transition temperatures up to 176 °C and good orientational stability. The dispersion of the linear and nonlinear optical properties of these polymers has been determined. The time-temperature behaviour of the chromophore relaxation has been modeled to obtain estimates of the lifetimes of electro-optic devices using these polymers.

Polymers and Other Advanced Materials: Emerging Technologies and Business Opportunities. Edited by P. N. Prasad *et al.*, Plenum Press, New York, 1995

POLYIMIDE SIDE AND POLYAMIDE MAIN CHAIN POLYMERS

Alkyl amino functionalized NLO azo chromophores were attached via a two or three carbon spacer linkage to an alternating styrene-maleic-anhydride copolymer in the case of the NLO side chain polymer.[6]

In the main chain polymer, the chromophores' dipole moments are fixed transverse to the polymer main chain, based on the assumption that in this arrangement the chromophores are easier to orient by an external field than in structures were the chromophores' dipole moments are pointing along the polymer main chain. The chosen NLO dye is based on DANS[7] [4-(dimethylamino)-4'-nitrostilbene]. It was modified by the introduction of two amino groups into one of the aromatic rings to achieve the new bifunctional chromophore DDANS [2',5'-diamino-4-(dimethylamino)-4'-nitrostilbene].

The chemical structure of the polymers described here together with the most important properties are given in Fig. 1 and Table 1. Details about the chemical synthesis are given elsewhere[8, 9].

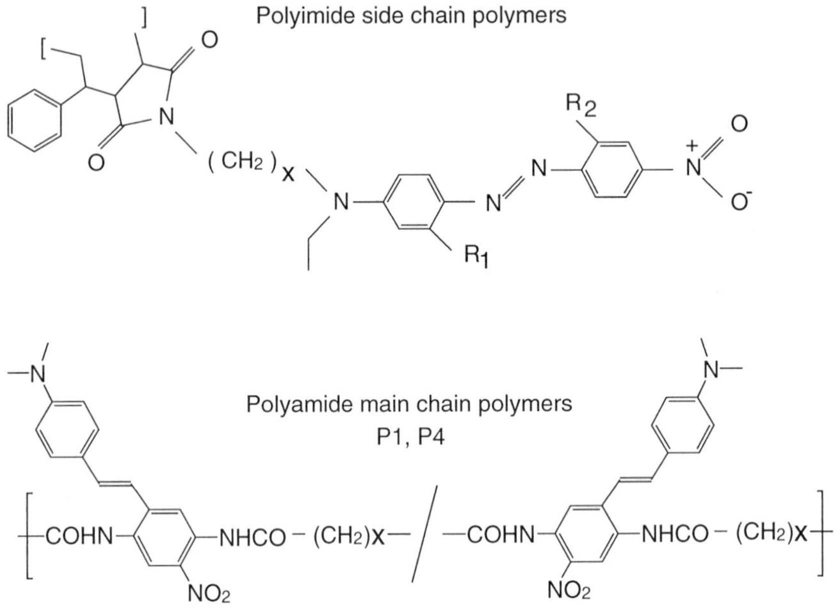

Polyimide side chain polymers

Polyamide main chain polymers
P1, P4

Figure 1. Chemical structure of the polyimide side and polyamide main chain polymers

Table 1. Nomenclature, azo dye substitution patterns, main and side chain spacer lengths and glass transition temperatures for the polyimide side and polyamide main chain polymers.

Type	Polymer	x	R_1	R_2	T_g [°C]
Polyimide	A-095.11	3	CH_3	H	137
side chain	A-097.07	3	CH_3	Cl	149
polymers	A-148.02	2	H	H	172
Polyamide main	P1	12	-	-	125
chain polymers	P4	6	-	-	176

LINEAR OPTICAL PROPERTIES

Linear optical properties of thin spin cast polymer films are most easily determined by transmission spectroscopy.[10] The difference between the refractive indices of the polymer and the substrate (fused silica) produces a modulation of the transmission properties due to optical interference in the thin (≈ 1 μm) polymeric films. The measured transmissivity is a function of the wavelength, film thickness, and the refractive indices of the substrate, the polymer film, and the covering medium. The real part of polymer refractive index was fitted to a Sellmeier-type dispersion formula,

$$n^2(\lambda)-1=\frac{q}{1/\lambda_o^2-1/\lambda^2}+A\ , \tag{1}$$

where λ_0 is the absorption wavelength of the dominant oscillator and A is a constant containing the sum of all the other oscillators. Fig. 2 shows a measured transmission curve and the calculated dispersion of the refractive indices. The Sellmeier parameters are given in Table 2.

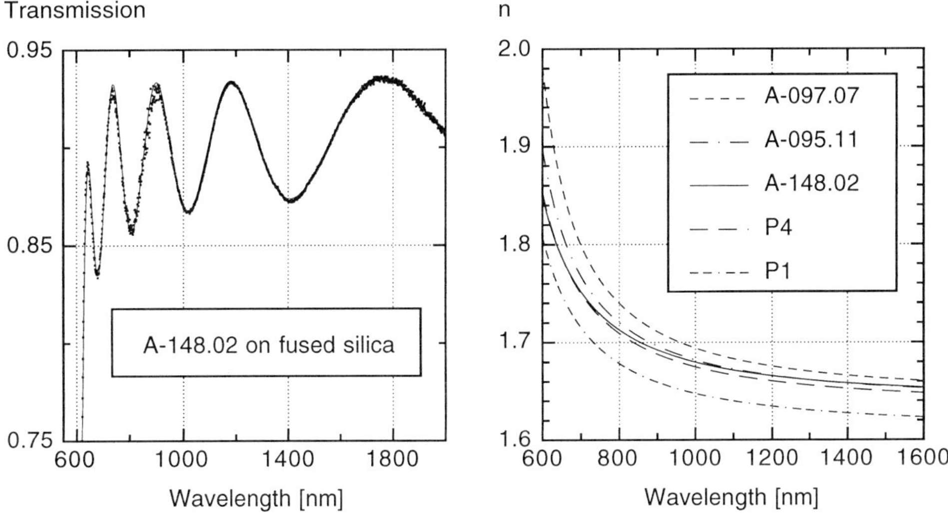

Figure 2. Left: transmission spectroscopy measurement for a 1060 nm thick film of polymer A-148.02 together with fitted transmission function (solid line). Right: dispersion of the refractive indices.

Table 2. Sellmeier parameters and refractive indices at selected wavelengths for the polyimide side and polyamide main chain polymers.

Polymer	A–095.11	A–097.07	A–148.02	P1	P4
λ_0 [nm]	491	510	470	471	473
q [10^{-6} nm^{-2}]	1.874	1.809	2.074	1.889	2.074
A	1.234	1.234	1.234	1.176	1.208
n @ 633 nm	1.84 ± 0.01	1.89 ± 0.01	1.80 ± 0.01	1.76 ± 0.01	1.81 ± 0.01
n @ 1313 nm	1.65 ± 0.01	1.67 ± 0.01	1.66 ± 0.01	1.63 ± 0.01	1.66 ± 0.01

NONLINEAR OPTICAL PROPERTIES

Nonlinear optical measurements were performed using a standard Maker-fringe technique.[11] Stimulated Raman scattering in compressed CH_4 (≈ 20 bar) in an approximately one meter long gas cells provided the fundamental wavelength of $\lambda = 1542$ nm. The Nd:YAG pump laser (BMI D.N.S 501) at $\lambda = 1064$ nm had pulse energies of 200 - 250 mJ/pulse and pulse durations of about 20 ns. This laser could be operated at $\lambda = 1338$ nm by changing the cavity mirrors.

Electro-optic measurements were performed for electric field frequencies much lower than acoustic and vibrational frequencies. At those low frequencies the free electro-optic coefficient r^T is the sum of the electronic contribution (r^e) and the effects from optic (r^o) and acoustic (r^a) phonons.[12] For polar inorganic materials the last two contributions are very important, however, for organic materials the electro-optic effect is assumed to be mainly of electronic origin. The electronic contribution to the electro-optic coefficient at low frequencies and the nonlinear optical susceptibilities are directly related by[13]

$$r_{ijk}^{e} = \frac{-4}{n_i^2 n_j^2} \cdot d_{ijk}^{EO} . \tag{2}$$

The superscript *EO* denotes the NLO susceptibility *d* extrapolated to zero frequency for the electric modulation field:

$$d_{ijk}^{EO} \equiv d_{ijk}^{-\omega;\omega,0} = \frac{f_i^{\omega} f_j^{\omega} f_k^{0}}{f_k^{2\omega'} f_i^{\omega'} f_j^{\omega'}} \cdot \frac{(3\omega_0^2 - \omega^2)\cdot(\omega_0^2 - \omega'^2)\cdot(\omega_0^2 - 4\omega'^2)}{3\cdot(\omega_0^2 - \omega^2)^2 \cdot \omega_0^2} \cdot d_{kij}^{-2\omega';\omega',\omega'} \tag{3}$$

ω' is the optical frequency at which $d_{kij}^{-2\omega';\omega',\omega'}$ is measured and ω the optical frequency involved in the electro-optic experiment. Therefore, second harmonic generation (SHG) experiments can be used to deduce the electronic part of the EO effect and provide a measure of the expected electro-optic response of a material. Table 3 contains these data for the polymers investigated in this work.

Table 3. Nonlinear optical susceptibilities d_{33} compared to a quartz reference (d_{11} = 0.4 pm/V) and electro-optic coefficients for the polyimide side and polyamide main chain polymers.

Polymer	d_{33} [pm/V] @ 1338 nm	d_{33} [pm/V] @ 1542 nm	r_{33}^{e} [pm/V] @ 1300 nm
A-095.11	69 ± 9	45 ± 6	18 ± 2
A-097.07	69 ± 21	45 ± 6	16 ± 5
A-148.02	57 ± 15	42 ± 1	17 ± 4
P1	42 ± 5	40 ± 4	16 ± 2
P4	31 ± 3	27 ± 3	10 ± 2

The electro-optic coefficients r_{33}^{e} for all the polymers have to be compared with the value r_{33} = 32 pm/V for $LiNbO_3$. It should be possible to reach this level within our new polymer concepts using an improved NLO dye with larger hyperpolarizability.

LONG TERM ORIENTATIONAL STABILITY

A useful functional representation for describing the non-exponentiality of structural relaxations in polymeric glasses in the time domain is the stretched exponential or Kohlrausch-Williams-Watts[14] (KWW) function.

$$\phi(t) = exp - (t/\tau)^b \ , \tag{4}$$

with $0 < b \leq 1$. The issue of nonlinearity of structural relaxations was addressed by Tool by introducing the concept of fictive temperature, T_f which is defined as the temperature at which the observed property would attain an equilibrium value. A calculation procedure for T_f was introduced by Narayanaswamy[15] and further developed by Moynihan[16] and Hodge[17].

The structural relaxation time τ of polymeric liquids, as interpreted by DiMarzio and Gibbs[18] involves the cooperative rearrangement of a number of molecular segments of the polymer molecule. Adams and Gibbs[19] subsequently obtained an expression for the structural relaxation time in terms of the configurational entropy S_c. Hodge has shown that the following expression can be derived for the relaxation time associated with the cooperative chain rearrangements:[20]

$$\tau = A\ exp\left(\frac{B}{T(1 - T_2/T_f)} \right) \tag{5}$$

where $B{\cdot}R$ is an activation energy with R the gas constant, T is the temperature, and A is a time parameter. We assume that the temperature T_2 in this formulation is equivalent to the "Kauzmann temperature",[21] T_K , which is described by $S_c = 0$.

Extended DSC (Differential Scanning Calorimetry) measurements on these polymers indicate the preexponential time factor A can be replace by

$$A = \tau_g\ exp\left(\frac{-B}{T_g - T_2} \right) \tag{6}$$

with τ_g the relaxation time at T_g, typically in the range of 10 - 100 s.

Below the glass transition T_f reaches a final limiting value in simple cooling processes, i.e. without sub-T_g annealing. Assuming the final fictive temperature to be T_g, we obtain the following temperature dependence relaxation time τ:

$$\tau = \tau_g \cdot exp\left(\frac{B}{T_g - T_2} \frac{T_g - T}{T} \right) = \tau_g \cdot exp\left(2.303\, C_1^g \frac{T_g - T}{T} \right) . \tag{7}$$

This result indicates that the primary or α-relaxation processes should have an Arrhenius temperature dependence below the glass transition. In addition, from the "universality" of the Williams-Landel-Ferry[22] (WLF) parameter C_1^g, one might expect the relaxation times to scale with the scaling parameter $(T_g - T)/T$, at least in the same class of glass forming liquids (e.g. polymers).

Second Harmonic Decay Measurements

Relaxation of the side-chain chromophores below the glass transition was investigated by the decay of the SHG signal from corona poled films at fixed temperature over several decades in time. The time dependence of the second harmonic decay was usually well represented by the stretched exponential function. An example of normalized decay traces for polymer P1 is shown in Fig. 3.

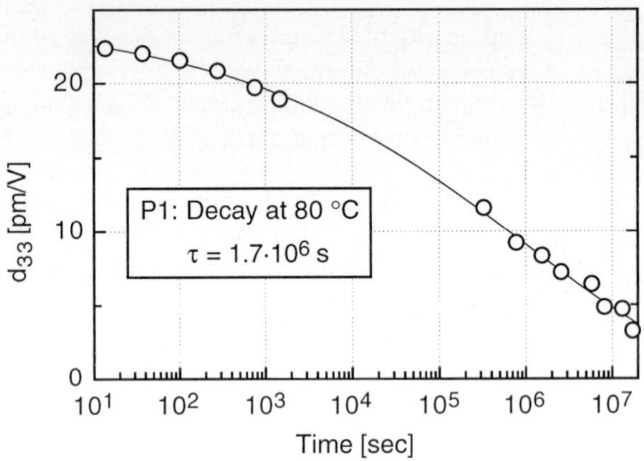

Figure 3. Second harmonic decay of P1 with stretched exponential decay function (solid line)

Scaling of Relaxation Times

Second harmonic decay times for all the polymers are displayed in Fig. 4 as a function of the scaling parameter $(T_g - T)/T$.

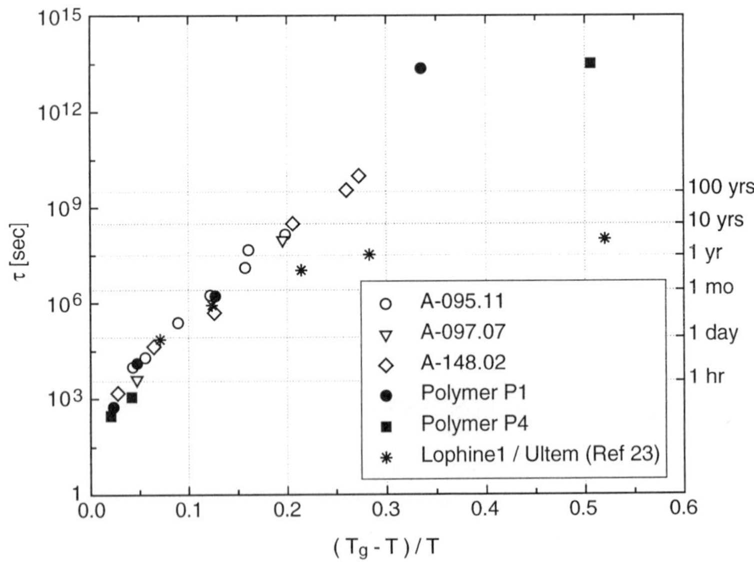

Figure 4. Scaling law for second harmonic decay times. For comparison, the results for the guest host system Lophine 1/ Ultem with a Tg of 180 °C from Stähelin et al.[23] are also shown in the figure.

Scaling for the polyimide side chain and polyamide main chain polymers is found over a large range in temperature. For the guest host system with a similar glass transition temperature of 180 °C a strong deviation from an Arrhenius behavior is found indicating that the orientational relaxation of side and main chain polymers correlate better with the main chain segmental motion than do guest host matrices.

CONCLUSIONS

Polyimide type side chain and polyamide main chain NLO polymers with glass transition temperature up to 176 °C have been prepared. They show nonlinear optical susceptibilities as high as 69 pm/V for d_{33} at a wavelength of 1.3 µm and electro-optic coefficients close to 20 pm/V for r_{33}. The structural relaxation in these polymers was modeled in terms of current phenomenological descriptions of the glass transition with the aid of a scaling relation. Relaxation times of several years at 80 °C make them superior to comparable guest host systems.

REFERENCES

(1) In *Nonlinear Optical Effects in Organic Polymers*; J. Messier, F. Kajzar, P. N. Prasad and D. R. Ulrich Ed; Kluwer Academic Publishers: Dordrecht, 1989.
(2) Meredith, G. R.; Dusen, J. G. V.; Williams, D. J., *Macromolecules* **1982**, *15*, 1385.
(3) Singer, K. D.; Sohn, J. E.; Lalama, S. J., *Appl. Phys. Lett.* **1986**, *49*, 248.
(4) Eich, M.; Reck, B.; Yoon, D. Y.; Willson, C. G.; Bjorklund, C. G., *J. Appl. Phys.* **1989**, *66*, 3241.
(5) Green, G. D.; Weinschenk, I., J.I.; Mulvaney, J. E.; Hall Jr., H. K., *Macromolecules* **1987**, *22*, 722.
(6) Prêtre, P.; Kaatz, P.; Bohren, A.; Günter, P.; Zysset, B.; Ahlheim, M.; Stähelin, M.; Lehr, F., *Macromolecules* **1994**, *27*, 5476.
(7) Oudar, J. L., *J. Chem. Phys* **1977**, *67*, 441.
(8) Ahlheim, M.; Lehr, F., *Makromol. Chem.* **1994**, *195*, 361.
(9) Weder, C.; Neuenschwander, P.; Suter, U. W.; Prêtre, P.; Kaatz, P.; Günter, P., *Macromolecules* **1994**, *27*, 2181.
(10) Manificier, J. C.; Gasiot, J.; Fillard, J. P., *J. Phys. E: Sci. Instrum.* **1976**, *9*, 1002.
(11) Jerphagnon, J.; Kurtz, S. K., *J. Appl. Phys.* **1970**, *41*, 1667.
(12) Bosshard, C.; Sutter, K.; Schlesser, R.; Günter, P., *J. Opt. Soc. Am.* **1993**, *B10*, 867.
(13) Boyd, G. D.; Kleinman, D. A., *J. Appl. Phys.* **1968**, *39*, 3597.
(14) Williams, G.; Watts, D. C., *Trans. Far. Soc.* **1970**, *66*, 80.
(15) Narayanaswamy, O. S., *J. Am. Cer. Soc.* **1971**, *54*, 471.
(16) Moynihan, C. T.; Crichton, S. N.; Opalka, S. M., *J. Non-Cryst. Solids* **1991**, *131-133*, 420.
(17) Hodge, I. M., *J. Non-Cryst. Solids* **1991**, *131-133*, 435.
(18) DiMarzio, E. A.; Gibbs, J. H., *J. Chem. Phys.* **1958**, *28*, 373.
(19) Adam, G.; Gibbs, J. H., *J. Chem. Phys.* **1965**, *43*, 139.
(20) Hodge, I. M., *Macromolecules* **1986**, *19*, 936.
(21) Kauzmann, W., *Chem. Revs.* **1948**, *43*, 219.
(22) Williams, M. L.; Landel, R. F.; Ferry, J. D., *J. Am. Chem. Soc.* **1955**, 3701.
(23) Stähelin, M.; Burland, D. M.; Ebert, M.; Miller., R. D.; Smith, B. A.; Twieg, R. J.; Volksen, W.; Walsh, C. A., *Appl. Phys. Lett.* **1992**, *61*, 1626.

BULK PHASE-MATCHED SECOND-HARMONIC GENERATION IN DRAWN AND POLED POLYMERS

X. T. Tao, H. Ukuda, T. Watanabe, and S. Miyata

Faculty of Technology, Tokyo University of Agriculture and Technology, 2-24-16 Nakamachi, Koganei-shi, Tokyo 184, JAPAN

Phase-matched second-harmonic generation (SHG) has been realized from a drawn and poled main-chain polyurea (PU) by end-fire coupling at the fundamental wavelength of 1.064 μm. The PU was synthesized from 4,4'-diphenylmethane diisocyanate and 4,4'-methylene bis(cyclohexylamine). The free-standing films were drawn and poled simultaneously by zone drawing and poling apparatus developed by ourself. By drawing, the birefringence between the in-plane and in the film normal directions can be controlled and such kind birefringence can be used for phase matching just like in biaxial single crystals. The type I phase-matching was realized in experiments. By this method, phase-matching at a number of interested wavelengths can be realized. Moreover, noncritical phase-matching in poled polymers at certain draw ratio is also possible.

INTRODUCTION

The development of compact, efficient, and reliable sources of coherent light in the visible region of spectrum is considerable interest. A possible approach to realize such a device is frequency doubling the output signal of a solid-state IR laser, e.g., a diode laser or a mini-YAG laser. In a SHG process, the fundamental waves interact in the nonlinear material induce a polarized wave. The polarization wave is forced by the fundamental wave, it travels with the same velocity, determined by $n(\omega)$, the refractive index at the fundamental wavelength. The polarization wave radiates a free second-harmonic wave that travels at a velocity determined by $n(2\omega)$, the refractive index at the harmonic wavelength. In normal dispersion materials, $n(2\omega) > n(\omega)$, as a result, the fundamental and second-harmonic waves travel at different phase velocities. To keep energy transferred constructively from fundamental wave into the second harmonic wave, the two waves should travel with the appropriate relative phase. If the harmonic wave and the source polarization wave travel at different phase velocities, then the efficiency of the interaction is reduced as energy is transferred alternately between the polarization wave and the free second harmonic wave.

To realize phase-matching, a general method is using birefringence in certain anisotropic crystals to compensate the refractive index dispersions between the fundamental and second harmonic wavelengths and realize $n^i_{\omega} = n^j_{2\omega}$ (i and j are different directions).[1-3] The merit of birefringence phase-matching is that one can realize this process very easily just by rotating the crystals. As a result of phase-matching, the second-harmonic light intensity will build up as the square of the interaction length. Now this method has been generally used in practical SHG single crystal devices. However, at present, it is not possible to design a crystal with the expected structure and birefringence for phase matching at the required wavelengths. Furthermore, the crystal growth, cutting and polishing are time consuming procedures and it is not easy to get crystals with large size and good optical quality, especially for organic materials.

On the other hand, in the last decade, a number of poled polymers have been studied for NLO applications since their good processibility and numerous variability through organic synthesis.[4-7] However, till now most of the researches were concentrated on the synthesis and characterization of NLO polymers and phase-matched SHG has not been extensively studied. The reported phase-matching methods in poled polymers include: modal dispersion PM,[8] Q-PM,[9] noncollinear light path PM,[10] Cerenkov radiation PM,[11] and abnormal dispersion PM.[12] It was believed that bulk phase-matching in poled polymers is not possible since that $n^i_{\omega} \neq n^j_{2\omega}$. Nevertheless, for Λ-type main-chain polymers, we found the refractive indices in the film plane are larger than those in the film normal direction. Such kind positive birefringence indicates the possibility of realizing bulk phase-matching in poled polymers.[13 (a)] In a previous paper, we reported phase-matched SHG in drawn and poled thin films by prism coupling.[13 (b)] Here we describe bulk phase-matching in thicker drawn and poled polymeric films through end-fire coupling at the fundamental wavelength of 1.064 μm.

EXPERIMENTAL SECTION

The used sample was a main-chain polyurea (U1). The synthesis and characterization have been described previously.[14] Films were prepared by casting of a N-methyl-pyrrolidone (NMP) solution on glass slides. The prepared films were baked at 140 °C in a vacuum oven to remove all the solvent. The films were characterized by an initial positive

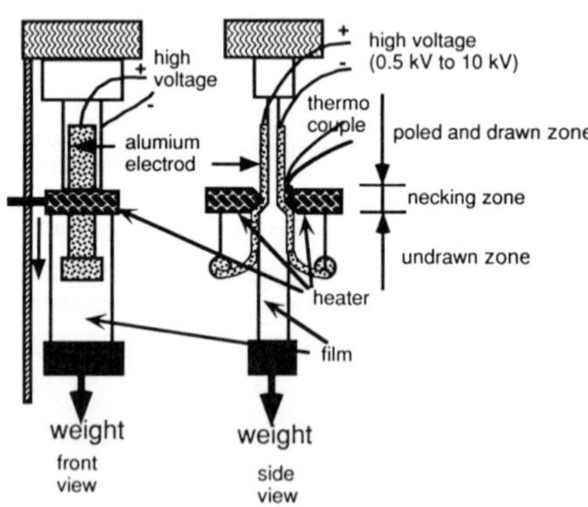

Figure 1. Schematic setup for drawing and poling experiment.

birefringence ($\Delta n = n_{TE} - n_{TM}$) and a molecular weight $M_n = 3.6 \times 10^4$ to $M_w = 12.0 \times 10^4$ as determined by gel permeation chromatography with polystyrene as a reference and NMP as solvent. The required noncentrosymmetric structure of polymer films for SHG was induced by contact poling. The poling temperature was selected at 170 °C, which is selected according to the in-situ SHG experiment as described previously.[14] To increase the positive birefringence, a drawing experiment was performed simultaneously. The schematical diagram of drawing and poling setup is shown in Figure 1. This self developed zone drawing and poling machine consists of three main parts, i.e., heating, drawing and poling parts. The free standing films have ~20 μm thick and with an area 100 x 10 mm were set in an upper holder and a tension was applied to the lower end of the strip by weighting. The temperature was controlled by a heater that can be moved up and down at a selected speed. In the same time, the films are poled by contact poling under 1000 voltage. At certain draw ratio, the temperature was decreased while keeping the polymer films under the stress of weighting and poling field till room temperature.

The phase-matched SHG experiment was realized on a sample with draw ratio of 1.18 by end-fire coupling. The samples were embodied in epoxy polymer holders and end face polished. The setup for phase-matched SHG measurement is shown in Figure 2. The sample was mounted on a stage which was computer controlled. The S-polarized beam of a Q-switched Nd:YAG laser (1064 nm, 10 ns/pulse, 10 Hz), was illuminated to the sample on the stage after being polarized and focused by a polarizer and a lens, respectively. The second harmonic (SH) signal was P-polarized and detected by photomultiplier tube and amplified.

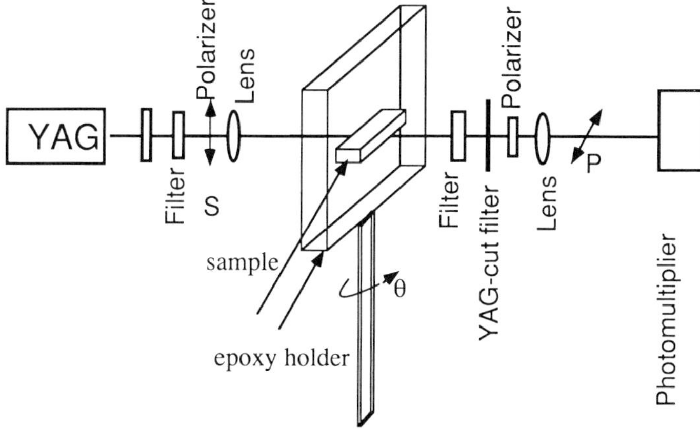

Figure 2. The experimental setup for the phase-matched second-harmonic generation . The incident light is S-polarized and the output light is P-polarized.

RESULTS AND DISCUSSIONS

Many drawing techniques have been developed for polymer films or fibers, such as drawing in an oven,[15] die drawing,[16] and solid-state extrusion[17] and so on. Compared to these methods, the zone drawing method has the following advantages: the films are only heated in a narrow zone for a short time, thermal degradation can be prevented; the heating zone is very narrow, films can be uniformly drawn with relative larger draw ratio without broken and the heat energy required to draw can be saved especially in the case of high temperature drawing.

The dielectric principle axes of the drawn and poled sample are defined as those in reference 13 (b). The refractive indices are defined as n_x, n_y, and n_z for the polarization vector in the film normal direction, the transverse direction in the plane of the film and parallel to the draw direction, respectively. The θ and ϕ are polar coordinates (referred to z and x, respectively). By drawing, the polymer chain will lie along the draw direction and hence a wave polarized parallel to this direction has larger refractive indices (n_z). On the other hand, a wave polarized perpendicular to the molecular plane will have relatively small refractive indices (n_x). The refractive indices along the transverse direction are expected to lie between those of n_x and n_z.

The range of transparency of the film spin coated on ITO was determined using a JASCO UV-visible spectrometer.[14] Results indicated PU has very wide transparency range. The cutoff wavelength was down to 0.3 μm, which represent the shortest transparency wavelength in poled NLO polymers. Both the spin-coated and drawn films are amorphous characterized by wide angle x-ray spectrum.

The refractive indices of the drawn and poled thin films have been measured by m-line method. From the refractive indices, phase-matching directions were calculated.[13(b)] But the m-line method has largely limited to thin films with the thickness in the range 0.5 ~ 10 μm. For a thick free standing polymer film, the refractive indices can be obtained by reflection and transmission method.[18] Figure 3 shows the refractive index dispersion of ~ 20 μm PU film with draw ratio of 1.18. Here only n_z and n_y were shown since n_x hasn't been exactly measured.

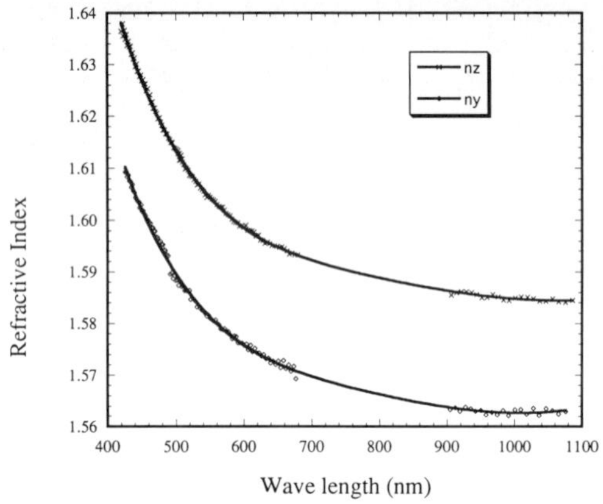

Figure 3. Refractive index dispersion of thick PU film with draw ratio of 1.18.

The poling field is along the film normal direction. By poling, the molecular dipoles will be aligned to this direction. For general spin-coated and poled polymers, it is isotropic in the film plane and such films have point symmetry of ∞mm,[4] whose symmetry operations include an infinite-fold rotation about the poling field direction and an infinity of mirror planes including this direction. In the present case, the poling field and the uniaxial drawing are perpendicular to each other and the in-plane becoming anisotropic. It was reported such kinds drawing and poling films have 2mm point symmetry.[19] The symmetry operations include a two-fold axis along the film normal direction and two mirror planes perpendiculars to each other and including the 2-fold axis.

For a system with 2mm point group, there is three independent non-zero nonlinear optical coefficients in assuming the Klainmen symmetry[20] is valid. The NLO coefficients were determined by Maker fringe method. The obtained NLO coefficients for drawn and corona poled thin film with draw ratio of 1.4 were d_{11} = 1.4, d_{12} =0.2, and d_{13} = 0.6 pm/V.[13 (b)]

To get high SHG conversion efficiency, phase matching between the polarized wave and free second-harmonic wave is necessary. This is because that the second harmonic conversion efficiency is generally related to the square of the propagation length. The phase matching condition in bulk single crystal can be deduced from the resolution of the propagation equation for the harmonic wave:[21]

$$\nabla^2 E=\mu_0 \sigma \frac{\partial E}{\partial t}+\mu_0 \varepsilon \frac{\partial^2 E}{\partial t^2}+\mu_0 \frac{\partial^2}{\partial^2 t} P_{NL} \tag{1}$$

were $(P_{NL})_i=d_{ijk} E_j E_k$

Resolution of Equation (1) leads to the following expression relating the harmonic optical intensity $I^{2\omega}$ to the fundamental one I^{ω} (the SH conversion efficiency):

$$\eta = \frac{I^{2\omega}}{I^{\omega}} = (2\omega^2 / c^3)[d^2 / n_{2\omega}(n_{\omega})^2](p_{\omega} / A)L^2 \\ [(\sin \Delta KL / 2) / (\Delta KL / 2)]^2 \tag{2}$$

where Δk is the phase mismatch factor. The reason for phase mismatch Δk between the pump and second-harmonic fields is due to the two waves travel with different velocity. The sign of power flow from one wave to the other is determined by the relative phase between the two waves. The continuous phase slip between these waves caused by their differing phase velocities leads to an alternation of the flow of power direction. The alternation of the sign of power leads to a repetitive growth and decay of the second harmonic intensity along the length of the interaction. The distance over which the relative phase of the two waves changes by π is the "coherence length" $l_c = \frac{\pi}{\Delta K} = \frac{\lambda}{4(n_2 - n_1)}$ which is the actual interaction length of the nonlinear optical process. Because of this phase mis-match, there is a reduction in second-harmonic determined by $n(2\omega) - n(\omega)$, and this reduction is expressed quantitatively by the factor

$$\frac{\sin^2(\Delta kL/2)}{(\Delta kL/2)^2} \tag{3}$$

Therefore, ensure $\Delta k = 0$ is necessary to obtain a maximal interaction length (i.e., equal to the physical length of the nonlinear material) and then maximize the corresponding nonlinear efficiency. In the PM case, the coherence length goes to infinity (not consider the walk off effect), the second-harmonic field grows linearly with distance in the medium and thus the intensity grows quadratically.

It was believed that bulk phase matching in poled polymers was not possible.[22] This is because poled polymer has a ∞mm symmetry,[4] and only three of the tensor elements are nonzero: d_{33}, d_{31}, and d_{15}. In considering the Kleinman symmetry, there are only two independent nonzero tensor elements. After poling, the dipole moment will be aligned along the film normal direction. Therefore, to realize bulk phase-matching in poled polymers, the second-harmonic wave should be polarized along this direction and the fundamental beam should polarized in the film plane. To this end, the refractive indices in the film plane must larger than those in the film normal direction and the refractive index dispersion between the fundamental and second harmonic waves must be offset by the positive birefringence. Figure 4 shows the schematic diagram of refractive index dispersions of side chain polymer and main-chain polymers with two dimensional charge-transfer chromophores. It is well known that the refractive indices of a polymer are determined by the orientations of chromophores and polymeric chains. For the general side-chain polymers, after poling, the chromophores will be aligned along the poling direction and the refractive indices in the film normal direction will be larger than those in the film plane. In this case, to realize bulk phase-matching, the fundamental wave should be polarized along the film normal direction and the second-harmonic wave should be polarized in the film plane. However, on average, there are no polarizable dipoles in the film plane and SHG can not be generated. On the other hand, for some main-chain polymers with Λ-type charge transfer chromophores, even after poling the refractive indices in the film plane are still large than those in the film normal direction. If we can realize $n_{\parallel}^{\omega(\theta)} = n_{\perp}^{2\omega}$, phase-matching can be realized.The relationship between SHG intensity and incident angle for end-fire coupling is shown in Figure 5. Phase-matching was realized at $\theta \sim 70°$, we express this direction as $\delta\theta = 0°$ in the figure. In this case, the phase-matching angle tolanence range is significantly larger than that films with draw ratio of 1.4.[13(b)]

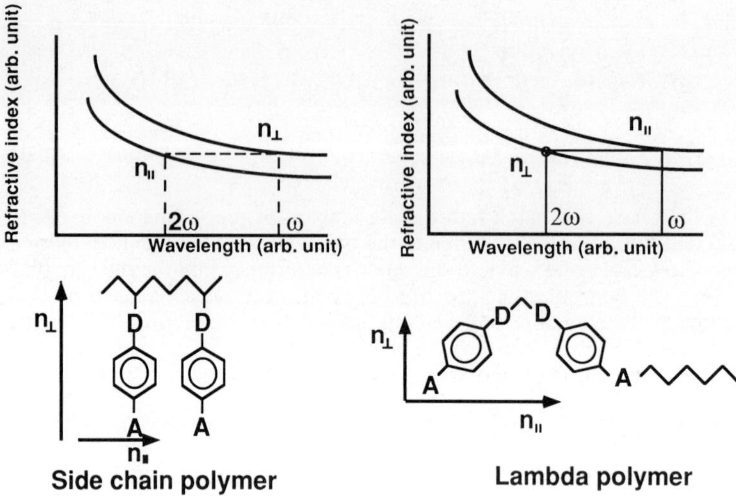

Figure 4. Schematic diagram of refractive index dispersion of side-chain polymer and main-chain polymer with two dimensional charge-transfer chromophores.

The phase-matching process described at above is a critical one. One of the mechanism which limit SHG in anisotropic medium is the walk-off of the extraordinary wave, a phenomenon also known as double refraction, both terms accounting for the different directions for phase (k wave vector) and energy flow (Pointing vector) propagation, which then make an angle (walk-off angle) inside the medium. The disadvantage of this double refraction is that it can prevent full use of the NLO properties of noncentrosymmetric birefringence materials. The interaction length is limited to [23]

$$l_a = w_0 \sqrt{\pi} / \rho \tag{4}$$

where ρ is the walk off angle and w_0 is the beam radius. The calculation of the walk-off angle in a general phase-matching direction is very complicated. However, for propagation in one of the principal planes, only the extraordinary wave departs from the original propagation direction and the calculation can be simplified. The calculated walk-off angle of type I phase-matching for the polyurea with draw ratio of 1.4 was 1.17°.

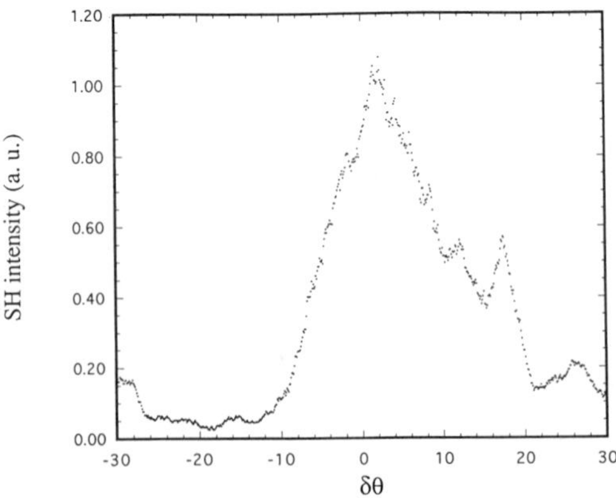

Figure 5. The relationship between second-harmonic intensity and incident angle.

Ideally, one would like to have propagation along one of the principal axes of the index ellipsoid. There are two advantages of using that direction. There is no walk-off affect and the angle tolerance range is extended. However, the refractive indices of single crystals are determined by their structures and constituents. It is difficult to adjust these data to realize noncritical phase-matching, although there are some experiments reported the temperature adjusted noncritical phase-matching in some special crystals and at special wavelengths.[24] Different from single crystals, in the drawn and poled polymers, it is possible to control the birefringence by draw ratio and poling field. Therefore, we can realize phase-matching at a number of required wavelengths. Furthermore, noncritical phase-matching is also possible.

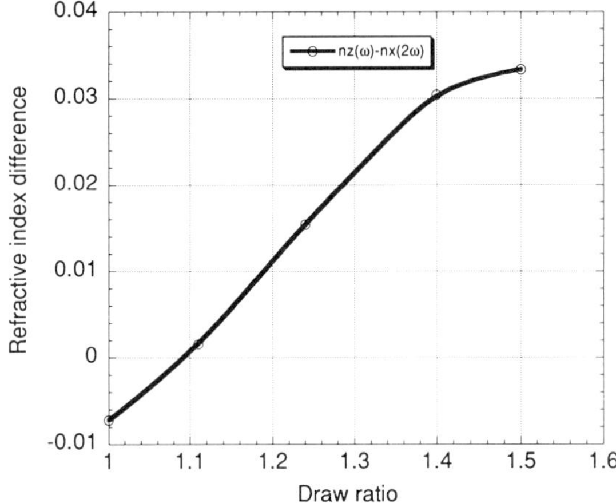

Figure 6. The relationship between refractive index difference and draw ratio.

The largest usable nonlinear coefficient in the drawn and poled polyurea is d_{13}, therefore, phase-matching using this component is necessary. In the case of noncritical phase-matching, we should have $n_3^{\omega} - n_1^{2\omega} = 0$. The relationship of refractive index difference with draw ratio for thin (~ 6 μm) films is shown in Figure 6. The results indicate the noncritical phase-matching will be realized at draw ratio near 1.09. The experimental confirmation of noncritical phase-matched second harmonic generation is under way.

SUMMARY

In this paper, we demonstrated bulk phase-matched SHG from a drawn and poled PU film by end-fire coupling. Visible green light can be easily observed when using a Q-switched Nd:YAG laser operated at 1.064 μm. Compared to other phase-matching methods used in poled polymers, the merits of using bulk phase matching include: (1) It is not need to control the film thickness or periodicity accurately as required in modal dispersion phase-matching or quasi-phase matching. Therefore, the preparation process is very simple. (2) For other phase matching methods used in polymer waveguides such as modal dispersion, Cherenkov radiation, and noncollinear light path, the conversion efficiency is not only related to the nonlinear coefficient, but also proportional to the square of spatial overlap integrals between the fundamental and SH waves. The overlap integrals between different modes are very small and will significantly decrease the obtainable conversion efficiency. In birefringence phase matching there is no such kind affection. (3) The refractive indices of single crystals are determined by their constituents and structures. Therefore, it is almost not possible to engineer a single crystal with the expected birefringence. However, the refractive indices of the poled polymers can be adjusted by varying draw ratio. Therefore, noncritical phase matching may be realized at a certain draw ratio. In this case a longer phase-matched interaction length can lead to a higher SH conversion efficiency.

REFERENCES

1. J. A. Giordmaine, " Mixing of light beams in crystals," Phys. Rev. Letters **8**, 19 (1962).
2. M. V. Hobden, "Phase-matched second-harmonic generation in biaxial crystals," J. Appl. Phys. **38**, 4365 (1967).
3. P. D. Maker, R. W. Terhune, M. Nisenoff, and C. M. Savage, "Effects of dispersion and focusing on the production of optical harmonics," Phys. Rev. Letters **8**, 21 (1962).
4. M. G. Kuzyk, K. D. Singer, H. E. Zahn, and L. A. King, " Second-order nonlinear-optical tensor properties of poled films under stress," J. Opt. Soc. Am. B **6**, 742 (1989).
5. G. R. Meredith, J. G. Vandusen, and D. J. Williams, "Optical and nonlinear optical characterization of molecularly doped thermotropic liquid crystalline polymers," Macromolecules **15**, 1385 (1982).
6. M. Eich, A. Sen, H. Looser, G. C. Bjorklund, J. D. Swalen, R. Twieg, and D. Y. Yoon, "Corona poling and real-time second-harmonic generation study of a novel covalently functionalized amorphous nonlinear optical polymer," J. Appl. Phys. **66**, 2559 (1989).
7. (a)Möhlmann, G. R., Ed. "Nonlinear Optical Properties of Organic Materials VI;" Proceedings of SPIE; SPIE Bellingham, WA 1993; Vol. 2025. (b) Miyata, S., Ed. "Nonlinear optics, Fundamentals, Materials and Devices" (North Holland, Amsterdam, 1992).
8. G. L. J. A. Rikken, C. J. E. Seppen, E. G. J. Staring and A. H. J. Venhuizen, " Efficient modal dispersion phase-matched frequency doubling in poled polymer waveguides," Appl. Phys. Lett. **62**, 2483 (1993).
9. G. Khanarian, R. A. Norwood, D. Haas, B. Feuer, and D. Karim, "Phase-matched second-harmonic generation in a polymer waveguide," Appl. Phys. Lett. **57**, 977 (1990).
10. Y. Shuto, H. Takara, M. Amano, T. Kaino, "Noncollinear phase-matched second-harmonic generation in stilbene-dye-attached polymer thin films," Jpn. J. Appl. Phys. **28**, 2508 (1989).
11 K. Clays, N. J. Armstrong, and T. L. Penner, "Blue and green Cerenkov-type second-harmonic generation in a polymeric Langmuir-Blodgett waveguide" J. Opt. Soc. Am. B **10**, 886 (1993).
12. J. R. Sounik, G. Khanarian, J. Popolo, and S. Meyer, "Zero Dispersion Nonlinear Optical polymeric Materials Containing Stilbene and Naphthalocyanine Dyes," Chem. Mater. **5**, 1253 (1993).
13. (a) X. T. Tao, T. Watanabe, S. Shimoda, D. C. Zou, H. Sato, and S. Miyata, "Λ-type main-chain polymers for second harmonic generation," Chem. Mater. **6**, 1961 (1994). (b) X. T. Tao, T. Watanabe, D. C. Zou, H. Ukuda and S. Miyata , " Bulk phase-matched second-harmonic generation in poled polymers," J. Opt. Soc. Am. B in the press.
14. X. T. Tao, T. Watanabe, D. C. Zou, S. Shimoda, H. Sato, and S. Miyata " Polyurea with large positive birefringence for second-harmonic generation," Macromolecules in the press.
15. J. N. Hay, D. J. Kemmish, J. I. Lanfold, and A. I. M. Rae, "The structure of crystalline PEEK," Polym. Commun. **25**, 175, (1984).
16. D. R. Rueda, F. Ania, A. Richardson., I. M. Ward, and F. J. Balta Calleja, "X-ray diffraction study of die drawn poly(aryletherketone) (PEEK)," Polym. Commun. **24**, 258 (1983).
17. A. E. Zachariades, E. S. Sherman, and R. S. Porter, "Ultradrawing of amorphous polymers by coextrusion illustrated with atactic polystyrene," J. Appl. Polym. Sci. **24**, 2137 (1979).
18. R. C. McPhedran, L. C. Botten, D. R. McKenzie, and R. P. Netterfield, "Unambiguous determination of optical constants of absorbing films by reflectance and transmittance measurements," Appl. Opt. **23**, 1197 (1984).

19. (a) J. B. Land, H. G. Olf, and A. Perterlin, "Nuclear magnetic resonance and X-ray determination of structure of poly (vinyllidene fluoride)," J. Polym. Sci. Part A-1, **4**, 941 (1966). (b) H. Kawai, "The piezoelectricity of poly (vinylidene fluoride)," Jpn. J. Appl. Phys. **8**, 975 (1969).
20. Kleinman, D. A. "Nonlinear dielectric polarization in optical media," Phy. Rev. **126**, 1977 (1962)
21. I. Ledoux, in "Organic Materials For Photonics, Science and Technology" G. Zerbi ed. Elsevier Science Publishers B. V., The netherlands, (1993).
22. D. M. Burland, R. D. Miller, and C. A. Walsh, "Second-order nonlinearity in poled-polymer systems," Chem. Rev. **94**, 31 (1994).
23. F. C. Zumsteg, J. D. Bierlein, and T. E. Gier, "$K_xRb_{(1-x)}TiOPO_4$: A new nonlinear optical material," J. Appl. Phys.**47**, 4980 (1976).
24. Ukachi, T.; Lane, R. J.; Bosenberg, W. R.; and Tang, C. L. "Phase-matched second-harmonic generation and growth of a LiB_3O_5 Crystal," J. Opt. Soc. Am. B**9**, 1128 (1992).

SYNTHESIS AND PROCESSING RELEVANT TO SYSTEMS INTEGRATION OF NONLINEAR OPTICAL MATERIALS

A. W. Harper, A. Hubbel, J. Laquindanum, Y. Ra, and L. R. Dalton

Loker Hydrocarbon Research Institute
Departments of Chemistry and of Materials Science and Engineering
University of Southern California
Los Angeles, California 90089-1661

INTRODUCTION

The various synthesis and processing steps required to fabricate prototype electro-optic modulators from polymeric materials containing nonlinear optical (NLO) chromophores are discussed. These include (1) synthesis of chromophores with large first hyperpolarizabilities, (2) covalent incorporation of these chromophores into polymer lattices which can be processed into optical quality thin films and poled by application of electric fields, (3) induction of lattice hardening reactions which lock-in poling-induced noncentrosymmetric order, (4) creation of buried channel, nonlinear optical waveguides by reactive ion etching, by electron cyclotron resonance etching, and/or by photochemical processing, (5) coupling of buried channel electro-optic modulators to fiber optic transmission lines by silicon V groove techniques, (6) deposition of metal drive electrodes (integration of semiconductor electronics and optical circuitry) and evaluation of prototype devices.

A variety of signal transduction and switching operations have been envisioned for electro-optic modulators[1] including signal transduction in CATV and real-time computer communication, ultrafast switching in local area networks (LANs), photonic sensing of electromagnetic radiation in radar and analytical instrumentation applications, radiofrequency (rf) power distribution, phased-array radar, remote sensing as in the electrical power industry, ultrafast analog to digital (A-D) conversion, ultrafast switching of electronic signals, full color flat panel displays, etc. Nonlinear optical organic materials afford a number of advantages for the fabrication of modulators for these various applications[1] including very high bandwidth, large optical nonlinearity, and ease of integration with semiconductor electronics. The large bandwidth of organic polymeric modulators derives from the fact that optical and radiofrequency waves co-propagate with nearly identical velocities in the polymer materials. That is, $\varepsilon \approx n^2$ which, in turn, derives from the fact that the π-electrons of the organic chromophore define both index of refraction, n, and dielectric constant ε. Because of the low dielectric constants of organic materials, electro-optic modulator bandwidths of 120-150 GHz are feasible for a 1 centimeter waveguide device. Organic chromophores can exhibit large, off-resonance (essentially nonresonant) first hyperpolarizabilities. For example, a number of chromophores have been synthesized[1] which exhibit optical nonlinearities, $\mu\beta$ (where μ is the dipole moment and β is the first hyperpolarizability), of 10^{-44} esu or greater. If such chromophores could be translated in high number density into highly ordered noncentrosymmetric macroscopic lattices, then electro-optic coefficients on the order of several hundred picometers per volt could be realized. Such values would be an order of

Polymers and Other Advanced Materials: Emerging Technologies and Business Opportunities. Edited by P. N. Prasad *et al.*, Plenum Press, New York, 1995

magnitude greater than that obtainable with inorganic materials such as lithium niobate or gallium arsenide. It has been generally assumed that nonlinear optical polymers, like conventional polymers, are amenable to processing appropriate for the integration with semiconductor electronics. For example, it should be possible to spin cast optical quality films onto semiconductor substrates. This is in contrast to crystalline inorganic materials such as lithium niobate which are not amenable to such processing although some efforts at epitaxial growth on semiconductor substrates have recently been attempted with $LiNbO_3$.

While the synthesis of chromophores with large first hyperpolarizabilities[1] appears to be well in hand, the preparation of macroscopic nonlinear optical materials and the processing of such materials to achieve integrated electro-optic devices appears to be a continuing challenge. These latter activities will be the focus of this article.

PREPARATION OF THERMALLY-STABLE NLO LATTICES

Three general approaches to realization of noncentrosymmetric lattices of NLO chromophores[1] can be defined: (1) Molecular self-assembly such as crystal growth, incorporation into inclusion compounds, exploitation of ordered domains such as found in asymmetric (e.g., ABA') block copolymer assemblies with an ordered B phase, etc. (2) Sequential synthesis exploiting Van der Waals, ionic, and/or covalent interactions as in Langmuir-Blodgett assembly, molecular beam epitaxial growth, Merrified-type covalent coupling, etc. (3) External force methods such as electric field poling or laser-assisted poling. For any of the above approaches to be successful, dipole-dipole repulsion between chromophores must be overcome. In this article, we focus upon external field methods as such methods are currently the most widely employed approach to the fabrication of prototype devices.

Electric field poling has been carried out both on chromophores dissolved in polymers to form composite materials and chromophores covalently attached to polymers to form new homogeneous polymeric materials. Although composite materials afford the advantage of simply mixing commercially available chromophores with commercially available polymers, a number of disadvantages are associated with the use of such materials including (1) poor solubility of chromophores in the polymer host resulting in poor loading or chromophore number density in composite material, (2) rapid relaxation of poling-induced order due to poor coupling of chromophores to the polymer lattice dynamics, (3) plasticization of the polymer with increasing chromophore concentration, (4) chromophore phase separation and aggregation with processing and with aging, (5) chromophore diffusion under electric field poling resulting in chromophore gradients and increased light scattering, (6) sublimation of chromophores at high processing temperatures, and (7) loss of chromophores with application of cladding layers due to chromophores dissolving in the spin casting solvent. These liabilities have, in general, prevented the realization of large electro-optic coefficients and have limited the optical quality which can be realized as heterogeneous distributions of chromophores can lead to serious light scattering when domain sizes begin to approach the wavelength of light.

In addition to covalent incorporation of chromophores to form a homogeneous polymer matrix, we have focused upon processing to achieve hardened noncentrosymmetric lattices characterized by glass transition temperatures in excess of 200°C. Our general approach has focused upon preparing a precursor to the final hardened material which can be spin cast (employing conventional solvents) into optical quality films and poled at modest temperatures. That is, every effort is made to fabricated precursor materials exhibiting good solubility in a variety of spin casting solvents and exhibiting glass transition temperatures below 200°C. Three general approaches have been made to achieve hardened lattices from a processible precursor material.

Asymmetric reactions employing double-end crosslinkable (DEC) chromophores

A DEC chromophores is typically one that is terminated by an acrylate functionality at one end and hydroxyl functionalities at the other end (see Figure 1). Polymerization of the acrylate functionality yields a precursor polymer which can be used to prepare optical quality films by spin casting. Thermally-induced condensation reactions involving the hydroxyl groups leads to a hardened lattice subsequent induction of noncentrosymmetric order by electric field poling.

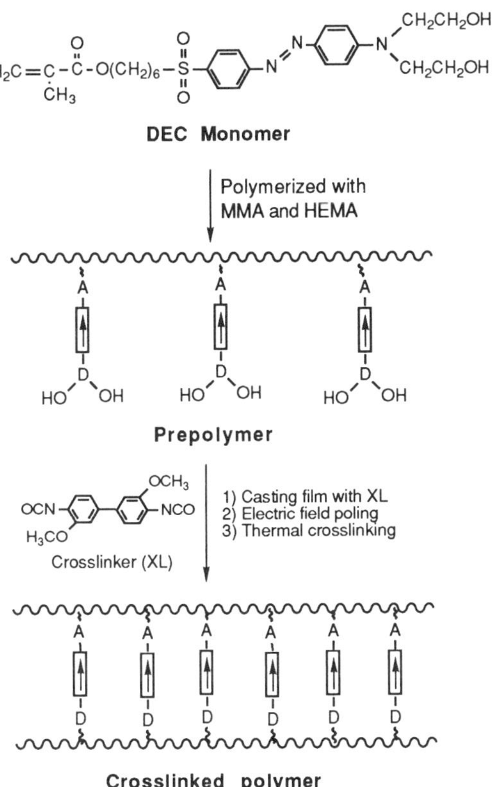

Figure 1. The production of a hardened NLO polymer lattices utilizing asymmetric reactivities associated with a DEC chromophore. Note that both ends of the chromophore are coupled to the polymer lattice by this two step scheme.

The hardening reaction elevates the thermal stability (defined by the temperature at which optical nonlinearity is first observed to decrease) of the optical nonlinearity (assayed by monitoring second harmonic generation efficiency while increasing the temperature at a rate of 10°C per minute) from 80°C to 170°C. The above procedure is readily adaptable to a wide range of chromophores including high $\mu\beta$ chromophores and further improvement in the thermal stability of optical nonlinearity has been achieved by increasing crosslink density and by reducing the extent of flexible chain segments. With azobenzene chromophores (such as shown in Figure 1) electro-optic coefficients have been obtained with values ranging from 12 to 20 pm/V (measured at 1.3 μm); polar order parameters as high as 0.4 have been obtained and optical nonlinearities have been found to be stable for periods of 1000 hours at 125°C. With high $\mu\beta$ chromophores, values have been obtained which range from 20 to 50 pm/V. Because of the flexibility of this scheme in permitting systematic variation of structure, good optical quality is typically obtained. At 1.3 μm, material optical loss on the order of 0.1 dB/cm is frequently obtained.

Lattice hardening by intramolecular condensation reactions

Marks and coworkers and Dalton and coworkers have employed the imidization reaction involving polyamic acids containing covalently incorporated NLO chromophores as pendants to produce hardened noncentrosymmetric NLO materials.[1] The thermal stability of optical nonlinearity (measured by increasing the temperature at a rate of 10°C per minute while monitoring second harmonic generation) was observed to increase from 80°C for the polyamic acid form to 175°C for the polyimide form.

Because of the limited solubility of polyamic acids in conventional spin casting solvents, greater difficulty is encountered in obtaining films of excellent optical quality. Liang and coworkers[2] have recently circumvented this problem by preparing modified polyamic acids containing more extended flexible chain segments which improve solubility. Both intra and intermolecular condensation reactions were employed to achieve lattice hardening with the result that hardened lattices competitive with traditional polyimides were obtained without the processing liabilities of the traditional materials. Yu and coworkers and Liang et al. have obtained modified polyimide materials exhibiting long term stability of optical nonlinearity to temperatures as high as 170°C.[1,2]

Lattice hardening by thermosetting reactions including sol-gel processing

In the thermosetting approach, chromophores are typically terminated at both ends by functionalities such as hydroxyl groups (see Figure 2)

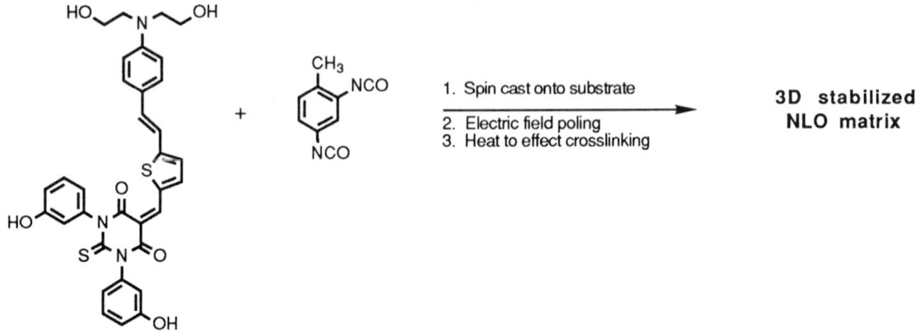

HO N OH

+ CH_3 NCO NCO

1. Spin cast onto substrate
2. Electric field poling
3. Heat to effect crosslinking

3D stabilized NLO matrix

Figure 2. Preparation of a hardened NLO lattice employing a thermosetting reaction.

With thermosetting schemes, the spin casting, poling, and lattice hardening steps are not cleanly separated. Typically, condensation is permitted to procedure to the point of producing a viscous solution appropriate for spin casting. Then, the temperature is elevated in a series of steps (with a poling field applied to the film). The polar order obtained is quite dependent on the exact protocol employed in increasing the temperature during the poling (and simultaneous lattice hardening procedure). To obtain optimum results, quite a bit more trial and error must be pursued than with the two proceeding lattice hardening schemes. Thermosetting reactions do have the advantage of permitting high chromophore number densities to be obtained.

A special case of thermosetting reactions is the sol-gel processing scheme. We[3-5] have synthesized chromophores with modified alkoxysilane functionalities at both ends. These materials lead to materials exhibiting thermal stability of optical nonlinearities approaching that obtained by other lattice hardening schemes.

Incorporation of high $\mu\beta$ chromophores into lattice hardening schemes and comparison of the optical and nonlinear optical characteristics of organic and inorganic materials

A major activity during the past year has been the incorporation of high $\mu\beta$ chromophores[1] (see Figure 3) into the aforementioned lattice hardening schemes. With cyanovinyl acceptors care must be taken to avoid nucleophilic reaction conditions. In general, high $\mu\beta$ chromophores are more difficult to pole because of their greater bulk and corresponding greater steric hindrance. Also, at the high chromophore concentrations, dipole-dipole repulsions appear to start to compete with electric field-dipolar ordering forces and reduced ordering is observed. Fortunately, trial and error modification of structure permits many of the problems encountered with these materials to be minimized and electro-optic coefficients in the range 20-60 pm/V appear to be a accessible (see Table 1).

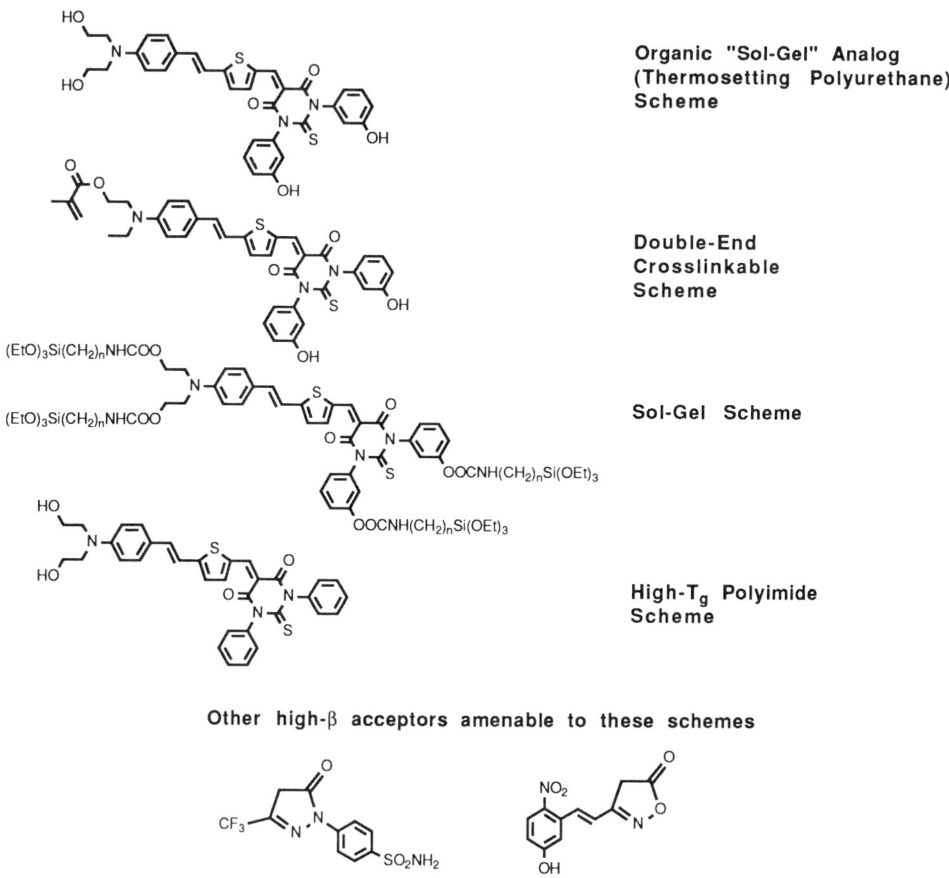

Figure 3. Several high $\mu\beta$ chromophores which have been synthesized for use with various lattice hardening schemes are shown.[1,6]

Table 1. Comparison of electro-optic materials performance (at 1.3 μm) for GaAs, $LiNbO_3$ and the best NLO chromophore-containing polymer exhibiting stability of optical nonlinearity for 1000 hours at 100°C.

Parameter	$LiNbO_3$	NLO Polymer	GaAs
r_{eff} (pm/V)	30	45	1.5
FOM (n^3r_{eff}/ε)	10	220	64
Space-Bandwidth Product (GHz-cm)	7	120	10
$V_\pi L$ (V-cm)	5	5	0.3
Optical Loss (dB/cm)	0.2	0.1-1.0	2

We have fabricated prototype Mach-Zehnder and birefringent modulators and have demonstrated modulation to 60 GHz with no signal attenuation attributable to the NLO polymer material. Currently, a test bed is under construction which will permit evaluation of prototype devices to 100 GHz.

PREPARATION OF BURIED CHANNEL NONLINEAR OPTICAL WAVEGUIDES

Buried channel nonlinear optical waveguides have been fabricated by reactive ion etching (RIE), electron cyclotron resonance (ECR) etching, and by photochemical processing[1]. ECR has afforded smoother waveguide channels and accompanying reduced optical loss relative to waveguides fabricated employing RIE. Recently, we have developed a photochemical processing scheme for the fabrication of buried channel waveguides and for producing graded index waveguides crucial for making transitions from large dimension silica fiber optic transmission lines to smaller dimension NLO polymer waveguides. The scheme is based on the realization that the penetration of electromagnetic radiation into a material is wavelength dependent (for a fixed exposure time at constant flux) in the vicinity of a resonance transition. A buried channel waveguide can be fabricated without deposition of a cladding layer by a two step process with change of mask and wavelength of radiation between steps. In the first step, a film of nonlinear optical polymer which undergoes a UV-Vis induced conformational change, is exposed to such radiation with a central channel of material protected by a mask. Irradiation is accomplished using a wavelength of light in the tail of the photochromic transition. Such radiation exhibits excellent penetration and exposure for a long period of time effects bleaching of regions of the nonlinear optical material not protected by the mask. In the second step, the mask is removed and the previously protected region is irradiated for a short period of time with radiation near the absorption maximum of the transition responsible for the index of refraction altering conformational change. Because the penetration depth is shallow and the exposure time is short, only the upper most region is bleached. Thus, this two step processing leads to a buried channel of high index of refraction NLO polymer surrounded by regions of lower index material. In like manner, controlled exposure using different wavelengths of light (or exposures for different periods of time using the same wavelength) can lead to graded index waveguides and tapered transitions. Currently, NLO polymeric waveguides exhibit total losses on the order of 1-2 dB/cm at 1.3 μm. Further refinement in processing protocols should reduce this figure.

COUPLING OF NLO WAVEGUIDES TO FIBER OPTIC LINES AND SYSTEM INTEGRATION

We have employed silicon V groove technology to achieve mechanically stable, low loss coupling of polymer waveguides to fiber optic transmission lines. The procedure involves (1) patterning thermally grown high optical quality SiO_2 as a lower cladding layer, (2) etching by RIE of V-grooves and patterning the lower cladding using a matched mask set, (3) placement of the silica fibers in the input and output V-grooves and bonding employing a UV-curable epoxy, and (4) spin casting the active polymer and the top cladding layer. In Figure 4, we show both a schematic representation of this coupling and an electron micrograph of fibers in V-grooves coupled to nonlinear optical polymer waveguides.

Recently, we have extended processing to full integration of a polymeric electro-optic modulator waveguide with semiconductor drive electronics. A film of nonlinear optical polymer was spin casted, poled, and processed into a buried channel waveguide on top of a semiconductor VLSI circuit board following the deposition of a planarizing polymer layer. This was accomplished without noticeable degradation in performance of the electronic circuitry. In Figure 5, we show transmission of light through the polymer waveguide utilizing prism coupling to in and out couple the light.

While these advances in integration of polymeric electro-optic modulators by no means resolve all the issues relevant to such integration and while optical loss remains a critical issue in the development of practical devices based upon polymeric electro-optic modulators, the demonstrations presented here attest to the ability to realize the putative advantages of polymeric materials in achieving sophisticated packaging of electronic and optical components and is an important step forward in the fabrication of actual devices.

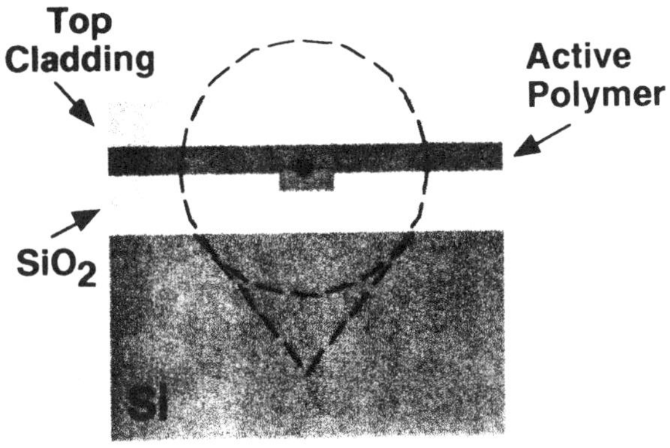

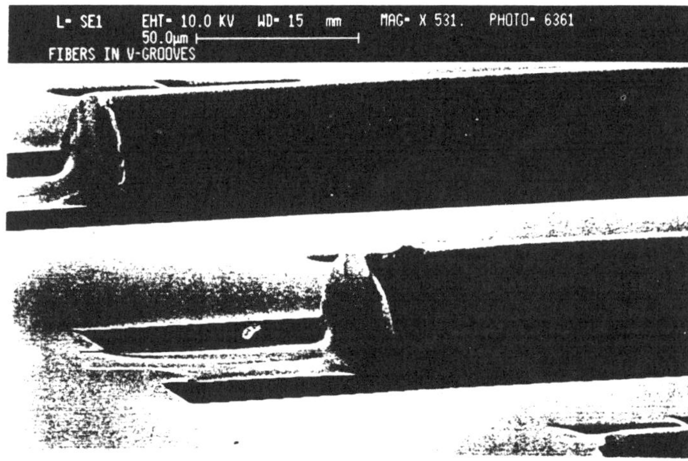

Figure 4. Top. A schematic representation of silicon V-groove coupling of a polymeric electro-optic waveguide to a silica optical fiber is shown. Bottom. An electron micrograph showing fiber-waveguide coupling is presented.

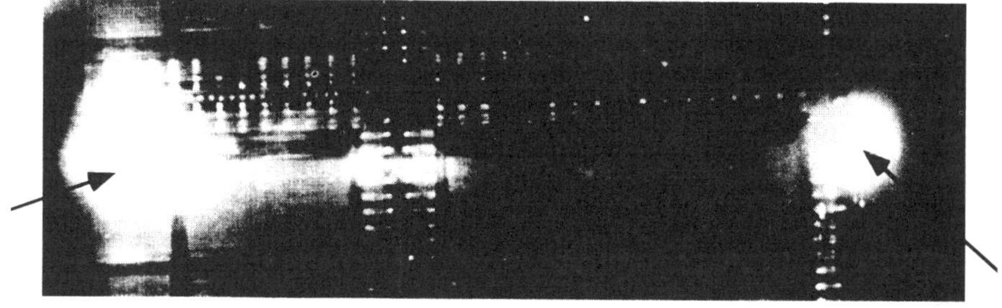

Figure 5. Waveguiding through a polymeric NLO waveguide integrated with a VLSI circuit is shown. The left arrow indicates in-coupled light while the right arrow indicates out-coupled light. Prism coupling is employed.

ACKNOWLEDGMENTS

The authors wish to thank Professors Steier and Fetterman for invaluable contributions to this work both in terms of ideas and execution of research. The authors gratefully acknowledge support from the National Science Foundation DMR-9107806 and from the Air Force Office of Scientific Research (Contracts F49620-94-1-0312 and F49620-94-1-0201).

REFERENCES

1. L. R. Dalton, A. W. Harper, R. Ghosn, W. H. Steier, M. Ziari, H. Fetterman, Y. Shi, R. V. Mustacich, A. K.-Y. Jen and K. J. Shea, Synthesis and processing of improved organic second-order nonlinear optical materials for applications in photonics, *Chem. Mater*.. 7:0000 (1995).
2. Z. Liang, L. R. Dalton, S. M. Garner, S. Kalluri, A. Chen, and W. H. Steier, Crosslinkable polyimides for second-order optical nonlinearities, *Chem. Mater*. 7:0000 (1995).
3. Z. Yang, C. Xu, B. Wu, L. R. Dalton, S. Kalluri, W. H. Steier, Y. Shi, and J. H. Bechtel, Anchoring both ends of chromophores into sol-gel networks for large and stable second-order optical nonlinearities,*Chem. Mater*. 6:1899 (1994).
4. S. Kalluri, Y. Shi, W. H. Steier, Z. Yang, C. Xu, B. Wu, and L. R. Dalton, Improved poling and thermal stability of sol-gel nonlinear optical polymers, *Appl. Phys. Lett*. 65:2651 (1994).
5. H. W. Oviatt, Jr., K. J. Shea, S. Kalluri, Y. Shi, W. H. Steier, and L. R. Dalton, Applications of organic bridged polysilsesquioxane xerogels to nonlinear optical materials by the sol-gel method, *Chem. Mater*. 7:493 (1995).
6. S. Gilmour, R. A. Montgomery, S. R. Marder, L.-T. Cheng, A. K.-Y. jen, Y. M. Cai, J. W. Perry, and L. R. Dalton, Synthesis of diarylthiobarbituric acid chromophores with enhanced second-order optical nonlinearities, *Chem. Mater*. 6:1603 (1994).

OPTICS AS A MEANS FOR STUDYING THE ORIENTATIONAL MOBILITY OF CHROMOPHORES AND FOR CREATING MACROSCOPIC ORDER IN AMORPHOUS FILMS

Michel Dumont

Institut d'Optique Théorique et Appliquée
Bât.5O3, B.P.147, F-914O3 ORSAY Cedex, France.
Fax: 33 (1) 6941 3192, E-mail: michel.dumont@iota.u-psud.fr

INTRODUCTION

It is no more necessary to demonstrate the interest of creating a macroscopic order in an ensemble of anisotropic optically active molecules dispersed in a polymer (or a sol-gel) film. If anisotropic chromophores are aligned the material is birefringent and eventually dichroic. This property can be used for patterning waveguides or gratings, or for information and image storage (by bits or holographically). If polar molecules are oriented in the same direction, the film becomes non centrosymmetric and first order optical nonlinearities appear ($\chi^{(2)}$): this is necessary for second harmonic generation (SHG) and for electro-optic modulation. For most of those applications the macroscopic order must be created once and one wishes a permanent stability of this order : a lot of job has been done everywhere to increase long term stability[13] (high Tg materials, integration of chromophores in the polymer chain, cross-linking...). Nevertheless some other applications, such as image processing by holographic methods, need a fast creation and a fast relaxation of the macroscopic order.

Many optical methods have been developed for the characterization of the macroscopic order of thin films. We will only emphasis dichroism measurement and Attenuated Total Reflection (ATR), for the measurement of birefringence and of electro-optic effects. But the main subject of this paper is the creation of macroscopic orders by optical means. We will discuss Photoinduced Anisotropy (PIA), Photoassisted Poling[1](PAP) and All Optical Poling[2] (AOP).

ORDER PARAMETERS.

The orientation of each molecule is defined by the three Euler angles, θ, φ, ξ, characterizing its individual axes with respect to the laboratory frame. The macroscopic properties depend on the angular distribution function $n(\theta,\varphi,\xi)$.

Most of the useful chromophores have an elongated shape, with conjugated π electrons, free to move along their principal axis Z. This structure gives them a much higher polarizability along the principal axis ($\alpha_{//}$) than in both transverse directions. If these molecules are polar, the permanent dipole moment, μ_0, is parallel to the axis Z and the first order hyperpolarizability tensor has one component, β_{ZZZ}, much larger than all the others. The main absorption band of such a molecule is a π-π transition with a transition dipole parallel to Z. For those reasons, in the following discussion, we will only consider ideal molecules with a cigar shape (both transverse directions are assumed to be equivalent with a polarizability $\alpha_{\perp}$. $\beta_{//} = \beta_{ZZZ} >> \beta_{ZXX} = \beta_{ZYY} = \beta_{\perp}$ are the only components of β). Furthermore, we assume that external fields impose an axial symmetry with respect to the laboratory axis 3. Then, the angular distribution of chromophores depends only on θ and it can be expressed as a development on Legendre polynomials[3]:

$$n(\theta,\varphi) = \frac{1}{2\pi} n(\theta) = \frac{1}{4\pi} \sum_{q=0}^{\infty} (2q+1)\ A_q P_q(\cos\theta) \tag{1}$$

$$\iint n(\theta,\varphi) \sin\theta\ d\theta d\varphi = A_0 = N \tag{2}$$

A_q are the order parameters which describe the angular distribution and N is the total number of chromophores per unit volume. One obtains (α and β include local field correction):

$$\chi_{33}^{(1)} = \left\langle \alpha_{//} \cos^2\theta + \alpha_{\perp} \sin^2\theta \right\rangle = \bar{\alpha}(A_0 + 2a\, A_2) \tag{3-a}$$

$$\chi_{11}^{(1)} = \chi_{22}^{(1)} = \left\langle \left(\alpha_{//} \sin^2\theta + \alpha_{\perp} \cos^2\theta\right) \sin^2\varphi + \alpha_{\perp} \cos^2\varphi \right\rangle = \bar{\alpha}(A_0 - a\, A_2) \tag{3-b}$$

$$\chi_{333}^{(2)} = \left\langle \cos^3\theta\, \beta_{//} + 3\sin^2\theta \cos\theta\, \beta_{\perp} \right\rangle = \tfrac{1}{5}\left[3A_1\left(\beta_{//} + 2\beta_{\perp}\right) + 2A_3\left(\beta_{//} - 3\beta_{\perp}\right)\right] \tag{4-a}$$

$$\begin{aligned} \chi_{311}^{(2)} = \chi_{322}^{(2)} &= \left\langle \cos\theta \left(\beta_{//} \sin^2\theta \cos^2\varphi + \beta_{\perp}\left((\cos^2\theta - 2\sin^2\theta) \cos^2\varphi + \sin^2\varphi\right)\right)\right\rangle \\ &= \tfrac{1}{5}\left[A_1\left(\beta_{//} + 2\beta_{\perp}\right) - A_3\left(\beta_{//} - 3\beta_{\perp}\right)\right] \end{aligned} \tag{4-b}$$

where $\beta_{\perp}$ is generally negligible and with the definitions:

$$\bar{\alpha} = \left(\alpha_{//} + 2\alpha_{\perp}\right)/3 \quad ; \quad a = \left(\alpha_{//} - \alpha_{\perp}\right)/\left(\alpha_{//} + 2\alpha_{\perp}\right) \tag{5}$$

As it can be seen on eqs. (3-4), only low orders have to be known (up to A_{n+1}, for $\chi^{(n)}$). Birefringence and dichroism are obtained from the real and the imaginary part of the tensor $\chi^{(1)}$: their measurement gives only parameters A_0, *i.e.* population, and A_2 which is often named "alignment". The Pockels electro-optic effect and SHG need a non centrosymmetric angular distribution for which at least A_1 ("orientation") and consequently $\chi^{(2)}$ are not vanishing.

Let us notice that many authors characterize the poling of films by measuring the variation of the absorption of a beam propagating perpendicularly to the sample (Parallel to the poling direction). They actually measure variations of absorbance $A_{\perp} \propto Im\left(\chi_{11}^{(1)}\right)$ which gives variations of A_2. This method is valid only if there is a thermal equilibrium (thermal poling) which establishes a temperature dependent relation between all A_q[note 1].

PROBING TECHNIQUES

If the symmetry axis of the angular distribution is in the plane of the film (for instance, when anisotropy is due to optical pumping by a polarized light or when it is due to a mechanical stretching of the film) the measurement of the absorption of light polarized parallel and perpendicular to the symmetry axis permits to follow the evolution of A_0 and A_2. $A_{//}$ and $A_{\perp}$ being absorbances of a film of thickness ℓ (Fig. 1-a), one has:

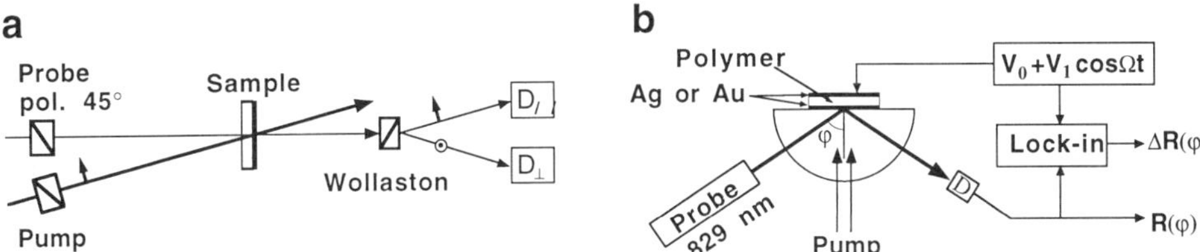

Figure 1. a: Photoinduced dichroism measurement. **b:** ATR experiment[4,5]. The polymer film is spin coated on a glass plate previously covered with a semi-transparent metallic layer. The glass plate is put in optical contact with a half sphere, through which the reflectivity of the sample is recorded as a function of the incidence angle, for both polarizations TM and TE. The position of reflectivity dips (Fabry-Perot rings) gives the three components of the index of refraction and the thickness of the film. With a modulated voltage applied to the top electrode (Ω) and a lock-in detection, one obtains Pockels and piezoelectric effects (detection at Ω) and Kerr and electrostriction effects[7] (at 2Ω). A DC voltage V_0can be applied for poling.

$$\tfrac{1}{3}(A_{//} + 2A_{\perp}) = \tfrac{2\pi\ell}{n\lambda} Im(\overline{\alpha})\, A_0 \quad ; \quad A_{//} - A_{\perp} = \tfrac{2\pi\ell}{n\lambda} Im\left(\alpha_{//} - \alpha_{\perp}\right) A_2 \tag{6}$$

The Attenuated Total Reflection (ATR)[4,5] (Fig. 1-b) is a powerful method, since it permits the measurement of the index of refraction in the three principal direction: n_z, perpendicular to the film; n_x and n_y in the plane of the film, respectively in the plane of incidence and perpendicularly to it.

$\chi^{(2)}$ (and consequently A_1 and A_3) can be measured by second harmonic generation. The SHG tensor, $d_{3j} = \frac{1}{2}\chi^{(2)}_{3jj}(-2\omega;\omega,\omega)$, is estimated by the measurement of the second harmonic intensity as a function of the angle of incidence of the fundamental beam (Maker fringes). This measurement is not very accurate for thin films and is often based on the assumption that $A_3 << A_1$ ($d_{33}=3d_{31}$), which is true for a low poling. The measurement of Pockels electro-optic effect by ATR[4,5] is more accurate since it is possible to measure independently Δn_x , Δn_y and Δn_z. As the DC field is applied along the normal direction z (which is assumed to be the symmetry axis), one has:

$$\begin{aligned} r_{33} &= -2n_z^{-4}\,\chi^{(2)}_{333}(-\omega;\omega,0) = -2n_z^{-3}\,\Delta n_z\, E_z^{-1} \\ r_{13} &= -2n_x^{-4}\,\chi^{(2)}_{113}(-\omega;\omega,0) = -2n_x^{-3}\,\Delta n_x\, E_z^{-1} \end{aligned} \tag{7}$$

The only problem of this method is the effect of piezoelectricity which produces an indirect electrooptic effect difficult to distinguish from the direct electronic Pockels effect[5]. But in many cases (for instance with DR1), the contribution of piezoelectricity is small. ATR permits also the measurement of Kerr and electrostriction effects proportional to $\chi^{(3)}E^2$ and sensitive to A_2 and A_4. Ellipsometry[12] gives the same kind of measurements. It is often easier experimentaly, but the analysis of experimental data is somewhat difficult if accuracy is needed

THERMAL DIFFUSION

In some materials, chromophores are free to rotate and spontaneously return to an isotropic distribution after any type of anisotropic perturbation. This is the case of many guest-host systems. This process is represented by a thermal diffusion equation,

$$\frac{\partial n(\theta,t)}{\partial t} = D\,\Delta n(\theta,t) = \frac{D}{\sin\theta}\frac{\partial}{\partial\theta}\left(\sin\theta\frac{\partial n(\theta,t)}{\partial\theta}\right) \tag{8}$$

which has a very simple expression in the order parameters formalism[8,9]:

$$\frac{dA_q}{dt} = -q(q+1)\,D\,A_q \tag{9}$$

D=kT/ζ is the angular diffusion constant in which k is the Boltzmann constant and ζ the angular viscosity, which depends on the free volume of active molecules in the polymer. This formalism, which leads to a simple exponential relaxation for each order parameter ($\tau_q=[D\,q(q+1)]^{-1}$), is known to fail to describe accurately experimental observations which are actually not mono-exponential. This behavior is usually explained by the inhomogeneity of the polymer in which each chromophore has a different free volume and consequently a different diffusion constant. The exponential behavior has to be averaged on the distribution of D values, f(D), but this distribution is unknown.

Many authors fit experimental relaxation curves with a phenomenological stretched exponential, $\exp(-(t/\tau)^{\beta})$, which is always better than a bi- or a multi-exponential approximation. In addition it has been proved[10] that the time constant, τ, varies with time after a perturbation which affects the structure of the polymer (such as a temperature variation or optical pumping switching off). In the following, we will assume exponential relaxation given by (9): this is the only way to write simple equations, but we will keep in mind the necessity of a final averaging on D values.

Materials with a high mobility of chromophores could be useful for real time holography and as a constituent of photorefractive materials[11], but for optical memories, waveguide drawing, SHG or electro-optic devices, one has to create a special order in a material which is able to keep this order indefinitely. The easiness of creation of orientational order is somewhat contradictory with the stability of this order: it is necessary to find processes which impose ordering forces and which simultaneously reduce temporary the rotational viscosity, allowing chromophores to respond to ordering forces within a reasonable time delay.

CREATION OF ORDER

Thermal poling

It is the usual method for creating non centrosymmetry for electro-optic devices or for SHG. In this method, the viscosity is reduced by heating the material above Tg, the glass transition temperature of the polymer, and the driving force is a DC field producing a torque, $\vec{\mu}_0 \times \vec{E}$, on polar molecules (we neglect the second order term αE^2). After that, the order is frozen by cooling the sample, in presence of the field which must be maintained until the viscosity recovers a high value. As show, in ref. 10, this recovery is not instantaneous after cooling and depends on cooling velocity. Thermal poling can be modelized by adding a torque in the diffusion equation (8) [8]:

$$\begin{aligned}\frac{\partial n(\theta,t)}{\partial t} &= \left(D\,\Delta n(\theta,t) + \frac{1}{\zeta}\vec{\nabla}\cdot\left(n(\theta,t)\vec{\nabla}(\mu_0 E\cos\theta)\right)\right)\\ &= \frac{D}{\sin\theta}\frac{\partial}{\partial\theta}\left[\sin\theta\left(\frac{\partial n(\theta,t)}{\partial\theta} + \frac{\mu_0 E}{kT}n(\theta,t)\frac{\partial\cos\theta}{\partial\theta}\right)\right]\end{aligned} \tag{10}$$

The new term expresses a viscous behavior in which velocity is proportional to the applied force. According to Benoit[8] the torque introduces a coupling between orders Δq=±1:

$$\frac{dA_q}{dt} = -q(q+1)\,D\left[A_q + \frac{X}{2q+1}\left(A_{q+1} - A_{q-1}\right)\right] \qquad \text{with: } X = \frac{\mu_0 E}{kT} \tag{11}$$

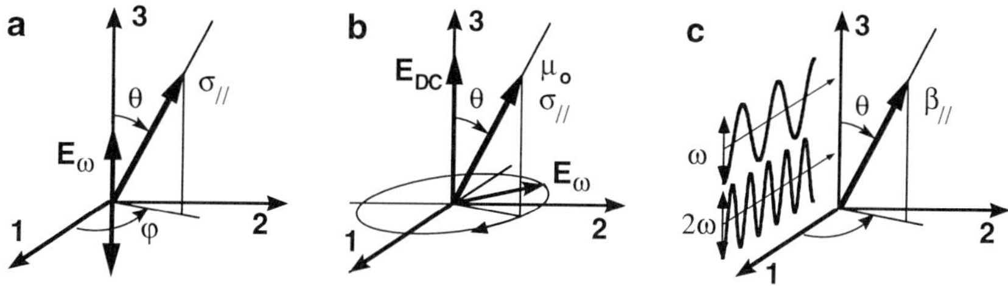

Figure 2. Fields geometry for PIA (a), PAP (b) and AOP (c).

The stationary solution of eq. (10) is the thermal equilibrium depending only on X ($n(\theta) \propto e^{-X\cos\theta}$). The different components of $\chi^{(1)}$, $\chi^{(2)}$, ... are given by Langevin functions[13,14] or expressed by Wu[9] in terms of order parameters (rigorous solution with the help of spherical modified Bessel functions[note 1]). The transient behavior of $A_q(t)$ cannot be expressed rigorously but a very good approximation can be obtained by neglecting all the A_q with q higher than q_{max}. Benoit[8] did this approximation with $q_{max}=2$ in order to get only two equations (A_1 and A_2), leading to a two-exponential solution. This approximation is reasonable for $X < 1$ [15] (which is true in many experimental cases). For higher values of X, higher values of q_{max} are necessary and a q_{max}-exponential solution can be obtained by numerical computation (Appendix).

Optical Ordering

Photoinduced Anisotropy (PIA)[16-18]**.** It is well known, for a long time (Weigert effect 1919) that optical pumping of anisotropic molecules by a polarized light beam (wavelength in the absorption band of molecules) produces a macroscopic anisotropy resulting in a birefringence and a dichroism of the sample. The first process is a selective optical pumping of molecules parallel to the pump polarization, also named "angular hole burning". More precisely, for linearly or for circularly polarized light, the pumping probability is, respectively :

$$Pr(\theta)_{lin} = I_p\left(\sigma_\perp + (\sigma_{//} - \sigma_\perp)\cos^2\theta\right) = \bar{\sigma} I\left(1 + 2eP_2(\cos\theta)\right) \qquad (12\text{-a})$$

$$Pr(\theta)_{circ} = I_p\left(\sigma_\perp + \frac{1}{2}(\sigma_{//} - \sigma_\perp)\sin^2\theta\right) = \bar{\sigma} I\left(1 - eP_2(\cos\theta)\right) \qquad (12\text{-b})$$

where I_p is the pump intensity, θ the angle of the principal axis of the molecule with the symmetry axis of the pumping, which is the polarization of light in (12-a) and its direction of propagation in (12-b) (fig. 2.a and 2.b). $\sigma_{//}$, $\sigma_\perp$ and $\bar{\sigma}$ are proportional to the imaginary part of $\alpha_{//}$, $\alpha_\perp$ and $\bar{\alpha}$ (eq. 5) and $e = (\sigma_{//}-\sigma_\perp)/(\sigma_{//}+2\sigma_\perp)$. If molecules are excited to a long living upper level or to a level relaxing toward a long living level (triplet state or isomer state) and if the saturation is moderate ($\sigma_{//}I_p\tau \sim 1$, where τ is of the order of the life time of the long living level) a large anisotropy (A_2) is created. For stronger pumping the relative anisotropy (A_2/A_0) remains large but the absolute anisotropy A_2 of the ground state, as measured by experiments described above, is reduced because of the depopulation (even if $\sigma_\perp I_p\tau \ll 1$, $\sigma_{//}I_p\tau\cos\theta$ can be of the order of 1 with large values of θ, near $\pi/2$). The angular distribution and order parameters can be rigorously calculated[16,18]

If hole burning is the only process, the anisotropy disappears when the pump is switched off, with the same time constant, τ, as the population. With some molecules, such as azobenzene derivatives (in particular, the famous DR1), experiments show that anisotropy

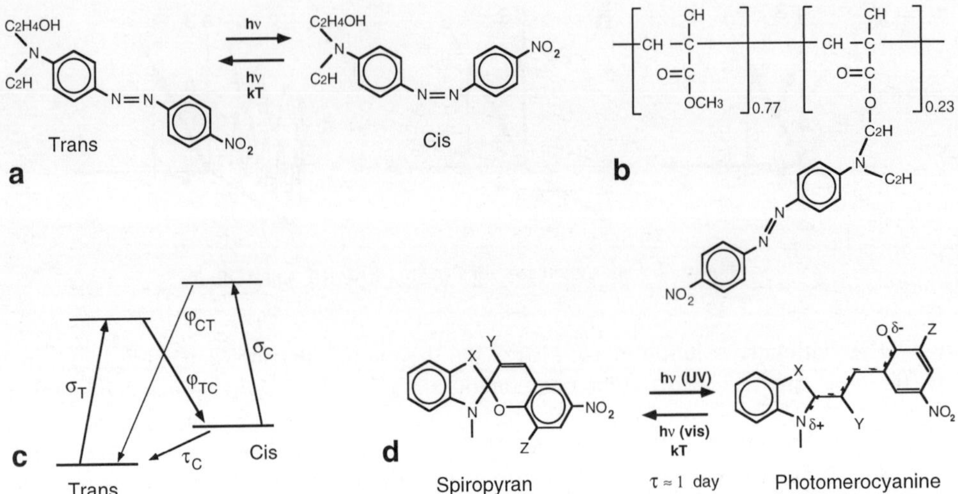

Figure 3. a: Trans-cis photoisomerization of DR1 (Dispersed Red 1). **b:** Copolymer of DR1 and MMA (Methyl Methacrylate). **c:** Simplified diagram of DR1 levels. The reverse optical pumping from cis to trans has been represented too. **d:** Spiropyran-Merocyanine photoisomerization. The diagram of fig. c can be applied also to this case. In experiments described in the text, $X=C(CH_3)_2$, $Y=Z=H$.

decays with a relaxation time much larger than τ (guest-host) or can be permanent (with DR1-PMMA copolymer of fig. 3b)[16]. This can be explained only by a rotation of molecules during the optical pumping cycle. With azobenzene derivatives this process has been studied in viscous solutions[19] or in polymer films[20] and has been attributed to trans-cis photoisomerization (fig. 3a). The lower level is the trans state which absorbs light in the blue-green spectral band. Through different possible cascades, the excited level has some probability to lead to the ground state of the cis shape which relaxes back to the trans shape (fig 3c), with a long lifetime, τ_c (few ms to several s, depending on the environment).

A random rotation may occur in the trans to cis photoisomerization and/or by thermal diffusion in the cis state (which is shorter and more compact than the trans state) and/or in the relaxation from cis to trans. With DR1, experiments cannot determine at which step rotation occurs, but they prove that the rotation angle is small at each cycle (smaller with copolymer than with guest-host) and that several optical pumping cycles for each molecule are necessary to change appreciably the macroscopic order. Equations have been written for the angular distribution of trans and cis states and they have been decomposed in a system of order parameters (T_q for trans and C_q for cis, according to the definition (1) of A_q):

$$\frac{dn_T(\theta)}{dt} = -Pr(\theta)\ n_T(\theta) + \frac{1}{\tau_c}\int R_{CT}(\theta' \rightarrow \theta)\ n_C(\theta')d\theta' + D_T\,\Delta n_T(\theta) \tag{13-a}$$

$$\frac{dn_C(\theta)}{dt} = \int P_{TC}(\theta' \rightarrow \theta)\ Pr(\theta')\ n_T(\theta')d\theta' - \frac{1}{\tau_c} n_C(\theta) + D_C\,\Delta n_C(\theta) \tag{13-b}$$

$$\frac{dT_q}{dt} = -I\sum_{q'} Pr(\varepsilon)_{qq'}\ T_{q'} + \frac{1}{\tau_c} R_q\, C_q - q(q+1)\, D_T T_q \tag{14-a}$$

$$\frac{dC_q}{dt} = I\,P_q \sum_{q'} Pr(\varepsilon)_{qq'}\ T_{q'} - \frac{1}{\tau_c} C_q - q(q+1) D_C\, C_q \tag{14-b}$$

with $I = \overline{\sigma} I_p \Phi_{TC}$ (Φ_{TC}: quantum yield) and with :

$$\begin{aligned} &\Pr(\theta) = \Phi_{TC} \Pr(\theta)_{lin} \qquad \text{or} \qquad \Pr(\theta) = \Phi_{TC} \Pr(\theta)_{circ} \\ &\Pr(\varepsilon)_{qq} = 1 + 2\varepsilon e q(q+1)(2q-1)^{-1}(2q+3)^{-1} \\ &(2q+1)\Pr(\varepsilon)_{q\,q\pm2} = 3\varepsilon e(q\pm1)(q+1\pm1)(2q+1\pm2)^{-1} \end{aligned} \tag{15}$$

ε=1, for linear polarization and ε=-1/2, for circular polarization. Angular diffusion has been introduced in cis (D_C) and in trans (D_T). Rotation produced by photoisomerization and relaxation are represented by the probabilities $P_{TC}(\theta'\rightarrow\theta)$ and $R_{CT}(\theta'\rightarrow\theta)$, in (13), and by factors P_q and R_q in (14). It is easy to take into account an eventual reverse cis-to-trans optical pumping (fig. 3c) by introducing a probability of leaving cis and a redistribution on the arrival in trans.

The $\cos^2\theta$, contained in optical pumping probability (12), couples orders $\Delta q = 0,\pm2$. As expected by symmetry considerations, it cannot create odd orders from an isotropic initial state[note 3]. The random rotation expressed by P_q, R_q and D_C doesn't couple different orders since they are isotropic processes. P_{TC} and R_{CT} depend only on the absolute value of the rotation angle but not on the initial or final orientation. P_q and R_q express the width of the angular spreading. P_q or R_q=1 represent the absence of rotation and P_q or R_q=0 a totally random redistribution (total loss of orientation in a single process). A phenomenological model for expressing P_q and R_q will be discussed of the next paragraph.

Photoassisted Poling (PAP)[1,16-18,21,note 2]**.** If a DC electric field is applied during the optical pumping, the diffusion in cis must be expressed like in thermal poling by (10) and (11) and introduces a coupling $\Delta q = \pm1$. In addition one can imagine that the rotation in both isomerization processes is no more isotropic around the initial orientation of the molecule but that the probability of rotation is higher with μ_0 in the direction of the field than in the opposite direction. Then eqs. (14) become:

$$\frac{dT_q}{dt} = -I\sum_{q'} \Pr(\varepsilon)_{qq'} T_{q'} + \frac{1}{\tau_c}\sum_{q'} R_{qq'} C_{q'} - q(q+1)D_T\left[T_q + \frac{X_T}{2q+1}\left(T_{q+1} - T_{q-1}\right)\right] \tag{16-a}$$

$$\frac{dC_q}{dt} = I\sum_{q'q''} P_{qq'} \Pr(\varepsilon)_{q'q''} T_{q''} - \frac{1}{\tau_c} C_q - q(q+1)D_C\left[C_q + \frac{X_C}{2q+1}\left(C_{q+1} - C_{q-1}\right)\right] \tag{16-b}$$

in which P_q and R_q are replaced by redistribution matrices $P_{qq'}$ and $R_{qq'}$ which couple different orders. In this case, we have chosen a non polarized (or circularly polarized; ε=-1/2) pumping beam, propagating in the direction of the DC field, in order to increase the efficiency (pumping of molecules perpendicular to E) and also to keep the uniaxial symmetry. A thermal poling term has been added in trans state too.

The determination of redistribution matrices $P_{qq'}$ and $R_{qq'}$ is a difficult problem which can be solved phenomenologically by assuming a virtual intermediate state in which an angular diffusion (like thermal poling) may occur[21]. These matrices depend only on two parameters, $X^* = \mu_0 E/kT^*$, in which T^* is a virtual temperature, which characterizes the competition between the ordering torque and the disordering thermal spreading of the distribution, and $D^*\tau^*$, the product of a virtual diffusion constant by the time spent in the virtual state. This last parameter measures the loss of memory of the initial orientation ($D^*\tau^* \ll 1$ means a very small rotation at each isomerization and $D^*\tau^* \gg 1$, a complete thermal poling in the virtual state in a single process). Of course, if E = 0, $P_{qq'}$ and $R_{qq'}$ are diagonal and one obtains the set of coefficients P_q and R_q of eqs. (14).

In order to solve eqs.(16) we first calculate matrices $P_{qq'}$ and $R_{qq'}$ by writing thermal poling equations (11) in the virtual intermediate state, by truncating then above $q_{max}+2$ and by solving then for all initial states $V_{q'} = \{A_{q'}=1; A_{q''}=0 \text{ for } q''\neq q'\}$,with the matricial method summarized in appendix. The same numerical method is then used for solving equations (16), truncated above q_{max}.

All Optical Poling (AOP)[2,22,23]**.** A few years ago, it has been demonstrated in inorganic materials (such as silica fibers), that the simultaneous illumination with a strong laser beam (ω) and with its own second harmonic (2ω) breaks the centrosymmetry[24]: a $\chi^{(2)}$ slowly grows and SHG becomes possible with an automatic phase matching. This technique is named "seeding" since it is necessary to seed with 2ω frequency, in order to produce 2ω after the growing of SHG ability. The mechanism in inorganic materials has been widely discussed and have been attributed to charges transfer. Recently seeding has been demonstrated in organic solutions[2] (by transient 4-wave mixing) and in chromophores-containing polymers[22].

In organic materials, the mechanism appears to be quite similar to photoinduced anisotropy. The pumping term only is different. As shown by Charra *et al.*[22], the amplitude of probability for finding a molecule in the excited state is given, at second order of perturbation, by (they assume two-level molecules with a transition, $a\rightarrow b$, almost resonant with the harmonic frequency, $\omega_0 \approx 2\omega$):

$$\langle b \mid \psi \rangle \propto \vec{\mu}_{ab} \cdot \vec{E}_{2\omega} + \frac{\left(\vec{\mu}_{ab} \cdot \vec{E}_{\omega}\right)\left(\left(\vec{\mu}_b - \vec{\mu}_a\right) \cdot \vec{E}_{\omega}\right)}{2\hbar\left(\omega_0 - \omega\right)} \tag{17}$$

The probability of transition from the ground state, a, to the excited state, b, is proportional to:

$$\Pr(a\rightarrow b) \propto \left|\langle b \mid \psi \rangle\right|^2 \propto \left|\vec{\mu}_{ab} \cdot \vec{E}_{2\omega}\right|^2 + Re\left\{\frac{\left(\vec{\mu}^*_{ab} \cdot \vec{E}^*_{2\omega}\right)\left(\vec{\mu}_{ab} \cdot \vec{E}_{\omega}\right)\left(\left(\vec{\mu}_b - \vec{\mu}_a\right) \cdot \vec{E}_{\omega}\right)}{\hbar\left(\omega_0 - \omega\right)}\right\} + \dots \tag{18}$$

The first term is linear absorption of 2ω and the second is a new cross term. By recombination of matrix elements μ_{ab} and μ_b-μ_a this term can be written $\beta \vdots \left(\vec{E}^*_{2\omega}\vec{E}_{\omega}\vec{E}_{\omega}\right)$ since $\beta \propto \mu^*_{ab} \otimes \mu_{ab} \otimes \Delta\mu$ Then it appears as an optical pumping depending on the hyperpolarizability β of molecules. If both beams are linearly polarized along the 3 axis (Fig. 1-c) this term can be written for rod like molecules ($\beta_\perp$=0) :

$$\Pr_{seeding} \propto \beta_{//} \left|E_{2\omega} E^2_{\omega}\right| \cos^3\theta \ \cos\varphi \tag{19}$$

where φ is the relative phase of ω and 2ω waves. The seeding term produces a selective pumping of molecules oriented upward or downward, according to the sign of $\cos\varphi$, as shown, in ref. 23. In thick materials, this $\cos\varphi$ insures automatically a quasi-phase-matching, with periodic reversal of orientation. This hole burning process may be followed by an angular redistribution, if some structural change occurs in the excited state or in the relaxation cascade. Then a permanent poling can be obtained[23]. For modelizing this process, eq. (13) can be used with the pumping probability (19), which is expressed by using the relation:

$$\cos^3\theta \ = \ \tfrac{1}{5}\left(3\,P_1(\cos\theta) + 2\,P_3(\cos\theta)\right) \tag{20}$$

This term introduces couplings $\Delta q=\pm1,\pm3$ (note 3), which has to be added to couplings $\Delta q=0,\pm2$ produced by linear absorption of 2ω (like in (15)) and to couplings $\Delta q=0,\pm2,\pm4$ from two-photon absorption of ω, since the laser beam at the fundamental frequency is strong (probability of pumping $\Pr_{2photon} \propto \cos^4\theta = \frac{1}{35}\left(7 + 20\,P_2(\cos\theta) + 8\,P_4(\cos\theta)\right)$). The optical pumping matrix in (14), becomes more complicated, but it is not a problem for the numerical solution of the appendix. The difficulty to simulate experiments, is the estimation of the ratios between the three optical pumping terms. Indeed experiments are performed with high power short pulses: saturation and transient processes could modified the ratios obtained from the stationary perturbation method, in (17) and (18).

The great advantage of AOP is that it does not need the existence of a permanent dipole: it is possible to orient octupolar molecules for which μ_0 is intrinsically zero[28].

SOME EXPERIMENTAL RESULTS.

The above theory has been used to simulate experiments performed with ATR (Fig.1-b) or photo induced dichroism set up (Fig.1-a). In this paragraph, we summarize the main experimental features which have been pointed out by theoretical simulations.

PIA. When the linearly polarized pump is switched on, $\Delta n_\perp$ and $\Delta A_\perp$ (index or absorbance variations for perpendicular polarization) always begins to decrease with a slope which is one third of that of $\Delta n_{//}$ and $\Delta A_{//}$, as predicted by hole burning theory. But after some time the angular redistribution tends to increase $\Delta n_\perp$ and $\Delta A_\perp$. Their final sign, as well as the time needed to reach the stationary state, depend on the pump intensity and on the average rotation at each pumping cycle. In DR1-doped PMMA we have shown[16,17] the change of shape of $\Delta A_\perp$ as a function of I_p. With DR1-PMMA copolymer the reorientation is slow (small rotation at each cycle) but a large permanent anisotropy is obtained[16]. When the pump is switched off, $\Delta n_{//}+2\Delta n_\perp$ (or $\Delta A_{//}+2\Delta A_\perp$) rapidly recovers its initial value, which proves the absence of photochemical degradation of DR1.

A better confirmation of the theory has been given by the measurement of dichroism variation in the UV range, where absorption is mostly sensitive to cis molecules[18,21] : at the beginning, during the hole burning period, cis is aligned parallel to the pump polarization since cis molecules are directly coming from the hole burnt in the trans angular distribution, but in a time of the order of 1 mn, this anisotropy disappears because of the slow angular redistribution : the theory shows that the equilibrium between the stationary state anisotropy, in trans, and the anisotropic optical pumping leads to an isotropic rate of arrival in cis (if $D_T=0$)

In a sol-gel doped with an azobenzene derivative[18,21], the dynamics of PIA has been shown to vary with aging of the sample: for the fresh sample, the angular redistribution was very rapidly dominant ($\Delta n_\perp>0$), while a few days later, hole burning remained dominant after an optical pumping of several minutes ($\Delta n_\perp<0$). Both transient curves have been precisely simulated with the same set of parameters, except the average rotation at each pumping cycle (*i.e.* the orientational memory), represented by the virtual $D^*\tau^*$ product in cis-to-trans relaxation, which had to be modified from 1 to 0.002. Let us notice that, in the absence of simultaneous measurement on the cis state, it is not possible to attribute rotation to one unique process in the trans-cis-trans cycle. The same result would have been obtained by attributing the rotation to trans-cis photoisomerization and/or to diffusion in cis.

PAP. PAP has been observed as well by ATR[1] as by SHG[25]. With guest-host DR1-PMMA, a partial orientation is obtained at room temperature by action of the DC field alone, but orientation is increased by optical pumping[1]. The relaxation after the removal of the field is much faster in the absence of optical pumping but in both case it is complete after a few hours. With the copolymer[1,16] there is almost no rotation without optical pumping and PAP produces a permanent poling. The general behavior has been correctly simulated[21] in both cases by varying only the diffusion constant in trans (D_T) but the precise shape of curves is not accurately represented since we use monoexponential relaxations. As explained previously , there is a wide dispersion of the diffusion constant from one chromophore to another.

With DR1, because of the short lifetime of the cis state, it is difficult to determine at which step of the cycle rotation does occur. On the opposite, with spiropyran, the photomerocyanine, obtained by UV pumping, has a long life time and ATR electro-optic measurements are essentially sensitive to merocyanine molecules which have been photoisomerized once only. By applying the DC field during or after the optical pumping period, a large difference of poling has been observed[21]: this proves that angular redistribution during photoisomerization is much more efficient than thermal diffusion in the photoisomer state. Let us notice that merocyanine is longer than spiropyran, while cis DR1 is shorter than trans: we must not generalize the above conclusion, but we have demonstrated that rotation can be associated with photoisomerization processes. The same experiments have also demonstrated that rotational mobility (for poling or relaxation) slowly decreases, one hours time scale, after the end of optical pumping. This is in agreement with results obtained by Bauer *et al.*[10] and with their interpretation by a slow relaxation of the polymeric environment, which has been perturbed by optical pumping of chromophores.

Different authors[10,26,27] have compared the efficiency of thermal poling and of optical poling. They found better results with thermal poling, as well for the final electro-optic coefficient as for its long term stability. The first limitation seems to be the small penetration depth of the pumping light, in high concentration films[10,26]. Indeed we obtained better poling with a pump wavelength on the side of the absorption band. The second limitation is probably that a proportion of molecules has a too small mobility to be oriented in a reasonable time: by heating the sample (but keeping $T<T_g$) the efficiency can be increased[27], as well as the long term stability[10]. Nevertheless we have some experimental proofs that the poling limit could be better by optical means, if these practical difficulties could be overcome.

AOP. It has not yet been possible to compare the above theory with AOP experiments, nevertheless the comparison of the dynamical behavior of DR1-PMMA guest-host with that of the copolymer, gives results quite similar to that obtained with PAP and PIA[23]: the rotation of chromophores is slower with the copolymer, but the final SHG coefficient is higher and much more stable.

APPENDIX

After truncature above order q_m, eqs. (11), (14) and (16) are of the form:

$$\frac{dX(t)}{dt} = -M\,X(t) + S \qquad (11)$$

where X(t) is a vector of dimension $p_m=2q_m+1$ (components: $T_1 \ldots T_{q_m}$, C_0, $C_1 \ldots C_{q_m}$), in (16), or $p_m=q_m+1$ (components: T_2, $T_4 \ldots T_{q_m}$, C_0, $C_2 \ldots C_{q_m}$), in (14), or $p_m=q_m$ (components: A_1, $A_2 \ldots A_{q_m}$), in (11). The vector S is built by the elimination of T_0 (or A_0) with the help of the conservation relation $C_0+T_0= N$ (or $A_0=N$) . The solution can be put in the following form:

$$X(t) = \sum_{p=1}^{p_m} \alpha_p V_p\, e^{-\lambda_p t} + V_\infty \qquad (12)$$

V_∞ is the steady state solution obtained by solving the system $M\,V_\infty = S$. V_p's are the p_m eigenvectors corresponding to the eigenvalues λ_p of M. The α_p are determined by solving the initial condition, $X(0) = \sum \alpha_p V_p + V_\infty$. As far as the truncature is valid, this method is rigorous, but it necessitates the use of a computer for matrix inversions and eigenvalues problems, except if it is possible to reduce the system to $p_m=2$ [8,17]. The minimum value of q_m necessary to obtain a good accuracy up to q=4 (if we are interested in $\chi^{(n)}$ up to n=3) depends on many parameters which will be discussed elsewhere. Let us say that for moderate values of E (X<10) and of I_p, $q_m= 6$ is sufficient in the case of PAP (except if e = 1). For thermal poling $p_m=q_m=2$ is enough if X<1 [15].

NOTES

note 1. At thermal equilibrium, according to Wu[9], $A_q/A_0 = i_q(\mu_0E/kT)/i_0(\mu_0E/kT)$, where i_q is the spherical modified Bessel function of order q.

note 2. We call this process PAP and not PIP, like many authors do[10,27], since light does not induce rotation but helps the action of the DC field.

note 3. If the pumping probability contains $\cos^p\theta$, this factor is expressed as a sum of $P_p(\cos\theta)$, $P_{p-2}(\cos\theta)$,... In projecting (13) on $P_{q'}(\cos\theta)$ one finds contributions to the matrix elements in (14) of the form: $\int P_q(u)\ P_p(u)\ P_{p'}(u)\ du = 2(2q'+1)^{-1}<p,0,q,0/p',0>^2$, where <./..> is a Clebsch-Gordan coefficient, not vanishing if q+p+q' is even and $/q-p/\leq q'\leq q+p$.

REFERENCES

1.a Z. Sekkat And M. Dumont, Photoassisted poling of azo-dye doped polymeric films at room temperature, Appl. Phys. B, 54, 486 (1992).

1.b Z. Sekkat And M. Dumont, Poling of polymer films by photoisomerisation of azo-dye chromophores, Mol. Chryst. liq. Chryst. Sci. Technol. secB: Nonlinear Optics, 2, 359 (1992).

2. F. Charra, F. Devaux, J.M. Nunzi and P. Raimond, Picosecond light-induced noncentrosymmetry in a dye solution, Phys. Rev. Lett. 68, 2440 (1992).

3.a J.D. Le Grange, M.G.Kuzyk and K.D.Singer, Effects of order on nonlinear optical processes in organic molecular materials, Mol. Cryst. Liq. Cryst. 150b, 567 (1987)

3.b K.D.Singer, M.G.Kuzyk, and J.E.Sohn, Second-order nonlinear optical processes in orientationally ordered materials: relationship between molecular and macroscopic properties, J. Opt. Soc. Am. B 4, 968 (1987).

4. M. Dumont and Y. Levy, Electro-optic properties of organic thin films by attenuated total reflection, in: "Nonlinear Optics of Organics and Semiconductors", T. Kobayashi, ed., Springer-Verlag, Berlin (1989) p.256

5. M. Dumont, Y. Levy and D. Morichière, Electrooptic organic waveguides : optical characterization, in: "Organic Molecules for Nonlinear Optics and Photonics", J.Messier, F.Kajzar and P.Prasad, ed., Nato ASI Series vol. E 194, Kluwer Academic Publishers, Dordrecht (1991), p461

6. M. Dumont, D. Morichère, Z. Sekkat and Y. Levy: Accurate measurement of thin polymeric films index variations. Application to elasto-optic effect and to photochromism, in "Photopolymer Device Physics, Chemistry, and Applications II", R. Lessard ed., SPIE proceedings Vol 1559 (1991), p 127.

7. D. Morichère, V. Dentan, F. Kajzar, P. Robin, Y. Levy and M. Dumont, Kerr effect and electrostriction in thin polymeric films. Application to polysilane films, Optics Comm., 74, 69, (1989).

8. H. Benoit, Contribution à l'étude de l'effet Kerr présenté par les solutions diluées de macromolécules rigides, Ann. Phys. (Paris) 6, 561 (1951).

9 J.W. Wu, Birefringence and electro-optic effects in polymer films: steady-state and transient properties, J. Opt. Soc. Am. B8, 142 (1991).

10. S. Bauer-Gogonea, S. Bauer, W. Wirges and R. Gerhard-Multhaupt, Pyroelectrical investigation of the dipole orientation in nonlinear optical polymers during and after photoinduced poling, J. Appl. Phys. 76, 2627 (1994).

11. W.E. Moerner and S.M. Silence, Polymeric photorefractive materials, Chem. Rev. 94, 127 (1994).

12.a C.C. Teng and H.T. Man, Simple reflection technique for measuring the electro-optic coefficient of poled polymers, Appl. Phys. Lett. 56, 1734 (1990).

12.b Y.Levy, M. Dumont, E. Chastaing, P. Robin, P.A. Chollet, G. Gadret and F. Kajzar, Reflection method for electro-optical coefficient determination in stratified film structures, Mol. Chryst. liq. Chryst. Sci. Technol. sec.B: Nonlinear optics, 4, 1 (1993)

13. D.M. Burland, R.D. Miller and C. A. Walsh, Second-order nonlinearity in poled-polymer systems,, Chem. Rev. 94, 31 (1994).

14. R.H. Page, M.C. Julrich, B. Reck, A. Sen, R.J. Twieg, J.D. Swalen, G.C.Bjoklund and C.G.Willson, Electrochromic and optical waveguide studies of corona-poled electro-optic polymer films, J. Opt. Soc. Am. B 7, 1239 (1990).

15. Z. Sekkat and W. Knoll, Stationary state and dynamics of birefringence and nonlinear optical properties induced by electric field poling in polymeric films, Ber. Bunsenges. Phys. Chem 98, 1231 (1994).

16. M. Dumont and Z. Sekkat, Dynamical study of photoinduced anisotropy and orientational relaxation of azo-dyes in polymeric films. Poling at room temperature, in: "Nonconducting Photopolymers and Applications", R. Lessard ed.,SPIE proceedings, Vol. 1774 (1992) p 188.

17. Z. Sekkat and M. Dumont, Photoinduced orientation of azo dyes in polymeric films. Characterization of molecules angular mobility, Synthetic metal, 54, 373 (1993)

18. M. Dumont,S. Hosotte, G. Froc and Z. Sekkat, Orientational manipulation of chromophores through photoisomerization, un: "Photopolymers and Applications In Holography, Optical Sensors and Interconnects", R. Lessard ed., SPIE proceedings Vol. 2042, (1993) p 2 .

19. H. Rau, Photoisomerization of azobezenes, in "Photochemistry and Photophysics, Vol. II", J.F. Rabek, ed., CRC press, Boca Raton, Florida (1990) p. 119.

B.S.Neporent, O.V.Stolbova: Opt. Spektrosk, 10, 287 (1961) [Opt. Spectrosc., 10, 146 (1961)]; Opt. Spektrosk, 14, 624 (1963) [Opt. Spectrosc., 14, 331 (1963)].

A.M.Makushenko, B.S. Neporent, O.V., Stolbova, Opt. Spektrosk, 31, 557 (1971) [Opt. Spectrosc. 31, 295 (1971)] ; Opt. Spektrosk, 31, 741 (1971) [Opt. Spectrosc. 31, 397 (1971)] ;

20. T.Todorov, L.Nikolova, N.Tomova, Polarization holography. 1:A new high-efficiency organic material with reversible photoinduced birefringence, Applied Optics, 23, 4309 (1984(.

21. M. Dumont, G.Froc and S.Hosotte, Alignment and orientation of chromophores by optical pumping, Nonlinear Optics , 9, 327 (1995)

22. F. Charra, F.Kajzar, J.M. Nunzi, P. Raimond and E. Idiart, Light-induced second-harmonic generation in azo-dye polymers,Optics Lett. 18, 941 (1993).

23.a C. Fiorini, F. Charra, J.M. Nunzi and P. Raimond, Photoinduced non centrosymmetry in azo-dye polymers, Nonlinear Optics 9 (1995) in press.

23.b J.M. Nunzi, C. Fiorini, F. Charra, F.Kajzar and P. Raimond, All-optical poling of polymers for phase-matched frequency doubling, in "Polymer Thin Films for Photonics", ACS symposium series (1994).

24. R.H. Stolen and H.W.K. Tom, Opt. Lett. 12, 585 (1987).

25. M. Dumont, Z. Sekkat, R. Loucif-Saibi, K. Nakatani and J.A. Delaire, Photoisomerization, photoinduced orientation and orientational relaxation of azo dyes in polymeric films, Mol. Chryst. liq. Chryst. Sci. Technol. secB: Nonlinear Optics, **5**, 395 (1993).

26. S. Yilmaz, S. Bauer and R. Gerhard-Multhaupt, Photothermal poling of nonlinear optical polymer films, Appl. Phys. Lett. 64, 2770 (1994).

27. P.M. Blanchard and G.R. Mitchell, A comparison of photoinduced poling and thermal poling of azo-dye-doped polymer films for second order nonlinear optical applications, Appl. Phys. Lett. 63, 2038 (1993)

28. J.M. Nunzi, F. Charra, C. Fiorini and J. Zyss, Transient optically induced non-centrosymmetry in a solution of octupolar molecules, Chem. Phys. Lett. 219, 349 (1994).

A THERMALLY STABLE ORGANIC LIGHT-EMITTING DIODE

P. Di Marco,[1] J. Kalinowski,[1,2] N. Camaioni,[1] V. Fattori[1] and G. Giro[1]

[1]Istituto di Fotochimica e Radiazioni di Alta Energia del C.N.R., Via P. Gobetti 101, 40 129 Bologna, Italy
[2]Department of Molecular Physics, Technical University of Gdańsk, Ul. G. Narutowicza 11/12, 80-952 Gdańsk, Poland

INTRODUCTION

The design of organic electroluminescence (EL) devices meets several barriers, thermal stability and injection ability of charge carriers from the electrodes being important examples. In spite of considerable efforts to surmount them,[1] they still create serious problems and require more experimental data and profounded knowledge on the device operation.

In this communication a thermally stable organic light - emitting diode (LED) is described and investigated in some detail.

EXPERIMENTAL

An EL diode based on Alq_3 as the light emitting material (EML)[2,3] and linear trans-quinacridone (QAC) as a hole-transporting layer (HTL) has been constructed following preliminary results of Tsutsui et al..[4] Schematic structure of the device is shown in Figure 1. The HTL is a polycrystalline QAC ($C_{20}H_{12}N_2O_2$) film and the EML consits of 8-hydroxyquinoline aluminum complex (Alq_3) polycrystalline layer standing in addition for an electron-transporting layer (ETL).

QAC | Alq_3

HOLE -TRANSPORTING MATERIAL | LIGHT-EMITTING MATERIAL

Both materials were deposited by vacuum-evaporation technique with a typical rate 0.2-0.4 nm/s on a room-temperature substrate. Injection of holes and electrons is provided from an indium-tin-oxide (ITO) anode (~30Ω/□) and Mg/Ag cathode, respectively. Electron-hole recombination near the organic interface region produces EL emission (hν) from Alq_3. Since the melting temperature of both Alq_3 and QAC (above 350°C) is much higher than softening

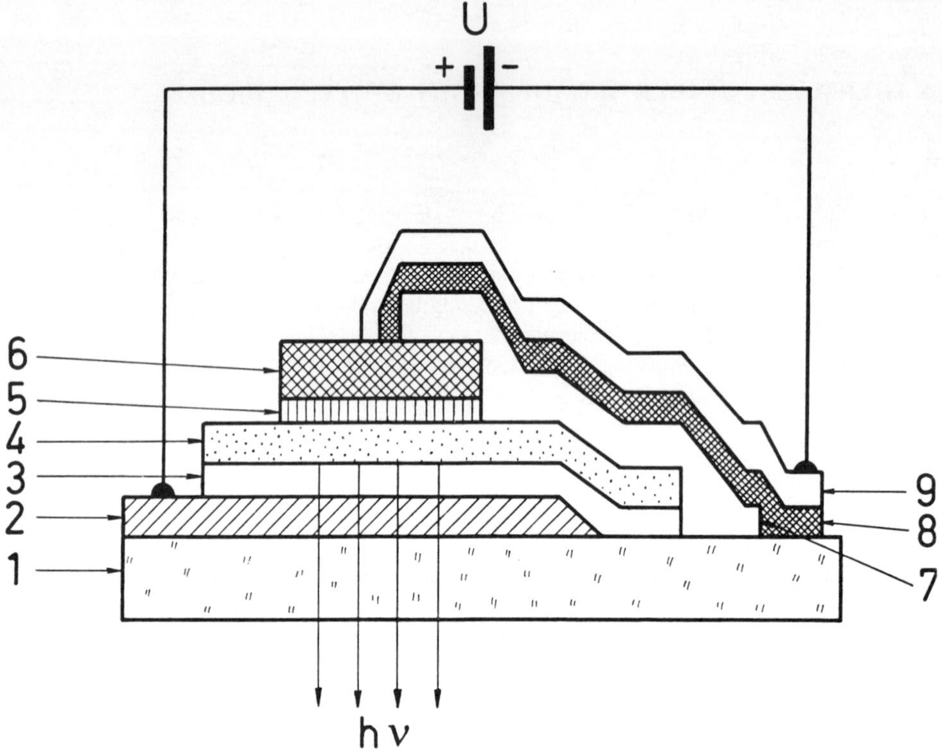

1 : Glass substrate
2 : Transparent anode (ITO ≈ 150 nm)
3 : Hole transporting layer (HTL) (QAC ≈ 80 nm)
4 : Light emitting layer (EML) (Alq_3 ≈ 90 nm)
5 : Cathode (Mg ≈ 20 nm)
6 : Protecting layer of Ag ≈ 200 nm
7 : Insulating layer of QAC ≈ 260 nm
8 : Ag contact film ≈ 100 nm
9 : Al contact strip ≈ 100 nm

Figure 1. Schematic structure of the EL device.

temperature (< 100 °C) of the commonly used hole-transporting aromatic amine N,N'-diphenyl-N,N'-bis(3-methylphenyl)-(1,1'-biphenyl)-4,4'-diamine (TPD) amorphous layers,[2-4] the device should show an improved thermal stability.[4] Indeed, some of our cells under current densities (j) exceeding 1A/cm^2 could survive near one hour. However, unusual behaviour of the EL cells under high-current conditions (voltage-independent current with concomitant strong increase in EL intensity) requires more careful studies of the origin of the emission within this current regime. To avoid this uncertainty and strong thermic effects we have investigated the cells under low-current conditions with j < 1 mA/cm^2 and U < 20V.

RESULTS AND DISCUSSION

The electric field characteristics of the current are shown in Figure 2, and EL intensity -current density relationship displayed in Figure 3. Assuming ITO to be a better injector of carriers into QAC than Mg into Alq_3, the results of Figure 2 suggest that the current flowing

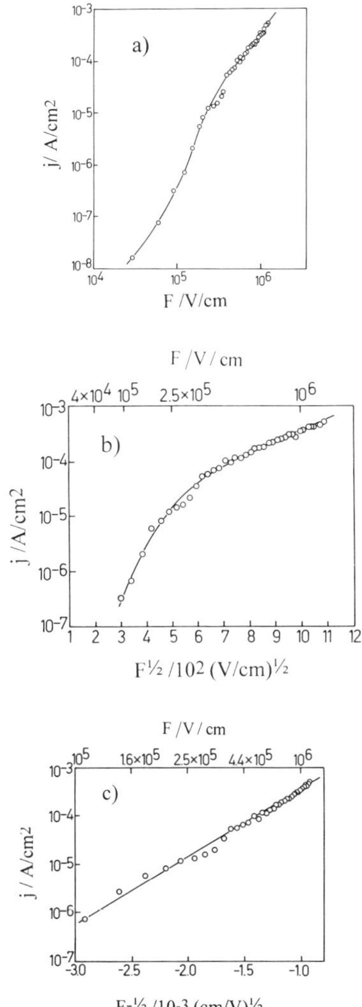

Figure 2. Field dependence of the current density in three different scales: (a) logj - lgF; (b) logj - $F^{1/2}$; (c) logj - $F^{-1/2}$.

through the cell is limited by injection from the cathode, following the one-dimensional Onsager model with the source current $j_{sc} = j_{oe} \exp(-\alpha_e x)$,[5]

$$j \approx j_{oe} \exp [-\alpha_e(\beta kT/eF)^{1/2}], \quad (1)$$

where $\beta = e^2/16\pi\varepsilon\varepsilon_0 kT$.

Such an expression holds at high fields when tunnelling constant

$$\alpha_e = (2\sqrt{2}/\hbar)(m^e_{eff} I_A)^{1/2} < eF/kT, \quad (2)$$

where m^e_{eff} is the effective mass of the electrons and I_A is the electron affinity of the material (Alq_3). This means that at an average field $\approx 5\times10^5$ V/cm the penetration depth of electrons $\alpha_e^{-1} > 5$Å, which seems to be a reasonable figure. Indeed, from the slope $S=(1/4)\alpha(e/\pi\varepsilon\varepsilon_o)^{1/2}$ of the straight-line relationship in Figure 2c, $\alpha^{-1} = 8.3$ Å is obtained, assuming $\varepsilon=4$. Thus, the injection current at x=0 can be evaluated as $j_{oc} \approx 10^{-2}$ A/cm^2.

The carrier concentrations at the QAC/Alq_3 interface are determined by the injection currents of holes and electrons flowing in the HTL and EML layers,

$$n_h = j_h / e\mu_h F_{HTL}, \qquad n_e = j_e / e\mu_e F_{EML}. \quad (3)$$

By definition, the EL intensity follows the relation[6,7]

$$\Phi_{EL} \sim n_h n_e \sim j_h j_e / \mu_h \mu_e F_{HTL} F_{EML} \quad (4)$$

with $F_{HTL} < F < F_{EML}$.

If the mobilities of holes, $\mu_h(F)$, and electrons, $\mu_e(F)$, are weakly varying functions of electric field, one would expect the high-field dependence of Φ_{EL} to be a power function of the electron injection current[6,7]

$$\Phi_{EL} \sim j_e^n \quad (5)$$

with

$$n = 1 + \alpha_h / \alpha_e = 1 + (m^h_{eff} E_g / m^e_{eff} I_A)^{1/2}, \quad (6)$$

where E_g is the energy gap of QAC. Indeed, Figure 3 shows $\Phi_{EL} \sim j^{5.7}$ for fields above 6.5×10^5 V/cm. From this value of the power n and from Equation (6) we conclude that the penetration depth of electrons into polycrystalline Alq_3 ($\alpha_e^{-1} \cong 8.3$Å) is about a factor 5 greater than that for holes into polycrystalline QAC ($\alpha_h^{-1} \cong 1.8$ Å). Then one can speculate on the relationship between effective masses of electrons and holes in respective materials if E_g and I_A are known. Assuming for QAC $E_g > 2.22$ eV, that is greater than the energy position of the first singlet level,[8] we find

$$m^h_{eff} = \frac{\hbar^2 \alpha_h^2}{8 E_g} < 0.12 m_e \quad (7)$$

and

$$m_{eff}^{e} < 0.012 m_e . \tag{8}$$

In both cases they are much smaller than the rest mass of an electron, m_e.

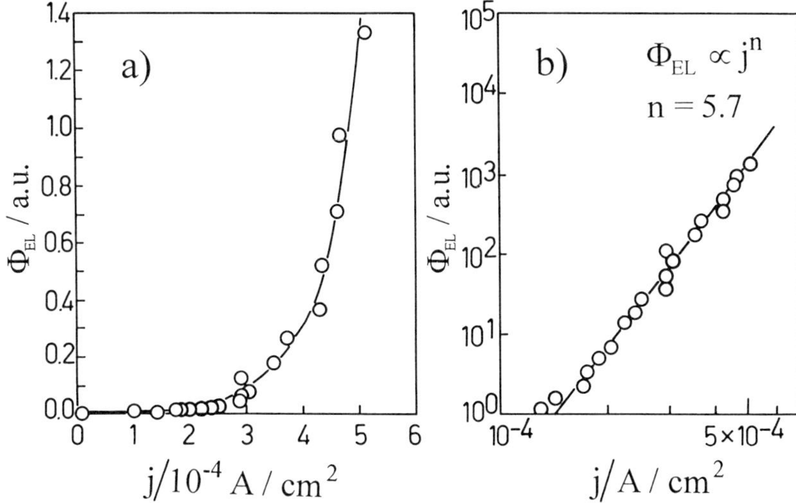

Figure 3. The relationship between EL intensity and current density in two scales: (a) linear, and (b) double logarithmic.

CONCLUSIONS

- An organic LED consisted of ITO/QAC/Alq_3/Mg-Ag shows to be thermal resistant (a current density higher than $1 A/cm^2$ can flow near one hour without breaking the cell).
- At lower voltages ($j<1 mA/cm^2$) the current flow is electron-injection limited, showing good correspondence to the one-dimensional Onsager model with the source current given by an exponential function of the coordinate along the electric field.
- The EL intensity - current density relationship is strongly superlinear, demonstrating different injection conditions at the interface between ITO and QAC, and Mg and Alq_3. The penetration depths of primary (tunnelling) electrons into Alq_3 and holes into QAC have been evaluated to be $\alpha_e^{-1} \cong 8.3$Å and $\alpha_h^{-1} \cong 1.8$Å, respectively. The effective masses of holes in QAC, $m_{eff}^{h} < 0.12\ m_e$, and electrons in Alq_3, $m_{eff}^{e} < 0.012\ m_e$, are shown to fall below the electron rest mass, m_e, to a greater degree as compared with light holes and electrons in silicon.[9]

ACKNOWLEDGEMENTS

This work has been done in the framework of the Program "Novel Materials" (Progetto Strategico "Materiali Innovativi") of the National Research Council (C.N.R.) of Italy. One of the authors (J.K.) wishes to thank Dr. C.W. Tang (Rochester, USA) and Dr. T. Tsutsui (Fukuoka, Japan) for interesting discussions on organic EL multilayer structures.

REFERENCES

1. "Electroluminescence Materials, Devices, and Large Screen Displays", Esther M. Conwell, Milan Stolka and M. Robert Miller, eds., *Proc. SPIE* 1910, USA (1993).
2. C. W. Tang and S. A. VanSlyke, Organic electroluminescent diodes, *Appl. Phys. Lett.* 51: 913 (1987).
3. C. W. Tang, S. A. VanSlyke and C. H. Chen, Electroluminescence of doped organic films, *J. Appl. Phys.* 65: 3610 (1989).
4. T. Tsutsui, E. Aminaka, Y. Fujita, Y. Hamada and S. Saito, Molecular design of organic dyes for thin-film electroluminescent diodes, *Synth. Metals* 57: 4157 (1993).
5. J. Godlewski and J. Kalinowski, Injection-limited currents in insulators, *Japan. J. Appl. Phys.* 28: 24 (1989).
6. J. Kalinowski, J. Godlewski and Z. Dreger, High-field recombination electroluminescence in vacuum-deposited anthracene and doped anthracene films, *Appl. Phys.* A37: 179 (1985).
7. J. Kalinowski, Limiting factors in operation and fabrication of organic electroluminescence devices, *in*: "Advanced New Materials and Emerging New Technologies", P.N. Prasad, E. Mark and J.F. Fung, eds., Plenum Press, New York (1995).
8. J. Kalinowski, W. Stampor, P. Di Marco and V. Fattori, Electroabsorption study of excited states in hydrogen-bonding solids: epindolidione and linear trans-quinacridone, *Chem. Phys.* 182: 341 (1994).
9. G. Yepifanov. " Physical Principles of Microelectronics ", Mir Publ., Moscow (1974).

NEW CONJUGATE POLYMERS - A MARRIAGE BETWEEN SYNTHETIC AND NATURAL POLYMERS

Naoya Ogata

Department of Chemistry, Sophia University
7-1 Kioi-Cho, Chiyoda-Ku, Tokyo 102, Japan

Abstract: Chemical modifications of enzymes such as Lipase or Tripsine to prepare novel enzyme conjugates were carried out by using telechelic oligo(N-isopropylacrylamide) which showed reversible dissolution-precipitation behaviors in water. The temperature-responsive enzyme conjugates retained the same enzymatic activities as native enzymes and recovery of the enzyme conjugates was possible by a simple temperature change to precipitate the enzyme conjugate out of the water solution and they were able to reuse in cycles without denaturation. This novel modulated systems is very useful for applications as a novel bioreactor system in comparison with immobilized enzyme systems.

INTRODUCTION

Stimuli-responsive polymers such as temperature-responsive polymers have been extensively studied with a common name as "Intelligent Polymers" which have three basic functionalities as sensors, actuators and controllers. Poly(N-isopropylacrylamide (IPAAm)) (PIPAAm) is a well-known water soluble polymer which shows unique, reversible hydration-dehydration changes in response to small temperature changes(1). An aqueous solution of PIPAAm demonstrates phase separation to precipitate the polymer out of the solution at a certain temperature, which is called as a lower critical solution temperature (LCST). The LCST of PIPAAm is known as 32°C. The LCST can be controlled from lower to higher temperatures by changing compositions of copolymers of PIPAAm and other comonomers such as butyl methacrylate (BMA) or N,N-dimethylacrylamide (DMA). Hydration-dehydration states of PIPAAm are schematically shown as scheme 1.

Temperature-responsive hydrogels derived from PIPAAm have been applied in novel approaches as intelligent polymers especially to biotechnological process controls such as drug-release systems (2-4), recovery of cultured cells (5,6), and immobilized enzymes (7,8). Immobilized enzymes which are commonly applied in biotechnology can be used to regulate enzymatic reactions, while the reaction takes place in a heterogeneous phase which may be considered as "insoluble conjugate", which may lead to reduce enzymatic activities, because of trapping enzymes within the matrix of hydrogels.

It was previously reported by my group that oligo-(IPAAm) (OIPAAm) with end reactive groups such as carboxylic or amine groups was synthesized by telomerization reactions of IPAAm (9). The OIPAAm with reactive end groups was then applied to prepare OIPAAm-biomolecule conjugates using bovine serum albumin and bovine plasma fibrinogen (10,11). These novel bioconjugates exhibited very interesting behaviors in terms of dissolution changes in water at LCST.

conjugates using bovine serum albumin and bovine plasma fibrinogen (10,11). These novel bioconjugates exhibited very interesting behaviors in terms of dissolution changes in water at LCST.

Polymers and Other Advanced Materials: Emerging Technologies and Business Opportunities. Edited by P. N. Prasad *et al.*, Plenum Press, New York, 1995

Poly(N-isopropylacrylamide) (P(IPAAm)) is a water -soluble polymer at room temperature. Aqueous solutions of this polymer exhibit a lower critical solution temperature (LCST) around 32°C.

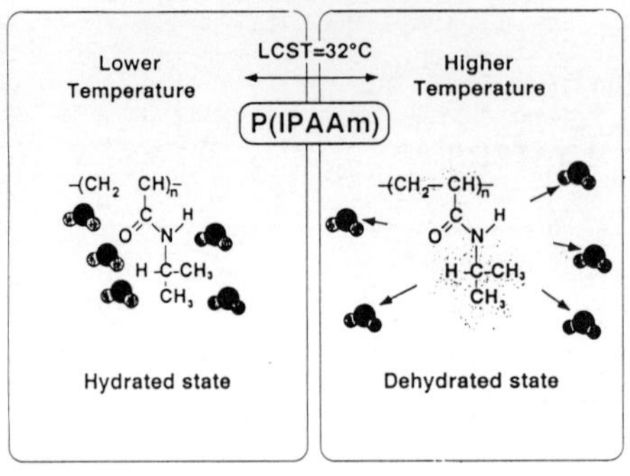

scheme 1

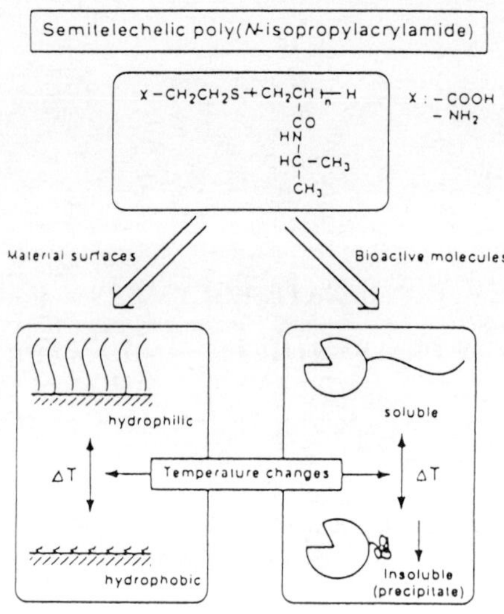

scheme 2

When enzymes are modified by the OIPAAm to prepare enzyme-OIPAAm conjugate, it is expected that the conjugates may exhibit a temperature-induced phase separation owing to the LCST behavior of the oligo(IPAAm), indicating reversible dissolution-precipitation behaviors at LCST. This reversible solubility changes of the enzyme conjugates may lead to the recovery of the enzyme conjugates after the enzymatic reactions finished. Moreover, the enzyme conjugates may retain the same enzymatic activities as native enzymes because the conjugates are dissolved in water in a free state which can not be attained by immobilized enzyme systems. The concept of the enzyme conjugate is shown as scheme 2. The modification of Lipase by the temperature-responsive OIPAAm was previously reported by my group(12).

This paper describes the novel enzyme conjugates by using Lipase and Tripsine to realize a happy marriage between synthetic and natural polymers.

RESULTS AND DISCUSSION

Synthesis of telechelic oligo(IPAAm) having end carboxylic group (IDc) was reported in a previous paper (9). Chemical modification of enzyme such as Lipase was also reported in a previous paper (12). Synthetic processes to prepare these enzyme conjugates can be refereed by these references. Trypsine was chemically modified by the same method as Lipase (12).

Figs. 1 and 2 indicate the LCST behaviors of the IDc-modified conjugates from Lipase and Trypsine, respectively. Native enzymes did not exhibit any changes in solubility, while the transmittance of the solution of the IDc conjugates remarkably decreased to zero at 37°C which is the same LCST as the original IDc. It is clear that the IDc conjugates of both Lipase and Trypsine are soluble in water below LCST, while they become insoluble above LCST to separate from the aqueous solution as droplets.

Enzymatic activities of the IDc conjugates of Lipase and Trypsine were investigated by using olive oil for Lipase and N-α-benzoyl-DL-arginine hydrochloride (BAPNA) as substrates for hydrolysis reactions and results are summarized in Figs. 3 and 4. The IDc-modified Lipase showed a slightly decreased activity in comparison with native Lipase, while the IDc-modified Trypsine indicated a rather improved activity. Presumably, hydrophobic isopropyl group of OIPAAm might have interactions with the hydrophobic amide compound so that the enzymatic activity of the IDc-Trypsine conjugate might be enhanced.

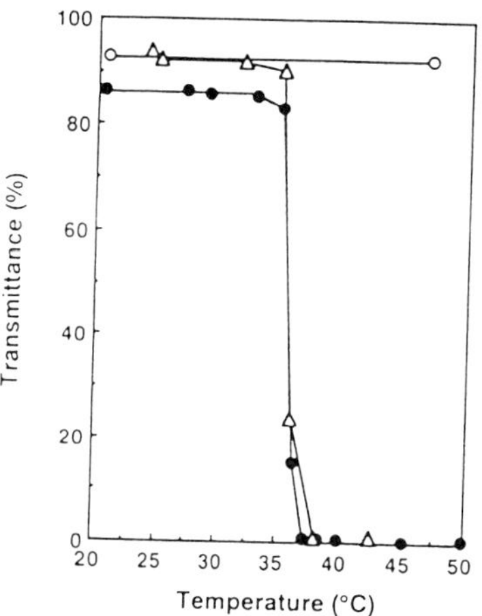

Fig. 1 LCST behavior of IDc-modified Lipase in terms of temperature dependence of optical transmittance of aqueous solutions of IDc-Lipase conjugate (1wt%,▵;IDc-10, ●; IDc-Lipase, ○; native Lipase, =500nm)

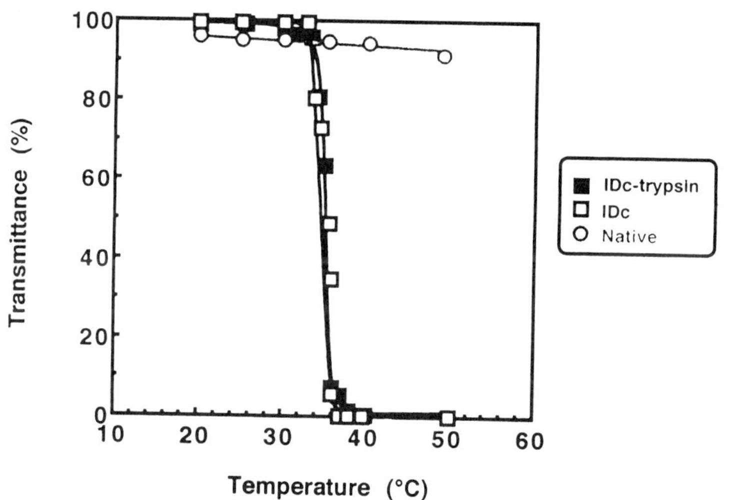

Fig. 2 LCST behavior of IDc-modified Trypsine in terms of temperature dependence of optical transmittance of aqueous solutions of IDc-Trypsine conjugate (1wt%, □;IDc-10, ■;IDc-Trypsine, o;native Trypsine, λ =500nm)

Effect of temperatures on enzymatic activities of the IDc-modified Lipase or Trypsine was investigated as shown in Fig.5. The enzymatic activity of the native Lipase increased with increasing reaction temperatures, while the IDc-Lipase conjugate decreased the enzymatic activity above 37°C which is the LCST of IDc because the IDc-Lipase conjugate was separated from the reaction phase owing to the precipitation as droplets. The IDc-Trypsine conjugates also decreased the enzymatic activities above LCST because of the phase separation out of the reaction media owing to the presipitation.

These results imply that the enzyme modification with temperature-responsive polymer enables the enzymes to halt their activities above the LCST of the conjugates and to be separated from the reaction phase. Since free mobile polymer chains having temperature responses are convalently grafted onto enzymes, their temperature-responsive behaviors reflect to instantaneous and simultaneous changes of solubilities of the conjugates. Thus, IDc telechelic co-oligomers play a critical role in the rapid switching of the conjugate solubility which is reversible.

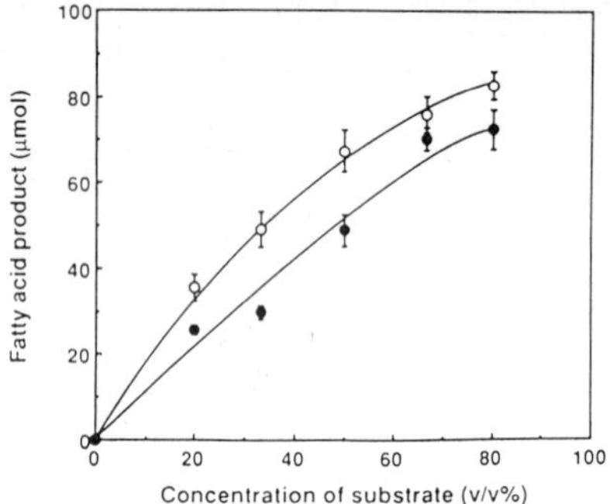

Fig. 3 Enzymatic activities of IDc-Lipase conjugate as functions of substrate concentration at 30°C (o; natice Lipase, •; IDc-Lipase. Lipase concentration was normalized to 80 μ g/ml , n=3, ± SEM)

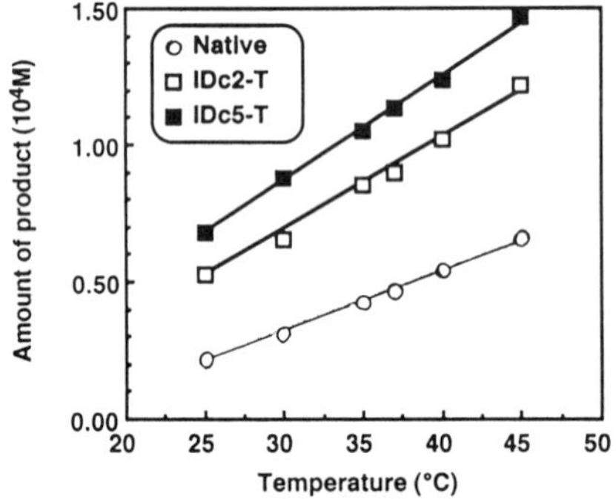

Fig. 4 Enzymatic activities of IDc-Trypsine as functions of temperatures (Try. concentration= 6.67μg/ml)

Lineweaver-Burk plots of the IDc-Trypsine conjugates were made at the Trypsine concentration of 38 μg/ml in a buffer solution of pH=8.15 containing 0.02 M of $CaCl_2$. Values of Km and Vmax are summarized in Table I. V_{max} values are almost similar, while Km of IDc-modified Trypsine was much smaller than that of native Trypsine, indicating that the IDc-modified Trypsine could more strongly attract substrates to enhance the enzymatic reactions, presumably owing to hydrophobic interactions between isopropyl moiety of IDc and substrates.

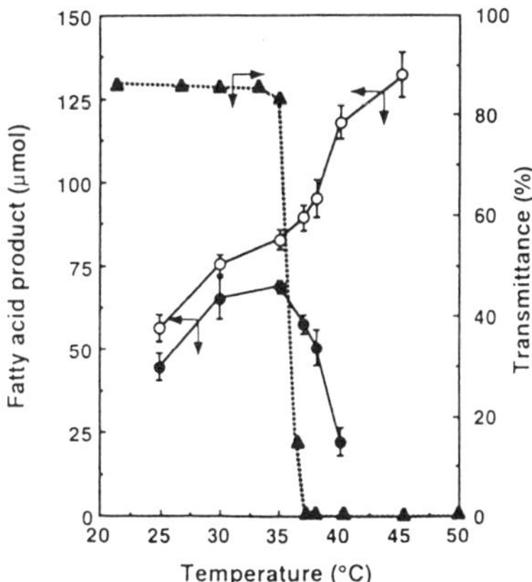

Fig. 5 Effect of temperature on Lipase activity (o; native Lipase, ●; IDc-Lipase. Substrate concentration; 50%(v/v), Lipase concentration was normalized to 100 μg/ml, n=6, ± SEM, ▲; transmittance of IDc-Lipase aquous solution at 500nm).

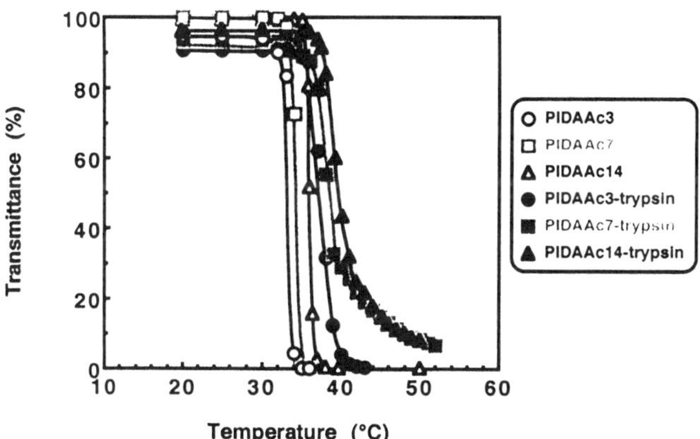

Fig. 6 LCST behaviors of PIDAAc and PIAAc-modified Trypsine as functions of temperatures (1 wt% in water).

The modification of enzymes as mentioned above was derived from singl end attachment of OIPAAm chains by telechelic oligo-IPAAm having one end carboxylic acid. Therefore, the grafted OIPAAm on enzyme surfases was free and flexible. It is interesting to incorporate PIPAAm chains as multi-point attachments on the surface of enzymes in order to compare enzymatic activities between single and multi-point attachments of the temperature-responsive PIPAAm. The concept of single- and multi-points attachment on enzymes is shown in scheme 3

Table I Kinetic constants of IDc-Trypsine conjugates

Sample	Km(10^3M)	V_{max} (10^4M/sec)
native	1.320	1.75
IDc-5T	0.278	1.77
IDc-2T	0.461	1.75

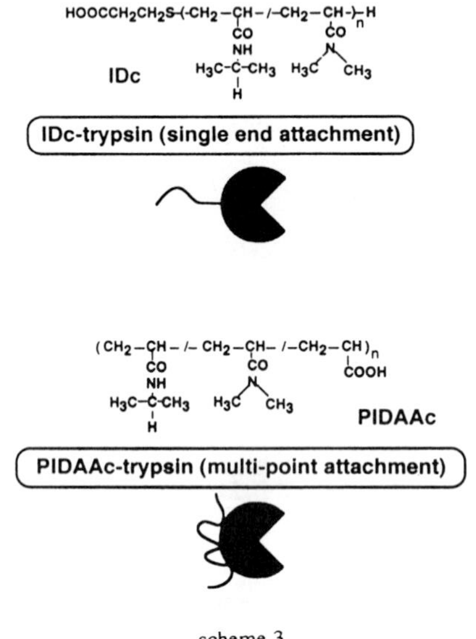

scheme 3

Terpolymers of IPAAm , DMA and acrylic acid (PIDAAc) were prepared by a common radical polymerization and they were bound to Trypsine by the same method as the case of single-point attachment. The amount of the incorporated acrylic acid moiety in PIDAAc was 2 mol% and the LCST of the PIDAAc is summarized in Table II, which indicates that the LCST increased slightly with increasing amount of DMA.

Trasmittance changes of the aqueous solution of PIDAAc having various amount of DMA units as functions of temperatures are indicated in Fig. 6, which also shows the LCST behaviors of the PIDAAc-modified Trypsine. It is seen in Fig. 6 that the LCST shifted toward higher temperature with increasing amount of DMA units in the terpolymers. Enzymatic stabilities of the modified Trypsine with either IDc or PIDAAc are investgated in Tris-HCl buffer (pH= 8.15) solution at 35°C and results are shown in Fig. 7. Comparing the stability of IDc- and PIDAAc-Trypsines, IDc-Trypsine was more stable than PIDAAc-Trypsine although both of them indicated much improved stabilities in comparison with native Trypsine. A single-point attachment of IDc has free and flexible chains on the surface of Trypsine, while multi-point attachment of PIPAAm on the surface of Trypsine might cause an enhancement of denaturation owing to a morphological change of Trypsine. Nevertheless, the modification of Trypsine by the temperature-responsive PIPAAm improved greatly the stability of Trypsine, as seen in Fig. 7.

Table II Properties of PIDAAc copolymers

Sample	IPAAm (mol%)	DMA (mol%)	AAc (mol%)	LCST
PIDAAc3	95	3	2	33.5
PIDAAc7	91	7	2	34.0
PIDAAc14	84	14	2	37.0

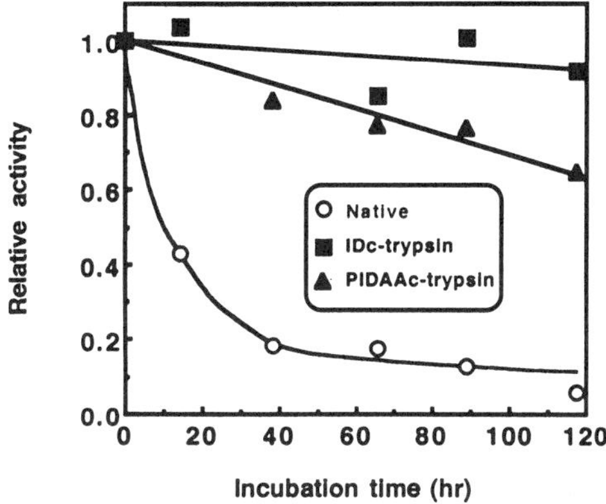

Fig. 7 Enzymatic stabilities of IDc- or PIDAAc-modified Trypsine as functions of temperatures (Tris-HCl buffer (pH 8.15) solution ar 30°C.

As mentioned before the solubility changes of the IDc-enzymes may lead to the novel recovery system of used enzymes. Since IDc enzyme conjugates precipitate as droplets above the LCST, these conjugates can be easily separated from the reaction phases by applying centrifugal separation followed by simple removal of supernatant after temperatures of the reaction phase raised above the LCST.

Enzymatic reactions of IDc conjugates of Lipase and Trypsine were carried out in the concentration of 100 μg/ml and substrate concentrations of 50%(v/v) at 30°C for 1 hr. After increasing solution temperature to 40°C to precipitate the conjugates, the precipitated conjugates were separated from the reaction products by a centrafuge. The recovered conjugates were reused in subsequent reaction after lyophilization. Specific enzymatic activities of the recovered enzymes in each cycle

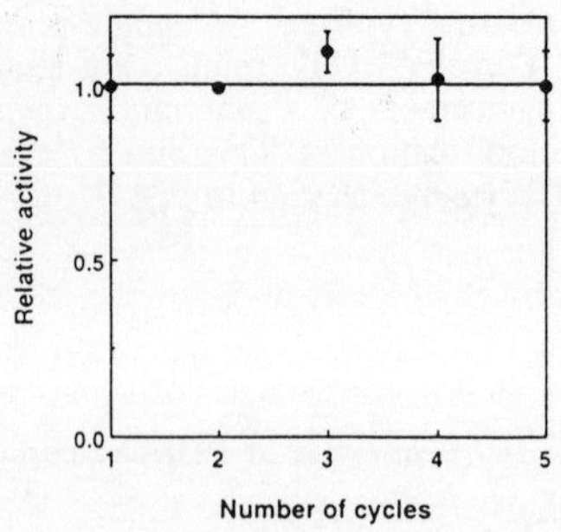

Fig. 8 Relative enzymatic actyivity of IDc-Lipase resued in repeated cyclic dissolution-precipitation lipolytic assays (n=3, +SEM).

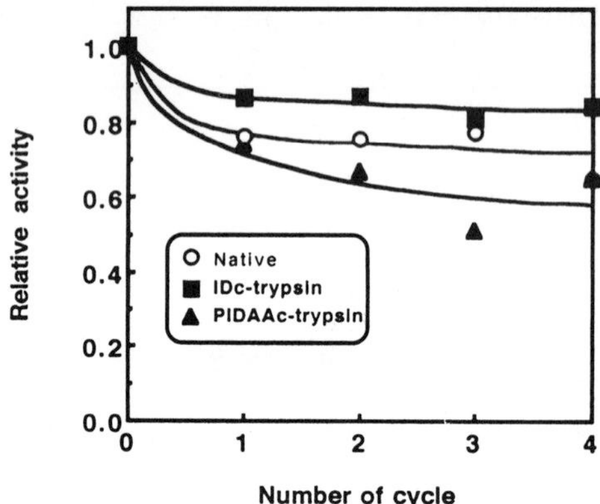

Fig. 9 Relative enzymatic activity of IDc- or PIAAc-modified Trypsines in repeated cyclic changes of temperarures below and above LCST(50°C - 4°C).

are shown in Fig. 8 for the IDc Lipase and Fig. 9 for the IDc Trypsine, respectively. It is seen in these Figs. 8 and 9 that the enzymatic activities of the IDc-modified Lipase or Trypsine remained almost constant and no significant denaturation of the enzymes occurred to decrease the enzymatic activities by repeated cycles of temperature changes.

Rapid hydration-dehydration changes of the incorporated OIPAAm onto the enzymes resulted in the solubility changes of the modified enzymes which did not change the enzymatic activities. Thus, novel recycling systems of enzymes were established through the happy marriage between synthetic and natural polymers.

CONCLUSION

Enzymes such as Lipase and Trypsine were chemically bound with temperature-responsive PIPAAm which showed hydration-dehydration phenomena at LCST, resulting in solubility changes of the modified enzymes. Thus, novel recycling systems of enzymes were established and new routes for novel bioreactors were open for further progress of biotechnology.

REFERENCES

1. M.Heskins, J.E.Guillet, E.James, J. Macromol. Sci., Chem., A2, 1441-1455(1968).
2. Y.H.Bae, T.Okano, R.Hsu, and S.W.Kim, Makromol. Chem., Rapid Commun., 8, 481-485(1987).
3. Y.H.Bae,T.Okano, R.Hsu and S.W.Kim, J. Polym.Sci., Polym.Phys., 28, 923-936(1990).
4. T.Okano, N.Yui, M.Yokoyama, and R.Yoshida, Japanese Technol. Reviews Section E: Biotechnology, Vol. 4, No. 1, Gordon and Breach Science Publishers SA, Yverdon, Switzland.
5. T. Okano, N.Yamada, H.Sakai, and Y.Sakurai, J. Miomed. Mater. Res., 27, 1243-1251(1993).
6. N.Yamada, T.Okano, T.Sakai, F.Karikusa, Y.Sawasaki, and Y.Sakurai, Makromol. Chem.,Rapid Commun., 11, 571-576(1990).
7. L.C.Dong and A.S.Hoffman, J. Controlled Release, 4, 223-227(1986).
8. E.Kokufuta and T.Tanaka, Macromol., 24, 1605-1607(1991).
9. Y.G.Takai, T.Aoki, K.Sanui, N.Ogata, T.Okano, and Y.Sakurai, Bioconjugate Chem., 4, 42-46(1993).
10. Y.G.Takai, T.Aoki, K.Sanui, N.Ogata, T.Okano, and Y. Sakurai, Bioconjugate Chem., 4, 341-346(1993).
11. Y.G.Takai, T.Aoki, K.Sanui, N.Ogata, Y.Sakurai, and T. Okano, Macromol., 27, 6163-6166(1994).
12. M.Matsukata, Y.Takai, T.Aoki, K.Sanui, N.Ogata, Y.Sakurai, and T.Okano, J. Biochem., 116,, 682-686(1994).

SOLID STATE BATTERIES FROM CHITOSAN BASED BIOPOLYMERS

A. K. Arof,[1] R. H. Y. Subban[2] and S. Radhakrishna[2]

[1]Physics Division
Centre for Foundation Studies in Science
[2]Institute of Advanced Studies
University of Malaya
59100 Kuala Lumpur
MALAYSIA

ABSTRACT

Chitosan is a biopolymer which can provide some ionic conductivity when dissolved in acetic acid. The conductivity is due to the presence of protons from the acetic acid solution. The conducting chitosan film was prepared by the solution cast technique. Several films were prepared and each film was complexed with different amounts of hydrated copper sulphate crystals. The undoped film is colourless and films with more than 1.5 g of hydrated copper sulphate crystals added to it contain whitish spots indicating the possibility of exceeding the solubility limit of the copper sulphate crystals into the chitosan matrix. The electrical conductivity at room temperature stops increasing after more than 2.0 g of copper sulphate crystals have been added to the chitosan-acetic acid solution. The copper doped film with the highest electrical conductivity of $\approx$ 7 μS/cm was used to fabricate some thin film solid state batteries using various anode materials and cathode intercalation compounds. In the ($Zn + CuSO_4$)/chitosan film/(PbO_2) battery, an open circuit voltage (OCV) of 1.668 V was obtained. From this OCV value, it can be deduced via the electromotive force method that the chitosan film is a mixed conductor. Internal resistance measurements and discharge characteristics of the batteries fabricated will be presented.

INTRODUCTION

Chitosan was first discovered when Hope-Seyler fused chitin with sodium hydroxide at 453 K in 1894. The product obtained contained less acetyl functional group and Hoppe-Seyler called this product chitosan[1]. This was followed by Araki in 1895, von Furth and Fusso in 1906 and Rigby in 1936. Chitosan also occurs in plant cell walls and in living systems such as mycellia and sporangiophores of Phycomyces blakesleeanus. The

Polymers and Other Advanced Materials: Emerging Technologies and Business Opportunities. Edited by P. N. Prasad *et al.*, Plenum Press, New York, 1995

first such observation was made by Kreger in 1954 and not long after this, it was discovered that a lot of chitosan was found in Mucor rouxii[2]. By the mid-seventies, it was established that chitosan refers to a material in which the acetyl content has been diminished. Nowdays, chitosan is generally referred to as β 1-4 linked glucosamine polymer.

Chitosan is insoluble in water. This poses a problem in the fabrication of solid state proton conducting polymer batteries because there will not be any water present in the chitosan which can act as a source of hydrogen ions. In other words, the proton conducting polymer needed for solid state battery application cannot be obtained from chitosan alone. Chitosan is soluble in formic acid, acetic acids and in 10% citric acid. The ability of, say, acetic acid to dissolve chitosan enables the formation of chitosan film by the solution cast technique which will contain protons contributed by the acid and water. We refer to this film as "acetylated chitosan" film and is by no means to be considered synonymous to the "acetylation of chitosan" where the acetyl functional group is incorporated into the chitosan matrix. The transport of these protons is thought to occur through the many microvoids in the polymer, since the dielectric constant values from piezoelectric property studies are small suggesting the polymer struture to contain many microvoids[3].

EXPERIMENTAL

Sample Preparation

The chitosan membranes were obtained from the Chemistry Department, National University of Malaysia. 1 g of the chitosan membrane was dissolved in 1% acetic acid solution (100 ml) and stirred continuously until all the chitosan has dissolved. This was then poured into a petri dish and the contents were left to cool and dry for about 7 to 10 days. In membranes which were doped, the dopant salt was added when the chitosan has compeletely dissolved in the acetic acid solution after which the solution was cast as before. Two types of chitosan membranes were used and they were differentiated by means of their colours; whitish and yellowish. This difference is probably due to their molecular weights.

Electrical Conductivity Measurement

The electrical conductivity of the samples was measured using the HIOKI 3250 LCR Hi-tester operated in the frequency range between 40 Hz to 100 kHz. The bridge was inter-faced to a microcomputer via an IEEE 488 multifunction card. The sample was placed between two stainless steel electrodes under a pressure of 100 bar.

Battery Fabrication

The dried blue coloured acetylated chitosan film was cut into pieces of about 1.5 cm X 1.5 cm. 0.03 g Zn powder and x g $CuSO_4.5H_2O$ (x = 0.01 to 0.09) was mixed with conductive carbon cement and used as the anode of the battery. The cathode consists of a mixture of conductive carbon cement and 0.05 g PbO_2 which is an intercalating compound.

Battery Characterization

The methods of measuring the open circuit voltage (OCV), internal resistance, transference number and discharge characteristics at constant current has been described elsewhere[4-6].

RESULTS AND DISCUSSION

The acetylated chitosan film used for battery fabrication in this work has an electrical conductivity greater than 1 μS/cm. The electrical conductivity is low and the addition of a divalent copper ion does not contribute much to the electrical conductivity. Moreover, chitosan has a high nitrogen content[2]. The nitrogen electrons play a prominent role in the copper ion fixation to the polymer. This can further reduce the ionic conductivity of the sample. However copper sulphate is preferred here over some of the colourless monovalent dopants available since the bluish colour of the doped acetylated chitosan film confirms the incorporation of copper inside the film.

The half cell reaction should be $H_2 + (1/2)O_2 \rightarrow H_2O$. If this is true then the theoretical OCV which is given by

$$OCV = \frac{-\Delta G}{zF}$$

where z is the valency of the conducting species, in this case, H^+ and F is Faraday's number will be 2.448 V since the Gibb's free energy of formation ΔG of H_2O is 56.43 kcal/mole. The measured OCV of 1.668 V therefore indicates that the ionic transference number is $1.668/2.448 \approx 0.68$ and that the doped acetylated chitosan film is a mixed conductor. Being a mixed conductor, the electronic conductivity must also be very high and the electronic transference number is $(1 - 0.68) \approx 0.32$.

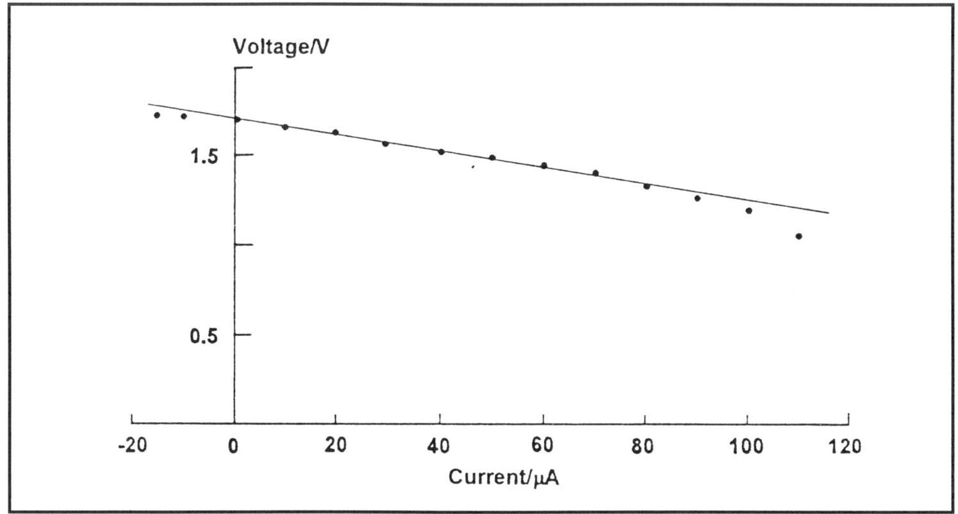

Figure 1. Voltage-current characteristics of the $(Zn + CuSO_4.5H_2O)$/doped acetylated chitosan film/PbO_2 battery.

The high electronic transference number should make the doped acetylated chitosan film less favourable for battery applications since internal short-circuiting can occur very easily. This can reduce the discharge life of the battery. Figure 1 shows voltage-current variation for the (Zn + $CuSO_4.5H_2O$)/doped acetylated chitosan film/PbO_2 battery. The intercept at the voltage axis will also give the OCV and the gradient of the graph will give the internal resistance of the battery. By the method of least squares, the OCV is 1.704 V and the internal resistance of the battery is ≈ 5 kΩ. Figure 2 displays the discharge characteristics of the cell when a current of 20 μA is drawn from the battery. The doped acetylated chitosan film was prepared from the whitish chitosan membrane.

It is to be noted that in our battery, we did not use any hydrogen storage material such palladium hydride. This could also explain why the discharge life is short.

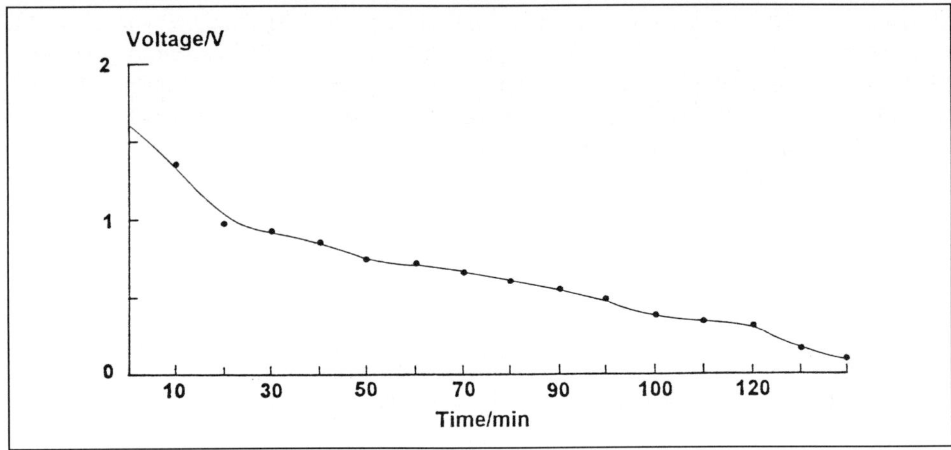

Figure 2. Discharge characteristics of chitosan battery at 20 μA (prepared from the whitish chitosan membrane).

The characteristics of batteries produced from the doped yellowish chitosan membranes together with their electrodes are as shown in Table 1.

Table 1. Batteries from acetylated chitosan films prepared from a more yellowish chitosan membrane.

$CuSO_4$ content (grams)	OCV (mV)	internal resistance (kΩ)	intercalation compound
0.01	767	45	PbO_2
0.03	830	58	PbO_2
0.05	750	95	PbO_2
0.07	630	138	PbO_2
0.09	815	89	PbO_2
0.03 *	632	125	PbO_2
0.03 *	5	-	V_2O_5

Zinc powder was used in the cathode of all the batteries in the above table except for the batteries marked *, copper powder was utilised. In the above investigation the battery with lowest internal resistance is that utilising 0.03 g Zn and 0.01 g $CuSO_4$ as the anode and 0.05 g PbO_2 as its cathode. The discharge characteristics curve is as shown in Figure 3.

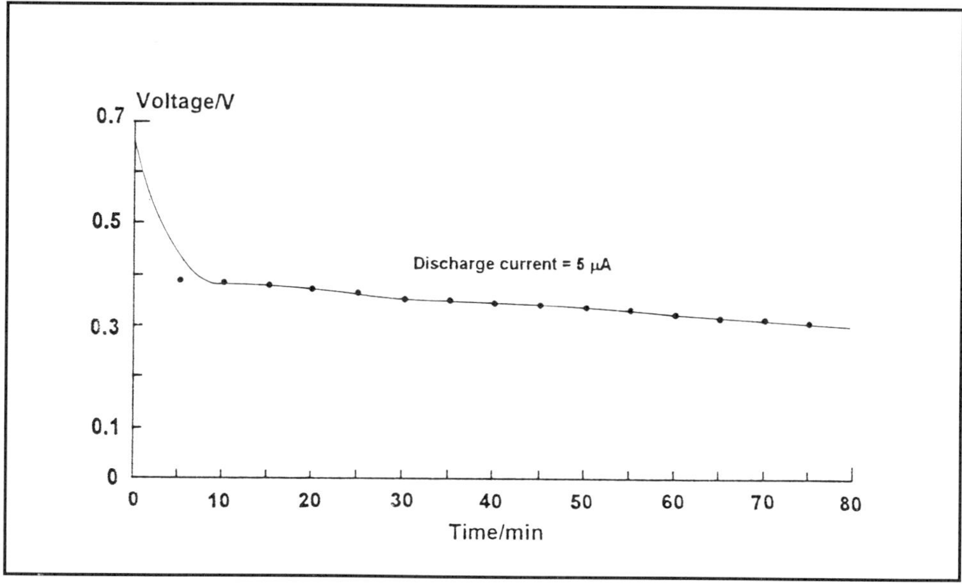

Figure 3. Discharge of battery with 0.03 g Zn and 0.01 g $CuSO_4.5H_2O$ crystals in the anode and 0.05 g PbO_2 in the cathode.

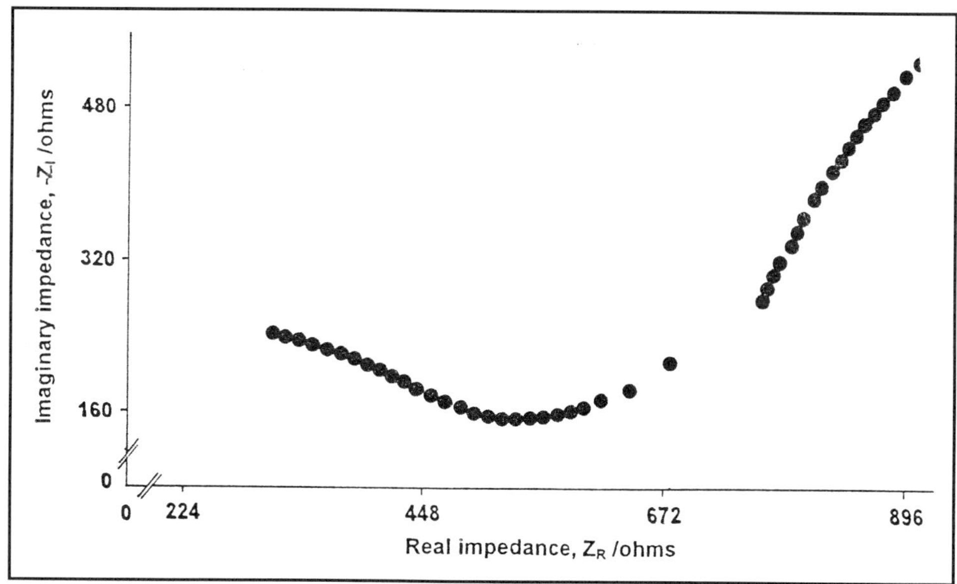

Figure 4. Complex impedance plot for a doped acetylated chitosan film.

The impedance plot showing the bulk resistance of the doped acetylated chitosan film is as depicted in Figure 4. As a means of comparison, the battery Na/(PEO + $NaYF_4$)/I_2 + C + electrolyte gives an OCV of 2.4 V[7]. This battery was discharged under a load of 10 MΩ. On discharge the OCV dropped to 1.25 V and the current drawn was 3.5 X 10^{-6} A. From first principles, the internal resistance of this battery is ≈ 329 kΩ which is comparable to that of the chitosan batteries in the present work.

CONCLUSION

We have shown that with some suitable means, we are able to fabricate solid state polymer batteries from chitosan. The choice of a more suitable electrode material may produce a better battery system.

Acknowledgements

R. H. Y. Subban thanks the MARA Institute of Technology for study leave which enabled this work to be carried out and to the University of Malaya for the vote PJP 242/94.

REFERENCES

1. R.A.A. Muzarelli. "Natural Chelating Polymers," Pergamon Press, Oxford (1973).
2. R.A.A. Muzarelli, A. Ferrero and M. Pizzoli, Talanta, 21:1137 (1974).
3. R.A.A. Muzarelli. "Chitin," Pergamon Press, Oxford (1977).
4. A.K. Arof, *J. Power Sources*. 45:250 (1993).
5. A.K. Arof, and S. Radhakrishna, Mater. Sci. and Eng. B20:251 (1993).
6. S. Chandra. "Superionic Solids: Principles and Application," North-Holland, Amsterdam (1981).
7. S.S. Rao, M.J. Reddy, K.N. Reddy, and U.V.S. Rao, Solid State Ionics, 74:225-228 (1994).

SYNTHESIS AND BIOLOGICAL ACTIVITIES OF ENDO-3,6-EPOXY-1,2,3,6-TETRAHYDROPHTHALIMIDE AND ITS POLYMERS

Gi Cheol Kim[1], Jin Gon Jeong[1], Neung Ju Lee[2], Chang Sik Ha[1] and Won Jei Cho[1]

[1]Department of Polymer Science and Engineering. Pusan National University, Pusan 609-735, Korea
[2]Department of Premedical Sciences, Kosin University, Pusan 602-735, Korea

INTRODUCTION

The high molecular weight antitumor compounds(HMAC) are attracting much interest from the standpoints of polymers synthesis as well as drug development, because of their higher specificity of action, lower toxic side effects and longer duration of drug action as compared with low molecular weight antitumor compounds(LMAC).[1-9]

It has been known that many polyanions exhibit antitumor activities. Among the polyanionic polymers, the 1:2 regularly alternating copolymer(DIVEMA) of divinyl ether(DVE) and maleic anhydride(MAH), first reported by Butler, has been extensively studied for its structure and its antitumor activity. DIVEMA has been shown to possess antitumor, antiviral, antibacterial, and antifungal activities as well as interferon inducing ability. However, it has also toxicity such as enlarged liver and spleen. Afterwards, many attempts have been made to obtain DIVEMA-like HMAC to reduce side effects as well as to enhance the antitumor activity. We have synthesized and characterized homopolymers and copolymers of exo-3,6-epoxy-1,2,3,6-tetrahydrophthalic anhydride(ETA)[10], exo-3,6-epoxy-1,2,3,6-tetrahydro phthalic glycinyl maleimide(ETGI)[11], N-glycinyl maleimide(GMI)[12,13].

In this study, monomeric endo-3,6-epoxy-1,2,3,6-tetrahydrophthalimide(ETPI) was synthesized by reacting maleimide and furan in diethylether. Poly(ETPI) and poly(ETPI-co-acrylic acid) were prepared by the photopolymerization. Synthesized ETPI and polymers were identified by IR and ^{1}H-NMR spectra. The composition of the copolymer containing ETPI was determined by elemental analysis. *In vitro* cytotoxicities of prepared ETPI and polymers containing ETPI moiety were evaluated against Fibrolast cells and K-562 human leukemia cells as a target cell.

In vivo antitumor activities of prepared ETPI and polymers containing ETPI moiety were investigated using mice bearing Sarcoma 180 tumor cells.

EXPERIMENTAL

Materials and instruments

Maleimide(MAI, Aldrich Co.), furan(Aldrich Co.), vinylacetate(VAc, Junsei chem.), acrylic acid(AA, Junsei chem.), maleic anhydride(MAH, Junsei chem.) and dimethoxy benzoin(DMB) were purified by conventional methods. 5-fluorouracil (5-FU, Aldrich Co.) was used without further purification. Fibroblast cells, K562 human leukemia cells, BALB/c mice and Sarcoma 180 as tumor cell were purchased from the Center of Genetic Engineering, Korea Institute of Science and Technology(KIST).

IR spectra were taken on a Jasco FT/IR-5300 spectrophotometer using KBr disk. ^{1}H-NMR spectra were recorded on a FT-300 MHz Bruker A-3000 spectrophotometer. Elemental analysis was made on a Carlo Erba Instruments Model EA 1108 elemental analyzer.

Synthesis of ETPI and polymers containing ETPI

Furan and MAI were dissolved in diethylether and stirred at room temperature for 2 days. The white precipitate was filtered and washed with diethyl ether. The melting point of ETPI was 122℃.

ETPI and dimethoxy benzoin(DMB) were dissolved in the 2:1(v/v) mixture of 2-butanone and acetone and introduced into a dry pyrex polymerization tube and placed in photochemical chamber reactor using λ_{max}=313nm lamps at 25±0.1℃ for 12 hours. The polymer solution was precipitated in diethylether. Precipitated poly(ETPI) was collected by filtration, washed with diethylether and dried under vacuum until it reached to a constant weight.

Poly(ETPI-co-AA) was prepared by the copolymerization of ETPI and AA. The copolymerizations was carried out by the same procedure as that of poly(ETPI).

In vitro cytotoxicity test

Fibroblast and K562 human leukemia cell line were used as target cell for *in vitro* cytotoxicity test. Cells were continuously cultured in RPMI1640 medium. 3×10^5 cells of target cell line and distilled water solutions of compounds were mixed, then placed in 96-well flat-bottomed plates and cultured for 60 hours in a CO_2 incubator at 37℃. After adding 0.1ml of trypan blue (0.4%), cell numbers were counted using a hemocytometer.

In vivo cytotoxicity test

To evaluate the antitumor activities of ETPI and polymers, mice bearing Sarcoma 180 tumor cells are used. BALB/c mice were first intraperitonearly(i.p.) implanted with Sarcoma 180 cells(2×10^5). The animals were then treated with the saline of sample on days 1-4. Three different dosages were tested : 0.8, 80 and 800 mg/kg.

For comparison, the antitumor activity of free 5-FU was also tested by the same method. A control group was divided into two groups : one group was treated with Sarcoma 180 cells along with the same volume of saline and the other was treated with only Sarcoma 180 cells. Each group was consisted of 10 animals.

RESULTS AND DISCUSSION

Characterization

The structure of ETPI was identified by IR and ^{1}H-NMR spectra. The IR spectrum of ETPI shows characteristic absorption bands at 3210cm^{-1}(NH), 2790 cm^{-1}(CH), 1716cm^{-1}(C=O), 1564cm^{-1}(CH=CH) and 1188cm^{-1}(C-O-C). In the ^{1}H-NMR spectrum of ETPI, the absorptions due to two methine protons of imide ring, two olefinic protons and two methine protons in cyclic ether group and imide hydrogen were shown at 3.4 ppm, 6.5 ppm, 5.2 ppm and 9.6 ppm, respectively.

^{1}H-NMR spectrum of poly(ETPI) shows methine protons of imide ring at 3.4 ppm, methine protons in polymer backbone at 2.9 ppm and methine protons of cyclic ether at 5.2 ppm and the absorption assingnable to the olefinic protons of ETPI monomer is not observed at 6.5 ppm.

The structure of poly(ETPI-co-AA) were also identified by its ^{1}H-NMR spectrum.

Solubility of ETPI and polymers containing ETPI moiety

Solubilities of ETPI, poly(ETPI) and poly(ETPI-co-AA) were listed in Table 1. ETPI and polymers containing ETPI moiety were soluble in polar solvents such as water, DMF, DMSO and methanol but insoluble in diethylether and chloroform.

Table 1. Solubility of ETPI and polymers containing ETPI moiety[1]

Solvent	ETPI	Poly(ETPI)	Poly(ETPI-co-AA)
Water	○	○	○
Dimethyl sulfoxide	○	○	○
N,N-Dimethylformamide	○	○	○
Methanol	○	△	○
Acetone	○	×	○
2-Butanone	△	×	×
Tetrahydrofuran	○	×	×
Ethylacetate	○	×	×
Chloroform	×	×	×
Diethyl ether	×	×	×
Benzene	×	×	×
n-Hexane	×	×	×

[1]○ : Soluble, △ : partially soluble, × : insoluble

Composition of polymers containing ETPI

The content of ETPI unit in poly(ETPI-co-AA) was determined 35% by elemental analysis. Composition of polymers containing ETPI was listed in Table 2.

Table 2. Composition of polymers containing ETPI.

Sample	Composition (%)			ETPI fraction (%)
	C	N	H	
ETPI	58.8	8.6	4.4	
Poly(ETPI)	57.6	6.9	5.7	
Poly(ETPI-co-AA)	51.0	4.4	5.4	35.4

In vitro cytotoxicity test

Cytotoxicities of ETPI and polymers containing ETPI moiety against Fibroblast cell(Normal Cell) and K562 human leukemia cell were shown in Figure 1 and 2. The cytotoxicities against Fibroblast cell(Normal Cell) and K562 human leukemia cells increased with the increase of the sample concentration. ETPI and polymers show considerable cytotoxicities at higher concentrations(5.0, 1.0, 0.1 mg/ml). The order of cytotoxicity against Fibroblast cells(Normal Cell) at 0.1 mg/ml is as follows : poly(ETPI-co-AA) > ETPI > poly(ETPI). The order of cytotoxicity against K562 human leukemia cells at 0.1mg/ml is as follows : ETPI >> poly(ETPI-co-AA) > poly(ETPI).

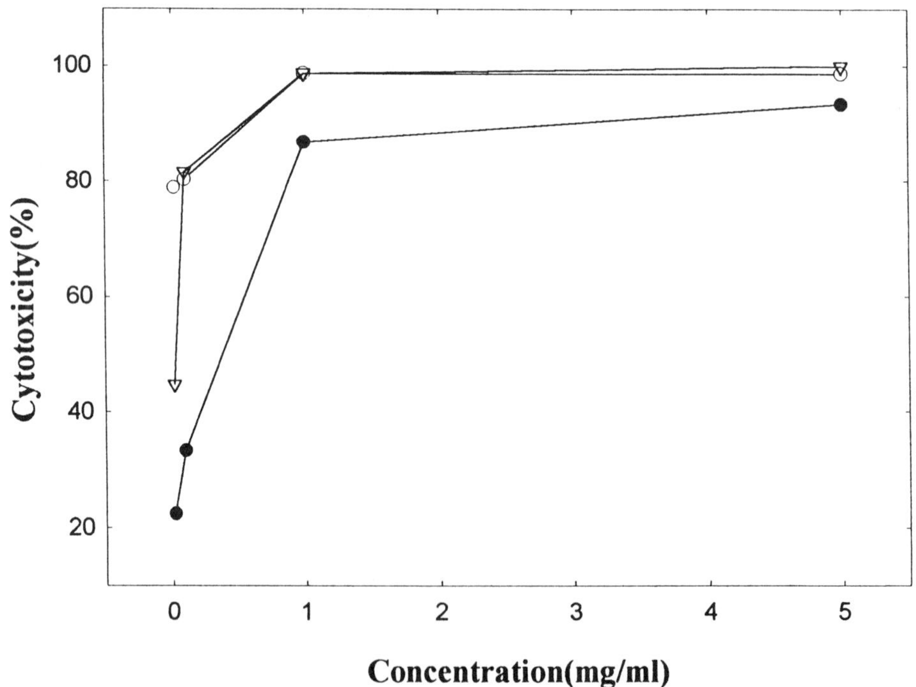

Figure 1. Cytotoxicity of ETPI and polymers containing ETPI moiety against Fibroblast cell(Normal Cell) ; ETPI[○], poly(ETPI)[●], poly(ETPI-co-AA)[▽].

In vivo cytotoxicity test

The results of antitumor activities of ETPI, polymers containing ETPI moiety, 5-FU and control group against Sarcoma 180 cells were showed in Figure 3. The life span of mice treated with poly(ETPI) at the concentrations of 800 and 80 mg/kg was increased by 229% and 195% as compared with the control group but 5-FU was increased by 50% and 140% in the same condition. The order of antitumor activity at a dosage of 80mg/kg is as follows : ETPI > poly(ETPI) > poly(ETPI-co-AA) > 5-FU > Control group.

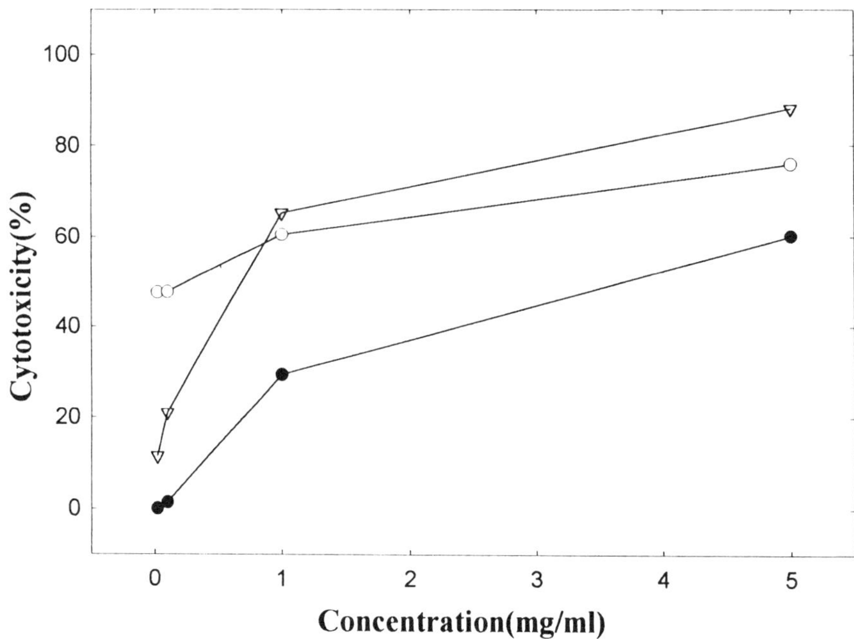

Figure 2. Cytotoxicity of ETPI and polymers containing ETPI moiety against K562 human leukemia cell line ; ETPI[○], poly(ETPI)[●], poly(ETPI-co-AA)[▽].

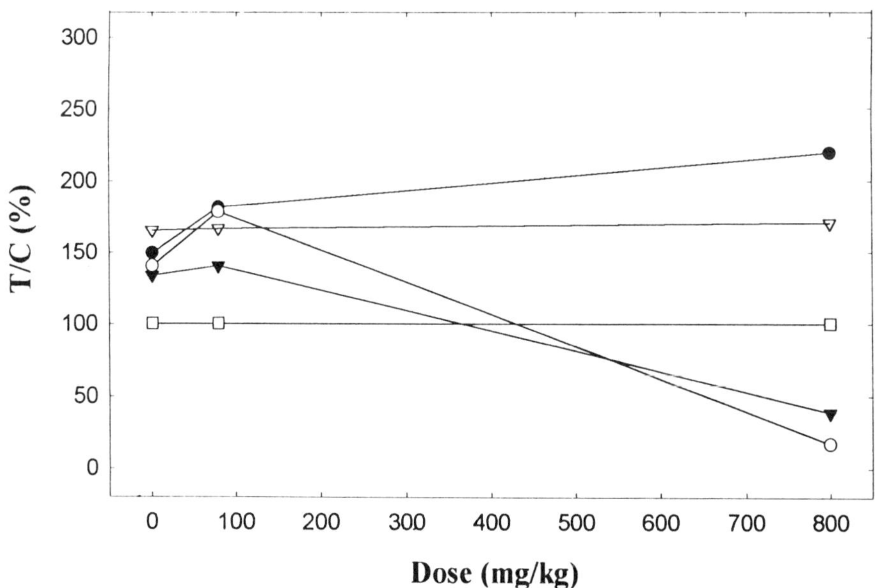

Figure 3. Antitumor activity of ETPI, polymers containing ETPI moiety, 5-FU and control against Sarcoma 180 ; ETPI[○], poly(ETPI)[●], poly(ETPI-co-AA)[▽], 5-FU[▼] and control group[□].

CONCLUSIONS

ETPI was synthesized by Diels-Alder reaction of furan and maleimide in diethylether at room temperature. Poly(ETPI) and poly(ETPI-co-AA) were prepared by photopolymerization at room temperature and their structures were identified by IR and ^{1}H-NMR spectroscopies. ETPI fraction in poly(ETPI-co-AA) was 35.4%. Solubilities of ETPI, poly(ETPI) and poly(ETPI-co-AA) are soluble in water, DMF, DMSO and methanol. The order of cytotoxicity against Fibroblast cells(Normal Cell) at a dosage of 0.1 mg/ml is as follows : poly(ETPI-co-AA) > ETPI > poly(ETPI). The order of cytotoxicity against K562 human leukemia cells at a dosage of 0.1mg/ml is as follows : ETPI >> poly(ETPI-co-AA) > poly(ETPI). *In vivo* antitumor activity of synthesized polymers was greater than that of 5-FU at a dosage of 80 mg/kg. The order of antitumor activity at a dosage of 80mg/kg is as follows : ETPI > poly(ETPI) > poly(ETPI-co-AA) > 5-FU > Control group.

REFERENCES

1. S.Ozaki,Y.Ike,H.Mizuno,K.Ishikawa and H.Mori, *Bull.Chem.Soc.Japan*,50,2406 (1977).
2. P.P.Umrigar,S.Ohashi and G.B.Butler, *J.Polym.Sci.,Polym.Chem.Ed.*,17,351(1979).
3. M.Akashi,K.BEPPU,I.Kikuchi. and N.Mizauchi, *J.Mcromol.Sci-Chem.ED.*,17,351 (1986).
4. R.M.Ottenbrite, *J.Mcromol.Sci-Chem.*,A22(57),819(1985).
5. T.Ouchi,H.Juzei,Y.Sakamoto,H.Chikachita and O Vogl, *J.Polym.Sci.,Polym.Chem.* Vol.23,2059(1986).
6. R.M.Ottenbrite,J.Sunamoto,T.Sato,K.Kojima,K.Sahara,K.Hara,M.Oka, *J.Bioact.& Compati.Polym.*3,184,(1988).
7. G.Butler, *J.Macromol.Sci-Chem.*,A13(3),351(1979).
8. N.J.Lee,C.S.Ha, and W.J.Cho, *J.Macromol.Sci-Chem.*,29,2,161(1992).
9. N.J.Lee,C.S.Ha, and W.J.Cho, *Bioact.& Compati.Polym.*7,39(1992).
10. J.G.Jeong,N.J.Lee,C.S.Ha and W.J.Cho, *Proc. 3rd Pacific Polymer Conf,Gold Coast, Australia*, Dec.13-17,pp.75-76(1993).
11. D.Y.Lee,N.J.Lee,J.G.Jeong,C.S.Ha and W.J.Cho, *The Society of Polymer Science, Japan*, 797page10/12-10/14.(1994).
12. G.T.Gam,J.G.Jeong,N.J.Lee,C.S.Ha and W.J.Cho, *J.Appl.Polym.Sci.* in print(1995).
13. G.T.Gam,J.G.Jeong,N.J.Lee,C.S.Ha and W.J.Cho, *IUPAC International Symposium on Functional and High Performance Polymers, Taipei, Taiwan*, Nov. 14-16,p.(1994).

ACKNOWLEDGEMENT

The authors are grateful for the financial support of the Korea Science and Engineering Foundation(Grant No. 931-0300-014-2).

BIOTECHNOLOGICAL APPLICATION OF ENZYMES FROM EXTREMOPHILIC ORGANISMS: SYNTHESIS OF MODIFIED MONOMERS

Gianfranco Peluso[1], Antonio Trincone[2], Francesco La Cara[1], Francesco Rosso[1], and Mosè Rossi[1]

[1]Institute of Protein Biochemistry and Enzymology - C.N.R.
[2]Institute for the Molecules of Biological Interest - C.N.R.

Via Toiano 6, 80072 Arco Felice, Naples, Italy

INTRODUCTION

In the last few years, the use of enzymes for industrial purpose has revealed a rapid growth owing to the advantages they confer to conventional chemical methods. For example biocatalysts are highly specific and efficient and are able to produce chiral compounds.

In particular, enzymes obtained from microorganisms already have numerous industrial applications, although till now their application fields are restricted by peculiar reaction conditions. In fact, these technological processes frequently involve exposure to extremes of temperature, pressure, ionic strenght, pH and organic solvents, and hence there is a continuing need to isolate microorganisms/enzymes which are able to function and catalyse specific reactions under these imposed conditions. Organisms able to growth in a diverse range of environments, considered extreme by man, are normally found in nature. The term extremophiles has been used to describe them and within this grouping are included thermophiles, psychrophiles, acidophiles, alkalophiles, halophiles, osmophiles, barophiles, radiation resistant and heavy metal tolerant organisms.

The object of this review is to summarize the potential biotechnological applications of thermophiles.

Extremophilic microorganisms

Temperature is probably the most important environmental factor affecting the activity and evolution of living microorganisms. Indeed the most part of microorganisms that have been studied exhaustively live at moderate temperatures and operate within a relatively narrow span of temperatures.

Polymers and Other Advanced Materials: Emerging Technologies and Business Opportunities. Edited by P. N. Prasad *et al.*, Plenum Press, New York, 1995

Generally, the term "thermophile" is used to describe those microorganisms with a growth temperature between 60 and 80 °C and the term "extreme thermophile" to describe those with T_{opt} higher than 80 °C[1,2].

In the recent year the general taxonomy of living organisms has been revised and now there is a general agreement in dividing living organisms in three groups: Bacteria, Archaea and Eukarya[3].

The new evolutionary line of Archaea is formed of three major phenotypes: extreme halophiles, methanogenes and thermophiles, each of which lives in environments that would normally kill all other organisms.

Thermophiles and extreme thermophiles can be either Bacteria or Archaebacteria but the archaebacterial phenotype seems to be the ancestral one.

In fact many of their niches, which appear extreme in terms of actual terrestrial conditions, might seem even normal in a warm primeval earth with a reducing atmosphere.

Life in such extreme environments is linked to the presence of exceptionally thermostable macromolecules such as lipids, proteins and nucleic acids. Such thermostability can be, however, intrinsic or gained by interactions with other macromolecules as in the case of DNA that is stabilized by the interaction with special DNA binding proteins[4].

Thermophilic organisms as a source of thermostable enzymes

Many attention has been focused on enzymes isolated from extreme thermophiles since we demonstrated that these enzymes are stable to temperature (thermostable) and are capable to act at high temperature (thermophilic)[5]. In addition they show enhanced stability in the presence of the common protein denaturants such as SDS, urea, guanidine hydrochloride, proteolytic enzymes and organic solvents.

Enzymes from extremophiles represent a new class of molecules with peculiar properties. Since the potential exploitation of proteins for practical purposes is one of the goal in many applications, the availability of extremophiles as a source of enzymes stable and active at high temperature and in adverse chemical environments, is an important development in the area.

However the general stability of these enzymes should confers a number of advantages in terms of industrial applications. In fact, in addition to longer lifetime (Table 1), the possibility to operate with enzyme reactors at high temperature will increase substrate solubility and diffusion rates and reduce viscosity of the medium and bacterial contamination. In the case of reactors and membrane reactors it would be easier to clean the enzyme by using, for example, urea at a certain concentration (2-4 M) and acetone.

Proteinases, cellulases, amylases, glycosidases, esterases, DNA polymerases, alcohol dehydrogenases, are the thermophilic enzymes more suitable for potential biotechnological exploitation. Cellulases, hemicellulases and amylases are a group of enzymes with several potential industrial applications since cellulose, hemicellulose and starch represent a large renewable resource from which many chemicals, including fuel and food, can be derived.

Table 1. Thermal stability of some thermophilic enzymes (adapted from Ref. 2).

Enzyme	Organism	Half Life
Proteinase	*Thermus sp.Rt41A*	13,5 h 80°C
Proteinase	*Thermococcus celer*	45' 95°C
Proteinase	*Desulfurococcus sp*	1.5 h 95°C
Proteinase	*Pyrococcus furiosus*	33 h 98°C
Endocellulase	*Clostridium thermocellum*	5 h 70°C
Endocellulase	*Acidothermus cellulolyticus*	2 h 85°C
Endocellulase	*Caldocellum saccharolyticum*	1.5 h 85°C
Exocellulase	*Thermotoga sp.*	11 h 95°C
Glycosidase	*Sulfolobus solfataricus*	2 h 95°C
Amylase	*Sulfolobus solfataricus*	> 1h 95°C
Amylase	*Desulfurococcus sp.*	> 1h 95°C
DNA polymerase	*Thermus aquaticus*	40' 95°C
DNA polymerase	*Thermococcus litoralis*	1 h 100°C
DNA polymerase	*Sulfolobus solfataricus*	30' 85°C
Esterase	*Sulfolobus acidocaldarius*	1 h 100°C
Esterase	*Bacillus acidocaldarius*	30' 85°C
Alcoholdehydrogenase	*Thermoanarobium broki*	2 h 85°C
Alcohol deydrogenase	*Sulfolobus solfataricus*	2 h 85°C

Several processes may employ thermostable cellulases, hemicellulases and amylases in the future. These include the production of liquid fuels from renewable biomass, food and fruit juice processing, production of HFS, paper manufacture and laundry detergency[6].

However, though enzymes from thermophiles seem well-suited for industrial applications, until now only a few of them have found application. Exceptions are the use, for example, of thermolysin in the industrial synthesis of the sweetener aspartame, and of DNA polymerase from *Thermus aquaticus* in the polymerase chain reaction[7]. The reluctance to change existing processes and the difficulty to economically produce thermophilic enzymes on industrial scale are the keys reasons of the few applications.

The only methodology for mass production of these enzymes lies in the cloning and overexpression of their genes in mesophilic hosts, which should be generally recognized as safe (GRAS) and characterized by high biomass yield and optimized fermentation conditions.

In the attempt to set up a general methodology for the production of archaebacterial enzymes, several enzymatic activities from *Sulfolobus solfataricus*, such as a β-glycosidase, a DNA polymerase, an alcohol dehydrogenase have been cloned and expressed either in *E.coli*, yeast and human cells[8-11].

The unique properties of thermophilic enzymes especially those regarding their specificity and/or stability are of interest in the field of biocatalysis where solution of specific problems are requested. Few examples are reported in literature for the application of thermophilic enzymes in this field[12,13].

Glycosidases for the synthesis of glycosidic bond

During the last two decades the scientific literature is growingly rich in examples of applications of biocatalysis for the construction of organic molecules especially regarding to those bearing chiral centers and/or functional groups of similar reactivity. The use of natural catalysts relies on their ability to promote very selective modifications difficult to obtain with chemical based methodologies. The dominance of only few biocatalytic systems, PLE (pig liver esterase), PPL (pig pancreatic lipase) CCL (*Candida cylindracea* lipase) and BY (baker's yeast), with respect to their number indicate a restriction of the potential of this field and show perspectives for the application of new enzymes. Moreover, very recently the use of molecular biology techniques spots the light for the application of modified enzymes and abzymes in biocatalysis opening opportunities for the application of a tailored biocatalyst for a particular modification[14,15].

Chemical syntheses of carbohydrate derivatives although well developed are complicated by the presence of multiple hydroxyl groups of similar reactivity in the starting molecules. However the need for well characterized compounds for biological studies, temporary anomeric protection, production of affinity adsorbent, synthesis of modified monomer for production of hydrophilic polymers etc. has made enzymatic synthesis of these molecules as an alternative and efficient choice.

Apart those enzymes used for direct functionalization of carbohydrates, such as lipases or proteases[16], the transferases are enzymes used by chemists as biocatalyst for the synthesis of glycosidic bond. One great disadvantage of these enzymes is however their limited availability in comparison to lower cost and wider availability of glycosidases. Infact, the biocatalyzed synthesis of glycosides, using different glycosidases, is of current interest due not only to the availability of these enzymes but also to their simple use (i.e. without the need of expensive cofactors).

Glycosidases are enzymes capable of hydrolyzing glycosidic bond, but they are also used to synthesize glycosides using two methods: reverse hydrolysis and transglycosylation. Both these techniques have been used in this field using enzymes from different sources. In the equilibrium controlled approach, a glucose molecule is attached to the acceptor in an inversion of catabolic route of the enzyme (reverse hydrolysis). In the kinetically controlled synthesis a reactive glycosylated enzyme is generated by the hydrolysis of a carbohydrate donor and can be intercepted by the glycosyl acceptor (transglycosylation) or by water (hydrolysis)[15].

The widespread application of this methodology suffers of some drawback such us poor stability of the mesophilic enzymes in the presence of high percentage of organic material in the reaction mixture, a common artifice used to improve yield of transglycosylation. In this context the use of thermophilic enzymes could be of some interest in that these biomolecules, other than thermophilic and thermoresistant are stable in conditions generally prohibitive for their mesophilic counterparts.

Different examples of functionalization of various hydroxyl group by the use of mesophilic glycosidases are reported in literature[15,17].

Obviously the primary goal for the enzymatic synthesis of glycosides is the selectivity of functionalization while keeping wide substrate accessibility into the active site of the enzyme. On the other hand the biological oriented methodology appears to be interesting because it consents the use of substrates with liable hydroxyl groups[15].

Applications of β-glycosidase from *Sulfolobus solfataricus*

Primary and secondary alcohols of different chain lenghts (C1-C8) can be used as acceptors in the transglycosylation reaction using the enzyme from *Sulfolobus solfataricus*. The yield of reaction depends upon chain lenght and OH position but a marked increase can be obtained increasing the molar excess of alcohol. Tertiary alcohols are not substrates[18].

Stereochemistry of β-glycosyl transfer from phenyl-β-galactoside, lactose, and phenyl-β-glucoside to various 1,2-, 1,3- and 1,4-diols, show a good enantioselection for the galactosyl transfer to the secondary hydroxyl group of different 1,2-diols[19]. When using phenyl-β-glucoside and -fucoside as carbohydrate donors this enantioselectivity was almost lost. However primary hydroxyl group is preferred (80%) and no enantioselection was observed for this position for all donors. The selectivity of reaction expressed for the case of 2-methyl pentane-2,4-diol, butane-1,4-diol and *trans*-cyclohexane-1,2-diol and other compounds is a valuable result from a synthetic point of view in that it permits the synthesis of pure β-anomeric form of monoglycosides of polyhydroxyaglycones while avoiding difficult chemical protection-deprotection procedures.

The study of these reactions is also useful to draw differences between the catalytic properties of mesophilic and thermophilic enzymes. Careful stereochemical analysis of products could lead to useful information about the interactions at level of active site of the enzyme (for a more complete discussion see ref. 20).

Immobilised cells has been used to synthesize n-hexyl-β-glucoside recycling the biocatalyst[19].

Examples of preparation of naturally occurring glycosides are the synthesis and stereochemical determination of aleppotrioloside[21] and the use of aromatic aglycones bearing aromatic and aliphatic hydroxyl groups. 3,4'-dihydroxypropiophenone 3-O-β-D-glucoside is a natural glycoside found in *Betula platyphylla* var. japoinica synthesized using various enzymatic sources and different reaction conditions[22,23]. The best yield was obtained using thermophilic β-glycosidase from *Sulfolobus solfataricus*[22].

Synthesis of modified monomers

Since glycosilated molecules are heavily involved in the biological recognition and the stereospecificity of glycosylation process is difficult to achieve with chemical metodologies, the enzymatic modification of polymers or monomers with the introduction of appropriate sugar residues, for their biomedical utilization, appears very promising[24,25].

In particular the use of enzymes isolated from extremophiles for their capability to act in organic solvents and at high temperature represents a new tool in this area. As an example we report on the synthesis of a derivative of hydroxyethylmethacrilate (HEMA) the GalEMA by using liophylized cells of the archeon *Sulfolobus solfataricus* or the purified or recombinant enzyme.

In order to establish the optimal temperature for the enzymatic HEMA galactosylation, the specific activity of the enzyme was evaluated in HEMA containing media at different temperatures. The activity of the cells increased up to 85 °C, as previously reported for the purified enzyme[26].

One of the main factor to be considered in transgalactosylation reactions is the dependence from water content. For this purpose we tested the degree of galactosylation at different HEMA/water ratios and an improvement of the product yield was found decreasing the water content as expected.

In a typical preparative enzymatic synthesis of GalEMA yields of 70-80% of isolated product was obtained.

Another interesting result is the cell recycling. In fact the enzymatic activity half life was detected after nine cycles of synthesis.

The data shown in this paper demonstrated that a high yield of GalEMA can be achieved by the catalyzing effect of the β-glycosidase activity of *S. solfataricus* cells. Studies are still in progress to extend the range of monomers to be modified by the thermophilic glycosidase activity. These efforts, as well as the utilization of other enzymes stable in organic solvents, can provide innovative solutions in the design of new chemical compounds.

REFERENCES

1. D.A. Cowan, Biochemistry and molecular biology of the extremely thermophilic Archaebacteria, *in*: "Molecular Biology and Biotechnology of Extremophiles," R.A. Herbert, and R.J. Sharp, eds., New York: Chapman and Hall (1992).
2. D.A. Cowan, Biotechnology of the Archaea,*TIBTECH* 10:315-323 (1992)
3. C.R. Woese, Bacterial Evolution, *Microbiol. Rev.* 51:221-271 (1987).
4. D.G. Searcy, The archaebacterial histone HTa, in: "Bacterial chromatin," C.O. Gualerzi, and C.L. Pow, eds., Heidelberg: Springer-Verlag (1986).
5. J.K. Kristjansson, Thermophilic organisms as sources of thermostable enzymes, *TIBTECH* 7: 349-353 (1989).
6. K. Peek, L.D. Ruttersmith, R.M. Daniel, H.W. Morgan, and P.L. Bergquist, Thermophilic enzymes as industrial catalysts? *Biotech Forum Europe* 9: 466-470 (1992).
7. R.A. Herbert, A perspective on the biotechnological potential of extremophiles, *TIBTECH* 10: 395-402 (1992).
8. M. Moracci, A. La Volpe, J.F. Pulitzer, M. Rossi, and M. Ciaramella, Expression of the thermostable ß-Galactosidase gene from the Archaebacterium *Sulfolobus solfataricus* in *Saccharomyces cerevisiae* and characterization of a new inducible promoter for heterologous expression, *J. Bacteriol.* 174: 873-882 (1992).
9. F.M. Pisani, C. De Martino, and M. Rossi, A DNA polymerase from the archaeon *Sulfolobus solfataricus* shows sequence similarity to family ß-DNA polymerases, *Nucleic Acids Res.* 20: 2711-2716 (1992).
10. S. Ammendola, C.A. Raia, C. Caruso, L. Camardella, S. D'Auria, M. De Rosa, and M. Rossi, Thermostable NAD^+-dependent alcohol dehydrogenase from *Sulfolobus solfataricus:* gene and protein sequence determination and relationship to other alcohol dehydrogenases, *Biochemistry* 31:12514-12523 (1992).
11. R. Cannio, D. de Pascale, M. Rossi, and S. Bartolucci, Gene expression of a thermostable ß-galactosidase in mammalian cells and its application in assays of eukaryotic promoter activity, *Biotechnol. Appl. Biochem.* 19: 233-244 (1994).
12. A. Trincone, B. Nicolaus, L. Lama, and A. Gambacorta, Potential application of *Sulfolobus solfataricus* as catalyst in organic synthesis, *Indian J. Chem. (Section B)* 32: 25-29 (1993).
13. E. Keinan, S. C. Sinha, and A. Sinha Bagchi, Thermostable enzymes in organic synthesis. 2. Asymmetric reduction of ketones with alcohol dehydrogenase from *Thermoanaerobium Brockii*, *J. Org. Chem.* 57: 3631-3636 (1992).
14. E. Santaniello, P. Ferraboschi, P. Grisenti, and A. Manzocchi, The biocatalytic approach to the preparation of enantiomerically pure chiral building blocks, *Chem. Rev.* 92: 1071-1140 (1992).
15. C-H. Wong, and G. M. Whitesides, Enzymes in synthetic organic chemistry Tetrahedron Organic series Vol. 12, J.E. Baldwin, F.R.S. Magnus and P.D. Magnus, eds., Great Britain Pergamon Press (1994).
16. H. Waldmann, and D. Sebastian, Enzymatic protecting group techniques, *Chem. Rev.* 94: 911-937 (1994).

17. E. J. Toone, E. S. Simon, M. D. Bednarski, and G. M. Whitesides, Enzyme-catalysed synthesis of carbohydrates, *Tetrahedron* 45: 5365-5422 (1989).
18. A. Trincone, B. Nicolaus, L. Lama, P. Morzillo, M. De Rosa, and A. Gambacorta, Enzyme-catalysed synthesis of alkyl-β-D-glycosides with crude omogenate of *Sulfolobus solfataricus*, *Biotechnol. Lett.* 13: 235-240 (1991).
19. A. Trincone, R. Improta, R. Nucci, M. Rossi, and A. Gambacorta, Enzymatic synthesis of carbohydrate derivatives using β-glycosidase of *Sulfolobus solfataricus*, *Biocatalysis* 10: 195-210 (1994).
20. A. Trincone, B. Nicolaus, L. Lama, and A. Gambacorta, Stereochemical studies of enzymatic transglycosilation using *Sulfolobus solfataricus*, *J. Chem. Soc. Perkin Trans I* 2841-2844 (1991).
21. A. Trincone, E. Pagnotta, and G. Sodano, Chemoenzymatic synthesis and stereochemisty of aleppotrioloside, a naturally occurring glycoside, *Tetrahedron Letters* 35: 1415-1416 (1994).
22. A. Trincone, E. Pagnotta, Efficient chemoselective synthesis of 3-4'-dihydroxypropiophenone 3-o-β-D-glucoside by thermophilic β-glycosidase from *Sulfolobus solfataricus*, *Biotechnol. Lett.* 17: 45-48 (1995).
23. K. Mori, Z-H. Qian, and S. Watanabe, Synthesis of 3-4'-dihydroxy-propiophenone 3-*p*-D-glucoside a constituent of *Betula platyphylla*, by enzymatic transglycosilation, *Liebigs Ann. Chem.* 485-487 (1992).
24. A. M. Blinkovski, and J. S. Dordick, Enzymatic derivatization of saccharides and their chemical polymerization,*Tetrahedron: Asymmetry* 6: 1221-1228 (1993).
25. J. S. Dordick, 1992, Enzymatic and chemoenzymatic approaches to polymer synthesis, *TIBTECH* 10: 287-293 (1992).
26. R. Nucci, M. Moracci, C. Vaccaro, N. Vespa, and M. Rossi, Exoglucosidase activity and substrate specificity of the β-glycosidase isolated from the extreme thermophile *Sulfolobus solfataricus*, *Biotechnol. Appl. Biochem.* 17: 239-250 (1993).

BIOLOGIC RESPONSES TO POLYMERS AND OTHER PROSTHETIC DEVICES

Anthony Paul Prezyna, M.D.
Clinical Associate Professor of Pathology and Ophthalmology
State University of New York at Buffalo
Buffalo, NY 14260

INTRODUCTION

Recently, the range of benefit to man and animal in alleviating bodily structural and functional defects has been greatly extended by innovative exogenous materials, chiefly polymers, select metals and ceramics.

The "alchemy" of the chemist continues to generate novel miracles of liquid, solid and gaseous synthetics in ever increasing variety and application. Currently, damaged structures within the body may be repaired or replaced, with significant improvement of function. Virtually every system in the body is now subject to study and/or amelioration by devices stemming from this applied science. Replacement or reinforcement of faulty structures is especially widespread in the cardiovascular, musculoskeletal, and ocular fields.

Attendant upon their use adverse sequelae of acute or chronic nature may ensue. There is heightened risk of infection, bleeding, blood coagulative phenomena, ischemic (low blood flow) damage to contiguous tissue, foreign body inflammatory response or excessive scarring. Some of these sequelae may have serious secondary implication in remote organs. The implant may undergo structural and subsequent functional failure. Finally, idiosyncratic reactions may give rise to allergic or suspect autoimmune phenomena, the latter with multisystem implications. The siliconE liquid implant is a possible example of the latter with sizable litigious consequences.

In the total framework of activity from creation to application of these materials, the shadow of adverse legal consequences will hover in spite of the best efforts of the investigator.

BIOMATERIAL APPLICATIONS AND MORBIDITY

The ever increasing variety of biomaterial applications, burgeoning in the last several decades, is spawning a morbidity and mortality that has been underestimated. Table I lists the general category of the biomaterial.

Table I

BIOMATERIALS
Polymers
Metals
Ceramics
Composites

Polymers and Other Advanced Materials: Emerging Technologies and Business Opportunities. Edited by P. N. Prasad *et al.*, Plenum Press, New York, 1995

The most widely used of these biomaterials is the silicone family of compounds. Recently, great controversy has been generated by claims of increased association with morbid phenomena with weighty litigious consequences. The uses of these materials in man and other animals have been designed for many purposes. They come in forms appropriate to their functional use (Table II).

Table II

FORMS OF BIOMATERIALS
Liquid, Solid, Gel Simple or Combined Resorbable, Removable or Permanent

The biomaterials serve to provide structural replacement or support in tissues such as the ocular lens, heart valves, myocardial (patches), accessory support for total heart replacements devices and widely to remedy osseous or joint disfunction. As scaffolds for tissue grafts they hold promise in the cornea, the parietes, and articulating joints. As conduits, they find use in replacing or by-passing damaged vessels, bile ducts, ureters, segments of upper respitory tract. Also, as conduits, they can provide egress or access to central nervous system, the eye, lacrimal structures and vicerol cavities.

Cosmetic needs are increasingly met by liquid, gel or solid components with pure silicone, silicone-encased or silicone-treated implants in a variety of surgical settings, most notably in breast reconfiguration efforts. Facial, orbital or testicular sites provide other examples.

As biosensors, silicone encase pacemakers, electrode leads and vascular access monitoring catheters. Some are incorporated into heart lung machines for gas-exchange or into renal dialysis units.

Recently, biomaterials serve as therapeutic drug carriers with a wide range of applications in managing infections, neoplasia or metabolic needs (Table III).

Table III

APPLICATIONS OF BIOMATERIALS
Structural Replacement or Support Conduits Cosmesis Biosensors Therapeutic Drug Carriers Heart Lung Machines/Dialysis Units

Adverse Reactions

With increasing use and time of residence of biomaterials in the host, adverse reactions are recognized and categorized. The expression of these will vary with the structure of pulative agent, the resident organ or tissue, time and host predisposition. Topical, regional or systemic manifestations may follow.

Structural failure may ensue due to material composition or mechanical properties tested by stress factors and the biological environment. Each of the three basic implant materials - polymers, metals and ceramics - exhibit different properties of tensile, compressive and yield strengths, corrosion, modulus, creep and wear that underline the requirements for clinical application. Methods of fabrication and of design may be the formula for either success or failure. Early silastic heart ball valves fragmented with systemic dissemination and occasional fatality. In orthopedics, metallic failure in the form of brittle, plastic or fatigue damage to implanted screws, wires, plates or protheses is a not uncommon problem.

Particulate Disease

Structural failure combined with host response has followed abundant expression in a form of pathology called particulate disease. Originally associated with the use of polymerized methyl methacrylate (PMMA) in bonding a prothesis to neighboring bone, thus it was initially dubbed "cement disease". It is dependant on a macrophagic response to particles for expression. This phenomenon now embraces reactions to polyethylene, the metals of stainless steel, chrome-cobalt, and titanium, the ceramics and barium sulfate (the latter incorporated into cements for purposes of radiography).

The frictional release of fragments under weight bearing stress initiates an aggressive granulomatas inflammatory response with macrophagic mobilization at interfaces. Release of high levels of acid phosphates, collognace, prostagledian E2 and interleukin-1 are implicated in bone resorbing processes (osteolysis). Aseptic loosening of prosthetic components leads to clinical disease expression within 4 to 10 years on average. Secondary phenomena include fibrosis, paraoseos interface separation, and hypotrophic reactive synovitis. Furthermore, osteo seletal instability may be attended by increased risk of infection and hemorrhage. Mobilization and migration of macrophage to regional nodes may give rise to lymphodemopathy and concern for neoplastic disease.

Host factors of age, genetic predisposition, nutritional and drug status (steroids, anticoagulates) etc. may greatly contribute to the outcome of a biomaterial application, either favorable or unfavorable. Astute clinical judgement assumes a preeminent role in the selection and timed use of biomaterial aids.

Table IV

SYSTEM CONSIDERATIONS OF BIOMATERIALS
Coagulopathy Sepsis Allergic-Toxic Autoimmune Disease Neoplasia

Foremost and commonest are problems of coagulopathy and infection. Abnormal bleeding and/or clotting predispositions should invoke consideration of all major host factors that may impact on tissue or organ biomaterial placement. Clotting of a heart valve prothesis or migration of fragmented clots to vital organs will defeat the most brilliant surgical placement. Similarily, careful appraisal of host factors may mitigate against disastrous septic complications.

Allergic-toxic effects of biomaterials have focused attention on metallic prothesis with their increasing use in young subjects and porous features of design that favor ion release. Nickel containing compounds have been associated with allergic dermatoses. While in high dose they are capable of systemic toxicity and mutagenicity - cobalt, aluminum and chromium also share dose-related potential for toxicity to various tissues. Minor elevations of nickel, cobalt, chrominium and titanium have been observed in the blood and urine of patients with methyl polyethylene implants. A variety of neoplasms have been rarely reported adjoining implants most notably the malignant fibrous histiosarcoma. Dissimilar metal composites favor increased corrosion and ion release and have greater potential for harm.

Autoimmune phenomena in relation to silicone use notably in breast augmentation mammoplasty have been mired in controversy. In the most serious setting it has been postulated that systemic autoimmune phenomena have been generated by silicone rupture, bleed, or free fragments with polyarthropathy in scleroderma-like presentation. Degeneration of interface fibronectin, collagen and fibrinogen macroneolecules has been advanced as a precursor to activate the immune system and generate humoral antibodies. The unequivocal link to disease expression may be host-generated as many asymptomatic individuals demonstrate there antibodies. Another face of silicone (and polyurebleave) products is excessive local fibrosis possibly related to cytokine activity with cellular, rather than humoral, factors playing a leading role. Extensive litigious activity has recently ensued.

Lastly, how will these biomaterials be impacted on or have an impact on the many diagnostic and therapeutic modalities, such as laser application. We conclude that with the rapid stride of new biomaterial development, we must be increasingly aware and knowledgeable of their potential for adverse reactions.

References

1. Jones, L.C., Hungerford, D.S., "Cement Disease", *Clin. Orthop.*, 1987, **225**, 192-206.
2. Kossoveky, N., Freiman, C.J., "Silicone Breast Implant Pathology", *Arch. Pathol. Lab. Med.*, 1994, **118**, 686-693.
3. Grunkemeier, G.L., Rahimtooca, S.H., "Artificial Heart Valves", *Annu. Rev. Med.*, 1990, **41**, 251-263.
4. Koval, K.J., et. al., Complications of Fracture Fixation Devices in Complications in Orthopedic Surgery - 3rd Edition, Charles H. Epps, Jr. Editor, J.B. Lejpiucott Co., 1994, Chapter 7, 131-154
5. Shaw, J.A., Green, R.B., Complications of Fracture Fixation Devices in Complications in Orthopedic Surgery - 3rd Edition, Charles H. Epps, Jr. Editor, J.B. Lejpiucott Co., 1994, Chapter 37, 1013-1026.
6. Warner, K.L., "Implications of Metallic Corrosion in Total Knee Arthroplasty", *Clin. Orthop.*, 1991, **271**, 12-20.

BIOARTIFICIAL POLYMERIC MATERIALS: NATURAL AND SYNTHETIC WORLD JOINING IN BIOMATERIALS RESEARCH

Paolo Giusti, Luigi Lazzeri, Maria Grazia Cascone, Niccoletta Barbani, Caterina Cristallini

Department of Chemical Engineering,
University of Pisa - via Diotisalvi 2,
56126 Pisa, Italy.

INTRODUCTION

The interactions of life science and material science, to use the great potentiality of the biological systems in realising both new materials and new industrial processes, have been successfully put into practice during the last few years, especially in the close related fields of biomaterials, bioengineering and biotechnologies.

Our main aim was to exploit these interactions to make new materials for applications in the biomedical field. The reason for this is that, in spite of the great improvement of the biological performances of many purposely designed new polymeric biomaterials, their interactions with living tissue components remain the major problem to be solved. Therefore, the idea was to smooth away the interactions between the synthetic and the biological systems by a material in which changes at a molecular level, due to synthetic-biological polymer interactions, have been already accomplished before getting in touch with the living tissue. Such a material, with already established molecular interactions, should behave macroscopically better than a fully synthetic material as far as the biological response of the host is concerned. In addition, the study of molecular level interactions would allow one both to select at a very early stage those synthetic polymers that will be potentially used in making new polymeric biomaterials, and to get information about the biocompatibility characteristics of the polymers.

Ten years ago, the design of new materials based on blends of biological and synthetic polymers started in our laboratories, the final purpose being the production of new processable polymeric materials that hopefully posses both good mechanical properties and biocompatibility. We also proposed that this class of materials be called "bioartificial polymeric materials"[1, 2, 3].

Bioartificial materials obtained by mixing synthetic polyurethanes with fibrinogen, and subsequently cross-linking the fibrinogen into fibrin by enzymatic treatment with thrombin, factor XIIa and Calcium ions, were studied for applications in the cardiovascular field[4].

Bioartificial materials based on the mixing of collagen with various commercial synthetic polymers were also examined[5,6,7].

Recently, new bioartificial materials have been prepared using hyaluronic acid (HA) or its ester derivatives (HYAFF) as biological components[3,8,9]. Different kind of materials such as hydrogels, films etc., were obtained by blending HA or HYAFF with poly(vinyl alcohol) or ethylene-vinyl alcohol copolymers. These materials were evaluated for applications as drug delivery systems[10,11] and dialysis membranes[12,13].

At the moment we are studying new methods to investigate synthetic-biological interactions by means of biological tests based on the use of enzymes. In particular, these methods could allow to evaluate both the behaviour of a biopolymer, as a substrate for a specific enzyme, when blended with a synthetic component, and the changes in the activity of an enzyme after its blending with a synthetic polymer.

Finally, as an alternative route to bioartificial materials made of pre-formed polymers, we are studying polymer blends prepared by template-polymerization of synthetic monomers on a biological matrix, in order to obtain both blends with higher miscibility degree and polymers with molecular recognition capability.

This paper will concern with bioartificial polymeric materials in which either collagen or hyaluronic acid is the biological component, whilst the synthetic component is poly-(vinyl alcohol) (PVA), poly(acrylic acid) (PAA) or poly(vinylsulphonic acid) (PVSA).

COLLAGEN-BASED BIOARTIFICIAL POLYMERIC MATERIALS

Collagen is one of the most important structural protein in the vertebrate body and is the principal extracellular protein in the connective tissues. Although several different types of collagen have been identified, the molecules of native collagen always consist of three polypeptidic chains arranged in a triple helix, and aggregate to form microfibrils and

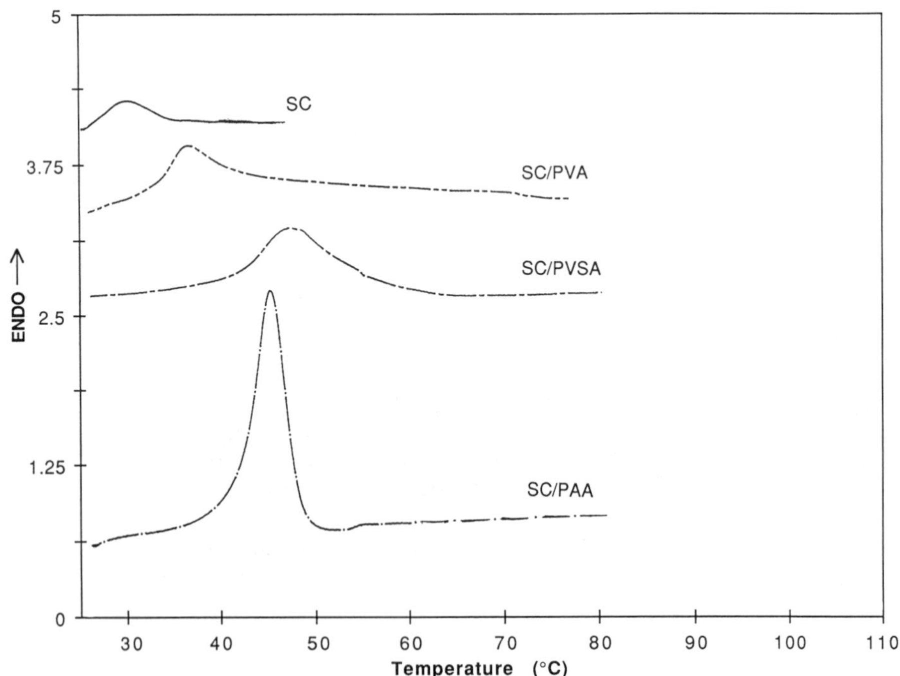

Figure 1. DSC thermograms of pure SC and SC/PVA, SC/PAA, SC/PVSA blends. The content of SC in the blends is 50% by weight. (Scan rate: 10 °C/min)

fibrils. The different properties of the collagen containing tissues, such as bone, tendon and skin, depend on the different organization of the collagen fibrils. Because of its important role in the living systems, collagen has been extensively studied as a polymer for use in manufactured materials and has attracted great attention for its potential applications in the biomedical field including dialysis membranes, heart valves, wound dressing and artificial skin.

In preparing bioartificial polymeric materials containing soluble collagen (SC) as biological component, we used water soluble polymers such as PVA, PAA and PVSA as synthetic components. In particular PVA has low production costs and has been largely used for different biomedical applications due to its good compatibility and versatility. PAA and PVSA also are easily available polymers, and have been mainly chosen in this work to study the molecular interactions between their specific functional groups and SC. Blends of SC and synthetic polymers, have been prepared by mixing their acqueous solutions. Films have been obtained by casting at room temperature and characterized by means of differential scanning calorimetry (DSC), dynamic mechanical thermal analysis (DMTA) and scanning electron microscopy (SEM).

Before the mixing of SC with PAA or PVSA, the pH of the solutions has been adjusted at a value just over the isoelettric point of SC[8].

After evaporation of the solvent, the films have been washed with an acetic acid solution at pH=2.5. Due to this treatment the composition of the SC/PVA and SC/PAA blends has not significantly varied while an enrichment in the SC content has been observed in the SC/PVSA blends.

The results of the DSC analysis reported in Fig. 1 show the denaturation temperature (T_d) of SC for blends in which the initial SC content was 50% by weight. It can be seen that the T_d value of SC in all the examined blends is higher than that of pure SC and follows the order: SC/PVA, SC/PAA, and SC/PVSA. No substantial differences from this trend have been observed in the case of blends composition other than 50% of SC.

The enhancement of the thermal stability of SC in the blends reveals that interactions between the biopolymer and the synthetic component occur. Among different factors which affect the strenght of these interactions, the nature of the functional groups belonging to the interacting polymers has been mainly considered in this work. The interactions in SC/PVA blends are mostly due to the relatively weak hydrogen bonding; while in the SC/PAA blends stronger ionic interactions take place due to the presence of partially ionized carboxyl groups on the PAA; even more stronger interactions occur in the SC/PVSA blends since the sulphonic groups of PVSA are completely ionized.

This could justify the observed dependence of T_d shift on the nature of the synthetic

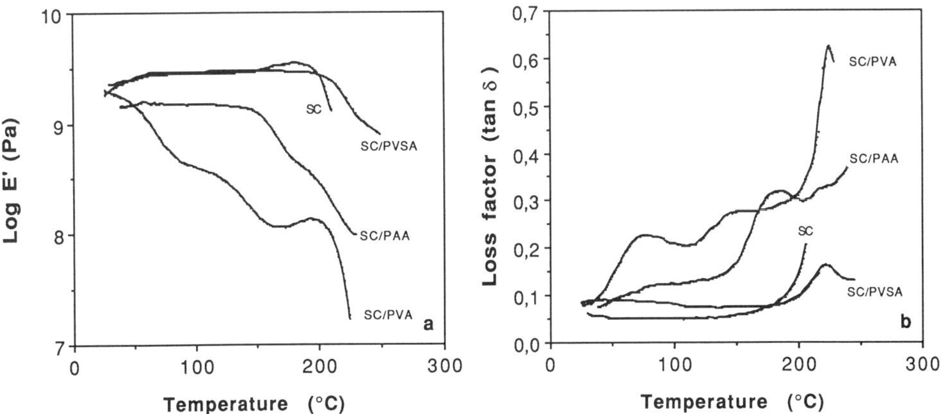

Figure 2. DMTA spectra of pure SC and SC/PVA, SC/PAA, SC/PVSA blends: (a) storage modulus, E'; (b) loss factor, tanδ. The weight content of SC in the blends is 80%. (Scan rate: 4 °C/min; frequency: 1 Hz)

component: the stronger interactions, the higher thermal stability of SC in the blend.

The nature and the extent of the interactions play also an important role in determining the thermodynamic miscibility of different polymers. Polymer systems in which stronger interactions occur would be expected to show a higher miscibility degree. This in turn would result in a material showing the mechanical behaviour of a single-phase system. This has been confirmed by the results of DMTA analysis carried out on SC based samples. In Fig. 2 a, b the trends of the storage modulus, E', and of the loss factor, tanδ, are shown for different blends in which the initial SC content was 50% by weight.

As expected, the SC/PVA blend shows two different transitions, before the final mechanical yield, related to the corresponding pure component transitions, that are the PVA glass transition at about 75°C and the SC denaturation at about 190°C, respectively. Conversely, both the SC/PAA and SC/PVSA blends show each a single transition at about 180°C and 225°C respectively.

These results also indicate an increase of the thermal stability as the extent of the molecular interactions increases.

HYALURONIC ACID-BASED BIOARTIFICIAL MATERIALS

Hyaluronic acid (HA) is a high molecular weight polysaccharide that can be found in many different animal tissues, often associated to proteins. The repeated unit of HA is constituted of a molecule of N- acetyl glucosamine and a molecule of D-glucuronic acid.

It is known that HA is able to influence the cellular migration, thus playing a special function during tissue growth and repair. Because of its high hydrophilicity, HA seems to control the tissue water content, the diffusion of other macromolecules and the lubrification of joints.

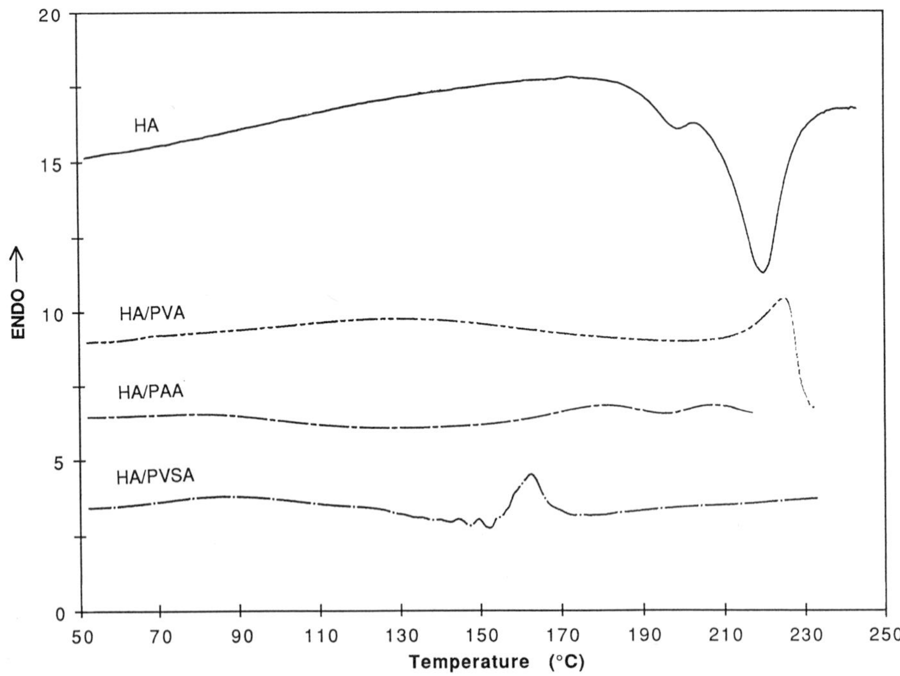

Figure 3. DSC thermograms of pure HA and HA/PVA, HA/PAA, HA/PVSA blends. The content of HA in the blends is 50%by weight. (Scan rate: 10 °C/min)

HA can be used in association with a great variety of synthetic polymers; it can be applied on the material surfaces or blended with the polymers solutions.

Also in preparing bioartificial polymeric materials containing HA as natural component, the polymers PVA, PAA and PVSA have been used as synthetic components. Aqueous solutions of the two components have been mixed and films have been obtained by solution casting at room temperature. The obtained films have been used in DSC and DMTA analysis.

The thermal behaviour of pure HA is characterized by an exothermal degradative process at about 200-220°C (Fig. 3). The mixing of HA with PVA results in a partial superimposition of the HA degradation peak with the melting of PVA, with no evident influence of one component on the thermal transition of the other. Substantially different results have been observed for the thermal behaviour of the HA/PAA and HA/PVSA blends. In the HA/PAA blends, no degradation event is seen by heating, while the HA/PVSA blends do not show substantial differences from the pure HA when the PVSA content is low (20 % by weight). At higher PVSA content, the HA degradation occurs at a temperature lower than that of the pure HA.

The results of the DMTA analysis are, in general, in agreement with those of the DSC analysis. The storage modulus and the loss factor show a similar trend for both the pure HA and the blends when the synthetic component content is not more than 20%. The HA/PAA and HA/PVA blends show a lower tensile strenght (Fig. 4). When the synthetic component content is about 50% the blends show a different behaviour; in particular the HA/PAA blend shows a single transition occurring at a temperature intermediate between those of the PAA glass transition and the HA degradation (Fig. 5), and indicating a miscibility close to the ideality.

These results are complex and controversial: in fact, in the case of the HA/PAA blends a stabilization of the HA can be observed, while in the case of the HA/PVSA blends, the HA degradation temperature shifts towards lower values as the PVSA content increases.

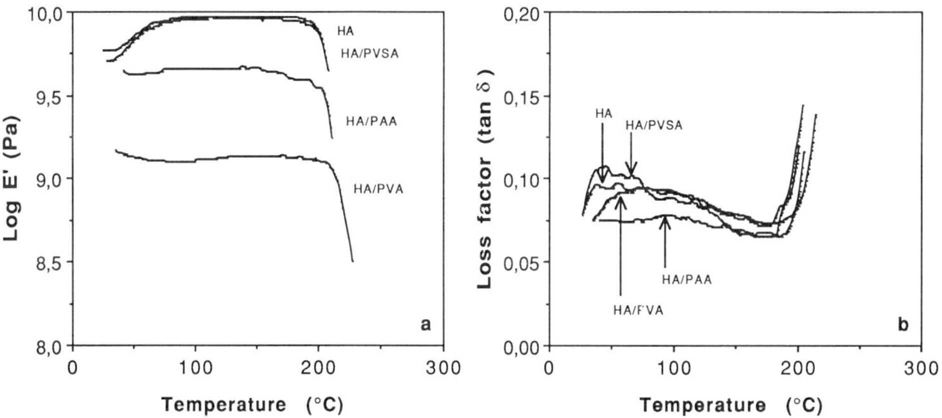

Figure 4. DMTA spectra of pure HA and HA/PVA, HA/PAA, HA/PVSA blends: (**a**) storage modulus, E'; (**b**) loss factor, tanδ. The weight content of HA in the blends is 80%. (Scan rate: 4 °C/min; frequency: 1 Hz)

CONCLUSIONS

The results of the study of the molecular interactions between SC and the examined synthetic polymers has shown not only that this joining is feasible, but also that it allows to improve the mechanical and biological properties so matching the goal expected for the

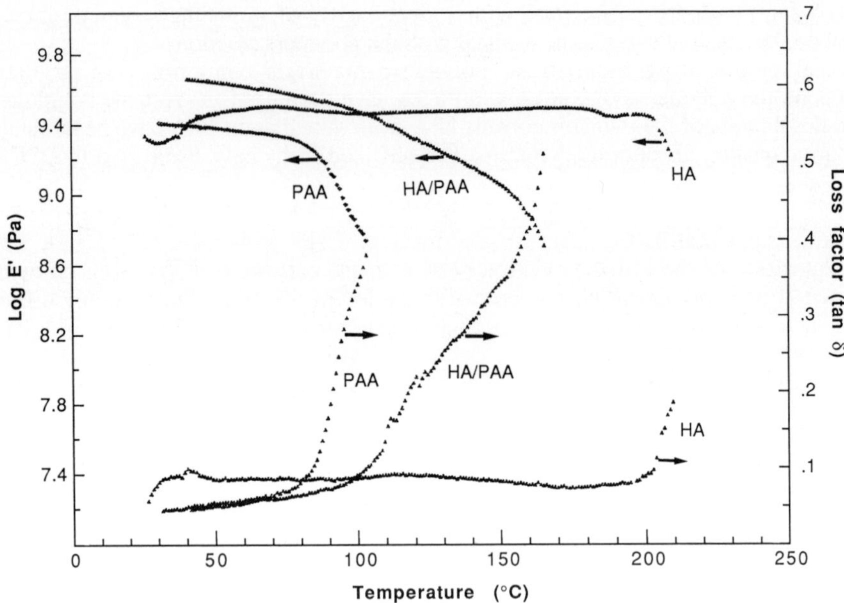

Figure 5. DMTA spectra of pure HA, pure PAA and HA/PAA blend (50/50 by weight). (Scan rate: 4 °C/min; frequency: 1 Hz)

bioartificial polymeric materials.

A miscibility close to the ideality has been found when SC was blended with PAA and PVSA, while in the case of PVA, interesting mechanical properties have been observed although the SC/PVA blends were less ideal.

These properties suggest important applications of these blends in preparing materials for devices as dialysis membranes and hollow fibers in the realization of which, the attempt to use pure SC has been unsuccessful mainly due to the poor mechanical properties of the riconstituted collagen.

The results of biocompatibility tests performed on these blends, reported elsewhere[14], confirm the improvement of the biological properties of the blends in comparison with the pure synthetic components.

In the case of HA the joining with the synthetic polymers has been less fruitful. But an interesting feature is represented by the low solubility of the HA-based blends, that allows the use of these blends for applications in HA releasing systems. In addition, these blends can be made fully insoluble by simple cross-linking methods. This is very important because allows the use of HA in those applications where its solubility represents an undesirable characteristic.

We can conclude that the idea to anticipate at a molecular level the interactions between the biological and the synthetic systems by the preparation of blends, has resulted successful because the new bioartificial polymeric materials realized in this work have shown interesting properties for applications in the biomedical field.

REFERENCES

1. P. Giusti, S. Soldani, R. Mercogliano, M. Pracella and N. Barbani, Applications of Advanced Polymers in Medicine: Bioartificial Polymeric Materials, Proc. 1st Mediterr. School on Science and Technology of Advanced Polymer-Based Materials, Vico Equense, Naples, 11-22 September 1989, p. 715.

2. P. Giusti, L. Lazzeri and L. Lelli, Bioartificial polymeric materials: a new method to design biomaterials by using both biological and synthetic polymers, *Trends in Polym. Sci.* 1:261 (1993).
3. P. Giusti, L. Lazzeri, N. Barbani, L. Lelli, S. De Petris and M.G. Cascone, Blends of natural and synthetic polymers: a new route to novel biomaterials, *Makromol. Chem. Macromol. Symp.* 78:285 (1994).
4. P. Giusti, G. Soldani, M. Palla and M. Paci, New fibrin containing thermoplastic elastomers for cardiovascular applications, *Adv. Biomat*, 6:221 (1986).
5. P. Giusti, L. Lazzeri, S. De Petris, M. Palla and M.G. Cascone, Collagen-based new bioartificial polymeric materials, *Biomaterials* 15:1229 (1994).
6. P. Giusti, L. Lazzeri, N. Barbani, P. Narducci, A. Bonaretti, M. Palla and L. Lelli, Hydrogels of poly(vinyl alcohol) and collagen as new bioartificial materials, *J. Mater. Sci. mater. in Med.* 4:538 (1993).
7. N. Barbani, L. Lazzeri, L. Lelli, A. Bonaretti, M. Seggiani, P. Narducci, G. Pizzirani and P. Giusti, Physical and biological stability of dehydro-thermally crosslinked collagen-poly(vinyl alcohol) blends, *J. Mater. Sci. mater. in Med.* 5:882 (1994).
8. L. Lazzeri, N. Barbani, M.G. Cascone, D. Lupinacci, P. Giusti and M. Laus, Physico-chemical and mechanical characterization of hydrogels of poly(vinyl alcohol) and hyaluronic acid, *J. Mater. Sci. mater. in Med.* 5:862 (1994).
9. R. Sbarbati Del Guerra, M.G. Cascone, N. Barbani and L. Lazzeri, Biological characterization of hydrogels of poly(vinyl alcohol) and hyaluronic acid, *J. Mater. Sci. mater. in Med.* 5:613 (1994).
10. M.G. Cascone, B. Sim and S. Downes, Blends of synthetic and natural polymers as drug delivery systems for growth hormone, *Biomaterials*, 16:569 (1995).
11. M.G. Cascone, L. Di Silvio, B. Sim and S. Downes, Collagen and hyaluronic acid based polymeric blends as drug delivery systems for the release of physiological concentrations of growth hormone, *J. Mater. Sci. mater. in Med.* 5:770 (1994).
12. M. Seggiani, L. Lazzeri, M.G. Cascone, N. Barbani, S. Vitolo and M. Palla, Collagen based bioartificial materials-evaluation as membranes for extracorporeal blood purification, *J. Mater. Sci. mater. in Med.* 5:868 (1994).
13. M. Seggiani, L. Lazzeri, P. Giusti, N. Barbani, L. Lelli, M. Palla and M. Taveri, Blends of hyaluronic acid derivatives with ethylene-vinyl alcohol copolymers as potential biomaterials, *J. Mater. Sci. mater. in Med.* 5:872 (1994).
14. M.G. Cascone, L. Lazzeri, N. Barbani, G. Polacco, A. Pollicino and P. Giusti, Dehydro-thermally crosslinked collagen-poly(vinyl alcohol) blends: mechanical, biological and surface properties, *J. Mater. Sci. mater. in Med.* (in press).

GUIDELINES FOR THE RE-STABILIZATION OF POST-CONSUMER RECYCLED PLASTICS

Karen Sun and Dr. Laurence Timothy Pearson

Additives Technical Centre
Ciba Geigy (Hong Kong) Ltd.
GPO Box 285, Hong Kong

ABSTRACT

Recycling of plastics is the preferred option for dealing with unavoidable plastic waste.The recycled polymer may have undergone degradation during its first use as virgin material. Such degradation can have two effects on recyclability. Firstly, it can lead to a change in physical and mechanical properties with respect to virgin material. Secondly, it can also render a polymer more susceptible to further degradation during re-use. Such susceptibility can be a result of the formation of oxidative by-products which initiate or otherwise take part in further degradation reactions.

This paper will first discuss the recycling situation in Hong Kong and a literature review on work that has been done in recyling plastics through re-stabilization. We will also discuss the degradation in Polypropylene (PP), Polyethyelene (PE), and Polypropylene Ethylene Propylene Diene Monomer (PP/EPDM) occuring during their first life in applications such as film, bumpers, cups, etc. and the fate of stabilizers that were added to the virgin material. We will focus on the use of stabilizers which can give processing, long term heat stability and light stability on recycled plastics. Some data charts on the advantages of using stabilizers in the mechanical and physical properties of the plastics will also be included.

INTRODUCTION

Recycling of plastics is the preferred option for dealing with unavoidable plastic waste. The recycled polymer may have undergone degradation during its first use as virgin material. Such degradation can have two effects on recyclability. Firstly, it can lead to a change in physical and mechanical properties with respect to virgin material. Secondly, it can also render a polymer more susceptible to further degradation during re-use. Such susceptibility can be a result of the formation of oxidative by-products which initiate or otherwise take part in further degradation reactions.

Polymers and Other Advanced Materials: Emerging Technologies and Business Opportunities. Edited by P. N. Prasad *et al.*, Plenum Press, New York, 1995

This paper will first discuss the recycling situation in Hong Kong and a literature review on work that has been done in recycling plastics through re-stabilization. We will also discuss the degradation in Polypropylene (PP), Polyethylene (PE), Polypropylene Ethylene Propylene Diene Monomer (PP/EPDM) and mixed polyolefins occurring during their first life in applications such as films, bumpers, etc. and the fate of stabilizers that were added to the virgin material. We will focus on the use of stabilizers which can give processing, long term heat stability and light stability on recycled plastics. Some data charts on the advantages of using stabilizers in the mechanical and physical properties of the plastics will also be included.

There are several ways of upgrading recycled plastics with a view to restoring or improving those physical properties that allow re-use. Some methods are physical and some are chemical. Physical methods would include processing steps such as washing, separation, compounding and melt filtration. Chemical methods include blending with various types of additives such as compatibilizers, polymer modifiers, fillers and stabilisers. They also include processes such as selective dissolution[1] , or feedstock recovery[2].

The problems inherent in physical methods are well exemplified by the situation in Hong Kong, where land is at a premium, and consumer awareness of the need for recycling is low. Based on a 1993 survey by the Waste and Water Policy Group of the Environmental Protection Department of Hong Kong, there had been 270,000 tons of plastics waste being recovered in Hong Kong. 4% of these are recycled locally and 96% are being exported[3]. All of these wastes are scrap and there are no post consumer recyclate (PCR) wastes. Some of the major constraints in recycling PCR in Hong Kong are the lack of space, high land premiums and labour cost, lack of markets for recovered materials and recycled products and the bulkiness and contaminated nature of these wastes may increase the processing cost and lower the value of the recycled products.

Chemical methods for upgrading, as opposed to chemical separation, could provide a way of reducing the stringency of requirements of physical methods and essentially involve the improvement of properties of recycled polymer as is, without the need to resort to, for example, separation techniques. The literature now contains many examples of physical property studies on blends from PCR. A paper by Hope et al. [4], for example, describes the use of compatibilizer in PP/HDPE blend. A 5 to10% blend of PP in HDPE will reduce the impact strength considerably if no compatibilizer is added. By adding 10 parts per hundred (pph) of compatibilizer, one can counter-act the property loss.

Another paper by J. Paul, et al. describes the advantages of blending PCR with virgin LLDPE/LDPE film grades [5]. As one increases the percentage of PCR, the impact and tear properties also decrease. By blending PCR HDPE with an LLDPE the properties are improved even with 25% PCR.

PCR can be inferior to virgin material in two distinct ways. The first is in terms of physical properties (in the solid state) and processability (melt properties). This effect is a result of degradation and hence molecular weight changes that have resulted from processing or use. The second way in which PCR can be inferior to virgin is in terms of stability. For example PCR HDPE can have a higher peroxide development than virgin materials[6]. A recent study on the effect of the number of re-processing cycles to the mechanical property[7] describes that depending on the number of re-processing cycles, PP film can be re-processed without re-stabilisation. But after eight re-processing cycles, the elongation to break and photo-oxidative stability decreased dramatically.

The subject of this paper is re-stabilization of PCR polymers with the objective of restoring primarily long term heat and light stability, but also in some cases process stability.

Re-stabilisation may be needed for two reasons. The first is that the initial stabilisation of a material may not be sufficient for its second life. If a light stabiliser has been added, for example, to give an expected stability of n years, and the polymer is taken out of service after m years, then after reprocessing, if no UV stabiliser is added, the remaining lifetime will be less than (n-m) years, taking into account degradation during re-processing and physical loss of stabiliser during processing and ageing. The second reason for re-stabilization is that the polymer after its first life may already have been damaged by shear, heat or ultraviolet (UV) light.

For re-stabilisation to be successful, several factors must be taken into account:

- the degree of the degradation which is already present,
- the amount of residual stabilisers,
- the type and quantity of the contaminant that exists,
- and the application of end-use

Knowing these factors will help judge how much and what stabilisers are to be added to the recycled polymer.

In this paper we will present a review of the results of several studies on the properties of re-stabilized materials, with particular regard to stability of physical properties after ageing and processing and will mainly focus on the re-stabilization of recycled PP, PE, PP/PE blends, and PP/EPDM in different applications.

1.1. PP SCRAP

When a polymer is processed, there are losses in both the phenolic antioxidant, and the phosphite. The phosphite suffers the greater losses, hence reduces the consumption of the hindered phenol during processing, thereby prolonging the long-term sevice life of the final product.[8] **Figure 1** for example shows the levels of residual process stabilizers found in a typical PP formulation during processing.

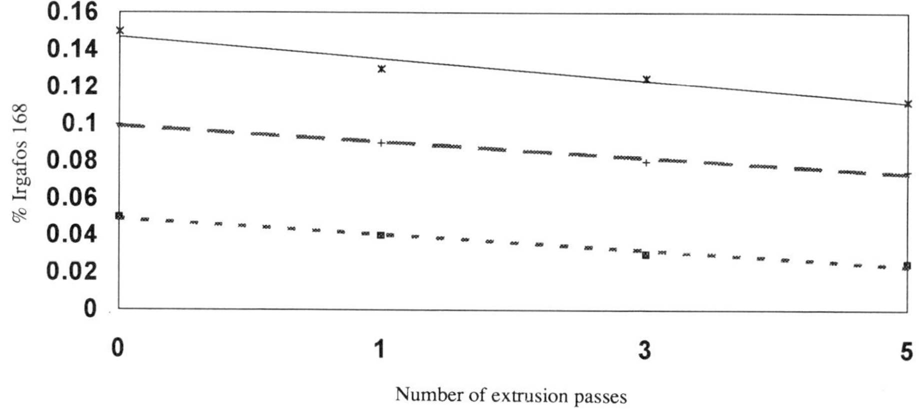

All samples contain 0.05% Ign 1010+0.1% CaSt2
Extrusion at 260°C

Figure 1. Consumption of Phosphite during Re-processing of PP.

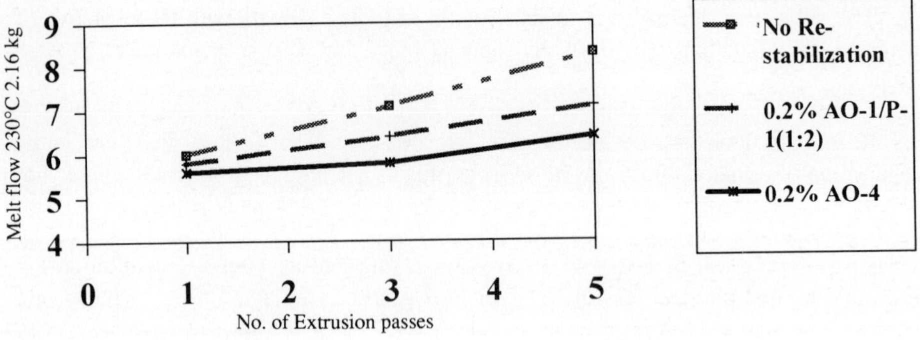

Extrusion: twin screw, max. 230°C
Approx. 0.1% residual process stabilizers

Figure 2. PP Production Scrap from Film Material Processing Stability.

Figure 2 shows the processing stability of agglomerated PP production scrap. The melt flow index (MFI) is measured after multiple passes. Although there are some residual stabilizers from the first use (0.07% phenolic anti-oxidant and 0.04% phosphite), the formulation that was not re-stabilized shows an increase in MFI which is an indication of molecular weight changes and deterioration of properties. By adding AO-1/PS-1 and AO-4 can reduce molecular weight changes.

The contribution of the phenolic anti-oxidant (AO) to long term thermal stability (LTTS) is shown in **Figure 3**. AO-4 can double the lifetime of the PP.

Figure 4 shows the light stability of agglomerated PP production scrap. Unstabilized PP would degrade in less than 1000 weather-o-meter (WOM) hours. When LS-1 is added to the polymer, the light stability is increased seven times. The tensile impact strength is measured against the weathering hours.

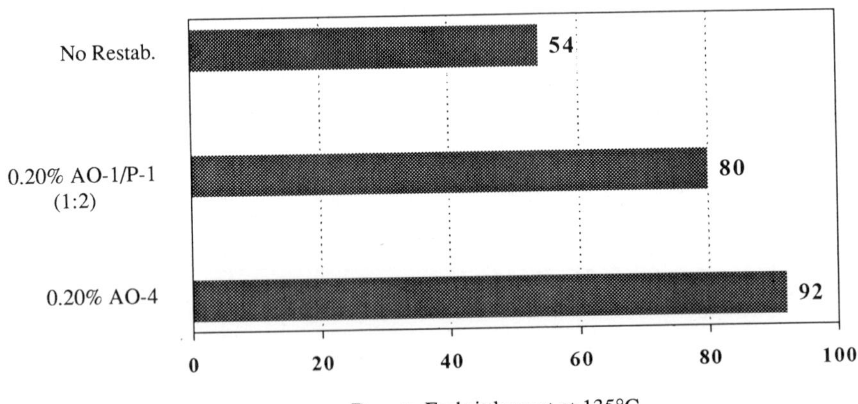

Extrusion: Twin screw, max. 230°C
Test specimens: 2-mm injection-molded plaques
Approx. 0.1% residual process stabilizers

Figure 3. PP Film Production Scrap Long Term Thermal Stability.

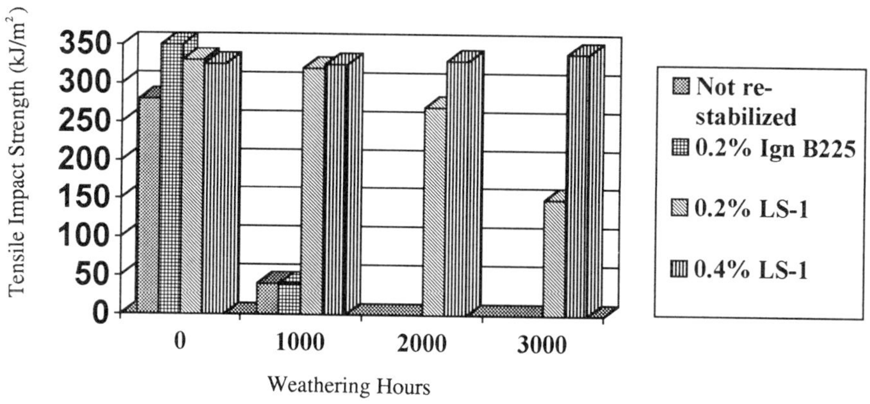

Figure 4. Light Stability of Agglomerated BOPP film Production Scrap.

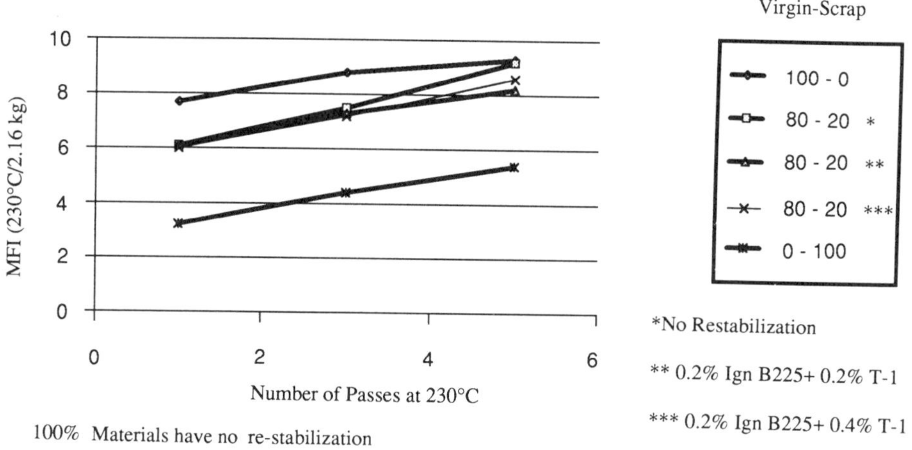

Figure 5. Effect of Painted Scrap on Process Stability of PP/EPDM Blends.

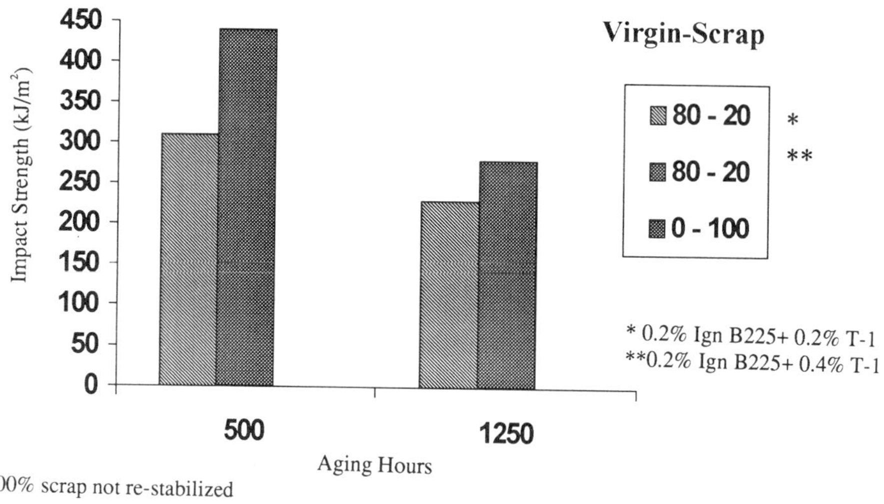

Figure 6. Effect of Painted Production Scrap on Impact Strength After Aging at 135°C.

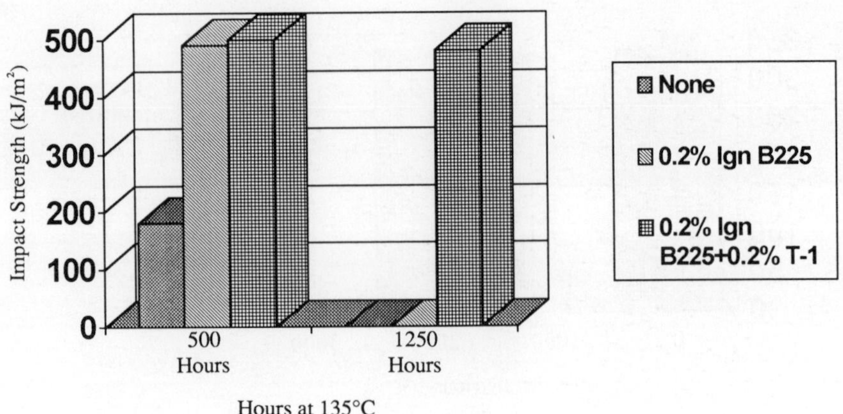

Figure 7. Effect of Restabilization of Impact Strength of a 10-year Old PCR Bumper Material.

1.2. PP/EPDM SCRAP

The simplest example of recyclate is post production scrap, which has not undergone any significant thermal or UV ageing, and is simply re-ground and re-processed. The fact that bumpers are painted can result in an effect on re-usability. **Figure 5** shows an example of how melt flow (MF) varies with extrusion number during re-processing. Pure scrap has undergone a significant decrease in melt flow. Blends of 80% virgin + 20% scrap show MF drift which is in between the virgin and the scrap, but basically re-stabilization (with phosphite + phenolic + thioester) does not seem to affect melt flow drift appreciably, and would seem in this case to be unnecessary.

However, **Figure 6** shows the long term heat stability of the scrap material and its blends with virgin. After 500 hours at 135°C, the scrap has no residual impact strength. Blends with virgin have better impact strength retention, and re-stabilization allows for further improvement in that property. **Figure 7** shows similar data and the LTTS is measured. Irganox B225/T-1 formulation can give equal performance as the virgin polymer.

Bumper material which has been in service for 10 years (i.e. bumper PCR) is shown in **Figures 8.** Pure un-re-stabilized PCR has some residual stability after 500 hours but

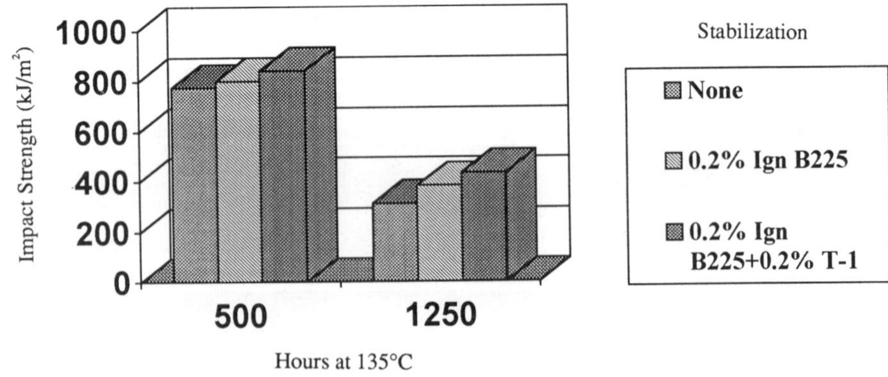

Figure 8. Effect of Restabilization on Impact Strength of 5% Recyclate/95% Virgin Blend.

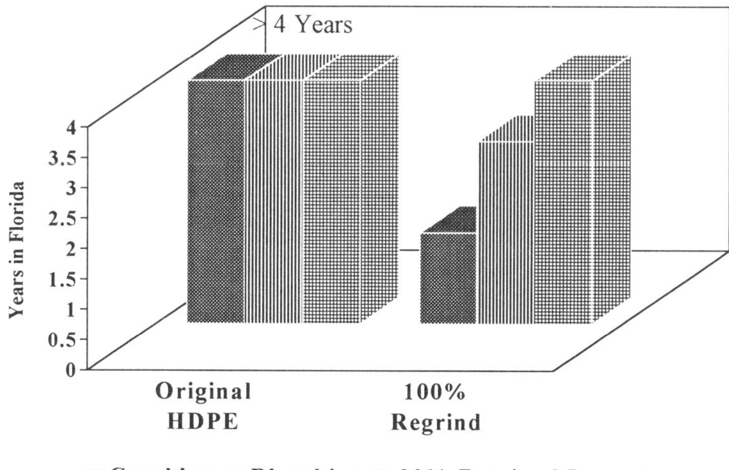

0.1% AO-1/P1+0.1% HALS-1+0.1% UV Absorber
Red pigmentation

Figure 9. Light Stabilization of HDPE Bottle Crates.

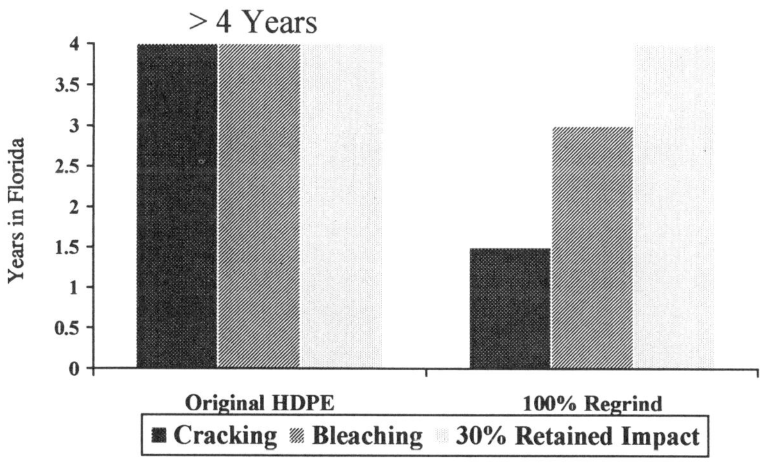

0.1% AO-1/P1+0.1% HALS-1+0.1% UV Absorber
Red pigmentation

Figure 10. Light Stabilization of HDPE Bottle Crates.

none after 1250 hours. Material that is re-stabilized with the phenol + phosphite blend (Irganox B225) has better retention at 500 hours, but for best long term heat stability a thioester can be used.

1.2.1. Post-consumer PP/EPDM

In **Figure 9**, a 100% PCR and a 50:50 used/virgin are compared in terms of their LTTS. The reference formulation is not re-stabilized and the LTTS is lower than that of the re-stabilized.

1.3. HDPE Scrap

Figure 10 the light stability of a virgin and a 100% recycled red pigmented HDPE bottle crates are compared. Both contain 0.1% HALS-1 + 0.1% UV absorber and after 4 years of outdoor exposure in Florida, the retained impact strength of the 100% recyclate is reduced to 30%. The concentration of light stabilizer package should be increased to achieve better light stability. In **Figure 11**, the effect of blending recyclates to virgin to the LTTS is shown. As you increase the percentage of recyclates, the LTTS is decreased.

1.3.1. PCR HDPE

Figure 12 compares the light stability of two pigmented PCR HDPE bottle crates. The samples are exposed by artificial and natural weathering and the hours to surface cracking is measured. The green pigment is more stable than the red pigment. The red pigment faded after 500 WOM Hours. Stabilizing PCR HDPE with HALS-2 increases the lifetime up to five times compared to the reference.

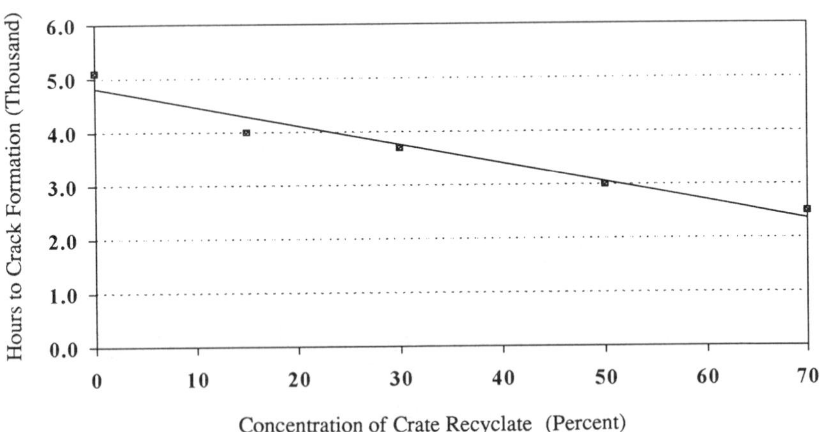

Xenon-arc weatherometer: 102' dry, 18' wet
All samples: 0.2% HALS-1 + 0.1% UV Absorber + 0.1% AO-1
Blue pigmentation, 2-mm thick

Figure 11. Light Stabilization of PCR HDPE Bottle Crates. Effect of Recyclate Content on UV Resistance.

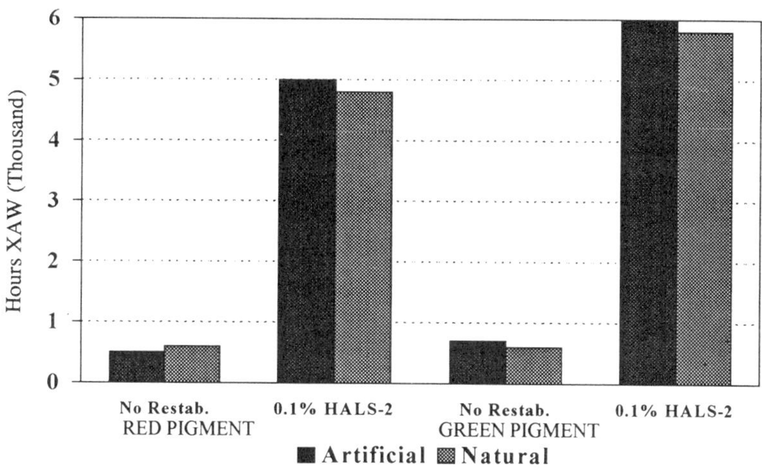

Note: Red pigment bleached after 500 weatherometer hours
and 6 months Florida exposure

Figure 12. Stabilization of PCR HDPE Crates.
Time to Surface Cracking.

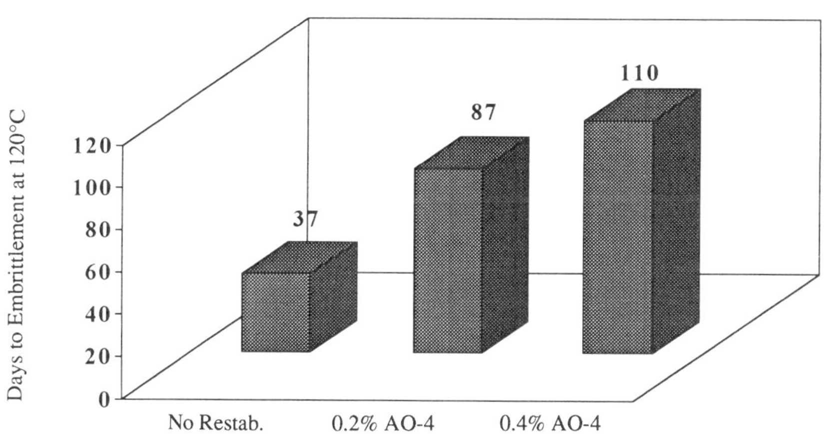

Extrusion: twin screw, max. 240°C
Reference recyclate contains approx. 0.02% residual stabilizer
Minimally stabilized HDPE virgin material: 100 days at 120°C

Figure 13. Mixed Polyolefin Recyclates (PE 90%, PP 10%).
Long Term Thermal Stability.

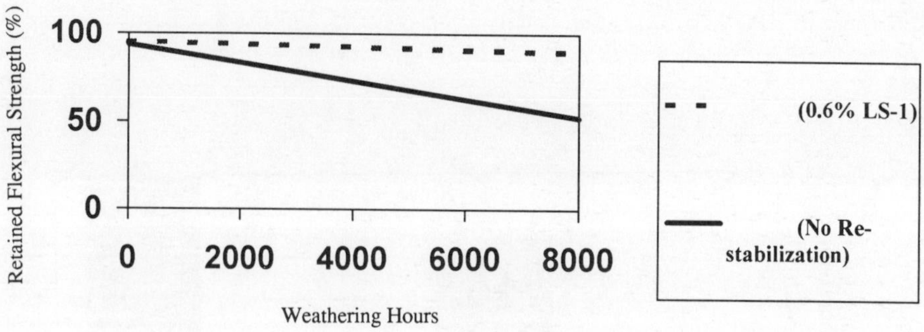

Polymer: 58% PE, 16% PP, 20% PS., 6% PVC
Artificial Weathering
Extrusion: 200°C

Figure 15. Light Stability of Mixed Plastics (Household waste collection).

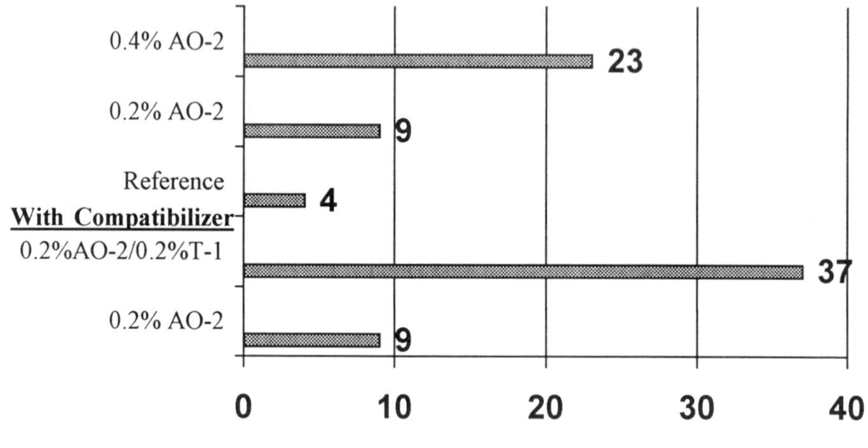

Polymer: 80%PO, 15% PS, 3% PVC, 2% PET
Samples: 2mm

Figure 14. Long Term Thermal Stability of Mixed Polyolefin Recyclate.

1.4. MIXED PLASTIC WASTE

Figure 13 shows that by using AO-4, the LTTS of mixed polyolefin recyclates is increased. Adding a compatibilizer and a thioester (T-1) gives synergistic effect to the LTTS.**(Figure 14).** LS-1 can also retain the flexural strength of the mixed plastic upon exposure. **(Figure 15)**

CONCLUSIONS

Re-stabilization is a necessary to be able to re-use recyclates, both scrap and PCR, especially in high value applications. There are many factors to consider for re-stabilization. One has to take into account the previous damage to the polymer and the residual stabilizer content. The additives to be used for re-stabilization must be specific to the polymer and application. Satisfactory service-life can be achieved by the use of appropriate stabilizers such as the following:

- combinations of phenolic anti-oxidant and phosphites are recommended to give processing and heat stability.
- combinations of ultraviolet absorbers (UVA) and hindered amine light stabilizers (HALS) for outdoor applications.

ACKNOWLEDGEMENTS

I would like to thank H. Herbst, K. Hoffman, R. Pfaender and F. Sitek from the Ciba Basle Venture Group, Additives for Plastics Recycling for their providing me with the data.

Also to the Waste and water policy group of the Hong Kong Government Environmental Protection Department for giving me an update on the recycling situation in Hong Kong.

And lastly, to my boss, Dr. Pearson, for all his support and help.

REFERENCES

[1] Lynch, Jerry and Nauman, Bruce E., "Plastics Recycling By Selective Dissolution", Technology Transfer Forum on Plastics Waste Management In Commonwealth Countries, March 1995.

[2] Troussier, Christian, "Feedstock Recycling: A European Initiative", Technology Transfer Forum on Plastics Waste Management In Commonwealth Countries, March 1995.

[3] Waste and Water Policy Group Environmental Protection Department, May 1994.

[4] Hope P.S., et. al., " A study of compatibilisers for recovering the mechnical properties of recycled polyethylenes, Plastics, Rubber and Composites Processing and Applications., Vol. 22, 1994, p.147-158.

[5] Paul J., et al., "Merits of blending stretch film PCR with Virgin LLDPE/LDPE film grades".

[6] Pearson, L. T., "Stabilization of Recycled HDPE" , Plastics Compounding, March /April, 1992.

[7] Saicu, D. et. al, "The photo-oxidative resistance of reprocessed Polypropylene", Polymer Degradation and stability, Vol. 46, 1994, p. 259-262.

[8] Drake, W.O. and Cooper, K.D., "Recent Advances in Processing Stabilization of Polyolefins", RETEC, 1993.

APPENDIX

List of Additives:

AO-1	Irganox® 1010 Pentaerythrityl tetrakis-3-(3,5-di-tert-butyl-4-hydroxy-phenyl)-propionate
AO-2	Irganox® 245 Triethyleneglycol-bis-3-(3-tert-butyl-4-hydroxy-5-methyl-phenyl)-proprionate
AO-4	Recyclostab® 411 (contains anti-oxidants and co-stabilizers)
T-1	Irganox® PS 802 Distearylthiodipropionate
Ign B225	Irganox® B225 (1:1 Irganox 1010/Irgafos 168)
P-1	Irgafos® 168 tris-(2,4,-di-tert-butyl-phenyl)-phosphite
LS-1	Recyclostab® 811 (contains anti-oxidants and light stabilizers)
HALS-1 -	Tinuvin® 770 (Bis-2,2,6,6-tetramethyl-4-piperidyl-sebacate)
HALS 2-	Chimassorb® 944 Poly-{[6,[1,1,3,3-tetramethylbutyl)-imino]-1,3,5,-triazine,2,4-diyl][2-(2,2,6,6-teteramethylpiperidyl)-amino]-hexamethylene-[4,(2,2,6,6-tetramethylpiperidyl)-imino]}

RECYCLING OF THERMOSET BASED COMPOSITES

M. Nasir, C. M. Ruzaidi and A. Halim

School of Industrial Technology
Universiti Sains Malaysia
11800, Minden
Penang, Malaysia.

INTRODUCTION

Recycling is now an important subject when dealing with most materials, in fact most often it is considered as an essential criterion in selecting a material for a particular application. In numerous engineering applications, thermoplastic composites have been advocated solely as the material in preference to thermosets in view of their recyclability factor alone. However thermosets are still preferred in areas in which elevated temperature and chemical resistance are critical. In other words the importance of thermoset matrices in composites are still necessary although with the availability of the so called engineering thermoplastics.

It is therefore imperative to investigate the recyclability of the thermosets so as to minimize the problems of disposal. The subject of recycling of such a system is not new, in fact many success stories have been reported (Endo, 1991; Wurtz, 1987), which include landfill use, energy regeneration etc. Here a simple process of regenerating and sorting of the thermoset based composites, specifically unsaturated polyester via a flocculation or cyclone technique combined with the usual grinding and breaking down of the material is presented. The segregated components viz. short and long fibres, the particulate matrices were then converted into ABS composites and their properties examined.

THE FLOCCULATION TECHNIQUE

The polyester composites supplied by the local industries comprising mainly of rejects, trimmers etc. were first cut into smaller pieces before grinding them using a powerful plastic grinder for more or less half an hour. They were than transferred into a drum having a stirrer and air was blown from the

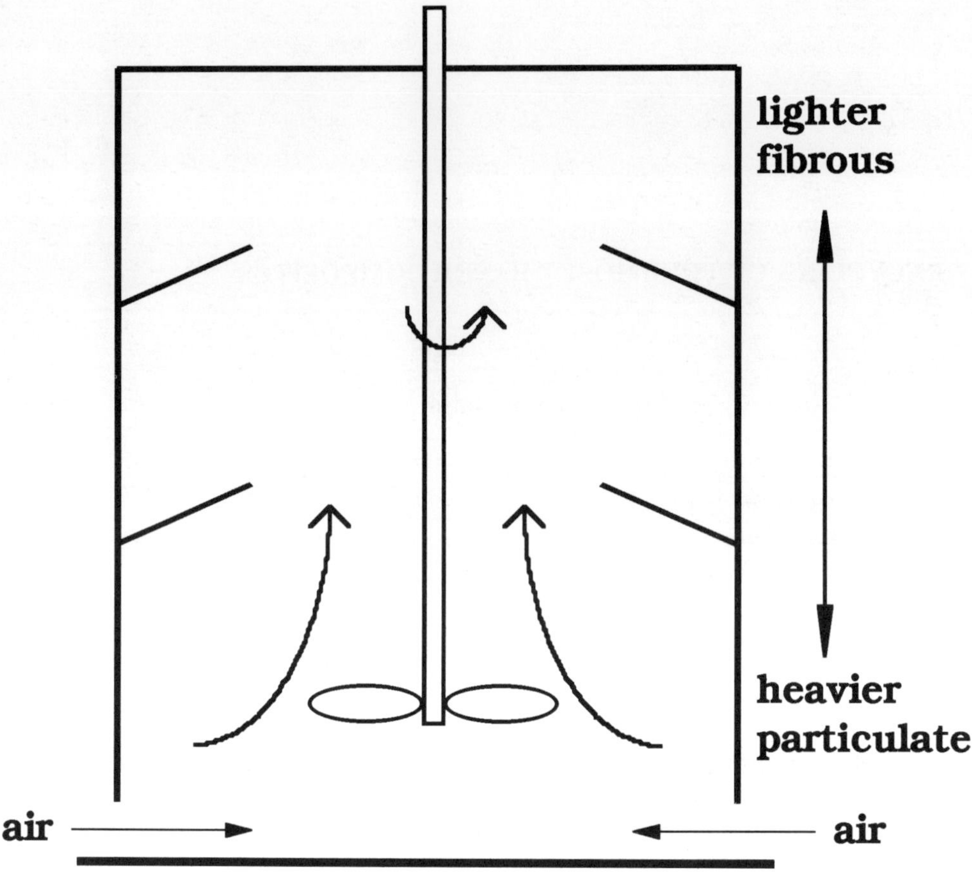

Figure 1. A schematic diagram showing the flocculation process of segregating the waste fibres.

bottom continuously until the various components are separated. The schematic diagram showing the segregation process is shown in Fig. 1. Since the segregation process works on the basis of density variation that are inherent with the components, its effectiveness depends very much on the grinding process. However the whole process can be repeated to ensure the desirable level of separation is achieved.

Fig. 2 shows the photograph of the segregated components, which were classified as types 1 and 2. The physical properties of both the type 1 and 2 which are basically different in the fibre weight fractions, W_f and also the resin weight fraction, W_r are given in Table 1. However it is obvious, both have dissimilar aspect ratios and particulate sizes of the fibres and the matrices, respectively, which will subsequently affect the composite properties viz. resin uptake, wettability, void content etc. (Crawford, 1987; Gachter and Muller, 1987)

ABS COMPOSITE PROPERTIES

ABS composite panels were prepared via a proprietary technique using both type 1 and 2 fibres/ fillers. Their subsequent properties viz. W_f and

Table 1. Type and characteristics of fibrous/ particulate fibres derived from the flocculation process

Type [a] of fillers derived by the flocculation process	Fibre weight fraction W_f %	Resin weight fraction [b] W_r %
1	44 - 58	42 - 56
2	32 - 38	62 - 68

[a] The type of fillers depend on the raw material feed, the grinding and the flocculation sequence
[b] Refers to the amount of polyester present

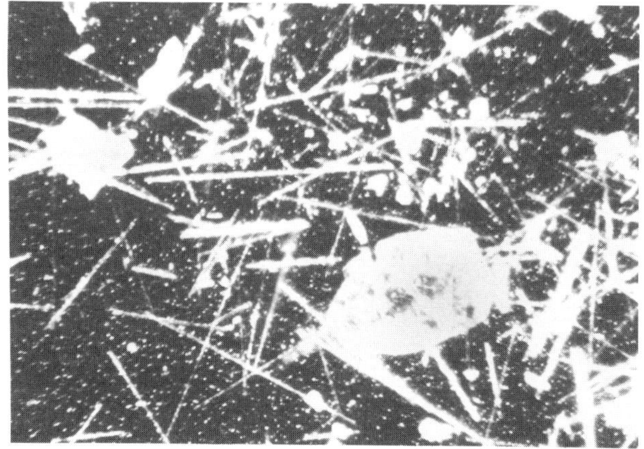

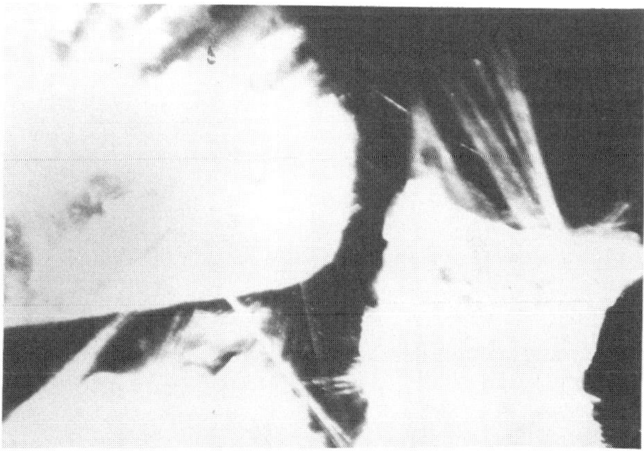

Figure 2. The photographs showing the type 1 and 2 fibres after the separation process.

compressive properties were evaluated, which are given in Table 2. It is apparent that type 2 gives higher compressive properties than type 1, which might be due to higher amount of fibre content present in its mixture than the type 1. Further, it well known (Gachter and Muller, 1987) that fibrous type fillers impart an overall better properties to those of a particulate type. In this case both type 1 and 2 fillers derived from the segregation process are basically a hybrid type, coming from the glass reinforcement and the polyester matrices. Further, the properties of the composite depend also the composition and the processing conditions; the latter in this case similar. The interphases and the matrices can be enhanced by the use of suitable coupling, which is not addressed yet.

Table 2. Properties of the moulded ABS thermoplastic composites

Type [a]	Composite			
	W_f %	W_f [b] %	Compressive strength (MPa)	Compressive modulus (MPa)
1	35 - 37	62 - 80	41.92	25.50
2	28 - 32	84 - 88	50.30	37.77

[a] Composite prepared with 40/60 w/w ratio via proprietary technique
[b] Values with ABS as matrices alone

Figure 3. The ABS composite panels demonstrating the thermoformability after moulding.

Since the composites are now thermoplastics, other advantages of such a system are easily obtainable. For example, these panels could be easily thermoformed after moulding as illustrated in Fig. 3. However the thermoformability would of course depends on the composite type.

CONCLUSIONS

The unsaturated polyester based composites can be converted into ABS composites after they have been broken down and segregated using the flocculation or cyclone technique. The subsequent ABS composites show properties typical of any thermoplastic composites such thermoformability, toughness, strength etc. which are somewhat governed by the segregation process composition and the moulding conditions.

Acknowledgement

This project is part of a composite realization scheme funded partly by the Government of Malaysia via the IRPA fund.

REFERENCES

Crawford, R. J., 1987, Plastics Engineering, Pergamon Press, Oxford, England.

Endo, M., 1991, Resources recovery and recycling of plastics., Kagaku Kugaku, 55(1) : **25**

Gacther, R. and Muller, H., 1987, Plastic Additives Handbook, Hanser Pub. Munich, Germany.

Wurtz, M. J., 1987, Plastic material recycling as part of scrap utilization possibilities and problems, Conversation Recycle, 10 : **177**.

BIODEGRADABLE FIBERS MADE FROM TRULY-BIODEGRADABLE THERMOPLASTICS

Masatsugu Mochizuki and Matsuo Hirami

Research & Development Center
Unitika Ltd.
23, Ujikozakura, Uji-Shi, Kyoto, 611, Japan

INTRODUCTION

The term " biodegradable fibers " has different meanings for different people. In this review, this term is used to refer widely to fibers which will degrade in biological environment, including any environment where *in vivo* or *in vitro* biological or biochemical processes are occurring, regardless of the degradation mechanisms.

According to this definition, the biodegradation of fibers involves non-enzyme-catalyzed chemical hydrolysis, especially bioabsorbable reactions *in vivo*, and enzyme-catalyzed chemical reactions caused by enzymes secreted by microorganisms under either aerobic or anaerobic conditions, as shown in Table 1.

Bioabsorbable sutures[1] made from poly(glycolic acid) (PGA), poly(lactic acid) (PLA), or their copolymers which decompose *in vivo* by chemical hydrolysis are well known since 25 years ago. They are, however, sensitive to atmospheric moisture which leads to hydrolysis in humid air alone, suggesting that they are not suitable for commodity or industrial uses.

An environmentally degradable fiber resistant to atmospheric moisture and primarily degraded by enzymes secreted by microorganisms in soil or sea water is required from the view point of industrial applications. Natural fibers or their regenerated fibers are stable during normal conditions of storage and use, but are degradable in microbially active environment. These fibers made from naturally occurring non-thermoplastic materials, however, have inherent disadvantages common to many natural substances, particularly variability in composition and properties, moderately poor mechanical properties, high water pickup, and incapability of thermal bonding, when

Table 1. Classification of biodegradable fiber

Type	Mechanism of degradation
Bioabsorbable	Non-enzymatic
Environmentally degradable	Enzymatic

Table 2. Classification of environmentally degradable fibers

Type	Thermoplastic	Non-thermoplastic
Fiber	Man-made	Natural or regenerated
Material	Aliphatic polyester	Polypeptide or polysaccaride
	Poly(ϵ-caprolactone)	Silk, Wool, Collagen
	Poly(butylene succinate)	Cotton, Linen, Rayon
	Poly(β-hydroxy butyrate)	Chitin, Ca alginate
Spinning method	Melt spinning	Solution spinning
Thermal bonding	Applicable	Not applicable
Hydrophilicity	Hydrophobic	Hydrophilic
Wet performance	Good	Poor

compared with man-made fibers made from biodegradable thermoplastics as shown in Table 2. The biodegradable man-made fibers made from truly-biodegradable thermoplastics, which are stable in humid air, but are degradable in the presence of enzymes secreted by microorganisms in compost, soil, sludge, lake or sea water, have not been commercially available yet. The authors have been developing this new type of fibers made from biodegradable thermoplastics such as poly (ϵ-caprolactone)(PCL), poly (butylene succinate)(PBS), poly (β-hydroxy butyrate) (PHB), and their copolymers.

The features of the biodegradation of biodegradable thermoplastic fibers are the principal subjects of this review, but important concepts related to this subject, especially types and mechanisms of biodegradation involved, types of polymers employed, and practical applications of the fibers, are also included.

MATERIALS AND FIBERS

As for truly-biodegradable structural materials for fiber making by melt spinning, aliphatic polyesters might be the only one that meets requirements regarding processability, physical properties and biodegradability of fibers obtained.

Most of aliphatic polyesters are biodegradable, while most of aromatic ones are not. An explanation that has been proposed for this difference is that aliphatic polyesters are flexible enough to fit into the active site of the enzyme, whereas aromatic polyesters are too rigid to permit this fit. Synthetic polymers like aliphatic polyesters, which have similar structures or functional groups to natural organic matter, are expected to be degraded non-specifically by some enzymes like lipase.

Commercially available aliphatic polyesters are classified into four types with respect to the primary chemical structure as shown in Table 3. It is suggested that the distance between ester groups in the polymer backbone is a factor in biodegradability as well as in the melting point and the physical property.

Poly(α-hydroxy acid)

In recent years, new applications of lactic acid as biocompatible and biodegradable plastics made from poly(lactic acid) (PLA), in other words polylactide or polylactate, have been highlighted because of its availability from renewable resources like corn. The most suitable method for the effective preparation of an optically active lactic acid is not chemical synthesis but rather fermentation using defined strains. The lactic acid can be polymerized to polyesters usually by the ring-opening polymerization of lactide, the cyclic dimer of lactic acid.

Table 3. Aliphatic polyesters

Compound／Chemical structure	Biodegradability	Examples
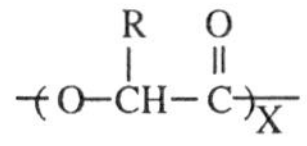 $-(O-CH(R)-C(=O))_X-$ Poly(α-hydroxy acid)	Chemical hydrolysis	R=H ; poly(glycolic acid)(PGA) R=CH_3 ; Poly(L-lactic acid)(PLLA) Lacty® / Shimadzu Corp.
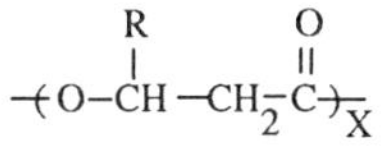 $-(O-CH(R)-CH_2-C(=O))_X-$ Poly (β-hydroxy alkanoate)	Enzymatic degradation	R=CH_3 ; Poly(β-hydroxy butyrate)(PHB) R=CH_3, C_2H_5 ; Poly(β-hydroxybutyrate valerate) (PHBV) Biopol® / Zeneca Ltd.
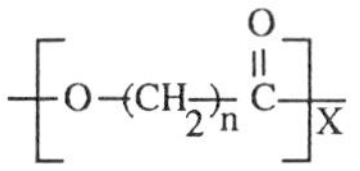 $[O-(CH_2)_n-C(=O)]_X$ Poly (ω-hydroxy alkanoate)	Enzymatic degradation	n=5 ; Poly(ε-caprolactone)(PCL) Tone® / Union Carbide Corp.
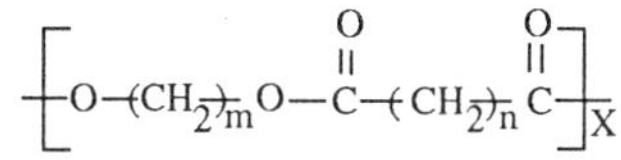 $[O-(CH_2)_m-O-C(=O)-(CH_2)_n-C(=O)]_X$ Poly(alkylene alkanoate)	Enzymatic degradation	m=4, n=2 ; Poly(butylene succinate)(PBS) Bionolle® / Showa Highpolymer Co., Ltd.

PLLA made from L-lactic acid is a hard, transparent and crystalline polymer having a melting of 170-180℃ and a glass transition temperature of about 53℃.

Poly(β-hydroxy alkanoate)

Poly(β-hydroxy alkanoate)s like poly(β-hydroxy butyrate) (PHB) are produced by the fermentation of renewable agricultural feedstocks by bacteria. PHB is naturally occurring material, accumulating in a wide range of microorganisms as a storage material for carbon and energy in the same way that humans accumulate fat. The microbial polyesters[3] are produced by a two-stage fermentation process, that is the cell growth and the subsequent polymer accumulation under nutrient limitation. The polymer is accumulated as discrete granules in the cell. Then the polymer is removed by an aqueous extraction process.

Commercially available poly(β-hydroxybutyrate valerate) (PHBV), namely " Biopol®" (Zeneca Ltd.), is composed of hydroxybutyrate (HB) units with between 0 and 30% of hydroxyvalerate(HV) units incorporated randomly throughout the polymer chain. The physical properties vary with the HV content of the polymer. For instance, as the HV content increases up to 20%, the melting point of the copolymer decreases from 177℃ similar to polypropylene down to 120℃ similar to polyethylene, and the flexibility and the toughness are improved.

It is possible to produce relatively strong fiber by in-line spin-draw technique using high molecular weight($\overline{M}w$>400,000) polymer with additives such as plasticizer and nucleating agent.

Poly(ω-hydroxy alkanoate)

Poly(ε-caprolactone) (PCL) is a synthetic polymer which is prepared from ε-caprolactone, a seven-membered ring compound, through an addition type of polymerization. The cyclic ester monomer opens its ring in the presence of an initiator that contains an active hydrogen atom. When a diol such as ethylene glycol is used as the initiator in conjunction with a catalyst such as stannous octanoate, dihydroxy-terminated polymer is formed. It exhibits both good hydrolytic and thermal stabilities since residual carboxylic acid end groups are basically absent.

High molecular weight PCL is a tough and partially crystalline polymer with a moderate low melting point of about 60℃, and is a semi-rigid material at room temperature having a modulus of between that of low-density and high-density polyethylene.

Using high molecular weight($\overline{M}n$=80,000) PCL, Tone® P-787(Union Carbide Corp.), tough and extremely high-strength fibers have been obtained by melt-spinning followed by cold-drawing. Both tensile and knot strengths of the fiber were observed to increase with increase in draw ratio, reaching tensile tenacity of over 8.5g/d and knot tenacity of 4.5g/d at draw ratio 9.

Poly(alkylene alkanoate)

Polycondensation process is a popular method of synthesizing polyesters like PET that are currently available in the market, while process was of no help in successfully developing high molecular weight aliphatic polyester utilizable as a biodegradable plastic. This is the reason why most of efforts to obtain high molecular weight aliphatic polyesters have been limited to ring-opening reaction of lactones or ring-opening polyaddtion of cyclic dimers.

It is possible to synthesize polyester up to a number-average molecular weight ranging from 15,000 to 18,000 by the conventional polycodensation process. As long as the molecular weight is around in this range, aromatic polyester such as PET is sufficiently suitable for practical use, but aliphatic polyester is not.

The combination of the discovery of a highly effective catalyst like alkyl titanate, that is not conductive to decomposition, with the addition of diisocyanate as coupling agent after the polycondensation reaction, is a key to successful synthesis of high molecular weight polyester.

Showa Highpolymer has developed a new type of aliphatic polyester(Bionolle®)[4] synthesized from aliphatic dicarboxylic acid and aliphatic diol by polycondensation process.

The standard grade(#1000), poly(butylene succinate)(PBS), composed of 1,4-butanediol and succinic acid, is reported to have a high molecular weight ($\overline{M}n$=30,000-40,000), and physical properties(Tm=119℃) comparable to polyethylene. Another grade(#3000) is a copolymer incorporating 20 mole % of butylene adipate(BA) as a comonomer, and has a melting point of 104℃.

Fibers

Biodegradable fibers of the above-mentioned biodegradable thermopastics were manufactured by conventional melt spinning process under the optimized spinning conditions.

Properties of monofilaments and multifilament yarns made from biodegradable thermoplastics are shown in Tables 4 and 5, respectively.

Table 4. Yarn properties of monofilaments

Material		Tensile tenacity (g/d)	Elongation at break (%)	Young's modulus (g/d)	Melting point (℃)
PCL	Poly(ε - caprolactone)	7.5 - 8.5	25 - 35	10 - 15	64
PBSA	Poly(BS -*co*-BA)	5.5 - 6.5	25 - 35	20 - 25	116
PHBV	Poly(3HB -*co*-3HV)	2.5 - 3.5	40 - 50	10 - 20	134

Table 5. Yarn properties of multifilaments

Material	Melting point (℃)	Glass transition (℃)	Tensile tenacity (g/d)	Elongation at break (%)	Young's modulus (g/d)
Poly(ε-caprolactone) (PCL)	64	-60	4.0 - 5.5	60 - 80	10 - 20
Poly (butylene succinate) (PBS)	116	-32	4.5 - 5.5	40 - 60	15 - 25
Poly(L-lactic acid) (PLLA)	178	53	3.0 - 5.0	20 - 30	55 - 65

BIODEGRADATION MECHANISM

There are discrete differences between chemical hydrolysis and enzymatic hydrolysis with regard to biodegradation mechanism[5].

Chemical Hydrolysis

The synthetic polyesters used in bioabsorbable suture, principally polyglycolide and poly(glycolide-*co*-lactide), are degraded in the body mainly by chemical hydrolysis. There is little evidence that the hydrolysis of these polyesters of α-hydroxy acids is catalyzed by hydrolase or depolymerase enzyme.

Poly(L-lactide) is biodegraded at a lower rate than polyglycolide as would be expected from the chemical structure having hydrophobic methyl side group. Figure 1 shows that polyglycolide suture is degraded in saline solution over a period of 2-3 weeks which is a period required for wound healing of soft tissue in the body. On the other hand, poly(L-lactide) is not degraded within 4-6 months which is required for recovery of hard tissue like bone[6].

The rate of biodegradation of poly(α-hydroxy acid)s such as PLA depends on the molecular weight distribution as well as the molecular weight of the polymer, and especially on the contents of monomer and oligomer. The monomer and the oligomer remained in the polymer accelerate the chemical hydrolysis not only in water but also in humid atmosphere. Both the tensile strength and the elongation at break of the fiber made from PLLA without purification decreased continuously with time in atmosphere, suggesting that the water in atmosphere could penetrate into the fiber and hydrolyzed the polymer. It is an evidence of the chemical hydrolysis that the molecular weight as monitored by relative viscosity of the polymer solution (0.5 wt% of tetrachloroethane/phenol(1/1) solution) decreased from 1.40 to 1.25 after six months.

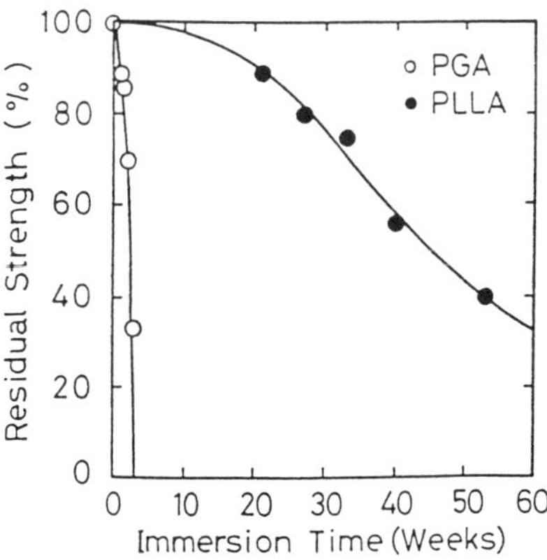

Figure 1. Change in the tensile strength of suture made from PGA and PLLA immersed in saline solution at 37 ℃[6].

Enzymatic Hydrolysis

The biodegradation behaviors of PHBV, PCL and PBS monofilaments were evaluated by field tests. The result[7] of the soil burial test of PCL monofilament showed that very rapid decreases in elongation at break and tensile strength of the sample occurred with time accompanying the weight loss, as shown in Figures 2(a) and 2(b). The tenacity of the monofilament decreased faster as the denier of the monofilament became finer, as might be expected due to the greater surface-to-volume ratio.

Figures 3(a), 3(b) and 3(c) are electron photomicrographs of the surface appearance of PCL monofilaments before and after the soil burial and the sea water immersion. The surface of the fiber before degradation is essentially smooth as seen in Figure 3(a). After partial degradation of the fiber in soil and in sea water, the surface has been deeply etched or eroded as seen in Figures 3(b)and 3(c).

This surface erosion mechanism is supported by two observations. The first observation is that the diameter of the fiber decreased gradually with time during the biodegradation test in soil and sea water. In addition, the molecular weight of the residual fiber in soil changed little . It is reasonable to think that enzymatic hydrolysis occurred only on the surface of the fiber, because the enzymes could not penetrate into the fibers.

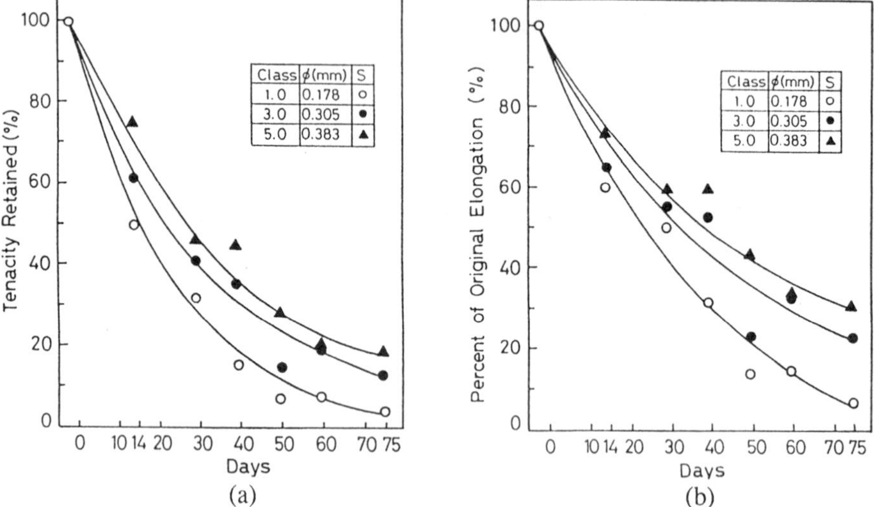

Figure 2. Change in the tenacity (a) and the elongation at break (b) of PCL monofilaments with different diameter which were buried in soil.

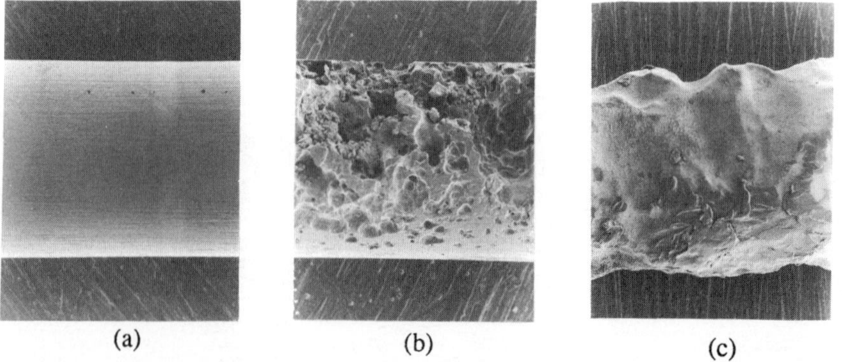

Figure 3. Electron photomicrographs of PCL monofilaments before(a) and after a month of soil burial(b) and sea water immersion(c).

Table 6. Primary features of biodegradation

Type	Chemical hydrolysis	Enzymatic hydrolysis
Polymer	PGA, PLLA	PCL, PBS, PHBV
Access point	Surface to inner	Surface only
Surface appearance	Smooth (Not eroded)	Eroded
Weight loss	No	Yes
Molecular weight reduction	Yes	No

Figure 3(c) shows the degradation of the PCL fiber immersed in sea water in Tokyo Bay. It is supposed that microorganisms in fresh sea water are responsible for the observed biodegradation, because it was observed that various buffers (pH 2-12) had no effect on PCL fiber, and that sea water stored in a flask in which no living microorganism also made no significant change in appearance of the fiber. It could be seen that very rapid reduction in tensile strength of PCL fibers occurred, indicating that PCL fibers were readily biodegraded when exposed to sea water as well as to soil.

These results confirm that biodegradation is initiated by the action of microorganisms such as bacteria or fungi, growing on the surface of the polymer and secreting extra cellular enzymes which degrade the polymer back to the individual molecular fragments. These soluble degradation products are then absorbed through the cell wall into the colonizing cells where they are used as a carbon source for growth.

Primary features of biodegradation in regard to the mechanism are shown in Table 6.

Effect of polymer morphology

The effect of solid state morphology on the course of biodegradation of semi-crystalline ploymer has been shown[8]. We have already reported the study[9] on hydrolysis of PCL fibers by lipase in regard to effects of the draw ratio on enzymatic degradation. The rate of biodegradation did depend adversely on the crystallinity of the fiber, because the enzymes degraded selectively amorphous or less-ordered regions which allowed for enzymes to penetrate into the fiber, although the crystalline regions were also degraded eventually.

In order to obtain an understanding of the effect of both material and morphology of fiber on biodegradation, soil burial test was carried out for four kinds of monofilaments with the same radius(around 285 μ m), but with or without drawing. The rate of weight loss in the soil burial test is shown in Figure 4. It can be seen that the rate of degradation is in the order of PHBV, PCL and PBS in respect of material, and that drawing reduced remarkably the rate of degradation of PBS fiber, probably due to higher crystalline content.

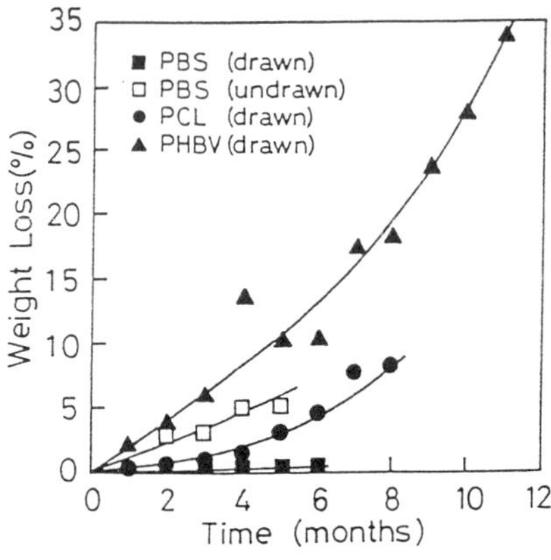

Figure 4. Weight loss of monofilaments with or without drawing in soil burial test.

APPLICATIONS

It is now widely accepted that there is no single solution to waste management of plastics, synthetic fibers or nonwovens. Under the circumstance where the reuse or recycling of a product is unlikely to be viable technically, economically, or logistically, the use of composting may provide a solution.

More than 50% of the waste stream today is degradable. If this is composted, it would not only relieve the mounting pressure on landfill but also make ppossible for this organic matters to be recycled biologically into a stable soil like product.

Biodegradability is of greatest benefit to those products which find their way into sewage systems, soil, compost or landfill sites which are managed for degradation. Obvious applications for this lie in disposable hygiene and agricultural sectors etc. as shown in Table 7.

Table 7 Applications of biodegradable fibers or nonwovens

Sector	Examples
Fishing	Fishing line, Fishing net
Agriculture	Seeds sheet, Horticulture sheet, Weed prevention sheet
Geotextile	Vertical drain sheet, Road-side protection sheet
Paper making	Fibrous binder
Hygiene	Disposable diaper, Personal care products
Medical device	Suture, Wound dressing
Commodity	Compost bag, Wiper, Filter

REFERENCES

1. E. J. Frazza and E.E.Schmitt, A new absorbable suture, *J.Biomed.Mater. Res. Symposium*, 1:43 (1971).
2. R. W. Lenz, Biodegradable polymers, *Advances in Polymer Science,* 107:1 (1993).
3. Y. Doi, "Microbial Polyesters", VCH Publishers, New York (1990).
4. E. Takiyama and T. Fujimaki, "BIONOLLE" biodegradable plastic through chemical synthesis, *in* : "Biodegradable Plastics and Polymers", Y. Doi and K. Fukuda, ed., Elsevier, The Netherlands (1994)
5. Y. Doi, Y. Kanesawa, M. Kunioka and T. Saito, Biodegration of microbial copolyesters: Poly(3-hydroxybutyrate-*co*-3-hydroxyvalerate) and poly(3-hydroxybutyrate-*co*-4-hydroxybutyrate), *Macromolecules*, 23:26 (1990).
6. T. Oka, Development of bioabsorbable medical devices, *Kinou Zairyo,* 13(12) :16(1993)
7. M. Mochizuki and K. Kudo, Polycaprolactone fiber, *in* : "Biodegradable Polymers", Y. Ikada ed., Kohbunshi Kankokai, Japan(1994).
8. S. J. Huang, Biodegradable polymers, *Encycl. Polym. Sci. Eng.*, 2:220(1985)
9. M. Mochizuki, M. Hirano, Y. Kanmuri, K.Kudo and Y. Tokiwa, Hydrolysis of polycaprolactone fiber by lipase : effects of draw ratio on enzymatic degradation, *J. Appl. Polym. Sci.*, 55:289(1995).

' ENVIRONMENTALLY FRIENDLY ' FATE OF PLASTIC WASTE ?.

Mohammed Yasin, Allan J. Amass and Brian J. Tighe

Biodegradable Polymers Unit, Speciality Materials Research Group, Department of Chemical Engineering & Applied Chemistry, Aston University, Aston Triangle, Birmingham B4 7ET, (UK).

INTRODUCTION

Over the last few years, there has been active debate on the options available in overcoming the increasing problems associated with plastic waste management, both from environmentalist and industrialist. The current interest and awareness in 'environmentally friendly' packaging has been brought on by both constantly changing legislation initiated by the politicians who need to be seen by the public that they are concerned about the environment and also through environmental conscience of the public and various sectors of the industry.

There has been 'evidence' that plastics cause environmental damage, are stable over a long period of time and those that degrade pollute water supplies from the degradation products. The current drive in the management of plastic waste is being directed towards plastic recycling and incineration schemes rather than disposal in landfill or efforts in the designing of degradable plastics. In this paper, the various options available for dealing with plastic waste management will be discussed together with the economic, social and environmental advantages and disadvantages presented by each option. The use of biodegradable plastics as part of a well-structured plastic waste management policy will be demonstrated.

WHY IS THERE A CURRENT INTEREST IN WASTE ISSUES ?.

Environmental issues are of topical concern on a global level. Governments, through pressure from environmentalist and the general public are starting to address the environmental problems that have not suddenly appeared, but have always existed. The changing life-styles of the consumer, and hence its need for items of convenience, whether as food or luxury, has had the effect of magnifying the problem of plastic packaging. This can be seen as litter and thus this has lead to demands for reduction of this visible waste.

Whereas the initiative for limiting emissions from automobiles came from the USA, the current influence on packaging minimisation and waste disposal, particularly in

Europe has come from Germany. The European Community Directive on packaging is currently being debated, industry is uncertain as to if it will be able to comply with the legislation in time. Politicians are eager to demand changes in waste disposal, to please the public not only because of 'Green Issues', but also of the "feels good" factor. They want to see sudden changes but do not realise how the mechanics of the actual changes will work. The Green Dot or the DSD system in Germany of collecting used plastic waste for recycling is an example of this; legislation was in place to the collect used plastic, but the infra-structure to recycle this plastic was not fully in place, so the German Government was paying companies to store this waste.

HOW BIG IS THE PROBLEM WITH PLASTIC WASTE ?

Various estimates have been made on the size of plastic waste, but if one looks at the types and amount of the differing materials that are present in municipal waste, plastic waste only accounts for about 8% (1). Most of this comes from plastic packaging and in the UK, this has been estimated as being equivalent to 7 kg of plastic per person per year. Although the major component of municipal waste is paper, plastic because of its low density (and hence large volume) has a larger environmental impact. Plastic packaging, particularly the polyethylene shopping bags, having found their way into the sewerage systems, have caused considerable drainage problems from total blockage of the effluent and build-up of potentially explosive levels of methane. Indeed, many countries have banned the use of plastic shopping bags in favour of the paper bags for this very reason. What are the various options available for the control or to decrease this plastic waste ?.

METHODS FOR PLASTIC DISPOSAL

Several options are available for plastic disposal and these are shown in Figure 1. The most obvious method of reducing plastic waste is to use less packaging and for this the manufacturers of the products will need to design the packaging. With the consumer market becoming more technological and the corresponding dramatic changes and needs to satisfy the customer, it was inevitable that there would be an explosion in the amount, quality and type of material being used for packaging of the goods for the sophisticated consumers. Several products, with 'environmentally conscious' redesigned packaging are already on the market and with increasing interest being shown by the consumer and the legislators, more no doubt will follow.

The reuse of packaging is another obvious method of reducing plastic waste, but this may not be practicable in many cases. Whereas the recycling of in-house plastic by the manufacturers has been practised for many years, the recycling of post-consumer plastic waste is an area that is receiving considerable attention currently. It is important to remember that at present very few successful schemes for the collection of plastic waste are available, so plastic will find its way into municipal waste. Hence, the major method of disposing municipal waste is through landfill. Organic matter is able to degrade in a well managed site, but most plastics will be left intact. Due to their low weight but high volume, plastics will have the tendency of occupying a very large space in the landfill, so this method is not favoured for their disposal. Certainly, recent increase in the charges for landfill sites has made this choice even more unpopular.

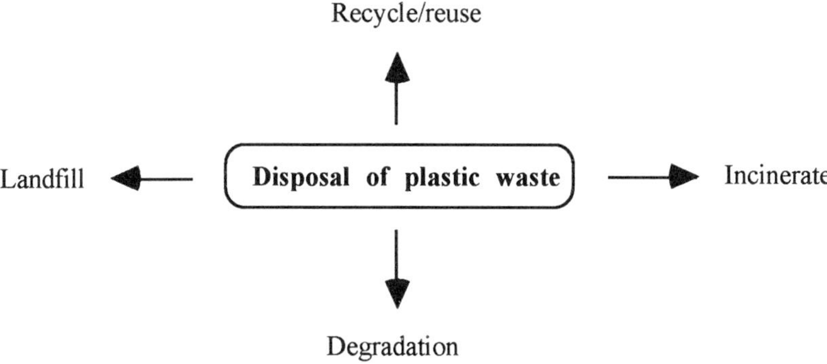

Figure 1. Methods for disposal of plastic waste.

Incineration or burning of municipal waste is another useful method for the disposal of plastic waste. Although the calorific value of plastics is higher than coal a number of considerations need to be taken in account for the operation of an 'environmentally friendly' incinerator. The emissions from the incinerators will have to be strictly monitored so that toxic gases are unable to escape into the atmosphere. This may involve redesigning of the incinerator plants or modifications to the operational procedures to increase the efficiency of the incineration process, at the same time reducing the environmental risks.

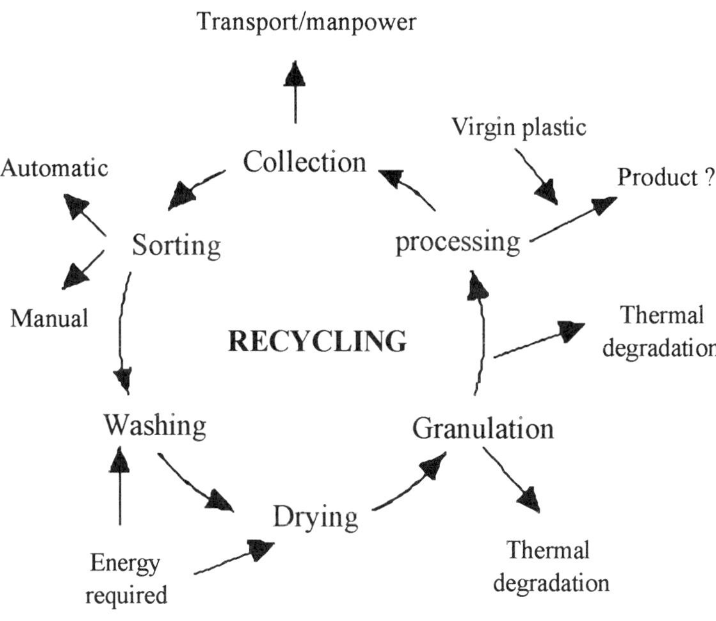

Figure 2. Various parameters for consideration in recycling of plastics.

Recycling is another viable option in theory but not every plastic article, and indeed not every plastic is recyclable. There is a limit to the number of times any plastic material can be recycled. Because during recycling and the re-fabrication process, plastics have to pass through several thermal processes these will affect their chemical and physical stability, and consequently will have drastic effects on their performance and usefulness. There is a shortage of post-consumer plastic waste due to limited collecting schemes and so the needs of the plastic converters are not fully met. As illustrated in Figure 2, collecting plastics for recycling is very labour intensive process and quite complex.

The ideal option is for the householders to separate the plastic out into its various types which could then be collected and sent to the recycling plant for processing. On its arrival the plastic has to be sorted out into the various types, most often manually because as yet a fully automated system for separation is not available. The plastic is then washed, dried and granulated into chips which can then be used for fabrication of the desired plastic product. It should be noted that the steps for the sorting, washing, drying and granulation require energy and time. The level of contamination of plastic should be the lowest possible to make it economically viable to wash and dry the plastic waste with the minimal costs being incurred. Both during the granulation and the fabrication process the plastic is subjected to heat and shear, hence both thermal and mechanical degradation may necessitate the need for the incorporation of a small amount of virgin material and/or stabilisers to be mixed with recycled plastic for achievement of acceptable levels of physical properties from the plastic article. This is another reason why articles made from recycled plastic are more expensive than those made from virgin material. Recycled plastic is commonly used to make down-grade products such as bins, dust bin liners, milk and beer crates, detergent and oil bottles, rather than for up-market, high value consumer products.

The fourth option is to design plastics that are either photodegradable or biodegradable. Ten years ago there was hardly any significant interest in degradable materials for packaging and indeed biodegradable polyesters such as polyhydroxybutyrate and polycaprolactone were either too expensive for packaging use or the materials available had low molecular weight and so inadequate physical properties[2].

BIODEGRADABLE POLYMERS FOR PACKAGING?

Biodegradable polymers can be classified as natural or synthetic. The four usual methods of designing new polymers are graft copolymerisation, polymerisation, alloying and blending Whereas both grafting and copolymerisation will need a new monomer and polymerisation technique, and hence there may be a long lead time for product development from chemical synthesis of the new polymer to product fabrication, both blending and alloy techniques are extremely useful techniques in this aspect. They allow the development of new materials, by mixing with existing available material in the shortest possible time. As the time needed to place new materials on to the market is reduced by using alloying and blending techniques, the development costs are also much reduced.

Several microbially produced or chemically synthesised polyesters are currently available commercially and are finding use as biodegradable material. Our work on developing materials for packaging applications has been involved with the bacterially produced biodegradable polyester, polyhydroxybutyrate (PHB) and its copolymers with hydroxyvalerate (HV). Applications of P(HB-HV) copolymers, such as for packaging has been limited because of their high cost of production in relation to commodity plastics.

One possibility of producing commercially viable biodegradable materials for applications where price is of importance is by blending P(HB-HV) copolymers with other polymers. Figure 3 illustrates the degradation of 20% HV P(HB-HV) copolymer and its blend with 30% amylose under hydrolytic (pH 7.4 and 37°C) and biodegradation in garden soil over a period of time.

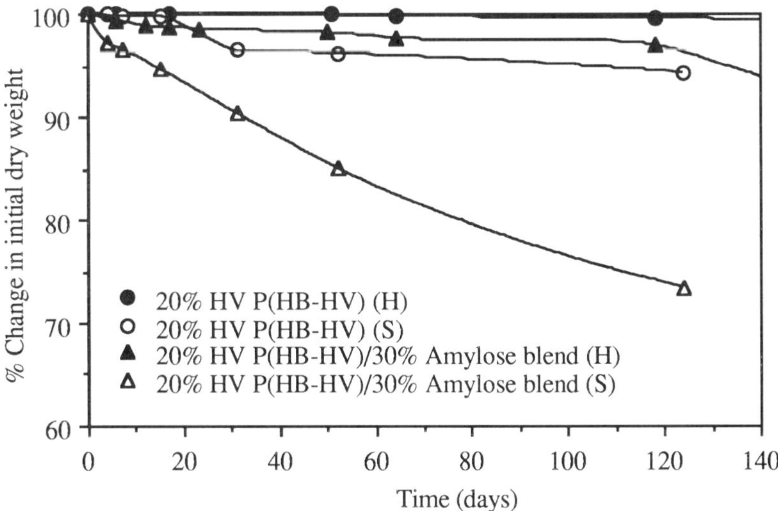

Figure 3. The change in the initial dry weight of 20% HV P(HB-HV) copolymer and 20% HV P(HB-HV)/30% amylose blend in pH 7.4 buffer at 37°C, (H), and in garden soil, (S), respectively. Injection moulded samples.

It is evident from Figure 3 the marked gulf between the rate of hydrolysis and true biodegradation for the P(HB-HV) copolymers. Whereas less than 5% weight loss occurs for the 20% HV P(HB-HV)/30% amylose blend in the buffer, in the soil nearly 25% weight loss occurs during the same time period. A detailed study of the degradation of these polymer blends under a range of conditions is to be published soon.

CONCLUSION

In the quest for the 'environmentally friendly' fate of plastic waste, a number of possible options are available. The best plastic waste management policy should use all the options available.

REFERENCES

1 F.E. Mark, The role of plastics in municipal solid waste combustion, PWMI Report.

2 M. Yasin and B.J. Tighe, Strategies for the design of biodegradable polymer systems: Manipulation of polyhydroxybutyrate-based blends, *Plastics, Rubber and Composites: Processing and Applications*. 19:15-27 (1993).

CHEMICAL RECYCLING OF WASTE PLASTICS: CATALYTIC CRACKING OF HEAVY OIL FROM WASTE PLASTICS OVER Ni-REY ZEOLITE

Ahmad Rahman Songip,[1] Takao Masuda,[2] Hiroshi Kuwahara,[3] and Kenji Hashimoto[2]

[1]Department of Chemical Engineering, Universiti Teknologi Malaysia, 54100 Kuala Lumpur (Malaysia)
[2]Department of Chemical Engineering, Kyoto University, Kyoto 606 (Japan)
[3]Sanwa Chemical Co. Ltd., Kyoto 613 (Japan)

INTRODUCTION

The increasing amount of waste plastics which causes serious pollution problems is a cheap and abundant source of chemicals and energy. Chemical recycling method which recovers hydrocarbons is recognized as the ideal approach in recycling the waste.[1] The economics and engineering constraints in the chemical method of a combined pyrolysis-reforming reactor has been discussed earlier.[2] The pyrolysis-reforming reactor was found to inhibit systematic study on the effects of various reaction conditions. A proposed process scheme of independently studying the pyrolysis and catalytic reforming reactions has been shown to alleviate the constraints of the combined pyrolysis-reforming scheme.[2]

The proposed scheme also offers lower costs of capital and transportation than that of the combined pyrolysis-reforming reactor.[2] We have also reported that rare-earth metal exchanged Y-type (REY) zeolite which has the proper pore size and acidic properties is the most effective catalyst from the catalytic screening test.[2]

High quality gasoline which has lower contents of aromatics and normal paraffins as well as high contents of iso-paraffins and cyclo-paraffins is sought in view of recent requirement for reformulation of gasoline pool.[3] High content of aromatics is undesirable since aromatics do not completely burn out, hence producing high carcinogen aromatics emission. While normal paraffins are known to reduce the research octane number.

We have reported the optimum reaction conditions and catalytic properties of REY zeolite that yield high gasoline yield and quality.[4] However, mere controlling the reaction conditions did not significantly reduce the aromatics and paraffins contents. Hence, a more effective method is desirable.

The objective of this paper is to examine preliminary efforts to increase the yield and quality of the gasoline product by reducing the undesirable components in the gasoline fraction, such as aromatics and paraffin products. These were performed by supporting nickel on REY zeolites and co-feeding steam.

Polymers and Other Advanced Materials: Emerging Technologies and Business Opportunities. Edited by P. N. Prasad *et al.*, Plenum Press, New York, 1995

REY zeolite, being the most effective solid acid catalyst for the reforming reaction,[2] was ion-exchanged with nickel in order to add the hydrogenation and dehydrogenation activities. Hydrogenation is required to reduce the aromatics products. On the other hand, dehydrogenation activity and the acidity of the zeolite support lead to the increase in iso-paraffins and cyclo-paraffins yields as well as to the decrease in n-paraffins yield. The performance of the prepared nickel supported REY (Ni-REY) zeolite was compared to other zeolite catalysts.

The main concern in the chemical recycling of waste plastics is on the economic viability of the process. The usage of nitrogen or hydrogen as the carrier gases would increase the production cost. Hence, a cheap and suitable carrier gas should be sought after. Steam, which is cheaply and abundantly available, is proposed to be used as a carrier gas. Also, the gaseous compounds, which are the by-products of the cracking reaction, can be utilized as fuel to generate steam from water.

Under steam atmosphere, the dealumination process of the zeolite catalyst is expected to occur which reduces the amount of strong acid sites and the activity of the catalyst. Also, REY zeolite is expected to show high stability under steam atmosphere as compared to HY and HZSM-5 zeolites, because the incorporation of rare earth metals in the zeolite stabilizes the aluminum atom in the zeolite frameworks.

Table 1. Chemical and Physical Properties of Catalysts.

Catalyst	ZSM-5	HY	REY	Ni-REY (0.5 wt%)	Ni-REY (1.0 wt%)	Ni-REY (3.0 wt%)
Si/Al	65	5.5	5.5	5.5	5.5	5.5
Pore Size (A)	5.3x5.6	7.4	7.4	7.4	7.4	7.4
Strong Acidity (mol/kg)	0.24	1.21	1.05	1.002	1.118	1.096
Weak Acidity (mol/kg)	0.122	0.44	0.375	0.2	0.268	0.255
Ni Loading (wt%)	-	-	-	0.52	0.95	1.12

EXPERIMENTAL

Nickel supported REY zeolite was prepared by ion-exchange method using the solution of nickel chloride. Table 1 summarizes the physical and chemical properties of catalysts used in this work. The amount of acid sites were determined using conventional TPD experiment. The measured amount of nickel loading for 0.5 and 1.0 wt% Ni-REY, as determined by using XRF, corresponds well to the preparation procedures.

The reaction was performed using a fixed bed reactor. The details of the experimental apparatus and procedures, analytical methods, lumping of reaction products as well as determination of research octane number (RON), and amount of coke loading were described elsewhere [2, 4]. The lumpings of the products are: gaseous compounds carbon number of 1 to 4: C1-C4, gasoline fraction (C5-C11), heavy oil (above C12), and coke.

H_2 and H_2O were employed as carrier gases. The nickel supported REY was first reduced at 723 K under flowing H_2 for at least 3 h. The feed oil was prepared by pyrolyzing polyethylene plastics at 723 K. The pyrolyzate was then distilled at 473 K to produce a residue having 95% heavy oil content and was used as the feed oil.

RESULTS AND DISCUSSION

Acidic Properties

Table 1 shows that the presence of nickel metal reduces both the amount of total and strong acid sites of the parent REY zeolite. Data for ZSM-5, HY, and REY zeolites have been published elsewhere.[2,4] The reduction in the amount of acid sites was due to the selectively adsorbed nickel on the strong acid sites, referred as nickel "blanketing".[5] Increasing the amount of nickel loading, however, resulted in the slight increased in the amount of acid sites to a maximum value at 1.0 wt% nickel loading.

Yield of Reaction Products

Table 2 shows the conversion (refers to the conversion of heavy oil) and yield of the reaction products at reaction temperature of 673 K, reaction time of 3 h and time factor of 1.0 kg-cat.h/kg for various catalysts and carrier gas atmosphere. These parameters were the optimum parameters from previous experiments.[2,4] The feed oil was effectively cracked to produce various reaction products: gasoline, gas and coke.

Although ZSM-5 zeolite has the highest conversion levels, the yield of gasoline is the lowest. The yield of coke on ZSM-5 zeolite is the lowest due to the non-existence of supercage in the structure of ZSM-5 zeolite.[2]

The levels of conversion for HY and REY zeolite were similar. However, the REY zeolite registered higher yield for gasoline and lower yield for gas and coke than that of HY zeolite. These results were due to the existence of rare-earth metals in REY zeolite which improves the performance and stability of the parent HY zeolite.[2]

By supporting nickel on REY zeolite (Ni-REY), the conversion decreased slightly. However, its yield for gasoline increased and yield for gaseous products decreased as compared with those for the REY zeolite. Ni-REY (0.5 wt%) registered the lowest yield for coke formation.

Table 2. Conversion of Heavy Oil and Yields of Reaction Products: Reaction Temperature, T= 673 K, Reaction Time, t = 3 h, Time Factor, W/F = 1.0 kg-cat.h/kg.,N2 = nitrogen atmosphere, H_2 = hydrogen atmosphere, (0.5 wt%) = 0.5 wt % of nickel loading.

Catalyst	ZSM-5	HY	REY	Ni-REY (0.5 wt%)	Ni-REY (1.0 wt%)	Ni-REY (3.0 wt%)
Carrier Gas	N_2	N_2	N_2	H_2	H_2	H_2
Conversion of Heavy Oil	86.17	77.34	78.57	75.89	72.87	66.06
Yield [wt%]						
Gasoline	30.36	37.83	47.52	50.53	48.85	46.32
Gas	55.77	38.38	30.00	24.48	23.37	19.11
Coke	0.04	1.13	1.05	0.88	0.66	0.63
Unreacted Heavy Oil	13.83	22.66	21.43	24.11	27.13	33.94

Gasoline Quality

The following equation was used to estimate the gasoline quality in terms of Research Octane Number (RON):[6]

$$RON = -1.0729\,Y_{NP2} + 0.7875\,Y_{IP1} + 0.0976\,Y_{IP2} + 0.3395\,Y_{CP} + 0.4049\,Y_{AR} + 69.0306 \quad (1)$$

where Y_i is the weight fraction of the i-th component in the gasoline fraction. Subscript NP2 is the n-paraffins without C_5, IP1 the total iso-paraffins from C_5 to C_7, IP2 the total isoparaffins without C_5 to C_7, CP to total cycloparaffins, and AR the total aromatics.

Eq (1) suggests that gasoline with a high RON value contains large amounts of iso-paraffins (IP1), aromatics (AR) and cyclo-paraffins (CP), and a small amount of n-paraffins (NP2). However, one of the basic requirements for the gasoline reformulation is the need to reduce the content of aromatics in the gasoline pool,[3] because aromatics are not easily burned out. Gasoline of high quality therefore contains large amounts of IP1 and CP and small amounts of NP2 and AR.

Figure 1 shows the RON values and the components in the gasoline fractions. Included also are the regular and high-octane commercial gasolines. The high RON values for gasoline obtained on ZSM-5 zeolite and commercial gasolines are due to the large amount of aromatics. The aromatics was expectedly produced on ZSM-5 zeolite due to the size of the pore channels of the zeolite which are nearly equal to that of benzene. The amount of environmental friendly iso-paraffins in those gasolines are the lowest.

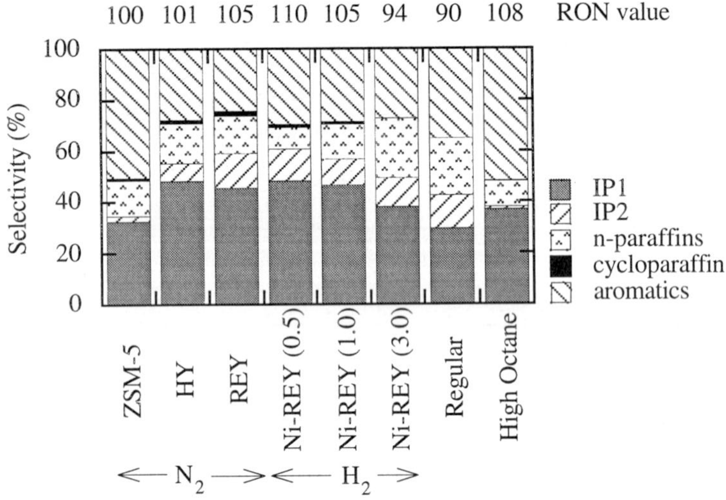

Figure 1. Gasoline RON Values and Distribution of Gasoline Components.

On the other hand, gasolines obtained on HY, REY and Ni-REY zeolites exhibit low amount of aromatics and high amount of iso-paraffins. Due to its proper pore size and proper acidic properties, REY zeolite is much more effective than HY zeolite.[2]

The introduction of nickel on REY zeolite exhibits further improvement on the quality of gasoline, especially at low amount of nickel loading. At 0.5 wt% nickel loading, the yields of iso-paraffins and n-paraffins are the highest and the lowest, respectively. The amount of aromatics, however, shows an increment over that of REY zeolite. Consequently, its RON value is the highest.

The increased in the amount of nickel loading resulted in two conflicting results: reduction of catalyst acidity due to the nickel "blanketing" and the the increased in the hydrogenation and dehydrogenation activities.

The reduction in the catalyst acidity lead to the reduction in cracking rate of n-paraffins and reduction in isomerization of n-paraffins. The increased in the hydrogenation and dehydrogenation rates lead to the decreased in the yields of cyclo-paraffins and aromatics, respectively. Because the former effect is much more pronounced than that of the latter, increasing the amount of nickel loading lead to the slight decreased in the quality of gasoline.

These results show that there is an interaction between the amount of strong acid sites and the amount of nickel loading. Figure 1 shows that the optimum amount of nickel loading on REY catalyst is about 0.5 wt % loading.

Effect of Time Factor, W/F

Figure 2 shows that the reaction rates were highest at lower time factor and then slowed to a steady-state values at higher time factor. The effects of time factor on the RON value, shown in Fig. 3, indicates that the optimum time factor is at about 1.0 kg-cat.h/kg.

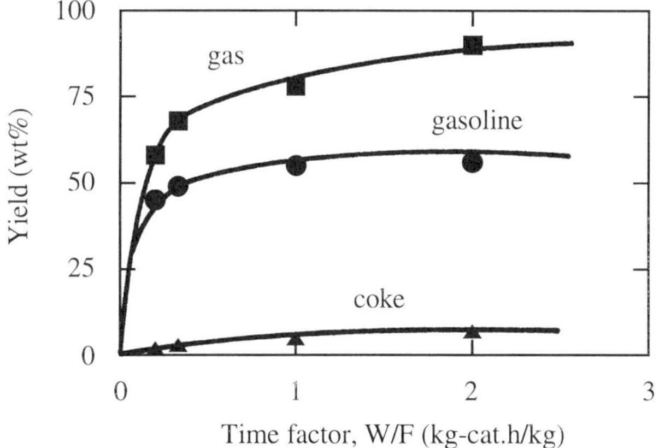

Figure 2. Effects of Time Factor on Product Yields.

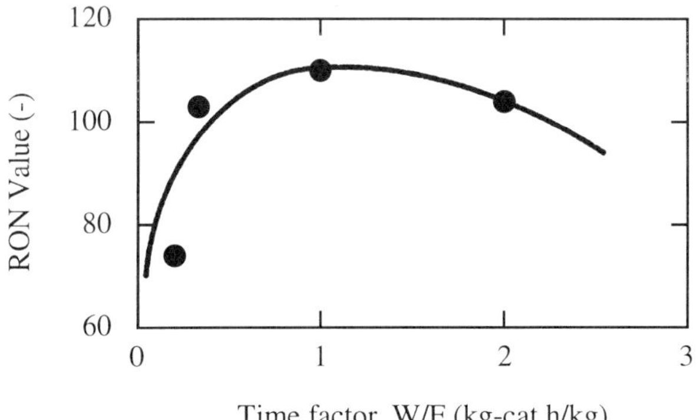

Figure 3. Effect of Time Factor on Gasoline Quality.

Table 3. Effects of Temperature and Carrier Gases on Conversion of Heavy Oil and Gasoline Yield using 0.5 wt% Nickel Loading and Time Factor, W/F, of 1.0 kg-cat.h/kg.

Temperature [K]	573	673	773	873
Conversion of Heavy Oil [%], under:				
H_2 atmosphere	42.3	78.9	64.8	60.0
H_2O atmosphere	47.2	75.8	75.2	70.1
Yield of Gasoline [wt%], under:				
H_2 atmosphere	39.4	50.0	30.3	24.8
H_2O atmosphere	37.1	55.1	41.7	30.0

Effect of Temperature and Steam Atmosphere

The effects of carrier gases were studied at optimum nickel loading of 0.5 wt% (Table 2 and Fig. 1) and optimum time factor of 1.0 kg-cat.h/kg (Figs. 2 and 3). The conversions of heavy oil and yields of gasoline under H_2 and H_2O atmospheres are shown in Table 3. The level of conversions and yields of gasoline increased to maximum values at 673 K and later decreased upon increasing the temperature.

The data at temperatures leading to 673 K are comparable for both carrier gases. However, upon increasing the temperature beyond 673 K, the conversions and yields of gasoline under steam atmosphere were higher than those under hydrogen atmosphere. The high activity of the catalyst under steam atmosphere at high temperature suggests the advantageous effects of dealumination reaction.

Dealumination reaction under steam atmosphere at temperature above 773 K[7] resulted in the lost of aluminum, in the form of $Al(OH)_3$, from the zeolite frame-work. The nickel "blanketing" that covers some of the stronger acid sites couples with the lost of aluminum atom leads to further reduction in the amount of stronger acid sites of the zeolite catalyst. Consequently, the deactivation of the catalyt is low and activity of the catalyst is high.

These results show that steam is suitable to be used as an effective carrier gases to replace the expensive hydrogen gas. Data shown in Table 3 indicates that the optimum temperature was found to be at about 673 K.

CONCLUSION

The quality of gasoline product was enhanced with the presence on nickel supported on REY zeolite, especially at low nickel loading. The enhancement was due to the slight reduction in the amount of acid sites and the addition of hydrogenation and dehydrogenation activities.

Gasoline obtained on Ni-REY (0.5 wt%) contains more environmental friendly components (iso-paraffins and cycloparaffins) and less undesirable aromatics and n-paraffins than those obtained on other catalysts and those of the commercial gasolines.

There is an interaction between the amount of strong acid sites and the amount of nickel loading. The optimum amount of nickel loading on REY catalyst and time factor were about 0.5 wt % and 1.0 kg-cat.h/kg, respectively.

Reaction under steam atmosphere was advantageous due to the dealumination reaction. It resulted in the reduction of stronger acid sites and catalyst deactivation, hence the activity of the catalyst was kept high. Steam was found to be a suitable carrier gas to replace the more expensive hydrogen gas. The optimum reaction was found to be at 673 K.

REFERENCES

1. *Chemical Engineering*, 30 (July 1992).
2. A. R. Songip, T. Masuda, H. Kuwahara and K. Hashimoto, *Appl. Catal. B: Environ.*, 2:153-164 (1993).
3. J. N. Armor, *Appl. Catal. B: Environ*, 1:221-256 (1992).
4. A. R. Songip, T. Masuda, H. Kuwahara, and K. Hashimoto, *Energy and Fuels*, 8(1):136-140 (1994).
5. O. A. Larson, D. S. Maciver, H. H. Tobin, and R. A. Flinn, *Ind. Eng. Chem. Process Des. Dev.*, 1(4):300 (1962).
6. G. P. Lovasic, N. Jambrec, D. D. Siftar, and M. V. Prostenik, *Fuel*, 69: 525-528 (1990).
7. P. S. Yarlagadda, H. Yaoliang, and N. N. Bakhshi, *Ind. Eng. Chem. Prod. Des. Dev.*, 25:257 (1986).

NANOCOMPOSITES FOR COATINGS AND OPTICS

Helmut K. Schmidt

Institut für Neue Materialien gem. GmbH
Im Stadtwald, Geb. 43 A
D-66 123 Saarbrücken, Germany

INTRODUCTION

The preparation of optical materials has to follow only one important requirement. The materials have to be transparent for the considered wavelength. In oder to obtain this, the phase dimension of these materials has to be considerably below the wavelength of the light to be used in order to avoid Rayleigh scattering. This simple requirement for a long time only could be fulfilled by glasses, single-crystals or some polymers the phase dimension of which is below 30 or 20 nm. With up-coming new technologies for the fabrication of dispersed materials with very small particle or pore sizes, a variety of new materials can be considered to be interesting for optics. One of the candidates are inorganic gels, for example produced by the sol-gel process [1, 2]. Depending on processing techniques, pores in gels can be made small enough to avoid light scattering, and if these pores are filled by components with optical functions, for example, lasing dyes [3], composite materials with high optical transparency and special functions can be produced.

Another example are solid particles, developed in inorganic or organic matrices [4, 5, 6]. Colloidal metals in glasses or organics, such as gold, silver or palladium can be of interest for reasons of intensive colours or non-linear effects. If the metal particle size is kept in the lower nano range, a transparency is not effected.

Summarizing, it is to say that if it is possible to keep the phase dimensions small, even composite materials can be of high optical transparency and, in addition to this, can show interesting optical effects. The focus of this paper is to show how to prepare various types of materials in small dimensions to prepare coatings, to incorporate them into optical transparent matrices and to fabricate passive or active optical components.

GENERAL PRINCIPLES

If small ceramic semiconductor or metal particles have to be fabricated, a variety of methods can be used. All these methods have to have in common that particles are

Polymers and Other Advanced Materials: Emerging Technologies and Business Opportunities. Edited by P. N. Prasad *et al.*, Plenum Press, New York, 1995

produced which are completely agglomerate-free since agglomerates enlarge the phase dimension and lead to undesired light scattering. Methods for preparing nano-scale particles are gas phase reaction, which, in most cases, produces heavily agglomerated particles if the yield of the fabrication method is increased into technically relevant dimensions. Precipitation processes from the solution lead to nano-scale particles, too, but also suffer from heavy agglomeration if one has to work in high concentrations in order to increase the volume yield. Agglomerate-free nano-scale particles of different shape can be obtained in very diluted solution [7]. Agglomeration in precipitation processes from solution takes place very rapidly if the surface of the growing particles is highly reactive, and, for example, covered by OH groups.

To solve these problems, the concept of the small organic molecule surface modification has been developed, as shown in fig. 1.

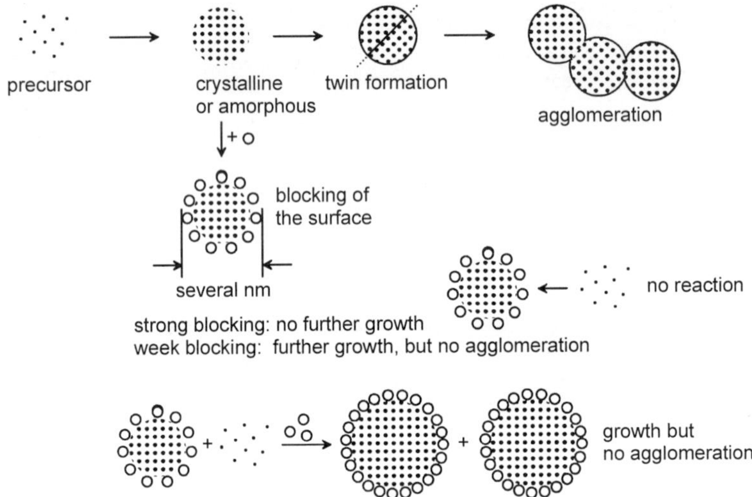

Figure 1. Concept of small organic molecule surface modification; • inorganic units; o small organic molecules.

Precipitation from solution takes place after nucleation. The crystal is growing due to the decrease of the free energy of the system as long as the system is supersaturated, a status which can be maintained by reactions producing insoluble monomers (for example, by sol-gel techniques, e.g. hydrolysis of alkoxides such as titania, alumina and others). Since the particles are very small at the beginning, the surface free energy of the system is rather high, and the particles tend to reduce their surface free energy by growing together (for example, twin formation or agglomeration). If at a special state (for example, in the nano range) additives are added which react with the surface and are able to reduce the surface free energy to a level below the level obtained by the growing particle, the reaction simply stops, and no further growth takes place. If the bond to the surface is rather weak, a growth can take place, but agglomeration can be avoided by proper choice of the additives.

Chemistry provides a large number of means to modify the surface of small particles and, as is has been shown elsewhere [8, 4], complexing ligands can be used which still remain fixed to the surface after colloid formation. In fig. 2 this principle is illustrated.

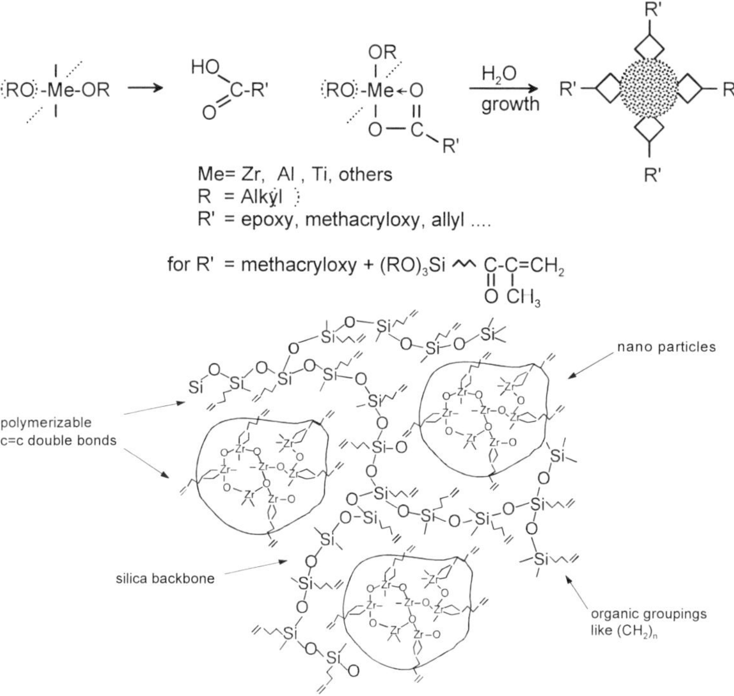

Figure 2. Schematics of the formation of ZrO_2 nanoparticles from Zr alkoxides using the controlled growth approach [after 4].

By proper choice of ligands, polymerization reactions can be used to cross-link the ligands to each other. If noble metals are used, for example, after crosslinking the metals can be reduced to atoms, and solid materials are obtained. In the case of ZrO_2, the particle size can be adjusted between 2 and 50 nm [9]. Similar results are obtained with metals from salts or complexes as far as they can be reduced. By heat treatment, for example, the nucleation process in $[AuX_4]^-$ complex containing ormocer films can be initiated, and metal clusters can be formed in polymeric matrices. In fig. 3, a gold colloid in an ormocer matrix [6] is shown. The particle diameter is about six nm.

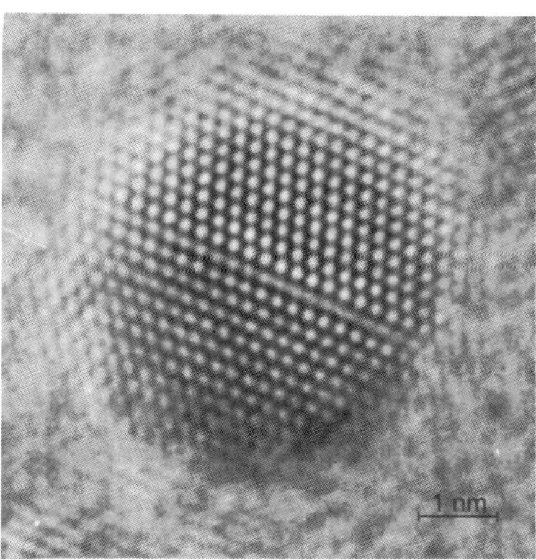

Figure 3. HRTEM micrograph from an Au cluster containing ormocer film [after 6].

What one can clearly see from this high-resolution TEM micrograph is the fact that close to the surface of the gold particle, the crystalline microstructure gets more and more disordered. In this case a twin formation has taken place.

The gold particles have been prepared from a gold amino complex where the amino group is linked to a silane (γ-ethylene aminopropyl triethoxysilane). The complex has been co-condensed with glycidyloxypropyl triethoxysilane, and the epoxide has been reacted to polyethylene oxide, forming a polyethylene oxide SiO_2 hybrid matrix. The only partially condensed system and polymerized to a viscous liquid was used for coatings. By heating up this film, the gold is reduced by organic molecules in the system (alcohols as solvents), and further heat treatments lead to nucleation and the gold colloid formation, and full condensation is obtained by this step, too. The gold colloid formation in ormocer films also has been observed by Mackenzie [5].

Another general principle is the use of sol-gel processes for small particle formation, as it has been shown with zirconia [10]. This principle can be used for other metal oxides, too, and, as it has been proved in the case of zirconia, crystalline particles are growing at temperatures around 120 °C, showing a monoclinic crystal structure. As shown in NMR and infrared studies, there are no indications that free methacrylic acid is present in the system. The CO frequency and the 13-C NMR signal show that methacrylic acid is still present in about the initial concentration. This result only can be interpreted that the methacrylic acid is bonded to the surface of the nanocrystalline particles. Similar results have been found with propionic acid for Al_2O_3 [8]. The particle size of the zirconia can be tailored by the concentration of the methacrylic acid and the reaction conditions like the water concentration. In other studies, it has been found that the colloidal particle size of gold or silver depends strongly on the ligand concentration, too. The influence of the ligand concentration on particle size depends strongly on the type of ligand.

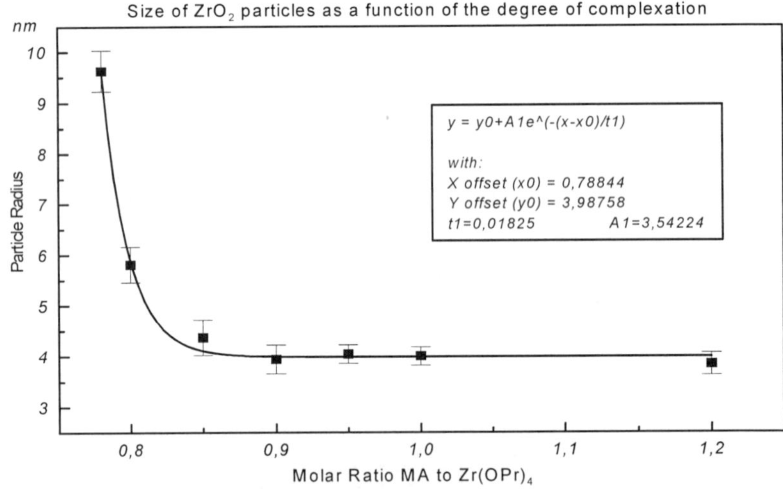

Figure 4. Dependence on colloidal particle size obtained by hydrolysis of Zr alkoxides in presence of methacrylic acid [after 9].

These results can be interpreted by a Gibbs free energy minimum, obtained by a specific surface coverage. It is assumed that through the interaction of the surface with the ligands a stabilization of the surface can be obtained and optimal surface coverage exists. If the optimal surface concentration is the driving force of the reaction, the particle size produced by a reduction process depends on the ligand concentration. This approach the first time has been demonstrated on sol-gel derived alumina with ß-diketones. In this case, ethyl acetoacetate has been used as complexing ligands [11]. It has been shown elsewhere that acetylacetone also can be used for particle size control of ZrO_2 in the µm range [12] as well as in the nm range [9].

In connection with bifunctional ligands, the synthesis principle can be used to produce tailored particle sizes and at the same time to establish special surface reaction possibilities on the colloidal systems. If groups with double bonds are used, polymerization reactions can be carried out, for example, with methacrylates, and photolithographic processing becomes possible. Similar reactivities can be obtained with epoxides. If reactive silanes are used as a second function, the particles can be used as a new type of sol-gel precursor. The different type of reactivities leads to elegant ways to incorporate the colloidal particles into a variety of matrix systems (sol-gel derived glass systems, ceramic systems or composite systems).

The third type of the tailoring of the reactivity of colloidal particles is shown in fig. 5. Using colloidal SiO_2 and modifying it with $\equiv Si\text{-}CH_3$ leads to sols with substantially reduced particle-to-particle interaction.

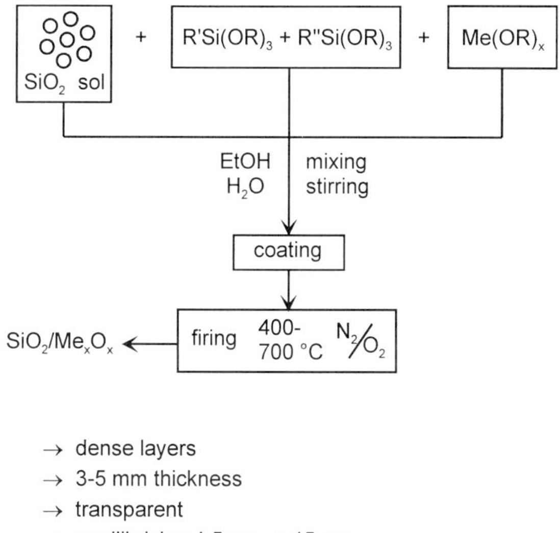

Figure 5. Schematics of the surface modification of aqueous or alcoholic SiO_2 sols [after 13].

It shows the surface modification of colloidal silica particles in order to control the particle-to-particle interaction. This becomes of importance if high solid systems should be prepared, which, in general, is one of the drawbacks of sol-gel processing. If methyl groups are used as surface modifiers, it has been shown that high solid content low viscosity clear sols can be obtained [13]. As it has been shown recently [14], thick layers up to 30 µm can be produced in a one-step coating process, and patterning can be carried out by embossing. Using this technique, SiO_2 waveguides can be produced in a very easy way. The embossed layers can be cured to pure SiO_2 at comparingly low temperatures (500 - 700 °C). This low curing temperature is not quite well understood, but it is attributed to nano or sub-nano pores, obtained by the burn-out of methyl groups. In fig. 6 the schematics of the layer and micropatterning process is shown. The progress of layer thickness is achieved by the increased solid content (> 50 wt. %), and the increased relaxation behavior due to the surface modification.

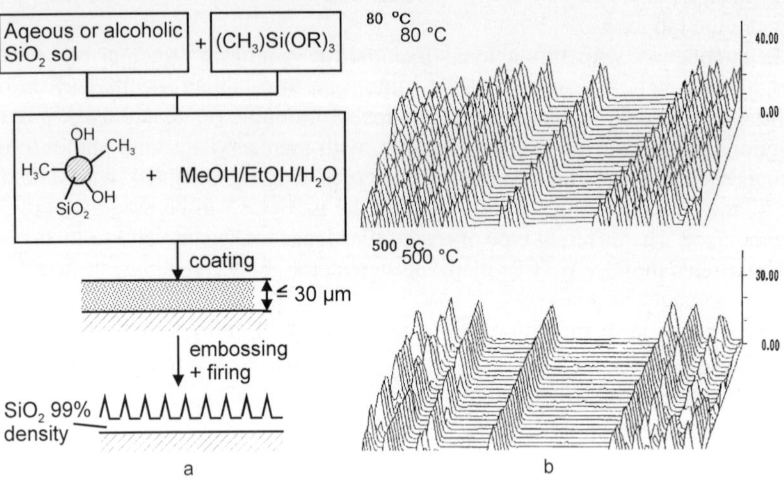

Figure 6. Schematics of the modification of the surface of SiO_2 particles and embossing of the non-cured films. After embossing the patterns can be cured to 99 % SiO_2 (according to the refractive index without changing their shape).

MATERIAL AND PROCESS DEVELOPMENT

The general approach of using interfacial thermodynamics to control the growth and the reactivity of colloidal particles can be used for many aspects. It can be used for controlling the particle-to-particle interaction, as already shown in the case of SiO_2. The extention of this principle to zirconia or titanium nitride has been investigated [15]. It could be shown that by appropriate surface modification, nano-scale titanium nitride (with particle sizes of about 30 nm) can be completely dispersed, and green densities of these "gels" by colloidal filtration of 60 % by volume after drying have been obtained. The titanium nitride surfaces covered by a very thin layer of titanium oxide, and for this reason, guanidino carbonate has been used as surface complex former forming strong bonds with Ti^{4+}. The surface modification with this short organic molecule firstly effects the complete dispersion of the agglomerated as received powder to the primary particle size and, secondly, reduces the particle-to-particle interaction to obtain high packing densities of 60 % of theory. These green compacts can be densified pressureless at temperatures of about 1250 - 1300 °C to almost complete density, and the most surprising effect is that almost no grain growth takes place. This seems to be contradictionary to the thermodynamics but has to be explained by a displacement mechanism of the atoms close to the surface, as known from nanocrystalline metals [16]. Due to the small particle size, the surface volume is remarkable and sufficient for a densification based on a displacement mechanism rather than on diffusion. By increasing the temperature above 1300 °C, rapid grain growth takes place, as to be expected. This leads to the possibility of separating "densification from crystal growth" and leads to an additional degree of freedom for processing ceramics. The principles of the surface modification of titanium nitride nanoparticles by guanidino carbonate is shown in fig. 7 as well as the effect of the surface modification on dispersion.

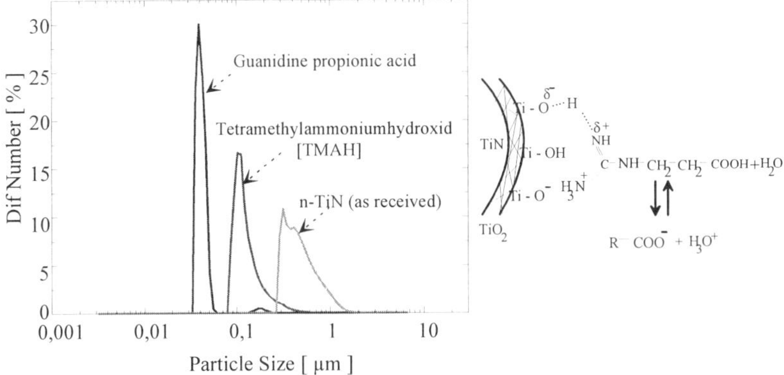

Figure 7. Surface modification of n-TiN by guanidino carbonate. The use of other ammines does not lead to complete dispersion [after 15].

The surface-modified sols also can be used for carrying out coatings, and in fig. 8 aluminum oxide coated by a titanium nitride sol is shown after densification in comparison to CVD coated Al_2O_3. The SEM micrograph shows the very fine crystal structure of the sol-gel coating compared to the CVD derived product. The irregular texture of the sol-gel coating is due to the aluminum oxide microstructure.

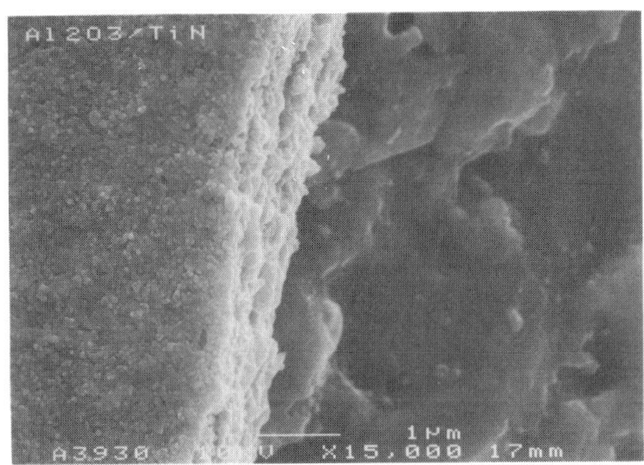

Figure 8. TEM micrograph of a sol-gel TiN coating on Al_2O_3 after 1,150 °C densification.

The described principle of handling nanoscale ceramic particles can be generalized and has been employed for zirconia [17] and alumina [18] processing.

As already described in the previous chapter, the surface modification of nano-scale particles also can be used for incorporating the particles into transparent matrices. Based on this principle, a variety of transparent coatings and optical materials have been developed. Depending on the desired properties, nano-scale particles can be used for a variety of purposes. For example, for the mechanical protection of optical components made from polymers, hard coatings are required. One of the major application areas are ophthalmics. Many investigations have been carried out to improve the scratch resistance of ophthalmic lenses. The deposition of inorganic coatings by vacuum or CVD techniques did not lead to satisfying results since the differences in the material properties between the coating (extremely brittle, but very hard) and matrix (ductile, but very soft) are not compatible. Liquid-phase hard coatings have been used for many years for the protection of ophthalmic lenses. It was of interest to investigate as far as nano-ceramic polymer matrix nanocomposites are able to improve the performance of liquid-phase hard coatings [19]. For these reasons, investigations are carried out to use alumina nanoparticles for the reinforcement of inorganic-organic composite matrix based on the ormocer principle. The results show that using this principle, a substantial improvement of the hard coatings could be obtained [20]. The effect is attributed to a very special structure of the material based on the fact that boehmite particles act as polymerization catalyst for the epoxide, forming a very special interface. Since the increase of hardness cannot be explained by the pure addition of the nano-scale filler, the effect is obtained by 5 % by volume only. The microstructure has been proved by IR and NMR analytics.

The development of new materials based on the principle of using nanoscale particles for maintaining the overall transparency of transparent matrices in composites can be extended by adding further components to the basic system. For example, if fluorinated silanes are added, a phase separation after coating takes place, leading to a gradient material with a high concentration of fluorinated groups at the surface. Using this approach, functional gradient materials have been prepared and components leading to a good adherence to substrates at the same time were obtained. This led to a surface similar to polytetrafluorethylene (PTFE), with surface free energies between 18 and 20 mJ/m^2 but with high transparency and good adherence to various types of substrates (e.g. glass, metals and polymers) [21, 22]. These systems have been successfully used for antisoiling coatings on glass, on metals and on metals and on plastics. Another interesting area of application are demoulding processes, which is a wide field in thermoplastic materials' moulding.

For passive optical application by using nano-scale particles the refractive index can be tailored in a wide range. For example, using zirconia, it can be varied between 1.51 and 1.53. If titania is added, coating materials up to 1.70 can be obtained. Together with polymerizable ligands, as shown in fig. 2, radical photopolymerization can be applied as well as radical thermopolymerization. Both techniques can be used for micropatterning techniques, for example, for embossing of microlenses, waveguides or for using photolithography or laser writing, for the fabrication of diffractive gratings or coupling systems. By the use of the Colburn-Haines-effect [23], holograms can be written into these layers by two-wave-mixing. Holographic gratings or microfresnel lenses can be produced. Figs. 9a and 9b give a survey about the techniques already developed for photolithography, holographic techniques and microembossing.

The described matrix materials also have been used as matrix materials for carrying additional optical functions, for example, semiconductor quantum dots or metal colloids for obtaining non-linear materials.

For the preparation of metal containing colloids the approach as described in the previous chapter has also been used. However, a reduction step has to be introduced, if one starts from salt solutions. This leads to several alternatives for the colloid synthesis. They can either be prepared in solution, by reduction, nucleation and growth in the presence of surface modifiers to stabilize the colloids and to control the particle size, or they can be introduced in an ionic complex form into a matrix system and then thermally be reduced in the presence of organic groupings, nucleated and grown in the already existing polymer or ormocer matrix. The different reaction routes are shown in figure 10.

This route has been investigated for gold colloids [6], silver colloids [29], palladium colloids [30] and for copper-colloids [31]. One application simply to be verified is the use of these materials as coatings on glass substrates since the coefficient of extinction of the colloids is in the range of 10^5 up to 10^6 $l \cdot mole^{-1} \cdot cm^{-1}$. This means that even in film thicknesses of 200 nm and colloid contents of some vol.-%, optical densities up to 3 can be obtained. The reaction path is shown in fig. 11.

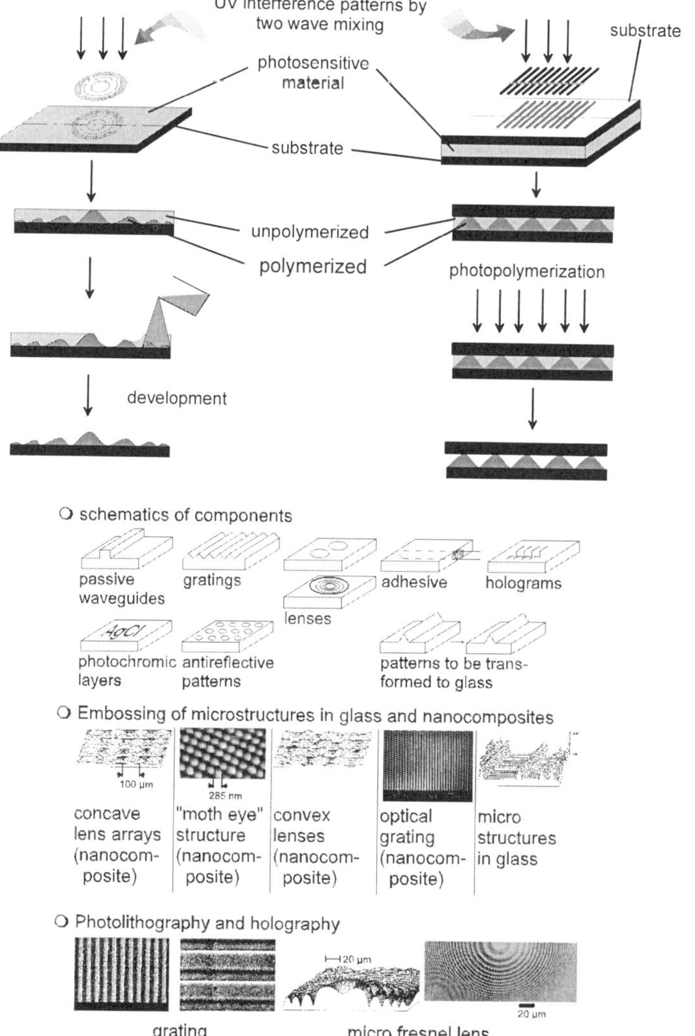

Figure 9a. Schematics and SEM micrographs of micropatterned optical materials by holographic techniques; **9b**: overview over components already fabricated by microtechniques [after 24 - 28].

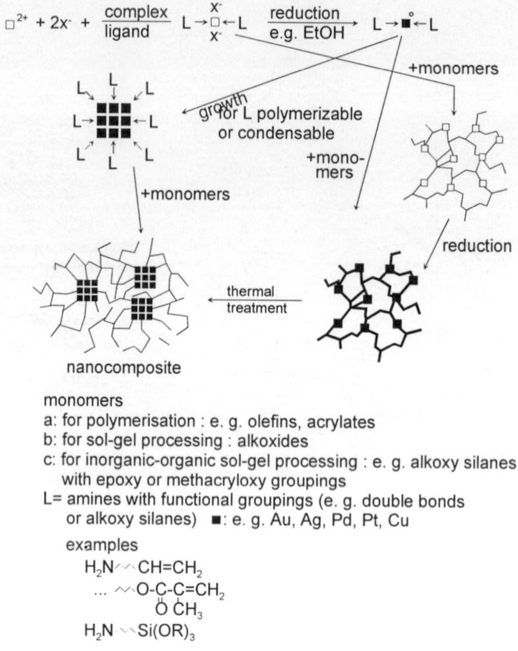

Figure 10. Reaction routes for the formation of metal-colloids in polymerizable matrices.

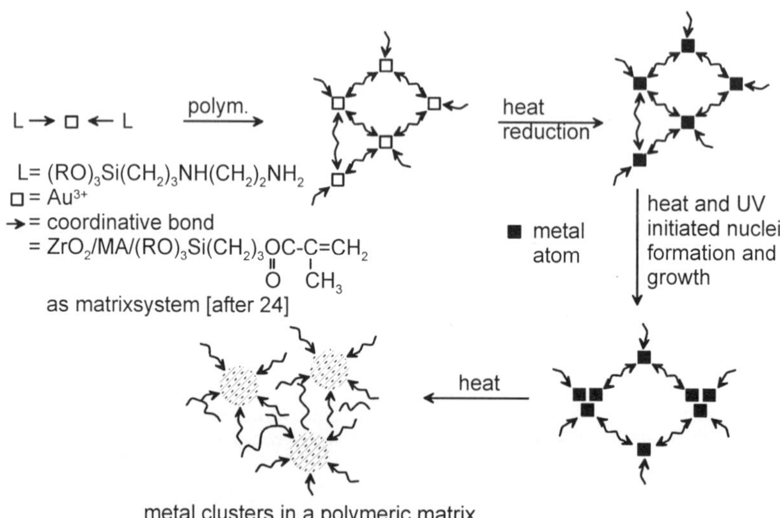

Figure 11. Reaction path for the formation of Au colloids in an ormocer matrix [after 6].

It is well known that nanoscale Au particles show interesting χ^3 susceptibilities [5]. For verifying high non-linear optical effects the tailoring of the so-called local electric field factor has been used as an approach (fig. 12).

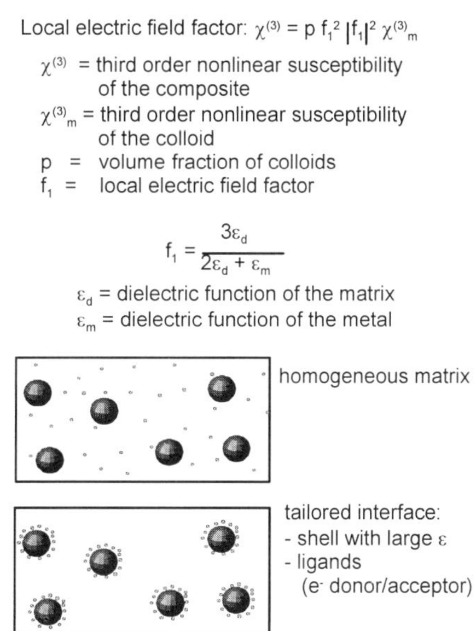

Figure 12. Schematics of the use of an enhanced local electric field factor to obtain high χ^3 susceptibilities [after 6, 32].

As pointed out in fig. 12, the local ecletric field factor is mainly defined by the dielectric function of the metal and the matrix. To increase the local electric field factor, the approach was followed to increase the electron density close to the metal colloid by appropriate ligands (amino, sulfido, cyano).

Investigations of the dependence of the χ^3 effect on the ligand show that amino ligands are much more effective than sulfido ligands. If the sulfido ligands are oxidized by heat treatment in air, resulting in sulfates, the χ^3 value decreases by several orders of magnitude. This is attributed to the fact that the sulfido ligands have a strongly decreased polarizability and do not contribute much to the local field factor. The highest χ^3 values have been obtained from amino-containing ligands. $\chi^3{}_m$ was up to $2 \cdot 6 \cdot 10^{-6}$, which is a rather high value. The χ^3 messurements have been carried out by a four-wave mixing process, as shown in figure 12, and in addition to this time depending experiments down to 7 ps have been carried out. In order to determine whether thermal effects are playing a role, the intensity dependence was investigated, showing no detectable dependence. Together with the time resolved experiments, it can be concluded that thermal effects do not play a role for the χ^3 susceptibilities.

Similar experiments have been carried out with Cu colloids [31]. By firing out the organics, two types of Cu colloids, depending on the atmosphere (reducing or oxidizing) are obtained: Cu clusters with and without oxide layers.

In fig. 14 the optical spectra and the χ^3 values of both types are shown.

The reduced form shows red and the oxidized form shows green colour. Green colour cannot be explained by the ionic absorbtion which would be far too less intensive and is attributed to the local electric field factor of the surrounding oxide.

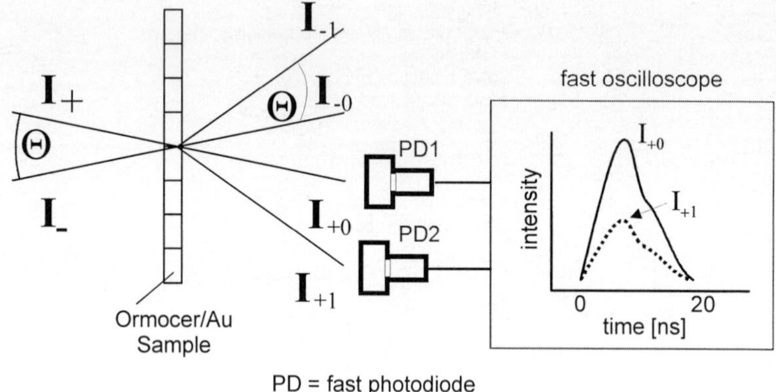

Figure 13. Schematics of the experimental χ^3 measurement set up and response curve.

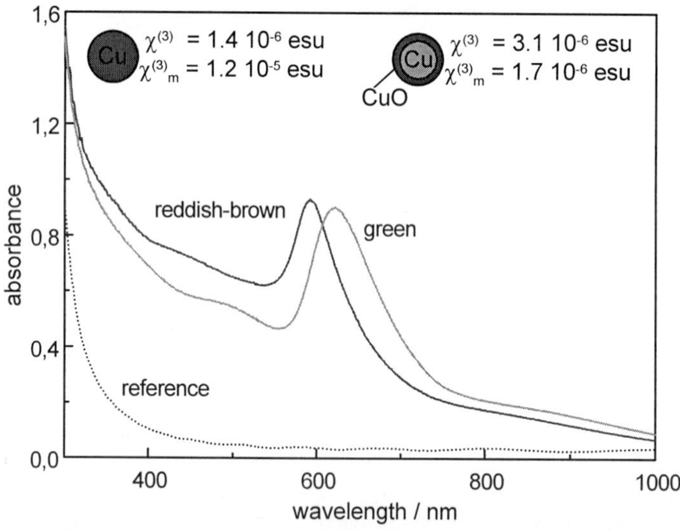

Figure 14. Optical spectra and χ^3 values of Cu colloids with and without oxide layer. $\chi^3_m = \chi^3$ metal.

Summarising one can say that using metal colloids together with selected ligands influencing the local electric field, non-linear optical properties can be increased remarkably.

One of the described effects of the tailoring of nano-sized particles is based on the reduction of interparticular interaction which leads to increased packing densities in compacts. It has been shown that increased packing densities can be used for monolithic processing, as in the case of ceramics. It can also be used for obtaining thick sol-gel films.

CONCLUSIONS

It could be shown that by using the surface thermodynamic approaches the control of growth and particle size of colloidal systems can be realized. In addition to this, it is possible to outfit the surface of the particles with organic groupings of different chemical properties and to immobilize them strongly on the surface by establishing chemical bonds. Based on these basic routes the colloidal particles can be used for advantageous processing of ceramics to films or inorganic-organic composites with interesting passive and active properties. With these techniques new routes for material synthesis have been opened, and the examples shown up to now only are scratching the surface.

REFERENCES

1. C. J. Brinker and G. W. Scherer. „Sol-Gel Science", Academic Press Inc., Boston (1990).
2. J. McKiernan, J. I. Zink, and B. S. Dunn, SPIE Vol. 1758:381 (1992).
3. R. Reisfeld, in: „Sol-Gel Science and Technology", M. A. Aegerter, M. Jafelici Jr., D. F. Souza, and E. D. Zanotto, eds., World Scientific Publishing Co. PTE Ltd., Singapore (1989) 322.
4. H. Schmidt, J. Non-Cryst. Solids 178:302 (1994).
5. J. Y. Tseng, C.-Y. Li, T. Takad, C. Lechner, and J. D. Mackenzie, SPIE Vol. 1758:612 (1992).
6. M. Mennig, M. Schmitt, U. Becker, G. Jung, and H. Schmidt, SPIE Vol. 2288:130 (1994).
7. E. Matijevic, in: Ultrastructure Processing of Advanced Ceramics, J. D. Mackenzie and D. R. Ulrich, eds., John Wiley & Sons, New York (1988) 429.
8. H. Schmidt, R. Naß, M. Aslan, K.-P. Schmitt, T. Benthien, and S. Albayrak, J. de Physique IV, Coll. C7, Vol. 3:1251 (1993).
9. V. Gerhard and S. Krämer, private communication, to be published later.
10. H. Krug, N. Merl, and H. Schmidt, in: Teubner-Texte zur Physik, Vol. 27: Integrated Optics and Micro-Optics with Polymers, W. Ehrfeld, G. Wagner, W. Karthe, H.-D. Bauer, and H. G. Moser, eds., B. G. Teubner Verlagsgesellschaft Stuttgart-Berlin (1993) 192.
11. R. Naß and H. Schmidt, in: Ceramic Powder Processing Science, H. Hausner, G.L. Messing, S. Hirano, eds., Deutsche Keramische Gesellschaft e.V., Köln (1989) 69.
12. G. Rinn and H. Schmidt, in: Ceramic Transactions, Vol. I, Part A: Ceramic Powder Science, American Ceramic Society, Westerville (1988) 23.
13. M. Mennig, G. Jonschker, and H. Schmidt, SPIE Vol. 1758:125 (1992).
14. M. Mennig and G. Jonschker, private communication.
15. R. Naß, S. Albayrak, M. Aslan, and H. Schmidt, in: Ceramic Transactions, Vol. 51: Ceramic Processing Science and Technology, H. Hausner, S. Hirano, and G. L. Messing, eds., American Ceramic Society, Westerville/Ohio (1995) 591.
16. H. Gleiter, Progress in Materials Science 33:223 (1989).

17. D. Burgard, Ch. Kropf, R. Naß, and H. Schmidt, in: Proc. 1994 MRS Spring Meeting, Symposium „Better Ceramics Through Chemistry VI", San Francisco, Mat. Res. Soc. Symp. Proc. (in print).
18. H. Schmidt, J. Sol-Gel Science and Technology 1:217 (1994).
19. G. Philipp and H. Schmidt, J. Non-Cryst. Solids 63:283 (1984).
20. R. Kasemann, H. Schmidt, and E. Wintrich, in: Proc. 1994 MRS Spring Meeting, Symposium „Better Ceramics Through Chemistry VI", San Francisco, Mat. Res. Soc. Symp. Proc. (in print).
21. R. Kasemann and H. Schmidt, oral presentation at „Fluorine in Coatings" Conference, September 1994, Salford, Great Britain.
22. H. Schmidt, R. Kasemann, T. Burkhart, G. Wagner, E. Arpac, and E. Geiter, in: ACS Symposium Series on Hybrid Organic-Inorganic Composites, J. E. Mark, P. A. Bianconi, and C. Y.-C. Lee, eds., American Chemical Society, Washington (in print).
23. C. Becker, M. Zahnhausen, H. Krug, and H. Schmidt, in: Extended Abstracts of the Fourth SAAR-LOR-LUX Meeting on Functional Advanced Materials. G. Kugel, ed., CLOES-SUPELEC, Technopole de Metz, November 1994.
24. H. Schmidt, H. Krug, R. Kasemann, and F. Tiefensee, SPIE Vol. 1590:36 (1991).
25. H. Schmidt, in: Proceedings NATO ARV on Applications of Organometallic Chemistry in the Preparation and Processing of Advanced Materials, Cap d'Agde/France, September 1994 (in print).
26. H. Schmidt and H. Krug, in: ACS Symposium Series 572 „Inorganic and Organometallic Polymers II: Advanced Materials and Intermediates, P. Wisian-Neilson, H. R. Allcock, K. J. Wynne, eds., American Chemical Society, Washington (1994) 183.
27. H. Krug, P. W. Oliveira, and H. Schmidt, in: Ceramic Transactions Volume 55: Sol-Gel Science and Technology, E. Pope, S. Sakka, and L. Klein, eds., American Ceramic Society (in print).
28. P. W. Oliveira, H. Krug, H. Künstle, and H. Schmidt, SPIE Vol. 2288:554 (1994).
29. M. Mennig, H. Schmidt, and C. Fink, SPIE Vol. 1590:152 (1994).
30. T. Burkhart, M. Mennig, H. Schmidt, and A. Licciulli, in: Proc. 1994 MRS Spring Meeting, Symposium „Better Ceramics Through Chemistry VI", San Francisco, Mat. Res. Soc. Symp. Proc. (in print).
31. M. Mennig, M. Schmitt, B. Kutsch, and H. Schmidt, SPIE Vol. 2288:120 (1994).
32. M. Mennig, U. Becker, M. Schmitt, and H. Schmidt, oral presentation at the American Society Annual Meeting, May 1995, Cincinnati/Ohio.

NANOCOMPOSITES OF GRAFTED POLYMERS

ONTO LAYERED SILICATE

Ahmed Akelah

Chemistry Department, UAE University,
P.O.Box 17551, Al-Ain United Arab Emirates

INTRODUCTION

Inorganic materials usually have high thermal stability, good optical properties and high modulus, but they are very brittle. On the other hand, polymeric materials have elastic properties. However, it is normally difficult to have in the same material both properties of high strength to sustain high loads and high toughness to absorb a large amount of energy during fracture that occurs by breaking of primary and secondary bonds, depending upon the structure of the material[1]. Inorganics are orginally introduced into polymers as fine solids to act as fillers or as reinforcing agents. The mineral fillers are used to dilute and hence to reduce the amount of the final polymers used in the shaped structures thereby lowering the economic high cost of the polymers. However, reinforced polymers with minerals show an increase in modulus, hardness, tensile strength, tear resistance, abrasion, and resistance to fatigue and cracking[2]. The reinforcing agent characteristics such as particle size, chemical composition and structure of the particle surface (degree of aggregation and agglomeration), level of hydration, surface acidity and adhesion between particle and polymer, play an important role in the improvement of the service life of the reinforced polymer articles.

Although some improvements of the lack of strength shown by many polymeric materials have taken place particularly by the incorporation of some inorganic additives, however, there is still significant need for producing polymeric materials with extremely high levels of stiffness and hardness in order to

Polymers and Other Advanced Materials: Emerging Technologies and Business Opportunities. Edited by P. N. Prasad *et al.*, Plenum Press, New York, 1995

become widely accepted as structural engineering materials. Blendings of polymers with finely divided inorganic additives result in a brittle composite material because of a very poor bond strength between the polymer matrix and the mineral, i.e. the bonding between them is not sufficient to provide the desired reinforced effect. In addition, the amount of the mineral that can be incorporated is limited. Consequently, the minerals are not uniformaly dispersed in the organic polymer. The efficiency of the mineral to modify the properties of the polymer is primarily determined by the degree of the mineral dispersion in the polymer matrix, which in turn depends on its particle size. However, the hydrophilic nature of the mineral surfaces impedes their homogeneous dispersion in the organic polymer phase. Thus, it is necessary to make the mineral surface hydrophobic in order to enhance its compatibility prior to compounding with the polymers. However, the spherical mineral additives have often coalesced into larger, irregularly shaped aggregates. Such aggregates are most frequently united into larger agglomerates by the atractive forces of the van der Waals type. The extent of aggregation and agglomeration has a marked influence on their reinforcing properties.

The key to the development of cracks is the stress distribution around indigenous flaws. The simplest way of using the filler particle as a retarder of crack propagation is by lengthening the crack's paths, since the crack has to move around the particle, thereby dissipating more energy. The aggregated particle chain is clearly more effective than single spherical filler particles. With fillers poorly bonded to the matrix, dewetting and vacuole formation occur upon a significant deformation, initiating cracks. Therefore, strong bonds between particle and polymer contribute to overall strength. Factors which absorb or dissipate energy by turning the potentially destructive stored free energy of an impact into a more harmless form as heat include, besides the matrix hystersis caused by uncoiling of chains (change in molecular conformation) and the frictional resistance to deformation, the break up of transient filler structures, alignment of polymer chains and particle aggregates, strain crystallization and stress relaxation.

Clays as most natural abundance minerals and available as inexpensive materials, have high mechanical strengths as well as high chemical resistance. Because of the small particle size and intercalation properties of clays, they afford an appreciable surface area for the adsorption of molecules. Polymer-clay materials have received considerable interest because the interactions between them have effect on the properties of both clay and polymer[3]. Polymers are added to clays in order to enhance the physical and colloidal properties of clays because of their agricultural potential as stabilizing and conditioning agents. The beneficial effect of polymer on clay and soils is related to the improvements in their structure and stability due to the aggregation and water stability of the aggregates formed.

The nanostructured materials with the ultrasmall size (<100 nm) often exhibit combinations of physical, optical, magnetic and mechanical properties that have characteristic lengths in this range and are not available in conventional materials[4]. Their special properties are determined by a complex interplays among the building blocks and interfaces between them. The properties are improved when the grains are reduced in size from 100μm to 100nm which are further improved by going to 1nm. Thus, nanocomposites offer a major opportunity for creating novel materials that offer new potentially useful combinations of properties. They are multilayered materials in which polymers are sandwiched between ultrathin of clay silicate sheets. The chemistry and crystal structure of each layer in a multilayer can be quite different from those a few atoms away, because the nanosized gaps that exist between the clay layers are infused with polymer chains. The confinement of the polymers in the two-dimensional spaces of the layers isolates them and forces them into a more orderly arrangement which has a major effect on the nanocomposite properties. When polymers are intercalated between silicate sheets, they do not behave the bulk polymers on heating and do not undergo the transitions of amorphous and crystalline polymers because the confinement of the molecules hinders their tranlational and rotational motions. The melting behaviour of crystallites requires more space to grow than is available in the nanosized gaps of the silicate lattice. Thus, the polymers intercalation in silicates increases their thermal and oxidative stabilities because the clay lattices protect the macromolecules confinement inside them from engaging in degradative behaviour. This approach allows the design of materials that combine a high strength and high thermal stability of the clay with the crack-deflecting properties and processibility of the polymer because of the frequent, periodic interlayer boundaries.

The principal objective of the present article is meant to be overview of the effort done on the synthesis of nanocomposite materials by the intercalation of organic polymers with the clay interlayers in which the polymers are molecularly dispersed within the inorganic phases.

PREVIOUS PREPARATIONS OF POLYMER-CLAY INTERACTIONS

Various attempts have been done to incorporate the polymers with clays depending on their physical, chemical or mechanical interactions[5]. The physical interaction leads to the formation of layers of immobilized polymers around the clay particles by adsorption, which can be desorbed on suspending the polymer-clay material in organic solvent, ie. it is reversible. However, the chemical interaction that leads to permanent bonding between macromolecules and clay particles, can be promoted by increasing the compatibility via the presence of reactive functionality in the polymer, that are capable of chemical reaction or H-bonding with the clay surface. In addition, the mechanical interaction has also been used for the occlusion of polymer molecules inside clay aggregates, i.e. hydrodynamic interaction. Each technique used for achieving the uniform dispersion of clay layers in the polymer medium to get the dimension of molecules has its own disadvantage.

Physical Interaction:

Polymerization of Impregnated Monomers. This technique has been used for the polymerization of the impregnated vinyl monomer as tetrafluoroethylene[6], butadiene, 2-butene[7], 4-vinylpyridine[7], acrylamide[7], methyl methacrylate[8], acrylonitrile[8-10] and styrene[11] into clays in the presence of a radical initiator. However, since the adsorption of monomer between the lamellae depends on the dipole moment of the monomer molecule, the yields of the physically inserted polymer were found to be low.

Direct Polymer Adsorption. The adsorption of uncharged linear polymers such as PVA[12], poly(vinyl pyrrolidone)[13], PEG[14], and PAm[15], onto clay surfaces has been used for preparing clay-polymer materials depending on the desorption of the hydration water from the surfaces. Since the gained entropy provides the driving force for this physical adsorption process,i.e.it is an entropy-driven process thus it does not lead to high polymer adsorption because the enthalpy change of the process is commonly very small.

Direct polymer melt absorptions have also been used for the interaction of an organically modified layered silicate [MMT-di-octadecyl-diMe-amm] with different polymers of varying degrees of polarity and crystallinity such as PS, PVF, PEO, poly(ϵ-capro lactone) poly(diMe-siloxane)[16]. This direct melt interaction of polymers in organo-silicates is primarily drived by enthalpic polymer-clay interactions. However, this process is completely reversible on using organic media.

Chemical Intercalation:

Chemical intercalation is the grafting of the organic polymers onto clay layers for creating new organic-inorganic materials which favour the increase of the assessibility of clay in polymer matrices by approaching the dimension of molecules under the microscopic observations. Incorporation of the polymer matrix into the space between layers of the clay mineral that has the higher hardness and rigidity leads to improvement of physico-mechanical properties of the polymer. But at the same time, such composites lose their rheological properties.

Grafting of Polymers via Coupling Agents. An indirect route for chemical bonding the polymers to the available hydroxyl groups on the surface of the clay particles has been carried out by either: 1) grafting of an unsaturated trialkoxysilane with the polyolefin, prior to compounding with the clay; or 2) pretreatment of the clay with the coupling agent for creating chemically active surface sites on the clay particles in which the vinyl pendants react with the radical sites generated on the polymer chains during compounding. The coupling agents are silane or titanate compounds that contain two reactive groups one suitable for reaction with the polymer and the other capable of reacting with the clay, i.e. they permit chemical bonding between the polymer and clay particles[17,18]. For example, vinyl-, amino- and mercapto-silanes have been used as coupling agents for kaolin clays modification to improve elastomers products[19-26], Scheme 1. Elastomers were grafted on the pretreated silica to create ester[27], urethane[28], or cationic grafting on clay surfaces[29]. This surface grafting increases the bond strength between the clay and the polymer, reduces dewetting and consequently increases the modulus. However, the interaction achieved by this chemical method is a surface modification process in which the polymers do not penetrate the interlayers of the clay and hence they are not homogeneously dispersed with each other.

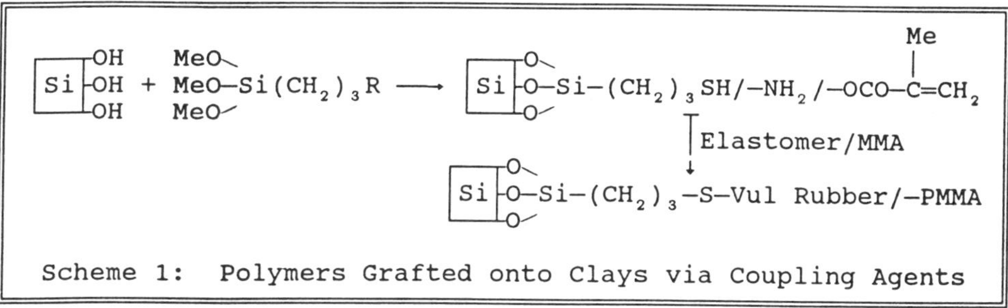

Scheme 1: Polymers Grafted onto Clays via Coupling Agents

Interactions of Polymer Functional Groups with Clays. The direct grafting of polymeric chains on clay surfaces has taken place through the interaction of polymer functional groups with the clay hydroxyl groups[29]. The compatibility of the clay with PE via the radical induced polymerization of maleic anhydride, has been carried out so that the PE is grafted by the anhydride moieties which concurrently react with the hydroxyl groups of the clay surface to form ester linkages and/or H-bonds[30].

Sol-Gel Processes. This synthesis is based on the hydrolysis and condensation of silicon alkoxides that lead to the formation of oxo-polymers which are then transformed into an oxide network upon calcination[31-37]. These reactions can be divided into three steps: 1) Nucleophilic addition of a negatively charged group on to the positively charged metal atom. 2) Proton transfer within the transition state toward the leaving group (OR). 3) Departure of the positively charged protonated leaving group ROH.

Three kinds of organic-silicon hybrid gels have been made by this process depending on the interactions between the oxide gel network and macromolecules: a) Polymers can be incorporated within oxide gel structure during nucleation and growth of the crystals, by mixing with silicon alkoxides in a common solvent, which become trapped into the oxide network during its formation b) Polymer can be incorporated into the layered silicate microstructure by dipping the gel into the polymer solution, i.e. by an exchange process with protons or water molecules of the gel. iii) Polymer can be intercalated into the layered silicate structure occurs through chemical modification of organically modified silicates which are synthesized from silicon alkoxides $Si(OR)_4$ and modified silicon alkoxides $R_2Si(OR)_2$ that act as bonding agent between the polymer and silica phases.

POLYMERS GRAFTING ONTO ORGANOPHILIC LAYERED SILICATES

Recently, the cation exchange grafting of polymers via chemical bonds to the layered clays to produce nanophase organic inorganic hybrids has been developed. This technique involve: a) intercalation of suitable monomers to the clay layers to produce organophilic calys, followed by subsequent condensation or free radical polymerizations of the monomers to grow the polymer chains in the interlayers space[38-47], or b) direct intercalation of the positively charged polymer chains from solution to the anionic clay layers[47-50].

Polymerizations of Monomers with Organophilic Monomer-Clays:

The intercalation technique of a suitable monomer to the clay interlayers to produce organophilic calys, has involved the attachment of monomer molecules, which contains a onium moiety to interact with the clay and a reactive group to react with the polymerizing system. However, two different techniques were used for achieving the polymerization of the monomers in the interlayers space of the clay. They depend upon the swelling of the organophilic modified clay either with the solvent or with the heat of the propagation reaction during the polymerizations.

Solution Polymerization. In this technique, the grafting of the polymers onto the clay interlayers depends on the swelling of the organophilic modified clay promoted by the solvent. The swelling is primarily due to the solvation of the interlayer organic cations which can be maximized by the use of appropriate solvents. The swelling of the clay is considered to be a result of a balance between the interlayers cohesive forces and the attractive forces between the solvent and intercalated cations. The interlayer distance can be increased remarkably in the solvents which have strong attractive forces toward the intercalated monomer molecules. Thus, the role of the solvent is to improve the wetting and the penetration of the monomer between the clay interlayers. The solvation of the monomer-clay complex with organic solvent depends on the nature of the intercalated monomer and also on the type of the solvent used, which must has attractive forces enough to increase the interlayer distance and must be good solvent for the monomer and the resulting polymer to facilitate the movement of the monomer molecules between the clay interlayers prior to the start of the polymerization.

For example, VM-g-MMT intercalate, I, which is able to swell and disperse in organic solvents, was prepared by exchanging the MMT meniral cations with vinylbenzyltrimethyl ammonium chloride. The resulting organophilic mineral has polymerizable moieties directly bonded to the layers surface of the mineral. Radical polymerizations of styrene between the interlayers of 5, 10, 25 and 50 wt% of I in selected solvent have resulted in PS-g-MMT materials[51-53], Scheme 2. The effect of MMT amounts on the formed PS as determined by the extraction with the solvents showed an increase in the grafted polymer formed (0.84-2.94g/1gMMT) and an decrease in the external PS with increasing the amount of MMT.

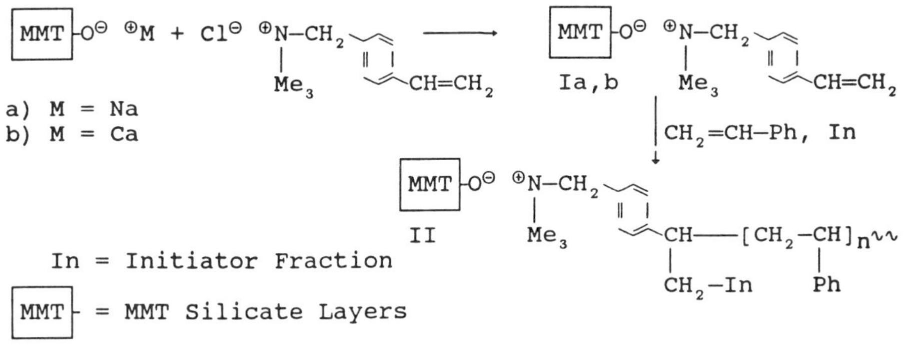

MMT–$O^{\ominus}$ $^{\oplus}M$ + $Cl^{\ominus}$ $^{\oplus}N(Me_3)–CH_2–C_6H_4–CH=CH_2$ ⟶ MMT–$O^{\ominus}$ $^{\oplus}N(Me_3)–CH_2–C_6H_4–CH=CH_2$ (Ia,b)

a) M = Na
b) M = Ca

↓ CH_2=CH–Ph, In

MMT–$O^{\ominus}$ $^{\oplus}N(Me_3)–CH_2–C_6H_4–CH(CH_2–In)–[CH_2–CH(Ph)]_n$∿∿ (II)

In = Initiator Fraction

MMT– = MMT Silicate Layers

Scheme 2: VM-MMT Intercalates & PS-MMT Nanocomposites

Melt Polymerization. In this technique, polymer interactions with the organophilic clay are driven by the heat of the propagation reactions during monomers polymerization. The penetration of monomers between the interlayers of the organophilic clay can be facilitated by the attractive forces between the modified clay interlayers and the free monomer molecules. The solvation energy of the monomer molecules is the driving force for the swelling of the clay with the monomer which is sufficient to overcome the attractive forces between the clay interlayers. Once the polymerization started in the clay interlayer space, the propagating chains can be swolled by the monomer molecules, due to the sufficiently attractive forces between them. In addition, the heat of the polymerization impart a degree of flexibility to the clay-polymer chains which in turn assists further insertion of monomers during polymerization process. The combination of these two factors is the driving force for the swelling of Org-clay complex, e.g. the bulk polymerization of acrylonitrile, 4-vinylpyridine, styrene-maleic anhydride in the presence of VM-g-MMT intercalate, I.

The polymcondensation type consists of polymerizing bifunctional monomers in the presence of organophilic-clay containing active groups capable of forming bonds between the clay and the formed polymer. The polycondensations of ϵ-caprolactam in the presence of clay-aminolauric, or clay-aminocaproic acid intercalates have been reported out to give nylon 6-clay hybrids[54-57].

A series of epoxyphilic-MMT derivatives prepared by the exchange of the inorganic cations of MMT with the ammonium ions of different organic derivatives have been used to eliminate the polymerization shrinkage problem that occurs during the curing of epoxy resins[58], Scheme 3. They have specific functional groups such as carboxylic, phenolic and amide which can react and absorb the epoxy formulation components giving rise to their swelling and separation of the silicate layers. This separation of the MMT interlayers as a result of their organophilicity towards the components of epoxy resins leads to the expansion of the cured epoxy resins by homogeneous dispersion of MMT in the epoxy components before curing. It was determined that the cured epoxy exists within the MMT interlayer by the observation of high interlayer spacings, even greater than 8 nm. Generally, epoxy resins containing MMT derivatives that had been swollen in the curing agent before curing exhibited larger interlayer spacings, especially among the nondispersed MMT layers. The maximum observed residual stress was reduced by greater than 50% in the epoxyphilic-MMT/epoxy resins which exhibited higher values of T_g, flexural modulus, and ultimate flexural strength than the pure epoxy.

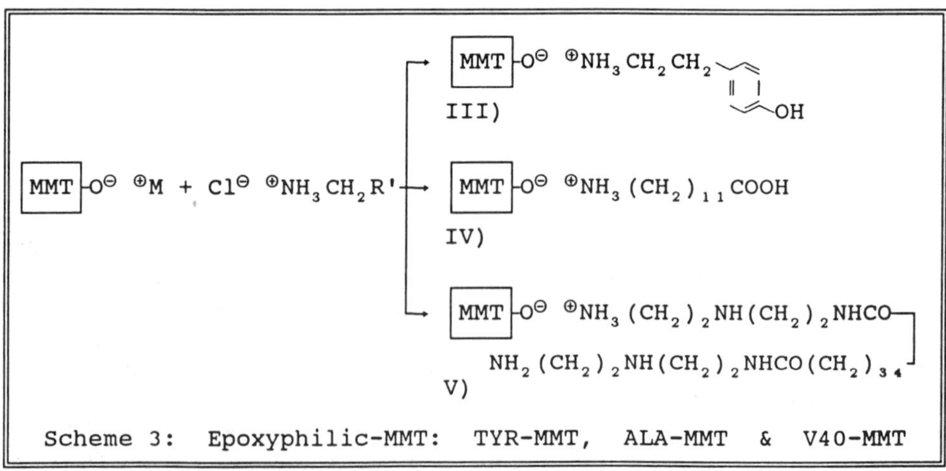

Scheme 3: Epoxyphilic-MMT: TYR-MMT, ALA-MMT & V40-MMT

Other epoxyphilic-MMT derivatives having various functional groups which can react with one or both of the components in amine-cured epoxy formulations such as anhydride, hydroxyl, amide or amine functional groups were prepared[59], Scheme 4. The intercalation of the organic cations to form epoxyphilic-MMT derivatives was confirmed by elemental analysis and FTIR study. The adsorption behavior of the modified MMT derivatives were compared with the unmodified MMT in an epoxy resin before and after the cure process.

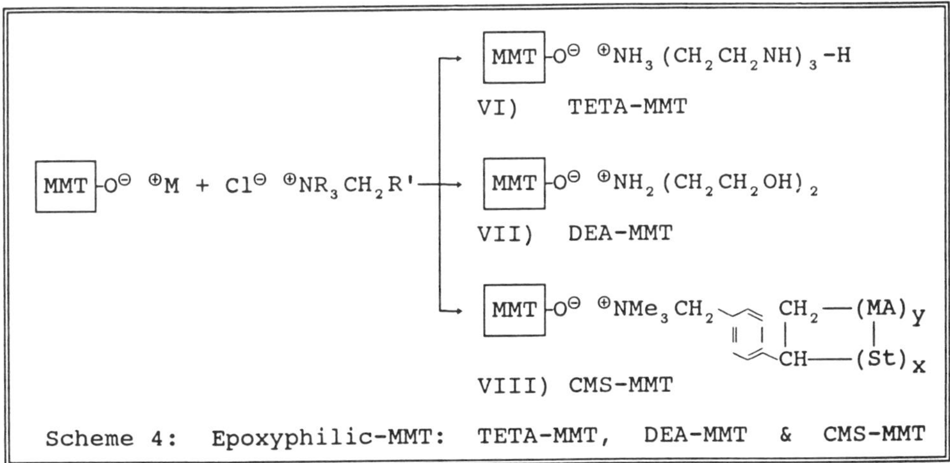

Scheme 4: Epoxyphilic-MMT: TETA-MMT, DEA-MMT & CMS-MMT

Chemical Modification of Clays with Polymers:

The chemical modification technique involves attachment of already preformed polymers containing positively charged onium moieties as intercalating groups such as pyridinium, ammonium and phosphonium salts, to the aqueous swolled clay. This modification technique is attractive from the simplicity of the procedure and the high molecular weight of the intercalated polymer. However, it is only depend upon the nature of the solvent mixtures used.

During polymer intercalation from solution, a relatively large number of solvent molecules need to be desorbed from the clay to accommodate the incoming polymer chains. Thus, the driving force for the polymer intercalation from solution is the entropy gained by desorption of solvent molecules due to the gain in translational freedom of desorbed molecules from the clay galleries. The resulting entropic gain compensates for the unfavorable loss of conformational entropy associated with the confined polymer and can be overcome by maximization of the number of the intercalated polymer. The intercalated chains must established numerous segment contacts to offset the large decrease in conformational energy which opposes intercalation.

These interactions impede the rotational and translational motions of the polymer. The specificity for the polymer can be enhanced by eliminating the competing clay-solvent and polymer-solvent intercalations.

Composites of amine terminated butadiene-acrylonitrile (ATBN) and MMT were prepared by ion exchange between the onium salt of the rubber and the interlamellar cations of the mineral in solvent mixtures of H_2O/Dioxane or H_2O/DMSO[59,62], Scheme 5.

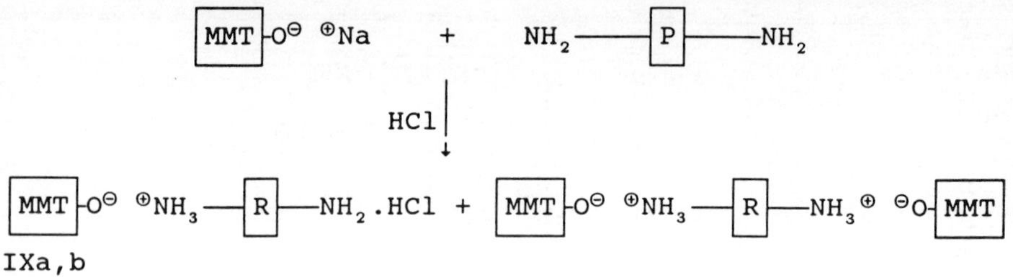

[MMT]-O⊖ ⊕Na + NH_2—[P]—NH_2

↓ HCl

[MMT]-O⊖ ⊕NH_3—[R]—NH_2.HCl + [MMT]-O⊖ ⊕NH_3—[R]—NH_3⊕ ⊖O-[MMT]

IXa,b

Scheme 5: Rubber-MMT Nanocomposites

Organophilic PS-MMT and PMMA-MMT supported catalysts have been prepared by grafting the copolymers of PMMA or PS with CMS that contain 2% ammonium salt groups, onto MMT interlayers via direct cation exchange process. The remained chloromethyl groups of the copolymer have been modified to produce onium salts[63], Scheme 6. The structural composition and catalytic activities of

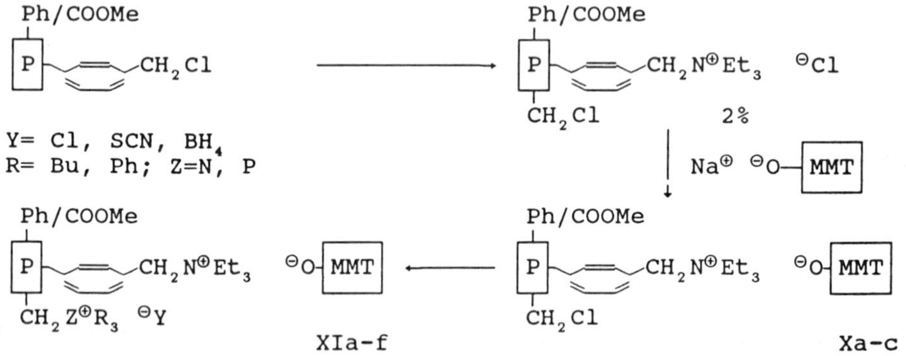

Ph/COOMe
[P]-C_6H_4-CH_2Cl ⟶ Ph/COOMe, [P]-C_6H_4-$CH_2N^{\oplus}Et_3$ ⊖Cl, CH_2Cl 2%

Y= Cl, SCN, BH_4
R= Bu, Ph; Z=N, P

↓ Na⊕ ⊖O-[MMT]

Ph/COOMe, [P]-C_6H_4-$CH_2N^{\oplus}Et_3$ ⊖O-[MMT], $CH_2Z^{\oplus}R_3$ ⊖Y (XIa-f) ⟵ Ph/COOMe, [P]-C_6H_4-$CH_2N^{\oplus}Et_3$ ⊖O-[MMT], CH_2Cl (Xa-c)

Scheme 6: PS-MMT & PMMA-MMT Supported Onium Catalysts

the onium moieties supported on PS-MMT and PMMA-MMT materials have been investigated in different nucleophilic substitutions under the effects of several factors governing the reactions rates, such as the composition of the polymer backbone, degree of loading, nature of the active moiety and solvent and reaction temperature.

CHARACTERIZATION OF POLYMER-MMT NANOCOMPOSITES

Characteristics of MMT:

Montmorillonite (MMT) is a hydrous alumina-silicate mineral whose lamellae are constructed from an octahedral alumina sheet sandwiched between two tetrahedral silica sheets. A net negative charge on the layers surface causes them to attach by sorbed Na/Ca cation. The layers of MMT are not stacked regularly enough to form true crystals but are grouped in small mumbers forming stacks, inside which the degree of organization is a parallelism sufficient to result in selective d_{001} reflections. The various organization levels in MMT are: a) Primary particles (ca. 1nm) consist of stacks of parallel elementary sheets (lamellae) with an average of 10 sheets per particle. b) Microaggregates (ca 8-10nm) formed by the association of several primary particles which are nearly parallel and joined together. c) Aggregates (0.10-10μm) comprise a large number of primary particles and microaggregates.

The irregular character of the distribution of the charges in MMT weakens the attraction between the interlayers enough for water to force them apart and center into the spaces, making the whole layer accessible to hydration and cations intercalation. The swelling and plasticizing of the MMT are due to increasing interparticle spaces rather than to the expansion of individual primary particles. Thus, MMT undergoes two interdependent processes during intercalation: swelling of the interlamellar spacing and disaggregation. This multifarious structure of the mineral furnishes abundant surface area (780 m^2/g) for the host molecules to interact by intercalation and otherwise.

Analytical Characteristics of Polymer-MMT Materials

Various analytical techniques were used to examine the nature of bonding between MMT and the polymer in the composite and to investigate the character of the spatial assembly of the composite and the evolving morphology. The observations of the analytical examination are categorized into three levels:

Interaction Bonding. Because polymer-MMT nanocomposites are multiphase materials, chemical continuity between the inter-phases which depends on the nature of the interaction bonding between them, plays a significant role in their durability. The chemical continuity of the phases has been established from elemental analysis, TGA and FTIR. They provided evidence of the formation and stability of the ionic level at which the polymer has accessed the interlayer charge sites and that the functional chain ends of the polymer have been ionically bonded to the mineral layers.

The inorganic ions (Na/Ca) contents in the composite as revealed by the elemental analysis give a clear indication of the efficiency of intercalation by the exchange of inorganic cations with the organic cations. The mineral contents (wt%) in crude dried polymer-MMT composites were evaluated by the thermogravimetric analysis. Additional evidence for the nature of the ionic bonding between the polymer and mineral was obtained from FTIR spectroscopy.

Elemental and FTIR analyses of VM-MMT intercalate (I) and PS-MMT (II) were used to ascertain their chemical structures and the stabilities of the ionic bonds[51]. The Na content (0.08%) in the rubber-MMT composite is equivalent to that arising from Na_2O implying that all ionic sites in MMT were exchanged with the onium salts of the ATBN. A ATBN-to-clay ratio of 1.5 (wt) was calculated from the C and N content. The ionic bonding of the terminal secondary amine (N-aminoethylpiperazine) salt of the ATBN with MMT shows four characteristic IR bands in the region 2700-2400 cm^{-1}, specifically at 2655, 2550, 2495 and 2440 cm^{-1}, Figure 1a. The higher bathochromic shift in the composite, Figure 1b, than the simple amine salt implies that the onium salt of the ATBN is intercalated with a more electronegative moiety of metal oxide. The broadening and the decline of the number of bands are suggested to be related to intermolecular interactions.

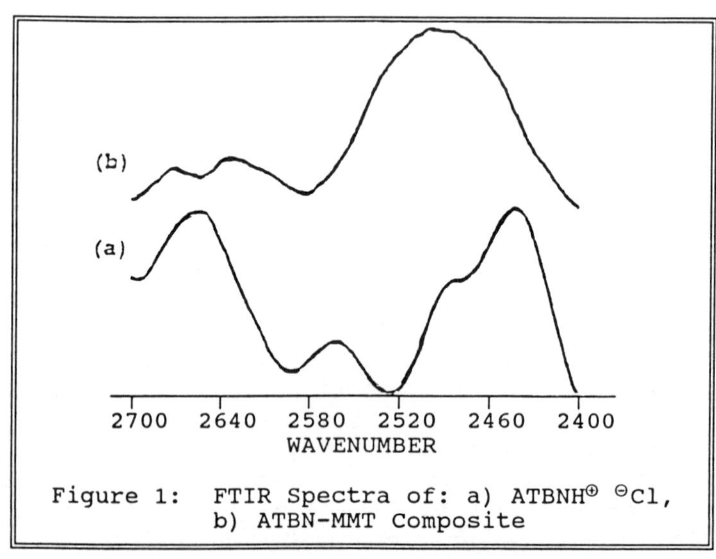

Figure 1: FTIR Spectra of: a) $ATBNH^{\oplus}\ {}^{\ominus}Cl$, b) ATBN-MMT Composite

Interlamellar Structure. WAXD measurements of the monomer-MMT and polymer-MMT gave rise to an interlamellar spacing. They were carried out to examine the interlayer activity in nanocomposites and to examine the polymer state within the interlayers before and after molding. The increase in the interlamellar spacing as determined by d_{001} reflections, amounts to a swelling in the interlayear spacing which give an indication about molecular packing of the monomer and the polymer between mineral layers. It also ascertain the complete exchange of inorganic cations and the amount of the monomer and polymer residing within the inter- layer spacing of MMT.

Typical WAXD behavior of PS-MMT composites containing 5% and 50% mineral are shown in Figure 2. Apparently, two diffracting species exist in these samples, one of which exhibit d-spacing of about 10 Å, is perhaps associated with unreacted mineral. The majority of the material exhibits interlayer diffraction of 2.1 nm and 1.7 nm for 5% and 50% composites, respectively.

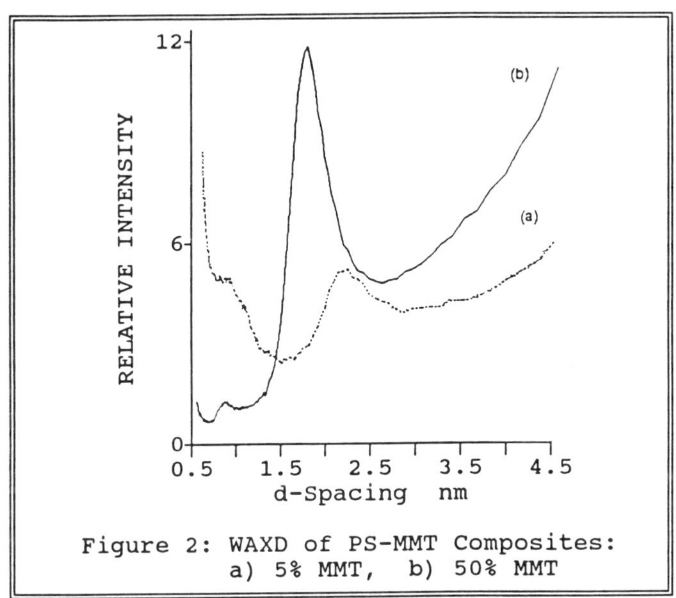

Figure 2: WAXD of PS-MMT Composites: a) 5% MMT, b) 50% MMT

The WAXD of MMT and compression molded ATBN-MMT samples are shown in Figure 3. The thickness of the polymer layer within the interlamellar span is determined by substracting the thickness of the silicate layer of MMT (9.3Å) from the observed d-spacing for ATBN-MMT (14.1Å). The corresponding value 4.9Å is slightly less than a bimolecular layer (2x2.8 Å) and might be considered to represent the mean thickness of a buckled and folded polymer chain, in which about half of the segments of each chain is in contact with the surface. Complete or partial collapse of the macromolecules at the surface represents a considerable degree of uncoiling. The amount of interlayer polymer remains unchanged for a number of treatments intended to explore significant chemical and thermal stability of the polymer-mineral bond at the molecular scale.

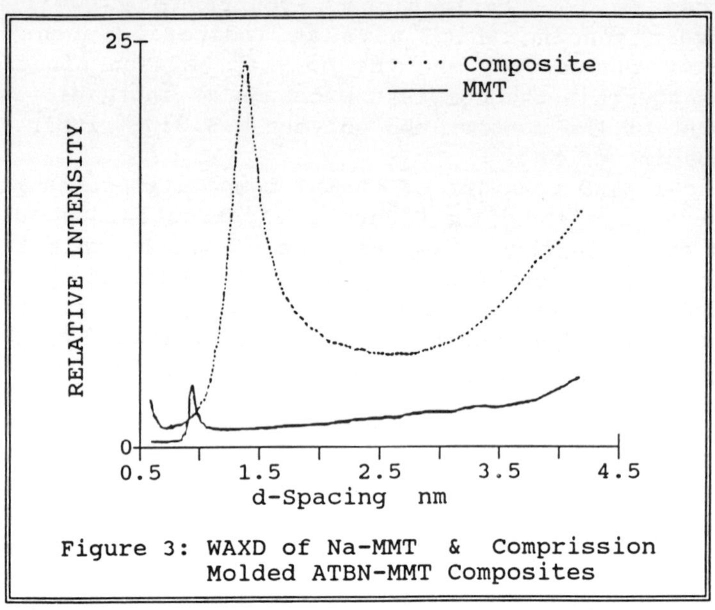

Figure 3: WAXD of Na-MMT & Comprission Molded ATBN-MMT Composites

Morphological Hierarchy. Microscopic evidence is presented to investigate the character of the assembly and to validate the morphological hierarchy of the produced nanocomposite. The elementary building block of the composite which is a pair of the mineral lamelae and the enclosed polymer may erect an agglomerate of a nanostructure which can be discerned by means of electron microscopic imaging or a matching technique. Nanostructures cannot simply be deduced from d-spacing measurements that detect lamellar activities. To identify how the building blocks of MMT (which exhibits a well recognized hierarchical morphology) disaggregate then reassemble to create the inorganic domain morphology in the composite, SEM and TEM evidences are used to unveil the morphological hierarchy of the polymer-MMT composites.

Quantitative analysis of the morphology is used to examine the swelling degree and disaggregation of MMT in the composite. Measurements of the thickness of the layered morphology reveals a distribution of sizes. The observed distribution of the thickness of the layered morphology implies that the primary building block of the nanodomains in this composite is, perhaps, an assembly of few to several lamellae (lamellar multiplets). The spacing between the multiplets was evaluated by means of an image analyzer.

TEM micrography of a thin film made from THF suspension of PS-g-MMT composite (50% MMT) is shown in Figure 4. The PS-g-MMT product appears to be comprised of mineral-rich spheres in which the polymer is bound to the mineral, evenly dispersed in a polymer matrix. The spherical mineral domains range in diameter from 150nm to 400nm and have basal spacings of 17.2-24.5Å. This observation leads into two worthwhile consequences. One is that the spherical clusters can not be perceived to undergo unlimted swelling upon molding. On the other hand, this morphology indicates the feasibility of extracting the mineral-rich domains from the polymer matrix by centrifuge tratment of the suspension to obtain a composite with higher MMT content.

Figure 4: TEM Micrograph of PS-MMT Composite

The domains of the compression molded PS-g-MMT were ca. 50nm which is much smaller than the mineral aggregate and comparable to that of the primary particle of the mineral. WAXD disclosed that the d-spacing of MMT in the composite ranged from 1.7 nm to 2.5 nm suggesting that the mineral aggregates (ca. 10 μm) were dissociated into individual layers then reassembled to lamellar nanoclusters.

SEM examination of a compression molded PS-MMT composite (50% MMT), Figure 5a, shows the presence of nodular submicroscopic domains (ca. 50 nm), which implies that the spherical clumps of Figure 4 are plausibly assemblies of smaller domains. Mapping the surface of the specimen imaged in Figure 5a for Si at the same magnification (Figure 5b) unveils that MMT domains in the molded sample probably maintain a wide size distribution, of which the nodular entities of Figure 5a are the largest, and that they are homogeneously dispersed.

Figure 5: a) SEM Micrograph, b) Si Mapping of PS-MMT Composite

SEM examination of the fracture surface of the compressed molded composite did not reveal the inorganic domains at the maximum possible magnification. Figure 6a shows a micrograph of THF etched fracture surface. At this magnification (20000) no mineral domains could be discerned. The search for mineral domains in this examination was aided by an energy dispersive X-ray probe. An image of elemental mapping for Si, Figure 6b, reveals the absence of large mineral aggregates. The submicron uniformity of the white dots, represent of Si, indicates that the mineral domains are submicron in size and homogeneously dispersed in the polymer matrix.

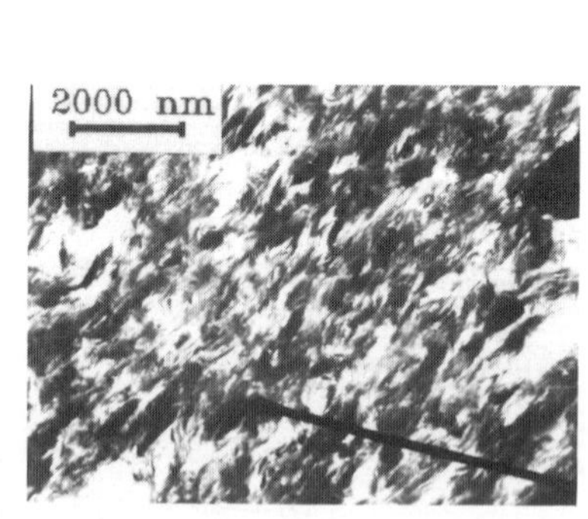

Figure 6: a) SEM Micrograph, b) Si Mapping of ATBN-MMT Composite

TEM observations of ultramicrotome sections of compression molded ATBN-MMT composites, Figure 6, display typical portraits of the nano mineral domains. It appears that the mineral domains are ellipsoids. Higher magnification was used to explore the primary blocks of the mineral nanodomains, Figure 7a,b. The photomicographs indicate that the layered structures do exhibit a wide range of thickness and spacing and the layers could assume extended configuration or bending respectively.

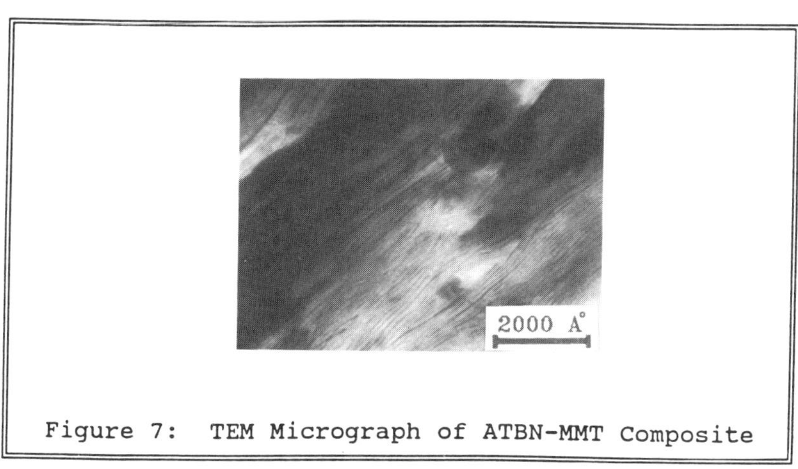

Figure 7: TEM Micrograph of ATBN-MMT Composite

Hierarchical Model for Nanocomposites

Nanostructure was concluded from WAXD measurements of the interlamellar spacing overlooking the fact that interlamellar expansion is possible without disaggregation of the MMT[6 7]. In view of the fact that the properties of structured materials arise from their morphological organization.The morphological hierarchy based on WAXD and quantitative analysis of TEM observations suggests that the MMT layers are assembled to laminated multiplets which are assembled to form mineral-rich clusters. The ability of the multiplets to diffract X-rays offers an alternate interpretation to the disappearance of the interlayer spacing from WAXD measurements.

A view illustrating the hierarchical morphology in the polymer-MMT composite is shown in Figure 8. The formation of the final nanoscale mineral domains involves a physicochemical reaction, which results in simultaneous disaggregation and swelling of the MMT. In the intercalating medium, the MMT must have been dispersed into its lamellar state, at which the ion exchange reaction took place. Short range interactions must have been maintained on the lamellar scale as evidenced by their presence as multiplets and their reassembly clusters. This view is augmented by the fact that MMT swollen in simple liquid then dried, reassembles into its nascent aggregates (0.1-10μm). The inability of the mineral to reaggregate is attributed to the influence of polymer via its dual function, macromolecular and cationic.

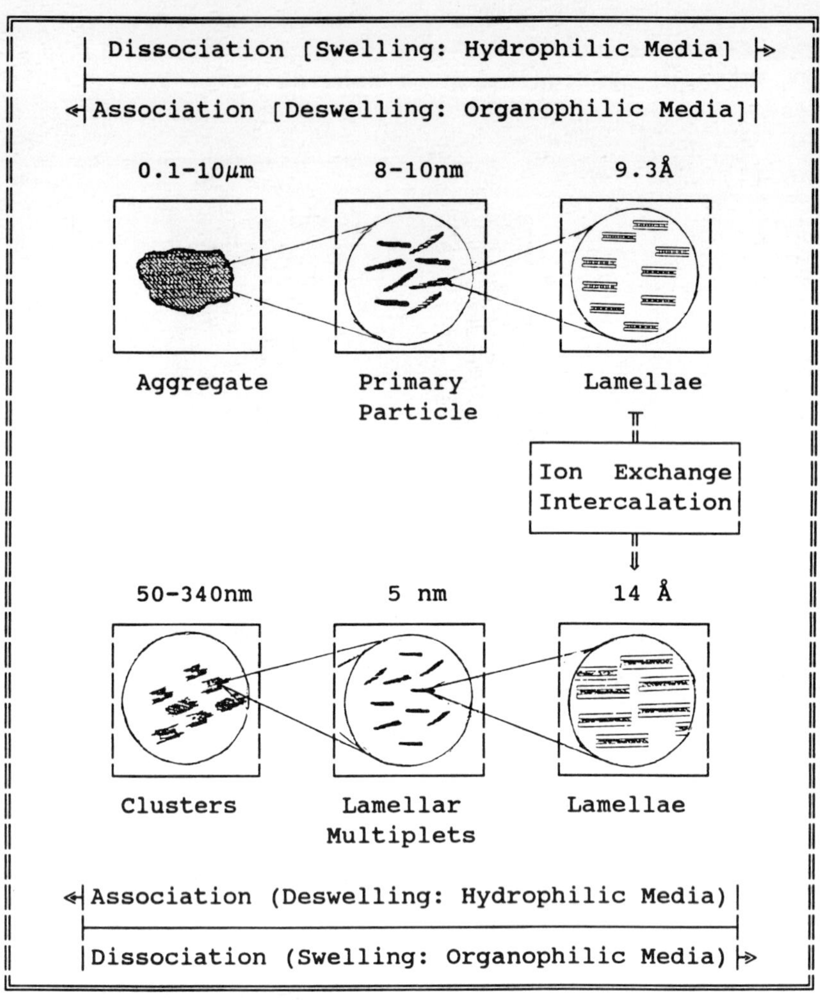

Figure 8: Hierarchical Model for Polymer-MMT Nanocomposites

REFERENCES

1- N.G.Gaylord, H.Ender, L.Davis & A.Takahashi, in "Modification of Polymers", C.E.Carraher & M.Tsuda, Eds, ACS Symp Ser 121, (1980)

2- A.Voet; J Polym Sci Macromol Rev 15: 327-373 (1980)

3- B.K.G.Theng, "Formation and Properties of Clay-Polymer Complexes", Elsevier, Amsterdam (1979)

4- L.E.Brus, W.L.Brown, R.P.Andres, R.S.Averback, W.A. Goddard, A.Kaldor, S.G.Louie, M.Moskovits, P.S.Peercy, S.J.Riley, R.W. Siegel, F.A.Spaepen & Y.Wang; J Mater Res 4: 704 (1989)
5- A.Akelah & A.Moet; J Appl Polym Sci Symp, 55: 153 (1994)
6- A.Wheeler; US Pat 2847391 (1958)
7- H.Z.Friedlander; Chem Eng News 42: 42 (1964) & J Polym Sci B-2: 475 (1964)
8- A.Blumstein; Bull Soc Chim Fr 899 (1961); J Polym Sci Part A-3: 2653 (1965) & Appl Polym Symp 25: 81 (1974)
9- Y.Sugahara, S.Satokawa, K.Kuroda & C.Kato; Clays Clay Miner 36: 343 (1988)
10- F.Bergaya & F.Kooli; Clay Miner 26: 33 (1991).
11- C.Kato, K.Kuroda & H.Takahara; Clays Clay Miner 29: 294 (1981)
12- D.J.Greenland; J Colloid Sci 18: 647 (1963)
13- C.W.Francis; Soil Sci 115: 40 (1973)
14- R.L.Parfilt & D.J.Greenland; Clay Miner 8: 305 (1970)
15- N.Schamp & J.Huylebroeck; J Polym Sci Symp 42:553 (1973)
16- R.A.Vaia, H.Ishii & E.P.Giannelis; Chem Mater 5: 1694 (1993)
17- E.P.Plueddemann, "Silane Coupling Agents", Plenum Press, NY, 1st edn, (1982) & 2nd edn, (1991)
18- E.Ruiz-Hitzky & J.Fripiat; Clys Clay Miner 24: 25 (1976)
19- R.L.Kaas & J.K.Kardos; Polym Eng Sci 11: 11 (1971)
20- J.D.Miller, K.Hoh & H.Ishida; Polym Compos 5: 18 (1984)
21- S.L.Savard, P.Blanchard, J.Leonard & R.E.Prud'homme; Polym Compos 5: 242 (1984)
22- B.D.Favis & P.Blanchard; Polym Compos 5: 11 (1984)
23- M.W.Ranney & C.A.Pagano; Rubber Chem Technol 44: 1080 (1971)
24- G.M.Cameron, M.W.Ranney & K.J.Sollman; Eur Rubber J 156: (3) 37 (1974)
25- G.W.MacDonald, Rubber Age April (1970)
26- T.A.Grillo, Rubber Age 37 (1971)
27- E.Papirer & V.T.Nguyen; J Polym Sci B-10, 167 (1972)
28- A.Vidal, E.Papirer & J.B.Donnet; Cour Polym J 12: 791 (1976)
29- E.Papirer, J.C.Morauski & A.Vidal; Angew Makromol Chem 42: 91 & 597 (1975); Angew Mol Chem 19: 65 (1971)
30- N.G.Gaylord; US Pat 3956230 (1976) & 4071494 (1978)
31- B.Tieke; Adv Polym Sci 71: 79 (1985)
32- L.C.Klein, "Sol-Gel Technology", Noyes Publ, Park Ridge, (1988); Ann Rev Mater Sci 15: 227 (1985)
33- L.G.Hubert-Pfalzgraf, New J Chem 11: 663 (1987)
34- J.Livage, M.Henry & C.Sanchez, Progr Solid State Chem 18: 259 (1988)
35- C.J.Brinker & G.W.Scherer,"Sol-Gel Science", Acad Press, San Diego (1990); J Non-cryst Solids 100: 31 (1988)
36- H.Schmidt; "Inorganic-Organic Composites by Sol-Gel Techniques", MRS Symp Proc 171: 3 (1990)
37- J.Livage, F.Babonneau & C.Sanchez, in "Inorganic and Organometallic Oligomers and Polymers", J.F.Harrod & R.M.Laine, Eds, Kluwer Axad Publ, Dordrecht, 217 (1991)
38- G.Cao & T.E.Mallouk; J Solid State Chem 94: 59 (1991)

39- J.E.Pillion & M.E.Thompson; Chem Mater 3: 777 (1991)
40- M.G.Kanatzidis, C.-G.Wu, H.O.Marcy, D.C.DeGroot & C.R. Kannewurf; Chem Mater 2: 222 (1990) & 3: 992 (1991)
41- M.G.Kanatzidis, C.-G.Wu, D.C.DeGroot, J.L.Schindler, M. Benz, E.LeGoff & C.R.Kannewurf; "NATO ASI, Chemical Physics of Intercalation II"; J.Fisher,Ed, Plenum Press, NY, (1993)
42- V.Mehrotra & E.P.Giannelis; in "Polymer Based Molecular Composites", D.W.Shaefer, J.E.Mark,Eds, MRS Pr 171: 1990
43- V.Mehrotra & E.P.Giannelis; Solid State Comm 77: 155 (1991) & 51: 115 (1992)
44- C.Kato, K.Kuroda & H.Takahara; Clays Clay Miner 29: 294 (1981)
45- M.Ogawa, K.Kuroda & C.Kato; Clay Sci 7: 243 (1989)
46- Y.Fukushima, A.Okada, M.Kawasumi, T.Kurauchi & O.Kamigaito; Clay Miner 23: 27 (1988)
47- E.P.Giannelis, V.Mehrotra, O.K.Tse, R.A.Vaia & T.C.Sung, in "Synthesis and Processing of Ceramics: Scientific Issues", E.W.Rhine, T.M.Shaw, R.J.Gottshall & Y.Chen, Eds, MRS Proc 249: 547 (1992)
48- P.Aranda & E.Ruiz-Hitzky; Chem Mater 4: 1395 (1992)
49- P.B.Messersmith & S.I.Stupp; J Mater Res 7: 2599 (1992)
50- L.F.Nazar, et al, in "Solid State Ionic II", MRS Proc; 210: (1991)
51- A.Moet & A.Akelah; Mater Lett 18: 97 (1993)
52- A.Moet, A.Akelah, A.Hiltner & E.Baer; Paper No. V-4.2, MRS Symp, San Francisco, Calif, 4-8 April (1994)
53- A.Moet, A.Akelah, A.Hiltner & E.Baer, in "Molecularly Designed Ultrafine/Nanostructured Materials", K.E.Gonsalves, G.M.Chow, T.D.Xiao and R.C.Cammarata, Eds, MRS Symp Proc, 351: 91 (1994)
54- C.Kato, K.Kuroda & M.Misawa; Clays Clay Miner 27: 129 (1979)
55- Y.Fukushima & S.Inagaki; J Incl Phenom 5: 473 (1987)
56- A.Okada, M.Kawasumi, A.U.suki, Y.Kojima, T.Kurauchi & O. Kamigaito; MRS Symp Proc 171: 45, (1990)
57- P.Kelly, A.Akelah, S.Qutubuddin & A.Moet; J Mater Sci, 29: 2274 (1994)
58- A.Akelah, P.Kelly, S.Qutubuddin & A.Moet; Clay Miner,29: 169 (1994)
59- A.Moet, A.Akelah, N.Salahuddin, A.Hiltner & E.Baer; Paper No V-6.1, MRS Symp, San Francisco, CA, 4-8 April (1994)
60- A.Akelah, N.Salahuddin, A.Hiltner, E.Baer, & A.Moet; Nanostr Mater, 4: (8), (1994)
61- A.Moet, A.Akelah, N.Salahuddin, A.Hiltner & E.Baer, in "Molecularly Designed Ultrafine/Nanostructured Materials K.E.Gonsalves, G.M. Chow, T.D.Xiao & R.C.Cammarata, Eds, MRS Symp Proc 351: 163 (1994)
62- A.Akelah, N.Salahuddin, A.Hiltner, E.Baer & A.Moet; Mater Lett, 22: 97 (1995)
63- A.Akelah, A.Rehab, A.Selim, & T.Agag; J Mol Catal 94: 311 (1994)

CdS NANOPARTICLES IN GEL FILM NETWORK : SYNTHESIS, STABILITY AND OPTICAL PROPERTIES

Arup K. Atta, Prasanta K. Biswas and Dibyendu Ganguli

Sol-Gel Laboratory
Central Glass and Ceramic Research Institute
Calcutta-700 032
India

INTRODUCTION

Development of quantum (Q) size metal and semiconductor nanoparticles[1-3] in various kinds of hosts and their optical and opto-electronic characterization have become an important area of research in recent times. The importance of this activity is derived from the possibility of application of such materials in photonics.Sol-gel processing has proved to be a very useful method for generating these new composites, and a large number of papers have been published [4-6] in the last few years. Much of this work deals with generation of sulphide semiconductor nanoparticles in monolithic silica or silicate bodies. In general the basic optical properties of such materials are characterized by an excitonic HOMO-LUMO transition at around 400-460 nm and a fluorescence band around 460 nm, both at 298 K. The transition depends on the size of the microcrystals confined in a suitable host. At the lowest excitonic state [7] the CdS particles should exhibit an absorption shoulder in the UV range (270 - 300 nm). This has also been observed experimentally.[8-10]

The present work reports the stable confinement of Q-CdS microcrystals at their lowest excitonic state in transparent baked zirconia gel thin film network by optical and fluorescence spectral studies. This also describes confinement of Q-CdS in zirconia-titania gel matrix for achieving a suitable host for attenuation of the impurity derived fluorescence band.

EXPERIMENTAL

The basic preparative principle of Q-CdS in gel film network is in-situ generation of CdS particles via a thiocomplex intermediate of cadmium within the gel films obtained from partially polymerized alkoxide based sols [11] containing dissolved nitrate or chloride of cadmium and thiourea. The equivalent weight ratio of ZrO_2 : CdS was maintained at 60:40 in the sols. Three sols were prepared with pH adjusted to 2.0, 4.0 and 4.8 (Jencons pH meter, model 3030) respectively.

The sols were used to deposit the films on ultrasonically cleaned silica and soda lime glass (SLG) substrates by the dipping and spinning techniques. The gel films were baked at temperatures 90°-300°C with long soaking time (2-4 h) at relatively low temperatures (90°-135°C) and short baking time (0.5 h) at relatively high temperatures (e.g. 210°C). The protective layer of silica (baking temperature, 100°C) deposited on top of the CdS doped zirconia gel films was prepared from an acid catalyzed 2.0 wt% equivalent silica sol of tetraethyl orthosilicate.

Optical spectra of the films were obtained on a Hitachi U3210 UV-visible spectrophotometer. The fluorescence properties of the films were studied with Fluorolog II, model F IIIA (298 K) and Perkin-Elmer MPF-44B(77K) fluorescence spectrophotometers. The size and nature of crystallites were characterized by TEM (JEOL, JEM-200 CX) at 70 and 100KV. For studying absorption and fluorescence properties, films of thickness in the range 800-1500 Å were prepared. For TEM studies, the thickness of the films was reduced to 200-300 Å. The thickness of protective silica layer was 400 $\pm$ 25 Å. The film thicknesses in all cases were measured with a surface profile analyser (DEKTAK II).

RESULTS AND DISCUSSION

Addtion of methanolic solution of cadmium nitrate and thiourea to a partially hydrolysed zirconia sol brought the pH value to 4.8 which was decreased to 4.0 and 2.0 by the addition of dilute nitric acid. The sols were deposited on silica glass

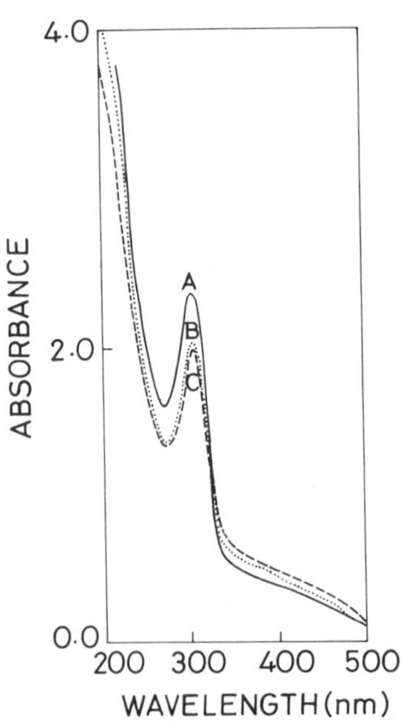

Figure 1. Absorption properties of the baked films (130°C) prepared from sols of different pH:A, pH 2.0 ; B,pH 4.0; C, pH 4.8

substrate and baked to 130°C. Characteristic UV absorption band and a shoulder appeared at 300 nm and 400-450 nm (Fig.1) respectively for CdS microcrystals. The film from the sol of pH 2.0 exhibited relatively high intensity at 300 nm band and hence most of the work was done with the sol of pH 2.0. As NO_3^- also exhibits a band at 300 nm for n —> π^* transition, hydrated cadmium chloride was also used for sol preparation and similarly developed films from such sols also exhibited a sharp peak at ~ 300 nm suggesting the generation of CdS nanoparticles in all cases, specially in baked films.

The effects of baking temperature on optical spectra are shown in Fig.2. With baking at 90°C, the UV peak was found to have grown without any absorption in the visible region indicating that the smallest particles of CdS just started to develop. On baking at 130°C, both UV peak and visible shoulder appeared. At 190°C the UV peak was at the threshold of vanishing probably due to coarsening of the CdS particles. It is interesting to note that with increasing baking temperature the UV peak did not shift but red shift was found in the visible shoulder at 400-450 nm (Fig.2).

The generation of the UV peak is due to the generation of Q-crystallites in the zirconia gel matrix with the size range 20-40 Å as predicted theoretically by Brus[7] and demonstrated experimentally by a number of workers.[9, 10, 12, 13]

Another important point for films baked at 300°C is that no UV peak was found at 298K while a strong peak was observed at 77K (Fig.3). This indicates

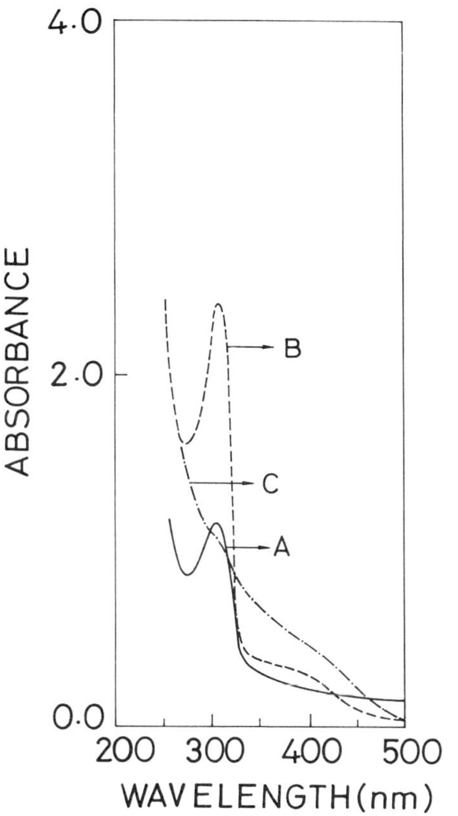

Figure 2. Absorption spectra (298K) of films baked at different temperatures : A, 90°C; B, 130°C; C, 190°C.

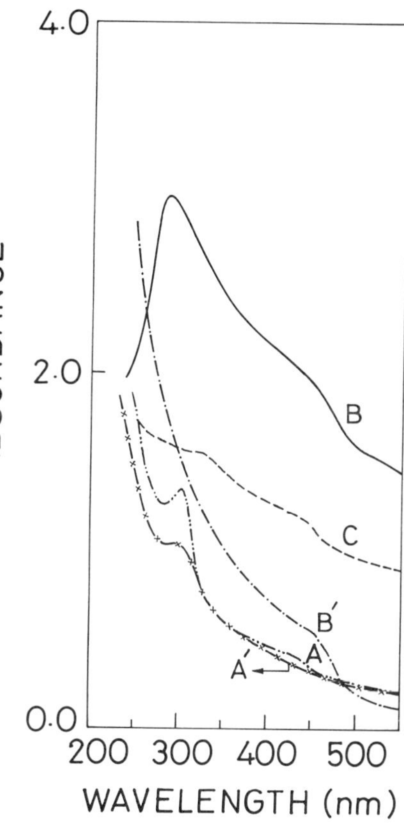

Figure 3. Absorption spectra (77K, 298K) of films baked at different temperatures A,90°C;B,210°C;C, 300°C (77K) & A' 90°C; B' 210°C (298K).

decrease in phonon interaction [14] at 77K, so that the excitonic effect was pronounced even with relatively low concentration of small size CdS.

Transmission electron microscopic pictures of the films (Fig.4) provide a tentative idea of size of the microcrystals. The 130°C and 210°C baked films suggested size range 20-40 Å and 80-120 Å respectively. The electron diffraction (ED) rings in the ED pattern in both cases were rather weak and sometimes diffuse. The coarse particles seemed to possess hexagonal structure. The diffuse ED pattern in 130°C baked films was possible due to change in symmetry from hexagonal to cubic due to reduction in size.[15]

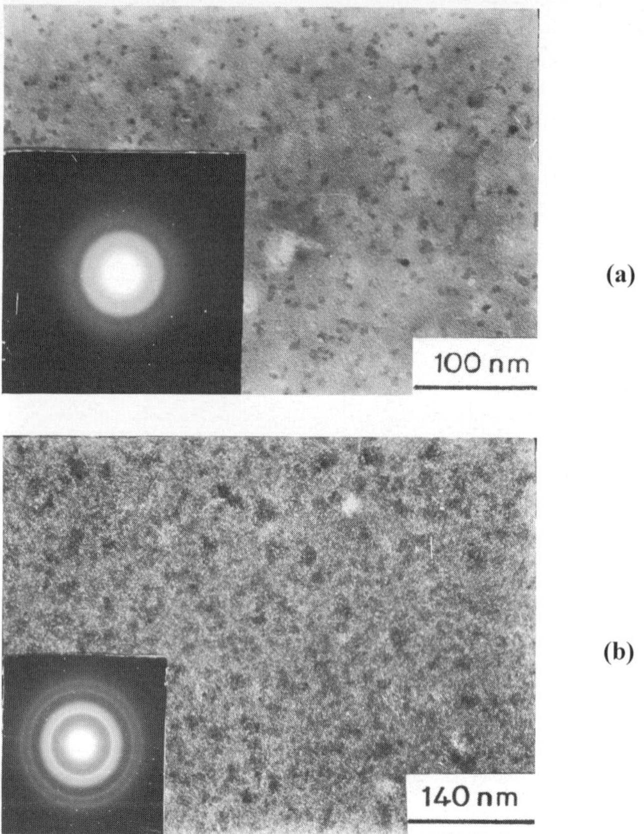

Figure 4. TEM micrographs of the films baked at different temperatures: (a) 130°C, (b) 210°C; Inset : electron diffraction pattern of the corresponding film.

Practically no degradation of the 300 nm peak was found for at least 2-3 weeks at low humidity (15-20%). However, at high humidity (70-90%), the peak progressively lost its intensity (Fig.5). Hence an overcoat of silica layer of 400 ± 25 Å thickness was deposited to increase the stability of the Q-size CdS microcrystals. It was observed (Fig.5) that the peak intensity still decreased at high humidity, but the rate of destabilization was considerably lower.

Fluorescence spectra (Fig.6) of 90°C baked films on SLG at 298K, derived from the sols of different pH exhibited three bands at 415, 470 and 520 nm which are due to glass substrate, CdS excitons and impurity effects respectively.

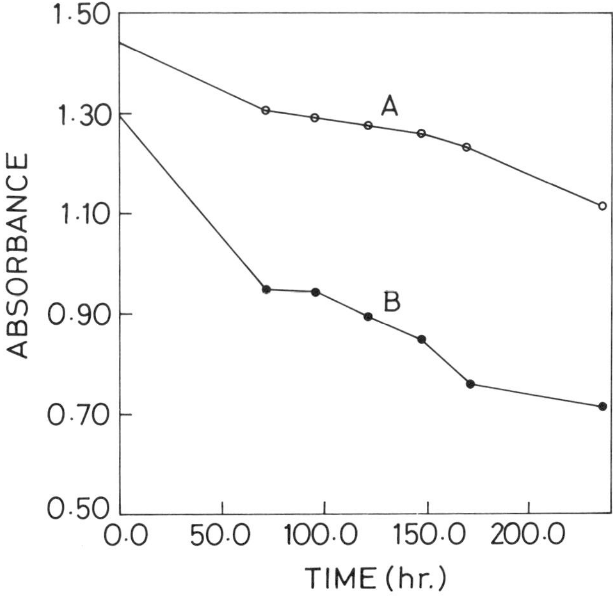

Figure 5. Degradation curve of lowest excitonic band intensity of CdS microcrystals in 130°C baked film with time : A, overcoated with a protective silica layer; (B), no overcoat.

With increasing pH, coarsening of the sulphide particles took place with creation of more defect centres resulting in prominent impurity bands. The film from sol of 2.0 pH shows a sharp band which stands out in comparison with the defect band.

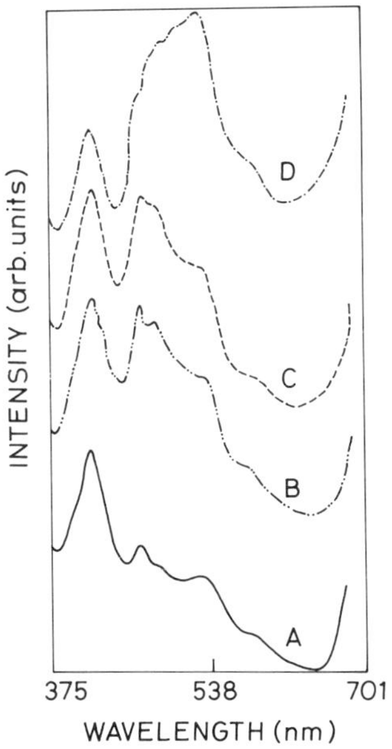

Figure 6. Fluorescence spectra (298 K, exc.380 nm) of the films (90 °C) : A, pH 0.8; B, pH 2.0; C, pH 4.0; D, pH 4.8.

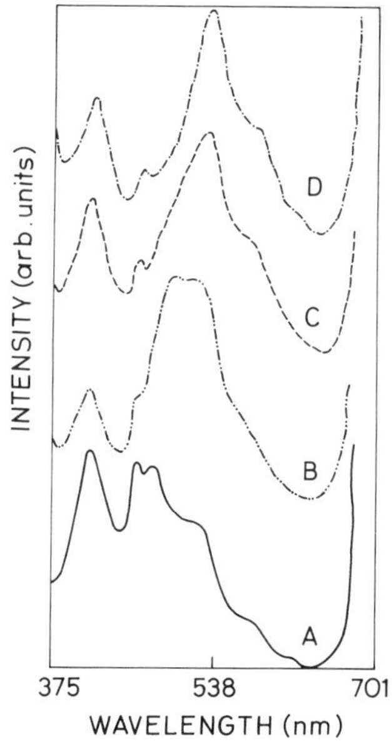

Figure 7. Fluorescence spectra (298 K, exc. 350 nm) of the films baked at different temperatures: A, 75 °C; B, 130 °C; C, 200 °C; D, 375 °C.

However, when the temperature of baking was raised to 130°C, the defect band became too prominent; at still higher temperature, the excitonic band became relatively very weak (Fig.7).

It was also observed (Fig.8) that at 77K, a wide shift of the defect band with broadening of the excitonic band occurred. A very interesting result was obtained when zirconia-titania mixed hosts were used. The impurity band showed almost total annihilation, and the excitonic band stood out in sharp contrast.The quenching effect of titania as observed in our earlier work [11] possibly caused annhilation of the defect band at 77K. The basic reason is possibly the shift of the electrons in excited CdS particles to the valence band of TiO_2 causing greater

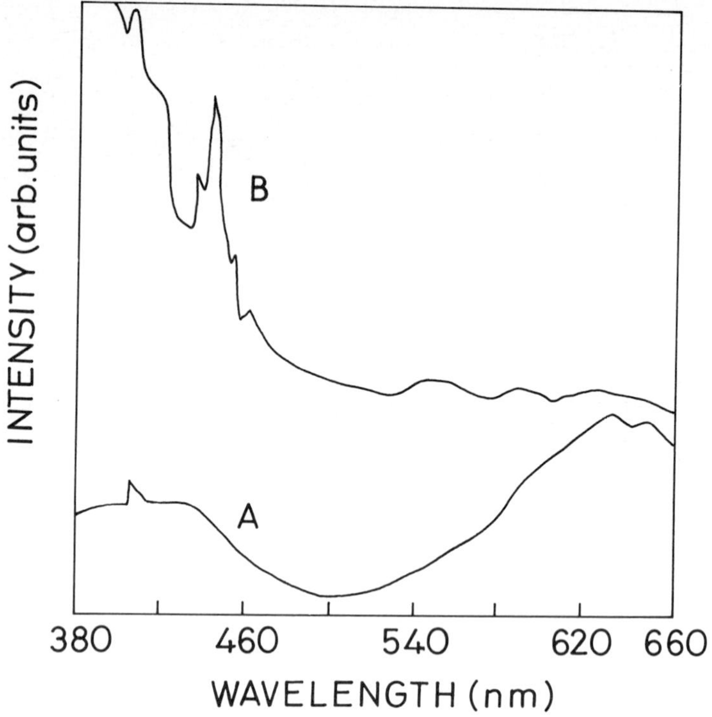

Figure 8. Fluorescence spectra (77K, exc. 350 nm) of the films (130°C) in different gel matrices : A, ZrO_2; B, ZrO_2 - TiO_2.

availability of such electrons in the defect sites also. For obtaining increased life and decreased response time, the defect band is undesirable and has to be annhilated by either high power laser irradiaition or addition of special chemicals like MV^{2+}.[16] Incorporation of titania in the gel matrix may be a better alternative, but more detailed studies are required to prove this.

Acknowledgment

The authors thank Dr. B.K. Sarkar, Director of this Institute for giving permission to publish this paper. Thanks are also due to Dr. S. Bhattacharya, Scientist, Glass Technology Section for obtaining the fluorescence spectra. One of the authors (A.K.A) thanks CSIR for providing Research Associateship.

REFERENCES

1. R.K. Jain and R.C. Lind, Degenerate four-wave mixing in semiconductor-doped glasses, *J. Opt. Soc. Am.* 73:647 (1983).
2. Anon., Techno Japan, 26: 63(1993).
3. B.J. Potter and J.H. Simmons, Quantum confinement effects in CdTe-glass composite films produced using r.f. magnetron sputtering, *J. Appl. Phys.* 68:1218 (1990).
4. M. Nogami and A. Nakamura, Emission spectra of CdS doped in sol-gel derived glasses, *Phys. Chem. Glasses*, 34:109(1993).
5. H. Nasu and J.D. Mackenzie, Non-linear optical properties of glasses and glasses- or gel-based composites, *Opt. Engg.* 26:102(1987).
6. T. Hasimoto, T. Yoko and S. Sakka, Third-order non-linear optical susceptibility of -Fe_2O_3 thin film prepared by the sol-gel method, *J. Ceram. Soc. Jpn.* 101:64(1993).
7. L.E. Brus, Electron-electron and electron-hole interactions in small semiconductor crystallites: The size dependence of the lowest excited electronic state, *J. Chem. Phys.* 80:4403(1984).
8. A. Henglein, Small-particle research: Physicochemical properties of extremely small colloidal metal and semiconductor particles, *Chem. Rev.* 89: 1861(1989).
9. L. Spanhel, E. Arpac and H. Schmidt, Semiconductor clusters in the sol-gel CdS nanocomposites, *J. Non-cryst. Solids*, 147 & 148:657(1992).
10. C.T. Dameron, R.N. Reese, R.K. Mehra, A.R. Kortan, P.J. Carroll, M.L. Steigerwald, L.E. Brus and D.R. Winge, Biosynthesis of cadmium sulphide quantum semiconductor crystallites, *Nature,* 338: 596(1989).
11. A.K. Atta, P.K. Biswas and D. Ganguli, Optical properties of CdS and (Cd, Zn)S doped sol-gel titania films, *Mat. Lett.* 15: 99-103(1993).
12. K. Osakada, A. Taniguchi, E. Kubota, S. Dev, K. Tanaka, K. Kubota and T. Yamamoto, New organosols of Cu_2S, CdS, ZnS, HgS, NiS and mixed metal sulfides in n-n-dimethyl formamide and dimethyl sulfoxide, preparation, characterisation and physical properties, *Chem. Mat.* 4: 562(1992).
13. L. Spanhel, M. Haase, H. Weller and A. Henglein, Photochemistry of colloidal semiconductors. 20. Surface modification and stability of strong luminescing CdS particles, *J. Am. Chem. Soc.* 109: 5649(1987).
14. Y. Fainman, J. Ma and S.H. Lee, Non-linear optical materials and applications, *Mat Sc. Reports*, 9: 53(1993).
15. A.I. Ekimov, I.A. Kudryavtsev, M.G. Ivanov and A.L. Efros, Spectra and Decay kinetics of radiative recombination in CdS microcrystals, *J. Lumin.*, 46: 83(1990).
16. J. Kuczynski and J.K. Thomas, Surface effects in the photochemistry of colloidal cadmium sulfide, *J. Phys. Chem.* 87: 5498(1983).

PREPARATION OF DENSE THIN FILMS BY SOL-GEL WITH VERY HIGH HUMIDITY SENSITIVITY

Gualtiero Gusmano,[1] Angelo Montenero,[2] and Enrico Traversa[1]

[1]Dipartimento di Scienze e Tecnologie Chimiche,
Universita' di Roma "Tor Vergata", 00133 Rome (Italy)
[2]Dipartimento di Chimica Generale e Inorganica, Chimica Analitica,
Chimica Fisica, Universita' di Parma, 43100 Parma (Italy)

INTRODUCTION

The currently available commercial humidity sensors are mainly based on porous sintered ceramics or on polymeric films.[1] Ceramics possess superior properties insofar chemical resistance and mechanical strength are considered,[2] but the conventional ceramic technology used for their production is not competitive with polymer sensor fabrication technology. However, the present trend in sensor research is towards the fabrication of integrated sensor, with miniaturization of sensing elements.[3] For this aim, the use of thin films is needed, and the recent development of ceramic deposition technologies was a great pulse to study ceramics in film form for chemical sensors. Recently, ceramic thin films fabricated with physical deposition methods, studied as humidity sensors, have shown sensitivities much lower than those measured for porous sintered ceramics, together with very large resistivities.[4,5]

Given that the sol-gel technology permits the production in an economical way of high purity ceramics in their final film shape (by dipping or spin-coating), the authors of this paper have decided to study, as humidity sensors, ceramic films prepared with this method. At first, the humidity-sensitive electrical properties of sol-gel processed TiO_2 films have been studied, but the results obtained were not entirely adequate.[6] Similar results were found also by other authors, which have improved the electrical response of sol-gel processed TiO_2 films in humid environments with the addition of SnO_2.[7] The addition of alkali ions was found to improve the humidity-sensitive characteristics of a number of ceramic materials,[8] and in particular of sintered TiO_2.[9,10] It was decided to study the effect of doping TiO_2 films with alkali during their processing with the sol-gel technique.[11]

In fact, the performance of titania films was dramatically improved by the addition of potassium, and results about their very high humidity sensitive electrical properties are reported in this paper. We report also the outstanding performances as active elements for humidity sensors recorded for sol-gel processed thin films based on iron oxide. Prototype humidity sensors were deposited by dip-coating the ceramic films onto Al_2O_3 substrates with comb-type Au electrodes. The humidity sensitive response of these materials was even larger than that reported for porous ceramics. It has been reported in the relevant literature that a large pore volume is necessary for high rh-sensitivity of ceramic oxides.[12] In our case, however, the high sensitivity was observed for films which were free of capillary pores, as revealed by the morphological SEM analysis.

Polymers and Other Advanced Materials: Emerging Technologies and Business Opportunities. Edited by P. N. Prasad *et al.*, Plenum Press, New York, 1995

EXPERIMENTAL PROCEDURE

TiO_2 containing different amounts of K ions was prepared by means of a sol-gel method.[13] The sol was prepared by mixing titanium isopropylalkoxide (TIPT) with iso-propyl alcohol, distilled water, and an acid catalyst (HCl 37%). K ions were incorporated in the titania sol by adding a suitable amount of CH_3COOK, respectively, to the TIPT mixed solution, and stirring for about 1 h. Four compositions were prepared, with an ion content of 1, 3, 6, and 10 at% (with respect to Ti).

The sol-gel preparation of iron oxide[14] was based on the synthesis of an iron alkoxide as the precursor.[15] The matrix sol was prepared by diluting the iron methoxyethoxide in methoxyethanol, with a final concentration of 4.4 M, which was then hydrolised by adding HNO_3 and water to a molar ratio equal to 1.

From the solutions, bulk gels and films were obtained for both materials. The bulk gels were dried in an oven for one day at 100°C, powdered, heated to various temperatures for 30 min in air, and then analysed. For electrical measurements, prototype sensors were prepared by depositing the ceramic films on Al_2O_3 substrates with comb-type Au electrodes. The films were deposited from the appropriate solutions by dipping, using a withdrawal speed of 25 cm/min for titania-based films and of 20 cm/min for iron oxide films. The films thus obtained were dried at 100°C for one day, and then heat-treated to various temperatures for 30 min.

Powdered bulk gels and films were analyzed using the following techniques: Fourier transform infrared (FTIR) spectroscopy, x-ray diffractometry (XRD), using Cu kα radiation ($\lambda = 0.154$ nm) for titania-based materials, and Co kα radiation ($\lambda = 0.179$ nm) for iron oxides, and scanning electron microscopy (SEM).

The humidity-sensitive electrical response of the specimens was evaluated by electrochemical impedance spectroscopy (EIS) measurements, using a frequency response analyzer (Solartron 1255) equipped with a home-made high impedance adaptor. EIS spectra were recorded in the frequency range 10^{-2} - 10^5 Hz, at 40°C and relative humidity (RH) values ranging from 4 to 85%, obtained by mixing dry and water-saturated air. The monitoring of RH within the test chamber was carried out using a hygrometric probe (Ultrakust mod. F3000), which gave results accurate to within ± 2%.

RESULTS

Sol-Gel Processed K-doped TiO_2

The FTIR spectra of the bulk gels of titania, heated to different temperatures, showed the presence of a large peak in the range 1000-400 cm^{-1}, which is attributable to the Ti-O bond. All the other peaks were assigned to the organic compounds remaining into the gel. These peaks decreased in intensity with increasing temperature, and they disappeared when the temperature reached 500°C. The higher the temperature, the narrower the peak attributed to the Ti-O bond.[6]

Figure 1 shows the XRD spectra of the TiO_2 bulk gels heated to various temperatures up to 1000°C. Below 200°C, the diffraction patterns showed the lack of any definite structure. The anatase structure started to appear above this temperature, and was maintained up to 750°C. The anatase lines were progressively narrowing with increasing temperature, showing that the degree of cristallinity of the specimens was increased. At higher temperatures, the anatase structure was progressively transformed into the rutile structure, which was the only phase present above 900°C. Standard XRD analysis performed on TiO_2 films heated up to 500°C did not reveal any definite structure. For films heated to 900°C, XRD measurements performed using very low scanning rate showed only the most intense peak of anatase, suggesting the presence of an oriented structure. This indicates a slower crystallization for TiO_2 films than for TiO_2 powdered bulks.

The addition of 1 at% K did not significantly change the thermal evolution of the titania gels. Some changes were evidenced for 6 at% K-doped TiO_2 powders: a small amount of rutile structure was found to be present also at low temperatures, at which only anatase was found for pure TiO_2. At these concentrations, no K-containing phases were detected, but only peaks attributable to anatase and rutile phases were observed.

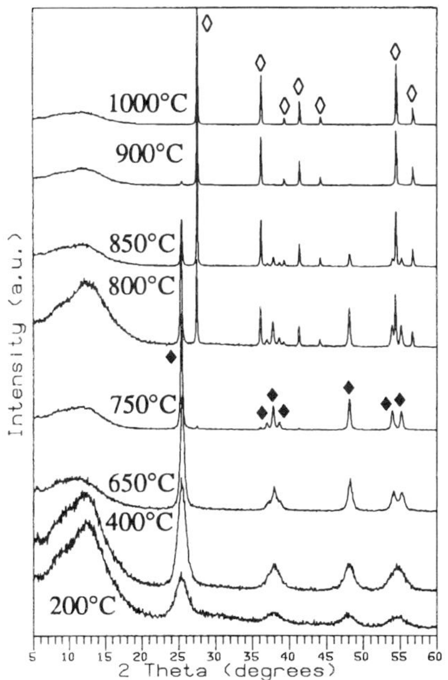

Figure 1. XRD spectra of TiO_2 powders heated to various temperatures (♦ = anatase, ◊ = rutile).

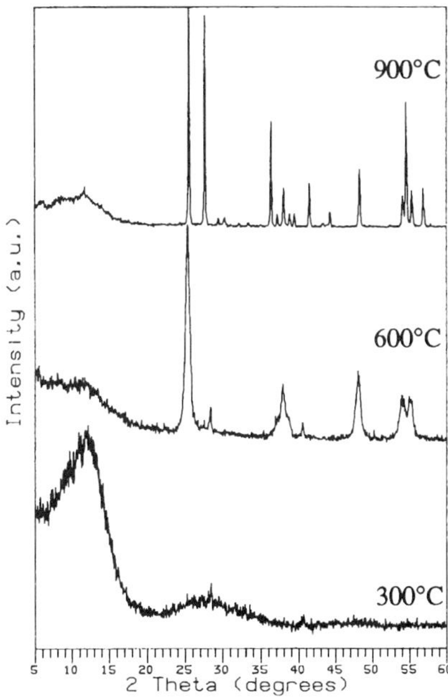

Figure 2. XRD spectra of 10 at% K-doped TiO_2 powders heated to various temperatures.

Figure 2 shows the XRD spectra of the 10 at% K-doped TiO_2 bulk gels heated to various temperatures up to 900°C. The XRD pattern of the powder heated to 300°C did not show any definite structure. The profile of the sample heated to 600°C showed mainly the presence of anatase with traces of rutile. The diffraction pattern recorded for the sample heated to 900°C showed clearly the presence of both anatase and rutile structures. Moreover, weaker signals centered at 2θ = 29.8°, 30.5°, and 33.5° were observed, which can be due to to K_4TiO_4. Another peak observed at 2θ near to 12°, not ascribable to TiO_2 phases, is very close to the position of the most intense peak of $K_2Ti_8O_{17}$. The presence of potassium titanates was observed only when the K content was as high as 10 at%.

The morphological SEM analysis of the undoped TiO_2 films showed that the films heated to 300 and 500°C were compact, free of porosity, and continuous, with no detectable grains. Some discontinuities were observed around the pores of the Au electrodes and of alumina. The addition of K did not modify the morphology of the films. A very thin, glassy layer covering the alumina substrate was observed, without the presence of pores.

The RH dependence of resistance for undoped and 10 at% K-doped TiO_2 films heated to 300°C is reported in Figure 3. The resistance was calculated on the complex plane plot from the intercepts of the semiarc of the EIS spectra, extrapolated where necessary, with the real axis. The undoped films showed a small resistance variation from 0% to about 50% RH, and a variation of 2 orders of magnitude in the RH range 50-85%. The behaviour of the TiO_2 films with concentrations up to 6 at% K was similar to that of undoped films. Films with 10 at% K showed a very large variation (7 orders of magnitude) of the resistance with RH, over the entire RH range tested. At the lowest RH values the variation is noticeable, with a decrease in resistance of nearly 3 orders of magnitude from 4% to 10% RH, while at higher RH values the resistance decreased less steeply.

It was observed that the RH sensitive electrical response of the 10 at% K-doped titania films depended on the frequency at which the resistance was measured.[16] At 50 Hz, a linear variation of the logarithm of the resistance with RH, of 4 orders of magnitude over the entire RH range tested, was observed. The large RH-sensitivity at low RH was measured at frequencies $\leq$ 1 Hz. Measurements carried out at different intervals of time without heat-cleaning treatments prior of a new measurement were very reproducible over time, demonstrating that for this material very little drift is observed even after a long time.[16]

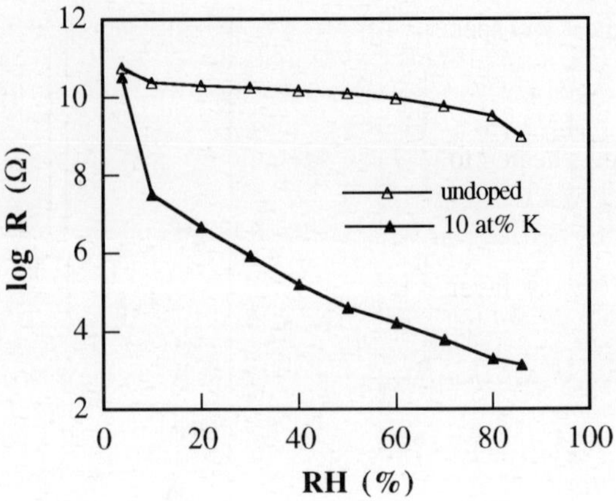

Figure 3. The RH dependence of the resistance of undoped and 10 at% Li- and K-doped TiO_2 films heated to 300°C, measured at 40°C.

Sol-Gel Processed Iron Oxide

The FTIR spectrum of the bulk gel precursor to iron oxide confirmed the successful synthesis of $Fe(OC_2H_4OCH_3)_3$, the desired product.[17] Most of the signals of the spectrum are related to the methoxyethoxy group. The formation of Fe-O bonds was proved by the signal in the low wavenumber region (around 590 cm^{-1}), which is typical for the ν(M-O) in metal alkoxides. Also the signal around 1000 cm^{-1} can be related to the formation of metal-oxygen bonds, being due to modes described both as ν(C-O) and ν(M-C-O).

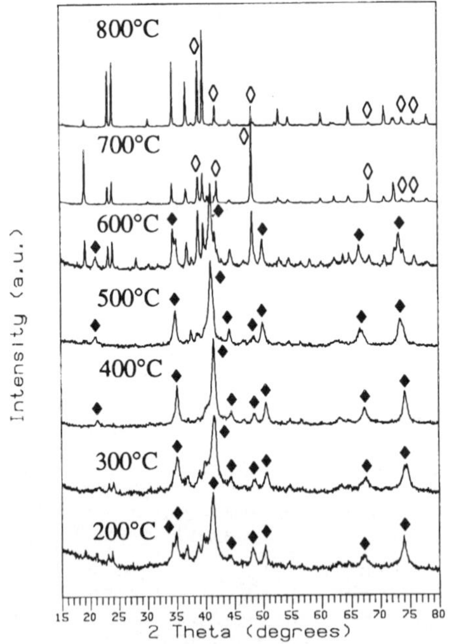

Figure 4. XRD spectra of iron oxide powders heated to various temperatures (♦ = anatase, ◊ = rutile).

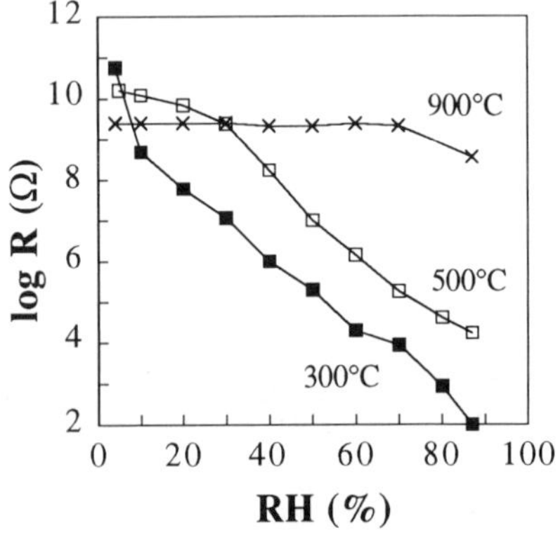

Figure 5. The RH dependence of the resistance of iron oxide films heated to various temperatures, measured at 40°C..

Figure 4 shows the XRD spectra of the iron oxide bulk gels heated to various temperatures. Crystallization occurred already for samples heated to 200°C. The XRD profile fits well the pattern of the magnetite phase, Fe_3O_4, which was the only phase present at temperatures < 500°C. At 600°C, the patterns of magnetite and α-hematite, Fe_2O_3, were detectable. For the samples heated to 700 and 800°C, the peaks of α-Fe_2O_3 were clear. Other peaks were present, but their identification was not possible by a simple comparison with data reported in the literature. Studies are currently in progress to clarify this point.

The morphology of the iron oxide films was similar to the one detected for the titania-based films. A thin, glassy layer covering the substrate was observed. The higher the heating temperature, the denser the films. The films heated to 900°C were fully dense.

Figure 5 shows the variation of the resistance, calculated from the EIS plots, with RH of the iron oxide films heated to different temperatures. The response to RH of the films heated to 300°C was even larger than the response of 10 at% K-doped, being of 9 orders of magnitude over the entire RH range tested. Also the films heated to 500°C showed a very large response, still as large as 6 orders of magnitude, but with a decrease in sensitivity for RH values lower than 30%. The films heated to 900°C were almost insensitive to humidity.

DISCUSSION

The sol-gel processed 10 at% K-doped TiO_2 and magnetite films showed very high RH-sensitivity, seldom observed even for porous sintered ceramics, although they were free of capillary pores. These outstanding results must be interpreted in terms of humidity sensitive conduction mechanisms which are different than the mechanism usually accepted for porous ceramics.[8]

For iron oxide films, the high RH-sensitivity was observed only when the oxide is present as Fe_3O_4 phase. With increasing the heating temperature of the films, an increase in the content of Fe^{+3} in the material can be observed, together with an increase in the crystallinity degree. This suggests that the simultaneous presence of Fe^{+2} and Fe^{+3} may cause polaronic conduction, greatly enhanced by RH, thereby resulting in the large humidity sensitivity. The films heated to 900°C, mainly made of hematite, are insensitive to RH.

For TiO_2 films doped with 10 at% K, in addition to the very high RH-sensitivity, other novel properties were observed. By choosing the frequency at which the resistance is measured, it is possible to modify their response to RH. When resistance is measured at frequencies ≤ 1 Hz, very high sensitivity was observed at the lowest RH values (4-10% RH). They showed also great stability over time. Heat-cleaning is probably not needed for practical uses. The addition of 10 at% K dramatically improved the performance of the sol-gel processed TiO_2 films, without affecting their morphology. This may be due to an increase in the number of water adsorption sites.[18] Recently, however, it has been reported that the RH-sensitivity of sintered ZnO compacts was improved by the addition of very small amounts of Li dopant, without affecting their microstructures.[19] Like in the present case, the addition of alkali ions, though in very different concentrations for different materials, led to a noticeable increase of the RH-sensitivity of the materials without microstructural changes. These results suggest that the humidity sensing mechanism of both materials may involve that alkali ions directly concur in the conduction in humid environments.

The interesting properties observed for the sol-gel processed 10 at% K-doped TiO_2 films can be ascribed to the features of the doped material: with the sol-gel process, very homogeneous and uniform materials can be obtained, with mixing at the atomic scale. The alkali-doping may be at nanoscale level, with the resulting formation of a nanostructured material which has integrated functions of insulator (titania) and ionic conductor (alkali-containing phase). These properties were not obtained for a K concentration lower than 10 at%. The increase in the alkali concentration led to the formation of a secondary phase, an alkaline titanate, which has ionic conductivity. In fact, although the films of 10 at% K-doped TiO_2 heated to 300°C did not show any XRD peaks, it is possible that they are nanocrystalline rather than amorphous: EXAFS tests performed on TiO_2 powders showed that the coordination shell of the powder heated to low temperatures, i.e. non-crystalline, was very similar to that of ordered TiO_2.[20] We may suppose that a defined order is present at nanoscale level on the TiO_2 films. By increasing the alkali concentration, it is possible that the environment of Ti ions was partly changed from titanium oxide to alkaline titanate.

CONCLUSIONS

The electrical characterization of sol-gel processed 10 at% K-doped TiO_2 films and of magnetite films showed that these materials possess a very high humidity sensitivity over the entire RH range tested, although they are free of capillary pores. Their properties cannot be explained in terms of the humidity sensing mechanism widely attributed to porous ceramics, but different conduction mechanisms must be involved. In the case of alkali-doped titania films, a direct participation of alkali ions to the conduction with humidity can be suggested, while for magnetite films, a polaronic mechanism can be enhanced in humid environments, due to the presence of Fe^{+2} and Fe^{+3} in the same material. The feasibility, using the sol-gel technology, of the fabrication of highly sensitive ceramic films without porous structure have been demonstrated. These materials are promising for use as active elements in integrated humidity sensors.

Acknowledgments

This work was supported by the National Research Council of Italy (CNR), under the auspices of the Targeted Project "Special Materials for Advanced Technologies".

REFERENCES

1. B.M. Kulwicki, Humidity sensors, *J. Am. Ceram. Soc.* 74:697 (1991).
2. J.G. Fagan and V.R.W. Amarakoon, Reliability and reproducibility of ceramic sensors: part III, humidity sensors, *Am. Ceram. Soc. Bull.* 72 [3]:119 (1993).
3. Q. Wu, K.M. Lee, and C.C. Liu, Development of chemical sensors using microfabrication and micromachining techniques, *Sensors and Actuators B* 13:1 (1993).
4. G. Gusmano, G. Montesperelli, E. Traversa, and G. Mattogno, Microstructure and electrical properties of $MgAl_2O_4$ thin films for humidity sensing, *J. Am. Ceram. Soc.* 76:743 (1993).
5. A. Bearzotti, A. Bianco, G. Montesperelli, and E. Traversa, Humidity sensitivity of sputtered TiO_2 thin films, *Sensors and Actuators B* 19:525 (1994).
6. G. Gusmano, G. Montesperelli, P. Nunziante, E. Traversa, A. Montenero, M. Braghini, G. Mattogno, and A. Bearzotti, Humidity-sensitive properties of titania films prepared using the sol-gel process, *J. Ceram. Soc. Jpn.* 101:1095 (1993).
7. N. Yoshimura, S. Sato, M. Itoi, and H. Taguchi, Electrical and humidity sensitive characteristics of TiO_2 and TiO_2-SnO_2 films prepared by sol-gel method, *Sozai Busseigaku Zasshi* 3:47 (1990).
8. E. Traversa, Ceramic sensors for humidity detection: the state-of-the-art and future developments, *Sensors and Actuators B* 23:135 (1995).
9. K. Katayama, K. Hasegawa, T. Noda, T. Akiba, and H. Yanagida, Effect of alkaline oxide addition on the humidity-sensitivity of Nb_2O_5-doped TiO_2, *Sensors and Actuators B* 2:143 (1990).
10. Y.C. Yeh, T.Y. Tseng, and D.A. Chang, Electrical properties of TiO_2-$K_2Ti_6O_{13}$ porous ceramic humidity sensor, *J. Am. Ceram. Soc.* 73:1992 (1990).
11. R. Zanoni, G. Righini, A. Montenero, G. Gnappi, G. Montesperelli, E.Traversa, and G. Gusmano, XPS analysis of sol-gel processed doped and undoped TiO_2 films for sensors, *Surf. Interface Anal.* 22:376 (1994).
12. Y. Shimizu, H. Arai, and T. Seiyama, Theoretical studies on the impedance-humidity characteristics of ceramic humidity sensors, *Sensors and Actuators* 7:11 (1985).
13. A. Montenero, L. Lan, G. Gusmano, and E. Traversa, *Italian Patent n° MI94A.001426*, 1994.
14. A. Montenero, J.D. Carrillo Heredero, G. Gusmano, and E. Traversa, *Ital. Pat. n° MI94A.001344*, 1994.
15. D.C. Bradley, R.C. Mehrotra, and D.P. Gaur, "Metal Alkoxides", Academic Press, London (1986).
16. G. Montesperelli, A. Pumo, E. Traversa, G. Gusmano, A. Bearzotti, A. Montenero, and G. Gnappi, Sol-gel processed TiO_2-based thin film as innovative humidity sensors, *Sensors and Actuators B* in press (1995).
17. J.D. Carrillo Heredero, G. Gnappi, A. Montenero, G. Montesperelli and E. Traversa, Properties of iron oxide films and powders by sol-gel, in P. Vincenzini (ed.), "Advances in Inorganic Films and Coatings", Techna, Faenza (1995).
18. F. Uchikawa and K. Shimamoto, Time variability of surface ionic conduction on humidity-sensitive SiO_2 films, *Am. Ceram. Soc. Bull.* 64:1137 (1985).
19. E. Joanni and J.L. Baptista, ZnO-Li_2O humidity sensors, *Sensors and Actuators B* 17:69 (1993).
20. I. Manzini, G. Antonioli, P.P. Lottici and A. Montenero, X-ray absorption spectroscopy study of crystallization processes in sol-gel derived TiO_2, *J. Non-cryst. Solids* in press (1995).

A NEW GENERATION OF COMPOSITE MATERIALS FROM AGRO-BASED FIBER

Roger M. Rowell

USDA Forest Service, Forest Products Laboratory,
One Gifford Pinchot Drive, Madison, WI 53705-2366
and Department of Forestry, University of Wisconsin,
1630 Linden Drive, Madison, WI 53706 USA

INTRODUCTION

Agro-based fibers are classified according to what part of the plant they come from. Five different fiber classifications will be used in this report: (1) bast or stem fibers, which are the fibrous bundles in the inner bark of the plant stem running the length of the stem; (2) leaf fibers, which run the length of leaves; (3) seed-hair fibers; (4) core, pith or stick fibers, which form the low density, spongy inner part of the stem of certain plants; and (5) all other plant fibers not included above. Examples of bast or stem fibers include jute, flax, hemp, kenaf, ramie, roselle, and urena. Leaf fibers include banana, sisal, henequen, abaca, pineapple, cantala, caroa, mauritius, and phormium. Seed-hair fibers include coir, cotton, kapok, and milk weed floss. Core fibers represent the center or pith fibers of such plants as kenaf and jute and can represent over 85 percent of the dry weight of these plants. The remaining fibers include roots, leaf segments, flower heads, seed hulls and short stem fiber.

While individual single fibers in all of these classes are quite short (except for flax, hemp, ramie, cotton, and kapok), fiber bundles can be quite long. For example, hemp, jute, and kenaf can have fiber bundles as long as 400 cm and abaca, mauritius, and phormium are about half this length. Considering all plant fiber types, there is a vast array of potential long and short fibers for composite production.

PLANT UTILIZATION FOR COMPOSITES

Scheme 1 shows possible processing pathways that lead to the composite products identified in this report that can come from each fraction of the plant. The entire plant (leaves, stock, pith, roots) can be used directly to produce structural and non-structural composites. By using the entire plant, processes such as retting, fiber separation, fraction purification, etc. can be eliminated which increases the total yield of plant material and reduces the costs associated with fraction isolation. This also gives the farmer a different option in their crop utilization. That is, bringing in the entire plant to a central processing center and not having to get involved in plant processing.

Polymers and Other Advanced Materials: Emerging Technologies and Business Opportunities. Edited by P. N. Prasad *et al.*, Plenum Press, New York, 1995

```
WHOLE PLANT -------->LONG FIBER ---->MATS
      |              |            |              |
      |              |            Combinations   Filters,
[Fiberize]           PITH         with other     Geotextiles,
      |              |            resources      Packaging,
      |              |                           Molded
      |              |                           composites,
Structural           Sorbents,
Non-structural       Packing,
composites           Light weight composites,
                     Insulation
```

Scheme 1 - Possible processing pathways from each plant fraction.

Another option is to separate the higher value long fiber from the other types of shorter fiber and use it in combination with other materials to make value added structural composites. When the long fiber is separated, the by-product is a large amount of short fiber and pith material that can be used for such products as sorbents, packing, light weight composites, and insulation. By utilizing the by-product from the long fiber isolation process, the overall cost of long fiber utilization is reduced.

The isolated long fiber can then be used to make mats which have value added applications in filters, geotextiles, packaging, molded composites, and structural and non-structural composites.

POTENTIAL COMPOSITES FROM AGRO-RESOURCES

For this report, a composite will be defined as any combination of two or more resources held together by some type of mastic or matrix. Composites can be classified in many ways: by their densities, by their uses, by their manufacturing methods, or other systems. For this report, they will be classified by their uses. Eight different use classes will be covered: (1) geotextiles, (2) filters, (3) sorbents, (4) structural composites, (5) non-structural composites, (6) molded products, (7) packaging, and (8) combinations with other materials Rowell 1993). There is some overlap between these areas. For example, once a fiber web has been made, it can be directly applied as a geotextile, filter, or sorbent, or can be further processed into a structural or non-structural composite, molded product, used in packaging, or combined with other resources. Within each composite made there are opportunities to improve the performance of that composite by improving the performance of the fiber going into the composite.

(1) Geotextiles

The long bast or leaf fibers can be formed into flexible fiber mats, which can be made by physical entanglement, nonwoven needling or thermoplastic fiber melt matrix technologies. The two most common types are carded and needle punched mats. In carding, the fibers are combed, mixed and physically entangled into a felted mat. These are usually of high density but can be made at almost any density. A needle punched mat is produced in a machine which passes a randomly formed machine made web through a needle board that produces a mat in which the fibers are mechanically entangled. The density of this type of mat can be controlled by the amount of fiber going through the needle board or by overlapping needled mats to give the desired density.

Numerous articles and technical papers have been written and several patents have been issued on both the manufacture and use of nonwoven fiber mats containing combinations of textile and lignocellulosic fibers or lignocellulosic fibers alone. This technology is particularly well-known in the consumer products industry. Medium- to high-density fiber mats can be used in several ways. One is for the use as a geotextile. Geotextiles derive their name from the two words geo and textile and, therefore, mean the use of fabrics in association with the earth.

Geotextiles have a large variety of uses. These can be used for mulch around newly planted seedlings. The mats provide the benefits of natural mulch; in addition, controlled-release fertilizers, repellents, insecticides, and herbicides can be added to the mats as needed. Research results on the combination of mulch and pesticides in agronomic crops have been promising.

The addition of such chemicals could be based on silvicultural prescriptions to ensure seedling survival and early development on planting sites where severe nutritional deficiencies, animal damage, insect attack, and weed problems are anticipated. Medium density fiber mats can also be used to replace dirt or sod for grass seeding around new homesites or along highway embankments. Grass or other type of seed can be incorporated in the fiber mat. Fiber mats promote seed germination and good moisture retention. Low and medium density fiber mats can be used for soil stabilization around new or existing construction sites. Steep slopes, without root stabilization, lead to erosion and loss of top soil.

Medium and high density fiber mats can also be used below ground in road and other types of construction as a natural separator between different materials in the layering of the back fill. It is important to restrain slippage and mixing of the different layers by placing separators between the various layers. Jute and kenaf geotextiles have been shown to work very well in these applications but the potential exists for any of the long agro-based fibers.

(2) Filters

Medium and high density fiber mats can be used for air filters. The density of the mats can be varied, depending on the size and quantity of material being filtered and the volume of air required to pass through the filter per unit of time. Air filters can be made to remove particulates and/or can be impregnated or reacted with various chemicals as an air freshener or cleanser.

Medium to high density mats can also be used as filtering aids to take particulates out of waste and drinking water or solvents.

(3) Sorbents

Tests are presently underway to to use agro-based sorbents to remove heavy metals, pesticides, and oil from rain water run off in several cities in the United States. Medium and high density mats can also be used for oil spill clean up pillows. It has been shown that agro-based core material from kenaf will preferentially sorb oil out of sea water when saturated with water. There are many other potential sorbent applications of agro-fiber/core resources such as removal of dyes, trace chemicals in solvents and in the purification of solvents.

It is also possible to use core materials as sorbents in cleaning aids such as floor sweep. While this is not a composite, as such, but it does represent another way agro-based resources can be used as sorbents.

(4) Structural Composites

A structural composite is defined as one that is required to carry a load in use. In the housing industry, for example, these represent load bearing walls, roof systems, subflooring, stairs, framing components, furniture, etc. In most, if not all cases, performance requirements of these composites are spelled out in codes and/or in specifications set forth by local or national organizations.

Structural composites can range widely in performance from high performance materials used in the aerospace industry down to wood-based composites which have lower performance requirements. Within the wood-based composites, performance varies from multi-layered plywood and laminated lumber to low cost particleboard. Structural wood based composites, intended for indoor use, are usually made with a low cost adhesive which is not stable to moisture while exterior grade composites use a thermosetting resin that is higher in cost but stable to moisture.

Performance can be improved in wood-based as well as agro-based composites by using chemical modification techniques, fire retardant and/or, decay control chemicals, etc.

(5) Non-Structural Composites

As the name implies, non-structural composites are not intended to carry a load in use. These can be made from a variety of materials such as thermoplastics, textiles, and wood particles, and are used for such products as doors, windows, furniture, gaskets, ceiling tiles, automotive interior parts, molding, etc. These are generally lower in cost than structural composites and have fewer codes and specifications associated with them.

(6) Molded Products

The present wood based composite industry mainly produces two dimensional (flat) sheet products. In some cases, these flat sheets are cut into pieces and glued/fastened together to make shaped products such as drawers, boxes, and packaging. Flat sheet wood fiber composite products are made by making a gravity formed mat of fibers with an adhesive and then pressing. If the final shape can be produced during the pressing step, then the secondary manufacturing profits can be realized by the primary board producer. Instead of making low cost flat sheet type composites, it is possible to make complex shaped composites directly using the long bast fiber.

In this technology, fiber mats are made similar to the ones described for use as geotextiles except during mat formation an adhesive is added by dipping or spraying of the fiber before mat formation or added as a powder during mat formation. The mat is then shaped and densified by a thermoforming step. Within certain limits, any size, shape, thickness, and density is possible.

These molded composites can be used for structural or non-structural applications as well as packaging, and can be combined with other materials to form new classes of composites. This technology will be described later.

(7) Packaging

"Gunny" bags made from jute have been used as sacking for products such as coffee, cocoa, nuts, cereals, dried fruits, and vegetables for many years. While there are still many applications for long fiber for sacking, most of the commodity goods are now shipped in containers. These containers are not made of agro-fibers today but there is no reason why they cannot be. Medium and high density agro-based fiber composites can be used for small containers, for example, in the tea industry and for large sea-going containers for commodity goods. These composites can be shaped to suit the product by using the molding technology described previously or made into low cost, flat sheets and made into containers.

Agro-based fiber composites can also be used in returnable containers where the product is reused several times. These containers can range from simple crease-fold types to more solid, even nestable, types. Long agro-fiber fabric and mats can be overlayed with thermoplastic films such as polyethylene or polypropylene to be used to package such products as concrete, foods, chemicals, and fertilizer. Corrosive chemicals require the plastic film to make them more water resistant and reduce degradation of the agro-based fiber. There are also many applications for agro-based fiber as paper sheet products for packaging. These vary from simple paper wrappers to corrugated, muti-folded, multi-layered packaging.

(8) Combinations with Other Resources

It is possible to make completely new types of composites by combining different resources. It is possible to combine, blend, or alloy leaf, bast and/or stick fiber with other materials such as glass, metals, plastics, and synthetics to produce new classes of materials. The objective will be to combine two or more materials in such a way that a synergism between the components results in a new material that is much better than the individual components.

Agro-based fiber/glass fiber composites can be made using the glass as a surface material or combined as a fiber with lignocellulosic fiber. Composites of this type can have a very high stiffness to weight ratio. The long bast fibers can also be used in place of glass fiber in resin injection molding (RIM) or used to replace, or in combination with, glass fiber in resin transfer molding (RTM) technologies. Problems of dimensional stability and compatibility with the resin must be addressed but this could also lead to new markets for property enhanced agro-based materials.

Metal films can be overlayed on to smooth, dimensionally stabilized fiber composite surfaces or applied through cold plasma technology to produce durable coatings. Such products could be used in exterior construction to replace all aluminum or vinyl siding, markets where agro-based resources have lost market share.

Metal fibers can also be combined with stabilized fiber in a matrix configuration in the same way metal fibers are added to rubber to produce wear-resistant aircraft tires. A metal matrix offers excellent temperature resistance and improved strength properties, and the ductility of the metal lends toughness to the resulting composite. Application for metal matrix composites could be in the cooler parts of the skin of ultra-high-speed aircraft. Technology also exists for making molded products using perforated metal plates embedded in a phenolic-coated fiber mat, which is then pressed into various shaped sections.

Bast or leaf fiber can also be combined in an inorganic matrix. Such composites are dimensionally and thermally stable, and they can be used as substitutes for asbestos composites. Inorganic bonded bast fiber composites can also be made with variable densities that can be used for structural applications.

One of the biggest new areas of research in the value added area is in combining natural fibers with thermoplastics (Sanadi et al 1994,a,b,c). Since prices for plastics have risen sharply over the past few years, adding a natural powder or fiber to plastics provides a cost reduction to the plastic industry (and in some cases increases performance as well) but to the agro-based industry, this represents an increased value for the agro-based component. Most of the research has concentrated on using a compatibilizer to make the hydrophobe (plastic) mix better with the hydrophil (lignocellulosic). The two components remain as separate phases, but if delamination and/or void formation can be avoided, properties can be improved over those of either separate phase. These types of materials are usually referred to as natural fiber/thermoplastic blends.

Another combination of fiber and thermoplastics is in products that can best be described as fiber-plastic alloys (Rowell et al 1994). In this case the fiber and plastic become one material and it is not possible to separate them. Fiber-plastic alloys are possible through fiber modification and grafting research. This can be done if you consider that agro-based fibers consist of a thermoset polymer (cellulose) in a thermoplastic matrix (lignin and the hemicelluloses). The glass transition temperature (Tg), however, of the dry natural thermoplastic matrix is higher than the decomposition temperature of the fiber. If the Tg were lowered through chemical modification, it should be possible to thermoplasticize the lignin and possibly the hemicelluloses to be able to do thermomolding at temperatures below decomposition. If a reactive thermoplastic is then reacted with the modified biobased fiber, it should be possible to form fiber/thermoplastic alloys. It is also possible to thermoplasticize cellulose but then the strong, stiff, reinforcing thermoset structure is lost. But, if the cellulose were also plasticized, it would be possible to produce agro-based films and foams.

CHEMICAL MODIFICATION FOR PROPERTY IMPROVEMENT

The performance of any lignocellulosic fiber composite is restricted by the properties of the fiber itself. Agro-based composites will change dimensions with changes in moisture content, be degraded by organisms, be degraded by ultraviolet radiation, and burn. If these negative properties of the natural fiber can be improved, all types of agro-based composites could have greatly improved performance. To understand how agro-based fiber can be used in property enhanced composites, it is important to understand the properties of the components of the cell wall and their contributions to fiber and fiber composite properties.

All agro-based fibers are three-dimensional, polymeric composites primarily made up of cellulose, hemicelluloses, lignin, and small amounts of extractives and ash. The cell wall polymers and their matrix make up the cell wall and in general are responsible for the physical and chemical properties of the agro-based fiber. Properties such as dimensional instability, flammability, biodegradability, and degradation caused by acids, bases, and ultraviolet radiation are a result of the environment trying to convert the natural composites back into their basic building blocks (carbon dioxide and water).

Composites made from agro-based resources change dimensions with changing moisture content because the cell wall polymers contain hydroxyl and other oxygen-containing groups that attract moisture through hydrogen bonding. The hemicelluloses are mainly responsible for moisture sorption, but the accessible cellulose, noncrystalline cellulose, lignin, and surface of crystalline cellulose also play major roles. Moisture swells the cell wall, and the fiber expands until the cell wall is saturated with water. Beyond this saturation point, moisture exists as free water in the void structure and does not contribute to further expansion. This process is reversible, and the fiber shrinks as it loses moisture.

Agro-based composites are degraded biologically because organisms recognize the carbohydrate polymers (mainly the hemicelluloses) in the cell wall and have very specific enzyme systems capable of hydrolyzing these polymers into digestible units. Biodegradation of the high molecular weight cellulose weakens the fiber cell wall because crystalline cellulose is primarily responsible for the strength of the cell wall . Strength is lost as the cellulose polymer undergoes degradation through oxidation, hydrolysis, and dehydration reactions. The same types of reactions take place in the presence of acids and bases.

Agro-based composites exposed outdoors undergo photochemical degradation caused by ultraviolet light. This degradation takes place primarily in the lignin component, which is responsible for the characteristic color changes. The lignin acts as an adhesive in the cell walls, holding the cellulose fibers together. The surface becomes richer in cellulose content as the lignin degrades. In comparison to lignin, cellulose is much less susceptible to ultraviolet light degradation. After the lignin has been degraded, the poorly bonded carbohydrate-rich fibers erode easily from the surface, which exposes new lignin to further degradative reactions. In time, this "weathering" process causes the surface of the composite to become rough and can account for a significant loss in surface fibers.

Agro-based composites burn because the cell wall polymers undergo pyrolysis reactions with increasing temperature to give off volatile, flammable gases. The hemicellulose and cellulose polymers are degraded by heat much before the lignin is. The lignin component contributes to char formation, and the charred layer helps insulate the composite from further thermal degradation.

Because the properties of the agro-based fiber result from the chemistry of the cell wall components, the basic properties of a fiber can be changed by modifying the basic chemistry of the cell wall polymers (Rowell 1983, 1990)

Dimensional stability can be greatly improved by bulking the fiber cell wall either with simple bonded chemicals or by impregnation with water soluble polymers. For example, acetylation of the cell wall polymers using acetic anhydride produces a fiber composite with greatly improved dimensional stability and biological resistance The same level of stabilization can also be achieved by using water-soluble phenol-formaldehyde polymers followed by curing.

Biological resistance of fiber-based composites can be improved by several methods. Bonding chemicals to the cell wall polymers increases resistance due to the lowering of the equilibrium moisture content point below that needed for microorganism attack and by changing the conformation and configuration requirements of the enzyme-substrate reactions. Toxic chemicals can also be added to the composite to stop biological attack. This is the basis for the wood preservation industry.

Resistance to ultraviolet radiation can be improved by bonding chemicals to the cell wall polymers, which reduces lignin degradation, or by adding polymers to the cell matrix to help hold the degraded fiber structure together so water leaching of the undegraded carbohydrate polymers cannot occur. Fire retardants can be bonded to the fiber cell wall to greatly improve the fire performance of composites. Soluble inorganic salts or polymers containing nitrogen and phosphorus can also be used. These chemicals are the basis of the fire retardant wood treating industry.

The strength properties of a fiber composite can be greatly improved in several ways. The finished composites can be impregnated with a monomer and polymerized in situ or impregnated with a preformed polymer. In most cases the polymer does not enter the cell wall and is located in the cell lumen. By using this technology, mechanical properties can be greatly enhanced. For example, composites impregnated with acrylates, methacrylate, epoxy, or melamine monomers and polymerized to weight gain levels of 60 to 100 percent, show increases (compared to untreated controls) in density of 60 to 150 percent, compression strength of 60 to 250 percent, and tangential hardness of 120 to 400 percent. Static bending tests show increases in modulus of elasticity of 25 percent, modulus of rupture of 80 percent, fiber stress at proportional limit of 80 percent, work to proportional limit of 150 percent, and work to maximum load of 80 percent, and at the same time a decrease in permeability of 200 to 1,200 percent.

CONCLUSIONS

Fiber from a wide variety of plants can be used to produce a wide variety of composites. These composites range from value added speciality products to very large volume commercial materials. High performance composite materials with uniform densities, durability in adverse environments, and high strength can be produced by using agro-based fiber, high performance adhesives, and fiber modification to overcome dimensional instability, biodegradability, flammability, and degradation caused by ultraviolet light, acids, and bases. Products with complex shapes can also be produced using flexible fiber mats, which can be made by nonwoven needling or thermoplastic fiber melt matrix technologies. Taking advantage of fiber cell wall modification chemistry and combining bast fiber with other materials provides a strategy for producing advanced composites and materials that take advantage of the enhanced properties of all types of materials, and it allows the scientist to design materials based on end use requirements within the framework of cost, availability, renewability, recyclability, sustainability, energy use, and environmental considerations.

REFERENCES

Rowell, R.M. 1983, Chemical modification of wood: A review. *Commonwealth Forestry Bureau*, Oxford, England, Vol. 6, N0. 12, p. 363-382.

Rowell, R.M. 1990, Chemical modification of lignocellulosic fibers to produce high performance composites. In E. Glass and G. Swift, eds., *Agriculture and synthetic polymers: Utilization and biodegradation*, Am. Chem. Soc. Advances in Chemistry Series 433, Washington, DC, p. 242-258

Rowell, R.M. 1993, Opportunities for composite materials from jute and kenaf. *International consultation on jute and the environment*, Food and Agricultural Organization of the United Nations, ESC:JU/IC 93/15, p 1-12.

Rowell, R.M., O'Dell, J.L., and Rials, T.G., 1994, Chemical modification of agro-fiber for thermoplasticization. In, Proceedings, *Pacific Rim Bio-Based Composite Symposium*, November, 1994 Vancouver, B.C. Canada, p. 144-152.

Sanadi, A.R., Caulfield, D.F., Jacobson, R.E., and Rowell, R.M., 1994a, Reinforcing polypropylene with agricultural fibers. Proceedings: *International Jute and Allied Fibre Symposium on Biocomposites and Blends*, New Delhi, India, p. 163-167.

Sanadi, A.R., Caulfield, D.F., and Rowell, R.M., 1994b, Reinforcing polypropylene with natural fibers. *Plastic Engineering*, Vol L(4): 27-28

Sanadi, A.R., Caulfield, D.F., Walz, K., Wieloch, L., Jacobson, R.E., and Rowell, R.M., 1994c, Kenaf fibers - Potentially outstanding reinforcing fillers in thermoplastics, In Proceedings, *Sixth Annual International Kenaf Conference*, New Orleans, LA, p. 155-160.

ADVANCED MATERIALS FROM ENZYMATIC POLYMERIZATION OF SUBSTITUTED PHENOLS IN ORDERED TEMPLATES

Madhu S. Ayyagari[1,2], Ferdinando F. Bruno[1,2], Sukant K. Tripathy[1], Kenneth A. Marx[1], David L. Kaplan[2], Joseph A. Akkara[2], D.V.G.L.N. Rao[3]

[1]Center for Advanced Materials, Department of Chemistry, University of Massachusetts Lowell, Lowell, MA 01854; [2]Biotechnology Division, US Army RD&E Center, Kansas Street, Natick, MA 01760; [3]Department of Physics, University of Massachusetts Boston, Boston, MA 02125.

ABSTRACT

Enzyme-catalyzed polymerization of *p*-substituted phenols and aromatic amines is described. The polymers were prepared as thin films at the air-water interface of a Langmuir trough, as spherical particles at the oil-water interface of reversed micelles or as amorphous material in bulk organic solvents. Horseradish peroxidase or laccase was used to catalyze the reactions. Molecular weight and size distribution of polyphenols formed with reversed micelles or in bulk solvents could be modulated by controlling solvent properties. For example, oligomers with fairly uniform size distribution could be prepared by varying the solvent polarity. *p*-Alkoxyphenols with long alkyl chains could be ordered into a monolayer on a Langmuir trough, and could be subsequently polymerized using enzymes. The polymers were analyzed for their thermal, mechanical, electrical and optical properties. The application potential of the polymers in electronics/photonics is discussed.

INTRODUCTION

Chemically synthesized phenol-formaldehyde resins find utility in a number of applications that include laminates, abrasives and other engineering materials. However, the extensive use of formaldehyde in the resin synthesis has been a subject of concern over the past few years due to its carcinogenic nature, and as a result, more benign procedures have been sought to produce polyphenols [Dordick *et al.,* 1987]. Enzyme-catalyzed polymerization of phenol, substituted phenols and aromatic amines was well studied in different reaction media [Dordick *et al.*, 1987; Akkara *et al.*, 1991; Rao *et al.*, 1993; Ayyagari *et al.*, 1995]. These studies indicated that the enzymatic production of polyphenols is feasible on a large scale, and that the process is a viable alternative to chemical synthesis of phenol-formaldehyde resins. While the enzymatic process offers advantages over chemical synthesis, it is unclear whether the lack of methylene bridges in the polymer structures formed by enzyme catalysis is a drawback for functional properties. The advantages in the enzymatic synthesis include: (a) avoiding the use of formaldehyde, (b) carrying out the reactions at ambient conditions of temperature and pressure, (c) avoiding or reducing the formation of by-products due to the specificity of the enzyme, (d) fast reaction kinetics, (e) ease of product recovery and (f) reuse of the biocatalyst in some cases. A number of substituted phenols were polymerized and their properties analyzed [Akkara *et al.*, 1991]. In general, the polymers were thermally stable. Thin films of the polymers could be solvent-cast to study their electrical, spectroscopic, electronic and/or photonic properties. Conductivity and third order nonlinear optical properties of the polymers were analyzed [Akkara *et al.*, 1994].

The present article describes reactions catalyzed by the enzyme horseradish peroxidase (HRP) involving oxidative coupling of the monomer units in the presence of hydrogen peroxide. The enzyme heme group undergoes oxidation during interaction with hydrogen peroxide and loses two electrons. The enzyme returns to its native state in two single electron steps during which two phenol radicals are produced that may combine to form polymer. The polymerization progresses in a step-wise manner in which one peroxide molecule is consumed per coupling [Rao *et al.*, 1993]. Polymerization of a number of substituted phenols or aromatic amines in aqueous media produced no more than dimers or trimers due to limitations of solubility as oligomers. Higher molecular weight oligomers and polymers could be obtained when the reaction was carried out in a mixture of 1,4-dioxane and water [Dordick *et al.*, 1987; Akkara *et al.*, 1991] or in isooctane-based microemulsions [Rao *et al.*, 1993].

Polyphenols have potential applications as advanced materials in the electronics/photonics industry. The resins are actively used in photoresists, batteries and other fields. However, electro-optical applications often require highly ordered structures. This ordering can be achieved by orienting the monomer molecules in a two or three dimensional template prior to their polymerization. The Langmuir trough is a useful system in this regard since surfactant molecules can be arranged in an ordered two dimensional monolayer at the air-water interface. Phenolic monomers, derivatized with long alkyl chains at the *para* position relative to the hydroxyl groups, can be spread on the aqueous subphase of the Langmuir trough and compressed into a highly ordered monolayer [Bruno *et al.*, 1992, 1994]. These oriented monomer molecules can be subsequently polymerized in the presence of hydrogen peroxide and HRP introduced in the aqueous subphase. A similar interfacial ordering of monomers was described by Rao *et al.* [1993] in which *p*-ethylphenol was positioned at the oil-water interface of reversed micelles prior to polymerization. Reversed micelles serve as a three dimensional template for *p*-ethylphenol to share the interface and act as a cosurfactant. This article provides details of polyphenol synthesis in these templates and the polymer characteristics.

MATERIALS AND METHODS

Isooctane (J.T. Baker, Phillipsburg, NJ) and chloroform (Fluka, Ronkonkoma, NY) were stored with molecular sieves to remove water from the solvents. Dioxane and N,N-dimethylformamide (American Bioanalytical, Natick, MA) were used as received. Dioctyl sodium sulfosuccinate (AOT) and 30% hydrogen peroxide (w/w) were supplied by Aldrich Chemicals Company (Milwaukee, WI), and the enzymes horseradish peroxidase (Type II) and laccase were purchased from Sigma Chemicals Company (St. Louis, MO). Polystyrene standards were purchased from Waters Associates (Milford, MA).

The reactions in reversed micelles, bulk solvents and Langmuir trough were previously described [Rao *et al.*, 1993; Akkara *et al.*, 1991; Bruno *et al.*, 1994]. Briefly, a reversed micellar solution was prepared by adding an aqueous preparation of HRP to a solution of the anionic surfactant AOT in isooctane. *p*-Ethylphenol was added to this transparent solution, and the reaction was initiated by adding hydrogen peroxide dropwise. Reactions in bulk solvents involved mixing the aqueous preparation of HRP with a solvent such as 1,4-dioxane or DMF and dissolving the monomer prior to the initiation of the reaction with hydrogen peroxide. Reactions at the air-water interface of a Lauda film balance (Langmuir Filmwaage, Model D, Lauda Brinkman, Westbury, NY) were carried out by first spreading a mixture of *p*-tetradecyloxyphenol (C14PP, synthesized by O-alkylation of hydroquinone with 1-bromotetradecane) and phenol on an aqueous subphase containing 0.85 mM HEPES buffer at pH 7.5 and 12-25 mg/L HRP. After compressing the monomer layer to a pressure of 15 mN/m, 0.3 ml 3% hydrogen peroxide was introduced into the subphase to initiate the reaction that was allowed to continue for 12 hours. A Perkin-Elmer 1760 FTIR-FTRaman spectrophotometer (Norwalk, CT) was used to record the spectral characteristics of the polymer.

Thermal characterization of the polymers was carried out on Du Pont thermal analyzers (TA Instruments, Inc., New Castle, DE). For differential scanning calorimetry (DSC) analysis, the polymers were hermetically sealed, and heated under a nitrogen atmosphere at $10^{o}C$ per minute from room temperature to $300^{o}C$. Thermogravimetric analysis (TGA) was carried out at the same temperature gradient and under nitrogen atmosphere, but heated to $600^{o}C$. Molecular weights were determined on a Waters LC Module I instrument (Millipore, Milford, MA) with an on-line gel permeation chromatography (GPC) column (GBR mixed bed linear column with a molecular weight range of 100 to over 20 million, Jordi Associates, Inc., Bellingham, MA). A UV detector at 270 nm was used to detect the polymer. The GPC data were collected and processed with Millenium GPC software supplied with the instrument. An eluent flow rate of 1 ml/min was maintained under isocratic conditions. Sample preparation for GPC analysis was described elsewhere [Ayyagari *et al.*, 1995]. Lithium bromide was used in DMF (0-1% w/v) to study the effect of molecular aggregation on the apparent molecular weights of polyphenols.

RESULTS

Synthesis in microemulsions and bulk solvents

Figure 1 is an illustration of the reaction scheme in different media discussed in this article. The repeat units in polyphenols synthesized enzymatically in solvent systems are linked

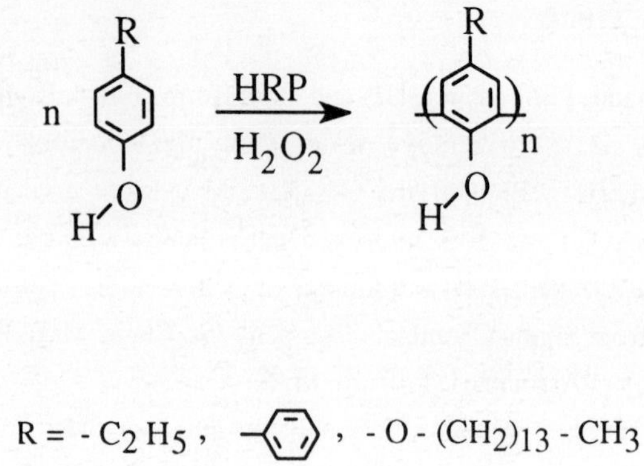

$$n \; \text{(R-C}_6\text{H}_4\text{-OH)} \xrightarrow[H_2O_2]{HRP} \text{-(C}_6\text{H}_3\text{(R)(OH))}_n\text{-}$$

R = $-C_2H_5$, $-C_6H_5$, $-O-(CH_2)_{13}-CH_3$

Figure 1. Generalized schematic of HRP-catalyzed polymerization of phenols.

Table 1. Molecular weight profiles of poly(*p*-ethylphenol) and poly(*p*-phenylphenol) prepared in reversed micelles and 85/15 dioxane/water.

Sample	Synthesized in...	M_w	M_w / M_n
Poly (*p*-ethylphenol)	AOT/isooctane reversed micelles	2700	1.4
Poly (*p*-ethylphenol)	85/15 dioxane/water	3200	> 2.0
Poly (*p*-phenylphenol)	85/15 dioxane/water	3200	> 2.0

exclusively through phenyl rings [Ayyagari *et al.*, 1995]. The free hydroxyls are therefore capable of forming polymer aggregates via inter-/intra-molecular hydrogen bonding. This phenomenon was found to overestimate the apparent molecular weight of the polymers in DMF. However, these aggregates could be easily dissociated in the presence of LiBr [Ayyagari *et al.*, 1995]. All GPC analyses were therefore carried out with 1% (w/v) LiBr/DMF as eluent.

It was demonstrated that *p*-ethylphenol could be rapidly polymerized in reversed micellar media with 100% monomer conversion [Rao *et al.*, 1993]. Table 1 compares the molecular weights of the polymer prepared in AOT/isooctane reversed micellar medium and the dioxane/water bulk polymerization system. The reversed micellar system is suitable where narrower dispersion is desired.

Figure 2 illustrates the DSC thermogram of poly(*p*-ethylphenol) prepared in reversed micelles. The polymer exhibited an exotherm at 110°C that indicated the possibility of crosslinking or crystallization. This exotherm was absent when the same sample was reheated. GPC analysis of the sample heated to 150°C showed that the average molecular weight of the polymer tripled on heating, presumably due to crosslinking, and the size distribution was broader. However, the polymer was thermally stable, as shown by TGA thermograms in Figure 3, with a weight loss of about 12% and 60% at 300°C and 600°C, respectively.

The morphology of poly(*p*-ethylphenol) was found to vary with the type of reaction medium [Ayyagari, 1992; John *et al.*, 1994]. When the polymer was prepared in reversed micellar medium, the surfactant molecules appeared to help the polymer chains aggregate into

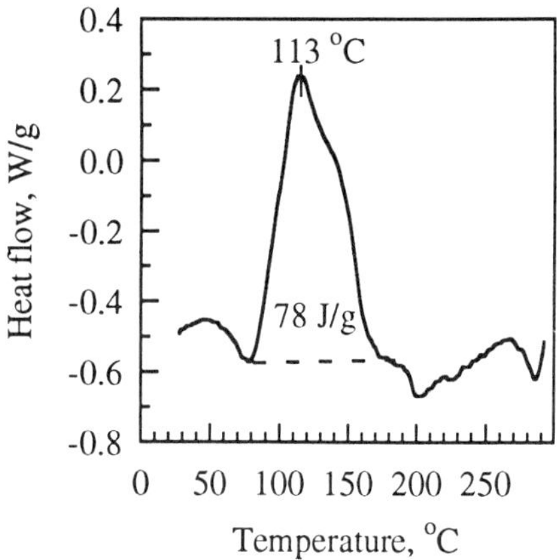

Figure 2. DSC thermogram of poly(*p*-ethylphenol) prepared in reversed micelles.

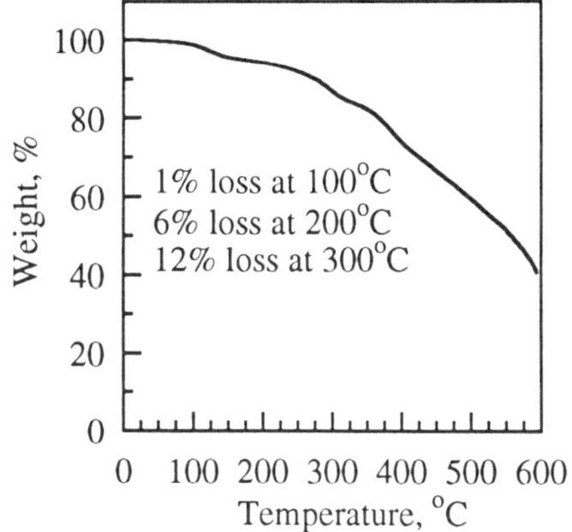

Figure 3. TGA thermogram of poly(*p*-ethylphenol) prepared in reversed micelles.

well defined spheres of an average diameter of 3-6 microns. In the absence of the surfactant, when the polymerization was carried out in dioxane/buffer system, the polymer was amorphous.

Some applications for polyphenols depend on molecular weight, therefore, tailorability of the polymer molecular weight is important. Studies involving the control of hydrogen peroxide concentration and reaction time showed little effect on the final polymer molecular weight. However, it was possible to control the solubility of the growing polymer chain by appropriate choice of the solvent system. The solvents (Table 2) were selected on the basis of the range of solubility parameters covered. The wide variation in solubility parameters and dielectric constants for each system was similar to that found in supercritical fluids as a function of pressure [Kamat *et al.*, 1993]. Polymerization of *p*-ethylphenol was carried out in these systems, and the results are summarized in Tables 3 and 4.

In the bulk DMF/water solvent system, the low molecular weight appears to be due to the combination of two factors: (1) the enzyme is catalytically inactivated at high DMF contents and (2) the solubility of oligomers in the solution decreases as the water content increases. Further

Table 2. Properties of solvent systems used in phenol polymerization reaction.

Solvent system	Solubility parameter ($MPa^{1/2}$)	Dielectric constant
Isooctane/chloroform	14 ~ 19	2 ~ 5
DMF/water	25 ~ 48	37 ~ 78
1,4-Dioxane/water	20 ~ 48	30 ~ 78

Table 3. Molecular weight profiles of poly(*p*-ethylphenol) prepared in DMF/water system.

Solvent composition	Monomer Conversion	Polymer Yield	M_w	M_w / M_n	Comments
100/0 DMF/water	0%	0%	-	-	no reaction
85/15 DMF/water	20%	10%	281	1.23	dimers soluble in 85% DMF
60/40 DMF/water	80%	75%	612	1.20	oligomers soluble in 60% DMF
40/60 DMF/water	80%	80%	675	1.05	oligomers soluble in 40% DMF
20/80 DMF/water	75%	75%	658	1.02	oligomers soluble in 20% DMF
0/100 DMF/water	50%	35%	400	1.90	oligomers soluble in 100% DMF

Table 4. Molecular weight profiles of poly(*p*-ethylphenol) prepared in $CHCl_3$/isooctane reversed micelles.

Solvent composition	Monomer Conversion	Polymer Yield	M_w	M_w / M_n	Comments
100% Isooctane	100%	100%	2700	1.36	$W_o = 15$
75/25 Isooctane/$CHCl_3$	100%	85%	1681	1.53	$W_o = 9$
50/50 Isooctane/$CHCl_3$	100%	75%	3461	1.85	$W_o = 9$
25/75 Isooctane/$CHCl_3$	75%	35%	3601	1.83	$W_o = 9$ phase separation
100% $CHCl_3$	20%	10%	1000	1.07	$W_o = 9$ phase separation

chain growth is prevented by the precipitation of the oligomers. Thus, as shown in Table 3, the product consisted predominantly of oligomers.

In chloroform/isooctane reversed micellar systems, higher molecular weights are realized because of better compatibility of the solvent system with the enzyme and better solubility of the growing poly(*p*-ethylphenol) chains. Table 4 gives the molecular weight and polydispersity of the polymer synthesized in this system. Higher molecular weights could not be obtained due to an unstable microemulsion system that did not support water droplets of larger size. As a result, the monomer conversion was poor, a phenomenon studied earlier [Rao *et al.*, 1993].

It was demonstrated that iron and gold particles could be encapsulated within the polymer spheres either during their growth or when the polymer was refolded in a suitable solvent in the presence of AOT and the metal particles [O'Connor *et al.*, 1994]. Thus, the polymer and its ordered morphology can be used to prepare nanoparticle composites that have potential applications in bioseparations, electronics and photonics.

Langmuir Trough Studies

Polymerization at the air-water interface of a Langmuir trough was carried out by HRP or laccase using a mixture of phenol derivatized phenols. The phenolic derivatives were chemically synthesized as described earlier [Bruno *et al.*, 1992]. Although a number of phenolic or aromatic amine monomers can be polymerized with different enzymes, HRP-catalyzed polymerization of *p*-tetradecyloxyphenol (C14PP) is discussed here. Pure C14PP could not be polymerized on a Langmuir trough presumably due to steric constraints between adjacent molecules. However, the reaction was possible when C14PP was premixed with phenol and spread on the trough [Bruno *et al.*, 1994]. The reaction rates were monitored as a function of the ratio between the monomers, and higher rates were realized at lower ratios of C14PP to phenol. A ratio of 1:10 was employed in this study. Figure 4 illustrates the pressure-area isotherms of the monomers at different

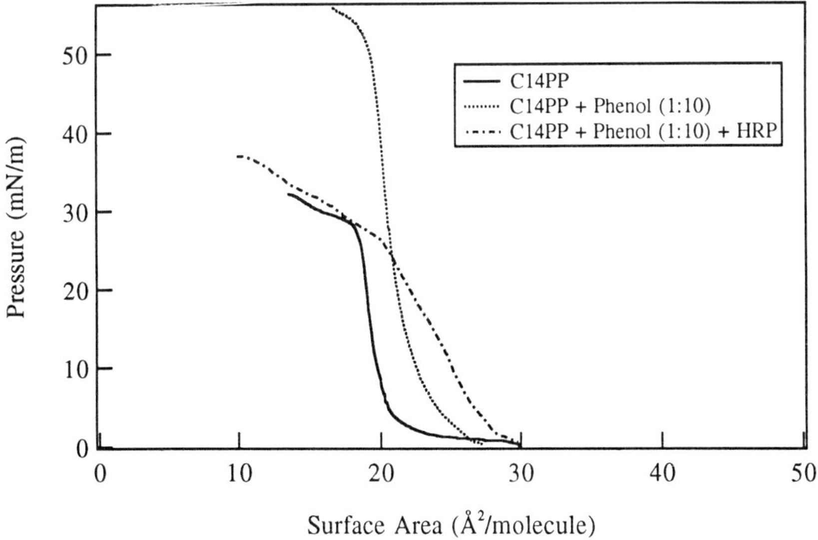

Figure 4. Pressure-area isotherms of C14PP, C14PP+phenol (1:10) and C14PP+phenol+HRP.

compositions and in the presence of HRP. After compressing the monomer layer to a pressure of 15 mN/m, polymerization was initiated by introducing hydrogen peroxide in the subphase. The evidence of reaction could be visibly seen as the monolayer developed color upon starting the reaction.

After 12 hours of reaction, the polymer was transferred onto a quartz substrate for spectroscopic studies. Figure 5 illustrates the FTIR-ATR spectra of the monomer and the polymer. The presence of polymer could be confirmed by characteristic changes in 1600-1700 cm^{-1} due to C=O attached to the delocalized conjugated backbone and around 3500 cm^{-1} due to rearranged hydrogen bonding among the hydroxyl groups. The mono/multilayer thickness and quality were confirmed by ellipsometry and atomic force microscopy [Akkara *et al.,* 1994].

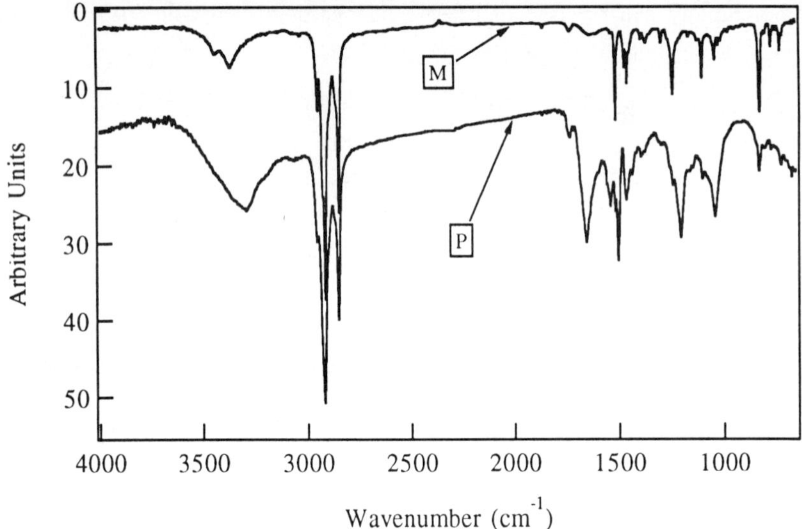

Figure 5. FTIR-ATR spectra of monomer (M) and polymer (P) of C14PP.

The two dimensional polymer networks, prepared enzymatically on the Langmuir trough, were thermally and temporally stable. Conductivity measurements on the doped polyphenols and polyanilines showed that the polymers fall in the semiconductor range. Third order nonlinear susceptibilities of the polymers prepared by different methods were measured, and in general, the $\chi(3)$ values of the polymers prepared from aromatic amines were two to three orders of magnitude higher than those prepared from phenols and substituted phenols [Akkara *et al.,* 1994].

CONCLUSIONS

Polyphenols were enzymatically synthesized in a two dimensional format of a Langmuir trough and a three dimensional format of water-in-oil microemulsions or reversed micelles. Poly(*p*-ethylphenol) synthesized in AOT/isooctane reversed micelles exhibited well defined spherical morphology and relatively narrow polydispersity. The polymers could be prepared with molecular weights ranging from 500-3000 by solvent engineering while maintaining low polydispersity. The polymer yields were generally high, and large scale production is possible.

Polymerization at the air-water interface was also demonstrated with *p*-alkoxyphenols spread on a Langmuir trough. An important feature of the methodology is the concurrent ordering of the monomer and its polymerization to form electrically conducting and optically active material.

REFERENCES

Akkara, J.A., Senecal, K.J., and Kaplan, D.L., 1991, Synthesis and characterization of polymers produced by horseradish peroxidase in dioxane, *J. Polym. Sci.*, 29:1561.

Akkara, J.A., Ayyagari, M., Bruno, F., Samuelson, L., John, V., Karayigitoglu, C., Tripathy, S., Marx, K.A., Rao, D., Kaplan, D.L., 1994, Biomimetic membrane and interface templates for enzyme-based polymerization reactions, ***Biomimetics*, 2(4):331.**

Ayyagari, M.S., 1992, Biocatalysis in water-in-oil microemulsions: Materials and chemicals synthesis in a microstructured environment, *Ph.D. Dissertation*, Tulane University, New Orleans, LA.

Ayyagari, M.S., Marx, K.A., Tripathy, S.K., Akkara, J.A., and Kaplan, D.L., 1995, Enzymatic polymerization of para-substituted phenols in organic media: Polymer synthesis and characterization, submitted to *Macromolecules.*

Bruno, F.F., Marx, K.A., Tripathy, S.K., Akkara, J.A., Samuelson, L.A., and Kaplan, D.L., 1994, Enzyme catalyzed polymerization of phenol and aniline derivatives on a Langmuir trough to form ordered 2-D polymer films, *J. Intel. Mat. Sys. Struct.*, 5(5):631.

Bruno, F.F, Akkara, J.A., Samuelson, L.A., Kaplan, D.L., Marx, K.A., and Tripathy, S.K., 1992, Method for synthesizing enzyme-catalyzed polymerized monolayer, US Patent, # 5143828.

Dordick, J.S., Marletta, M.A., and Klibanov, A.M.,1987, Polymerization of phenols catalyzed by peroxidase in nonaqueous media, *Biotechnol. Bioeng.*, 30:31.

John, V.T., Akkara, J.A., and Kaplan, D.L., 1994, Use of hydrate formation to control membrane mimetic systems, US Patent, # 5324436.

Kamat, S.V., Iwaskewycz, B., Beckman, E.J., and Russell, A.J., 1993, Biocatalytic synthesis of acrylates in supercritical fluids: Tuning enzyme activity by changing pressure, *Proc. Natl. Acad. Sci. USA*, 90:2940.

O'Connor, C.J., Lee, Y.S., Tang, J., John, V.T., Kommareddi, N.S., Tata, M., McPherson, G.L., Akkara, J.A., and Kaplan, D.L., Supermagnetism of ferrite particles dispersed in spherical polymeric materials, IEEE Trans. Magnetics, 30(6):4954.

Rao, A.M., John, V.T., Gonzalez, R.D., Akkara, J.A., and Kaplan, D.L., 1993, Catalytic and interfacial aspects of enzymatic polymer synthesis in reversed micellar systems, *Biotechnol. Bioeng.*, 41:531.

HUMIC ACIDS ARE VERSATILE NATURAL POLYMERS

Geoffrey Davies,[1] Elham A. Ghabbour,[2] Susan Jansen[3] and James Varnum[3]

[1]Chemistry Department, Northeastern University, Boston, MA 02115
[2]Soil Salinity Laboratory, Bacos, Alexandria, Egypt
[3]Chemistry Department, Temple University, Philadelphia, PA 19122

INTRODUCTION

Life depends on exquisitely designed enzymes, nucleic acids, polysaccharides and structural proteins. These materials return to the earth, where they are *humified* to make organic polymers called humic substances. These living polymers interact with air, earth and water and have many environmental functions. Natural humification makes efficient use of life's waste.

This paper discusses a subset of humic substances (HS) called *humic acids* (HA), which are versatile and smart materials. Recent work suggests that HA may have a limited number of building blocks and shows how they can be processed to give materials of even greater utility.

Humic Substances

Humic substances are living brown organic polymers with average molar mass $<M_w>$ from 1500 to >100,000 Da. HS result from natural polymerization of smaller organic molecules to give lignins, HS, peats and coals, whose structures are affected by the chemical feedstock and polymerization conditions at the source.[1]

HS solubilities classify four main subsets.

Fulvic acids (FA) are soluble in natural waters or at high or low pH. Their highest $<M_w>$ is about 5000 Da.

Hymatomelanic acids (HY) differ from HA by being soluble in small alcohols like methanol. Their $<M_w>$ is in the range 5000-11,000 Da.

Humic acids form as gels from aqueous alkaline solutions on reducing the pH to 1. HA are insoluble in small alcohols or in water below pH 8. Their lowest $<M_w>$ is about 12,000 Da.

Humin is insoluble in water at any pH, has much lower adsorptive power than HA and $<M_w>$ is at least 70,000 Da. Humin is further along the progression toward peats, coals and graphite.[1]

Humic Acids

HA are immobilized in the environment by their insolubility in water. They are

Polymers and Other Advanced Materials: Emerging Technologies and Business Opportunities. Edited by P. N. Prasad *et al.*, Plenum Press, New York, 1995

found throughout nature and especially in soils and water sediments. These living polymers regulate environmental processes. They impart fluidity and porosity and act as adsorbents and reservoirs of water, plant nutrients, metals, herbicides, pesticides and other soil and sediment components.[1] As such, HA are crucial environmental materials.

HA also play roles in the gastro-intestinal tract of humans and animals and are absorbed by animals.[2] They circulate with the blood[3] and are metabolized in the liver.[4] Oral doses of HA reduce heavy metal adsorption in animals[5] and decrease pesticide toxicity.[6] HA can be administered prophylactically and therapeutically in animals, including pregnant animals, without apparent risk.[7] Some HA control uterine cancer in rats[8] and HA markedly reduce the mutagenic effects of benzo-pyrene, 3-aminoanthracene, 2-nitrofluorene and 1-nitro-pyrene.[9] Desmutagenic effects depend on adsorption of dangerous chemicals on the HA surface.[10] A recent comprehensive volume clearly links HA properties with human health.[11]

Isolation, Composition and Structure. HA gels are isolated from compost and other sources by sequential solvent and aqueous base extraction followed by treatment with mineral acids and dialysis.[12] Acid-treated HA are essentially metal-free[13] and can be identified with a range of physico-chemical techniques.[14]

Until recently, HA were thought only to result from humification. However, we have isolated HA from a living plant (*Pilayella littoralis*) for the first time[14] and are searching for HA in other live plants. Plants that contain HA and biomass that is easily composted are of enormous value for soil creation, improvement and remediation.

A central question is whether HA building blocks are numerous or limited. Does humification break down waste molecules partially or extensively? Partial breakdown would give an infinite number of HA building blocks and structures, while synthesis from fewer, smaller building blocks would limit the possibilities. The state of knowledge on this question resembles that concerning proteins before they were shown to be made by condensation of amino acids. Finding HA in live plants[14] and animals[2-10] where synthesis is under biochemical control suggests a limited number of HA building blocks. There are clues in the compositions of HA that suggest this may be the case.

The C, H, N and O element contents of extensively purified HA[13] and the type (amine, carbonyl, carboxylic acid, phenol, quinone), amounts and relative amounts of their functional groups depend on the source but do not vary greatly (Table 1).[1,11,15-25] Table 2 lists the analytical properties of proposed average HA building blocks, which are covalent structures with pendant functional groups.[11,21-25] HA grow by oxidation and condensation of their functional groups but the nature and number of these reactions remain to be settled.[1,11,15-25]

Solutes like (-)nicotine[26] solubilize HA by breaking the internal condensation linkages and "capping" the surface and liberated functional groups. The stoichiometric, solubilizing reaction of compost-derived HA with (-)nicotine consumes 17.0 mmoles nicotine/gm HA.[26] The structure in Figure 1 (last entry in Table 2) has 13 functional groups and a molar mass of 765 Da. Reaction of (-)nicotine with every functional group consumes 13,000 mmoles/765 gm HA = 17.0 mmoles/gm HA. Agreement with the experimental data[26] recommends this structure, which fits well with the average analytical data in Tables 1 and 2, as an average HA building block. Our specific modifications of the previously proposed structure $C_{40}H_{45}O_{17}$ (Table 2)[25] replace one quinone with an amino group, two methyl groups with two hydrogen atoms and two alcohol groups with two aldehyde functions.[27]

We are working to identify the building blocks of HA from different sources with microbiological approaches and through HA reactions with capping amines. The proposed **BB** in Figure 1 appears to represent their anticipated properties.[25] It has seven chiral centers, 64 conformations and 64 mirror images. We have computed the energy of all unique conformations of this structure and probed the complete conformational space.

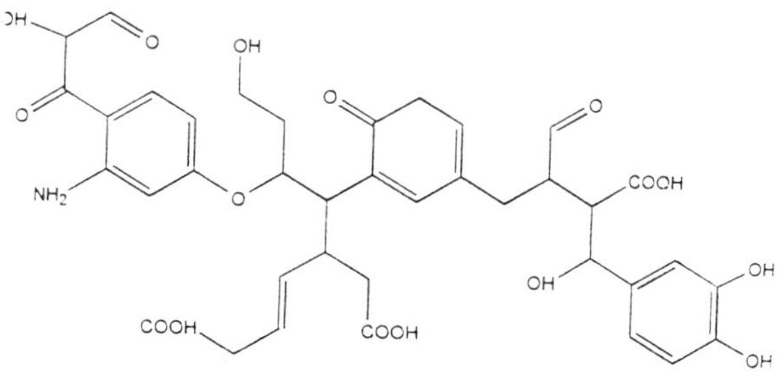

Figure 1. Proposed HA Building Block.[27]

Table 1. Acidity and Total Functional Group Data (meq/g) for Humic Acids

Source	Acidity	Functional Group[a] COOH	Ar-OH	C=O	Ar=O	Ref
Marine	2.8-5.8	1.8-3.9	0.9-1.9	1.2-3.0	0.7-2.1	15
Marine	3.0-7.0	2.0-5.0	0.5-2.5	3.0-6.0		16
Soil	6.1-6.7	2.7-3.3	3.5-3.9	3.8-5.2	1.4	17
Soil	6.0-9.7	3.9-5.7	2.1-4.4			18
Soil	7.3	3.8	3.5			19
Soil	6.7	3.6	3.9			20
Average[b]	6.3	3.6	2.7	3.1	1.4	

[a] Data are the total content (meq/gm) of each functional group after complete HA hydrolysis with strong mineral acid; [b] Average calculated by taking the mid range values from rows 1 - 4 and combining them with the values in rows 5 and 6.

This work is preliminary to study of experimentally isolated building blocks and gives a possible description of functional group involvement in HA growth.

The **BB** in Figure 1 has the lowest energy conformation shown in Figure 2. This conformation creates metal binding regions I - III that appear to match the requirements of metals like Cu(II), Fe(II) and Fe(III) that are selectively bound by HA. The conformation can be linked by condensation of its amine and carboxylic acid functional groups to give the helical octamer in Figure 3.[30] The Cotton effect in the circular dichroism (CD) spectrum of Figure 4a shows that compost-derived HA contains an α-helix[31, 32] and the NMR spectra in Figure 4b indicate an amide linkage.[32] These findings encourage a re-invigorated search for actual HA structures as a means of understanding their functions.

Adsorptive Properties. HA are adsorbents, ion-exchangers, biochemical regulators and reservoirs. Their functions and uses depend on these properties.[1,11] HA adsorb and react with a wide range of solutes.[1,11] However, very few data for adsorption on HA have been analyzed with models that generate thermodynamic parameters for comparison of different systems.[26,33-40]

This situation has been improved by a thorough study of aqueous adsorption of nucleic acid constituents adenine (**1**), adenosine (**2**), adenosine-5'-monophosphate (**3**), guanosine (**4**), guanosine-5'-monophosphate (**5**), cytosine (**6**), cytidine (**7**), cytidine-5'-

Table 2. Proposed and Predicted Properties of Humic Acid Building Blocks

	Calcd. Anal. (%)[b]				Functional Groups[c]			
MF[a]	C	H	N	O	COOH	Ar-OH	Ar=OC=O	
$C_9H_6O_4$(178)[21]	61	3.4	0	36	5	6	0	1
					(3)	(2)	(1)	(3)
$C_{32}H_{32}O_{13}N_2$(652)[22]	59	5.0	4.3	32	1	3	60	
					(5)	(4)	(2)	(5)
$C_{70}H_{58}O_{28}N$(1360)[23]	62	4.3	1.0	33	2	1	3	5
					(5)	(4)	(5)	(2)
$C_{72}H_{47}O_{36}N_4$(1543)[24]	56	3.0	3.6	37	5	8	25	
					(6)	(4)	(6)	(2)
$C_{40}H_{45}O_{17}$(797)[25]	60	5.6	0	34	3	2	2	1
					(3)	(2)	(1)	(3)
$C_{38}H_{39}O_{16}N$(765)[27]	60	5.1	1.8	33	3	2	1	3
					(3)	(2)	(1)	(3)
Average[d]	56	4.5	2.5	36				

[a] Proposed molecular formula of BB (calculated formula weight, Da); [b] calculated elemental percentages of the proposed BB structure; [c] the number of functional groups in the proposed structure; numbers in parentheses are the respective numbers calculated from the formula weight and the average meq/gm data in Table 1, as in this example: humic acids contain an average 3.6 meq/gm of -COOH (Table 1). This predicts the number 3.6 x 10^{-3} eq/gm x 765 gm = three carboxylic acid groups in the proposed HA structure[27] with FW 765 Da: hence the entry (3) in column 6, row 12; [d] median averages from refs. 1, 19, 28, 29.

monophosphate (**8**), thymine (**9**), thymidine (**10**), thymidine-5'-monophosphate (**11**), uracil (**12**), uridine (**13**), uridine-5'-monophosphate (**14**) and of RNA on a fully characterized compost-derived HA.[41] Solutes **1**, **6**, **9** and **12** are nucleobases, **2**, **4**, **7**, **10** and **13** are nucleosides and **3**, **5**, **8**, **11** and **14** are nucleotides. Nucleosides are the 3'-ribopentose derivatives of the respective nucleobase and nucleotides are the corresponding 5'-monophosphoric acids.

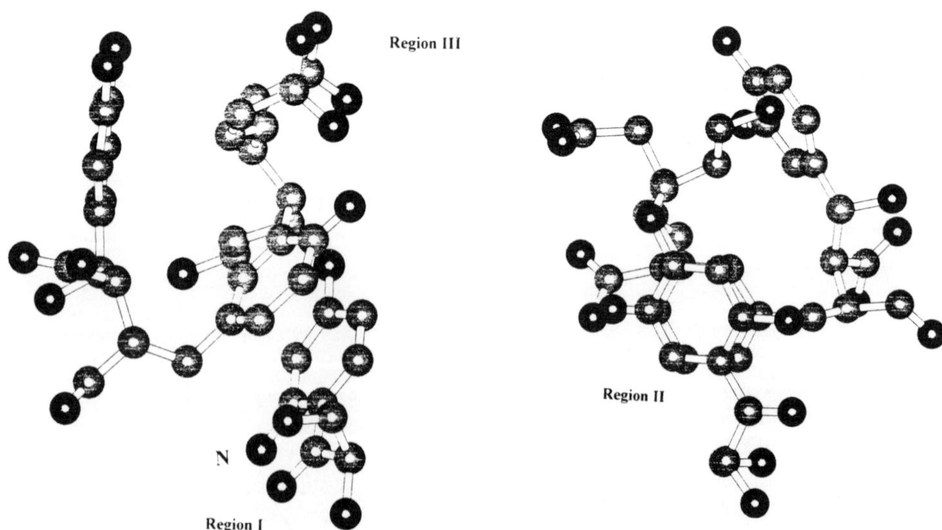

Figure 2. Views of the lowest energy conformation of the HA building block in Figure 1.[30]

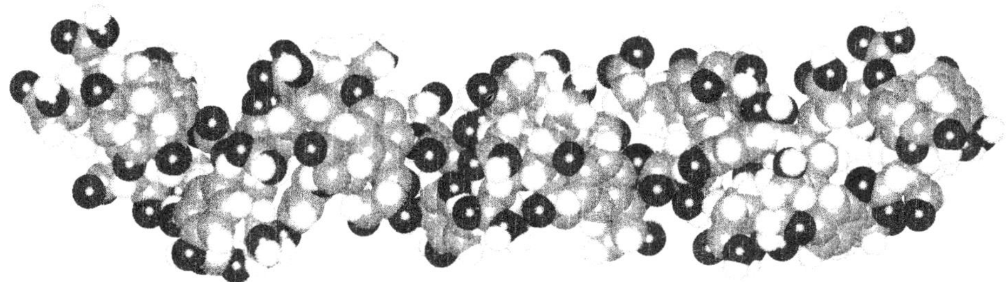

Figure 3. A helical octamer from condensation of the conformation in Figure 2.[30]

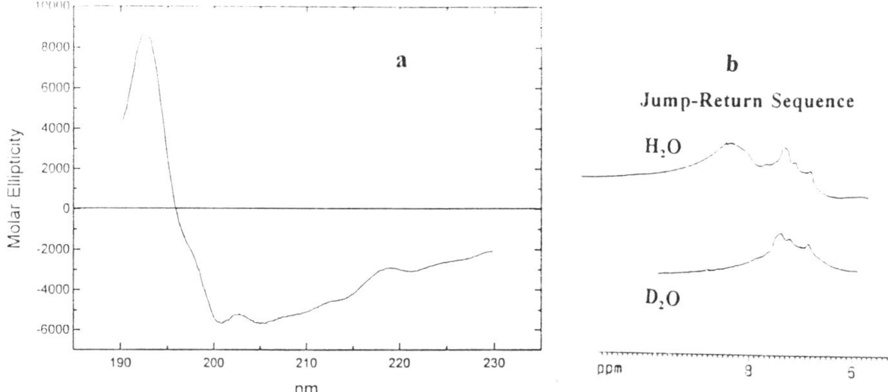

Figure 4. (a) CD spectrum and (b) NMR spectra of compost-derived HA in alkaline solution.[32]

Individual samples of HA (m = 10 mg) were equilibrated with 20 known aqueous concentrations (0 to 600 μM) of each solute at fixed temperature. The pH range was 4.6 to 6.6 for nucleobase- or nucleoside-HA solutions and 3.2 to 3.7 for nucleotide-HA solutions. Nucleobases and nucleotides are neutral under these conditions and the nucleotides are monoanionic. Equilibrium solute concentration c and the amount of solute adsorbed (x, mmoles) in each experiment were used to construct an isotherm (x/m *vs.* c) for each solute at fixed temperature. Isotherms for each solute at six temperatures in the range 10.0 to 40.0°C (i.e a total of 15 x 6 = 90 isotherms) were obtained.[41]

The isotherms indicate two (**A** and **C**) and often three HA adsorption sites **A**, **B** and **C**, depending on the solute and experimental conditions (Figure 5). Extensive data enabled a thorough test of the Langmuir model for reversible adsorption through plots of m/x *vs* 1/c for each adsorption step. Most of the data fit this model, which generates the site capacity v and the equilibrium constant K for each adsorption step. The few data giving negative v, inconsistent v or K at different temperatures or non-Langmuir isotherms are attributed to occasional HA sample inhomogeneity and discarded.[41]

Site capacity v for a given solute and HA site is independent of temperature, as required by the Langmuir model; v for solute families like thymine, thymidine and thymidine-5'-monophosphate are independent of solute molar mass, indicating interaction of the nucleobase component of nucleosides and nucleotides with the HA surface.[41]

Except for adenine adsorption, the site capacities increase in the order $\langle v_A \rangle < \langle v_B \rangle < \langle v_C \rangle$, where $\langle\ \rangle$ is the average value over the experimental temperature range. Close comparison of $\langle v_A \rangle$, $\langle v_B \rangle$ and $\langle v_C \rangle$ indicates that many solutes

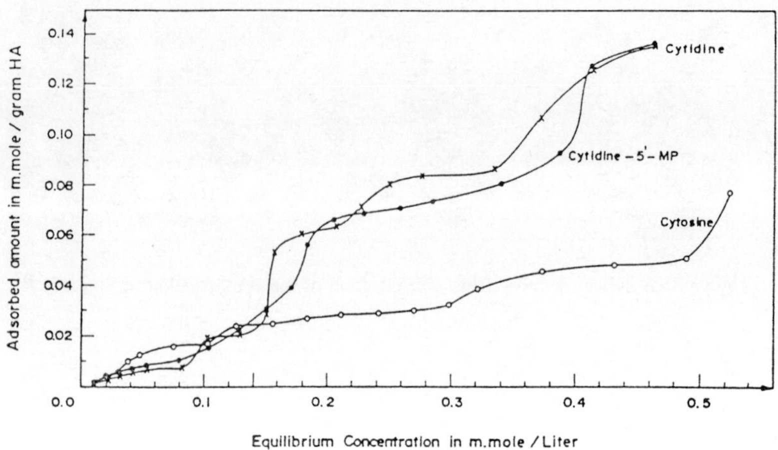

Figure 5. Adsorption isotherms for compost-derived HA at 40°C.[41]

interact with a given site with the same stoichiometry. However, some are decidedly different. For example, $<v_C>$ is 0.28 ± 0.07 mmole solute/gm HA for most of the solutes but $<v_C>$ for thymine is smaller than average and $<v_C>$ for cytosine, guanosine and adenosine are much larger than expected. The solutes that have or do not have average $<v>$ for adsorption at a particular HA site are different and thus HA sites **A**, **B** and **C** are different. Comparison of total site capacities $\Sigma<v> = <v_A> + <v_B> + <v_C>$ confirms some unusual adsorptive behavior.

The temperature dependences of K for adsorption of different solutes on HA sites **A**, **B** and **C** indicate exothermic, thermoneutral or endothermic adsorption, depending on the solute and the adsorption site. Taken together, the v and K data indicate the following.[41]

1. Agreement is observed with limited data for adsorption of solutes on HA from different sources where Langmuir behavior and analysis of the results are reported.[26,33-40]

2. The average humic acid molecule ($<M_w>$ 33,000 Da) has one **A**, two **B** and eight **C** adsorption sites accessible to an average solute. The surface functional group ratios differ from those in Table 2 because condensation of building block functional groups involves some functional groups but not others. Only about 2% of the total number of functional groups in an average HA molecule are accessible to an average solute. This fraction increases with decreasing HA molar mass.[27,41]

Site Creation on Adsorption. HA are "smart" materials: some solutes create new sites as a result of primary adsorption on HA. For example, adenine has an enormous capacity to create new **B** sites while it has average properties when it adsorbs on site **C**. Site creation ability depends on K for adsorption of a given solute on a given HA site. Site creation is negligible for adsorption of uridine, which serves as a standard solute for the three HA sites.

Variations of K and site creation ability are responsible for marked solute selectivity at sites **A**, **B** and **C**. For example, site **A** cannot discriminate between adenosine, thymine and thymidine while site **C** has a marked preference for thymine.

A remarkable correlation of adsorption enthalpies and entropies (Figure 6) points to a common, underlying adsorption mechanism for all the solutes. Solute adsorption could be chemical or physical. It definitely is reversible.[41]

HA molecules consist of a covalent backbone that is strengthened and likely propagated by condensation reactions between amine, carbonyl, carboxylic acid, phenol

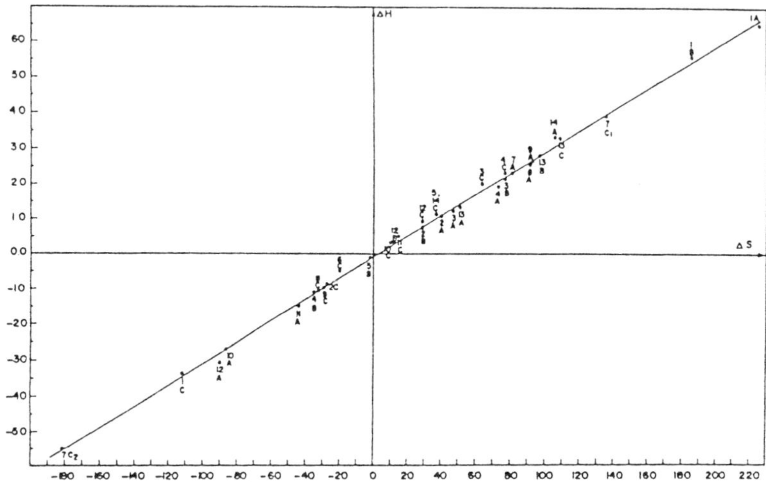

Figure 6. Correlation of thermodynamic data for nucleic acid constituent adsorption on compost-derived HA (Example: 3A refers to adsorption of solute **3** on site **A**).[41]

or quinone functional groups.[1,11,15-25] Our proposed site creation model has four major components: a) primary solute adsorption; b) creation by some solutes of new sites near the primary adsorption site through breaking of the bonds that link HA building blocks; c) adsorption of the solute on these new sites; d) adsorption of water on the other "half" of the broken HA building block linkages. Each of these steps has its own associated enthalpy and entropy changes.[41]

Site creation and water adsorption enthalpy/entropy changes are negative and strongly correlated. Site creation and water adsorption at HA site **B** are most exothermic and exhibit the largest entropy decreases. Facile site creation by solutes like adenine at higher temperatures parallels the appearance of S-curves for HA adsorption of herbicides and pesticides.[27,33-40]

HA sites **A**, **B** and **C** need chemical identification. Bases adenine, cytosine and (-)nicotine all cause HA to dissolve at pH $<$ 7 because of their extreme ability to create new adsorption sites. They break condensation linkages that hold HA building blocks together and cap the functional groups to prevent them from rejoining. The soluble product spectra clearly indicate solute capping of carboxylic acid and phenol functions in HA. This is the most comprehensive description of adsorption on HA to date.[41]

A Humic Acid Aerogel. Aerogels are materials with very low density and very high surface areas that are made by supercritical fluid drying.[42-44] We have found that the surface area of an HA extracted from *ascophyllum nodosum* kelp can be increased to 230 m^2/gm by supercritical fluid CO_2 drying of the gel from the last HA extraction step. The product is a fractal material.[45] Important keys to successful aerogel production are replacing the water in the gel with a solvent like acetone and use of SCF-CO_2, whose critical temperature is below the HA decomposition temperature. Preliminary data indicate that the water retention and adsorptive properties of such high surface area HA are much better than those for HA isolated by conventional drying methods. HA aerogels look promising for soil creation, improvement and remediation and have many other potential applications.

Acknowledgements

This work was supported in part by a grant from the National Agricultural Research Project administered by the US Agency for International Development. We thank our co-authors for their contributions to the work presented here.

References

1. W. Ziechmann, "Humic Substances", BI Wissenschaftsverlag, Mannheim (1993).
2. A. H. Khairy, *Acta Medica Empirica*, 11:898 (1981); *De Natura Rerum.*, 3:229 (1989); A. H. Khairy, S. S. El-Gendi, and H. H. Bhagdadi, *De Natura Rerum.*, 5:76 (1991).
3. R. Klockling, U. Eichhorn, and T. Blumohr, *Fres. Z. Anal. Chem.*, 292:408 (1978).
4. S. A. Visser, *Acta Biol. Med. Germ.*, 31:569 (1973).
5. F. Ridwan, S. Molnar, and W. Rochus, *Naturwiss.*, 65:539 (1978).
6. S. Golbs, M. Kuchnert, and V. Fuchs, *Z. Ges. Hyg.*, 30:720 (1984).
7. S. Golbs, F. Volkhard, and M. Kuchnert, *Arch. Exp. Veternaermed.*, 36:179 (1982).
8. F. Miawu, Z. Lisheng, C. Lauping, Z. Ping, and X. Hongji, *Xaoxue Tongbao.*, 16:502 (1981).
9. T. Sato, Y. Ose, and H. Nagase, *Mutation Res.*, 162:173 (1986); T. Sato, Y. Ose, H. Nagase, and K. Hayase, *Sci. Total Envir.*, 62:305 (1987).
10. T. Sato, Y. Ose, H. Nagase, and K. Hayase, *Mutation Res.*, 176:199 (1987).
11. "Humic Substances in the Global Environment: Implications for Human Health," N. Senesi, and T. M. Miano, eds., Elsevier, Amsterdam (1994). For recent work on HA structure see F. H. Frimmel, H. M. Abdel-Fattah, and G. Abbt-Braun, *Modell., Meas. Control*, 44:43 (1994); M. V. Cheshire, D. B. McPhail, and M. L. Berrow, *Sci. Total Environ.*, 152:63 (1994); N. Senesi, *Environ. Int.*, 20:3 (1994); G. Ricca, and F. Severini, *Geoderma*, 58:233 (1993); A. Jankowska, T. Siemieniewska, K. Tomkov, and M. Jasienko-Halat, *Carbon*, 31:871 (1993).
12. F. Scheffer, W. Zeichmann, and G. Pawelke, *Pflanzenernahr. Dung. Bodenkd.*, 90:58 (1960); R. H. Pierce, Jr., and G. T. Felbeck, Jr., *Proc. Int. Mtg. Humic Subst.*, Nieuwersluis, 1972, p. 217.
13. M. Schnitzer, and S. U. Khan, "Humic Substances in the Environment," Dekker, New York, (1972), p. 288.
14. E. A. Ghabbour, A. H. Khairy, D. P. Cheney, V Gross, G. Davies, T. R. Gilbert, and X. Zhang, *J. Appl. Phycol.*, 6:459 (1994).
15. A. H. Gillam, and J. P. Riley, *Analyt. Chim. Acta*, 141:287 (1982).
16. M. A. Rashid, and L. H. King, *Geochim. Cosmochim. Acta*, 34:193 (1970).
17. A. Piccolo, *Environ. Soil Sci.*, 146:418 (1988).
18. M. B. Holder, and S. M. Griffith, *Can. J. Soil Sci.*, 63:151 (1983).
19. M. Schnitzer, Recent findings on the characterization of humic substances extracted from soils from widely differing climatic zones, *in*: "Soil Organic Matter Studies," International Atomic Energy Agency, Vienna, Vol II (1977), p. 117-132.
20. Y. Chen, N. Senesi, and M. Schnitzer, *Geoderma*, 20:87 (1978).
21. W. Fuchs, *Kolloid* Z., 52:124 (1930): see also "Surface and Colloid Chemistry in Natural Water and Water Treatment," R. Beckett, ed., Plenum, New York (1990) p. 3-20.
22. C. C. Dragunov, H. H. Zhelokhovtseva, and E. J. Strelkova, *Pochvreddenie*, 7:409 (1948).
23. W. Flaig, *Sci. Proc. Roy. Dublin Soc.*, 4:49 (1960).
24. F. J. Stevenson, "Humus Chemistry: Genesis, Composition, Reactions," Wiley, New York (1982), pp. 26-39, 258-263.
25. C. Steelink, Implications of elemental characteristics of humic substances, *in*: "Humic Substances in Soil, Sediments and Water," G. R. Aiken, D. M. McKnight, R. L. Wershaw, and P. MacCarthy, eds.: Wiley-Interscience, New York (1985) pp. 457-476.
26. A. H. Khairy, H. H. Baghdadi, and E. A. Ghabbour, *Z. Pflanzenernahr. Bodenkd.*, 153:33 (1990).
27. G. Davies, E. A. Ghabbour, A. H. Khairy and H. Z. Ibrahim, submitted to *J. Phys. Chem.*
28. W. Flaig, H. Beutelspacher, and E. Rietz, Chemical composition and physical properties of humic substances, *in*: "Soil Components. VI. Organic Components," J. E. Gieseking, ed., Springer-Verlag, New York (1975) pp. 1-211.
29. F. J. Stevenson, and J. A. H. Butler, Chemistry of humic acids and related substances, *in*: "Organic Geochemistry," G. Eglinton, and M. T. J. Murphy, eds., Springer, New York

(1969), pp. 534-557.
30. S. A. Jansen, J. Varnum, M. Malaty, E. A. Ghabbour and G. Davies, in preparation.
31. R. W. Woody, *in*: "The Peptides," Academic, New York (1985), Vol. 2, Ch.2 and references therein.
32. J. Varnum, S. A. Jansen, and G. Davies, to be published: for jump-return sequence used to obtain the data in Figure 4b, see J. P. Hoare, *J. Magn. Res.*, 55:283 (1983); G. M. Clore, B. J. Kimber, and A. M. Gronenborn, *J. Magn. Res.*, 68:526 (1986).
33. P. Ruggiero, C. Crecchio, R. Mininni, and M. D. R. Pizzigallo, *Sci. Total Environ.*, 123/124:93 (1992).
34. M. Adhikari, and P. Sen, *Indian Agric.*, 26:107 (1982).
35. S. U. Khan, *Can. J. Soil Sci.*, 53:199 (1973).
36. Y. A. Madhhun, V. H. Freed, and J. L. Young, *Soil Sci. Soc. Amer. J.*, 50:319 (1986).
37. A. Piccolo, G. Celano, and G. Pietramellara, *Sci. Total Environ.*, 123/124:77 (1992).
38. U. Muller-Wegener, *Z. Planzenernach. Bodenk.*, 145:411 (1982).
39. G. C. Li, and G. T. Felbeck, Jr., *Soil Sci.*, 113:140 (1972).
40. C. Maqueda, J. L. Prez-Rodriguez, and M. C. Hermosin, *Soil Sci.*, 136:75 (1983).
41. A. H. Khairy, H. Z. Ibrahim, E. A. Ghabbour, and G. Davies, manuscripts in preparation; E. A. Ghabbour, Doctoral dissertation, Alexandria University (1994).
42. R. W. Pekala, *J. Matl. Sci.* 24:3221 (1989).
43. L. W. Hrubesh, *Chem. & Ind*, 24:824 (1990).
44. W. J. Schmitt, R. A. Greiger -Block, and T. W. Chapman, *in*: "Chemical Engineering at Supercritical Fluid Conditions," M. E. Paulitis, ed., Ann Arbor Science, Ann Arbor (1983), pp.445-460.
45. A. Radwan, R. J. Willey, G. Davies, A. Fataftah, and S. J. Rouanet, to be published.

PALM OIL BASED POLYURETHANE FOAMS

Salmiah Ahmad,[1] Parthiban Siwayanan,[1]
Ooi Tian Lye,[1] Rohana Ali,[1]
Azman Rafaei,[1] Zuhairi Zainuddin,[1]
Dieter Wiese,[2] and Chua Meng Choo[2]

[1]Palm Oil Research Institute of Malaysia
[2]Intermed Sdn. Bhd.

ABSTRACT

The current raw materials for the production of polyurethanes (PUs) are polyols and isocyanates. These are mostly petrochemical-based. The objective of this paper is to evaluate the technical and economic viability of producing polyols and polyurethanes from palm oil and palm oil products (POP) and to discuss the characteristics of POP-based PU foams (PUFs).

Earlier study has indicated the technical viability of producing polyols from POP. In this second phase of our study, it was observed that the hydroxyl values of the polyols produced could be varied by changing the types of palm oil and palm oil products (POP) used or by changing the various formulations.

It was also demonstrated that, the production of polyurethanes from these polyols are technically viable. The polyurethanes formed ranges from semi-rigid to rigid types. The properties (in terms of density, foam structures & flammability) of these polyurethanes could be further varied by changing the types and concentrations of the additives used.

Our initial calculation indicate that polyols derived from POP could be economically competitive to those derived from petrochemicals. Since the isocyanates used are the same, the cost of the PUFs are expected to be similar. But PUs derived from POP are expected to be more environmentally friendly.

Keywords: Palm oil, epoxidized palm oil, polyols and polyurethanes

INTRODUCTION

A urethane group is formed by the exothermic chemical reaction between an alcohol and an isocyanate. Urethane polymers or polyurethanes (PUs) result from the reactions between alcohols with two or more reactive hydroxyl groups per molecule (diols or polyols) with isocyanates that have more than one reactive isocyanate group per molecule (polyisocyanates).

Isocyanates also react with water to form unstable carbamic acids, which decompose with the evolution of carbon dioxide and amines. The released carbon dioxide if trapped within the polymers result in the formation of cell-structures, better known as foams.

Relatively few basic isocyanates and a range of polyols of different molecular weights and functionalities are used to produce a whole spectrum of polyurethane materials. Polyurethane foams (PUFs) are classified in many ways. Examples are:-

1) flexible, semi-rigid and rigid,
2) closed-cell, opened-cell and mixed type and
3) block, in-situ foamed and moulded & sprayed type.

The most popular classification is the flexible, semi-rigid and the rigid type. The rigidity depends on the density or the degree of cross-linking which in turn depends on the types of polyols and isocyanates used. The borderline between rigid and flexible foams is therefore diffused due to the wide overlapping range of properties and the intermediate class, semi-rigid was introduced for convenience.

PUs are very versatile materials and they play vital role in several industries such as ship building, footware, construction of cars, furniture and mattresses, adhesives and sealants, packaging, coating, paints, encapsulation, refrigeration etc.

The reaction between these two raw materials can be modified by adding various types of additives. These in turn would change either the rates of reaction, the structures, appearances and the properties of the products thus increasing their areas of applications. In 1991, the total world PUs market share by industry is as follows:-

Industry	Percentage
Furniture	31.31
Transport	16.39
Coatings/Adhesive/Sealant	17.65
Construction	14.95
Appliance	6.37
Footware	3.34
Other Industries	9.99

Source: Hazimah *et.al,* 1993

THE CHEMISTRY & MATERIALS FOR POLYURETHANES

Polyisocyanates

Reaction of isocyanates with polyols

Polyurethanes are addition polymers formed by the reaction of polyisocyanates with polyols. The reaction between polyol and isocyanate to form the urethane group is the most important reaction for the preparation of polyurethane.

$$HO\text{-}R\text{-}OH + O{=}C{=}N\text{-}R'\text{-}N{=}C{=}O \longrightarrow \left| \begin{array}{l} C\text{-}N\text{-}R'\text{-}N\text{-}C\text{-}O\text{-}R\text{-}O\text{-} \\ \| \;\;\; | \qquad\;\; | \;\;\; \| \\ O \;\; H \qquad H \;\; O \end{array} \right|_n$$

Reaction of isocyanates with water

Reaction with water is the next most important reaction of isocyanates. Isocyanate reacts with water to produce amine and carbon dioxide. This reaction provides the principle source of gas for "blowing" in the manufacture of PUFs.

$$2H_2O + O{=}C{=}N\text{-}R'\text{-}N{=}C{=}O \longrightarrow \begin{array}{l} H_2N\text{-}R'\text{-}NH_2 \\ + CO_2 \end{array}$$

Reaction of isocyanates with amines

The reaction of isocyanate with primary or secondary diamines is the third most important reaction of isocyanates. Diamines are used as chain-extending and curing agents in PUs since they increase the reactivity of the reaction mixture.

$$OCN\text{-}R'\text{-}NCO + H_2N\text{-}R'\text{-}NH_2 \longrightarrow \begin{array}{l} OCN\text{-}R'\text{-}N\text{-}C\text{-}N\text{-}R'\text{-}N\text{-}C\text{-}N\text{-}R'\text{-}NCO \\ \qquad\qquad\;\; | \;\; \| \;\; | \qquad\;\; | \;\; \| \;\; | \\ \qquad\qquad\;\; H \; O \; H \qquad H \; O \; H \end{array}$$

Owing to the above three chemical properties of isocyanates, chemical reactions between a polyol and a polyisocyanate start spontaneously when they are blended together in the presence of suitable catalysts.

Several aromatic and aliphatic isocyanates are available. However, more than 95% of the PUs produced are based on two of them i.e. toluene diisocyanate (TDI) and diisocyanate-diphenylmethane (MDI). Both materials are derived from readily available petrochemical intermediates. The volatility of TDI had caused some severe problems when it was sprayed. This made the development and production of MDI a necessity.

POLYOLS

A wide range of polyols is available, but about 90% of the polyols used in making polyurethanes are polyethers with terminal hydroxyl groups. Hydroxyl terminated polyesters are also used to obtain polyurethanes with special properties but they are usually more expensive.

The structures, molecular weights and functionalities of the polyols play vital roles in determining the properties of the final urethane polymers. The characteristics of the polyols used to make flexible and rigid PUs are as given below:-

Similar to isocyanates, these hydroxyl rich compounds are usually obtained from petrochemicals (or carbohydrates).

Characteristics of Polyols	Flexible PU foams	Rigid PU foams
Molecular weight (range)	1000 - 6500	150 - 1600
Functionality (range)	2.0 - 3.0	3 - 8
Hydroxyl value (range) mg KOH/g	28 - 160	250 - 1000

Hydroxyl value = a measure of the concentration of cyanate-reactive hydroxyl group per unit weight of the polyol

Source: Woods, 1987.

ADDITIVES

Apart from polyisocyanates and polyols, a wide variety of auxiliary chemicals may be added in order to control and modify both the polyurethane reaction itself and the properties of the final polymer. These additives include; catalysts, cross-linking agents, chain extending agents, blowing agents, surfactants, colouring materials, fillers, smoke depressant and flame retardants.

Catalysts

A number of catalysts can be used for the reaction of isocyanates with water and polyols. These include aliphatic and aromatic tertiary amines, organometallic compounds, especially tin compounds. Tertiary amines are widely used. They function by donating electrons to carbonyl carbon of isocyanates group thus forming complexes.

Organo-metallic catalysts are used to accelerate the urethane reaction. The most popular are stannous octoate and dibutyltin dilaurate. An organometallic catalyst forms an intermediate complex with isocyanate group and the hydroxyl group of the polyol. Water blown PU foams are usually catalysed with synergistic mixture of one or more amine and an organotin catalyst.

Chain Extenders

Chain extenders or cross linking agents are used to increase the level of covalent bonding in rigid PUs.

Blowing Agents

Cellular PUs or PUFs are manufactured by using blowing agents to form gas bubbles in the polymerising reaction mixtures. They are usually made using the carbon dioxide formed during the reaction of isocyanate with water either as the sole blowing agent or as the principle blowing agent in association with chlorofluorocarbons (CFC's). When

CFC's are used and they are retained in the closed cells, their low thermal conductivity characteristics, render the resulting products good insulating properties. However, due to the possible environmentally harmful effect of CFC's, they need to be replaced. Todate cyclopentane and pentane are being actively evaluated as possible substitutes.

Surfactants

Surface active agents are essential ingredients in the manufacture of most PUs. Selected surfactants or mixtures, help in mixing the incompatible components of the reaction mixtures.

They are useful to help control the size of the foam cells by stabilising the gas bubbles formed during nucleation. The most common surfactants are organosiloxanes or silicone based surfactants.

Colouring Materials

PUs to be used for specialised applications may be highly coloured. The usual method of colouring is by adding pigment pastes to the foam reaction mixtures. The most widely used colouring material is carbon black. They are used at a level of 0.1 to 100 parts of the polyols.

Fillers

Particulate or fibrous material fillers are added in PUs for various reasons which include – to reduce flammability, to increase weight, to increase resistance to compression, to increase stiffness and to increase the range of operating temperatures of the foams. Some of the fillers commonly used include calcium carbonate, clays, glass flakes, silicates, cements and short fibres.

Flame Retardants

Common to all organic polymers, PUs will burn given sufficient time, heat and oxygen. Flame retardants are added to reduce ignitability and the burning rate of PUs thus reducing their flammability. The most widely used flame retardants are the chlorinated phosphate esters.

OBJECTIVES OF THIS STUDY

As indicated above, majority of the raw materials used for the production of PUs or PUFs are petrochemical-based. It was of interest to carry out an exploratory study to check on the technical and economic feasibilities of producing any of these raw materials for PUs or PUFs production, from palm oil and palm oil products (POP). Once the technical and economic feasibilities have been proven, it is hoped that the PUs and PUFs produced from POP would be more environmentally friendly.

The first phase of our work (Hazimah, *et.al*, 1993) indicated that it was technically feasible to prepare polyols from POP. In this second phase of the study, the technical feasibility of producing polyurethanes foams from palm oil and palm oil products was investigated. This paper shall report our findings.

EXPERIMENTAL PROCEDURES

MATERIALS

Commercially available MDI and additives were purchased. POP based polyols were prepared using the procedure described by Hazimah *et.al* (1993). These POP-based polyols were then reacted with MDI in the presence of commercially available blowing agent and additives.

PROCEDURES

Polyurethanes Foams

The polyols were prepared according to the procedure described by Hazimah et.al (1993). The additives to be studied (types and concentration of catalysts, types and concentration of flame retardants, types and amount of blowing agents etc.) were added into the polyol and stirred. When all additives were throughly mixed, MDI (amount to be decided based on the formulation to be studied) was added and the mixture was again stirred. When the formation of foams could be seen to be slowly taking place, the stirring was stopped and the reaction mixture was allowed to stand. The foams would rise with time and after about five hours, the foams would not be sticky to touch upon which they could be taken out and evaluated.

The density, foam characteristics and heat loss of the foams prepared were evaluated.

a) **Density**
Density of the foams were calculated by weighing the PUFs cut into cubes.

b) **Heat loss**
A simple method to evaluate the heat loss through the PUFs was attempted. The PUFs blocks were put on a piece of ice and the time taken for the PUFs block to reach lowest temperature within certain amount of time was recorded.

c) **Flammability**
The flammability was checked by actually igniting the foams. Only foams that did not burn at all were further evaluated.

d) **Foam-structures**
The foam structure was observed with the naked eyes.

RESULTS & DISCUSSION

Hydroxyl Values

Various palm oil and palm oil products (POP) were used for the production of polyols in this study, in contrast to earlier study where only one type of oil/fat had been used. It was observed that the hydroxyl value of the polyols could be varied within wider range by using various types of POP as well as by varying reaction conditions. This is an interesting finding since it implies that different types of PUFs can be prepared by varying the types of POP used as well as the reaction conditons. In Table 1, POP-3/GUN13 gave a rigid foam instead of the flexible foam predicted based on woods (1987) classification. The rest of the formulations in Table 1 gave foams with characteristics similiar to that predicted.

Table 1. Polyols Derived From Palm Oil and Palm Oil Products

POP	FORMULATIONS	OHV (mgKOH/g)	(PREDICTED) POSSIBLE TYPES PUF
POP-1	GUN11	105	flexible
	GUN12	93	flexible
	GUN13	95	flexible
	GUN21	107	flexible
	GUN31	111	flexible
POP-2	GUN11	325	semi ⟵ rigid
	GUN12	515	semi ⟶ rigid
	GUN13	481	semi ⟶ rigid
	GUN21	242	semi ⟵ rigid
	GUN31	285	semi ⟵ rigid
POP-3	GUN13	124	flex ⟵ semi-rigid
	GUN31	521	semi ⟶ rigid

Note: the Hydroxyl value was determined based on FCC method (Food Chemicals Codex of FDA, USA)

EFFECT OF BLOWING AGENTS ON DENSITY & HEAT LOSS

Earlier in our study, three types of blowing agents (BA1 to BA3) were investigated. It was found that BA1 was the best among the three and this was concentrated on. The effect of adding various amounts of blowing agent was investigated and this is as summarised in Tables 2A, 2B and 2C.

Generally, it was observed that increasing the amount of BA1 would result in the increase in the foam volume or decrease in density. With increasing BA1, the density of the foam would pass through a minimum after which it would increase again.

Similarly the structures of the PUFs would change from being very fine and condensed to "good volume". After the maximum amount of BA1 has been added, the structure of the foam would become more condensed with increasing amount of BA1.

EFFECT OF FLAME RETARDANTS ON FOAMS, DENSITY & HEAT LOSS

Two types of flame retardants were investigated, however one (FR1) was found to be better than the other and only this shall be discussed. The minimum amount of FR1

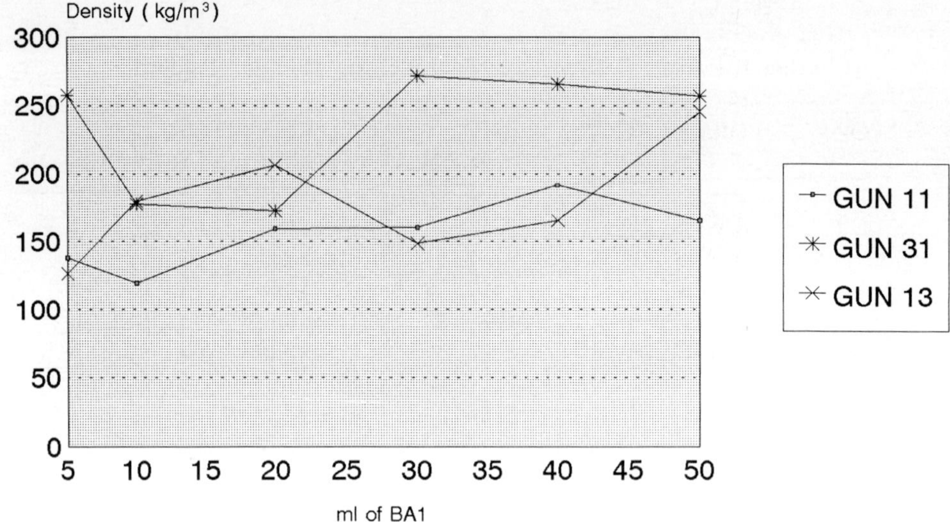

Figure 1. Effect of BA1 on the Density of PUF (without flame retardant)

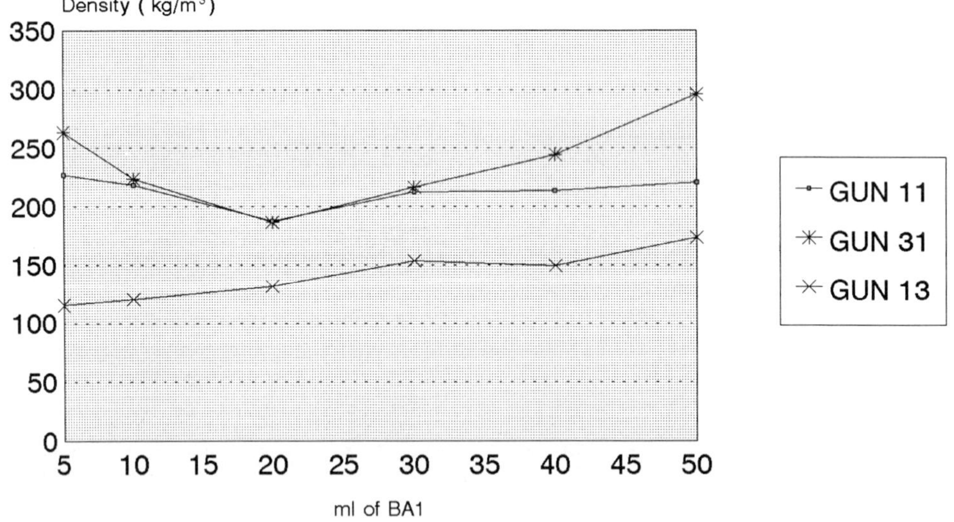

Figure 2. Effect of BA1 on the Density of PUF (with FR1)

Table 2A. Effects of Flame Retardant (FR1), Blowing Agent (BA1) on Density, Structure and Heat Loss

PUFs CHARACTERISTICS - POP-2								
CODE	BA1	FR1	Height	Density	Heat Loss	Rigidity	Structure	Remarks
GUN 11	5	w.o.	82	136-140	0.2364 @ 22	semi rigid	fine	tiny pores
		w.	64	218-236	0.2500 @ 24	rigid	very fine	
	10	w.o.	110	103-136	0.2750 @ 24	semi rigid	coarse	small pores
		w.	73	218	0.1925 @ 27	rigid	very fine	
	20	w.o.	101	146-173	0.4037 @ 27	semi rigid	coarse	small pores
		w.	90	186-188	0.3400 @ 20	rigid	very fine	
	30	w.o.	104	137-184	0.3308 @ 26	semi rigid	coarse	small pores
		w.	71	212-213	0.3200 @ 25	rigid	very fine	
	40	w.o.	95	169-214	0.2538 @ 26	semi rigid	coarse	small pores
		w.	81	211-215	0.3182 @ 22	rigid	very fine	
	50	w.o.	94	159-172	0.5158 @ 19	semi rigid	coarse	small pores
		w.	80	216-224	0.3500 @ 20	rigid	very fine	

Height ... (mm.)
Water ... (ml.)
Density ... (kg/m^3)

Heat Loss ... Temp.(°C)/time(min.)
(temperature drop per minute)

w.o. ... without flame retardant
w. ... with 10g. FR1

Table 2B. Effects of Flame Retardant (FR1), Blowing Agent (BA1) on Density, Structure and Heat Loss

PUFs CHARACTERISTICS - POP-2								
CODE	BA1	FR1	Height	Density	Heat Loss	Rigidity	Structure	Remarks
GUN 13	5	w.o.	107	110-143	0.2286 @ 28	semi rigid	fine	tiny pores
		w.	117	111-120	0.2357 @ 28	semi rigid	fine	tiny pores
	10	w.o.	102	178-181	0.2750 @ 24	semi rigid	fine	tiny pores
		w.	115	115-126	0.2320 @ 25	semi rigid	fine	tiny pores
	20	w.o.	90	201-212	0.4100 @ 20	semi rigid	course	small pores
		w.	118	123-140	0.3143 @ 21	semi rigid	fine	tiny pores
	30	w.o.	83	145-152	0.3619 @ 21	semi rigid	coarse	large pores
		w.	97	152-154	0.2609 @ 23	semi rigid	fine	tiny pores
	40	w.o.	82	127-204	0.2913 @ 23	semi rigid	coarse	large pores
		w.	100	145-154	0.2591 @ 22	semi rigid	course	small pores
	50	w.o.	96	145-146	0.4211 @ 19	semi rigid	coarse	large pores
		w.	92	171-176	0.4333 @ 18	semi rigid	course	small pores

Height ... (mm.)
Water ... (ml.)
Density ... (kg/m^3)

Heat Loss ... Temp.(°C)/time(min.)
(temperature drop per minute)

w.o. ... without flame retardant
w. ... with 10g. FR1

Table 2C. Effects of Flame Retardant (FR1), Blowing Agent (BA1) on Density, Structure and Heat Loss

PUFs CHARACTERISTICS - POP-2								
CODE	BA1	FR1	Height	Density	Heat Loss	Rigidity	Structure	Remarks
GUN 31	5	w.o.	52	250-265	0.4231 @ 26	rigid	coarse	small pores
		w.	55	263-264	0.2083 @ 24	semi rigid	very fine	
	10	w.o.	70	153-202	0.2640 @ 25	semi rigid	coarse	small pores
		w.	62	219-228	0.2190 @ 21	semi rigid	fine	tiny pores
	20	w.o.	76	161-184	0.3130 @ 23	semi rigid	coarse	small pores
		w.	80	172-200	0.2667 @ 21	semi rigid	fine	tiny pores
	30	w.o.	59	262-281	0.2371 @ 28	semi rigid	coarse	large pores
		w.	73	215-217	0.2071 @ 28	semi rigid	fine	tiny pores
	40	w.o.	63	260-271	0.2519 @ 27	semi rigid	coarse	large pores
		w.	70	236-242	0.2870 @ 23	semi rigid	fine	tiny pores
	50	w.o.	61	245-269	0.2348 @ 23	semi rigid	coarse	large pores
		w.	64	292-300	0.2182 @ 22	rigid	fine	tiny pores

Height ... (mm.)
Water ... (ml.)
Density ... (kg/m^3)

Heat Loss ... Temp.(°C)/time(min.)
(temperature drop per minute)

w.o. ... without flame retardant
w. ... with 10g. FR1

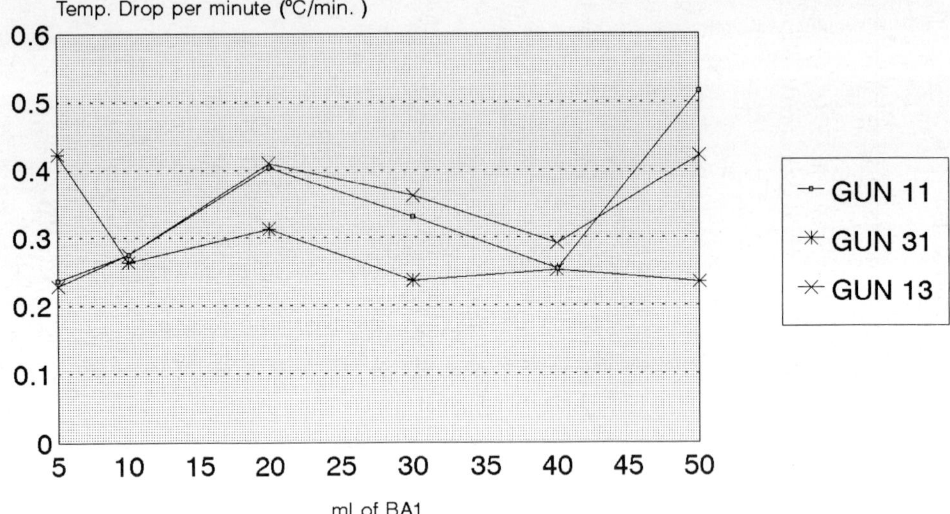

Figure 3. Heat Loss Characteristics of PUF (without flame retardant)

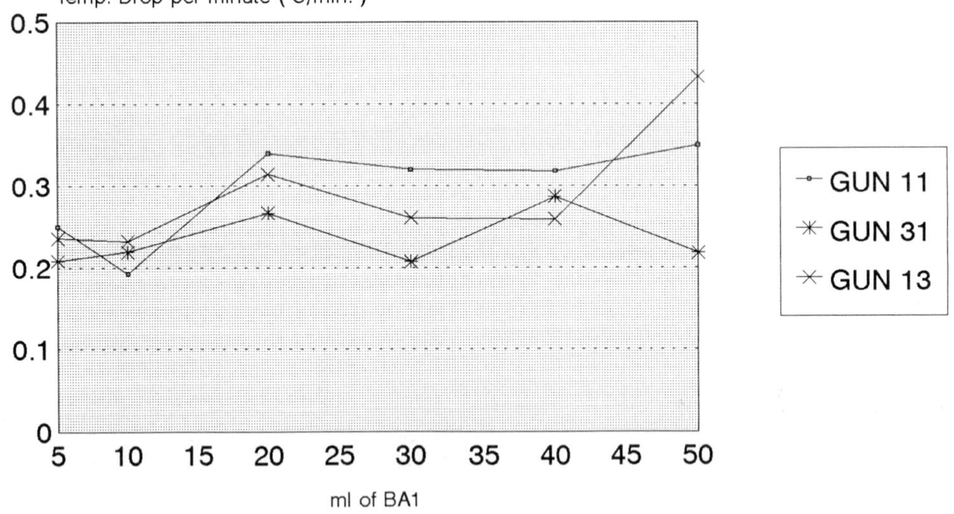

Figure 4. Heat Loss Characteristics of PUF (with FR1)

needed to ensure non-flammability was found to be 10g per 100g of sample. This ratio was used throughout the various formulations.

In general FR1 was found to increase the hardness or the rigidity of foams as compared to those with BA1 but without FR1. The structures of the foams were also found to be changed in the presence of FR1. The cell structure decreased and became more regular with the addition of FR1.

The additions of FR1 to the various formulations resulted in a decreased in the heat loss compared to those without FR1. These results are better illustrated in Figures 1 to 4.

COST OF PALM OIL-BASED POLYOLS VS. PETROLEUM-BASED POLYOLS

The price of the polyols produced commercially would vary depending on the availablity of the petrochemicals. The prices of the petrochemical-based products are expected to increase in the future and so would the prices of the polyols. Our initial calculation indicated that the price of palm based polyols were very competitive to petrochemical based polyols. Since MDI is used in both formulations (petrochemicals and POP) any change in price would be irrelevant to the cost comparison. There is therefore a possibility that polyols from POP could be economically feasible in the future.

CONCLUSION

These studies indicate that palm oil and palm oil products can be used to prepare polyols. The hydroxyl values of these polyols could be varied by varying the formulations used. The range of hydroxyl values obtained indicate that various types of polyurethanes foams ranging from flexible to rigid could be prepared from these polyols. These indications were technically confirmed.

When these polyols were used to manufacture polyurethane foams, various types of foams were produced. The characteristics of the foams were found to be influenced by the types of additives used, such as blowing agent, flame retardants and others, allowing tailor-made products to be manufactured.

ACKNOWLEDGEMENT

The authors of this paper would like to thank the Director General of PORIM for his permission to present this paper at the 3rd. International Conference on Frontiers Polymers and Advance Materials, 16 to 20 January, 1995. Shangri-La Hotel Kuala Lumpur.

REFERENCES

1. George Woods (1987). The ICI Polyurethanes Book. Publ: John Wiley & Sons, New York.
2. Hazimah A. Hassan, Yeong Shoot Kian & Salmiah Ahmad (1993). Palm-based polyols for polyurethanes foams. Proceedings, PORIM's International Palm Oil Conference, 1993, Hotel Istana, Kuala Lumpur.

RADIATION VULCANISATION NATURAL RUBBER LATEX AND ITS PROPERTIES[1]

Wan Manshol bin W. Zin

Malaysian Institute For Nuclear Technology Research (MINT)
Bangi, 43000 Kajang, Malaysia

ABSTRACT

In natural rubber latex vulcanisation, the use of gamma radiation helps to simplify the process, produces cleaner vulcanised natural rubber latex and hence dipped latex products. The vulcanisates of radiation vulcanised natural rubber latex (RVNRL), meet the current specifications required for dipped products. RVNRL films vulcanisates have the tensile strength of about 28 MPa and an elongation at break of about 1000%. Incineration of the film vulcanisates gave a detectable level of sulphur of about 0.11 ppm and an ash residue of less than 1%. Products from RVNRL are free from chemical-induced allergies, hence user-friendly and environment-friendly.

INTRODUCTION

Radiation vulcanisation of natural rubber latex (RVNRL) is a technology of radiation induced crosslinking of cis-polyisoprene in an aqueous dispersed phase. Evaporation of

[1] *Keywords : Radiation, Vulcanisation, Latex*

water from the irradiated latex brings about adhesion of rubber particles and consequently the formation of homogeneously crosslinked rubber films.

A schematic representation of natural rubber latex vulcanisation by gamma irradiation is given in Figure 1. The attraction lies in the simplicity of the process. It requires no curatives such as sulphur, zinc oxide and accelerators. The use of a sensitiser in the preparation helps to increase the efficiency of the vulcanisation process and makes it more attractive to users. No special latex is required, the normal high ammonia latex concentrate can be used for the preparation. Moreover, RVNRL can be processed into dipped products according to the standard technique available. RVNRL does not only provides an alternative in natural rubber latex vulcanisation, but also , the results of the research studies carried out this far show that RVNRL produces prevulcanised natural rubber latex with the key characteristics which make it particularly suitable for latex dipped products manufacturing at this point in time. It is most suitable for use in the relevent industries wanting to comply with regulations or to overcome the problem related to the use of latex dipped products as listed below:

* nitrosamines regulations
* accelerator induced allergies
* protein induced allergies
* zinc in factory effluents
* disposal of used products - acid combustion gases.

RVNRL having the advantages[1,2,3] it has to offer may promote it as a frontier material to related industries. The present studies involved in the scale up preparations of RVNRL and its properties, plus other advantages possessed by RVNRL which may be exploited by latex dipped products manufacturers.

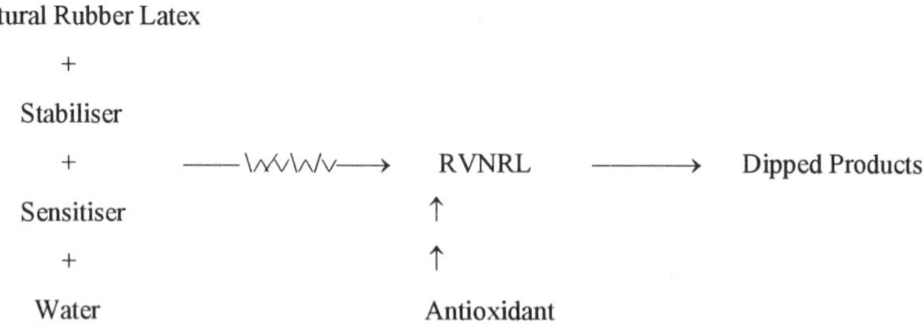

Figure 1. Schematic Representation of RVNRL Preparation

MATERIALS AND METHOD

Natural Rubber Latex

Locally available concentrated natural rubber latices of high ammonia type were used in carrying out the research work. Before further work, the latices were analysed for the total solid content (tsc) and dry rubber content (drc). The mechanical stability test, pH reading and viscosity reading were recorded.

Sensitiser

A sensitiser was used to increase the efficiency of the vulcanisation process and to reduce the dose required for producing RVNRL of maximum tensile strength. Among the sensitisers studied, n-butyl acrylate (n-BA) was found to be the most effective in reducing the dose required for producing RVNRL of optimum tensile strength[4].

Stabiliser

Certain sensitisers were found to affect the stability of certain latices. These effects were more pronounced when the sensitiser used was n-BA. To overcome the problem, the use of a stabiliser was thus a prerequisite. Potassium hydroxide solution of 10% concentration, and potassium laurylic acid were employed for the purpose.

Formulation

A typical latex formulation for RVNRL preparation is given in Table 1. The sensitiser, stabiliser and water were first prepared into an emulsion before slowly adding into the latex with continous stirring. Once the addition of the emulsified materials were completed, the latex mixture was left stirring for a few hours. It was then transferred into 25 litres capacity jerry-can plastic containers and irradiated with gamma rays from a cobalt-60 source to a dose of 12.5 kGy[4]. After irradiation, at which stage the latex is now known as RVNRL, was tested and evaluated for its properties.

Table 1. Latex Formulation

Materials	Amount, pphr	Part by Weight, g
NR Latex (60% DRC)	100.00	167.00
Stabiliser	0.02	0.02
Sensitiser	5.00	5.00
Water	Add to 50% DRC	26.40

PROPERTIES OF RVNRL

The RVNRL was evaluated for its viscosity, mechanical stability test (MST), pH reading, ammonia content and total solid content.

Viscosity measurements were carried out at 25°C using a Brookfield Digital Viscometer Model LVTD, and a spindle number 18 operated at 30 rpm was employed for the purpose.

The mechanical stability tests (MST) were carried out using a Klaxon Natural Rubber Latices Mechanical Stability apparatus, at a speed of 14 000 +/- 200 rpm.

The pH readings were determined using a Schott Gerate pH-Meter Model CG820. Other properties such as total solid content and ammonia content were determined using the standard techniques available.

PROPERTIES OF RVNRL FILMS VULCANISATES

Physical Properties

The physical properties of RVNRL were determined from cast film vulcanisates and coagulant dipped film vulcanisates.

RVNRL dipped films were prepared by using glass plates. Clean glass plates were first dipped into a 20% solution of calcium nitrate in industrial spirit, dried in an air convection oven at 70°C before dipping into the RVNRL at room temperature, followed by drying again in an air convection oven at 70°C until transparent.

For the preparation of cast films, clear glass plates with strips of glass glued to all the sides were cleaned. A known quantity of RVNRL was then strained onto the plates. These were then left to dry at room temperature.

All films were subjected to leaching in distilled water overnight at ambient temperature, followed by drying at room temperature before the films were given heat treatment in an air convection oven at 70°C for one hour. Five dumbbell-shaped test pieces were cut from each sample according to BS 6746. The physical properties of the vulcanisates were determined as per relevent ASTM standards.

Ash Content

The ash contents were determined by heating a known weight of RVNRL film vulcanisate prepared by coagulant dipping method. A known weight of film was cut into small milligram pieces, placed in a porcelain crucible and then heated in a furnace until a constant weight of ash was obtained. The percentage of the weight of the sample before and after heating was taken as the ash content. For comparison the experiment was repeated on five commercially available gloves produced from sulphur prevulcanised natural rubber latex.

Sulphur Content

The sulphur content was determined by a private laboratory, Core Laboratories (Malaysian) Sdn. Bhd. For the purpose of this study, the sulphur contents analysis were done on both the finger cots made from RVNRL and a commercially available finger cots made from sulphur prevulcanised natural rubber latex.

Soluble Protein Content

Soluble protein analysis of RVNRL film vulcanisates were carried out using modified Bradford-Lowry Method with precipitation of proteins using biorad DC protein assay. Absorption was measured at 750nm using a spectrophotometer. The results were calibrated against Bovine Serum Albumin (BSA) standard. For the purpose of this study, the level of soluble protein content in the RVNRL vulcanisates subjected to different leaching times were determined.

RESULTS AND DISCUSSIONS

To date the preparation of RVNRL is being upgraded from a laboratory scale of 75 kg per batch to a semi pilot scale of 500 kg per batch. The RVNRL obtained was found to be of consistent properties. These are given in Table 2. An increase of about 3% in the total solid content of the latex after the irradiation stage was determined and this was probably due to the polimerisation of the sensitiser. The ammonia content of RVNRL was determined to be lower than its latex in the unvulcanised stage, i.e. approximately 0.7%. This, however, did not seem to give any negative effect either to its alkalinity or to its mechanical stability time (MST). In fact, the MST of RVNRL was always on the high side, being greater than 2000 seconds. The reason for the high MST is not known for certain, as similar MST increase was observed upon an irradiation of unformulated latex. Although RVNRL was found to possess high MST, however, in dipped films preparations, as well as laboratory scale products preparations, it was found that a solution of 20% calcium nitrate in industrial methylated spirit was suitable to be used as a coagulant. The physical properties of examination and surgical gloves made from RVNRL meets the international standards. The tensile strength, modulus and elongation at break were found similar to the values obtained from dipped films prepared on glass plates, these are given in Table 3.

Table 2. Properties of RVNRL

Total Solid Content,%	Ammonia Content,%	pH Reading	MST secs.	Brookfield Viscosity,Spindle 18/30rpm/25°C
54	0.4	9-10	>2000	30-35

The low ammonia level is certainly an advantage to the latex dipped products industries wishing to use RVNRL. Disposed effluent will not only be free of zinc oxide free but also low in ammonia concentration.

Table 3. Properties of RVNRL Film Vulcanisates

Films Vulcanisates	Tensile Strength,MPa	Modulus @ 500%	Modulus @ 700%	Elongation @ Break,%
Coagulant Dipped	28.0	2.1	6.6	1050
Cast	25.0	1.7	4.7	1000

From the results given in Table 3, it was learned that the tensile properties of RVNRL film vulcanisates are influenced by the methods of films preparations. Higher tensile properties are shown by the film vulcanisates prepared by coagulant dipping method compared to the films prepared by cast film method treated under similar conditions.

The higher tensile strength exhibited by coagulant dipped film vulcanisates was said to be attributed by the inter particles crosslinking through the calcium ions[5]. The low moduli exhibited by RVNRL film vulcanisates make RVNRL useful for the production of latex dipped goods such as examination and surgical gloves, balloons and finger cots of a better quality than those available in the market today. For example, the gloves made from RVNRL are softer. Therefore wearing the gloves for an extended period may be less tiring to the fingers especially when attending to a more delicate work. Moreover, the balloons have improved air retention property and are easier to blow to the intended sizes. The use of any suitable pigment produces products of brighter and more attractive colours.

Table 4 and Table 5 show the results of the sulphur content and ash content analysis respectively. The sulphate and sulphide contents of RVNRL vulcanisates are lower than that of sulphur prevulcanised latex vulcanisates. The low level of sulphate and sulphide detected from RVNRL vulcanisates probably come from natural resources.

Table 4. The Concentration of Sulphate and Sulphide in RVNRL Vulcanisate

Sample	RVNRL	SPVL
Sulphate as SO4, ppm	0.11	0.21
Sulphide as S, ppm	< 0.001	0.007

Note: SPVL is sulphur prevulcanised natural rubber latex.

The low sulphur content exhibited by RVNRL vulcanisates is an advantage not only in the production of medical products, but also other products such as finger cots for the electronic components assembling plants operators, since any sulphur from the finger cots may be the source of corrosion of the copper parts of electronic components.

Table 5. The Percentage Ash Contents of Commercial and RVNRL Gloves

Gloves Samples	Ash Content,%
Commercial Gloves A	2.5
Commercial Gloves B	2.0
Commercial Gloves C	2.7
Commercial Gloves D	2.2
Commercial Gloves E	2.0
RVNRL Gloves	0.7

As shown in Table 5, the commercial gloves made from conventionally vulcanised natural rubber latex have an average ash content of about 2.4% whereas an ash content of less than 1% determined from the RVNRL gloves. The low level of ash and the low level of toxic gases liberated during its incineration allow for the safer disposal of dipped products from RVNRL.

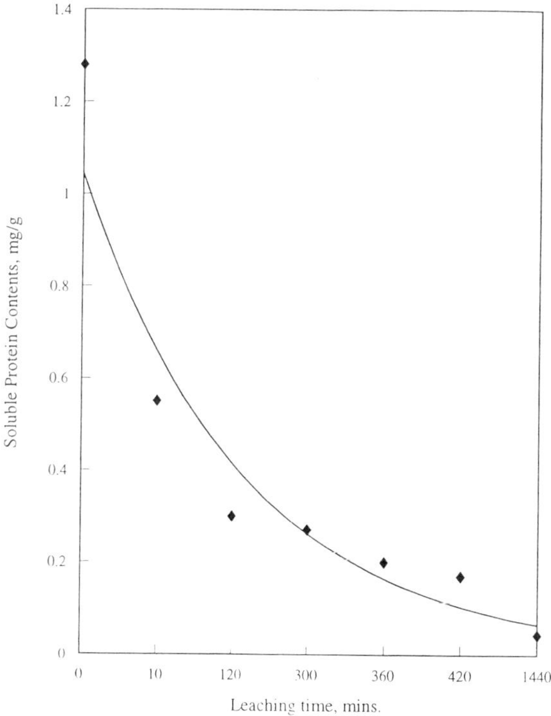

Figure 2. Soluble Protein Content of RVNRL Film Vulcanisate.

At present there is a lot of fuss especially in the USA on the role of proteins in latex as the main cause of on increasing number of cases of anaphylactic shock and other responses elicited by contact with rubber gloves. In view of this, studies have been initiated to determine the amount of soluble proteins from RVNRL dipped film vulcanizate.

The results of the studies carried out this far indicate that the amount of soluble protein determined from the RVNRL film vulcanizate was influenced by the leaching time at a maximum leaching time of 24 hours, refer to Figure 2. The longer the leaching time the lower the soluble protein detected.

Clinical tests are being carried out to identify any possible allergy that may be caused by the proteins detected in RVNRL vulcanisates. The results will be reported in the near future.

CONCLUSIONS

Gamma irradiation technology can be used in natural rubber latex vulcanisation. The prevulcanised natural rubber latex has the required properties and can be used in dipped latex goods manufacturing. RVNRL uses no toxic chemical in its preparation, possess the advantages that other prevulcanised latex may not be able to offer and are ready to be fully exploited by latex dipped products manufacturers.

REFERENCES

1. Nakamura, A., Ikarashi, Y., Tsuchiya, T. and Kaniwa, M. (1990). Proc. of the Int. Sym. on RVNRL, JAERI-M, Takasaki, Japan 89-228.

2. Pendle, T. D. (1993). A paper presented at the National Executive Managemant Seminar on RVNRL, Kompleks PUSPATI, Bangi, Malaysia.

3. Makuuchi, K. (1993). A paper presented at the National Executive Management Seminar on RVNRL, Kompleks PUSPATI, Bangi, Malaysia.

4. Wan Manshol, W.Z., Norjanah, M., Jusoh, H., Wan Khairul Amalina, W.M.N. and Zulkifli, J. (1993). The Preparation of RVNRL Using Malaysian-Produced Latices.Radiat. Phys. Chem.,42, 101-105.

5. Saito, T., Yoshii, F., Makuuchi, K and Ishigaki, I. (1990). Proc. of the Int. Sym. on RVNRL, JAERI-M., 89-228.

COMMUNICATING WITH RESPONSIVE INTELLIGENT MEMBRANES

A. Mirmohseni, W. E. Price, C. J. Small, C. O. Too, G. G. Wallace and H. Zhao

Intelligent Polymer Research Laboratory,
Chemistry Department,
University of Wollongong,
Northfields Avenue,
Wollongong,
NSW 2522,
Australia.

INTRODUCTION

At present, membrane technology is being applied in a wide range of industries from food and chemical processing to biotechnology and waste management. The current (1995) global market has been estimated[1] to be in excess of US$6-9 billion; of which 30% is attributed to membrane materials alone. These traditional membrane products are quite sophisticated. For example, polysulfone asymmetric membranes[2] have a dense thin skin side that is supported by a more porous open structure; thus high selectivity can be achieved without sacrificing high flux.

Current commercial membranes, however, are manufactured from materials that have properties which are permanently set during their production. Thus these membranes are "non-intelligent" in that they cannot be made to respond to their environment. This can lead to inefficiencies such as loss of selectivity, or loss of flux due to concentration polarisation and/or fouling.

This paper presents new and exciting membrane materials based on conducting electroactive polymers (CEPs) which have the adaptive properties which enable them to meet the above challenges and form a new generation of products. These materials can be fabricated into responsive intelligent membranes. The necessary hardware, applications and mechanisms are also discussed.

EXPERIMENTAL

All reagents mentioned in the text were used as received except for pyrrole which was redistilled prior to use. The 3-carboxy-4-methylpyrrole was synthesised in our laboratories.

A PAR 173 galvanostat/potentiostat (Princeton Applied Research) was used for membrane electrosynthesis; where the constant current method was used in general except when constant potential was used for electro-copolymerisation. Transport studies were performed using the PAR 173 in conjunction with in-house built potential pulse signal

Polymers and Other Advanced Materials: Emerging Technologies and Business Opportunities. Edited by P. N. Prasad *et al.*, Plenum Press, New York, 1995

generator and electrochemical cells and systems (Figure 1). Quantitative chemical analysis of inorganic cations was carried out using a Perkin-Elmer Model AA/AZ 551 Atomic Absorption spectrometer. Organic anions were quantified by HPLC using a Beckman 114 solvent delivery module, a variable wavelength spectrophotometric detector (Linear) and a Waters μBondapak C_{18} column (3.9 x 300 mm).

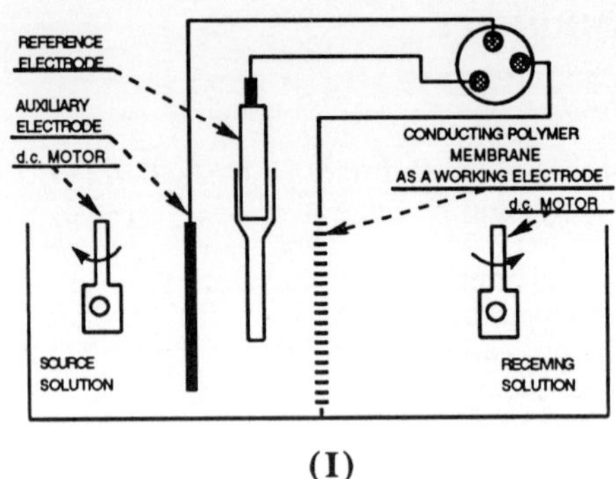

(I)

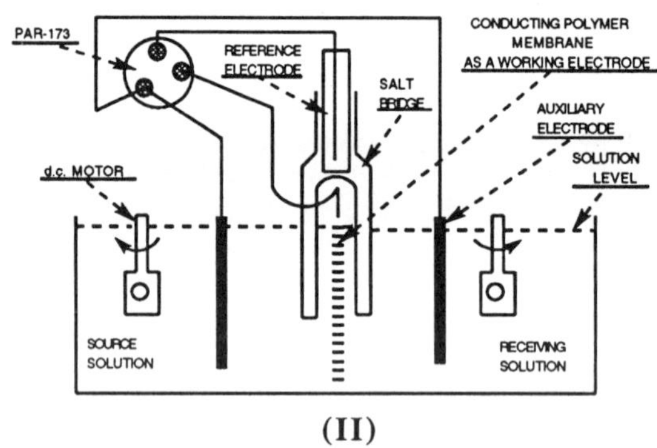

(II)

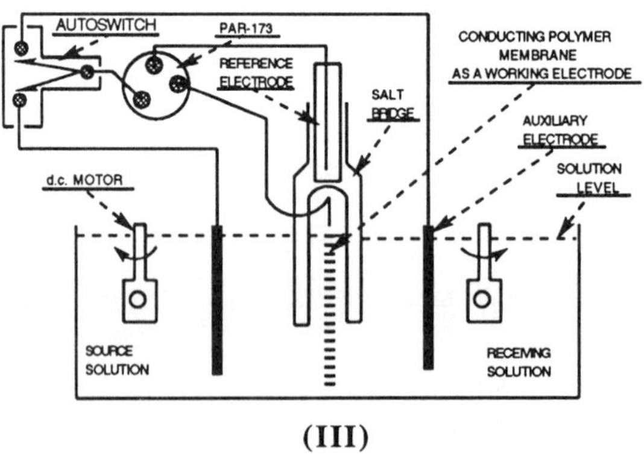

(III)

RESULTS AND DISCUSSION

The electrochemical synthesis of a CEP such as polypyrroles can be depicted by:

$$\text{pyrrole}(R_1, R_2)\ \xrightarrow[A^-]{-e}\ \left[\text{pyrrole}(R_1, R_2)\right]_n^{+}\ A^- \qquad (1)$$

where A^- is the counterion to be incorporated into the polymer. Therefore, the properties of the CEP can be tailored to intended applications by varying the nature of the substituents R_1 and/or R_2, and/or varying the nature of the counterion A^-. Electro-copolymerisation can also be used to achieve desired CEP characteristics in the merging of the desired properties of the two co-monomers together to form new CEPs, for example:

$$\text{pyrrole}(R_1, R_2) + \text{pyrrole}\ \xrightarrow[A^-]{-e}\ \left[\left(\text{pyrrole}(R_1, R_2)\right)_n \left(\text{pyrrole}\right)_m \right]^{+} A^- \qquad (2)$$

Similar considerations are also valid in the synthesis of CEP-hydrogel composites where the CEP is electrosynthesised within the hydrogel matrix.

The addressable properties of these new membrane materials can be accessed in-situ by using suitable electrochemical transport cells and systems. The following results and discussion will demonstrate the feasibility and applicability of these responsive intelligent membranes. The synthesised membranes that will be discussed are given in Table 1.

As stated above, communication with responsive intelligent membranes will depend upon the electrochemical cells and systems as well as the electrical stimuli (such as pulsed potentials) used. When a more efficient system is used, then better selectivity can be achieved even at higher flux because transport of species across these stand alone membranes is electrochemically facilitated. This will be evident from later discussions on

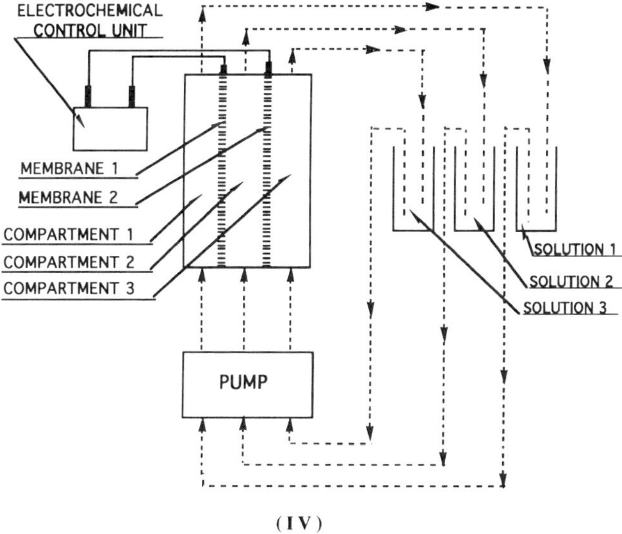

Figure 1. Electrochemical cells and systems for membrane separations

Table 1. Responsive Intelligent Membranes

Membrane	Abbreviation	Dry state conductivity ($S\ cm^{-1}$)	Tensile strength (MPa)
Polypyrrole-benzenesulfonate[3]	PPy-BS	19-20	17-23
Polypyrrole-paratoluenesulfonate[3]	PPy-pTS	90-110	70-80
Polypyrrole-4-ethylbenzene sulfonate[3]	PPy-EBS	90-110	60-70
Polypyrrole-2-mesitylenesulfonate[3]	PPy-MS	50-70	36-47
Polypyrrole-1,3-benzenedisulfonate[3]	PPy-BDS	47-70	40-55
Polypyrrole-1,5-naphthalene disulfonate[3]	PPy-NDS	50-70	40-50
Polypyrrole-poly(vinylsulfonate)[4]	PPy-PVS	14-15	70-90
Poly(3-carboxy-4-methylpyrrole)-paratoluenesulfonate	PCMP-pTS	2	6
Copolymer of pyrrole with 3-carboxy-4-methylpyrrole	PPy/PCMP-pTS	95	65
Composite of poly-acrylamide gel/polypyrrole-nitrate[5]	PAG/PPy-NO_3	6	n.a.

Notes: n.a. = not available

mechanisms of transport. The range of hardware under consideration here is given in Figure 1. The main feature for systems I to III is that the CEP membrane is installed as the working electrode which divides the source solution from the receiving solution. The main differences between systems I to III are discussed below.

(i) System I has the auxiliary and reference electrodes only in the source side of the cell. The electrodes are connected to the potential pulse generator in a three-electrode configuration. This directs electrochemical control of the membrane to the source side.

(ii) System II has auxiliary and reference electrodes in both the source side and receiving side of the cell. In this case the two auxiliary electrodes are connected simultaneously to the potential pulse generator. In addition, the reference electrode is in contact with both compartments of the cell via a two-pronged salt bridge. This configuration directs electrochemical control of the membrane simultaneously to both compartments of the cell.

(iii) System III is similar to system II except that the automatic switch for the two auxiliary electrodes allows control of the membrane to be effected independently between the two compartments. For example, the anodic pulse can always be directed to the source side whereas the cathodic pulse can always be directed to the receiver side, or vice versa.

A novel system[6] which is more electrochemically efficient and easier to scale up is shown as Figure 1(IV). The main features are that this is a flow through system with two CEP membranes acting as electrodes in a two-electrode configuration. Thus feed solution is pumped through the middle compartment of the cell between the two membranes whilst pulsed potentials are applied across the two membrane electrodes. In this way electrochemically facilitated transport occurs across both membranes because, when potential pulses are applied, one membrane acts as the anode whilst the other acts as the cathode in an alternate fashion. This results in two receiver streams.

Electrochemical control[7] of these responsive intelligent membranes can be summarised as:-

$$\left[\text{polypyrrole}(R_1,R_2)\right]_n^{+}\,A^- \underset{-e}{\overset{+e}{\rightleftharpoons}} \left[\text{polypyrrole}(R_1,R_2)\right]_n^{0} + A^- \qquad (3)$$

where the counterion (A^-) is mobile, and/or

$$\left[\text{polypyrrole}(R_1,R_2)\right]_n^{+}\,A^- + M^+ \underset{-e}{\overset{+e}{\rightleftharpoons}} \left[\text{polypyrrole}(R_1,R_2)\right]_n^{0}\,A^-M^+ \qquad (4)$$

where the counterion (A^-) is less mobile or immobile, and M^+ is a cation from the electrolyte solution.

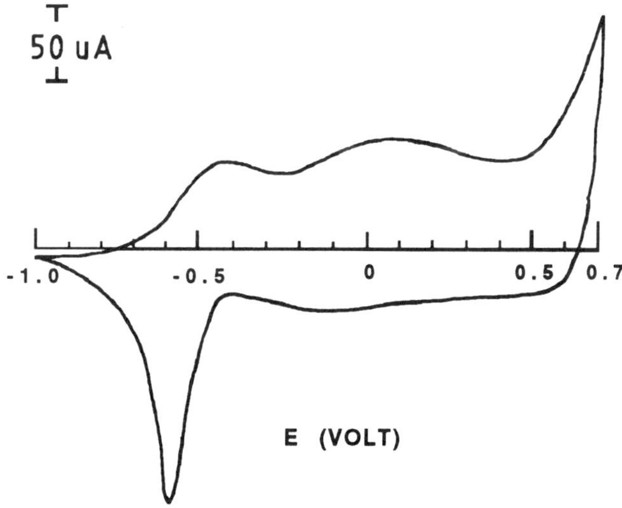

Figure 2. Cyclic voltammogram of polypyrrole-paratoluenesulfonate in KCl. Supporting electrolyte = 0.2 M KCl. Scan rate = 100 mV s^{-1}

Thus Equation 3 shows the anion exchange character of the CEP whereas Equation 4 shows the cation exchange character. Furthermore, in general, the CEP becomes more hydrophobic when it is reduced and less hydrophobic when it is oxidised.

PPy-pTS is a typical CEP which shows cyclic voltammetric responses (Figure 2) of the mechanisms above. Its qualities as a selective membrane are demonstrated[7] in Figure 3 using electrochemical system I. The notable feature here is that no transport of ions is evident until the pulsed potential is applied, and transport ceases when the potential is switched off. Furthermore, K^+ and Na^+ are selectively transported across the membrane. In this case the selectivity factor, flux of K^+ : flux of Na^+, is 3.1.

The nature of the CEP membrane can also be modified by incorporating different counterions during synthesis of the membrane[3]. The effect on membrane transport properties was tested using a K^+/Ca^{2+} mixture. The results are tabulated in Table 2 which shows that a selectivity factor, flux of K^+ : flux of Ca^{2+}, as high as >240 was obtainable but that it decreases with increasing size of the counterion incorporated at membrane synthesis. High selectivities are obtained because K^+ and Ca^{2+} compete for ion exchange sites and the incorporation of the smaller K^+ is favoured over the larger Ca^{2+}.

In the case of a PPy-PVS membrane, no transport at all was obtained using electrochemical system I. This was possibly due to the lower conductivity of PPy-PVS leading to lack of efficient electrochemical control. Furthermore, PPy-PVS may have a stronger affinity for cations (Equation 4) and so transport is inhibited. However, transport was obtained by use of electrochemical system III. The results[4] for the K^+/Na^+ system show that higher fluxes (as high as 5.1×10^{-9} mol cm^{-2} s^{-1}) are obtained (Figure 4); when compared to Figure 2 which was obtained from electrochemical system I. The selectivity factor, flux of K^+ : flux of Na^+, in this instance, was lower at 1.6.

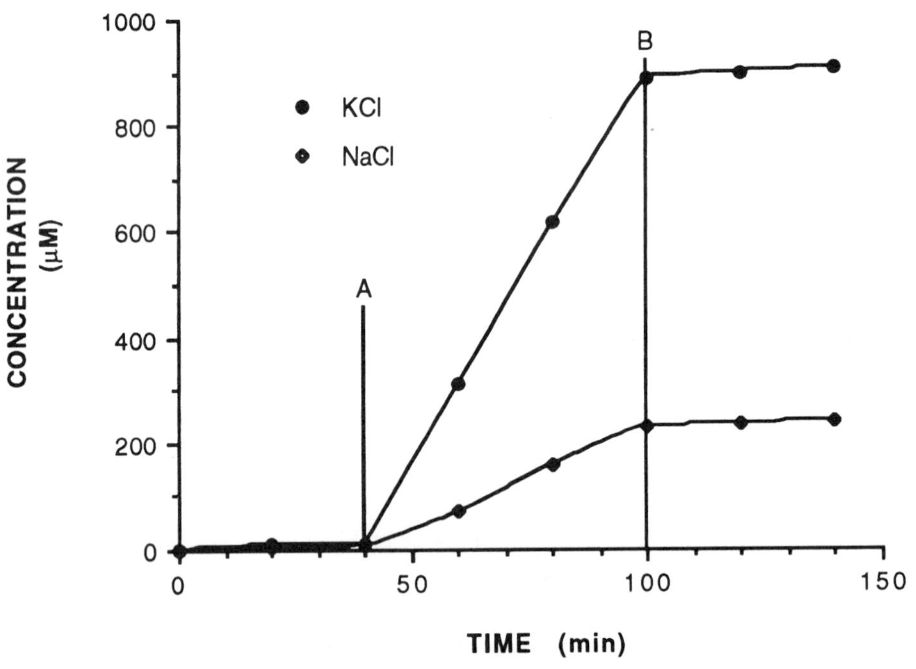

Figure 3. Selective separation of K^+ and Na^+ across a PPy-pTS membrane. Source solution = 0.1 M KCl + 0.1 M NaCl. Receiver solution = Milli-Q deionised water Pulsed potential limits = +0.70 V and -1.00 V (vs Ag/AgCl). Pulse width = 50 s Pulsed potentials applied during period labelled A to B.

Table 2. Effect of the counterion employed at synthesis on the membrane separation characteristics[3]

Membranes:	PPy-BS	PPy-pTS	PPy-EBS	PPy-MS	PPy-BDS	PPy-NDS
Flux K^+ (mol cm^{-2} s^{-1})	$2.4x10^{-10}$	$1.9x10^{-9}$	$2.9x10^{-9}$	$3.1x10^{-9}$	$2.6x10^{-9}$	$3.5x10^{-9}$
Flux Ca^{2+} (mol cm^{-2} s^{-1})	$<10^{-12}$	$2.0x10^{-11}$	$4.9x10^{-11}$	$5.4x10^{-11}$	$6.0x10^{-11}$	$9.4x10^{-11}$
Flux K^+/ flux Ca^{2+}	>240	94	59	57	43	37

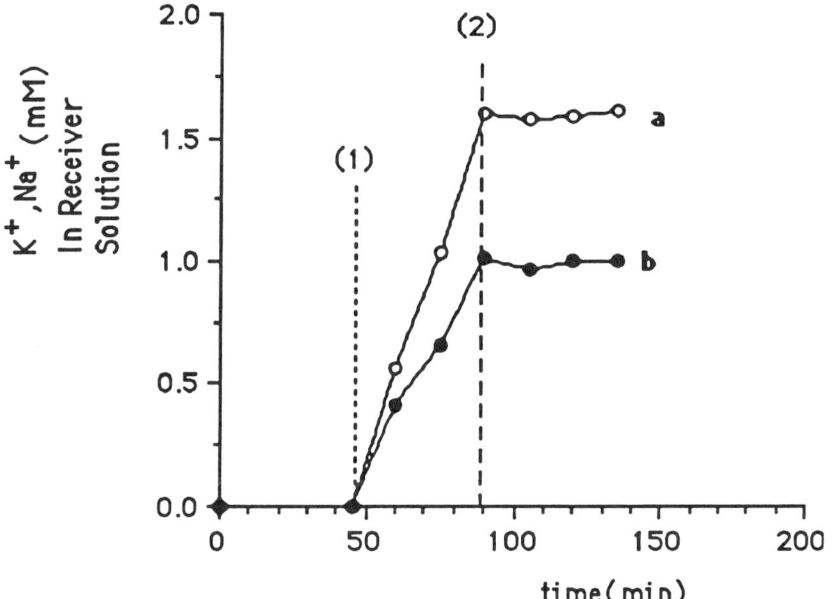

Figure 4. Selective separation of K^+ and Na^+ across a PPy-PVS membrane. Source solution = 0.1 M KCl + 0.1 M NaCl. Receiver solution supporting electrolyte = 0.01 M $MgCl_2$. A square wave potential was applied during period labelled 1 to 2: Source side = -0.60 V, Receiver side = +0.70 V, Pulse width = 30 s

Apart from varying the counterion for CEP synthesis in order to vary the resultant dynamic properties, the nature of the monomer itself can be varied by substitution. Thus PCMP-pTS can be electrosynthesised. Reference to Table 1, however, will quickly show that this particular CEP has low conductivity and low mechanical strength which renders it unsuitable as a stand alone membrane material. Consequently the copolymer of pyrrole with 3-carboxy-4-methylpyrrole was electrosynthesised. This copolymer has suitable properties required of responsive intelligent membranes (Table 1). It was tested on an industrially significant system where K^+ was present in a large excess of Mn^{2+}; for example as might occur in the production of electrolytic manganese dioxide. As can be seen from the results (Figure 5), selective electrochemically controlled transport of K^+ (100 ppm) in Mn^{2+} (1 M) is achievable. In this case, the concentration ratio of Mn^{2+} : K^+ was reduced from 54.9:1 in the source side to 2.9:1 in the receiver side. The flux of Mn^{2+} transport was 3.36 x 10^{-10} mol cm^{-2} s^{-1} and that of K^+ transport was 1.61 x 10^{-10} mol cm^{-2} s^{-1}.

Apart from the electrochemically facilitated transport of inorganic species, organic species can also be selectively transported[8]. The results in Table 3 were obtained using electrochemical system III. This shows the selective transport of small organic anions, compared with pTS^-, across a PPy-pTS membrane. As can be seen, selectivity factors as high as 27.5 can be obtained. The membrane also gave different selectivity factors for 4-sulfobenzoic acid as compared with 3-sulfobenzoic acid; thus showing that it was sensitive even to substituent position on the benzene ring of the species to be transported.

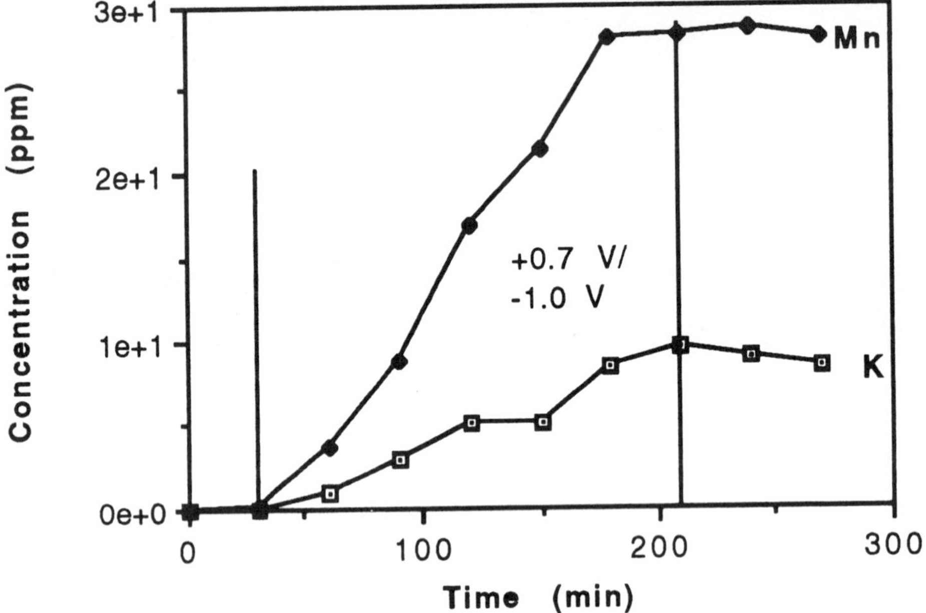

Figure 5. Selective separation of K^+ and Mn^{2+} across a PPy/PCMP-pTS copolymer membrane. Source solution = 1 M Mn^{2+} + 100 ppm K^+. Receiver solution = Milli-Q deionised water. Square wave potential limits = +0.70 V and -1.00 V (vs Ag/AgCl). Pulse width = 50 s

Finally, a demonstration of a novel dual CEP membrane system (electrochemical system IV)[6] is shown in Figure 6. Once again, no transport of K^+ species is observed without the application of pulsed potential stimuli. During electrical stimuli K^+ is transported across both membranes (1 and 2) at similar fluxes (15.5 x 10^{-9} mol cm^{-2} s^{-1} compared with 16.5 x 10^{-9} mol cm^{-2} s^{-1}). Thus permeate throughput is doubled and the electrochemical efficiency of the system is enhanced.

Table 3. Separation of a variety of organic anions from pTS^- across PPy-pTS membranes[8]

Anion	Average fluxes (10^{-9} mol cm^{-2} s^{-1}) Anion	pTS^-	Selectivity factor (Flux pTS^-/Flux anion)
Benzene sulfonate	0.76	0.87	1.14
4-hydroxy benzene sulfonate	0.17	0.47	2.76
4-sulfo- benzoic acid	0.08	0.22	2.75
3-sulfo- benzoic acid	0.07	0.33	4.71
1,3 benzene disulfonate	0.02	0.55	27.50

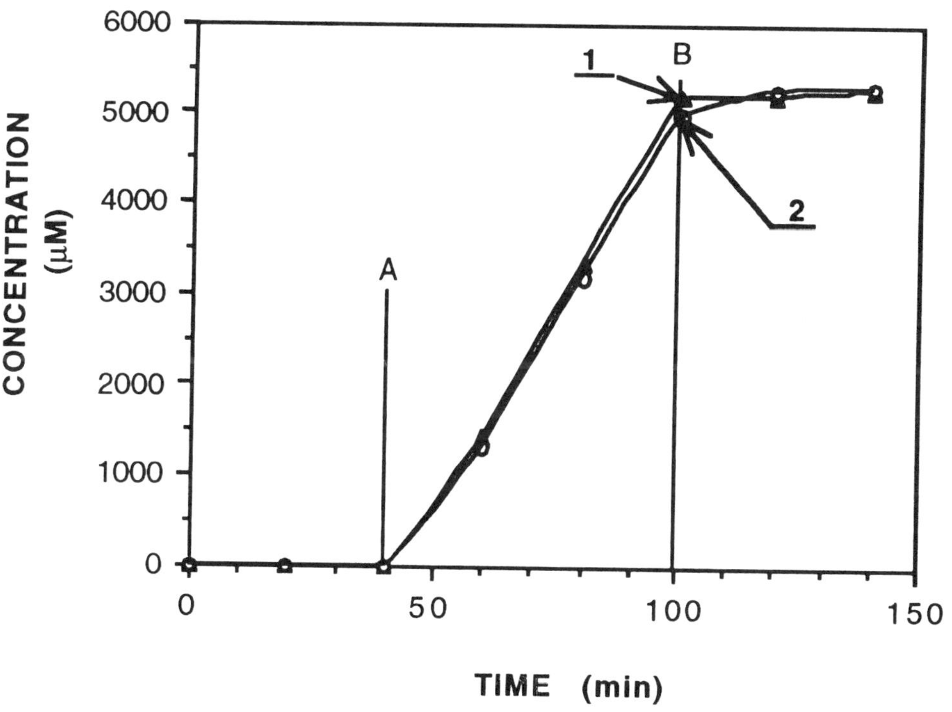

Figure 6. Electrochemically facilitated transport of K^+ across PPy-pTS membranes in a dual membrane flow-through cell. Source solution = 0.2 M KCl. Receiving solution = Milli-Q deionised water. Flow rate = 1.0 ml min^{-1}. Pulsed potential limits = +/- 1.5 V. Pulse width = 20 s. Pulsed potentials applied during the period labelled A to B. Curve 1 for membrane 1, and Curve 2 for membrane 2

CONCLUSIONS

It has been clearly demonstrated that with the proper cell design, systems and instrumentation it is possible to communicate with our responsive intelligent membranes. This communication allows reversible control of the membrane characteristics thus allowing the membrane to adapt to the separations environment and even impose conditions upon the feed species so that good separations can be effected. Selective separations of inorganic and organic species have been achieved. Furthermore, the potential for scale up, especially in industrial applications, has been realised by use of the dual CEP membrane system.

ACKNOWLEDGMENTS

Parts of the work presented in this paper were funded by an Australian Department of Industry, Technology and Commerce GIRD grant, and an Australian Government Targetted Institutional Links Program grant.

REFERENCES

1. A.J. Jones, Membrane and Separation Technology: the Australian Perspective 1986, Department of Industry, Technology and Commerce, Australian Government Publishing Service, Canberra (1987).

2. S. Loeb and S. Sourirajan, Sea water demineralisation by means of an osmotic membrane, saline water conversion II, Adv. Chemistry Ser. 38:117(1963).

3. H. Zhao, W.E. Price and G.G. Wallace, Effect of the counterion employed during synthesis on the properties of polypyrrole membranes, J. Membrane Sci. 87:47(1994).

4. A. Mirmohseni, W.E. Price and G.G. Wallace, Electrochemically controlled transport across conducting polymer composites - basis of smart membrane materials, Polymer Gels and Networks 1:61(1993).

5. K. Gilmore, A.J. Hodgson, B. Luan, C.J. Small and G.G. Wallace, Preparation of hydrogel/conducting polymer composites, Polymer Gels and Networks 2:135(1994).

6. D. Zhou, H. Zhao, W.E. Price and G.G. Wallace, Electrochemically controlled transport in a dual conducting polymer membrane system, J. Membrane Sci., in press (1994).

7. A. Mirmohseni, W.E. Price, G.G. Wallace, and H. Zhao, Adaptive membrane systems based on conductive electroactive polymers, J. Int. Mat. Syst. Struct. 4:43(1993).

8. A. Mirmohseni, W.E. Price and G.G. Wallace, Electrochemically controlled transport of small charged organic molecules, J. Membrane Sci., in press(1994).

DESIGN OF CERAMIC MATERIALS FOR CHEMICAL SENSORS WITH INTELLIGENT PROPERTIES

Enrico Traversa

Dipartimento di Scienze e Tecnologie Chimiche,
Universita' di Roma "Tor Vergata", 00133 Rome (Italy)

INTRODUCTION

Research in the field of materials has always been oriented towards the improvement of their performance and of their reliability. Recently, however, materials R&D has taken a new direction, searching for novel functions.[1] The development of materials with intelligent functions is a key point for the development of novel technologies, which should be environment and user friendly.[2] Intelligent materials are those materials which are able to modify and to adapt themselves to external changes, mimicking the behaviour of living organisms.[3] One class of materials which has outstanding potential for application as intelligent materials is ceramics, because they may have completely different properties.

Intelligent ceramics have potential applications in a wide range of engineering fields.[4] One of the most appealing uses of ceramics with intelligent functions is in the chemical sensor field, because of the problems still to be solved for the materials conventionally used in this field. For the maintenance of a safe environment, the easy and accurate detection of the minor components in the gas phase is fundamental. At present, very expensive and bulky analytical methods are mostly available for air pollutants' detection. The conductivity changes of semiconducting oxides, caused by the reaction of gases with the oxygen adsorbed on the oxide surface, offer an inexpensive method for monitoring hazardous gases.[5] The use of semiconducting oxides has been exploited for gas monitoring since more than 30 years,[6] but, for environmental monitoring, they have the disadvantages of insufficient gas selectivity, inability to detect very low gas concentrations, and changes in sensing properties caused by surface contamination.

One of the major problems also for the stability and reliability of humidity sensors is surface contamination.[7] In the case of humidity sensors made of porous ceramics, their surface conductivity decreases during prolonged contact with humid environments because of the formation of stable surface hydroxyl ions.[8] This leads to a drift in sensor resistance over time, which must be recovered through a so-called heat-cleaning treatment: the sensor is heated to a high temperature (about 400°C) to remove the adsorbed molecules.[9]

Different approaches have been attempted in order to overcome these problems.[10,11] One possible solution is the design of materials with novel detection mechanisms and intelligent properties. Interfaces between p- and n-type semiconducting oxides, expected to have novel functions different from those of each individual material, have been tested for gas[12] and humidity[13] sensors. Ceramic heterocontacts have shown interesting properties of selectivity towards CO[14] and characteristics which allow us to describe them as intelligent materials.[1] For humidity detection, p-n contacts have shown a different mechanism from those known for single-phase ceramics.[7]

Polymers and Other Advanced Materials: Emerging Technologies and Business Opportunities. Edited by P. N. Prasad *et al.*, Plenum Press, New York, 1995

The design of intelligent mechanisms can be achieved by the hybridization and/or the integration in simple structures of materials with differing properties. The interaction in the multiphase structures leads in general to the development of novel functions which are not possessed by the single phase materials. The materials in the ceramics class may have completely different properties, and thus offer outstanding potential for the development of intelligent functions.

In this paper, examples of intelligent ceramic materials for humidity and gas sensors studied by the author are reported, i.e. La_2CuO_4/ZnO p-n heterocontacts (for humidity and reducing gases), Au/ZnO Schottky barriers (for humidity), and sol-gel processed thin films of TiO_2 with 10 at% of K (for humidity), discussed in terms of their functions as intelligent materials. The intelligent functions of these systems are based on the multiphase interaction of materials having different properties, such as p-type/n-type semiconducting oxides, metal/ceramic, and conductor/insulator.

EXPERIMENTAL PROCEDURE

Materials

The heterocontacts were prepared by mechanically pressing sintered discs of the two oxides, La_2CuO_4 (p-type) and ZnO (n-type semiconductor). The compacts were prepared by pressing the powders at 20 MPa into discs 10 mm in diameter and 2 mm in thickness. Discs were sintered at 1100°C (at 1000°C for ZnO) for 5 h in air.[15] The relative density was 75% for La_2CuO_4 discs and 98% for ZnO discs. For electrical measurements, silver paste for La_2CuO_4 and aluminium paste for ZnO were used as ohmic electrodes.

ZnO pellets prepared in the same way were used to obtain Au/ZnO Schottky barriers. The pellets had an ohmic contact on one side, made of fired Al paste. On the other side, a gold electrode was mechanically pressed by a spring to the ZnO surface in a home-made test cell.[16] In both cases, open metal/oxide or p-type/n-type oxide interfaces were obtained, accessible to gases and to moisture.

TiO_2-based films with 10 at% of K (with respect to Ti) were prepared by means of a sol-gel method.[17] The matrix sol was prepared by mixing titanium isopropylalkoxide (TIPT) with isopropyl alcohol, distilled water, and an acid catalyst (HCl 37%). K ions were incorporated by adding to the titania sol an amount of CH_3COOK to a composition of 10 at% of K (with respect to Ti). For electrical measurements, prototype sensors were prepared by depositing the ceramic films on Al_2O_3 substrates with comb-type Au electrodes. The films were deposited by dipping in the above solution, using a withdrawal speed of 25 cm/min. The films obtained were dried at 100°C for one day, and then heat-treated at 300°C for 30 min.

Measurements

The humidity-sensitive electrical response of the specimens was analyzed in a chamber where both temperature and humidity were controlled. Different relative humidity (RH) values were obtained by controlling flow rates of dry and water-saturated air or N_2. Relative humidity was monitored using a commercial sensor (Mod. PCRC-11, Phys-Chem Research Co.), which gave results accurate to within ± 2%. For the open interfaces, dark current-voltage (I-V) characteristics were measured using a digital multimeter and a programmable d.c. source, controlled by a computer. High frequency (100 kHz) dark capacitance-voltage (C-V) characteristics were measured at various RH values. The electrical properties of the TiO_2-based films were evaluated at 40°C using electrochemical impedance spectroscopy (EIS). EIS spectra were collected in the frequency range 10^{-2} to 10^5 Hz by means of a frequency response analyzer.

For the sensing of reducing gases (CO and H_2), the heterocontacts were placed in a tube furnace at temperatures in the range 200-400°C, in environments at different gas concentrations, obtained by controlling the flow rates of dry air and of the gases tested. The I-V curves were carried out using a digital electrometer and a programmable d.c. source, controlled by a computer.

RESULTS AND DISCUSSION

La_2CuO_4/ZnO Heterocontacts

Figure 1 shows the I-V characteristics of La_2CuO_4/ZnO heterocontacts at room temperature in a nearly dry environment. The curve shows typical p-n diode behaviour, with a rectifying current under forward direction ($La_2CuO_4^+/ZnO^-$). Under high humidity, the forward current increased markedly, while reverse current remained at low levels, showing thus a highly rectifying character (Figure 2). The RH-sensitivity was bias-dependent, as shown in Figure 3, and thus externally tunable by changing the applied voltage. Very high RH-sensitivity, expressed as I_{wet}/I_{dry}, where I_{dry} is the current recorded in a nearly dry environment, was measured for the contacts. This was up to about 4 orders of magnitude at 2.5 V d.c. and 95% RH.

The RH-sensitivity of the heterocontacts was due to phenomena occurring at the p-n interfaces. Measurements carried out on the single pellets of La_2CuO_4 and ZnO confirmed that single oxides were insensitive to RH.[18] In fact, the RH-sensitivity of porous ceramics is due to water adsorption processes on their surface. Ionic-type porous ceramics react to moisture by decreasing their impedance due to (chemical and physical) water adsorption and/or capillary condensation. Dense pellets were used for the heterocontacts in order to hinder the ionic-type humidity-sensitive conduction mechanism of the ceramic materials.

The humidity detection mechanism of the heterocontacts is thus different from that of single-phase porous ceramics, and can be described as follows.[19] In the proximity of the p-n junction sites electron holes are injected from the p-type semiconductor into the adsorbed water molecules, giving rise to protons in the adsorbed water phase. The positive charge is liberated at the surface of the n-type semiconductor, resulting in electrolysis of adsorbed water, with current increases only in the forward bias. The amount of hydrolysis increases at higher RH, because the higher the humidity, the higher the amount of water adsorbed. Electrolysis provides cleaning during working with a self-cleaning/self-recovery mechanism. In this respect, p-n junctions can be regarded as intelligent materials; heat-cleaning is not needed, because the sensing mechanism itself is the cleaning treatment.

Figure 4 shows the the bias-dependence in the reverse bias region of $1/C^2$, measured at high-frequency (100 kHz), in a nearly dry environment and at 95% RH. The intersection of the extrapolated line of $1/C^2$, evaluated in the reverse bias region, with the bias axis usually indicates the diffusion potential at the p-n junction interface. However, in this case, the diffusion potential value obtained would be unrealistically large both with and without moisture. The interface barrier at the La_2CuO_4/ZnO heterocontacts is, thus, not a physical diffusion barrier, but a chemical-like barrier, due to the presence of a high density of interface states derived from chemical adsorbates. These adsorbates are likely to be derived from oxygen and/or water adsorption. The importance of original interface states for the sensitivity of thin-film CuO/ZnO heterojunctions has also been reported.[20]

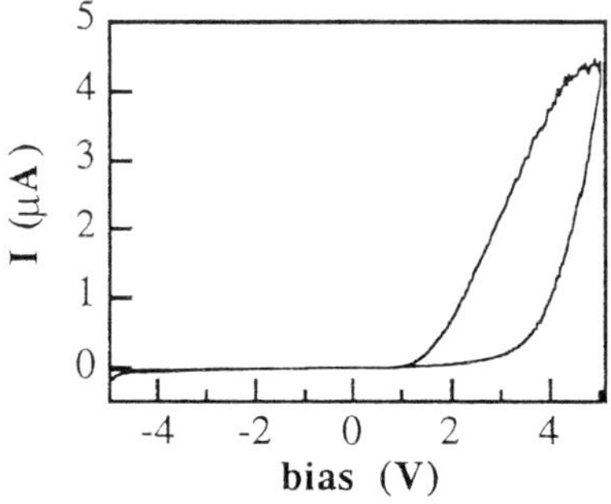

Figure 1. Dark I-V curve of La_2CuO_4/ZnO heterocontacts at room temperature and 0% RH.

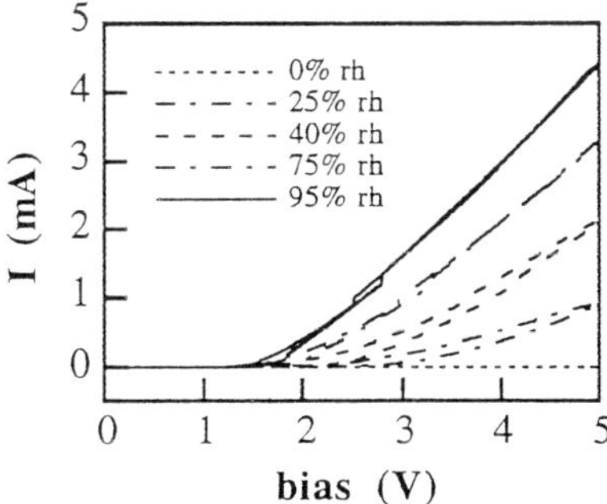

Figure 2. Dark I-V curves of La_2CuO_4/ZnO heterocontacts at room temperature, at various RH values.

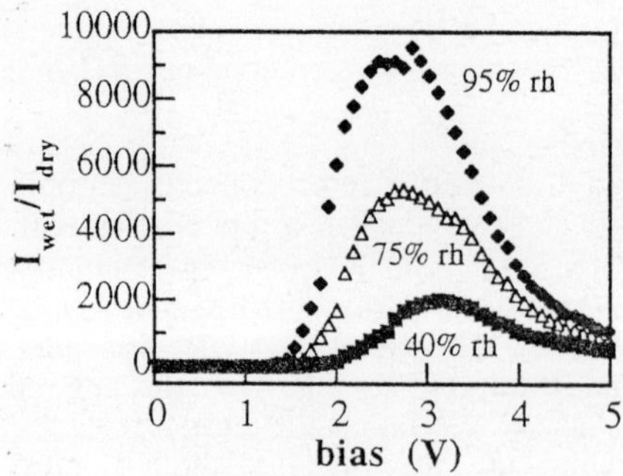

Figure 3. The voltage dependence of the RH-sensitivity of La_2CuO_4/ZnO heterocontacts at various RH values, evaluated from the data in figure 2.

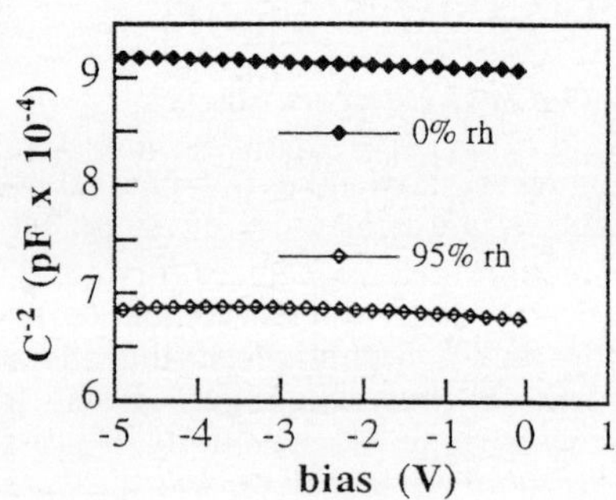

Figure 4. The voltage dependence in the reverse bias region of $1/C^2$ for La_2CuO_4/ZnO heterocontacts at 0% and 95% RH, at room temperature.

The high RH-sensitivity can thus be explained in terms of the saturation of the original interface states by physisorbed water, which, with increasing RH, leads to the release of trapped electrons, resulting in an increase in the forward current.[21] This mechanism explains the influence of the external electric field on the reactions with water. The presence of interface states modifies the potential barrier at the p-n junction, and an applied electric field can control the barrier height, and thereby the RH-sensitivity of the device. This is another function of p-n heterocontacts as intelligent materials for chemical sensors.

At high temperatures, in the range from 200°C to 400°C, the La_2CuO_4/ZnO heterocontacts continued to show I-V curves typical of p-n diode behaviour when measured in dry air.[15] The rectifying character of I-V curves was enhanced upon contact with increasing gas concentrations, as shown in Figure 5. The highest sensitivities, expressed as I/I_0, where I is the current flowing with gases and I_0 the current flowing in air, were measured at 400°C. For the La_2CuO_4/ZnO heterocontacts, no distinctive selectivity to CO was observed, as is the case for the CuO/ZnO system.[14] Much higher gas sensitivities were however measured. The gas sensitivity was bias-dependent, as shown in Figure 6. For bias values higher than 0.5 V, the greatest hydrogen sensitivity (above 10 for 2000 ppm) was observed. For CO, a different trend for sensitivity with bias was observed. At 0.5 V and 2000 ppm CO the sensitivity was about 3.5.

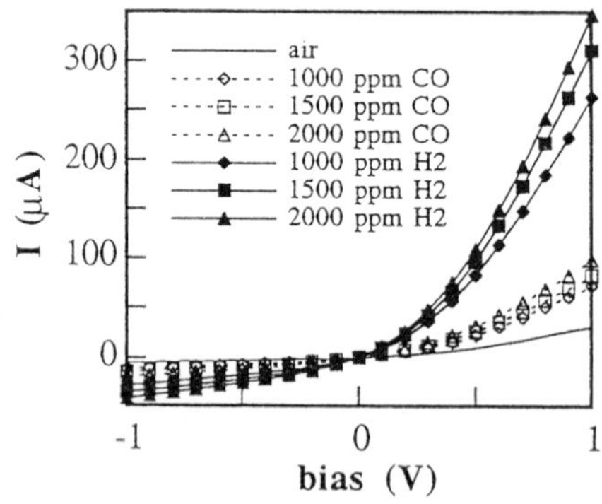

Figure 5. Dark I-V curves of La_2CuO_4/ZnO heterocontacts at 400°C in environments at various gas concentrations.

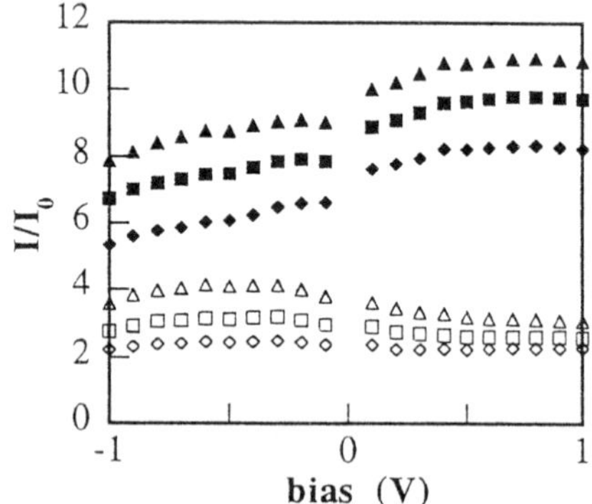

Figure 6 The voltage dependence of the gas sensitivity of La_2CuO_4/ZnO heterocontacts at various gas concentrations, evaluated from the data in figure 5. The symbols are the same as in fig. 5.

The predictable reactions between gas molecules and the oxygen originally adsorbed at the interface are the basis of the gas sensing mechanism of these heterocontacts. CO and hydrogen are ionized by electron holes injected from the p-type oxide and react with negatively charged oxygen ion adsorbates to form CO_2 or vapour. This results in an increase of the forward current across the p-n junction. These chemical reactions are enhanced when the forward bias is greater than 0.5 V, for hydrogen. The external electric field may control the chemical reactions, and thus the gas sensitivity, giving rise to the possibility of tuning the gas sensitivity of the heterocontacts by externally applied bias. If the p-type oxide has a preference for the adsorption of CO instead of H_2, as does CuO, CO selectivity may result, but this is not the case for La_2CuO_4.

Au/ZnO Schottky Barriers

Figure 7 shows the I-V curves of dense ZnO pellets with one Al and one Au electrode at 2% and 85% RH, at room temperature. As for the p-n heterocontacts, a diode behaviour was observed, enhanced in wet environments, due to the formation of Schottky barriers at the chemically abrupt interfaces between Au (a metal with a high work function) and ZnO, at contact points between the mechanically pressed ZnO disc and the Au sheet. The RH-sensitivity of the Au/ZnO Schottky barriers was bias-dependent (Figure 8). With forward bias (Au^+/Al^-) the specimens responded to RH, with a maximum sensitivity of 2.6, at 85% RH upon application of 10 V. The humidity sensitive electrical properties are given by the Au/ZnO interface. In fact, as already stated above, the use of dense ZnO samples hindered the ionic-type conduction mechanism in the presence of humidity. Dense ZnO pellets with two Al contacts showed linear I-V curves which were independent of the RH level.[24]

The bias-dependent, RH-sensitive resistivity of Au/ZnO interfaces may be determined by variations of the Schottky barrier height, with a mechanism similar to that shown by Schottky barrier gas sensors.[23] Semiconducting oxides usually possess surface states associated with adsorbed oxygen or water-decomposed species. At low RH, the barrier height is affected by the presence of extrinsic surface states which can be formed by water chemisorption. The shape of the I-V curves in the forward-bias region is due to the band-bending at the ZnO surface, which is in turn due to the charge of the intrinsic states and the charge of the extrinsic states by chemisorbed hydroxyl ions. At high RH, layers of water molecules physisorb on the chemisorbed layer. Therefore, exposure of the specimen to humidity leads to the complete saturation of hydroxyl ions through hydrogen bonds, resulting in the neutralization of the charge due to the extrinsic surface states. The result is a further decrease in the height of the barrier at high RH, resulting in an increase in forward current.

These results show that the RH-sensitivity of dense ZnO pellets with one Au electrode is due to a sensing mechanism which is close to that of p-n heterocontacts. The presence of a Schottky barrier gives a rectifying character, which in the case of heterocontacts is given by p-n junctions. The RH-sensitivity is related to the variation of the barrier height because of the presence of surface/interface states. The neutralization of these states by means of physisorbed water causes the current enhancement with RH in the forward-bias region. This mechanism, not yet fully clear, is different from those already proposed for ceramics.

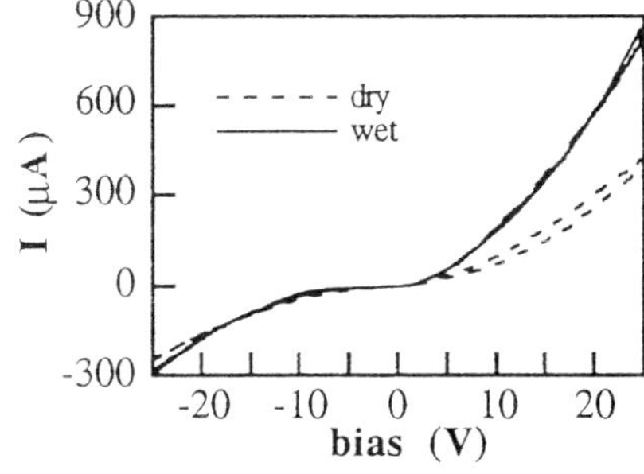

Figure 7. Dark I-V curves of ZnO discs with one Al and one Au electrode, in dry (2% RH) and wet (85% RH) environments, at room temperature.

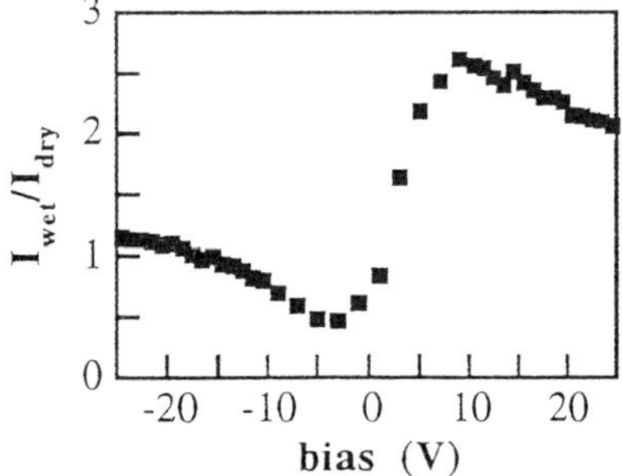

Figure 8. The voltage dependence of the RH-sensitivity of Au/ZnO Schottky barriers at 85% RH and room temperature, evaluated from the data in fig. 7.

Sol-Gel Processed TiO_2 Films with 10 at% of K

Figure 9 shows the RH-dependence of the resistance of sol-gel processed 10 at% K-doped TiO_2 films measured at different frequencies. The RH-sensitivity depends on the frequency at which the resistance was measured. In particular, at 50 Hz the variation of the logarithm of the resistance, linear with RH, was of 4 orders of magnitude over the entire RH range tested (4-85% RH). At the lowest RH values, a noticeable RH-sensitivity of about 3 orders of magnitude in the RH range from 4% to 10% was measured only at frequencies $\leq$ 1 Hz, as shown in figure 10 for resistance measured at 100 mHz. Another interesting feature of these films is their stability over time: for this material very little drift was observed even after a long time of testing.[17]

These results show that the 10 at% K-doped TiO_2 films possess some unique properties. For ceramic materials, it has been reported that a large pore volume is needed for high RH-sensitivity.[24] The sol-gel processed K-doped TiO_2 films, although they were free of capillary pores, showed very high RH-sensitivity, seldom observed even for porous sintered ceramics. No morphological features were detectable by SEM observations.[17] The undoped TiO_2 films, prepared in the same way by sol-gel, showed very limited humidity-sensitivity.[25] The addition of alkali ions has been found to be effective in increasing the RH-sensitivity of several ceramic oxides.[8] The increase in sensitivity observed for doped-TiO_2 films with respect to the undoped films may be explained in terms of an increased number of water adsorption sites. The EIS measurements, however, lead us to suppose that there is a humidity sensing mechanism which is different from that generally accepted for porous ceramic oxides. This mechanism may involve the direct participation of alkali ions in conduction in wet environments. Recently, it has been reported that the RH-sensitivity of sintered ZnO was dramatically increased by adding very small amounts of Li dopant, which did not affect the microstructure of the porous compacts.[26] As in the present case, the addition of alkali ions, though in very different concentrations for different materials, led to a noticeable increase in the RH-sensitivity of the materials, and also in this case it is possible to hypothesize the direct participation of alkali ions in the conduction mechanism.

Another interesting feature is the possibility of measuring very low RH values with great sensitivity using 10 at% K-doped TiO_2 films, by measuring the resistance in a.c. at low frequencies. Moreover, the high RH-sensitivity is stable over time, and heat-cleaning is probably not necessary for practical applications. Both these characteristics are extremely interesting for practical application. By choosing the frequency at which the resistance is measured, it is possible to modify the response of the alkali-doped films. Therefore, the RH sensing properties of the films are tunable with the frequency. This property is among those of intelligent materials. The structure of the material may explain their intelligent properties: with the sol-gel process, very homogeneous and uniform materials can be obtained, with intimate mixing at the atomic scale. The resulting material has integrated functions of insulator (titania) and ionic conductor (K-containing phase). Also in this case the integration of materials with different functions (insulator/conductor) led to an improved material with novel characteristics, which are not possessed by the single phases.

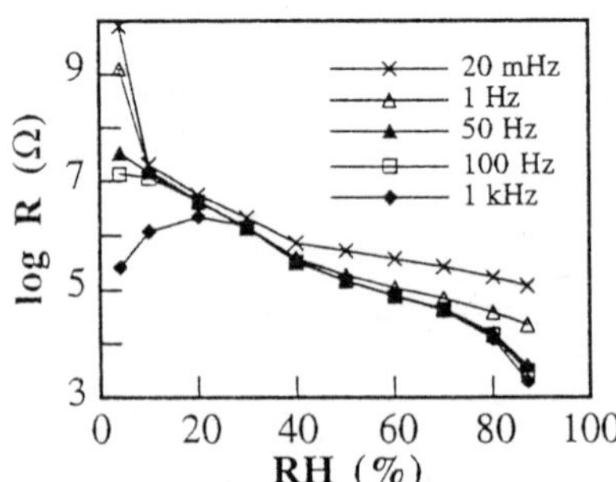

Figure 9. The RH dependence of the resistance of sol-gel processed 10 at% K-doped TiO_2 films at 40°C, measured at various frequencies.

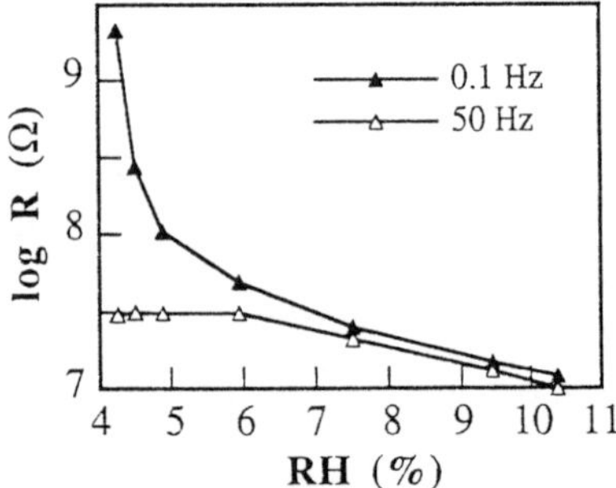

Figure 10. Log resistance versus low values of RH curves for sol-gel processed 10 at% K-doped TiO_2 films at 40°C, measured at various frequencies.

Acknowledgments

First of all, I would like to thank Prof. Hiroaki Yanagida and Prof. Masaru Miyayama of the University of Tokyo, who introduced me to the study of intelligent materials. I wish to thank Prof. Angelo Montenero (Univ. of Parma) for his fundamental collaboration in the preparation of materials by sol-gel. Very special thanks to Dr. Andrea Bearzotti (Inst. Solid State Electronics, CNR, Rome) for his continuous help in providing experimental results and useful criticisms. Last but not least, I heartly thank Prof. Gualtiero Gusmano (Univ. of Rome "Tor Vergata") for his encouragement.

REFERENCES

1. H. Yanagida, Intelligent materials - a new frontier, *Angew. Chem.* 100:1443 (1988).
2. H. Yanagida, Intelligent ceramics, *Ferroelectrics* 102:251 (1990).
3. C.A. Rogers, From the editor, *J. Intelligent Mater. Systems and Structures* 1:3 (1990).
4. R.E. Newnham and G.R. Ruschau, Smart electroceramics, *J. Am. Ceram. Soc.* 74:463 (1991).
5. N. Yamazoe and N. Miura, Environmental gas sensing, *in*: "Techn. Digest of the 7th Intern. Conf. on Solid-State Sensors and Actuators (Transducers 93)," Yokohama, Japan (1993).
6. T. Seiyama, A. Kato, K. Fujiishi, and M. Nagatani, A new detector for gaseous components using semiconductive thin films, *Anal. Chem.* 34:1502 (1962).
7. J.G. Fagan and V.R.W. Amarakoon, Reliability and reproducibility of ceramic sensors: part III, humidity sensors, *Am. Ceram. Soc. Bull.* 72 [3]:119 (1993).
8. E. Traversa, Ceramic sensors for humidity detection: the state-of-the-art and future developments, *Sensors and Actuators B* 23:135 (1995).
9. B.M. Kulwicki, Humidity sensors, *J. Am. Ceram. Soc.* 74:697 (1991).
10. N. Yamazoe, New approaches for improving semiconductor gas sensors, *Sensors and Actuators B* 5:7 (1991).
11. K.D. Schierbaum, Engineering of oxide surfaces and metal/oxide interfaces for chemical sensors: recent trends, *Sensors and Actuators B* in press (1995).
12. Y. Nakamura, T. Tsurutani, M. Miyayama, O. Okada, K. Koumoto, and H. Yanagida, The detection of carbon monoxide by the oxide-semiconductor heterocontacts, *J. Chem. Soc. Jpn.* 1987:477 (1987).
13. K. Kawakami and H. Yanagida, Effects of water vapor on the electrical conductivity of the interface of semiconductor ceramic-ceramic contacts, *J. Ceram. Soc. Jpn.* 87:112 (1979).
14. Y. Nakamura, H. Yoshioka, M. Miyayama, H. Yanagida, T. Tsurutani, and Y. Nakamura, Selective CO gas sensing mechanism with CuO/ZnO heterocontact, *J. Electrochem. Soc.* 137:940 (1990).
15. E. Traversa, M. Miyayama, and H. Yanagida, Gas sensitivity of ZnO/La_2CuO_4 heterocontacts, *Sensors and Actuators B* 17:257 (1994).
16. E. Traversa and A. Bearzotti, Humidity sensitive electrical properties of dense ZnO with non-ohmic electrode, *J. Ceram. Soc. Jpn.* 103:11 (1995).
17. G. Montesperelli, A. Pumo, E. Traversa, G. Gusmano, A. Bearzotti, A. Montenero, and G. Gnappi, Sol-gel processed TiO_2-based thin film as innovative humidity sensors, *Sensors and Actuators B* in press (1995).
18. E. Traversa, A. Bianco, G. Montesperelli, G. Gusmano, A. Bearzotti, M. Miyayama, and H. Yanagida, ZnO/La_2CuO_4 hetero-contacts as humidity sensors, *in*: "Ferroic Materials: Design, Preparation, and Characteristics," A.S. Bhalla, K.M. Nair, I.K. Lloyd, H. Yanagida, and D.A. Payne, eds., The Am. Ceram. Soc., Westerville (1994).
19. Y. Nakamura, M. Ikejiri, M. Miyayama, K. Koumoto, and H. Yanagida, The current-voltage characteristics of CuO/ZnO heterojunctions, *J. Chem. Soc. Jpn.* 1985:1154 (1985).
20. Y. Ushio, M. Miyayama, and H. Yanagida, Effects of interface states on gas sensing properties of a CuO/ZnO thin film heterojunction, *Sensors and Actuators B* 17:221 (1994).
21. E. Traversa, A. Bearzotti, M. Miyayama, and H. Yanagida, Study of the conduction mechanism of La_2CuO_4/ZnO heterocontacts at different relative humidities, *Sensors and Actuators B* in press (1995).
22. E. Traversa and A. Bearzotti, A novel humidity detection mechanism for ZnO dense pellets, *Sensors and Actuators B* 23:181 (1995).
23. K.D. Schierbaum, U.K. Kirner, J.F. Geiger, and W. Göpel, Schottky-barrier and conductivity gas sensors based upon Pd/SnO_2 and Pt/TiO_2, *Sensors and Actuators B* 4:87 (1991).
24. Y. Shimizu, H. Arai, and T. Seiyama, Theoretical studies on the impedance-humidity characteristics of ceramic humidity sensors, *Sensors and Actuators* 7:11 (1985).
25. G. Gusmano, G. Montesperelli, P. Nunziante, E. Traversa, A. Montenero, M. Braghini, G. Mattogno, and A. Bearzotti, "Humidity-sensitive properties of titania films prepared using the sol-gel process", *J. Ceram. Soc. Jpn.* 101:1095 (1993).
26. E. Joanni and J.L. Baptista, $ZnO-Li_2O$ humidity sensors, *Sensors and Actuators B* 17:69 (1993).

THE USE OF SOL-GEL FILM AS A MATRIX FOR ENTRAPMENT OF FLUORIMETRIC REAGENTS

Musa Ahmad[1] and Ramaier Narayanaswamy [2]

[1]Chemistry Department
Faculty of Physical and Applied Science
Universiti Kebangsaan Malaysia
43600 Bangi, Selangor D.E. Malaysia

[2]Department of Instrumentation and Analytical
Science, UMIST, P.O. Box 88
Sacville Street
Manchester M60 1QD, England

ABSTRACT

Sol-gel process was used in this study to prepare a reagent phase for used in the development of an optical fibre Al(III) sensor. This include the preparation of a sol-gel film using a dip-coating technique and incorporated fluorimetric reagent into the film. This paper will basically discussed the details of the technique used for the entrapment of the reagent, the results obtained and some fundamental problems faced during the study.

INTRODUCTION

Generally, the sol-gel process consists of three distinct stages[1] i.e. mixing of the reaction component (mainly metal oxide source, water, solvent and catalyst) in solution to achieve intimacy on a molecular scale, gelling of the solution in a manner that will retain the chemical homogeneity and thermal treatment of the gel to convert it into a glass-like matrix.

Tetraethylorthosilicate (TEOS) is the most common of the alkoxy compounds used for sol-gel process[2], the preparation of which can be described by two types of reactions, as follows:

Polymers and Other Advanced Materials: Emerging Technologies and Business Opportunities. Edited by P. N. Prasad *et al.*, Plenum Press, New York, 1995

$$Si(OEt)_4 + xH_2O \underset{\text{esterification}}{\overset{\text{hydrolysis}}{\rightleftharpoons}} Si(OH)_x(OEt)_{4-x} + xEtOH \qquad (1)$$

$$\text{-SiOEt} + \text{HOSi-} \underset{\text{alcoholysis}}{\overset{\text{alcohol condensation}}{\rightleftharpoons}} \text{-Si-O-Si-} + EtOH \qquad (2a)$$

$$\text{-SiOH} + \text{HOSi-} \underset{\text{hydrolysis}}{\overset{\text{water condensation}}{\rightleftharpoons}} \text{-Si-O-Si-} + H_2O \qquad (2b)$$

If the gel is not subjected to extensive thermal treatment, it is envisaged that the final gel will consist of a three-dimensional polysiloxane structure with some unreacted ethoxy and hydroxyl groups.

$$\text{- Si - O - Si(OEt) - O - Si - O - [Si - O -]}_n\text{ Si(OH) -}$$

The hydrolysis and polycondensation reactions occur simultaneously. The relative rates of these reactions, determined by processing factor such as solvent, pH, catalyst, aging, temperature and time, and concentration of water, influence the growth and subsequent physical properties of gel so that the overall process is complex and difficult to model. After limited thermal treatment, the gel structure becomes glass-like but still retains an element of its original porosity, the extent of which is largely dependent on the degree of thermal treatment. Thus, the sol-gel process can lead to the preparation of a porous glass-like material and it is this feature which suggest that it will be suitable in preparing a reagent matrix for optical fibre chemical sensors.

The ability to trap organic and bioorganic molecules in organic oxides through the sol-gel process was first introduced in 1984[3], which opened the road a whole new class of materials and intensive activity in many laboratories. In the area of chemical sensing, sol-gel technique was widely used as a matrix for the preparation of the reagent phase. While most of the published works concentrated on the preparation of reagent phase for pH[2,4,5] and gas sensing such as oxygen[6], flammable gases[7] and ammonia vapour[8], none of them reported the possibility of using sol-gel film in preparing a reagent phase for metal sensing. In this study, we highlight the results obtained for entrapment of two Al(III) sensitive reagents i.e. 3-hydroxy-2-napthoic acid (HNA) and purpurin. In solution, HNA[9] and purpurin[10] were reported to form complexes which fluoresces at wavelength of 510 nm and 550 nm when excited at wavelength of 383 nm and 480 nm, respectively.

EXPERIMENTAL

(i) Reagent:

Standard Al(III) solution of 6.4×10^{-3} M was prepared by dissolving required amount of aluminium potassium sulphate (Aldrich) in distilled deionised water. Standard solutions of

1.0×10^{-3} M purpurin and HNA was prepared by dissolving appropriate amount of the reagent in ethanol. A mixture of equal volume of saturated fluoride solution and distilled deionised water was used as regenerating solution in this study. Saturated fluoride solution was prepared by dissolving excess amount of sodium fluoride (Aldrich) in distilled deionised water.

Some of the solid supports used in this study include filter paper, melinex 505, acetate sheet, cover glass and microscope slide. All these materials which provide a support for the sol-gel film were rinsed with ethanol before used. Tetraethylorthosilicate (Aldrich), ethanol (Aldrich), distilled deionised water, nitric acid (BDH) and triton X-100 (BDH) were the precursors used for sol-gel preparation.

(ii) Instrumentation

LS-5 Luminescence spectrometer was used for all fluorescence measurement using fibre optic accessory. The fabrication of the accessory has been described elsewhere[11] and its probe was designed to accommodate the sample in the form of 5 mm diameter disk.

(iii) Procedure

For preparation of a blank sol-gel the procedure as described by Chernyak *et al.*[8] was followed. The sol-gel mixture contained of TEOS, water, ethanol and nitric acid at the mole ratio of 1:4.2:9.4:0.5. Triton X-100 (15 drops) was finally added to 25 ml of the above mixture. The mixture was stirred for a few minutes for homogeneity.

The sol-gel films were deposited on the support materials by dip coating. In its simplest manifestation, Brinker *et.al*[12] explained that sol-gel dip coating consists of the withdrawal of a support material from a fluid sol: gravitational draining and solvent evaporation, accompanied by further condensation reactions, results in the deposition of a solid film. Dip-coating technique is preferred in this study because it requires considerably less equipment and is potentially less expensive than conventional thin film forming processes such as sputtering and evaporation. The support materials (3.0 cm x 1.5 cm) were mechanically dipped into the blank sol-gel mixture and recovered slowly and steadily to provide even coating and was allowed to dry for 24 hours in ambient temperature.

The Al(III) sensitive fluorescent reagent was impregnated into an already formed sol-gel film by immersing the sol-gel film in the reagent solution for 24 hours. A sample disk was prepared by circularly cutting the sample (entrapped reagent in sol-gel film supported by support material) with a diameter of 5 mm. This was done for ease of coupling the sample at the distal end of the optical fibre. Nylon mesh was used to retain the sample disk in its position during the measurement. Regeneration of the entrapped reagent after reaction with Al(III) was carried out by dipping the tip of the fibre probe into fluoride solution and washing the probe in distilled deionised water before the next measurement was taken.

The study of the effect of HNO_3 was carried out by introducing different amounts of HNO_3 during preparation of sol-gel solution, while maintaining the total volume of HNO_3-water a constant. The effect of the concentration of HNA and Al(III) were studied using the same procedure by maintaining the total volume of HNA-ethanol and Al(III)-water a constant, respectively. An attempt was also made to use hydrochloric acid as a catalyst during the sol-gel formation instead of HNO_3 acid.

RESULT AND DISCUSSION

Filter paper as support material produced the best result of all the support materials used. The entrapped Al(III)-HNA complex was observed to fluoresce at a wavelength of 514 nm when excited at a wavelength of 350 nm (Figure 1). As can be seen, the entrapped HNA alone produced a very high fluorescence signal and produces a low difference in the fluorescence signal of the reagent and its Al(III) complex.

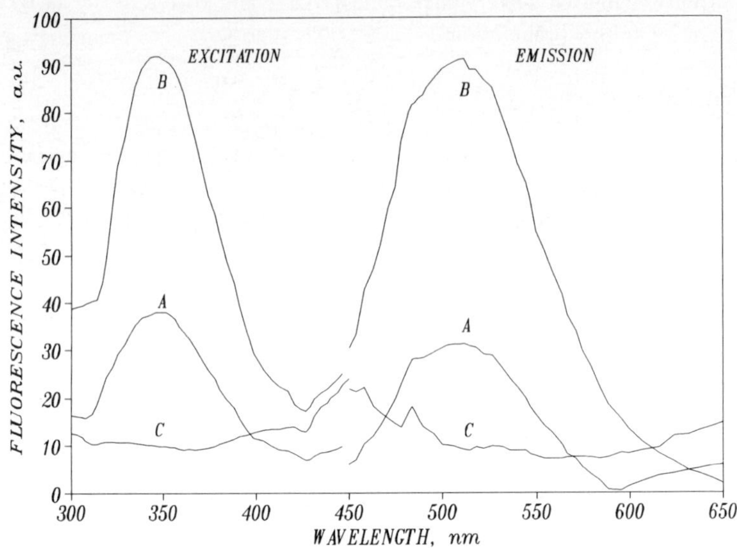

Figure 1: Luminescence spectra of the blank sol-gel (C), entrapped HNA (A) and its Al(III) complex (B).

Figure 2 on the shows solid surface fluorescence spectra of the entrapped purpurin in the sol-gel film before and after reaction with Al(III). Not as with HNA, the results obtained here appear to be very encouraging. A big changes in fluorescence intensity of the entrapped purpurin upon reaction was observed at wavelength of 600 nm.

Figure 3 shows the response of the entrapped purpurin in sol-gel film towards same concentration of Al(III) at a single wavelength. This result was unexpected because the reproducibility of the probe seems to be very poor. This observation indicates that the entrapped purpurin has been washed out from the sol-gel film or some photobleaching of the reagent had taken place since the fluorescence signal of the complex was continuously decreasing with time when the probe was continuously immersed in the solution. The photostability study carried out for the period of more than 48 hours however shows that the entrapped purpurin was fairly stable when exposed to light. There was also no indication that the reagent had been washed out. The fluoride solution used in this study was previously shown not to quench the fluorescence signal[11]. An optical fibre sensor for fluoride ion based on enhanced fluorescence measurement has also been reported[13] without any problem of quenching from the ion. Therefore, an attempt was made to study the effect of varying the composition of the sol-gel prepared. The study was concentrated only to HNA.

(i) The Effect of HNA Concentration.

The effect of HNA concentration on the Al(III)-HNA complex formation was studied both in sol-gel solution and with the sol-gel immobilised on the filter paper. The results

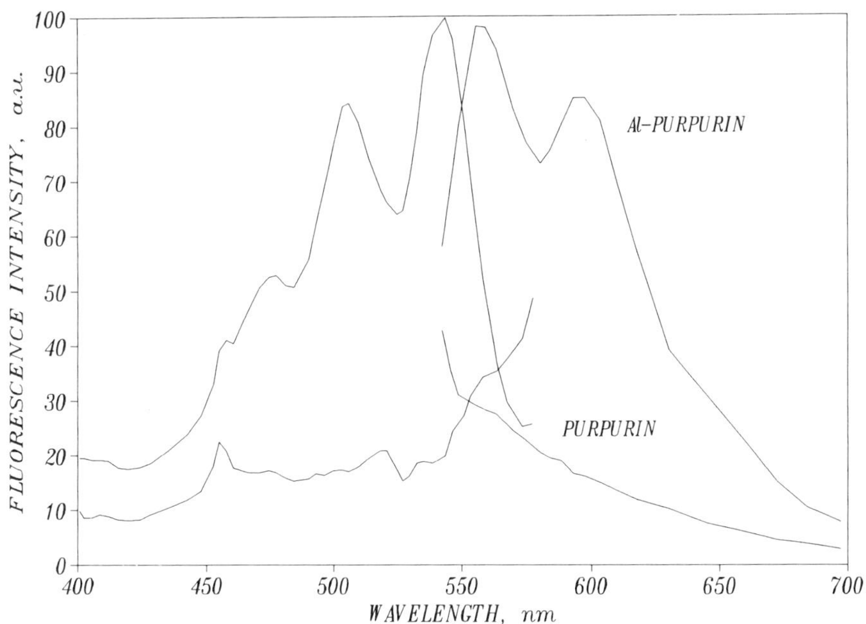

Figure 2: Luminescence spectra of entrapped purpurin in sol-gel film and after reaction with Al(III).

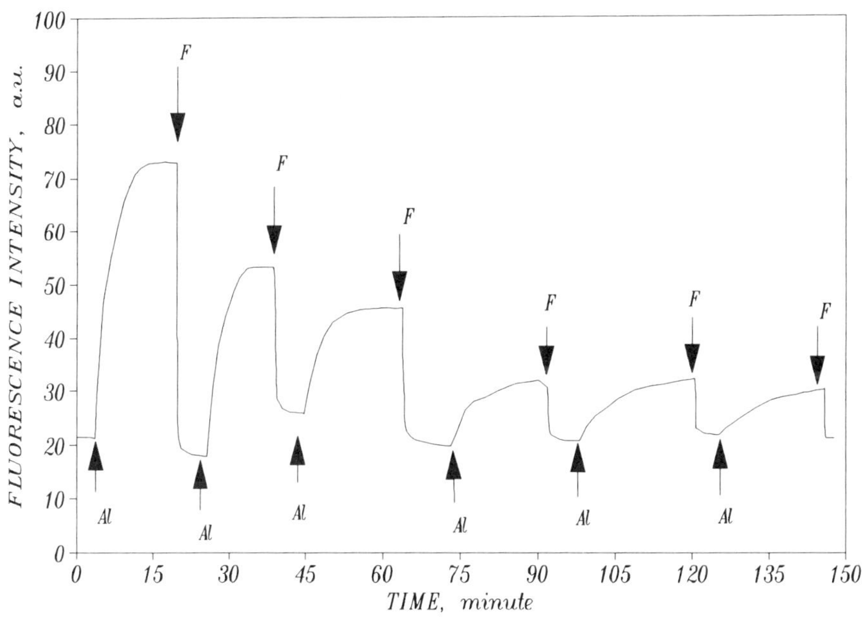

Figure 3: The response of the entrapped purpurin in sol-gel film towards same concentration of Al(III).

obtained recorded at six different HNA concentrations are shown in Figure 4. As the concentration of HNA in sol-gel decreased, the fluorescence intensity of the complex was found to decrease. At low concentrations of HNA, low fluorescence signals are observed due to the formation of small amounts of Al(III)-HNA, complex. Similar results were observed with the sol-gel immobilised on the filter paper. As the concentration of HNA decreased, the fluorescence intensity of both the complex and the reagent were also decreased and emission was hardly detectable at very low concentration of HNA, because the amount of the complex that had been formed became insignificant.

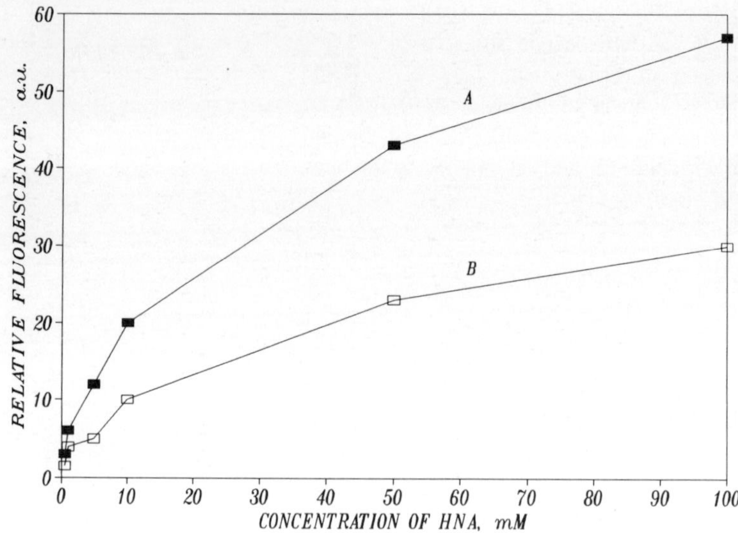

Figure 4: The effect of HNA concentration to the fluorescence intensity of the complex in sol-gel solution (A) and sol-gel film (B).

(ii) The Effect of Al(III) Concentration.

In the sol-gel solution, it was found that as the Al(III) concentration decreased, the fluorescence signal of the complex was also decreased and finally coincided with that of the reagent alone (Figure 5). The effect of Al(III) concentration was also studied with the sol-gel immobilised on a filter paper, by measuring the fluorescence of the entrapped complex. The results showed a slight increases in the fluorescence signal with increasing Al(III) concentration.

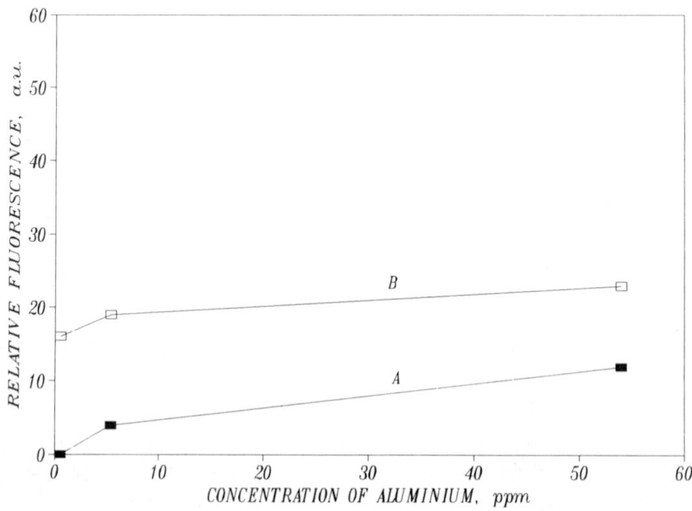

Figure 5: The effect of Al(III) concentration to the fluorescence intensity of the complex in sol-gel solution (A) and in sol-gel film.

(iii) The Effect of HNO_3 in Sol-Gel Preparation.

In this study, it was found that the use of the sol-gel technique for reagent entrapment, give rise to a high fluorescence signal of the reagent alone. Concentrated nitric acid is reported to render aluminium metal passive[14] and therefore may have the same effect to this metal ion when present in solution during sol-gel preparation. An experiment was therefore carried out to study the effect of HNO_3 on the fluorescence intensity of the complex in ethanol, sol-gel solution and in sol-gel immobilised on filter paper. The results are summarised in Figure 6.

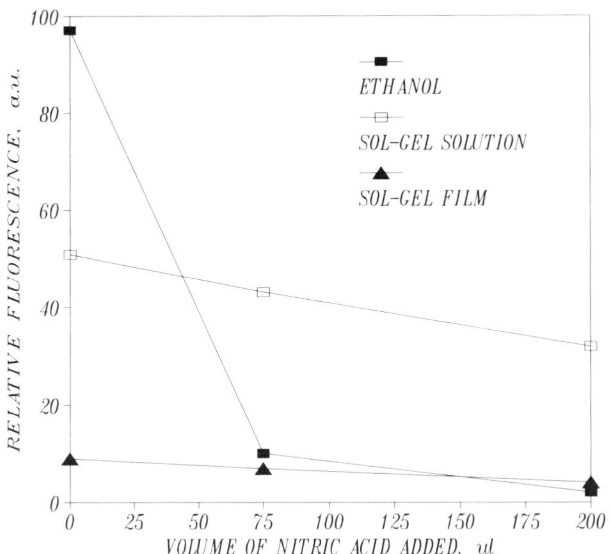

Figure 6: The effect of addition of HNO_3 acid to the fluorescence intensity of HNA and its Al(III) complex in ethanol, sol-gel solution and sol-gel film.

Addition of HNO_3 was found to decrease the fluorescence intensity of the complex in ethanol Addition of about 200 μl of concentrated HNO_3 caused the relative fluorescence intensity of the complex to approach zero value. Addition of HNO_3 was also found to decrease the relative fluorescence intensity of the complex in sol-gel solution. However, the degree of reduction is found to be less in sol-gel solution than in ethanol. With the sol-gel immobilised on a filter paper, the fluorescence signal of the entrapped complex decreased only very slightly with addition of HNO_3. An attempt to use hydrochloric acid instead of HNO_3 also gave similar results (Figure 7). Since the reaction between aluminium and HNA is pH dependent, the decrease in the fluorescence intensity of the complex with addition of HNO_3 or HCl would be due to the decrease in pH of the solution. In ethanol, this complex was found to give strong fluorescence signal at pH 5.8[9].

(iv) The Effect of pH

The effect of the pH on the fluorescence intensity of the entrapped complex in the sol-gel film was also studied by dipping the filter paper (containing sol-gel film with entrapped complexes) into the buffer solution of different pH value. However the variation in the fluorescence intensities with pH was found to be very small.

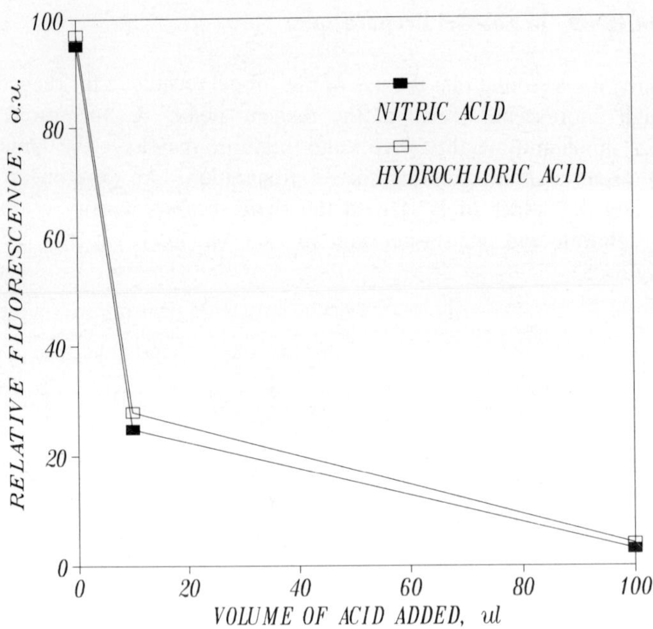

Figure 7: The effect of addition of acids to the fluorescence intensity of HNA and its Al(III) complex in ethanol.

From the studies which have been carried out above, none of the them shows any indication that they were responsible for the observation shows in Figure 3. After one month, the same sample of the entrapped purpurin which gave spectra shown in Figure 2 was recorded for its fluorescence spectra and the results obtained is shown in Figure 8. Comparing this spectra with its corresponding spectra (Figure 2),it was noted that the former gave a very smooth spectra with a very low gain used in the instrument. However, the change in the fluorescence signal between the entrapped reagent and its Al(III) complex was found to be very small. The same observation was also made for HNA. Therefore it was postulated (at least based on observation in this study) that a period of one month was probably sufficient for the sol-gel film to dry out and increase in the rigidity of its structure as well that of the reagent and hence, observed increase in the fluorescence signal of the reagent. Since formation of Al(III) complex has already contributed towards the rigidity of the structure, no further increase in fluorescence would be expected for the complex by drying the sol-gel film. As a results, the change in the fluorescence signal between entrapped reagent and its Al(III) complex would be very low.

CONCLUSION

It is possible to entrap HNA and purpurin in sol-gel film with filter paper as the support material. The entrapment of the reagent in sol-gel film can be carried out easily by immersing immobilised sol-gel film (on filter paper) in the reagent solution. Immobilised reagent however shows a very high fluorescence signal (nearly equal to that of the Al(III)-reagent complex). Although all these reagents can be successfully entrapped in sol-gel film, the fluorescence intensity of the entrapped reagent increased after a period of time. This made the difference in the fluorescence intensity of the reagent and the aluminium complex become very small.

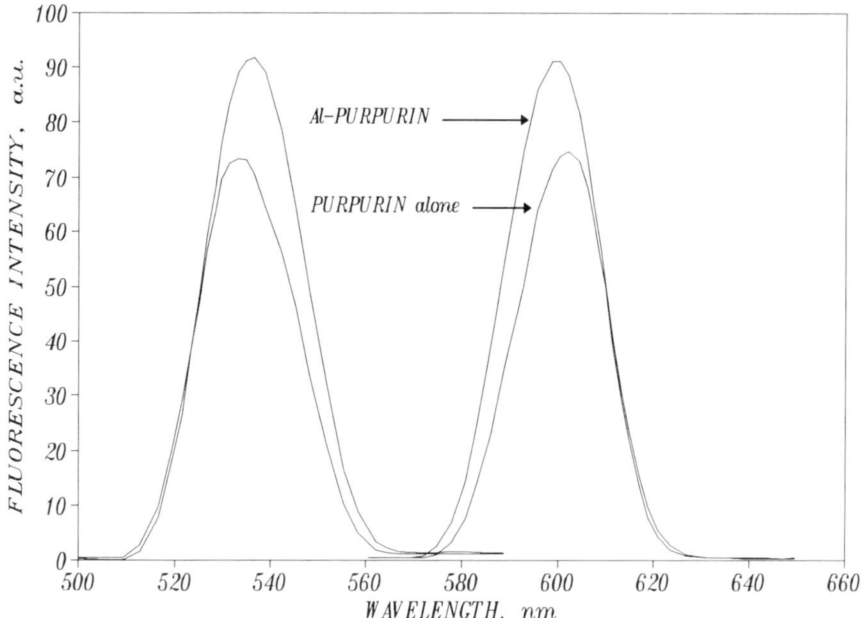

Figure 8: Solid surface fluorescence spectra of the entrapped purpurin in sol-gel film and its Al(III) complex after the sample was kept in room temperature for one month.

ACKNOWLEDGMENT

Association of commonwealth Universities and Universiti Kebangsaan Malaysia are gratefully acknowledged for financial support which enabled this research to be carried out.

REFERENCES

1. K.T.V. Grattan, G.T. Badini, A.W. Palmer and A.C.C. Tseung, Use of sol-gel techniques for fibre optic sensor applications, *Sensors and Actuators A*, 25-27:483 (1991).

2. G.T. Badini, K.T.V. Grattan, A.W. Palmer and A.C.C. Tseung, Development of pH-sensitive substrate for optical sensor applications, *in:* "Springer Proceedings in Physics vol. 44 : 6th. Int. Conf. Optical Fibre Sensors," H.J Arditty., J.P Dakin. and R.T Kersten, eds., Springer-Verlag , Berlin (1989).

3. D. Avnir, S. Braun and Ottolenghim, Encapsulation of organic molecules and enzymes in sol-gel glasses: a review of novel photoactive optical sensing and bioactive materials, *ACS Symposium Series*, 499:384 (1992).

4. B.D. MacGraith., Ruddy, C. Potter, J.F. McGilp and B. O'Kelly, Fibre optic fluorescence sensors based on sol-gel entrapped dyes, *SPIE-Chemical and Medical Sensors*, 1510:101 (1991).

5. B.D. MacGraith, Enhanced evanescent wave sensors based on sol-gel derived porous glass coatings, *Sensors and Actuators B*, 11:29 (1993).

6. B.D. MacGraith, C.M. McDonagh, G. O'Keeffe, E.T. Keyes, J.G. Vos, B. O'Kelly and J.F. McGilp., Fibre optic oxygen sensors based on fluorescence quenching of evanescent wave excited ruthenium complexes in sol-gel derived porous coating, *Analyst*, 118(4):385 (1993).

7. A. Wilson, J.D. Wright., J.J. Murphy, M.A.M. Stroud. and S.C. Thorpe, Sol-gel materials for gas-sensing applications, *Sensors and Actuators B*, 18-19:506 (1994).

8. V. Chernyak, R. Reisfeld, R. Gvishi and D. Venezky, Oxazine-170 in sol-gel glass and PMMA films as a reversible optical waveguide sensor for ammonia and acids, *Sensors and Materials (Tokyo)*, 2:117 (1990).

9. G.F. Kirkbrigth, T.S. West and C. Woodward, Spectrofluorimetric determination of sub-microgram amounts of aluminium and beryllium with 2-hydroxy-3-napthoic acid, *Anal. Chem.*, 37(1):137 (1965).

10. M. Ahmad and R. Narayanaswamy, Direct fluorimetric determination of aqueous aluminium (III) with purpurin, paper submitted to *Science International.*

11. M Ahmad and R Narayanaswamy, Optical fibre aluminium (III) based on solid surface fluorescence measurement, paper submitted to *Analyst.*

12. C.J. Brinker., G.C. Frye., A.J. Hurd. and C.S. Ashley, Fundamentals of sol-gel dip coating, *Thin Solid Films*, 201:97 (1991).

13. D.A. Russell. and R. Narayanaswamy, An optical fibre sensor for fluoride, *Anal. Chim. Acta.*, 220:75 (1989).

14. G. Svehla., "Vogel's Qualitative Inorganic Analysis," John Wiley and Sons, New York (1987).

MULTIFUNCTIONAL CONJUGATED ORGANIC MATERIALS INCORPORATING STABLE CHARGE STATES

Charles W. Spangler and Mingqian He

Department of Chemistry
Northern Illinois University
DeKalb, IL 60115-2862

INTRODUCTION

Electroactive materials have been studied extensively for several years due to their unique chemical and photophysical properties. Originally of interest due to the dramatic increases in conductivity when chemically or electrochemically doped,[1,2] it has more recently been recognized that these materials are truly multifunctional in nature. In most of these emerging applications, the introduction of stable charge states into a conjugated π-electron sequence seems to play a key role.[3] In this paper we will review how well-defined stabilized bipolaronic charge states can be incorporated into both oligomeric and copolymeric conjugation sequences and review their potential application.

MULTIFUNCTIONALITY IN CONJUGATED ORGANIC MATERIALS

When conjugated π-electron polymers are oxidatively doped to metallic conductivity, it has been recognized that spinless bipolarons dominate the charge carrier population,[4] as illustrated below for polythiophene:

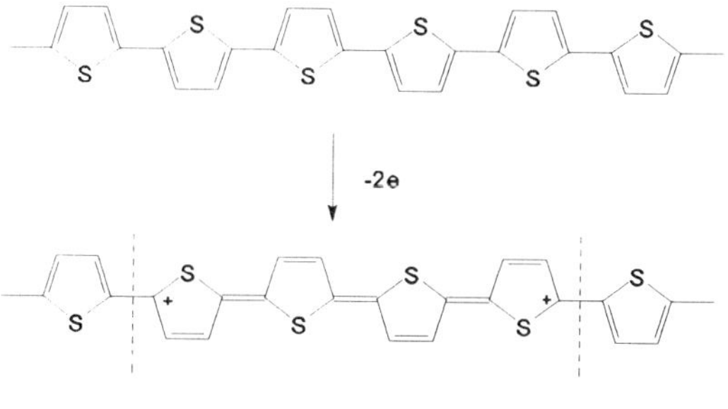

S S S S S S

-2e

S S S S S S

Polymers and Other Advanced Materials: Emerging Technologies and Business Opportunities. Edited by P. N. Prasad *et al.*, Plenum Press, New York, 1995

These charge carriers are usually localized over 4-6 repeat units. Bipolaron formation in conjugated polymers is accompanied by dramatic changes in their absorption spectra, with new bands appearing at lower energy.[3] Thus doping produces two dramatic changes: an increase in conductivity of more than 10 orders of magnitude, coupled with shifts in oscillator strength of 1-2 eV to lower energy. However, it has recently been shown that these materials also have large nonresonant third-order nonlinearities,[5] and that introduction of charge states into the conjugation sequence can lead to $\chi^{(3)}$ enhancement.[6] Within the last five years, charge state injection in conjugated polymers such as poly[p-phenylene vinylene] (PPV) has been shown to lead to light emission.[7] Although the mechanism of photon emission from these materials is still a hotly debated topic, it has been suggested that polaron excitons are intimately involved in the process. It is also of some interest that the electroluminescence (EL) and photoluminescence (PL) in polymers such as PPV are almost identical, indicating a common mechanism. Thus the production, propagation and stabilization of charge states in conjugated materials is of critical importance in polymer conductivity, nonlinear optical properties and in applications involving either PL or EL.

CHARGE STATE MODELING IN SMALL MOLECULES

Over the past 10 years we have studied the formation of polaron-like radical cations and bipolaron-like dications in a variety of model systems:[8-10]

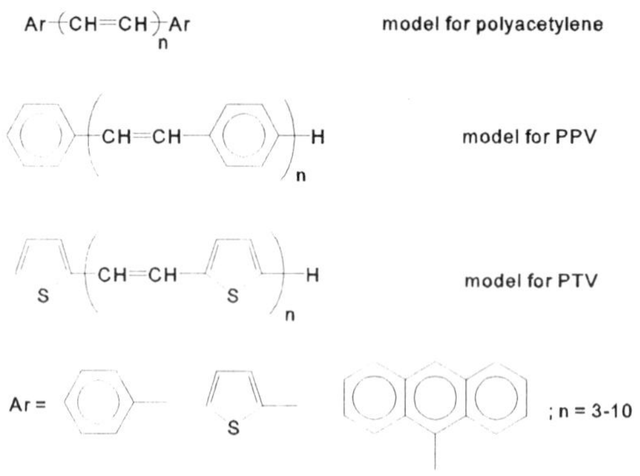

$Ar\!-\!(CH{=}CH)_n\!-\!Ar$ model for polyacetylene

model for PPV

model for PTV

Ar = ; n = 3-10

In all cases, stable bipolaron-like charge states can be produced in CH_2Cl_2 solution by chemical oxidation with $SbCl_5$. These charge states are quite stable and long-lived under ambient laboratory conditions, either in solution or as polycarbonate composites. For example, α,ω-bis-anthracenyl polyene dications are stable for up to six months as 1% polycarbonate composites with no apparent decay in optical properties.[6] However, one difficulty inherent in all of the above oligomeric models is the lack of solubility, particularly for the larger conjugation sequences ($n > 4$). This lack of processibility is a serious material defect which would preclude using these multifunctional materials in a wide variety of potential applications. The formation of these charge states is illustrated below for bis-thienyl polyenes (Figure 1).

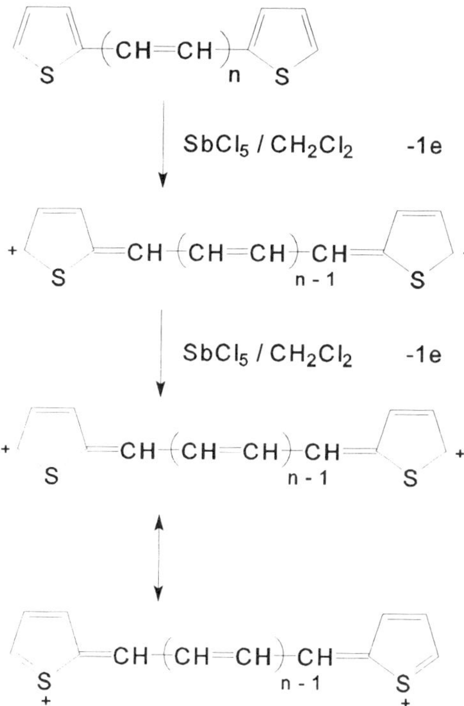

Figure 1. Formation of polaron-like radical cations and bipolaron-like dications in bis-thienyl polyenes.

SYNTHESIS OF BIS-THIENYL POLYENES INCORPORATING SOLUBILIZING SUBSTITUENTS

We have previously described the synthesis and formation of bipolaron-like species in both diphenyl and bis-thienyl polyenes.[8-12] Bipolaron formation and stabilization is greatly enhanced by mesomerically interactive electron-donating substituents. However, in these cases, lack of solubility precluded spectroscopic study of those materials beyond the hexaene in all cases. This lack of solubility also made evaluation of these materials for nonlinear optics and photonics application *via* degenerate four wave mixing (DFWM) studies extremely difficult. We have recently made use of the solubility enhancement attending alkyl group substitution to synthesize a series of 3,4-dibutylthienyl polyenes.[13,14] These materials, in contrast to the unsubstituted bis-

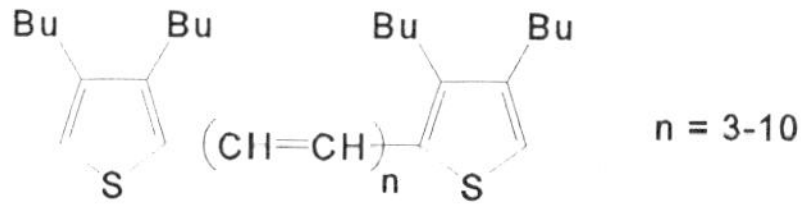

n = 3-10

thienyl polyene series, are soluble in a wide variety of solvents up to the decaene. Longer polyene sequences have been synthesized (n = 11, 12), but their stability in air becomes a problem at these longer lengths. Upon doping, dramatic shifts in oscillator strength are observed, with the BP absorption shifting from the VIS to NIR portion of the spectrum. This is shown in Table 1.

The third-order optical nonlinearity of the bis-thienyl polyene series has been studied *via* DFWM in collaboration with Hellwarth's group at both 532 and 1064 nm.[14,15] In all cases in which the resonance enhancement of the neutral and bipolaron species was minimal, a $\chi^{(3)}$ enhancement of a factor of 3 was observed. For the larger sequences, γ values approaching 10^{-29} esu were obtained, indicating that these are among the most nonlinear small molecules yet studied.

Table 1. Absorption spectra of bis-thienyl polyenes and polyene bipolaron-like dications.

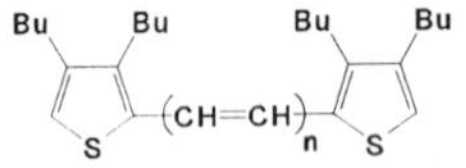

n	λ_{max} neutral (nm)[a]	λ_{max} bipolaron (nm)[a]
3	422, $\underline{399}$, 380	$\underline{655}$, 593
4	443, $\underline{418}$, 396	$\underline{661}$, 600
5	462, $\underline{435}$, 412	$\underline{715}$, 679
6	480, $\underline{450}$, 426	809, $\underline{719}$
7	496, $\underline{464}$, 439	$\underline{849}$, 790
8	510, $\underline{477}$, 450	$\underline{914}$, 855
9	521, $\underline{489}$, 461	$\underline{971}$, 892
10	534, 499, 471	$\underline{1022}$, 950

[a] 10^{-5} M solutions in CH_2Cl_2; underlined peak equals peak of maximum absorption.

INCORPORATION OF BIS-THIENYL POLYENES AS FORMAL COPOLYMER REPEAT UNITS

During the past decade the emerging field of photonics has shown dramatic progress in the potential application of π-conjugated organic polymers in the areas of frequency modulation and optical switching.[5] The bis-thienyl polyene series represents an intriguing opportunity to control the absorption characteristics in both the UV-VIS and NIR for a variety of photonics applications if they could be incorporated as formal repeat units in processible copolymers. Using an approach developed by Hadziioannou's group in the Netherlands for the synthesis of polymers for light-emitting diodes,[16] we have been able to synthesize a series of copolymers which incorporate the bis-thienyl polyene sequences, as outlined below:

Br–(Bu, Bu thiophene)–(CH=CH)$_n$–CHO + $Bu_3\overset{+}{P}CH_2(CH{=}CH)_mCH_2\overset{+}{P}Bu_3$, $2X^-$

NaOEt / EtOH / DMF / Δ

Br–(Bu, Bu thiophene)–(CH=CH)$_p$–(Bu, Bu thiophene)–Br p = 2n + m + 2

1. 4 eq *tert*-BuLi; 2. Me_2SiCl_2

[–(Bu, Bu thiophene)–(CH=CH)$_p$–(Bu, Bu thiophene)–SiR_2–]$_n$

These polymers are all obtained in good yields (> 90%), and are soluble in common solvents (CH_2Cl_2, THF). An important question for copolymer formulations in which alternating repeat units are to be oxidized to bipolaron domains, is whether all of the repeat units can indeed be oxidized. This is illustrated in Figure 2.

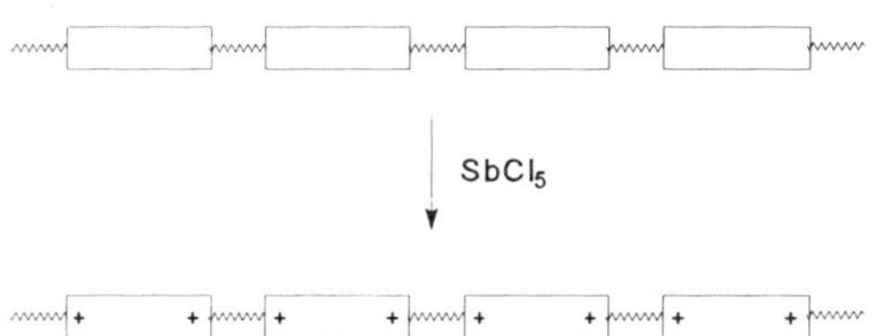

Figure 2. Oxidation of copolymers incorporating electroactive repeat units to copolymer bipolarons.

The copolymers where p = 4, 5 and 6 were oxidized in CH_2Cl_2 solution with $SbCl_5$, as previously described for the bis-thienyl polyene series. In all cases, a complete bleaching of the neutral π-π^* transition is observed within a few seconds. A new red-shifted band is observed which is quite stable in solution, showing no decay after 24 hours. The new absorption parallels that of the bis-thienyl polyenes listed in Table 1. In Table 2, these absorptions are compared.

Table 2. Comparative bipolaron spectra for bis-thienyl polyenes and bis-thienyl polyene copolymers.

n	λ_{max} polyene BP (nm)[a]	λ_{max} polymer BP (nm)[a]
4	$\underline{661}$, 600	660
5	$\underline{715}$, 679	752, $\underline{694}$
6	$\underline{809}$, 719	796, $\underline{739}$

[a]Underlined peak equals peak of maximum absorption.

Since there is *no* residual absorption from the original neutral polyene repeat unit, it would appear that *all* of the bis-thienyl polyene repeat units have been oxidized to the dication form. We are currently studying how best to cast or spin cast the oxidized polymers as optical quality films for DFWM studies. We anticipate that these materials will have large nonresonant $\chi^{(3)}$ values.

CONCLUSIONS

For the first time, long polyene sequences stabilized by thienyl substituents have been incorporated in processible copolymer formulation. Bipolaronic charge states can be incorporated in these materials by chemical oxidation. These new materials may have many potential applications in nonlinear optics, as conducting copolymers or polymeric LEDs, and possibly as optical limiters and sensor protectors. Since both the design approach and the fine control of both conjugation length and absorption characteristics offer great flexibility, these new materials should prove quite attractive as true multifunctional materials.

ACKNOWLEDGMENTS

This research was supported, in part, by the Air Force Office of Scientific Research.

REFERENCES

1. T.J. Skotheim, ed. "Handbook of Conducting Polymers," Vol. 1 and 2, Marcel Dekker, Inc., New York (1986).
2. H.G. Kiess, ed. "Conjugated Conducting Polymers," Springer-Verlag, Heidelberg (1992).
3. C.W. Spangler, K.O. Havelka, and P.-K. Liu, The formation of charge states in organic molecules, oligomers and polymers for applications in molecular electronics and photonic devices, *in*: "Molecular Electronic Devices: Science and Technology," Vol. III, K. Sienicki, ed., CRC Press, Boca Raton (1994).
4. J.I. Kroschwitz, ed. "Electrical and Electronic Properties of Polymers: A State-of-the-Art Compendium," John Wiley & Sons, New York (1988).
5. P.N. Prasad and D.J. Williams, "Introduction to Nonlinear Optical Effects in Molecules and Polymers," John Wiley & Sons, New York (1991).
6. E.G. Nickel, C.W. Spangler, N. Tang, R. Hellwarth, and L. Dalton, Third-order non-linearity in polymer models and composites containing stabilized bipolarons, *in*: "Organic Materials for Non-Linear Optics III," G.J. Ashwell and D. Bloor, eds., Royal Society of Chemistry, Cambridge (1993).
7. J.H. Burroughes, D.D.C. Bradley, A.R. Brown, R.N. Marks, K. Mackay, R.H. Friend, P.L. Burn, and A.B. Holmes, Light-emitting diodes based on conjugated polymers, *Nature* 347:539 (1990).
8. C.W. Spangler, L.S. Sapochak, and B.D. Gates, Polaron and bipolaron formation in model extended π-electron systems: potential non-linear optics applications, *in*: "Organic Materials for Non-Linear Optics," R.A. Hann and D. Bloor, eds., Royal Society of Chemistry, Cambridge (1989).
9. C.W. Spangler and T.J. Hall, Oxidative doping studies of PPV oligomers, *Synthet. Met.* 44:85 (1991).

10. C.W. Spangler and P.-K. Liu, Oxidative doping studies of PTV oligomers, *Synthet. Met.* 44:259 (1991).
11. C.W. Spangler, P.-K. Liu, A. Dembek, and K.O. Havelka, Preparation and oxidative doping of α,ω-dithienyl polyenes, *J. Chem. Soc. Perkin Trans. 1* 799 (1991).
12. C.W. Spangler and P.-K. Liu, Preparation and oxidative doping studies of α,ω-dithienyl polyenes stabilized by long chain alkylthio substituents, *J. Chem. Soc. Perkin Trans. 2* 1959 (1992).
13. C.W. Spangler and M.Q. He, Preparation and oxidative doping studies of bis-thienyl polyenes stabilized by alkyl group substitution, in press, *J. Chem. Soc., Perkin Trans. 2* (1995).
14. C.W. Spangler, M.Q. He, J. Laquindanum, L.R. Dalton, N. Tang, J.P. Partanen, and R. Hellwarth, Bipolaron formation and nonlinear optical properties in bis-thienyl polyenes, *Mat. Res. Soc. Symp. Proc.* 328:655 (1994).
15. N. Tang, J. Partanen, R. Hellwarth, J. Laquindanum, L. Dalton, C.W. Spangler, and M.Q. He, Studies of optical nonlinearity in bis-thienyl polyenes, *Proc. SPIE* 2285:186 (1994).
16. G.G. Malliaras, J.K. Herrema, J. Wildeman, R.H. Wieringa, R.E. Gill, S.S. Lampoura, and G. Hadziioannou, Tuning of the photo- and electroluminescence in multi-block copolymers of poly[(silanylene)-thiophenes]s via exciton confinement, *Adv. Mater.* 5:721 (1993).

SABIC AND THE SAUDI ARABIAN PETROCHEMICAL INDUSTRY: ANATOMY OF A LEADER

Moayyed A. Al-Qurtas
Director General, Research & Development
Saudi Basic Industries Corporation (SABIC)

Mr. Chairman
Conference Delegates
Ladies and Gentlemen

On behalf of Saudi Basic Industries Corporation -- more commonly known as SABIC -- I want to express appreciation to the Conference organizers for providing SABIC the opportunity to participate in this Third International Conference on Polymers, Malaysia.

Thank you for your welcome to Kuala Lumpur. I am honored to be with you today.

When one discusses the petrochemical industry in the Middle East, the point of reference is SABIC.

In just a year from now SABIC will celebrate the 20th Anniversary of the issuance of the Royal Decree that transformed our company from a mere vision to the reality that SABIC is today.

In 1993 we completed our first decade as a major chemical producer, and one of the world's fastest growing industrial conglomerates.

It has been a relatively brief, but very busy period for us. Petrochemicals is a tough and expensive business to break into. There are few new players.

The fact that SABIC's petrochemicals hold about a 5 percent market share reflects the respect and attention SABIC has gained in the global petrochemical industry over these two decades of development and growth.

It also reflects industry awareness to the economic maturity that the Kingdom of Saudi Arabia has attained.

Polymers and Other Advanced Materials: Emerging Technologies and Business Opportunities. Edited by P. N. Prasad *et al.*, Plenum Press, New York, 1995

However, before getting into the SABIC story, I would like to take a moment to familiarize you with the country where SABIC is based.

The Kingdom of Saudi Arabia is a big country. It has a land area of some 2.5 million square kilometers, or nearly a million square miles -- largest of the six nations which occupy the Arabian Peninsula, which also include: Kuwait, Qatar, the United Arab Emirates, Oman and Yemen.

Saudi Arabia is characterized by three salient features:

1. A geographic position astride historical trade routes between East and West.
2. Immense hydrocarbon reserves.
3. Financial strength.

With the advent of oil exploration and discovery in Saudi Arabia during the 1930s, Saudi Arabia has been, and continues to be, the major supplier of oil to the world.

In more recent times, the value-added concept took hold. It manifested itself into the conversion of oil into oil products, and later into capturing formerly flared natural gas associated with crude oil production and converting this into value added production, as well.

First: ammonia and derivatives, and later conversion of methane and ethane into a variety of petrochemicals, fertilizers and plastics, and as an energy source for metal industries.

The industrialization of Saudi Arabia over the past 20 years has had few parallels in history. SABIC has been a key factor in this achievement.

In the early 1970s the government of Saudi Arabia drafted plans to establish a domestic petrochemical industry as the foundation for a massive industrialization effort.

Large quantities of hydrocarbon gases produced in association with crude oil -- particularly methane and ethane -- were to be harnessed as feedstock for the manufacture of petrochemicals. To achieve this, SABIC was established in 1976.

Implementation of the industrialization plan was a monumental effort, financed with the Kingdom's oil income, for the purpose of reducing the nation's dependence on oil.

It was based on a three pronged approach:

First: Construct and operate a Master Gas System -- the MGS. This collects the associated gas produced with crude oil, processes it, and distributes it for use as feedstock and industrial fuel.

Second: Develop an industrial infrastructure to support world scale manufacturing complexes which would be supplied methane and ethane as feedstocks by the Master Gas System.

Third: Establish SABIC and build basic industries to be the commercial catalyst that would encourage further downstream diversification of the Kingdom's industrial base.

The goals SABIC set for itself were certainly challenging.

The company would broaden and deepen its presence in world markets, both in terms of products offered and in geographic reach. It would:

- Add value to Saudi Arabia's hydrocarbon and mineral resources.
- Expand and diversify production capacities.
- Develop Saudi human resources into a highly trained workforce.
- Provide raw materials to expand downstream product manufacture. And it would diversify and increase Saudi Arabia's export earning capability. The goals would be accomplished through the building of capital intensive world scale manufacturing projects incorporating the transfer of the newest and best technology available. This would result in the production of large volume commodity products to be marketed worldwide.

It was logical to choose petrochemicals to spearhead the industrialization of Saudi Arabia. They are comprised largely of hydrocarbons. They also consume considerable energy in their manufacture.

In other words, they depend on the one resource that the Kingdom has in abundance. Thus, an industry was created for large volume commodity petrochemicals where it made the most sense to do it -- at the world's largest source of hydrocarbons.

However, in spite of the historical precedent, most chemical producers and industry "experts" thought that the establishment of a large petrochemical industry in Saudi Arabia was not feasible. Or at best, it would offer no advantages over facilities built elsewhere.

The audacity of the company's plan centered on building three world scale ethane crackers and derivatives plants, plus methane derivatives plants, plus steel and support industries, with a total of nearly 8 million tons of product capacity.

The scale of this undertaking was met with considerable skepticism.

Well, the experts were proven wrong. SABIC turned out to be an astonishing industrial success story.

AR-RAZI, The Saudi Methanol Company, was the first of our new industries to come on stream in 1983, It was quickly followed by our introduction of urea fertilizer and steel production.

In quick succession, SABIC established a string of production affiliates in the Kingdom. Nine of these affiliates manufacture petrochemicals.

SABIC played a crucial role in developing Saudi Arabia's agricultural sector by providing a wide range of fertilizers to farmers. Three SABIC affiliates produce fertilizers widely used in Saudi Arabia. Surplus production is exported to international market. New production is being added.

One of the key planks of SABIC's industry-building strategy was the stress on providing the national economy with basic raw materials that support or create other industries. The Saudi construction industry has been among the beneficiaries of this strategy. SABIC's steel industry produces more than 2 million tons of steel bar and wire coil for concrete reinforcement.

A stand-alone SABIC affiliate provides nitrogen and oxygen industrial gases to SABIC's industries at Al-Jubail Industrial City.

Today, most of SABIC's industries are located in the industrial cities of Al-Jubail on the Gulf coast, and Yanbu on the Red Sea. These 15 world class manufacturing affiliates in Saudi Arabia produce a total of some 20 million metric tons of high quality SABIC products annually.

Additionally, SABIC is a partner in four companies based in Bahrain that manufacture and market methanol, ammonia, and aluminium.

A global marketing network supplies SABIC's petrochemicals fertilizers and plastic resins to some 75 countries.

SABIC Marketing Limited, one of the wholly-owned SABIC affiliates, is charged with marketing domestically and worldwide SABIC-produced petrochemicals, fertilizers and plastic resins.

In addition to being based close to the world's biggest oil and gas fields, SABIC's marketing effort benefits from the Kingdom's access to major shipping lanes serving Europe and the Far East.

China, Japan, South Korea and Taiwan are already among our most important customers.

Malaysia, Thailand, Indonesia, India and Pakistan are strong growth markets for SABIC's products.

And SABIC's rapidly growing Research and Development capability is a vital element in Saudi Arabia's emergence as a principal player on the global petrochemical stage.

The means to develop advanced technologies in-house is the ambition of every industry seeking to reinforce its place among major international manufacturers and product marketers.

SABIC created its Industrial Complex for Research and Development in 1992 to serve this ambitious goal. The SABIC R & D Complex occupies a 33 hectare science campus in Riyadh.

It is one of the largest industrial R & D centers in the Middle East, with the latest facilities and equipment, and with staff expertise recruited from all over the world.

Since R & D is my particular responsibility at SABIC, I would like to explain one especially unique feature about our operation. And at the end of my remarks I will be happy to discuss any questions you have about SABIC, and especially our R & D activities.

Let me simply note that a primary goal of SABIC's R & D activities is to develop technologies especially tailored to Saudi Arabia's industrial growth. There are many technologies that are of interest to SABIC.

Unfortunately, however, many of these are out of reach, either because they are monopolized and controlled, or because the conditions stipulated for their licensing are not acceptable to us.

Since it is not easy to start from scratch, the best alternative is to work with the Kingdom's universities, research centers and scientific organizations to identify mature research projects available in the market that are well suited to Saudi Arabia's industrial growth.

We can also indulge in cooperative research and development with other companies which have similar interests. We share the risk, the effort and the cost and arrive at a common goal.

An early result of this cooperative R & D effort has been the development of butene-1 technology in cooperation with I.F.P. -- the Institute Francais du Petrol. SABIC shares the licensing rights for worldwide marketing of butene-1 technology. Seven overseas licenses for this technology have thus far been granted.

Maximizing SABIC's technological capabilities is a vital key to Saudi Arabia's emergence as a principal player on the global industrial stage.

SABIC has also made a significant contribution to further the adoption of the TQM -- Total Quality Management concept -- throughout the corporation. As a result, 11 of SABIC manufacturing affiliates have been awarded ISO 9002 certification for quality production. It is expected that all of the affiliates will soon be ISO certified, further expanding the marketing horizons for SABIC's products.

From the beginning SABIC's growth strategy has been based on the concept of shared participation in the utilization of the Kingdom's resources for chemical production.

Most of SABIC's industries are joint venture partnerships with some of the world's leading chemical companies.

The partners were chosen on the basis of their proven record in the industry and on their willingness to provide state-of-the-art technology and hands' on employee training.

In return, SABIC's partners would share in SABIC's success, taking advantage of plentiful feedstocks, a healthy economic climate, a modern and highly developed industrial infrastructure and easy access to thriving global markets.

From the start of production in 1983, SABIC has established itself as a leading international petrochemical company through a series of expansion and diversification projects, which are continuing.

Specifically, SABIC is in the process of enhancing considerably its position in three principal and long-established product areas -- Methanol, Ethylene and Ammonia -- and their immediate derivatives.

This is an on-going and ambitious program, and part of SABIC's strategic long-term goal to anticipate future growth opportunities. Most of the expansion projects are being developed within the existing infrastructural frame-work of SABIC's industries.

For example, world demand for methanol is forecast to rise by about 6 percent a year, and will continue to do so through the rest of this decade. SABIC has grown to become a leading global producer of chemical grade methanol. Expansion projects now underway will boost methanol capacity to more than 3 million metric tons by 1997.

This tonnage represents about one-third of the product required to manufacture the environmentally-friendly gasoline additive MTBE (methyl tertiary butyl ether). Demand for MTBE is growing at an annual rate of 20 percent, or better.

SABIC's MTBE production capacity will increase from 1.9 million tons, to more than 2.5 million tons per year by 1996.

Here is another example: Plant construction of SABIC's 16th industrial complex is well underway at Yanbu Industrial City on the Red Sea. This is IBN RUSHD -- The Arabian Industrial Fibers Company -- SABIC's first polyester manufacturing facility, which is scheduled to launch production next year.

The new plant is sited near SABIC's YANPET affiliate, which will supply it with ethylene glycol, a feedstock for the manufacture of polyester.

The IBN RUSHD complex will be integrated further by 1997 with the addition of two major process plants.

One plant will manufacture 350,000 tons per year of PTA (purified terephthalic acid). This is the other primary feedstock needed to produce polyester.

A second production unit will manufacture aromatics, to include 300,000 tons of benzene, and 380,000 tons of xylenes.

SABIC's total output of petrochemicals, chemicals, fertilizers, thermoplastic resins, metals and industrial gases totaled just over 20 million metric tons in 1994, an increase of 3 million tons from 1993.

While SABIC's industries were created world-scale in size with the intent of exporting a substantial part of production, the market which has grown far beyond our expectations has been in Saudi Arabia.

SABIC's greatest achievement has been to foster the development of new domestic downstream industries.

The growth of the plastics industry in Saudi Arabia is a good example of downstream industrial progress. Today there are 266 plastic processing plants in the Kingdom, versus only a handful a decade ago.

At that time, the market for plastic resins was about 15,000 tons per year. Today that market is approaching 300,000 tons and this should nearly double by the year 2000.

SABIC now produces and supplies all five of the major thermoplastic resins, along with Melamine, a thermosetting resin, all under SABIC's LADENE trademark -- the Arabic word for plastic.

Polyester staple fiber and POY (partially oriented yarn), plus Polyethylene Terephthalate (PET) bottle grade resin, will be available late next year.

As we look forward to the year 2000, there is no doubt that SABIC has quite a bit of activity planned.

We forecast the addition of a further 10 million metric tons of production through horizontal expansion, vertical integration and product diversification.

- Many new projects are planned including:
- Basic Products: such as ethylene, propylene, aromatics and methanol.
- Intermediate Products: propylene oxide, ethylene dichloride, styrene and PTA.
- Final Products including: MTBE, urea and polyester.
- New investment will reach about 12 billion U.S. dollars.
- We are looking to form new global business alliances in manufacturing joint ventures, in research and development, in marketing, and in technology cross-licensing.
- These new strategic alliances will be based on a cooperation model that has proven its practical efficiency and sustained viability with our global partners.

Well, there you have it -- the anatomy of SABIC:

Where we've come from;
What we are today; and
Where we're going.

I want to conclude my remarks by simply noting this: Saudi Arabia is investing heavily in the future of petrochemicals -- not only as a means of developing a strong domestic downstream industrial base -- but also to maintain and solidify SABIC's growing share of the global marketplace for petrochemicals.

As we look to the future, we see major expansion opportunities and new projects that will provide further growth opportunities for SABIC.

We also see these growth opportunities being supported by SABIC-developed science and technology, leading to new product development, as well as expanding the horizon of applications for our current products.

I want to leave you with one other thought.

SABIC is an environmentally aware, responsible and committed enterprise. The company is confident that additions to our manufacturing facilities, new product introductions, and increased production will not lessen our sensitivity to the environmental concerns -- not only of Saudi Arabia and its citizens -- but of the world we all share.

It is hoped that the awareness SABIC has demonstrated and the procedures the company has incorporated will set precedents for other developing nations to follow.

Thank you for your interest and attention.

-- END --

POLYMER TECHNOLOGY AND EDUCATION FRONTIERS

A CONTRIBUTION FROM AUSTRALIA

Dr. Alastair N. McKee,
Director,
Cooperative Research Centre for Polymer Blends
32 Business Park Drive,
Notting Hill VIC 3168
Australia

INTRODUCTION

The Cooperative Research Centre for Polymer Blends (CRCPB) is one of sixty organisations funded under an Australian Government programme to encourage greater collaboration between researchers in industry, government and academia. The CRCPB, which began operation in July 1992, is one of eight in the manufacturing technology area, and the sole CRC in the chemicals sector.

BACKGROUND

The challenge and cost of developing new polymers has become so great that existing polymers are frequently blended to improve or to balance properties such as impact strength, stiffness, heat deflection under load, chemical and solvent resistance, flame retardancy or barrier properties, and processability. The performance of low cost polymers can be improved to meet certain performance capabilities of engineering resins, and to tune a material to meet customer specifications: in the ideal case, synergy may result in unexpected improvements.

THE CONCEPT

The CRCPB adopted a mission to provide a focus in Australia for the development of advanced polymer blends, processing technologies and skills encompassing the whole product life cycle, and to provide a resource to Australian industry to enhance international competitiveness.

The scope was interpreted to incorporate an integration of:

- blend component design and synthesis
- blend processing and implications of processing on melt and solid state structure and property relationships
- product design, engineering and production questions, and
- resource utilisation, recycling and reprocessing of plastics materials, including mixed materials to achieve products of high value

THE SCIENTIFIC CHALLENGES

The work area brings together many fields of science and engineering, directed towards well defined targets. Included are thermodynamics, chemistry, physics, rheology and process control.

Key scientific areas include:

- understanding molecular structure, architecture and mechanisms of compatibilisation
- conceiving, designing and synthesising new and improved compatibilisers
- altering functionality by reactive processing
- studying and controlling the rheology of blends
- influencing morphology by optimising processing conditions (and time), leading to targeted structure-property relationships
- researching the unmixing or separation problem
- improving the economics of blends by re-introducing polymers to valuable second and subsequent uses

The goals of the CRCPB differ from those of a scientific institute in that the full measure of success is gained only when scientific and engineering advances are taken up commercially. This adds an important marketing emphasis to its work.

THE FIELD OF INTEREST OF THE CRCPB

Since blends are included in the general field of polymers, by definition we are interested in one sector of this, and the CRCPB research providers can and do operate in the wider field of polymer blends. The CRCPB industrial participants are interested in separate sub fields, primarily polyolefins on the one hand, and ABS and vinyl esters on the other. There are many areas in the blends field not of interest to the current industrial participants, leading to the opportunity being pursued for the CRCPB to develop contract relationships with other companies.

The participants in the CRCPB are shown in Table 1. The skills, experience and resources of these groups allows the CRCPB to offer expertise from computer simulation to predict the behaviour of compatibilisers and polymer blends through synthesis and manufacture of trial quantities of product for evaluation across the complete range of relevant parameters. Formulations, composition and economics can be tuned to ensure optimum in-service performance of materials in their intended products.

When the participants of the CRCPB cannot provide the required range of capabilities to meet a customer need, we are able to draw in additional resources from affiliates such as the Casey College of TAFE Plastics Technology Centre, or other academic groups through the developing Australian Polymer Network. With this we can offer from a single "shopfront", access to a tremendous breadth of knowledge for problem solving, development and research work.

The CRCPB programme comprises a number of core activities in which underpinning technologies are developed to support work in shorter term, more industrially focussed projects. In the second group in particular, the support and close involvement of people from the industrial partner are crucial to ensuring that the investigations undertaken are directed to solving problems of practical importance, which if solved will lead to a commercial benefit by way of exploiting the outcomes.

In addition to industry people, staff for each project are drawn from a number of participants, leading to closer personal and institutional ties, while ensuring an effective cross-organisational transfer of information and access to equipment and other resources. It is the policy of the CRCPB to augment existing laboratory and other facilities established by our participants, by funding additional items of equipment and people.

Table 1: Participants in the CRCPB

Type	Organisation	Division
Government	Commonwealth Scientific and Industrial Research Organisation (CSIRO) Defence Science and Technology Organisation (DSTO)	Division of Chemicals and Polymers Aeronautical and Maritime Research Laboratories
Industrial	Huntsman Chemical Company Australia Limited ICI Plastics Australia	
Academic	Monash University Royal Melbourne Institute of Technology (RMIT)	Department of Materials Engineering • Polymer Technology Centre • Rheology and Materials Processing Centre • Department of Applied Chemistry
Associates	Cray Research (Australia) Pty Ltd Plastics and Chemicals Industry Association of Australia	

Table 2: Current CRCPB Projects

Short Title	Partner	Leader	Node
1. Melt Strength Enhancement of Polyolefins	ICI	Prof. S.N. Bhattacharya	RMIT
2. Polyolefin Blends	ICI	Mr. E.I. Kosior	RMIT
3. Toughened Polypropylene	ICI	A/Prof. R.A. Shanks	RMIT
4. Blends of Engineering Thermoplastics	Huntsman	Dr. W.D. Cook	Monash
5. Blends of Polyolefins with Elastomers		Dr. R.Y. Wu	CSIRO
6. Tough Thermoset Resin Blends	Huntsman	Dr. C.E.M. Morris	DSTO
7. Design & Synthesis of Compatibilizers & Polymer Blend Components	Core Activity	Dr. E. Rizzardo	CSIRO
8. Computer-aided Technologies in the Design of Polymer Blends	Core Activity	Dr. G. Moad	CSIRO
9. Education	Core Activity	Dr. A. Goodwin Mr. E. Kosior	Monash RMIT

EDUCATION AND TRAINING

Monash and RMIT are very well established in a wide range of teaching and training activities, with students coming from many countries to study in Australia. In recent times they have paid particular attention to offering their products in increasingly diverse geographical areas, and the CRCPB in its area will also seek to participate in selected regional markets.

A feature of the educational approach is to develop and offer courses with close industrial linkages, seeking to prepare people to make a rapid, practical contribution in a workplace by ensuring relevance of the course work to the plastics and rubber processing industries. A sound theoretical knowledge provides a basis for clear and full understanding of the principles and processes involved, with emphasis also on problem-solving techniques and continuous improvement practices. Project work is selected wherever possible from a range collated with industry input, and involvement with companies includes vacation and other student placement periods.

Training is provided across the complete range of activities of importance to the polymers industry.

Award courses are offered at diploma, bachelor, masters and doctoral level in Polymer Engineering, Materials Engineering, Chemistry and other branches of science. The CRCPB sponsors about a dozen, mostly PhD candidates who may be co-supervised by academics from more than one institution, and by government - and industry-based - people. Topics are selected to study in detail aspects of the projects being undertaken within the CRCPB: in this way, students gain an early understanding of the needs and constraints of industrial research.

Short Courses are designed to combine state of the art lectures supplemented by laboratory or other experience illustrating the applicability of research effort to industrial activity. In other cases, a particular subject may be treated at levels appropriate to operators, to supervisors or to management, so that knowledge across the company can be disseminated to support adoption of new initiatives. Courses can be tailored to meet specific industry needs, and can be offered through industry associations or other groupings.

Seminars and Workshops may be based on visiting experts in key areas of interest across a variety of companies, institutions or government agencies. Specialised workshops may be organised at scientific meetings or industry exhibitions, to help people remain current with international progress and standards. Individual, private company workshops may be arranged in addition to the open sessions, and consulting organised.

BUSINESS DEVELOPMENT

The Board of the CRCPB is comprised of nominees of each participant, an independent chairman and two individuals representing major customer industries, namely automotive and packaging. The Board has decided that the CRCPB should seek to prosper beyond mid-1999, the date to which funding is contracted in the Centre Agreements. Incidentally, the Australian Government matches inputs contributed by the participants.

To develop the required income stream, the CRCPB is obliged to build a separate character from that of its participants, by earning a reputation for its ability to negotiate contracts based on agreed work programmes, meet time and deliverable commitments, and to add value to its customers' operations. Marketing has commenced in Australia, and we are now starting to sign our first substantial contracts. We recognise a need to work in a broader sphere, with larger industry and other opportunities than exist in the relatively small if quite sophisticated economy of Australia, and attendance at the 3rd International Conference on Frontiers in Polymers and Advanced Materials and participation in its Exhibition represents a first step. We would like to develop on-going relationships with firms and others in countries such as Malaysia, Singapore, Thailand, Indonesia and ASEAN generally, as well as in countries further north (such as Vietnam, Japan and Korea) and west (such as India). We believe that the range of contributions which the CRCPB can draw together could add to existing resources of large and small companies, and perhaps there may be opportunity in consulting to governments and regional organisations in collaboration with local partners.

The Australian Polymer Network has already been mentioned and, in conjunction with the Plastics and Chemicals Industry Association, effort has been directed in recent months towards collaboration within the Commonwealth Plastics Network (CPN). We are building on the CPN membership of the RMIT Polymer Technology Centre, and are preparing for an Australian contribution to the Plastics Waste Management Forum being organised by the CPN, the Standards and Industrial Research Institute of Malaysia and the Malaysian Plastics Manufacturers' Association in Kuala Lumpur in March, 1995. At this meeting, a very exciting extension of scope of the CPN will result in the formation of a Global Plastics Network, with membership open to all nations. We believe that this offers great potential for regional cooperation, and it is our strong desire that the Cooperative Research Centre for Polymer Blends plays a part.

FUTURE OF SILICON SCIENCE AND TECHNOLOGY

F. W. Gordon Fearon
Central R&D
Dow Corning Corporation
Midland, MI 48686

Silicon is the most abundant solid element in the earth's crust and probably in the universe as a whole as far as we can tell. The earth's crust contains almost 26% of silicon by weight combined with oxygen and other metals as sand or silicates. Mankind has worked with silicon based materials for thousands of years, ceramics, glasses, bricks and mortar, many useful materials which have been discovered, developed and utilized are based on silicon. However, our scientific understanding of these materials is relatively still in its infancy. It was not until the late 19th and early 20th century that scientists began to study silicon materials intensively, and it was not until the 1940's that synthetic materials based on silicon were produced and employed. In the last fifty years the discovery, development and application of new silicon based materials has expanded rapidly so that today over 8000 distinct products can be identified, accounting for sales of over $6 billion worldwide, business which provides employment for at least 20,000 people. Will this growth continue?

Figure I shows the number of publications covering synthetic silicon based materials in recognized scientific journals for five year periods starting in 1930. Data was collected by an information specialist with many years of experience in this field, so whereas the absolute numbers can be challenged, the relative trends are real. Figure II shows similar data for patents covering the same material classes. Between 1950 and 1955 approximately 100 publications were issued, between 1985 and 1990 the number is greater than 20,000. During the same time period, the number of patents increased from less than 200 to greater than 50,000. The publications provide a measure of scientific knowledge, patents provide a measure of the conversion of this knowledge to potentially useful technology. These data all suggest that the field of silicon science and technology is growing very rapidly and

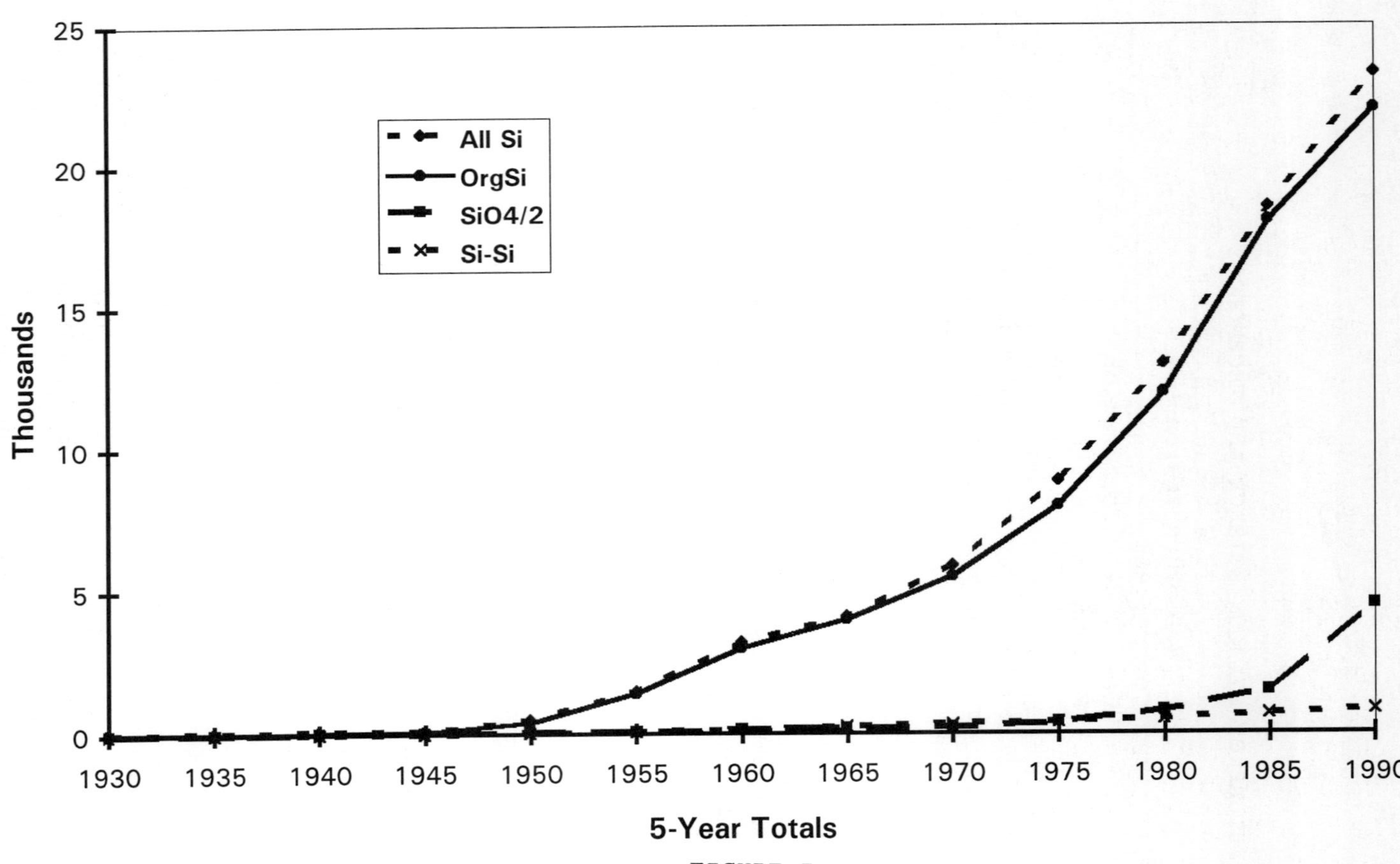
Silicon-containing Publications
Thousands
25
20
15
10
5
0
1930
1935
1940
1945
1950
1955
1960
1965
1970
1975
1980
1985
1990
5-Year Totals
All Si
OrgSi
SiO4/2
Si-Si

FIGURE I

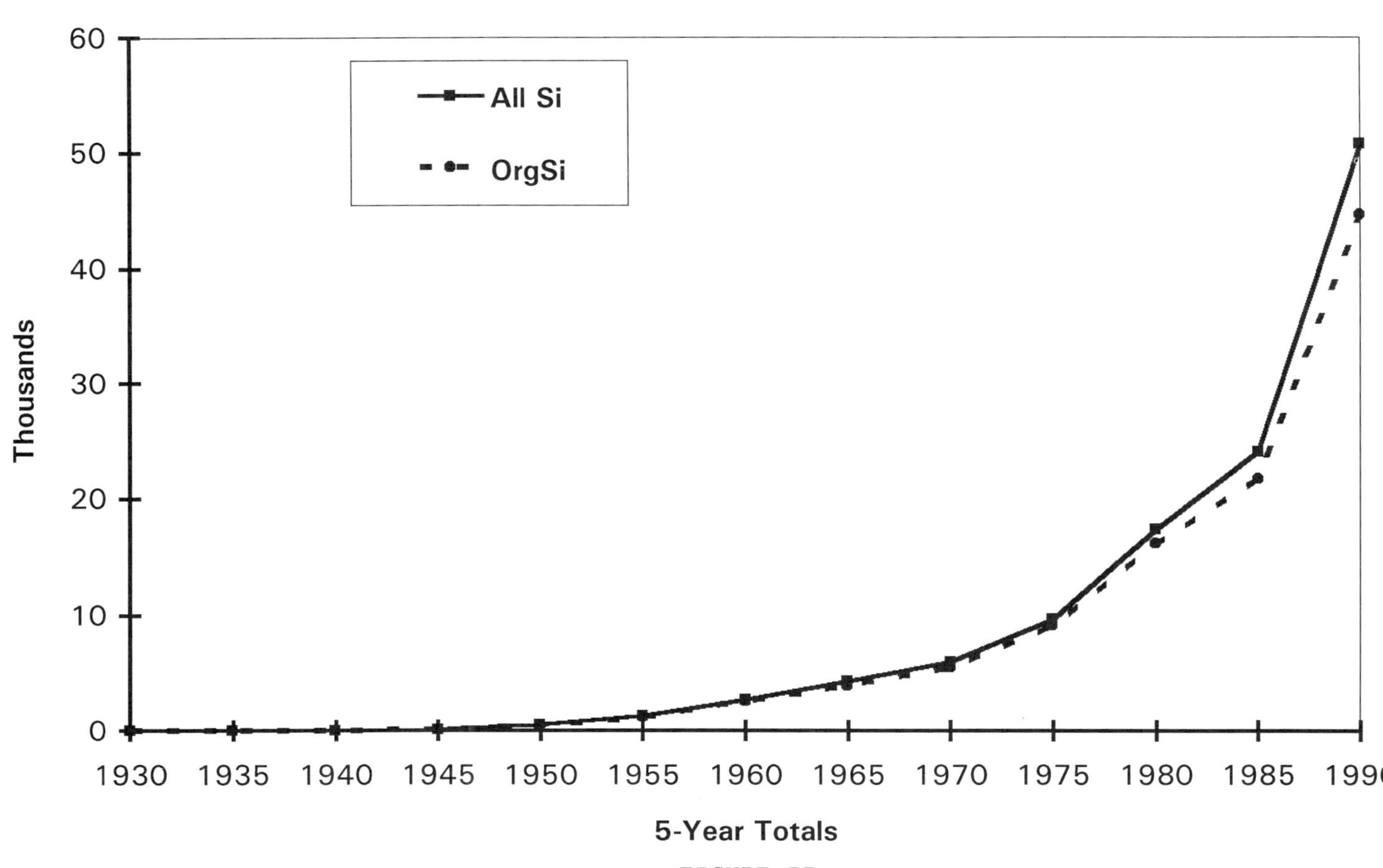
Silicon-containing Patents
All Si
OrgSi
Thousands
60
50
40
30
20
10
0
1930
1935
1940
1945
1950
1955
1960
1965
1970
1975
1980
1985
1990
5-Year Totals

FIGURE II

will continue to grow. In addition to the total number of publications describing silicon based materials including semiconductor grade silicon Figure I shows the total number of publications covering organosilicon polymers, combinations of silicon polymers and organic polymers, sol gel systems and polysilanes. Each of these classes will be dealt with in detail later.

In this brief overview, I will give one perspective from industry on the future for this field of materials science and technology. I will focus on specific applications which are, or are likely to be commercially significant. I will deal with these in two categories those materials which are already commercially significant and which appear set for continued growth, and those materials which I believe will be commercially significant in the future. For those interested in learning more about this field, I recommend several general texts and reviews which are listed in the bibliography.

Silicon, silanes, polysiloxanes, polysilsesquioxanes and silicates are certainly the materials which form the basis for today's business, and they will continue to be very important in the future. In addition, I believe that new families of materials based on polysilanes, polycarbosilanes, polysilazanes and combinations of these materials with other organic or inorganic materials will add valuable new dimensions to the silicon material base.

Several rapidly growing applications for silicon based materials have emerged because material suppliers have learned to produce silicon chemicals and materials based on these, to a controlled and very high level of purity. Let us now look at these in some detail.

The semiconductor industry is based on silicon. The bulk of all semiconductor devices manufactured today are based on silicon which is produced synthetically by chemical vapor deposition of trichlorosilane. This trichlorosilane is purified to a level of parts per billion impurity, and this level of purity is maintained during conversion to silicon. Semiconductors can be based on materials other than silicon, gallium arsenide for instance, and recent work shows that semiconductors can be made from organic materials. However, all evidence suggests that silicon will continue to be the most important basic material for the semiconductor industry into the foreseeable future.

Photovoltaic devices are a second major application for high purity silicon. Today photovoltaic cells based on silicon power watches, calculators and a whole variety of systems which will require electrical power remote from the normal electrical distributor networks. Several countries today support long range programs to develop photovoltaics as a viable source of power generation. Fundamental to these programs is continuous improvement of the purity and control of the structure of the silicon. This is turn is based on continuous development of the silicon chemicals used to produce these materials and deeper understanding of the structure, property, process relationship of the derived silicon.

Many learned sources describe our emerging information based society, and rapid access to information is now judged to be the pacing step in many industries. The use of optical fibers to transfer large amounts of data is now well established. These optical fibers are made from high purity silica, and this in turn is made from high purity chlorosilanes or siloxanes. Once again, the ability of the silicon material suppliers to provide high purity intermediates has opened up a major new application for silicon based materials which is still in its infancy.

One of the most useful features of organosilicon materials is as the names implies, the ability to combine organic, carbon based, with inorganic, silicon based properties. The benefit of this feature is readily apparent in silane coupling agents. These materials combine an organic functional group with a silicon functional group. Some examples of this broad family of materials are listed in Figure III. The primary application for these materials is to interface between an organic and inorganic material, for instance, to promote adhesion between organic materials and inorganic substrates. The choice of organic and inorganic functionality is large, so that these coupling agents can in principle be tailored to specific systems.

Synthons, intermediates designed to steer the course of chemical reactions, are key to the synthesis of high value molecules, especially pharmaceuticals. Use of silicon chemicals in this type of application is growing rapidly and is expected to continue to expand into the foreseeable future.

During the last decade great advances have been made in defining polymerization processes based on metallocene catalysts designed to carry out very specific polymerization. Silicon is a key building block in these catalysts.

Silicon chemicals are key building blocks and intermediates for many systems but by far the most important today are polymeric and of these, dimethylpolysiloxane is the most important. Families of products based on this polymer provide the bulk of today's business. Molecular weights of these polymers can be controlled from hundreds to millions providing access to a wide variety of material forms from low viscosity liquids to ultra high viscosity, but still flowable, gums. Functional groups can be incorporated to add value by enhancing high temperature or low temperature performance, solvent resistance, interfacial activity, dielectric or rheological properties. Reactive sites can be incorporated to facilitate crosslinking or to promote interaction with other materials. The systems can be readily reinforced with various silicas to yield elastomeric systems.

These materials provide the basis for families of fluids, gels, sealants, elastomers and coatings which find application in many fields too numerous to list here. But when all is said and done, these materials are used either because they resist harsh environments or because they have high interfacial activity. Dimethylpolysiloxane can resist harsh environments because it is able to absorb and dissipate large quantities of energy without rupture of the polymer chain. Compared with most common organic materials, this polymer has a high resistance to radiation, heat and dielectric breakdown - features which are the basis for families of protective coatings, sealants and adhesives, solid and liquid insulating systems, mechanical fluids, lubricants and many families of fire resistant materials. The polymer chain is one of the most flexible known, a characteristic which is primarily responsible for the high interfacial activity. This feature is widely utilized in families of release agents and adhesives, profoamers, antifoams and cosmetics.

The above is, of course, a very simplistic overview of a very complex material, but in the context of this whole subject, probably appropriate coverage. Key points to remember, the polydimethylsiloxane system forms the basis of most of today's silicone business. The polymer system is very versatile, and its properties can be easily tailored for specific applications. Applications for the system continue to grow rapidly and will continue to grow into the foreseeable future.

The second large class of commercially significant silicone polymers are families of branched materials generally referred to as resins, based on tri- as well as di-monofunctional monomers. Historically, these represent some of the first commercial silicon products that were ever made, and many are still very valuable today in specific applications. But, relative to the linear polymers, we know little about the detailed structure property relationships of these complex formulations. A typical material may be formed by the cohydrolysis and condensation of several mono-, di- and tri-functional monomers in processes which are still quite artistic. The true structure of the final materials is often merely speculation. Understanding the structure, property, process relationship of resins is one of the challenging frontiers of silicon materials research.

Even more complex and valuable materials are based on tetrafunctional monomers. Hydrolysis and condensation of a tetraethoxysilane, for instance, gives silica in various forms. Incorporation of a monofunctional monomer into the hydrolysis can yield families of discrete, soluble siloxylated silicas which are very valuable for reinforcement and surface modification of siloxane systems. Recent work by several groups is showing that these materials are comprised of well defined cage structures. Once again, the understanding and control of these materials provides a significant challenge for material scientists.

Silicon materials are based on sand, but in most of the polymeric materials this is not apparent, most silicon polymers look very similar to their carbon based counterparts. A very high molecular weight polysiloxane looks very similar to a low molecular weight polyolefin, but a very simple way to tell the difference is to burn the material. Pure polyethylene burns to give carbon dioxide and water, pure dimethylpolysiloxane burns to give carbon dioxide, water and silica. Heat release from polyethylene is approximately 1000kW/M^2. Heat release from dimethylpolysiloxane is approximately 150 kW/M^2. The silicone material burns with difficulty because the silica forms a protective skin on the surface which effectively reduces pyrolysis. The intensity of a polyethylene fire depends on the size, larger fires feed back more energy to the polymer, accelerate pyrolysis, which provides more fuel so the fire burns more intensely. The silica skin which forms on the silicone reduces pyrolysis so that the intensity of the fire is almost independent of size.

During the last few years the fire science of silicon based materials has been under intense study. In addition to the characteristics detailed above, evidence is now emerging that silicon based materials are more difficult to ignite than their carbon based analogs and that flame spread is much slower for silicon based than carbon based materials. Likely these characteristics are linked to the ability of silicon based materials to absorb energy as was mentioned before, but this is speculation.

The subject of fire science makes a nice bridge to the future and to emerging materials. Certainly silicon based materials are used today in applications where their reduced flammability is of advantage, as liquid dielectrics in transformers and in insulation for high voltage cables, to name two. As society continues to demand safer materials for structures and furnishings, especially for transportation, it is reasonable to expect that use of silicon based materials will become even more important.

Continuing with this theme, I will now briefly review some families of material which are emerging as the silicon based materials for the future. In my opinion, these will augment, not replace, the established polysiloxanes to provide a basis for continued growth of the business. Good candidates are polysilanes, preceramic polymers based on polycarbosilanes and similar precursors, blends and copolymers of silicone and organic polymers, sol gel materials and related cage structures and composites, plus multilayer structures based on these.

The polysilane backbone imparts some potentially useful features to this family of material, such as conductivity and photosensitivity. At the present time, wide application of these materials is hindered by the lack of good, cost effective manufacturing technology. Most synthetic routes involve metal reduction of halosilanes in some way or another. These routes inherently produce much waste and are high cost. Approaches via the polymerization of silane intermediates show more promise. Once satisfactory process technology is developed, these materials will likely find broad application.

Silicon carbide, silicon nitride and other families of silicon based ceramics are engineering materials with outstanding resistance to high temperatures, corrosion and wear. However, processing these materials into useful forms in a controlled manner has always been difficult. Approaches to these valuable materials through careful structured preceramic polymers has demonstrated potential to yield fibers, matrices, and derived composites which resist very harsh environments. For example, silicon carbide composite components for next generation turbojets are routinely resisting temperatures of greater than 1200°C in operating prototypes.

Publications, Figure I, suggests that sol gel materials are the fastest growing family of silicon based materials today. This is a broad family of materials, not all are based on silicon, but the silicon based versions are probably the most common and easiest to handle. In general, these materials are produced by cohydrolysis of an alkoxysilane with an inorganic colloid such as silica. The number of tri- or tetrafunctional alkoxysilane that can be used is large. Colloids can be of various sizes and compositions, and are often combined with other organic and silicon functionality, so the number of chemical compositions that can be prepared are very large indeed. If publications are any evidence, several groups are well on their way to synthesizing all available combinations!

To date, the number of commercially significant products based on sol gel systems is quite small. Families of coatings have been used for some time as hardcoats for plastics, particularly transparent plastics used in opthalmics and glazing. Similar materials show promise in corrosion protective coatings for non-ferrous metals such as aluminum. Other fields of application which are presently being explored including use as matrix materials for solid lubricants, catalysts, and optical elements.

Recent discoveries suggest that silicon will continue to form the heart of the semiconductor into the foreseeable future. New designs have been shown to greatly speed up devices, many of the insulating components of devices are made from silica or silicon nitride. Silicon carbide is emerging as a most effective high temperature (600°C+) semiconductor and a host of silicon chemicals and materials are being used to process semiconductors.

Through the last two decades many workers have explored the combination of silicon material with organic materials to form silicon organic hybrids. Blends, copolymers, silicon functional organic materials, and organic functional silicon materials all have been investigated, patented and widely reported.

Mechanical blends are simple to produce. For instance, it is easy to mix dimethylpolysiloxane with polystyrene or a host of other thermoplastic materials. Thermosetting materials are somewhat more challenging but various techniques are known to compatibilize these blends. Simplistically, it is possible to mix most silicon polymers with most organic polymers. What use are these mixtures? Introduction of small quantities of siloxane, particularly polydimethylsiloxane into an organic system will significantly change surface properties such as release and coefficient of friction. Sometimes changes in bulk properties such as rheology or modulus result, but these are often very process dependent so commercial application is not common.

Block or graft copolymers of silicon and organic systems are slowly being developed as important items of commerce. Such combinations often have unique surface or mechanical properties which are valuable. For instance, block and graft copolymers of polysiloxanes and polyethers are very active surfactants, which have been used to stabilize polyurethane and other organic foams for many years. Similar copolymers are now finding application in aqueous systems where powerful surfactants are needed.

Though much effort has been expended, there are as yet few if any examples where silicon organic block or graft copolymers are being used for their bulk properties. Many interesting materials have been investigated, and have shown useful properties compared with materials presently being used but have ultimately been judged not to be cost effective.

For example, multiphase composites of poly(styrene-g-dimethylsiloxane-g-acrylonitrile) have much higher impact strength and better weather resistance than ABS, poly(acrylonitrile-g-butadiene-g-styrene)(-ABS). To date, engineers have not demanded these superior properties.

The use of small amounts of silicon functionality to enhance the properties of an organic polymer is probably the most successful and fastest growing example of silicon modified organic material today. For example, organic chemicals shown in Figure III are easily combined with organic polymers to provide adhesion or to provide crosslinking sites that can be activated and crosslinked at room temperature. These small amounts of silicon functionality are the critical enabling technology for families of composites, adhesives, coatings, and sealants. All evidence suggests that application for these types of materials will continue to grow rapidly.

Based on material availability, breadth of chemistry, derived material compositions and possible application, the future for silicon based materials seems secure. But challenges exist.

In this day and age, all materials are examined carefully for impact on health and the environment. Silicon based materials are no exception. To prove that something is safe is impossible, it is only possible to show that different materials are relatively more or less safe or have relatively more or less impact in specific applications compared with known, accepted materials which can be used as standards for acceptable performance.

Today all evidence suggests that dimethylpolysiloxane is one of the most inert polymeric materials known. In the environment the material appears to degrade simply to silica carbon dioxide and water. Based on this evidence, it may be expected that materials based on dimethylpolysiloxane will ultimately prove to be some of the most acceptable materials to mankind from a health and safety standpoint. But not all silicon based materials are inert. For instance, families of very physiologically active silicon based materials are known today and active research programs seek effective silicon based pharmaceuticals. So another major challenge is to educate the public that all silicon based materials are not the same and certainly all of them are not inert!

Silicon based materials are expensive compared with organic counterparts so they are only used where their high value can be justified. Today the first step in the production of most silicon chemicals and derived polymers involves the carbothermic reduction of silica to silicon, an energy intensive and expensive process. The full potential for silicon based materials will not be realized until synthetic processes are discovered to bypass this step and find a less expensive route to useful silicon based materials.

Some Silicon Functional Materials

- $CH_2 = CH\ Si(OCH_3)_3$
- $ClCH_2CH_2CH_2Si(OCH_3)_3$
- $\overset{O}{\widehat{CH_2CH}}CH_2OCH_2CH_2Si(OCH_3)_3$
- $CH_2 = \overset{CH_3}{\overset{|}{C}}\text{-}COOCH_2CH_2CH_2Si(OCH_3)_3$
- $H_2NCH_2CH_2CH_2\ Si(O\ C_2H_5)_3$
- $HS\ CH_2CH_2CH_2CH_2Si\ (OCH_3)_3$

FIGURE III

To summarize, today, early 1995, just over 50 years since the first silicon based polymers were introduced as items of commerce, the field of silicon science and technology is dynamic and appears to be expanding rapidly in many parts of the world. Continued increase in the number of publications and patents each year suggest that the discovery and application process are working well so one may reasonably expect that the field of silicon science and technology will continue to grow in scope and value into the foreseeable future. Indeed, as many have suggested, silicon based materials are Materials for the 21st Century.

gf22095.1

REFERENCES

Allcock, H., Mark, J., and West, R. eds. "Inorganic Polymers," New Jersey: Prentice Hall, 1982.

Auner, N., Weis, J. eds. "Organosilicon Chemistry - From Molecules to Materials," New York: Weinher, 1994.

Chojnowski, J., Marciniec, B. eds. "Progress in Organosilicon Chemistry," Amsterdam: Overseas Publishers Association, 1995.

Clarson, S.J., Semlyen, J. Anthony eds. "Siloxane Polymers," New Jersey: Prentice Hall, 1993.

Fearon, G. and Ziegler J. eds. "Silicon Based Polymer Science," American Chemical Society, 1990.

Rochow, E. "Silicon & Silicones," Berlin: Springer Verlag, 1987.

ADVANCED MATERIALS-COMPETITIVENESS BY DESIGN

Björn Thundal

Celsius Materialteknik
Linköping, Sweden

INTRODUCTION

The purpose of this paper is to emphasise and demonstrate that **to exploit advanced materials, materials technology has to be integrated with product development and manufacturing.**

Furthermore, this means that **the responsibility for advanced materials commercial development to a very large degree is moved from material suppliers to component and system producers.**

BASE OR KEY TECHNOLOGY?

The purpose of this paper is to emphasise and demonstrate that **to exploit advanced materials, materials technology has to be integrated with product development and manufacturing.**

Furthermore, this means that **the responsibility for advanced materials commercial development to a very large degree is moved from material suppliers to component and system producers.**

Isn´t this obvious?
No it is not. People still don´t recognise materials technology as the key technology it is.
First, why is it so important to state this?
Basically it depends on the observation that key technologies are characterised by rapid development and that large differences in knowledge between different companies are common.
Thus, key technologies are technologies which can give large competitive advantages even with limited resources.
Why don´t people recognise advanced materials as a key technology?
To understand this, I think it is necessary to take a closer look into the field of materials technology.

Polymers and Other Advanced Materials: Emerging Technologies and Business Opportunities. Edited by P. N. Prasad *et al.*, Plenum Press, New York, 1995

CLASSIFICATION OF MATERIALS

There is - at least - three ways to classify materials:

* by structure
* by volumes used
* by properties used

Structure

Metals:	Fe, Al, Cu, Mg, W, Pb,........
Plastics:	PVC, PE, PET, PEEK, epoxi,.....
Ceramics:	Al2O3, TiB2, B4C, SiC, Si3N4,.....
Intermetallics:	Ni3Al, Ti3Al, TiAl, FeAl, Mo2Si,......
Composites:	PMC, MMC, CMC, IMC
Others:	Wood, concrete, stone, rubbers,......

Volume

Bulk materials	wood, concrete, steel, alumimium, plastics, rubber,.......
Advanced mtrl´s	superalloys, heavy alloys, engineering ceramics, semiconductors, advanced plastics, PMC, MMC,RSM,....
Developing mtrl´s	Existing materials but with no or very small commercial consumptions like ceramic superconductors, nanocrystalline materials, fullerenes, diamond coatings,.....

Properties used

Structural materials:	Mechanical properties
Functional materials	Electrical properties
	Magnetic properties
	Optical properties
	Thermal properties
	Piezoelectric properties

Naturally the limits between these three groups are not sharp. Roughly, I would say that in the first group official price notations for the materials as raw materials are common.
This is not so in the second group. Here prices normally relate to different components and products produced in the different materials.
The third group represents materials which do physically exist, but where commercial usage does not exist or is at very low levels.

The listing above directly shows that the field of materials technology both is vast and complex. This complexity normally is not realised.
People generally associate materials technology with structural materials, and actually only those structural materials which are bulk materials.
This is very natural. Structural bulk materials are the materials that historically by far have been of the greatest importance, regardless if we measure in tonnes, dollars or employment.
However, structural bulk materials rarely are key technologies - normally they represent base technologies.
These are well known and have a slow development. Thus, it is difficult to get competitive advantages with limited resources.

CHARACTERISTICS OF ADVANCED MATERIALS

Now, advanced materials, regardless of structure and properties used, differs from bulk materials in many regards:

- **Advanced materials often are produced in the same process as the components themselves. Material and component production are fully integrated processes.** Examples are filament winding of advanced PMCs, HIPing of advanced PM-parts and casting of single crystal turbine blades.

- **As material and component production are fully integrated processes, material development has to be an integrated part of product design and development.** With material development as integrated part of product design and development, it immidiately becomes a key technology.

  Let me take some very well-known exaples:

  The most obvious example is **information industry**. Information technology to a large degree is based on advanced materials technology: ultra-pure semi-conducting materials like silicon and gallium-arsenide doped with minute quantities of specifically selected elements, complex integrated circuits tailor-made by the most advanced technology in surface treatment, and optical cables made from non-crystalline glass in stead of metallic copper wires.

  Another, more every-day example, is the **sporting goods industry**. Look at raquets for tennis, squash and badminton. Here new generations of products are based on advanced materials like carbon-reinforced polymers in-stead of earlier wood and other natural materials. Equipment for golfing, skiing, fishing and surfing is made from imposing combinations of advaced materials.

- The fact that material production and component production are integrated processes, means that fabrication of semis disappears in the production chain. This, together with the fact that materials development becomes a part of product design and development, means that **the responsibility for materials development moves down-stream the product chain.**

- While the economic value of bulk materials is calculated as tonnages x material prices, this is certainly not true for advanced materials. Instead the economic value of advanced materials is associated with better system properties, new product generations, decreased over-all productions costs and increased market shares. With other words
  Economic value of advanced materials is shown by the competitiveness and profitability of the components and products involved.

EXAMPLES

To end up, I would like to show some more product examples based on advanced materials:

- Naval Drive Shafts (PMC)
- Swivel Dampers for Hoover-Crafts (PMC)
- Launching Tubes (PMC)
- KE-penetrators (advanced PM materials)
- Engine Case (MMC)

BUSINESS OPPORTUNITIES FOR A NEW TYPE OF LCP AND ITS BLENDS

Christer H. Bergström

Corporate Technology
Neste Oy, Keilaniemi
Espoo, FIN-02151, Finland

INTRODUCTION

Liquid Chrystalline Polymers, LCP, have received a lot of attention in the recent years because of their very special properties like good heat resistance, flowability, dimensional stability, modulus and strength etc. Because of high prices and anisotropical behaviour their end-uses are, however, still rather limited. LCP is mostly used in the electrical / electronics field and the predominant processing method is injection moulding.

In order to obtain a better price / performance ratio for the end-products Neste started 5 years ago to develop LCP / PE- and LCP/ PP-blends and related extruded and coextruded products. Also a new family of LCP, "Type 4", with a heat resistance comparable to Type 2 and a processing temperature comparable to Type 3 was developed as well as several compatibilization concepts for LCP / Polyolefins blends.

The properties of the above mentioned polymers, blends and end-products will be presented and the potential for new applications, business opportunities and cooperative arrangements will be discussed.

PREPARING A NEW TYPE OF LCP

Since 5 years Neste Oy has worked in the field of thermotropic mainchain LCP development with the aim of using it as a blend component in polyolefins. Because of this rather special aim of upgrading polyolefins some special requirements were placed on the LCP. We wanted to have an LCP which could be processed at the same temperature as PE and PP but still have enough heat resistance and stiffness. A low processing temperature could be obtained with a type 3 LCP like Unitika's Rodrun but, then the other requirements were not achieved. Also the compatibilization of the type 3 LCP:s was not successful. The type 1 and type 2 LCP:s which were stiff and heat resistant needed, however, too high temperatures in compounding and processing. Also for the type 1 and type 2 LCP:s the compatibilization was unsuccessful.

By combining many types of monomers a successful LCP for blending purposes was finally obtained. This LCP we call a type 4 LCP and its processing temperature and HDT are seen in Table 1 in comparison with other types of LPC:s. The type 4 LCP contains 5 different monomers 2 of which are obtained from very small amounts of PET. The type 4 LCP has no crystalline melting point but a Tg of 147°C. Because of the feature of one of the monomers, the compatibilization with functionalized polyolefins is also significantly improved.

Table 1. Types of LPC:s.

Thermotropic mainchain LPC:s are stiff, fully or partly aromatic polyesters or ester-copolymers, which can be crystalline or amorphous. Such thermotropic mainchain LCP:s have until now been devided into 3 categories due to their thermal properties. Now Neste is introducing a new "type 4 LCP".

LCP-TYPES	SOME LCP-ALTERNATIVES	PROCESSING TEMPERATURE	HDT/A
1	Amoco, Xydar SRT-300	400 - 450 °C	355°C
2	Hoechst, Vectra A950	280 - 320 °C	180°C
3	Unitika, Rodrun LC3000	210 - 240 °C	64°C
"4"	Neste, "Type 4 LCP"	210 - 260 °C	130°C

PREPARATION OF A NEW TYPE OF LCP

In Table 2 also the mechanical properties of the type 4 LCP are presented in comparison with Rodrun LC-3000 and Vectra B950 which has a higher HDT than Vectra A950 presented in Table 1. One can see that the tensile and flexural modulus and the flexural strength of the type 4 LCP are on the same level as for Vectra B 950 and significantly better than for the type 3 LCP Rodrun LC-3000. These properties are of great significance when using the type 4 LCP as reinforcement fibres (possibly made in-situ in the compounding process) in polymer blends. The notched Charpy is however lower for the type 4 LCP. In Table 2 the HDT values are made with method B in order to be able to compare with the HDT values of the LCP blends.

PROPERTIES OF LCP-BLENDS

In Table 3 are seen the properties of 40% LCP-blends with HDPE. These results are presented only to show the basic effects of LCP-type and compatibilizer. Since the dispersed LCP-phase is composed of droplets (no in-situ fibre formation) the improvement on the mechanical propertis is very insignificant. However, it can be seen that also for blends with HDPE the HDT is higher for the type 4 LCP compared to Rodrun LC-3000. Also the tensile and

Table 2. Properties of Neste's new "Type 4 LCP".

PROPERTIES	"Type 4 LCP"	RODRUN LC-3000	VECTRA B950
TENSILE MODULUS (GPa)	18	9	21
FLEXURAL STRENGTH (MPa)	216	114	262
FLEXURAL MODULUS (GPa)	17	9	17
NOTCHED CHARPY AT 23 °C (kJ/m²)	7	60	28
UNNOTCHED CHARPY AT 23 °C (kJ/m²)	23	nb	31
HDT/B (°C)	148	83	230

flexural modulus as well as the tensile strength are higher. The Charpy impact properties are however lower when using compatibilizer M. Regarding the blends of Rodrun LC-3000 with HDPE one can see that compatibilizer M is mainly improving the Charpy impact properties.

In Table 4 are presented LCP / HDPE-blends with three different compatibilizers, M, E, and C. Here one can see that by replacing compatibilizer M with compatibilizer E for the type 4 LCP-blends softer blends are obtained. The stiffness and strength properties as well as HDT decrease whereas the impact and elongation properties increase. The unnotched Charpy and elongation are actually better for the type 4 LCP-blends compared to the Rodrun LC-3000-blends when using compatibilizer E. Using compatibilizer C in the Rodrun LC-3000-blends the Charpy impact properties can, however, be further improved.

The results presented here show that the properties of LCP / HDPE -blends depend very much on the type of LCP used as well as the compatibilizer. In order to obtain the optimum combination of properties for each application it is therefore very essential to choose the right combination of LCP and compatibilizer.

In order to obtain barrier properties also LCP / LDPE -blends with LCP as the continuous phase (70%) were developed (Table 5). Also in this case the right combination of type of LCP and type of compatibilizer is essential. We have developed two types of LCP / LDPE -blends for coextrusion coating for barrier application. One blend contain our type 4 LCP and the other a competitor's grade. In this application it is essential that the melt is elastic enough so that an even layer is obtained in coextrusion. Also the anisotropic properties as well as the impact

Table 3. Properties of LCP / HDPE -blends.

Blends of 40% LCP and 60% HDPE (and compatibilizer) were made with a corotating twin-screw extruder and injection moulded into ISO test specimens. Optimum compounding and injection molding temperatures for each LCP-grade were used.

PROPERTIES	"Type 4 LCP" and 10 % compatibilizer M	RODRUN LC-3000 and 10 % compatibilizer M	RODRUN LC-3000 without compatibilizer
TENSILE STRENGTH (MPa)	55	36	39
TENSILE MODULUS (GPa)	6	3	3
TENSILE ELONGATION (%)	1	2	3
FLEXURAL MODULUS (GPa)	5	3	3
NOTCHED CHARPY AT 23°C (kJ/m²)	3	19	13
UNNOTCHED CHARPY AT 23°C (kJ/m²)	9	21	16
HDT/B (°C)	135	71	68

Table 4. Influence of compatibilizer on the properties of LCP / HDPE-blends.

Neste has developed 3 fundamentally different compatibilization concepts for LCP / polyolefins blends. Presented below are blends of 40% LCP and 60% HDPE (and compatibilizer) made with 10 % of compatibilizer M, E, and C.

Properties	"Type 4 LCP" compat. M	"Type 4 LCP" compat. E	Rodrun LC-3000 compat. E	Rodrun LC-3000 compat. C
Tensile strength (MPa)	55	23	35	37
Tensle modulus (GPa)	6	1	3	3
Tensile elongation (%)	1	16	2	2
Flexural modulus (GPa)	5	1	3	3
Notched Charpy at 23°C (kJ/m^2)	3	3	8	21
Unnotched Charpy at 23°C (kJ/m^2)	9	25	13	22
HDT/B (°C)	135	94	68	70

Table 5. The use of LCP / polyolefins -blends in barrier applications.

- Neste has developed technology for the coextrusion coating of LCP / LDPE -blends (> 70% LCP continuous phase) for barrier applications.

- The LCP / LDPE -blends used for barrier applications need a special compatibilizer in combination with a special type of LCP (Neste's "Type 4 LCP" and a few other LCP-grades can be used).

- These LCP / LDPE -blends can be coextruded with LDPE or HDPE without tie-layers and if used as a core layer the thickness can be below 5 μm depending on the die design and the extrusion conditions.

- Oxygen transmission rates less than 0.1 cm^3/m^2. 24 h. bar (normalized to a 25 μm thick layer) have been obtained at 85% relative humidity, which is less than low ethylene content (32%) EVOH at dry conditions.

- There is no interaction between the gas or the fluid and the LCP, which means that these LCP / polyolefins -blends have good barrier properties also for other media like water vapour, gasoline, solvents, flavours, and other gases.

- Coextrusion coated LCP / LDPE -blends can be used for the replacement of EVOH (sensitive to water vapour) or aluminium-foil (micro-cracks, leakage at the corners) in orange juice cartons etc.

Table 6. Other potential applications for LCP/polyolefins -blends.

- Multilayer pipes for radiator heating (oxygen barrier), gas transmission (gas barrier) etc.

- Structures in optical cables (dimensional stability etc.)

- Thermoformed multilayer PP-sheet for retortable food containers (good oxygen barrier under autoclaving conditions)

- Coextrusion blow-moulded gasoline tanks for cars (3-layers instead of 5-layers, easy flow allowing complicated design etc.)

- Multilayer films for special barrier purposes (for replacing aluminium in lids, covers etc.)

- Electronics' components (connectors etc.)

properties need to be good enough. All these properties can be obtained by using the right combination of type of LCP and compatibilizer. Most LCP-types available on the market were unsuccessful in this application.

Finally we have listed some very interesting potential applications for LCP / polyolefins -blends, which are presented in Table 6., and we are looking for partners in order to develop these applications together.

INDEX